ICAZ

Behaviour Behind Bones

The zooarchaeology of ritual, religion, status and identity

Proceedings of the 9th Conference of the International Council of Archaeozoology, Durham, August 2002
Series Editors: Peter Rowley-Conwy, Umberto Albarella and Keith Dobney

Behaviour Behind Bones

The zooarchaeology of ritual, religion, status and identity

Edited by
Sharyn Jones O'Day, Wim Van Neer
and Anton Ervynck

Oxbow Books

First published in the United Kingdom in 2004. Reprinted in 2017 by
OXBOW BOOKS
The Old Music Hall, 106–108 Cowley Road, Oxford OX4 1JE

and in the United States by
OXBOW BOOKS
1950 Lawrence Road, Havertown, PA 19083

Paperback Edition: ISBN 978-1-78570-571-7
Digital Edition: ISBN 978-1-78297-911-1

A CIP record for this book is available from the British Library

For a complete list of Oxbow titles, please contact:

UNITED KINGDOM
Oxbow Books
Telephone (01865) 241249, Fax (01865) 794449
Email: oxbow@oxbowbooks.com
www.oxbowbooks.com

UNITED STATES OF AMERICA
Oxbow Books
Telephone (800) 791-9354, Fax (610) 853-9146
Email: queries@casemateacademic.com
www.casemateacademic.com/oxbow

Oxbow Books is part of the Casemate Group

Contents

Preface

Peter Rowley-Conwy, Umberto Albarella and Keith Dobney

This book is one of several volumes which form the published proceedings of the 9th meeting of the International Council of Archaeozoology (ICAZ), which was held in Durham (UK) 23rd–28th August 2002. ICAZ was founded in the early '70s and has ever since acted as the main international organisation for the study of animal remains from archaeological sites. The main international conferences are held every four years, and the Durham meeting – the largest ever – follows those in Hungary, the Netherlands, Poland, England (London), France, USA, Germany and Canada. The next meeting will be held in Mexico in 2006. The Durham conference – which was attended by about 500 delegates from 46 countries – was organised in 23 thematic sessions, which attracted, in addition to zooarchaeologists, scholars from related disciplines such as palaeoanthropology, archaeobotany, bone chemistry, genetics, mainstream archaeology etc.

The publication structure reflects that of the conference, each volume dealing with a different topic, be it methodological, ecological, palaeoeconomic, sociological, historical or anthropological (or a combination of these). This organisation by theme rather than by chronology or region, was chosen for two main reasons. The first is that we wanted to take the opportunity presented by such a large gathering of researchers from across the world to encourage international communication, and we thought that this could more easily be achieved through themes with world-wide relevance. The second is that we thought that, by tackling broad questions, zooarchaeologists would be more inclined to take a holistic approach and integrate their information with other sources of evidence. This also had the potential of attracting other specialists who shared an interest in that particular topic. We believe that our choice turned out to be correct for the conference, and helped substantially towards its success. For the publication there is the added benefit of having a series of volumes that will be of interest far beyond the restricted circle of specialists on faunal remains. Readers from many different backgrounds, ranging from history to zoology, will certainly be interested in many of the fourteen volumes that will be published.

Due to the large number of sessions it would have been impractical to publish each as a separate volume, so some that had a common theme have been combined. Far from losing their main thematic focus, these volumes have the potential to attract a particularly wide and diverse readership. Because of these combinations (and because two other sessions will be published outside this series) it was therefore possible to reduce the original 24 sessions to 14 volumes. Publication of such a series is a remarkable undertaking, and we are very grateful to David Brown and Oxbow Books for agreeing to produce the volumes.

We would also like to take this opportunity to thank the University of Durham and the ICAZ Executive Committee for their support during the preparation of the conference, and all session organisers – now book editors – for all their hard work. Some of the conference administrative costs were covered by a generous grant provided by the British Academy. Further financial help came from the following sources: English Heritage, Rijksdienst voor het Oudheidkundig Bodemonderzoek (ROB), County Durham Development Office, University College Durham, Palaeoecology Research Services, Northern Archaeological Associates, Archaeological Services University of Durham (ASUD), and NYS Corporate Travel. Finally we are extremely grateful for the continued support of the Wellcome Trust and Arts and Humanities Research Board (AHRB) who, through their provision of Research Fellowships for Keith Dobney and Umberto Albarella, enabled us to undertake such a challenge.

Two of the 24 sessions organised included the themes of "Ritual and Religion" and "Status and Identity". Since these broad study areas (as aspects of human behaviour that clearly go beyond sheer economic and ecological frameworks) cover much common ground, it was obvious that these sessions should be conjoined for the purposes of publication.

Introduction

Sharyn Jones O'Day, Wim Van Neer and Anton Ervynck

Behaviour behind bones

Amongst those who study animal bones from archaeological sites, there exists a smouldering debate regarding whether this scientific discipline should be called archaeozoology or zooarchaeology. The first name suggests zoological research based on material derived from archaeo logical sites and indicates a close link with fields such as palaeontology, natural history, ecology or even veterinary medicine. In the Old World, the study of animal bones from archaeological sites indeed grew as 'archaeozoology' and focused upon the quaternary evolution of the animal realm, the history of domestic animals, the economy of former human settlements and the ecology of the landscape around it. In the New World, however, juggling with bones has been called 'zooarchaeology', the study of past human life through the animal remains found at their sites. Through a close link with anthropology and sociology, the emphasis is on the reconstruction and understanding of human behaviour. Of course, both the archaeozoological and zooarchaeological viewpoints are equally valuable, but the dichotomy may yield the unfortunate effect of dividing schools of researchers and inhibiting communication between each other.

Over the last decades in the Old World, more and more published animal bone reports deal with sites from historical periods (*e.g.*, from Roman, medieval or post-medieval times). In this temporal frame, the evolution of the animal world or domestication history are no longer the main research themes, being replaced by more compelling issues such as urbanisation, colonialism, industrialisation, social uprise. Through the development towards the analysis of historical contexts, it also became clear that economy and ecology alone were not sufficient to explain what had happened at these sites. Historians had, long ago, already provided the data and insights to indicate that people's behaviour in the past was not only directed by economic and ecological constraints but was equally influenced by social mechanisms, including religion and ideology in general. Taking this historical approach back to pre-historical periods, Old World archaeozoologists are now increasingly coming into the influence spheres of anthropology, the four field discipline of most New World archaeologists. On the other hand, New World zooarchaeologists, even those working within the field of the New World 'Historical Archaeology', have a growing interest in the synergy between the rich corpus of written sources and the animal remains studies in the Old World. In this way, anthropology, history, zooarchaeology and archaeozoology are now coming together.

The present volume is a result of the recent trends described. Herein we bring together studies of animal remains from different parts of the world, each representing specific research lines, combining history and prehistory, ethnologic comparisons and anthropological insights. Importantly, however, all contributions deal with behavioural aspects of former human lifeways that go beyond economy and ecology. Contributors to this volume focus on humans, and not merely on the animals they ate, sharing the common viewpoint that we study people and not just bones; hence the title of this publication, 'Behaviour behind Bones', a phrase taken from a popular textbook (Thomas 1998, 319).

The zooarchaeology of ritual, religion, status, and identity

Throughout the history of their discipline, anthropologists and archaeologists have concerned themselves with heuristic categories of religion, ritual, and social status. These categories serve to separate specific aspects of human culture and behaviour into approachable divisions that may be understood through our modes of investigation and analysis. In addition, both the ways that cultures are viewed by anthropologists and classified in a comparative sense are fundamentally influenced by our interpretations of indigenous systems, which often hinge on relative degrees of social stratification or hierarchy and the "type" of religion practised (*e.g*, shamanism, animism, totemism, polytheism, monotheism). The integration of these topics into one volume is appropriate since they overlap considerably in theory (etic conceptions) and in practice (emic concepts and actual human experience).

Recently, theoretical and methodological approaches in archaeology have explicitly directed investigations toward deciphering the role of identity, ritual, and religion in social systems through the material remains of the past (see Joyce 2000; Dietler and Hayden 2001; Gazin-Schwartz 2001; Insoll 2001; Wills 2001; Ashmore 2002; Lee and Zhu 2002). In step with this development, we present a wide range of case studies, both spatially and temporally, examining these social issues and the contribution archaezoological/zooarchaeological studies can make.

Beyond calories: the zooarchaeology of ritual and religion

Ritual and religion are incredibly broad concepts as used by anthropologists. These issues have been defined in countless ways, thus to attempt an overview is certainly beyond the scope of a single volume such as this. Recently, a number of books have focused on ritual and religion, providing historical reviews and critiques of the concepts and their applications in anthropology and/or archaeology (*e.g.*, Bell 1997; Rappaport 1999; Insoll 1999; Bowie 2000; Doty 2000; Insoll 2001); the reader is referred to these texts for an in-depth treatment of the subjects. A working definition for religion that is deliberately broad and thus applicable to the multiple ways the contributors use it here is: the institutionalised expression of "...the belief in, worship of, or obedience to a supernatural power or powers, considered to be divine or to have control of human destiny" (Scarre 1996, 590). While religion is considered to be fundamental, if not *the* universal features of human societies, archaeologically it is often considered as the most elusive of not entirely inaccessible components of culture (Parker Pearson 2001, 204). The related concept of ritual is even less concrete than religion. In anthropology, ritual is traditionally distinguished from other forms of human action by its supposed non-utilitarian and potentially non-rational qualities (*e.g.*, Tylor 1873); thus, even in modern anthropology, ritualised behaviours are often contrasted to ordinary and logical modes of behaviour (Bell 1997, 46). This interpretation leaves little room for a more encompassing view of ritual, which includes patterned or formalised secular activities and modes of everyday life that are directed by belief systems (for example, norms of public behaviour or the spatial organisation of houses and villages). Ritual can be generally defined as, "a complex sociocultural medium, variously constructed of tradition, exigency, and self-expression; it is understood to play a wide variety of roles and to communicate a rich density of over determined messages and attitudes" (Bell 1997, xi).

Contributors to this volume have primarily examined ritual and religion using archaeological indicators. Our interpretations are also often supplemented with or driven by ethnographic and ethnohistorical data (Cooke, *this volume*; Emery, *this volume*; Millerstrom, *this volume*; Muir and Driver, *this volume*; O'Day, *this volume*; Rofes, *this volume*). Some of the authors identify religious rituals through association with non-secular activities (such as offerings to gods or ceremonial feasting) and sacred places, including temples and funerary monuments (Corrado *et al.*, *this volume*; Ikram, *this volume*; Lauwerier, *this volume*; Stanc and Bejenaru, *this volume*; Warman, *this volume*; Wilkens, *this volume*). These interpretations are often bolstered by textual evidence as well. Most of the papers focus on religious rituals. Secular rituals that are tied to religious beliefs or cosmology are discussed, either directly or indirectly, in papers by Cooke, Lauwerier, Millerstrom, O'Day, Kansa and Campbell, and Rofes. Crabtree (*this volume*) discusses appropriate theoretical approaches to investigate ritual and religion through the faunal record.

Behavioural phenomena referred to as religious practices are identifiable in an archaeological material context, which is distinct from the "everyday" (Crabtree, *this volume*; Galik, *this volume*; Kansa and Campbell, *this volume*; Lentacker *et al.*, *this volume*; Muir and Driver, *this volume*). An assemblage associated with religious activities may include unique faunal markers or evidence of particular processing techniques (Cope, *this volume*; Weissbrod and Bar-Oz, *this volume*), but is not limited to these types of remains (Lauwerier, *this volume*). Archaeological phenomena such as pottery, religious icons, and petroglyphs may display visual representations of animals that have symbolic and religious connotations (Cooke, *this volume*; Lentacker *et al.*, *this volume*; Millerstrom, *this volume*). Thus, material culture depicting and/or associated with animals provides clues about ideology, religious practices and the role of animals within spiritual systems.

These chapters present historical, or deep archaeological, perspectives that have great potential to inform about the development of religions regionally and temporally. Social interactions (*e.g.*, communication, diffusion, conflict, and exchange) between populations have given rise to what Timothy Insoll (2001) refers to as World Religions, including Buddhism, Christianity, Hinduism, Islam, and Judaism. Indeed, these same social interactions may have given rise to all forms of religion throughout human history. By practising a particular religion and regulating eating practices, ethnic or social groups may separate themselves from their neighbours and/or enemies (*e.g.*, Judaism's kosher diet; see, Cope, *this volume*; Soler 1997 [1973]; Harris 1985). Contributions by Cope, Stanc and Bejenaru, and Wilkens all explore rituals, and food-related religious practices that at one level serve as social markers, differentiating groups of people from surrounding populations.

Relatively few zooarchaeological publications have

explicitly focused on ritual or religion. The paucity of studies on these issues is likely due to many factors, four primary explanations are offered. First, in some parts of the world, sacred sites are frequently inaccessible to archaeologists; for example, many sacred sites in North America and Oceania are protected from excavations by law, or simply remain unidentified or invisible archaeologically. Second, some zooarchaeologists argue that ideological issues are either unimportant ("noise") or are not subjects that our studies can reliably contribute to. Third, as a sub-discipline zooarchaeology (or archaeozoology) tends to operate somewhat autonomously from other areas of anthropology; researchers often specialise exclusively in faunal analysis and may not be involved in a dialogue with researchers outside this speciality. Fourth, when large religious monuments are excavated, as in the case of some Buddhist temples (Coningham 2001), faunal remains are rarely collected and/or analysed; in other situations, archaeologists may focus on grand architecture or human burials rather than fauna. In the succeeding chapters we hope that readers will gain a broader appreciation for the contribution faunal remains can make to the study of ritual and religion. Ritual and religion are expressed through a multiplicity of ways in the archaeological record generally and the faunal record specifically; each chapter displays this diversity, interpreting it differently through a variety of theoretical and methodological approaches. Further, with this volume the authors aim to open lines of communication between anthropologists, archaeologists, and faunal specialists. The case studies presented here demonstrate that ideology is accessible from archaeological materials, and more specifically from the faunal record when multiple lines of evidence are combined. Through the careful study of all material remains (*e.g.*, faunal, architectural, textual, visual), archaeologists and anthropologists will come to better understand ritual and religion and changes that these social phenomena have undergone through time.

Equations for inequality: the zooarchaeology of identity, status and other forms of social differentiation

Archaeological and anthropological research illustrates that virtually all societies throughout human history showed some degree, and kind, of inequality (Béteille 1994). This inequality not only manifests itself in the social relations between individuals and groups (or the absence thereof), but especially in the distribution of material and other resources. Food, for the living or the dead, for people or for their gods, certainly is an important parameter within this context. As Claude Lévi-Strauss (1968) stated: "a society's cuisine acts as a language through which it unconsciously expresses its structure". Indeed, many social mechanisms associated with nutrition can be conceived that, on a daily basis or on special occasions, confirmed or re-established the structure of society and thereby illustrated its internal inequalities. Examples include foraging peck orders, mechanisms of exchange, hunting rights, differences in purchasing power, access to traded goods, feasts, food offerings, grave gifts, food rules, food taboos, culinary fashions, etc. The main question, however, is whether zooarchaeological research is able to investigate, analyse and explain these social mechanisms? Are the differences between zooarchaeological contexts easy to relate to the social mechanisms described? In other words: are we able to find equations describing inequality?

Of course, differentiation within a society, or between societies within a certain part of the world, is not exclusively related to socio-economic status. Religion can act as a differentiating factor, which may for example manifest itself through food taboos. Different ethnic traditions or origins can play the same role. Even general ideas or beliefs about health and healthy food may cause the emergence of different consumption patterns within groups. Some of these patterns, such as vegetarianism, also might invoke food taboos. Thus, when zooarchaeology records differences in food consumption within a former society, this does not mean that an explanation for the patterns found is straight forward. From the bones themselves, it will almost never be possible to tell whether socio-economic factors, the environment or ideology was the differentiating factor. Putting the data into an anthropological, historical and ecological framework of course helps but even then the ultimate questions *why* often remains unanswered. In-depth insights about the structure of society (any society, in general) will only appear from zooarchaeological research when broad comparative studies can be undertaken, searching for recurrent patterns that may be related to similar evolutionary scenarios. Certainly, we have not yet come that far, but the corpus of evidence is building up.

The contributions to this volume provide data and raise questions, which often cannot be answered, but that stimulate further research. How, for example, must the different consumption patterns attested for the Neolithic village of Arbon-Bleiche be explained (Marti-Grädel *et al., this volume*)? Why, at Arbon-Bleiche, did ones neighbours eat differently although they lived in the same type of houses and shared the same material culture? Perhaps the ethnic traditions of immigrants in a local population provide the answer? The emphasis is certainly also laid upon ethnicity in a study of human-animal relationships in the Bronze Age of Hungary (Choyke *et al., this volume*). Both studies, however, deal with groups classified as 'tribes' and the situation becomes even more complex when 'states' appear.

In a Neolithic case study, from eastern Turkey, dating back to roughly the same period, it is suggested that the rise of centralised states explains why animal breeding and meat consumption no longer differed between two geographically distinct areas (Siracusano, *this volume*). Inequality on a regional level (within society) thus created equality upon a supra-regional level (between societies in different landscapes). The well-structured society of ancient Egypt provides another example of social differentiation within a centralised state (Rossel, *this volume*), but here the presence of a clear historical context (written sources, iconography) supports the interpretations.

Historical interpretations are also possible when studying animal remains from medieval Europe. Ervynck (*this volume*) introduces a synthesis explaining how different consumption patterns within medieval society can be interpreted within the model of a society divided in three orders. Importantly, support for this interpretation was written down by medieval people themselves although real life was of course much more diverse than the model can account for. There were general characteristics in the monastic diet, but there were also differences between monastic orders or even abbeys, an example of which is given by the material from a Carthusian monastery from Austria (Galik and Kunst, *this volume*). On the other hand, it appears that people have always tried to escape the rigid framework of a structured society, for example when urban households tried to imitate the consumption at a noble court (Pigière *et al., this volume*). Moreover, the medieval philosophical model of society was an ideological one that did not take into account people they did not want to recognize. Hence, it is interesting to study, through their food remains, the Jewish population of late medieval Buda (Daróczi-Szabó, *this volume*) or the late medieval to early post-medieval 'nomadic' Cumanian population of the Hungarian Plain (Nyerges, *this volume*).

At the other side of the ocean, an even more diverse spectrum of social structures may be found. The complexity of the relationships between people and animals within pre-Columbian Panama is well illustrated by the analysis of animal depictions and artefacts made from animal products found in a number of cemeteries (Cooke, *this volume*), illustrating that the link between animals and status does not need to be studied through consumption patterns alone. The importance of hunting in social differentiation is analysed in a case study from the pre-Columbian southwest of North America (Potter, *this volume*). A more northern example touches the hitherto undiscussed aspect of colonialism, causing socio-economic differences between tribes that interacted with the occupiers and those that did not (Lapham, *this volume*). Less choice was given to the slave populations in 19th-century southern USA. Nevertheless, a comparison between plantations in the coastal region and upland sites shows that different labour regimes resulted in different forms of inequality (Lev-Tov, *this volume*).

The two African studies within this volume highlight the importance of historical and ethnographical information. This is the case for a discussion of the interaction in West Africa of several cultural groups with different attitudes towards dog eating (Linseele, *this volume*) and the study of the role played by hunting among rural pastoral and farming communities in the western Rif mountains of Morocco (Moreno-García, *this volume*).

Finally, despite the wealth of information provided, scientific deontology forces us to end with a warning finger. It is indeed true that researchers should never stop asking whether their data set is reliable, for instance to make sure that depositional factors, or taphonomy in general, have not biased the zooarchaeological records of the contexts studied (Driver, *this volume*). Moreover, when differences within societies are expressed in terms of status, how trustworthy are the indicators, and how often are we trapped into the pitfall of circular reasoning (Van Neer and Ervynck, *this volume*)?

Towards a broader view

In conclusion, this compilation of essays lends insights into the lives of humans, thus transcending the notion that zooarchaeology illuminates purely economic and/or environmental issues. We aim to explore how the material remains of behaviours and lifeways inform archaeologists and anthropologists about the social issues of religion, status, and identity in the past. Moreover, although zooarchaeological data are the central focus of all the papers in this volume, most authors have attempted to explore the animal remains broadly and within the context of multiple lines of evidence. In particular, the contributors draw from other forms of archaeological evidence (botanical remains, tool technologies, ornamental artifacts, visual representations, and mortuary evidence), architecture, texts, and ethnohistorical and ethnographic data. The authors' interpretations come from a variety of perspectives including functional, ecological, symbolic, and historical. Without doubt, such broader approaches provide a better understanding of how people believed, acted, differentiated and identified themselves in the past.

Acknowledgements
This volume originally took shape during the 9th International Council of Archaeozoology Conference held at the University of Durham, in August 2002. We would like to thank the conference organisers, Umberto Albarella, Keith Dobney, Jacqui Huntley, and Peter Rowley-Conwy for sponsoring the conference and making this volume possible.

The editors are indebted to our colleagues for their helpful comments and suggestions on the papers contained in this volume. We thank the many people who provided insightful reviews on these papers, including: Pierre Bonte (Ecole des Hautes Etudes en Sciences Sociales, Paris), James Brady (California State University, Los Angeles), Susan deFrance (University of Florida), Jacopo De Grossi Mazzorin (Universita degli Studi di Lecce), Jonathan Driver (Simon Fraser University, Burnaby), Kitty Emery (Florida Museum of Natural History, Gainesville), David Grove (University of Florida), Yannis Hamilakis (University of Southampton), Michael Heckenberger (University of Florida), Stan Hendrickx (Katholieke

Universiteit Leuven), Richard Hoffmann (York University, Ontario), Daniel Kaufman (University of Haifa), Omri Lernau (Mevaseret Zion), Arturo Morales (Universidad Autonóma de Madrid), Patrick O'Day (University of Florida), Wietske Prummel (Groningen Institute of Archaeology), Elizabeth Reitz (Museum of Natural History, University of Georgia, Athens), John Speth (University of Michigan), William Walker (New Mexico State University), Heather Walsh-Haney (University of Florida), and James Woollett (Hunter College, CUNY). Sheila Hamilton-Dyer (Southampton) is acknowledged for the linguistic corrections of some of the articles. Help with the editing was provided by Wim Tavernier and Alain Reygel (Royal Museum for Central Africa, Tervuren). Artwork for the cover was kindly provided by Elvira Servaes (Ename Center for Public Archaeology and Heritage Presentation).

References

Ashmore, W. 2002. Decisions and dispositions: socializing spatial archaeology. *American Anthropologist* **104**, 1172–84.

Bell, C. 1997. *Ritual: Perspectives and Dimensions*. Oxford: Oxford University Press.

Béteille, A. 1994. Inequality and equality, pp. 1010–39 in Ingold T. (ed.), *Companion Encyclopedia of Anthropology*. London/ New York: Routledge.

Bowie, F. 2000. *The Anthropology of Religion*. Oxford: Blackwell.

Coningham, R. 2001. The archaeology of Buddhism, pp. 61–95 in Insoll, T. (ed.), *Archaeology and World Religion*. London: Routledge.

Dietler, M. and Hayden, B. (eds) 2001. *Feasts: Archaeological and Ethnographic Perspectives on Food, Politics, and Power*. Washington DC: Smithsonian Institute Press.

Doty, W.G. 2000. *Mythography: The Study of Myths and Rituals* (Second Edition). London: The University of Alabama Press.

Gazin-Schwartz, A. 2001. Archaeology and folklore of material culture, ritual and everyday life. *International Journal of Historical Archaeology* **5**, 263–80.

Harris, M. 1985. *Good to Eat: Riddles of Food and Culture*. New York: Simon Schuster.

Insoll. T. (ed.) 1999. *Case Studies in Archaeology and World Religion* (BAR International Series S755). Oxford: Archaeopress.

Insoll, T. (ed.) 2001. *Archaeology and World Religion*. London: Routledge.

Joyce, R. A. 2000. Girling the girl and boying the boy: the production of adulthood in ancient Mesoamerica. *World Archaeology* **31**, 473–83.

Lee, Y.K. and Zhu, N. 2002. Social integration of religion and ritual in prehistoric China. *Antiquity* **(76) 293**, 719–24.

Lévi-Strauss, C. 1968. *Mythologiques III. L'origine des manières de table*. Paris: Plon.

Parker Pearson, M. 2001. Death, being, and time: the historical context of the world religions, pp. 203–19 in Insoll, T. (ed.), *Archaeology and World Religion*. London: Routledge.

Rappaport, R.A. 1999. *Ritual and Religion in the Making of Humanity* (Cambridge Studies in Social and Cultural Anthropology 110). Cambridge: Cambridge University Press.

Scarre, C. 1996. Religion, pp. 590–3 in Fagan, B.M. (ed.), *The Oxford Companion to Archaeology*. Oxford: Oxford University Press.

Soler, J. 1997 [1973]. The semiotics of food in the Bible, pp. 55–66 in Counihan, C. and Van Esterik, P. (eds), *Food and Culture: A Reader*. London: Routledge.

Thomas, D.H. 1998. *Archaeology* (3rd edition). Fort Worth: Harcourt Brace College Publishers.

Tylor, E.B. 1873. *Primitive Culture* (2nd edition). London: Johan Murray.

Wills, W.H. 2001. Ritual and mound formation during the Bonito phase in Chaco Canyon. *American Antiquity* **66**, 433–51.

Part 1

Beyond calories: the zooarchaeology of ritual and religion

edited by Sharyn Jones O'Day

9th ICAZ Conference, Durham 2002
Behaviour Behind Bones, (ed. Sharyn Jones O'Day et al.) pp. 2–13

1. Feasting with the dead? – a ritual bone deposit at Domuztepe, south eastern Turkey (*c*. 5550 cal BC)

Sarah Whitcher Kansa and Stuart Campbell

*The Halaf period in the Near East (*c. *6000–5200 cal BC) saw important developments in social complexity illustrated by elaborate craft production, use of seals, and the establishment of long-distance exchange networks. This period laid the foundations for the later rise of state institutions in the Near East but also saw the continuation of earlier forms of organization. The site of Domuztepe is located at the north eastern extreme of the Halaf cultural phenomenon in south eastern Turkey. Domuztepe is the first large Halaf site to be extensively excavated, offering a unique window on Halaf settlement in this region.*

A unique feature of the settlement at Domuztepe is an extensive and complex funerary deposit, focused around a pit in Operation I in which portions of at least 40 individuals were recovered. Animal bones were also found associated with the human bones in the pit. This paper presents results of a detailed, albeit preliminary, examination of the animal remains in this ritual deposit. The results of this study are critical to a better understanding of both the overall ritual activity itself and the social strategies that integrated an early large-scale society.

The identified portion of the Domuztepe faunal assemblage consists of 8,000 fragments, about 2,000 of which come from this pit. The nature of the faunal remains from this pit are presented and then compared with those from the rest of the site. Differences in body part representation and in relative abundance of taxa between the two areas reflect both human choice and preservation biases. The role played by cattle in forming this special assemblage is highlighted. The results from Domuztepe are then compared with archaeological and ethnographic parallels as we attempt to understand the nature of the (likely) ritual feasting that formed this assemblage.

Introduction

This paper presents the initial results of faunal analysis from an extensive funerary deposit at Domuztepe, a 6th millennium site in south eastern Turkey. This pit, generally referred to as the "Death Pit" contained highly fragmented and jumbled human and animal bones, as well as a few complete and articulated bones. This study asks how zooarchaeological analysis adds to our understanding of this pit and the ritual associated with its creation.

First, we will present an overview of the site of Domuztepe, with specific commentary on the Death Pit. Then, we will delineate some of the criteria that have been used in the past to identify "ritual" in the zooarchaeological record. We will then discuss the results of zooarchaeological analysis, and how these results can help to understand ritual behavior at this site. Finally, we will draw upon one ethnographic example of ritual feasting that helps demonstrate the variability and unpredictability of human actions that can result in assemblages such as theDomuztepe Death Pit assemblage.

The Domuztepe excavations

The site of Domuztepe is located to the south of the modern city of Kahramanmaras in south eastern Turkey (Fig. 1). Although it has distinctive local characteristics, its most obvious links are with the Halaf phenomenon in north Mesopotamia. Domuztepe is the first large Halaf site to be excavated, offering a unique window on prehistoric settlement in this region. The site itself covers at least 20 hectares, making it one of the largest prehistoric sites in the Near East. Most of this area was probably occupied simultaneously in the latter stages of occupation.

Although several areas at Domuztepe have been examined, the main concentration has been on Operation I, a large exposure on the southern part of the site, which now extends to over 1000 square meters (Fig. 2). The periods excavated are confined to about 200–250 years, dating to the mid-6th millennium BC. The architecture includes both rectangular and circular buildings. The ceramics from the site show strong links with the Halaf tradition in north Mesopotamia but also with areas to the south down the

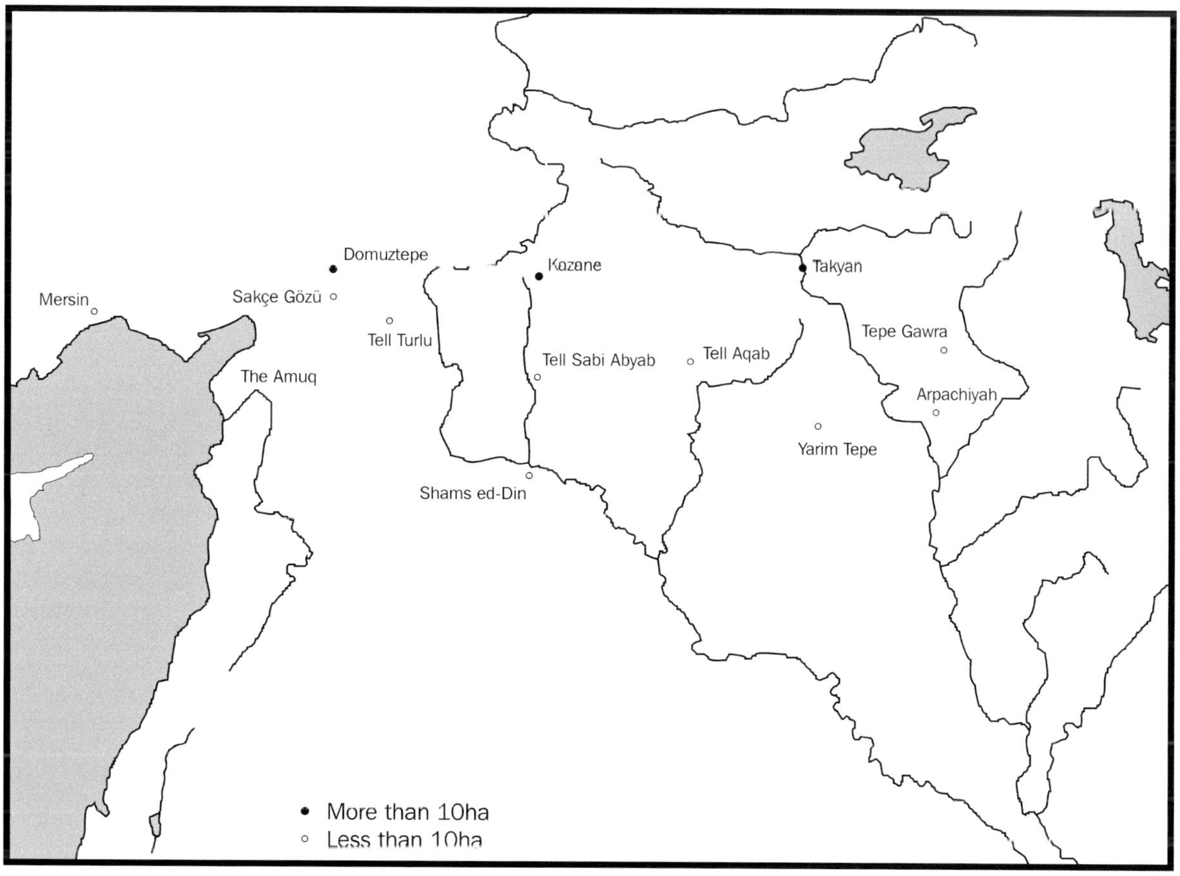

Fig. 1. Map showing the location of Domuztepe in south eastern Turkey.

Levantine coast. Both the faunal and botanical assemblages are dominated by domestic species, although wild plants, in particular, were also extensively exploited.

The Death Pit

A unique feature of the settlement at Domuztepe is an extensive and complex funerary deposit, focused around a pit in Operation I in which portions of at least 40 individuals were recovered. Significant quantities of animal bones were also found associated with the human bones in the pit. Four radiocarbon dates from the Death Pit, all from cereal grains, are very consistent and a Bayesian calibration gives an estimated calibrated date of 5582–5562 BC (at 1 standard deviation) or 5592–5557 BC and 5555–5537 BC (at 2 standard deviations).

Fig. 3 illustrates the main deposits in the Death Pit. These deposits form a complex sequence, but their actual deposition almost certainly took place over a short time, possibly days rather than weeks. (Note: The description here is based on the state of knowledge at the start of the 2002 season, at the time this paper was written. Further excavation during summer 2002 clarified and amplified our knowledge of the early events in the development of the Death Pit, although without invalidating the information upon which this paper is based. The fill designations used here are provisional and may be revised.) A pit was dug into the edge of a terrace on the southern part of the site. An initial series of deposits, mainly of animal bones, in the bottom of the pit was followed by an episode in which the pit was flooded leading to thick silt deposits (Fills A and B). As soon as this had dried a dense mix of fragmented bones, almost entirely human, together with packed pisé and patches of ash were used to construct a raised hollow in the northern half of the Death Pit (Fill C). In parallel, deposits were placed to the immediate south which were almost devoid of human remains but superficially have a strong resemblance to domestic midden deposits with quantities of broken pottery, animal bones and lithics (Fill D). After the raised hollow had been created, several more human skulls and intact long bones were placed within it and the hollow filled with a dense ash deposit which spills out into the surrounding area (Fill E). The area remained of significance, being marked by later pits filled with identical ashy deposits, two large posts, and a series of deposits of human remains, generally in very fragmentary form.

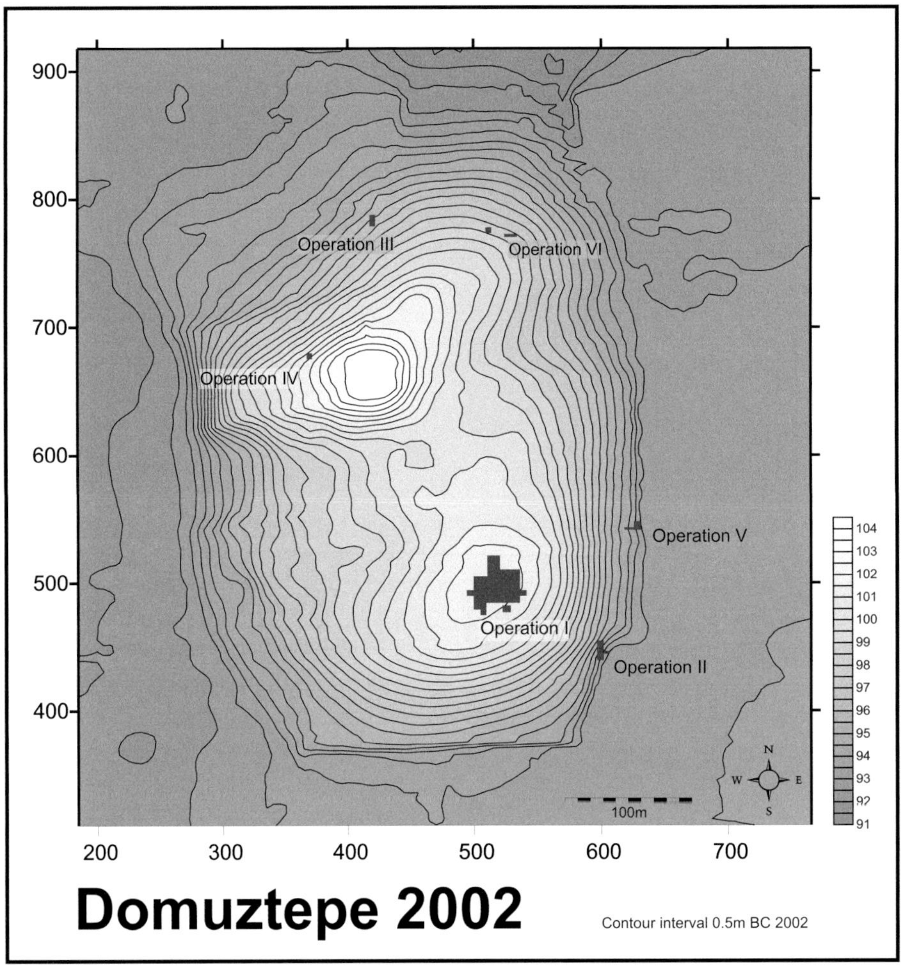

Fig. 2. Overview of the Domuztepe excavations.

Assumptions

The analytical results presented here are based on observations made comparing the Death Pit and the other deposits from the site. While the site certainly has distinctive deposits that merit close examination and comparisons of their own, for the purpose of this study, we are assuming that the site reflects an overall pattern of daily subsistence refuse. By contrasting these two areas, we hope to elucidate differences that will help us understand the extent to which the Death Pit contains an assemblage that can be described as reflecting daily subsistence refuse, or if it holds a different kind of assemblage that might be more closely related to distinctive ritual activities.

Identifying ritual feasting in the zooarchaeological record

There are inherent complications in examining ritual using zooarchaeological remains since, as Grant (1991, 110) has pointed out, "ritual and economic [behaviors] are inextricably entwined". Rituals are frequently accompanied by food and drink, small meals and large feasts. Indeed, the gathering of people around a daily meal almost invariably has ritual aspects. The frequent association of food with ritual makes it a difficult task to distinguish ritual activities involving food from daily or mundane meal refuse. In general, scholars detect ritual using any number of criteria that mark a deposit as distinctive, where their location, context, and modification add symbolic or ritual significance, such as: (1) the presence of whole, unbutchered animals or articulated portions of animals; (2) the presence of very young animals or very old animals; (3) a selection of specific parts (horn cores, heads, etc.); (4) an abundance of one sex; (5) an abundance of a particular taxon; (6) the presence of rare taxa; (7) association with human remains; (8) association with grave goods (criteria based on Horwitz 1987).

While any one of these characteristics can often be attributed to non-ritual behavior, the co-incidence of a number of these characteristics taken together may point to some kind of activity that can be seen as out of the ordinary. Other non-zooarchaeological lines of evidence can also help to identify special, distinctive deposits. The

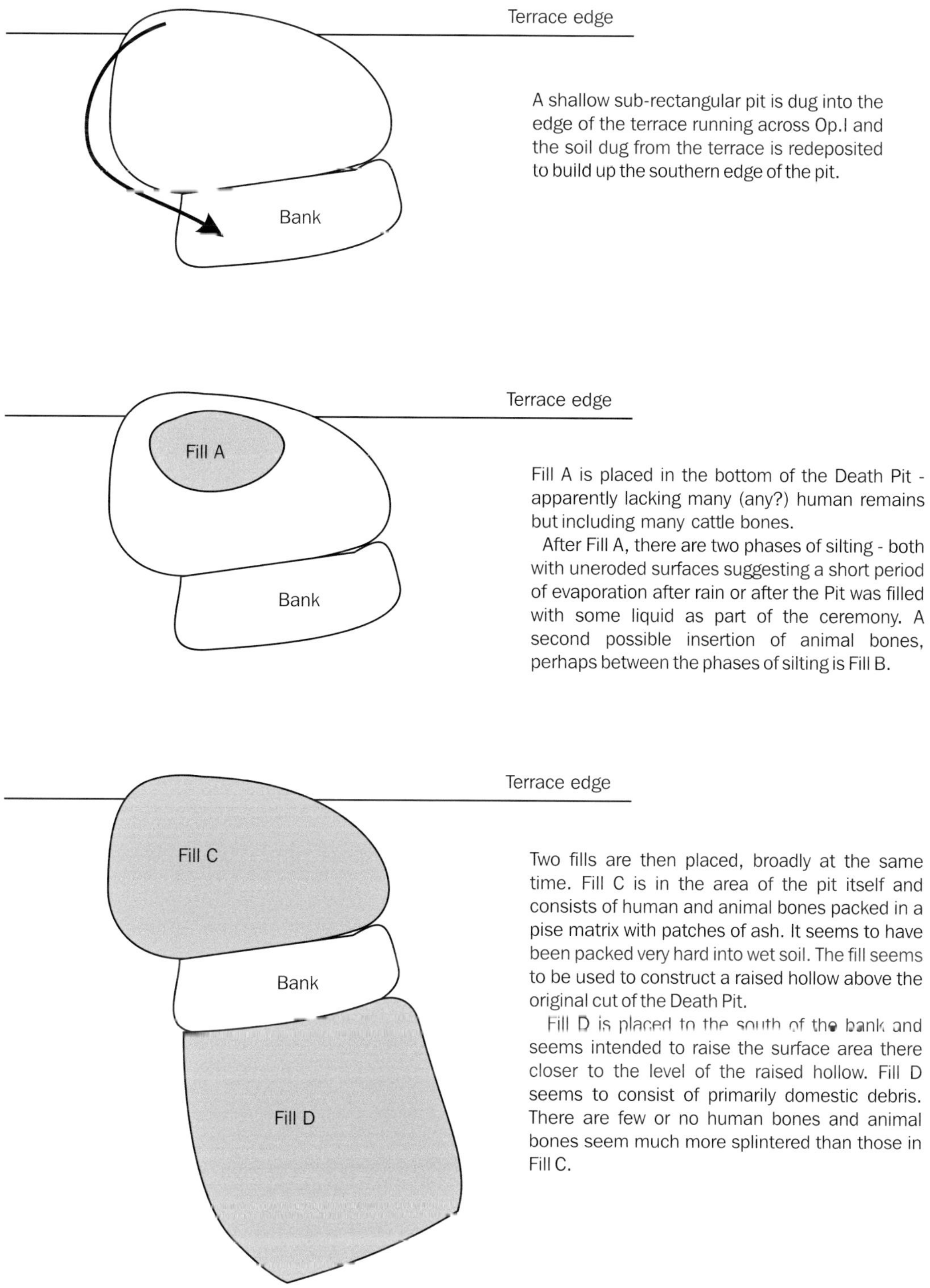

Fig. 3. The main deposits in the Death Pit.

Domuztepe Death Pit assemblage meets a number of these criteria, the most obvious of which is association with human remains. An entire series of ritual steps and associations can be articulated within the Death Pit. These associations range from the details of the carefully executed sequence of procedures involved, to the large and highly visible fire that produced the ash that covered the area, to the way in which the location of the Death Pit was marked and respected for a considerable period. This, together with other characteristics explained below, point to a special use of animals in this particular assemblage.

The Domuztepe animal bone assemblage: observations

The first five years of excavation at Domuztepe recovered over 100,000 bone, tooth, and antler fragments. During five field seasons Kansa drew from these 100,000 fragments a portion equaling approximately 20% of the entire assemblage, that she determined to be identifiable; that is, fragments that could be identified to genus and element. Upon export of the identifiable portion, Kansa undertook full analysis on the assemblage in the Zooarchaeology Laboratory of the Peabody Museum, Harvard University, under the auspices of Dr. Richard Meadow. To date, Kansa has completed analysis on nearly half of the identifiable portion of the assemblage, or 8,000 bones. 25% of the analyzed portion of the assemblage (or about 2,000 bones) comes from the Death Pit.

Taxonomic variation

The Domuztepe bone assemblage is dominated by domestic animals: cattle, sheep, goat, and pig (Fig. 4). Dog bones also occur in low numbers. Wild taxa include very small proportions of equid, gazelle, deer, bear, fox, hare, bird, fish, and reptile, among others. While the spectrum of taxa represented in the Death Pit generally reflects that from the site, there is a striking difference in the relative percentages of taxa in the two areas. Specifically, the Death Pit contains 30% more cattle bones and half as many pig bones than the rest of the site (Fig. 5).

When we consider this variation in taxa by context in the Death Pit, clear differences emerge, particularly in the relative percentage of cattle and pig (Fig. 6). The lowest layer (Fills A and B) has the highest number of cattle bones and the fewest pig bones. The number of pigs remains low, even though in the last layer of ash (Fill E) it is closer to the level of the rest of the site. Cattle bones are distinctly high in Fills A, B and C, relating to the initial deposits in the Death Pit and the deposits within which the majority of the human remains were concentrated. Fill D, which was placed at the same time as and to the south of Fill C, is markedly different. It is like Fills A-C in that it has very low quantities of pig, but the proportions of cattle and sheep/goat are very similar to the rest of the site. Fill D is not associated with significant quantities of human remains. The animal bones from Fill E most closely parallel to the rest of the site. This fill also contains very few human remains, although it was placed directly over human remains resting on the base of the hollow constructed as the upper surface of Fill C. In summary, cattle bones are particularly associated with the initial phase of deposition, without human bones, and with the next phase of deposition when they were distinctively associated with human remains. Pig were largely excluded not only from these fills but also from Fill D which was being placed at the same time as human remains were being placed, albeit in an adjacent zone. It is only when the final deposits of the Death Pit are placed, after all the human remains have been deposited in this area, that proportions of animal bones resemble the same mix from elsewhere on the site.

Another interesting characteristic of the Death Pit assemblage is the high proportion of dog bones. There are nearly five times more dog bones in Death Pit than in the rest of the site, and 80% of these bones are nearly complete. Additionally, all but two of the dog bones from the Death Pit were found in Fill C, a layer of the Death Pit where human bones predominate. These dog bones represent at least three individuals and mainly come from the skull: complete crania, maxillary bones, and mandibles. The absence of corresponding post-cranial bones for dogs is a broadly similar situation to that of human bones in the Death Pit. While this is intriguing, it does not differ greatly from the dog remains from the site, where, while having far fewer dog bones overall, nearly half of those from the site are skull bones as well (specifically, mandibles from at least 3 individuals).

Taxon		**Death Pit (%)**	**Site (%)**
Domestic animals			
Bos taurus	Cattle	31.6	22.7
Ovis aries / Capra hircus	Sheep / Goat	52.3	49.1
Sus scrofa	Pig	13.3	25.7
Canis familiaris	Dog	1.4	0.3
Wild animals			
Cervus elaphus / Dama dama	Red deer / Fallow deer	0.5	0.9
Equus sp.	Equid	0.1	<0.1
Vulpes vulpes	Fox	<0.1	0.2
Lepus sp.	Hare	<0.1	0.1
Ursus sp.	Bear	<0.1	0.1
Other wild		0.4	0.6
Bird		0.3	0.2
Fish		0.1	0.1
Total # of Bones		*1511*	*5314*

Fig. 4. Relative frequencies of taxa represented in the Domuztepe assemblage.

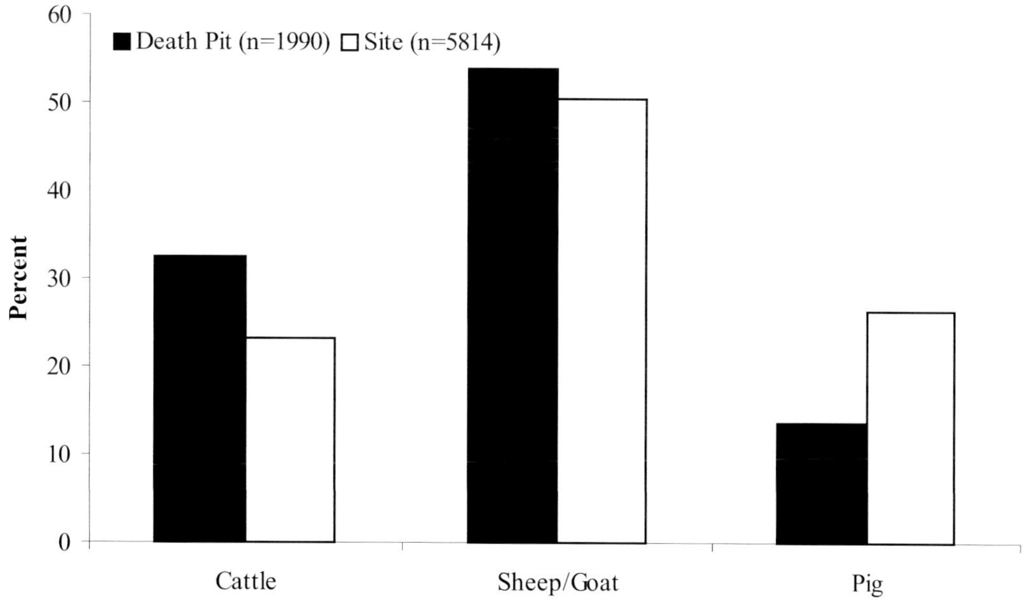

Fig. 5. Relative percentages of predominant taxa at Domuztepe.

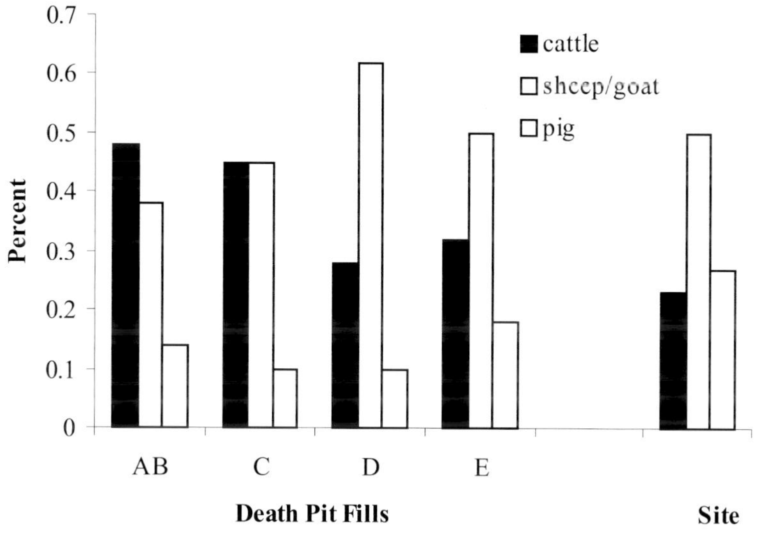

Fig. 6. Distribution of predominant taxa in the Death Pit and the site.

A general absence of cut marks on dog bones and a high percentage of complete bones throughout the Domuztepe assemblage indicates that dogs were not a normal part of the diet at Domuztepe. Only two dog bones in the pit have cut marks, right and left mandibles that appear both to have been crushed behind the third molar. The much higher occurrence of dogs in the pit suggests that dogs were seen as different and were intentionally interred with humans as part of this ritual.

Selection of animals: age and sex

Sheep and goat are represented by a ratio of 1:1 in both the Death Pit and the rest of the site. Kill-off patterns for sheep and goats were determined based on mandibular tooth eruption and wear (following the method of Payne 1973). Results show that sheep and goat mortality is nearly identical for the Death Pit and the site (Fig. 7). In both cases, most animals survive the first year (up to stage "D"), after which there is a steady kill-off throughout adulthood. Similarly, cattle kill-off in the Death Pit resembles that in the rest of the site, with both areas showing about 60% of cattle surviving to maturity (based

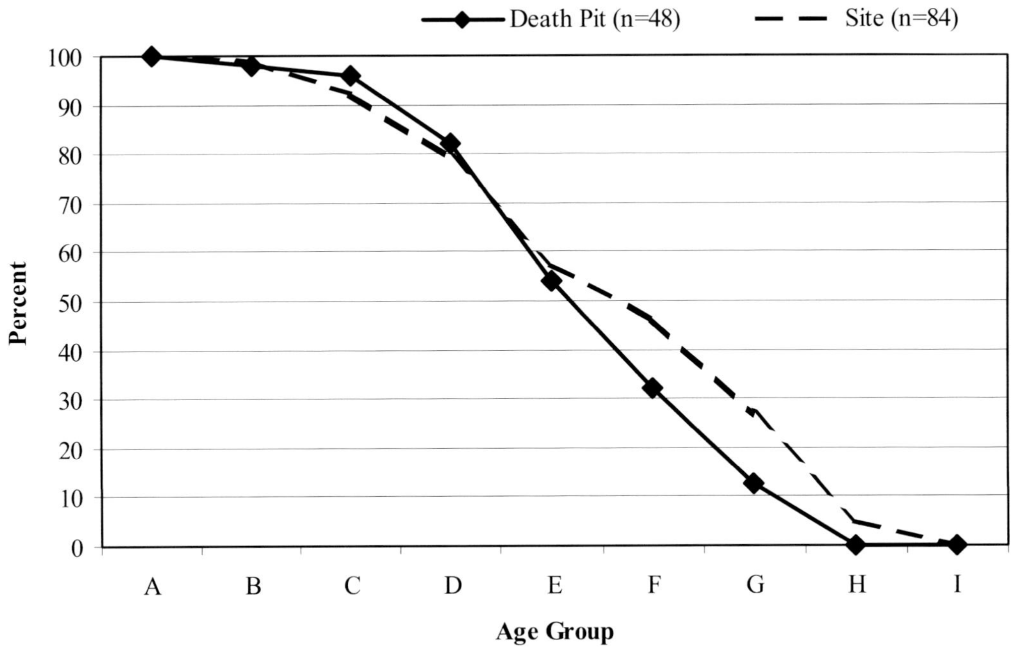

Fig. 7. Sheep/goat survivorship at Domuztepe.

on epiphyseal fusion). This similarity in kill-off patterns between the Death Pit and the site for the two predominant taxa indicates no preference for animals of a particular age for use in the formation of the Death Pit assemblage.

Pelves provided the largest and most reliable body of data for determining the sex ratio for sheep/goat and cattle (based on criteria laid out in Boessneck 1969). For both cattle and sheep/goat, the ratio in the Death Pit is an average of four females to every one male, while the rest of the site shows a ratio of one to one (based on a total of ten cattle and 78 sheep/goat bones from the site, and a much smaller sample of four cattle and sixteen sheep/goat bones from the Death Pit). Considering that these are mainly prime-age animals, as we have seen from the kill-off patterns, the predominance of females in the Death Pit reflects special selection of valuable animals (adult females for milking and breeding) for the ritual associated with the Death Pit's formation. The fact that a high ratio of females to males is found for both cattle and sheep/goat suggests that the people who created the Death Pit assemblage were following a specific selection strategy for both species. In sum, the feasting activities associated with the Death Pit formation placed more emphasis on the sex of the animals chosen rather than the age. While the age is consistent with everyday food preparation practices, the intentional selection of females over males points to a considerable contribution of valuable resources.

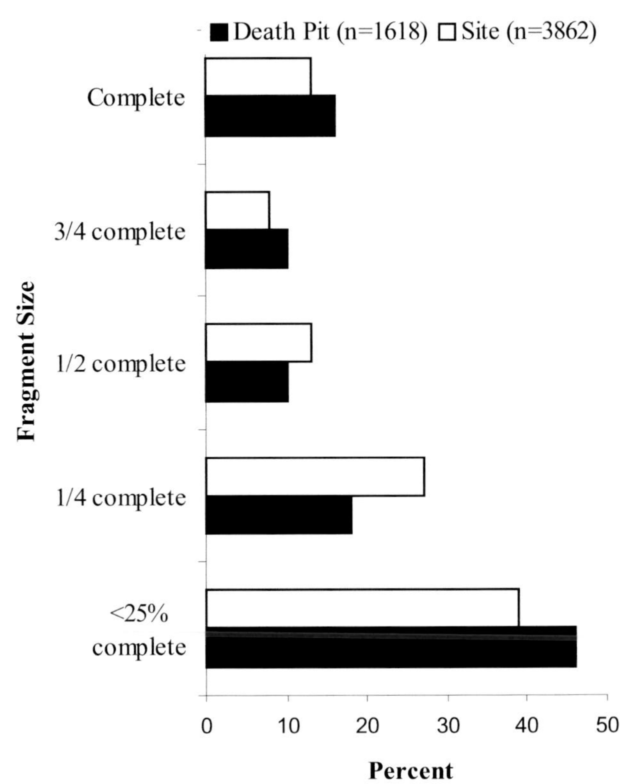

Fig. 8. Fragmentation at Domuztepe.

Animal processing: fragmentation and butchery

An overall comparison of animal bone fragmentation in the Death Pit and the site (Fig. 8) shows that the Death Pit contains a higher number of complete or nearly complete animal bones than the rest of the site. This observation

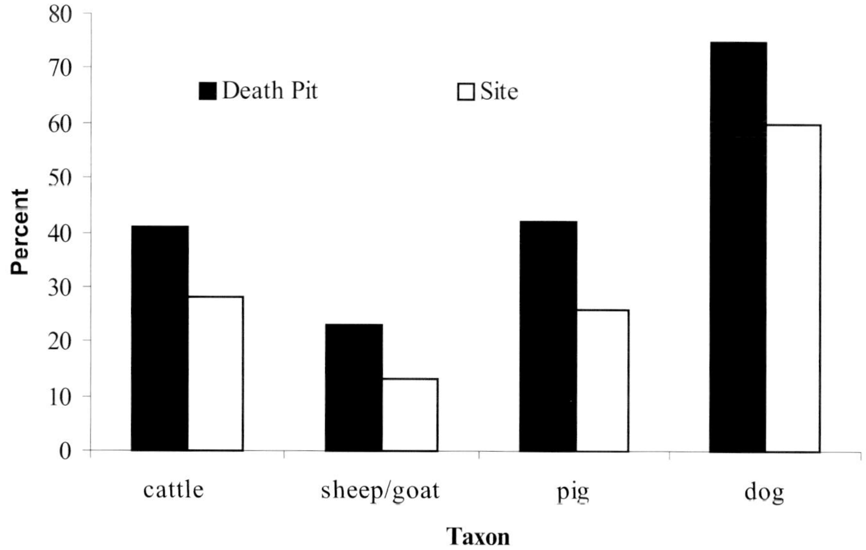

Fig. 9. Occurrence of complete bones for each major taxon at Domuztepe.

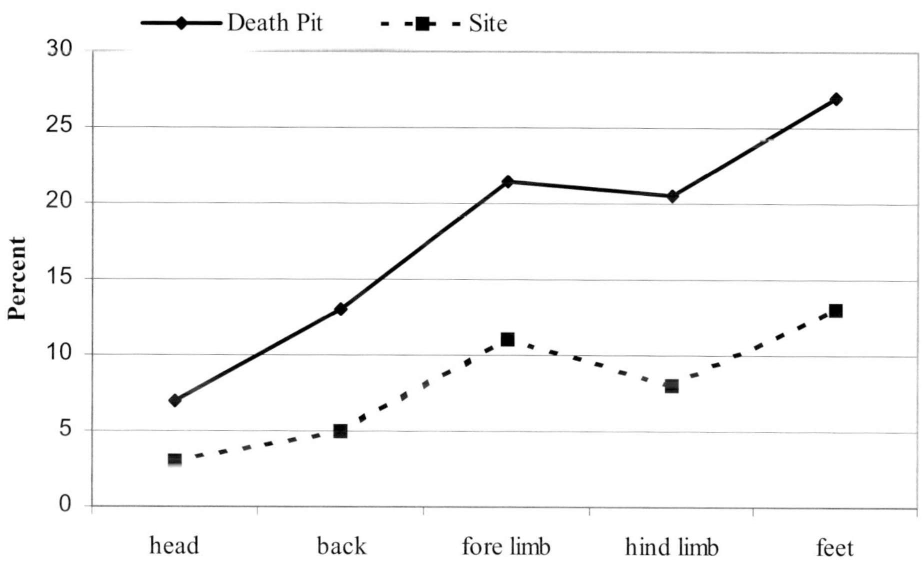

Fig. 10. Percent burned bones per skeletal area for all taxa at Domuztepe.

holds true for each of the major taxa (Fig. 9), where most taxa have nearly twice as many complete bones in the Death Pit as the rest of the site. In the case of dogs, the percentage of complete bones is high in both contexts, again reflecting the fact that dog bones appear *not* to have been butchered in the same way as the major food animals. While the presence of more complete bones in the Death Pit is likely due to the protected nature of the deposit and its speedy deposition (so that the contents were not subjected to weather, trampling, and scavenging animals in the same way that household refuse would be), it also may reflect the nature of the meal or feast that accompanied the formation of the Death Pit. In such a situation, where a large number of animals are killed in a relatively short amount of time, their bones might not be processed quite as thoroughly as in daily life (such as being boiled and crushed for marrow).

The Death Pit contains twice as many burned bones as the rest of the site; however, there is no disproportionate burning on any one species or bones from a particular skeletal area (Fig. 10). In fact, the ratio of burning between skeletal areas is the same in the Death Pit and the site, suggesting that no special animal processing activity resulted in the higher proportion of burned bone in the Death Pit. The higher proportion appears to be a result simply of deposition in a context in which hot ash was

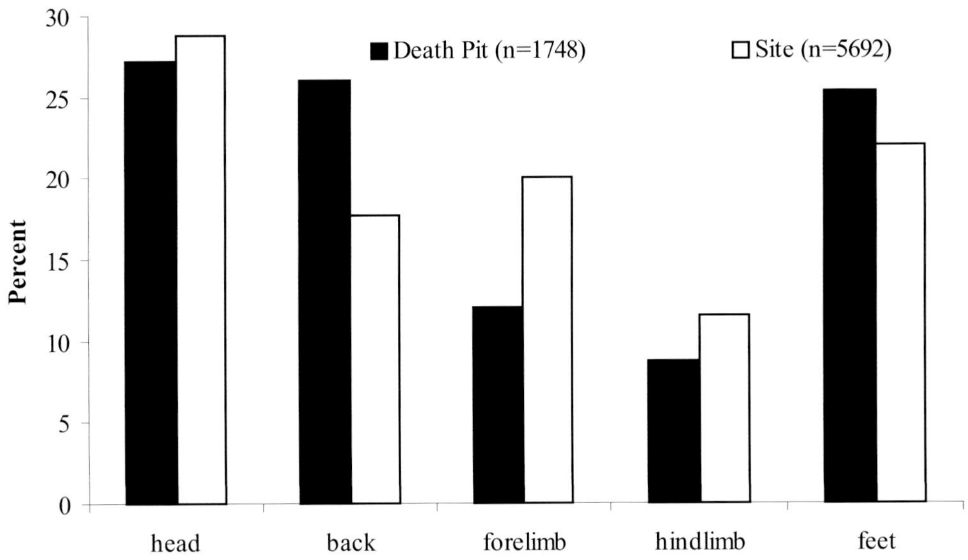

Fig. 11. Distribution of body parts for all taxa at Domuztepe.

	Death Pit				Rest of Site
	Fill AB	*Fill C*	*Fill D*	*Fill E*	
Cattle					
Head	9%	19%	17%	30%	23%
Back	23%	34%	22%	26%	19%
Fore limb	0%	8%	6%	4%	13%
Hind limb	11%	9%	3%	11%	11%
Feet	57%	30%	52%	29%	34%
Total Cattle	*35*	*184*	*102*	*27*	*1192*
Sheep / Goat					
Head	44%	40%	44%	42%	35%
Back	19%	12%	13%	12%	16%
Fore limb	11%	14%	9%	21%	21%
Hind limb	7%	13%	14%	15%	14%
Feet	19%	21%	19%	9%	14%
Total Sheep / Goat	*28*	*198*	*247*	*41*	*2601*
Pig					
Head	55%	36%	30%	20%	34%
Back	11%	2%	12%	7%	10%
Fore limb	11%	24%	23%	27%	21%
Hind limb	0%	2%	10%	27%	11%
Feet	22%	36%	25%	20%	24%
Total Pig	*9*	*42*	*40*	*15*	*1362*

Fig. 12. Body part representation for each major taxon in the various fills of the Death Pit and in the rest of the site. Key to body parts: Head: cranial bones, mandibles, loose teeth; Back: ribs, vertebrae; Fore limb: scapula, humerus, radius, ulna; Hind limb: pelvis, femur, tibia, fibula; Feet: carpals, tarsals, metapodials, phalanges

present. Indeed, burning on Death Pit bones is light and ashy, likely resulting from association with ash rather than direct burning from processing.

Further evidence for the mundane nature of animal processing for the Death Pit deposits comes from an examination of body part representation. All taxa show few differences in the carcass parts present in both the Death Pit and the site (Fig. 11). There are more back bones (that is, ribs and vertebrae) in the pit, in particular a high number of cattle ribs. This is most likely due to the protected nature of the deposit, which would favor fragile ribs and vertebrae better than the exposed contexts of the rest of the site. Considering body part representation in the various fills of the Death Pit (Fig. 12), the one outstanding fact is the great number of cattle foot bones in the earliest fills. Half of Fill AB comprises cattle bones, and 60% of these are foot bones, perhaps reflecting the initial butchery refuse before the cattle carcasses were processed further. The remaining 40% of the cattle bones in this context are bones from the meaty back (ribs and vertebrae) and hind limb. A mere 9% of the cattle bones in Fill AB are head bones. The expected number for a complete skeleton would be much higher than this. Fills C through E have a more typical body part representation, one that would be expected as if whole carcasses were being processed and disposed in this context.

Thus, it appears that there is little preference for carcass parts in the Death Pit, except for an initial concentration of foot bones in the lowest layer of the pit. On the whole, body part use is similar to the rest of the site and indicates that whole animals were butchered and disposed of in the vicinity. Cut marks on bones provide further support for "normal" carcass processing. While there are over twice as many cut marks in the Death Pit as in the site, this is likely due to preservation since accretions obscure marks on many of the bones from elsewhere in the site. 80% of the cut marks in both the Death Pit and the site are on cattle and sheep/goat bones. Only 3% of the cut marks in the pit are on pig bones, as opposed to 11% on pig bones in the site, a difference largely explained by the paucity of pig bones in the pit. Similar cut marks on the Death Pit and site bones indicate that these were probably not sacrificed animals. While there are more complete and articulating bones in the pit, they nevertheless appear to have been butchered and generally fragmented in a similar way to the bones from the site.

The secure nature of the Death Pit is also reflected in the number of articulations, where twice as many of the bones in the Death Pit articulate or pair than those in the site. The presence and types of cut marks indicate that animals were, indeed, butchered for consumption; that is, butchery marks indicate a spectrum of activities, from skinning to disarticulation to chopping into smaller parts. Therefore, the articulating bones are most likely refuse from butchery, where the discarded bones were still held together by skin and tendons. Indeed, all of the articulating cattle bones from the Death Pit are foot bones (non-meaty parts that would have been discarded in the initial butchery process). Many of the sheep/goat articulations are foot bones and vertebrae, but with a number of fore limb articulations as well (specifically, the radius and ulna). A particularly outstanding context is Fill AB, where 9 of the 35 bones articulate or pair (left and right elements of the same animal), and represent a minimum number of three cattle. This high occurrence of articulating or pairing elements is proof that Fill AB was left more or less undisturbed (untouched by scavengers) and was covered over rather quickly so that one fourth of the bones remained in articulation. There is no doubt the rest of the site would show similar numbers of articulating elements with a similarly protected context. Indeed, most of the cattle foot bones from the Death Pit and from the rest of the site are complete, suggesting that they were normally discarded in both areas (or, if processed, they were not processed to the extent that it was destructive to the bone). In sum, the animals involved in the creation of the Death Pit were selected primarily for their value; specifically, cattle and prime-age females were chosen whose milk and breeding would be sorely missed. Numerous lines of evidence suggest that, after this specific choice of animals, the Death Pit and the rest of the site show comparable carcass processing activities.

Discussion: evidence for ritual at Domuztepe

Zooarchaeological analyses on the Domuztepe Death Pit compared to the rest of the site indicate the following:

- Fragmentation and butchery evidence show that people generally processed the animals in the same way as usual, indicating food preparation and consumption rather than strictly sacrifice of animals. A higher number of butchery marks as well as a higher number of complete bones in the Death Pit are likely a result of better preservation in the pit. However, the presence of more complete bones and more articulations might also reflect different cooking and/or consumption processes for this special event. Indications such as these would accord with ethnographic parallels where ritual consumption is often distinguished from the mundane by special preparation or consumption practices.
- People preferred certain taxa over others for the food activities associated with the formation of the Death Pit than they did in their day to day lives.
- People selected adult female sheep/goat and cattle four times as frequently as males, indicating a special emphasis on valuable animals (prime-age milking and breeding animals).
- In the case of cattle it appears that people butchered whole animals and put all the bone refuse in the Death Pit. Whether this is a result of better preservation in the pit remains to be seen.

North section		South section	
Fill E	*Burning: Obscuration* Few Human: Some Cattle: Many Sheep/goat: Some Pig *Burial* Many Human		
Fill C	*Consumption: Burial: Construction* Many Human: Many Cattle: Many dog: Some Sheep/goat: Few Pig	Fill D	*Consumption: Mundane: Construction* No Human: Some Cattle: Many Sheep/goat: Few Pig
Fill A/B	*Butchering: Signification of place* No Human: Many Cattle: Some Sheep/goat: Few Pig		

Fig. 13. Possible functional and ritual associations of animal remains and zones and processes within the Death Pit.

- Of special note is the role of cattle in the formation of the Death Pit assemblage. Cattle (probably mostly adult females) were preferentially selected for this feast. They appear to have had a special significance to the people of Domuztepe, an observation that may be strengthened by the numerous sherds with bucrania as a central motif.
- Cattle butchery debris was substantially placed in the Death Pit amongst the earliest deposits.
- A high proportion of dog bones in the pit, as well as their concentration in one single layer of the pit, suggests a different treatment of dogs than other food taxa. The association with human remains may be significant.
- Pig were largely excluded from particular contexts within the Death Pit, specifically those in which cattle were highly prominent and those laid down when human remains were being placed within the Death Pit.

In sum, zooarchaeological analysis on the Domuztepe Death Pit assemblage indicates special selection of particular taxa and sexes, with fewer differences in butchery and consumption patterns compared to everyday life. When we consider the very large number of animal and human bones in the pit, together with a short depositional history, these results largely point to feasting in association with whatever burial rite took place to create this deposit.

Parker Pearson (2001, 194) recently stated that, "Ritual is not something altogether removed from the everyday, but is situated in constant reference to other daily and seasonal activities… drawing on aspects of everyday life, but through more formal conventions". Indeed, it appears that food was a significant part of the ritual associated with the pit at Domuztepe, and that everyday methods were used to prepare this food which was then used in a special way, laden with different meanings than the daily meal. Conversely, some of the special meanings attached to the consumption of food within the Death Pit ritual were probably also present, although perhaps in less formalized form, in much more mundane but still very meaningful contexts of consumption.

Different episodes within the Death Pit were marked by the use of different species and parts of animals. In each case, a set of activities, ritual meanings and distinctive selection and use of animals are identified with each other. The animal remains do not simply represent food being consumed within a ritual but show animals being situated within a much wider ritually constructed view of the world. Although greater analysis is needed, and in particular more detailed integration with other aspects of the Death Pit, it is possible to suggest some of the possible associations (Fig. 13).

Conclusions

We would like to suggest one possible interpretation for the Domuztepe death pit, with the ethnographic parallel from North America, documented by French missionaries in the 17th century. The Huron "Feast of the Dead" in southern Ontario involved a large gathering of many villages once every 8–10 years. For this gathering, people would disinter their dead from the past decade, deflesh them, burn the flesh, and wash the bones. They would then bring the bones in bundles to a central place where the Feast of the Dead was to take place. Among much feasting and dancing over several days, a large pit was prepared, a scaffold built, and the bundles of bodies hung from the scaffold. From this scaffold the contents of the bundles were eventually emptied into the communal burial pit, the bones mixed up, and buried. This ritual of communal burial was seen as reinforcing social ties and notions of identity among people of the same tribe and neighboring regions (Trigger 1969, 108). Tooker (1967, 139–40) explains, "By means of this ceremony the Indians confirmed their friendships, saying that as the bones of their deceased relatives and friends were united in one place, so they would live together in the same unity and harmony".

This is just one example of numerous and unpredictable ways people treat their dead and, whether consciously or not, use this treatment of the dead to strengthen ties, legitimize power, or emphasize distinctions or status

(Pearson 2001, 194). It seems clear that the Death Pit was far more than simply concerned with the disposal of human remains. It was the culmination of a much more extensive process through which social relations were defined, renegotiated and restated. In this, the role of animals, both symbolically and as food, would appear to have been an important one. At a very large settlement such as Domuztepe, the importance of communal ritual events may have been particularly important as a means of integrating large numbers of people. We may, however, move further. The Death Pit is not simply funerary; it also integrates a wider range of ways of seeing and rationalizing the living and supernatural worlds. The general exclusion of sheep/goat from the main Death Pit contrasts strongly with the appearance of sheep/goat in large numbers along with cow in the Fill D to the south, which was placed as part of the same ritual process. This might suggest a conscious binary divide in which 'domestic' and 'mundane' material (the terms are used loosely) is juxtaposed with use of cow and dog remains within the area of human burial, albeit in highly specific positions. We might tentatively propose this as evidence for conceptual divisions within a Domuztepe worldview. The limited inclusion of pig in the Death Pit until the final Fill E which obscured the whole area with ash is equally significant in indicating the way in which different animals were appropriate within particular ritual zones and activities. The Death Pit does not simply allow us to examine funerary processes but also allows us to approach the ways in which animals and their appropriate roles fit into a much wider cosmology. The role of animals as a suitable subsistence resource is something which is a cultural as well as a nutritional construct, and the Death Pit at Domuztepe provides one exciting avenue into a broader understanding of the position of animals in the prehistoric world.

References

Boessneck, J. 1969. Osteological differences between sheep (*Ovis aries* Linné) and goats (*Capra hircus* Linné), pp. 331–58 in Brothwell, D. and Higgs, E. S. (eds), *Science in Archaeology* (2nd edition). London: Thames and Hudson.

Grant, A. 1991. Economic or symbolic? Animals and ritual behaviour, pp. 109–14 in Garwood, P., Jennings, D., Skeates, R., and Toms, J. (eds), *Sacred and Profane: Proceedings of a Conference on Archaeology, Ritual, and Religion.* (Oxford University Committee for Archaeology, Monograph 32). Oxford: Oxbow Books.

Horwitz, L. K. 1987. Animal offerings from two Middle Bronze Age tombs. *Israel Exploration Journal* **37**, 251–5.

Pearson, M. P. 2000. *The Archaeology of Death and Burial*. College Station: Texas A&M University Press.

Payne, S. 1973. Kill-off patterns in sheep and goats: the mandibles from Asvan Kale. *Anatolian Studies* **23**, 281–303.

Tooker, E. 1967. *An Ethnography of the Huron Indians, 1615–1649*. Midland, Ontario: Huronia Historical Development Council.

Trigger, B. G. 1969. *The Huron: Farmers of the North*. New York: Holt, Rinehart, and Winston.

Sarah Whitcher Kansa
Peabody Museum, Harvard University,
11 Divinity Ave., Cambridge, MA 02138, USA
E-mail: skansa@alexandriaarchive.org

Stuart Campbell
Department of Art History and Archaeology,
University of Manchester, Oxford Road,
Manchester M13 9PL, UK
E-mail: stuart.campbell@man.ac.uk

9th ICAZ Conference, Durham 2002
Behaviour Behind Bones, (ed. Sharyn Jones O'Day et al.) pp. 14–19

2. Animal offerings found in Necropoleis belonging to Santana of Mures-Cerniahov culture from the east and the south extra-Carpathian Zones of Romania

Simina Stanc and Luminita Bejenaru

Our paper discusses the frequency of the burial rites, both inhumation and cremation, and the types and characteristics of the animal offerings found in the inhumation tombs from the twelve Santana of Mures -Cerniahov necropoleis in Romania. The Santana of Mures-Cerniahov culture spread over an area of more than one million square km, including the eastern and southern extra-Carpathian regions of Romania. The expansion of this culture over such a massive area was likely related to the invasion and extension of the powerful Goths during the 4th century AD.

Introduction

Our paper discusses animal offerings in the Santana of Mures-Cerniahov culture during the 4th century AD in Romania's history. Ethnically, this culture is a heterogenous one, a mixture of Germanic, Geto-Dacian and Sarmatic populations. Geto-Dacians populations occupied the Carpatho-Danubian basin, living in the territory named Dacia. They belonged to the North-Thracian branch of the Thracian tribes living east of the North Balkan Peninsula. The Geto-Dacians' separation into the north branch of the Thracian tribes took place at the beginning of the Iron Age. The Geto-Dacians had a state form of social organization as early as the 1st century BC, which was characterized by a tributary type of exploitation (Tarcan and Bejenaru 2001).

In the year AD 106 part of Dacia (the Transylvania, the Oltenia and the Banat regions) became various Roman provinces when the area was conquered by the emperor Traian. The three other regions, Muntenia, Moldova and Maramures, were not occupied by the Romans. In these areas the Dacians lived beyond Roman political control. During the following two centuries, the relations between the free Dacians and the Romans was characterized by both peaceful periods, when commercial exchanges took place, and times of social conflict. The Dacians generally led a sedentary life focused on agricultural production.

The free Dacians were influenced from neighboring Sarmatic and Germanic population (the Vandals and the Goths). During the entire period of the Roman occupation of the Lower North Danube territories, the Sarmatians infiltrated themselves in the territories of the free Dacians, especially in the steppe and silvosteppe zones. The Sarmatians belonged to the western branch of the Iranian populations and their existence continued for about eleven centuries (from the 7th century BC until the 4th century AD). At first they spread over and occupied southern regions of the Ural mountains and areas around the Ural and Volga rivers. These people were nomadic and are known to have been stock-breeders. Architecturally, necropoleis are the only archaeological evidence of the Sarmatians in Romania.

After the withdrawal of the Roman army and administration in AD 274–275 from Dacia, the new political force that asserted itself in the Lower North Danube region was that of the population of Goths. The Goths belonged to the eastern branch of the Germanic tribes, to which the vandals also belonged, and they were settlers in a region south of the Baltic Sea (300 BC – 200 AD). Their domination all over Eastern Europe caused major transformations, which brought about a considerably extended process of acculturation. The Santana of Mures-Cerniahov Culture is an example of the outcome of these interactions. This culture's name comes from discoveries at the sites of Santana of Mures, Transylvania and Cerniahov, in the Ukraine (middle Dniper). This culture is extremely widespread; it includes vast territories in Romania and Ukraine. Its most southern limit reaches the Lower Danube and the Black Sea shore. In the east it occupied the Lower Dniper

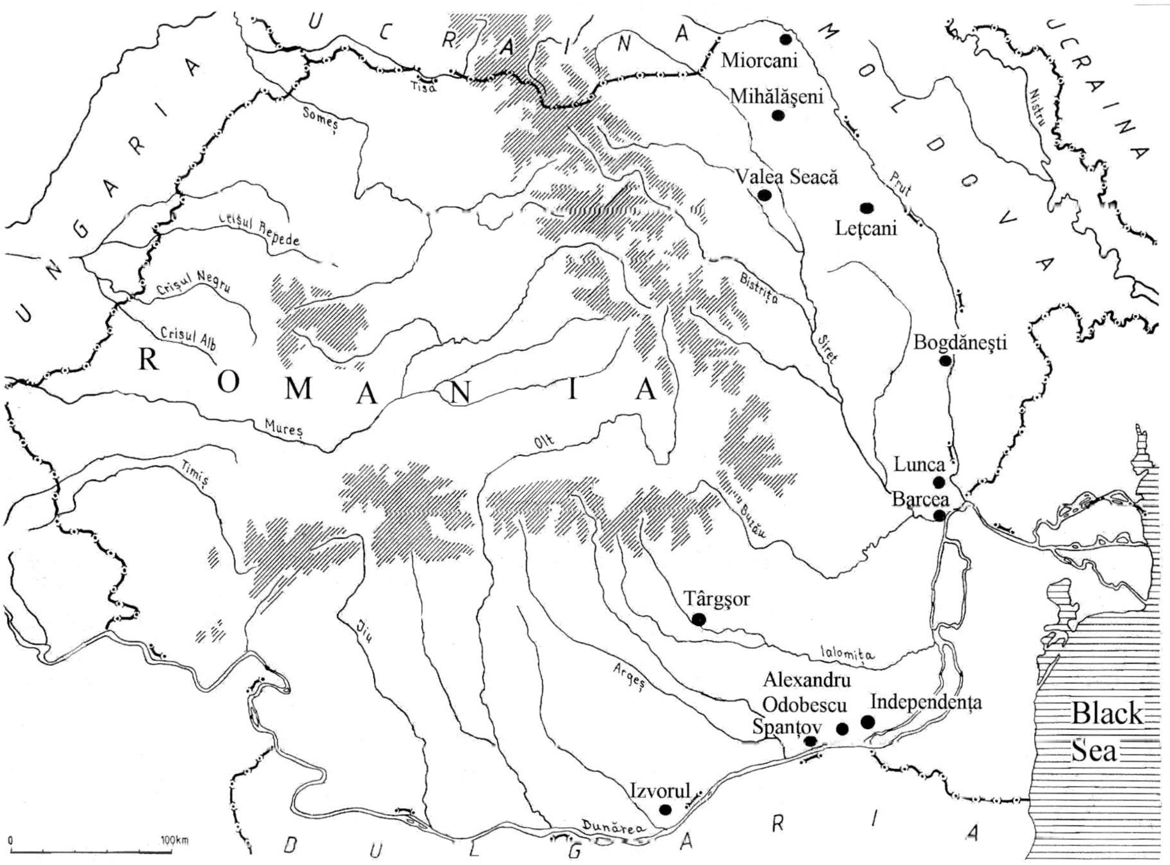

Fig. 1. Map of Romania showing the Santana of Mures-Cerniahov necropoleis.

region, while in the north-east it occupied the whole of middle Dniper basin and the upper part of the north Donet. Its northern limit reaches the Sejm and Desna rivers, up to the surroundings of Kiev City. To the west, it reaches west of the Bug river after which it spreads southwards to the Carpathians (the eastern and southern Carpathians) and to the Danube.

The Santana of Mures-Cerniahov material culture remains present astonishing unitary characteristics in spite of the culture's spreads over such an extended area. There are, of course, local variations; one of the most frequently used criteria in distinguishing the local variants is the hand-made ceramics. Variations in ceramics are argued to evidence particular ethnic groups within the Santana of Mures-Cerniahov culture. The presence of the Sarmatians within this culture is particularly evident by specific burial practices and objects of Sarmatic tradition. Finally, pressed by the invasion of the Huns, the Goths moved to the south of the Danube and the Santana of Mures-Cerniahov culture ceased to exist at the beginning of the 5th century AD (Ionita 2001).

Our research explores zooarchaeological samples extracted from twelve necropoleis of the Santana of Mures-Cerniahov culture studied by different zooarchaeologists. Seven of the assemblages are from the east (Moldova) of Romania and five are from the south (Muntenia) (Fig.1). The bone assemblage recovered from Miorcani were studied by the authors.

Rites of burial

The necropoleis of Santana of Mures-Cerniahov culture were placed close to residential areas although some of these structures were very large and contained many tombs (*e.g.*, Valea Seaca, Targsor, Mihalaseni). Both incineration and inhumation were practiced in all the necropoleis. In the early periods of the culture, incineration was practiced on a larger scale than inhumation, but gradually the latter practice became more frequently used, especially towards the final stages of the culture.

Incineration took various forms, depending on the shape of the grave, the quantity of remains, the state and the position of the earthen urns, and the type of lid used. Usually, the incineration graves containing urns, with or without lids, are by far more frequent than the calcinated remains burned directly in simple graves (Ionita 2001).

Three variants of inhumation burials have been discovered. Inhumation type depends on the form and the orientation of the graves and on the position of the

Necropolis	References	Inhumation		Cremation		Cenotaph	Total
		No	%	No	%		
Izvorul	Mitrea and Preda 1966	31	96.88	1	3.12	–	32
Independenta	Mitrea and Preda 1966	27	77.14	8	22.86	–	35
Al. Odobescu	Mitrea and Preda 1966	15	68.18	7	31.82	–	22
Spantov	Mitrea and Preda 1966	56	81.16	10	14.49	3	69
Targsor	Diaconu 1965	157	59.02	109	40.98	–	266
Letcani	Blosiu 1975 Enciclopedia (II) 1996	32	66.67	17*	33.33	–	48
Valea Seaca	Palade 1986	252	46.07	295	53.93	–	547
Bogdanesti	Enciclopedia (I) 1994	92	63	54	37	–	146
Miorcani	Ionita 1974	121	90.3	13	9.7	–	134
Mihalaseni **	Haimovici 1988 Haimovici 1989	–	–	–	–	3	516
Barcea	Enciclopedia (I) 1994	119	71.26	48	28.74	–	167

*Fig. 2. Comparison of the two burial rites for some necropoleis in the south and east of Romania. * tomb 37b, a cremation, was found inside the inhumation tomb 37a; ** 516 is the total number of excavated tombs, from which 140 contained animal offering (123 inhumation tombs, 14 cremation tombs and 3 cenotaph).*

skeleton. The first type includes graves that are oriented in the general north-south direction and containing flexed skeletons. The second group includes graves that differ only in the position of the hands and of the legs, which are straight (non-flexed). The third variant, which is less frequently found, contains west-east oriented graves, with non-flexed burials (legs and the arms straight along the body or with arms on the chest or on the pelvis). All three categories of inhumation graves are usually found together inside the same necropolis and only rarely occur in separate areas (Ionita 2001).

The diversity in the practice of burial rites may be explained by the presence of different ethnic groups within the larger cultural population. Variations are expected to begin with acculturation process (from the end of the 3rd century to the beginning of the 4th century AD). Basic differences between the funerary practices of the free Dacians and the Sarmatians populations certainly exist. While the Sarmatians inhumed their dead, the Dacians incinerated their dead in most instances (they practiced inhumation only for some children and adolescents). Sarmatians occasionally continued to practice, even in the new territories, artificial distortion of the skull as a sign of ethnic distinction, which they considered to be beautiful; this practice was not acquired by the free Dacians (Ionita 2001).

The inhumation rite is by far the most common rite. There is an exception in the Valea Seaca necropolis only, where the number of cremation burials is larger than the number of inhumation burials (Fig. 2). Unfortunately, we do not know the total number of excavated tombs at Lunca necropolis.

Animal offerings

The analysis of the animal remains extracted from seven inhumation burials in Miorcani necropolis provided valuable data: the species used in the offerings included *Ovis aries/Capra hircus* (in three tombs), *Gallus domesticus* (in one tomb) and *Sus domesticus* (in three tombs). All of the assemblages of sacrificial material that contained remains of only one species are referred to as simple offerings. Six of the assemblages were simple offerings. One assemblage consists of bones of two individuals of *Ovis/Capra*. Generally, offerings consist of remains of young animals.

The inventory of the inhumation burials of Santana of Mures Cerniahov necropoleis is diverse. Under the title "animal offerings" we included both offerings of sacrificed animals and of objects of animal origin (artifacts of bone and shell).

Not all the tombs contain sacrificed animal offerings. The number of inhumation graves with sacrificed animal offering represents 37.5% of the total inhumation graves in Letcani, 33.3% in Alexandru Odobescu, 29% in Izvorul, 20.6% in Bogdanesti (Fig. 3). We do not know the total number of inhumation tombs for Lunca and Mihalaseni necropoleis.

We determined NISP and MNI from the bone assemblages of Spantov, Lunca, Targsor, Letcani, and Miorcani. For the bone assemblages extracted from Valea Seaca, Bogdanesti, Barcea, Al. Odobescu, Independenta, Izvorul and Mihalaseni, only MNI data was available. Unfortunately, the initial research on these assemblages did not specify the exact number of remains for each species from each tomb (Fig. 4). However, some patterns are apparent. For example, all burial tombs with sacrificed animal offerings contained an archaeological inventory of artifacts. Also, the domestic animals that were offered in this burial custom are: *Bos taurus, Ovis aries/Capra hircus, Sus domesticus, Equus caballus, Canis familiaris, Gallus domesticus, Anser* sp. The sacrificed remains from wild mammals include: *Bos primigenius, Cervus elaphus, Sus scrofa, Lepus europaeus,* and fish (*Cyprinus carpio,* and *Carassius carassius*). In all necropoleis, the most

Necropolis	References	Total inhumation tombs	Tombs with animal offering	
Independenta	Mitrea and Preda 1966	27	7	25.9%
Al. Odobescu	Mitrea and Preda 1966	15	5	33.3%
Izvorul	Mitrea and Preda 1966	31	9	29%
Spantov	Mitrea and Preda 1966 Bolomey 1967	56	13*	23.2%
Targsor	Bolomey 1967 Diaconu 1965	157	22**	14%
Lunca	Bolomey 1967	–	8	–
Mihalaseni	Haimovici 1988 Haimovici 1989	–	123	–
Letcani	Blosiu 1975 Haimovici 1975	32	12	37.5%
Valea Seaca	Haimovici 1994 Palade 1986	252	35	13.8%
Bogdanesti	Haimovici 1994 Enciclopedia (I) 1996	92	19	20.6%
Barcea	Enciclopedia (I) 1994 Haimovici 1988	119	13	10.9%
Miorcani	Ionita 1974 Stanc and Bejenaru, unpublished data	121	7	5.7%

*Fig. 3. Frequency of inhumation burials with sacrificed animal offerings. * Mitrea and Preda (1966) reported thirteen tombs with animal offerings; Bolomey (1967) analyzed 14 assemblages of animal offerings. ** The animal offerings from 6 tombs were not analyzed due to poor preservation.*

Species	Spantov		Lunca		Targsor		Letcani		Miorcani		Barcea	Valea Seaca	Bogdanesti	Alexandru Odobescu	Izvorul	Independenta	Mihalaseni
	NISP	MNI	NISP	MNI	NISP	MNI	NISP	MNI	NISP	MNI	MNI	MNI	MNI	MNI	MNI	MNI	MNI
Ovis aries	128	7	77	3	112	8	97	8	–	–	6	16	4	–	–	–	43
Capra hircus	–	–	–	–	–	–	–				1	–	–	–	–	–	3
Ovis/Capra	1	1	–	–	16	3	?	2	14	3	3	12	9	3	4	6	21
Sus domesticus	23	2	11	2	12	1	–	–	31	3	–	2	2	–	–	–	18
Sus domesticus ?	–	–	–	–	–	–	–	–	7	–	–	–	–	1	–	–	–
Bos taurus	1	1	8	2	6	3	–	–	–	–	1	2	6	1	1	–	29
Bos taurus ?	–	–	–	–	–	–	–	–	–	–	–	–	–	–	–	–	1
Equus caballus	–	–	–	–	–	–	–	–	–	–	–	–	–	–	–	–	6
Sus scrofa	–	–	–	–	–	–	–	–	–	–	–	–	–	–	–	–	1
Sus dom/Sus scrofa	–	–	–	–	–	–	–	–	–	–	–	–	–	–	–	–	1
Bos t./Bos primig.	–	–	–	–	–	–	–	–	–	–	1	–	–	–	–	–	2
Cervus elaphus	–	–	–	–	–	–	–	–	–	–	–	1	–	–	–	–	–
Lepus europaeus	1	1	–	–	–	–	–	–	–	–	–	–	–	–	–	–	–
Gallus domesticus	–	–	–	–	?	2	20	3	5	1	4	13	2	–	–	–	37
Gallus domesticus ?	–	–	–	–	–	–	–	–	25					–	–	–	–
Anser sp.	?	?						–	–	–	–	–	–	–	–	–	–
Aves	?	2	?	2	?	1	–	–	–	–	–	–	–	2	4	2	–
Cyprinus carpio	–	–	–	–	–	–	–	–	–	–	–	–	–	–	–	–	1
Carassius carassius	–	–	–	–	–	–	–	–	–	–	–	–	–	–	–	–	1
Pisces	–	–	–	–	–	–	–	–	–	–	–	1	–	–	–	–	–
Bone indet.	1	–	–	–	–	–	–	–	–	–	–	–	–	1	1	–	–
Eggs shell	–	–	–	–	–	–	*	–	–	–	–	–	–	–	–	–	*
Total	155	15	96	9	146	18	117	13	82	7	15	47	23	8	10	8	164

Fig. 4. Quantification of the animal remains from the necropoleis. ? – unknown number of remains.

frequently represented animals are *Ovis/Capra*, based on estimates of minimum number of individuals (MNI). Following ovicaprines in abundance are birds; birds were found in most of the necropoleis (Spantov, Letcani, Barcea, Alexandru Odobescu, Iyvorul, Independenta).

Remains from pigs were not found in the samples from Letcani, Barcea, Independenta, Izvorul. In the assemblages that contain pig bones (Spantov, Lunca, Letcani, Targsor, Bogdanesti, Mihalaseni), these occur in a small proportion compared with remains of *Ovis/Capra*. Cattle bones (*Bos taurus*) were not identified among the samples from Independenta, Miorcani and Letcani. *Bos taurus* offerings are generally identified less frequently than ovicaprines; when cattle bones do occur they are represented by isolated pieces, as a result it is unclear if certain parts of the animal body were preferred.

Some graves contained the bones from two individuals of the same species. For example, two MNI of *Ovis aries* were identified in one tomb at Spantov and in three tombs at Mihalaseni; *Ovis/Capra* was found in one tomb at Miorcani. One tomb at Mihalaseni contained multiple individuals of *Bos taurus*, as did two tombs at Mihalaseni.

Age	Targsor	Spantov	Bogdanesti	Letcani	Barcea	Valea Seaca	Mihalaseni
0–1 year	7	5	5	5	4	14	38
1–2 years	–	–	2	1	1	5	3
2–3 years	–	–	–	–	–	–	1
3–4 years	–	–	–	–	–	–	2
4–5 years	–	–	–	–	2	1	–
8–10 years	–	–	–	–	–	–	1

Fig. 5. Age of sacrifice (dental age) for Ovis/Capra *(MNI).*

Two individuals of *Gallus domesticus* occurred in four tombs at Mihalaseni.

Simple offerings predominate in all necropoleis, with the exception of Alexandru Odobescu. These simple offerings represent 100% at Miorcani, 91.6% at Letcani, 88.8% at Izvorul, 87.5% at Lunca and Targsor, 78.9 % at Bogdanesti, 71.4% at Valea Seaca, 70.7% at Mihalaseni. Second, are the double offerings, which represent 23% at Barcea, 22.7% at Mihalaseni, 22.8% at Valea Seaca, 21% at Bogdanesti, 14.3% at Independenta, 8.3% at Letcani.

In most parts of the necropoleis, the offerings were of a simple and double type, as we found at Independenta, Izvorul, Targsor, Lunca, Letcani, Bogdanesti and Barcea necropoleis (Fig. 3). Three kinds of offerings (simple, double and triple) were found in the tombs in Valea Seaca (5.7% of the offering tombs were triple offerings) and at Spantov (7.7% of all the tombs with animal offerings were triple). The Mihalaseni necropolis had the most diverse offerings, concerning the kind of offerings (there were found simple, double, triple and even four species content) and of the variety of the identified species.

In all cases the preferred offerings were sheep/goat, especially the young ones, below one year of age (Fig. 5). The studies for Lunca, Izvorul, Al. Odobescu, Independent bone assemblages do not offer data about the age of the sacrificed animals. Alexandra Bolomey (1967), studying the offerings from the cemeteries in the south of Romania, argues that there was a predilection of the Santana of Mures-Cerniahov populations to use young sheep (under one year of age). This might have been based on a religious belief, from a practical point of view it would have been much more convenient to sacrifice older or sick animals (Bolomey 1967). Pieces of all parts of the body were offered, but those with little meat on them predominated. Among the sheep/goat bones, the skulls and the extremities were found in a much greater numbers. Remains from wild species were identified at Mihalaseni (*Cyprinus carpio*, *Carassius carassius*, *Sus scrofa* and probably *Bos primigenius*), at Valea Seaca (*Cervus elaphus* and *Cyprinus* sp.) and at Spantov (*Lepus europaeus*).

Offerings consisting of objects of animal origin

Objects in this category include domestic items, such as combs, awls, bone tubes, and ornaments including beads and pendants. We have no data concerning the objects of animal origin found in tombs at Mihalaseni. From the tombs where we do have information, combs made of red deer antlers were the most common bone objects found (Fig. 6). All necropoleis contained one or more combs. Usually, the combs consist of two main parts: the segments with teeth and two semicircular plates forming the handle. The segments with teeth are five or eight in number and they are compressed to the two semicircular plates by bronze or iron rivets (Mitrea and Preda 1966). Metal fibulas and buckles are also common objects found in the inventories of the tombs of this culture.

Comb offerings were found in children's, women's and in men's tombs. There was no standard place where combs were placed within the tombs. At the necropoleis of Valea Seaca, Bogdanesti, Targsor, Barcea, and Lunca, we do not know the exact number of the tombs with combs, but the inventories included these objects. Generally, just one comb was deposited in a tomb, but there are cases when two combs were found in a tomb (*e.g.*, in four tombs at the Alexandru Odobescu).

Bone tubes for needles were found. They were made of long bird bones, generally the femur. They are oblong in shape and circular in section, their length being of about 10–12 cm. This kind of tubes were found at the necropoleis in Izvorul, Independenta, Alexandru Odobescu, Targsor, Lunca and Letcani. Pendants were usually manufactured from wild boar fangs. An ornamented and pierced canine was recovered from Alexandru Odobescu. At Letcani a small tubular bone connected with a bronze pendant was found, and at Targsor and Al. Odobescu prismatic pendants made of antlers were inventoried (Diaconu 1965). Chain links made of burrs of red deer antlers were found in the tombs of the necropoleis in Letcani, Targsor, Barcea and Valea Seaca.

In addition to the objects we have already described, in some necropoleis a variety of shells (both bivalves and gastropods) were recovered. Some of the shells were perforated and for this reason, they were considered pendants. Bivalves of *Cardium* sp., and gastropods of *Cypraea* sp. were discovered in the necropoleis in Letcani, Bogdanesti, Spantov, Independenta and Alexandru Odobescu. The Sarmatic populations commonly wore cowry shell belts (Blosiu 1975).

The position of different types of offerings in the tombs exhibits patterning. Combs were placed around the head and the meat offerings were positioned at the end of the legs while bone tubes were found between the legs (around

Necropolis	References	Tombs with domestic objects			Tombs with body ornaments			
		comb	bone tube	awl	shell	pendant	chain link	bone bead
Independenta	Mitrea and Preda 1966	11	2	–	3	–	–	1
Al. Odobescu	Mitrea and Preda 1966	4	2	–	2	3	–	–
Izvorul	Mitrea and Preda 1966	7	1	–	–	–	–	–
Spantov	Mitrea and Preda 1966	10	–	–	3	–	–	–
Targsor	Diaconu 1965	+	+	–	+	+	+	–
Letcani	Blosiu 1975	18	1	1	2	1	1	1
Valea Seaca	Palade 1986	+	+	–	–	–	+	–
Bogdanesti	Enciclopedia (I) 1994	+	–	–	+	+	–	–
Barcea	Enciclopedia (I) 1994	+	–	–	+	+	+	+
Lunca	Enciclopedia (II) 1996	+	+	–	+	+	–	–
Miorcani	Ionita 1974	32	–	–	–	+	–	–

Fig. 6. Objects of animal origin found in the inhumation tombs. + included in the inventory; – not found in the tomb.

the femurs and knees) at some necropoleis in the south of Romania (Mitrea and Preda 1966). At the Letcani cemetery, meat offerings were situated in the head and leg regions; three child tombs were an exception to this pattern.

Conclusions

Our analysis have yielded a number of interesting findings:

- In the Santana of Mures-Cerniahov necropoleis the inhumation burial custom is better represented then the cremation burial custom.
- Most of the animal offerings consisted of a single species, but there were also double and triple offerings.
- The most commonly offered species were sheep, especially the young ones (below one year of age).
- No correlation was found between the type and the quantity of the animal offerings and the age or sex of the deceased.

References

Blosiu, C. 1975. Necropola din secolul al IV–lea e.n. de la Letcani (jud. Iasi). *Arheologia Moldovei* **8**, 203–80.

Bolomey, A. 1967. Ofrande animale in necropole din secolul al IV-lea e.n. *Studii si cercetari de antropologie* **4** (**1**), 25–35.

Diaconu, G. 1965. *Targsor, necropola din secolele III–IV e.n.*. Bucuresti. Editura Academiei Romane.

Haimovici, S. 1975. Studiul resturilor faunistice depuse ca ofranda in cimitirul de la Letcani (jud. Iasi). *Arheologia Moldovei* **8**, 287–91.

Haimovici, S. 1988a. L'étude des restes animaliers déposés comme offrande dans les tombes de la nécropole de la Barcea (IV – V siecle n.è.) appartenant à la culture Santana de Mures – Cerneahov. *Analele Stiintifice ale Universitatii "Al. I. Cuza"* Iasi **34, II**, 59–60.

Haimovici, S. 1988b. Studiul materialului paleofaunistic depus ca ofrande in mormintele din necropola birituala de la Mihalaseni (jud. Botosani) apartinand culturii Santana de Mures (sec. IV e.n.). *Hierasus* **7-8**, 235–58.

Haimovici, S. 1989. Studiul materialului paleofaunistic depus ca ofrande in mormintele din necropola birituala de la Mihalaseni (jud. Botosani) apartinand culturii Santana de Mures (sec. IV e.n.) (partea a II – a). *Hierasus* **9**, 195–228.

Haimovici, S. 1994. Studiul resturilor faunistice depuse ca ofrande in mormintele din doua necropole – Valea Seaca si Bogdanesti (jud. Vaslui) – apartinand culturii Santana de Mures. *Carpica* **25**, 141–58.

Ionita, I. 1974. Necropola din secolul IV e.n. de la Miorcani (jud. Botosani). *Cercetari istorice* **5**, 81–92.

Ionita, I. 2001. Populatia locala in secolul al IV–lea in regiunile extracarpatice. *Istoria Romanilor* **2**, 617–37.

Mitrea, B. and Preda C. 1966. *Necropole din secolul al IV–lea e.n. in Muntenia*. Bucuresti: Editura Academiei Romane.

Palade, V. 1973. Sapaturile arheologice de la Bogdanesti-Falciu, jud. Vaslui (1967–1968). *Materiale si cercetari arheologice* **10**, 169–89.

Palade, V. 1986. Nécropole du IVe et commencement du Ve siècle de n.è. à Barlad-Valea Seaca. *Inventaria Archaeologica* **12**, 72–92.

Tarcan, C. and Bejenaru, L. 2001. The importance of domestic mammals during the La Tène Period in Romania. *Archaeofauna* **10**, 7–23.

Enciclopedia arheologici si istoriei vechi a Romaniei I. 1994. Bucuresti: Editura Enciclopedica.

Enciclopedia arheologici si istoriei vechi a Romaniei II. 1996. Bucuresti: Editura Enciclopedica.

Simina Stanc and Luminita Bejenaru
Al.I.Cuza University, Faculty of Biology,
11 Carol I Bul., Iasi, 6600, Romania
E-mails: siminams@yahoo.com
lumib@uaic.ro

9th ICAZ Conference, Durham 2002
Behaviour Behind Bones, (ed. Sharyn Jones O'Day et al.) pp. 20–24

3. Caprines and toads: taphonomic patterning of animal offering practices in a Late Bronze Age burial assemblage

Lior Weissbrod and Guy Bar-Oz

*The faunal assemblage from the Late Bronze Age (*c. *1550–1130 BCE) Ara burial cave, Lower Galilee, Israel, consists of large and small animal remains. The small faunal remains were mainly retrieved through fine screening the fill of the ceramic vessels. Sheep/goat remains dominate the large size fauna, with an age structure revealing selection for young individuals. This conforms to the generally accepted interpretation of ritual offerings. However, non-selective body-part representation, together with an absence of articulated bones and their non-association with the vessels suggest* in situ *preparation and consumption. This observation accords with the alternative Bronze Age mortuary practice of honorary ritual banquets. The small fauna is comprised predominantly of anuran remains. Bronze Age small faunal assemblages are typically considered either as intrusive or the product of predator accumulation. This is indicated for the Ara assemblage by the good state of bone preservation and nearly complete skeletal part representation. However, the context of a sealed burial cave, the presence of the material within the vessels, and the anomalous absence of anuran cranial elements, suggest their preparation as ritual food offerings.*

Introduction

The analysis of animal bones from burial sites provides a valuable contribution to our understanding of mortuary ritual practices of past societies. Zooarchaeologists largely recognize the significance of animal bones from such contexts for reconstructing ancient burial customs. Animal bones found in tombs may represent the remains of funerary feasts, food offerings for the deceased (for the afterlife), or gifts to appease the gods (see review and references in Horwitz 2001a). In this paper we explore the possible mortuary ritual significance of the archaeofaunal assemblage recovered from Late Bronze Age (2nd millennium BCE) Ara burial cave, Lower Galilee, Israel. Systematic collection of large faunal remains was complemented with restricted retrieval of small faunal remains from within ceramic vessels. This was accomplished through fine sieving of their sedimentary fills in the laboratory. The assemblage yielded large and small mammals, amphibians, reptiles and fish bone fragments. The presence of animal remains, in association with human remains, raises the issue of their possible mortuary ritual significance. The complete bone report is detailed in Bar-Oz and Weissbrod (*in press*). Here we present a general summary of the fauna and suggest tentative explanations, raised by the analysis of the faunal material, for the established interpretive framework of this type of assemblage.

The tomb is located adjacent to Tel Ara, the site of the modern village of Kfar Ara in the Yiron Valley. This corridor connects Israel's central coastal plain, in the west, with the Jezreel Valley further inland and to the northeast. Tel Ara is situated on the historic international trade route *Via Maris* between Egypt and Syria. Several surveys of the Tel, which extends over approximately four hectares, revealed remnants from most of the Near-Eastern historical periods, beginning with the Middle Bronze Age (*c.* 2000–1550 BCE).

The salvage excavation (directed by Yuval Gadot and Yotam Tepper, Ramot Archaeology of Tel Aviv University), uncovered two adjoined multi-chambered burial caves forming part of the Tel Ara cemetery (for site plan and excavation details see http://www.tau.ac.il/~archpubs/projects/ara.html). Only the faunal material from tomb one was made available to us. The tomb was quarried into a narrow outcrop of soft Senonian chalk, in which all other burial caves in this region were also uncovered. The tomb was discovered sealed. It consists of a large central chamber and five subsidiary rooms. The entrance to the tomb was through a squared shaft continuing into a narrow corridor. Inside the tomb an unusually dense concentration of finds was recovered comprised mainly of a large variety of local and imported ceramic vessels. Other finds included bronze artifacts such as arrow heads, a dagger and fastening pins; various types of beads and bone tools; a female ceramic figurine; and several scarab seals, some of which are inscribed with the name of the Pharaoh Amenhotep III. The human osteological material represents a minimum number of 26 individuals, including male and female adults as well as adolescents (see the above web site for figures and details). The context of the multitude of cluttered finds in the tomb is the result of centuries of its recurrent use. Most likely, this tomb served as a family burial estate being suitable for multiple burials.

Based on the artifact typology the chronology of the tomb's use spans the periods: Intermediate Bronze Age (*c.* 2300–2000 BCE) when the cave was quarried, the Middle Bronze Age I (*c.* 2000–1800 BCE), and the Late Bronze Age II (*c.* 1300–1200 BCE) to which the majority of the finds are attributed (Yuval Gadot, *pers. comm.*). In the course of this time the tomb contents were cleared aside more than once in preparation for its reuse. Consequently, most of the artifacts were discovered piled up in several heaps inside the tomb. The context of the tomb as a whole clearly demonstrates its burial function. Presumably, the dead were laid to rest with various burial offerings placed beside the bodies. These would have included several ceramic vessels containing food and drink in addition to jewelry and other artifacts. The offerings could have been intended either for the use of the dead in the afterlife or for the living paying tribute to the dead.

The number of studied and published burial faunal assemblages from the near-eastern biblical periods is rather limited, to date. Consequently, the faunal database of mortuary contexts is restricted (see, for example Horwitz 2001a and Horwitz 2001b for reviews of the Middle Bronze and the Iron Age, respectively). This is especially true for the Late Bronze Age. The Ara tomb faunal assemblage is analyzed according to the research design outlined by the excavators. The principal questions thus concern the role of the tomb's contents as burial offerings.

The zooarchaeological and taphonomic coding and analysis procedures used in recording the data for this research, as well as the complete zooarchaeological report are detailed in Bar-Oz and Weissbrod (*in press*). The abundances of the different taxa were quantified using NISP (number of identified specimens) and MNI (mini mum number of individuals). These values were calculated based on the assumptions described in Klein and Cruz-Uribe (1984). Anuran skeletal element representation values were computed with the relative abundance formula as presented by Andrews (1990).

Evaluation of ritual patterning from the large faunal assemblage

A small assemblage of 423 complete and fragmentary large faunal remains was recovered from the tomb fills. All bones were dispersed over the entire area of the tomb and inter-mingled with the human remains. Burnt bones are absent from the assemblage. The assemblage is clearly dominated by caprines (sheep *Ovis aries* and goat *Capra hircus*), which make up 92% of the identifiable bones (Fig. 1). Separation of sheep and goats was based on morphological criteria following Boessneck (1969). It seems that both species are present in equal numbers. Other identified species included cattle (*Bos taurus*) and equid (most probably *Equus asinus*).

Skeletal part representation of the caprines reveals that all body parts are represented at Ara tomb (Fig. 2), although none were discovered in anatomical articulation. No preference for particular side of the animal carcass is evident. The caprine remains probably represent a mixture of both primary butchery (slaughter and carcass division) and secondary butchery (food processing and consumption) (Hesse and Wapnish 1985). The remains of the ass and the cattle are represented by both cranial and post-cranial elements.

The assemblage contains high proportions of immature animals (based on epiphyseal closure, following Silver 1969). Approximately 15% of the caprines were slaughtered before the age of 10 months. All other caprines represented are younger than 36 months but older than 12 months. The results suggest that these individuals were raised primarily for their meat, and the fairly young optimal butchering age is therefore represented (Payne 1973; Grigson 1987). The remains of the ass and cattle (MNI = 1 for both species) are also of immature individuals. The sex of the animals could not be determined on the basis of the available materials.

Horwitz (1987) established a suite of criteria that can be used to differentiate between faunal assemblages from mortuary and domestic contexts. This is based on her extensive analysis of Middle Bronze Age faunal assemblages (see also Horwitz 2001a). According to these criteria, mortuary assemblage are mainly characterized by a narrow range of species with a typical dominance of caprines, an abundance of young and particularly immature animals, differential representation of body parts and bone modifications.

The assemblage from the Ara burial cave is compatible with the first two of these characteristics. However, it exhibits a more or less uniform representation of body parts. The general pattern of body part representation in Middle and Late Bronze Age burial assemblages is that of differential representation (Horwitz 2001; Lev-Tov and Maher 2001). However, this pattern is not always consistent and suggests the existence of varying modes of selection of body parts. Only in a few isolated cases were whole articulated carcasses found (see review in Horwitz

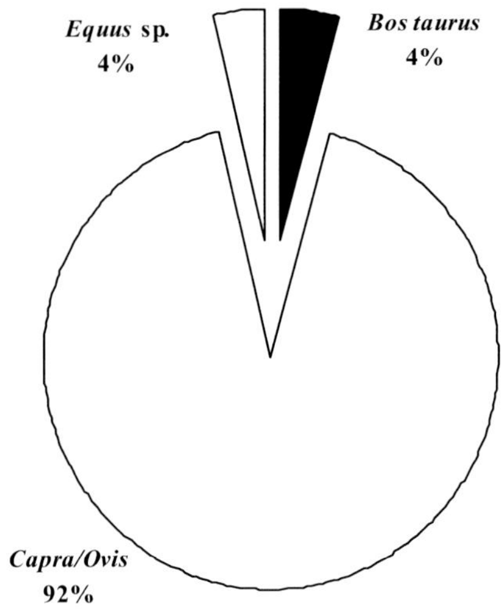

Fig. 1. Relative abundance of the large fauna from Ara burial cave (n = 423, based on NISP).

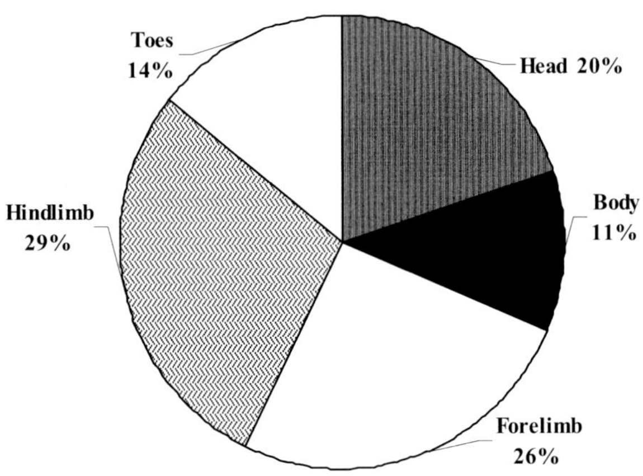

Fig. 2. Relative abundance of sheep/goat body parts (based on NISP).

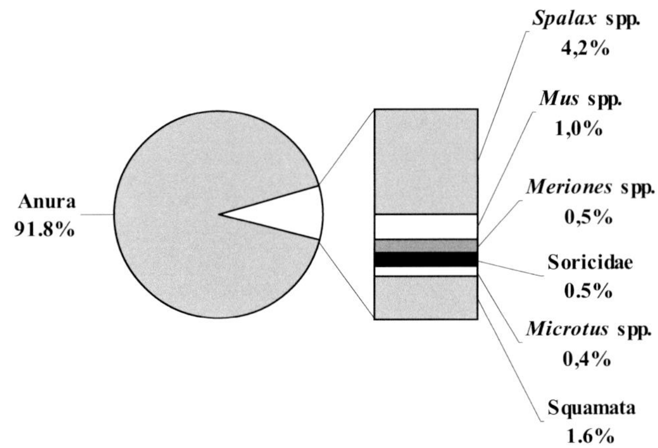

Fig. 3. Relative abundance of the small fauna from Ara burial cave (n = 970 based on NISP).

2001). The occurrence of skeletal elements from all parts of the body suggests *in situ* butchering and possibly also consumption of the slaughtered animals. Such a practice is reminiscent of the mortuary tradition of honorary ritual banquets. This type of ritual is one of several mortuary traditions pertaining to animal offerings that have been identified in the ancient literature of the Ugaritic texts and the Old Testament, as outlined recently by Lev-Tov and Maher (2001). So far, the record of archaeofaunal assemblages from contemporaneous burial contexts in Israel has not demonstrated unequivocal evidence for this type of ritual.

Evaluation of ritual patterning from the small faunal assemblage

The small faunal assemblage collected primarily from the sifted fill of the ceramic vessels contains a total of 970 identified skeletal elements. The majority of the small fauna (*c*. 92%) are amphibian remains (order Anura), while the remainder includes five species of micro-mammals: mole rat (*Spalax* spp.), common mouse (*Mus* spp.), vole (*Microtus* spp.), jird (*Meriones* spp.), and shrew (family Soricidae), and a few squamata vertebrae (Fig. 3).

Taphonomic observations on the small faunal remains indicate a good state of preservation as attested by a relatively high average representation of the anuran major limb bones (*c*. 76%; Fig. 4). In one instance, two almost complete skeletons of mole rats were retrieved from one vessel. Significant anatomical interconnectedness is also indicated by the restriction of the 16 reptilian vertebrae to one vessel. This indicates that the contents of some of these vessels remained largely undisturbed for an extensive period of time. A large number of the limb bones are complete, while the rest are broken only in half. In marked contrast to this pattern of favorable preservation there is a total absence of anuran cranial elements (both skull and dentaries) and under-representation of axial elements (low rate of vertebrae, the urostyle, and pelves and absence of scapulae) and extremities (phalanges and metapodials) (Fig. 4).

In view of the taphonomic characteristics of the small faunal assemblage its potential origins can be outlined. The micro-mammals, which make up only a minor component of this assemblage, are most likely commensals. The five species present in the assemblage represent the most consistently found micro-mammal species in archaeofaunal assemblages from Israel since prehistoric times (*e.g.*, Haas 1952; Bar-Yosef and Tchernov 1966; Weissbrod 2002). In the case of the anuran remains the

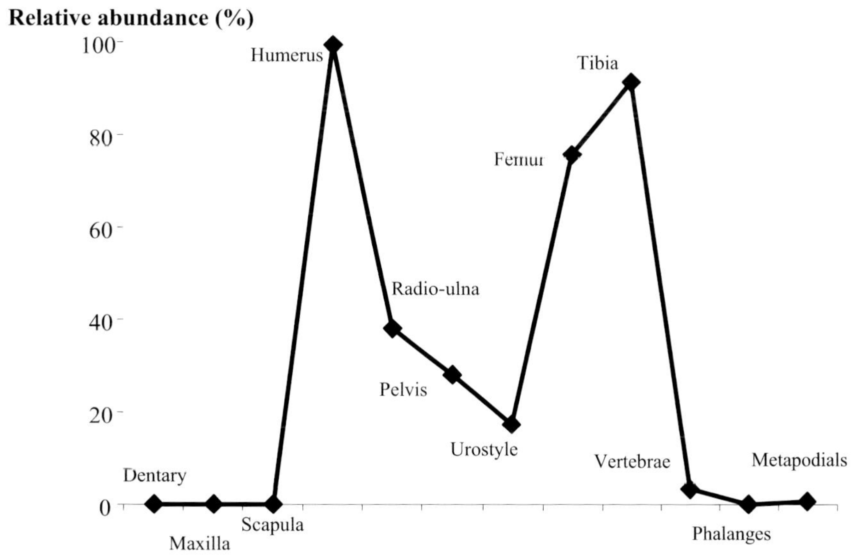

Fig. 4. Relative abundance of anuran skeletal elements (expressed as %MNE)

implications are not as straightforward. Various predators such as owls are known as accumulators of anuran skeletal material (*e.g.*, Llona and Andrews 1999). Therefore, one may speculate that Ara tomb served as a nesting or roosting site for certain raptors (see Wapnish and Hesse 2002 for similar conclusions). However, extant barn owl prey assemblages from Israel are in fact typically dominated by micro-mammals (Dor 1982), keeping in mind that the barn owl is one of the most opportunistic raptors. In addition, a more complete pattern of skeletal element representation including cranial, axial and extremity elements is expected for owl anuran prey as these raptors are known to cause a minimum amount of damage to skeletal material (*e.g.*, Andrews 1990).

Another possibility is the intrusion of anuran material into archaeological deposits that can result from either the death of aestivating toads (Andrews 1990) or entrapment (Whyte 1991; Shaffer and Neely 1992). One of the local toad species, the Syrian spadefoot (*Pelobates syriacus*), is known to burrow into the ground during the dry season (Dgani and Mendelssohn 1990). According to Llona and Andrews (1999), who studied a Middle Pleistocene cave assemblage from Spain, anuran skull elements are generally not preserved due to their decreased durability. However, anuran cranial elements have been retrieved from another Middle Pleistocene cave assemblage from Gibraltar (Gleed-Owen 2001). Moreover, given that the tomb was discovered sealed the case for either predation or intrusion is weakened.

Alternatively, the association with the vessels and the complete absence of the cranial and certain axial and foot elements hints at a cultural origin of the anuran material. This absence can be ascribed to methods of preparation of anuran species as food items. A Native American recipe for preparing a toad instructs as follows: "Catch early frogs, twist off heads, peel off skin…" (Whisler 1973). Indeed, a toad butchering experiment determined that skinning of the animal was facilitated by initially pulling off its head (Cooke 1989). Cooke also ascertained that pulling the skin down over the toad's legs tends to damage the foot bones, while snapping of the legs invariably damages the vertebrae, urostyle, and pelves. Thus, decapitation and skinning of anurans as part of their preparation as food offerings could account for the observed pattern of skeletal representation.

The religious and medicinal significance of anurans in the Greek and Roman worlds (Wissowa 1894), and more importantly in the Egyptian world, is well documented (Houlihan 1996). The inclusion in burials of figurines and vessels in the shape of frogs is known from Egypt. At Tel Dan, in Northern Israel, a scarab seal depicting a squatting frog was found in a Late Bronze Age burial context (Brandl, *in press*). Numerous other parallels of this seal are known from around the Near East.

Summary

We have demonstrated the possible mortuary ritual significance of both the large and small faunal assemblages from the Ara burial Cave. The possible representation of both retrospective and prospective mortuary practices of food offerings, for the large and small fauna respectively, may correspond to the extended temporal use of the tomb. Otherwise, these two types of ritual could have been practiced concurrently. These alternative explanations contribute to the understanding of the observed variability in Late Bronze Age mortuary practices. In view of the great explanatory potential demonstrated by the careful taphonomic analysis of small faunal remains it is highly recommended that systematic collection of small material be regularly implemented in the excavation of historic contexts. Furthermore, in order to substantiate the suggestions presented here more assemblages of this type that are retrieved under carefully controlled methods of excavation are required for study. We believe the suggestions presented call for further study and merit consideration.

Acknowledgements
We thank Yuval Gadot and Yotam Teper for providing us with the bone assemblage, N. Lev-Tov for sorting the faunal remains, and Daniel Kaufman and Mina Weinstein-Evron for their comment on an earlier draft of this manuscript.

References

Andrews, P. 1990. *Owls, Caves and Fossils: Predation, Preservation and Accumulation of Small Mammals Bones in Caves*. Chicago: University of Chicago Press.

Bar-Oz, G. and Weissbrod, L. in press. Faunal remains from the Late Bronze Age burial cave of Ara. *Tel-Aviv*.

Bar-Yosef, O. and Tchernov, E. 1966. Archaeological finds and the fossil faunas of the Natufian and microlithic industries at Hayonim cave (Western Galilee, Israel). *Israel Journal of Zoology* **15**, 104–40.

Boessneck, J. 1969. Osteological differences between sheep (*Ovis aries* Linné) and goats (*Capra hircus* Linné), pp. 331–58 in Brothwell, D. and Higgs, E. S. (eds), *Science in Archaeology* (2nd edition). London: Thames and Hudson.

Brandl, B. A. in press. Frog-scaraboid from tomb 387 at Tel Dan.

Cooke, R. G. 1989. Anurans as human food in tropical America: ethnographic, ethnohistoric and archaeological evidence. *Archaeozoologia* **3**, 123–42.

Dgani, G. and Mendelssohn, H. 1990. Anura, pp. 206–7 in Arbel, A. (ed.), *Plants and Animals of the Land of Israel, Reptiles and Amphibians*. Vol. 5. Israel: Ministry of Defense (in Hebrew).

Dor, M. 1982. The barn owl as an "aid" to science. *Ha-oznia* **5**, 29–35 (in Hebrew).

Gleed-Owen, C. P. 2001. A preliminary report on the late Pleistocene amphibians and reptiles from Gorham's cave and Vanguard cave, Gibraltar. *Herpetological Journal* **11**, 167–70.

Grigson, C. 1987. Different herding strategies for sheep and goat in the Chalcolithic of Beersheva. *Archaeozoologia* **12**, 115–26.

Haas, G. 1952. The fauna of layer B of the Abu Usba Cave. *Israel Exploration Journal* **2**, 35–47.

Hesse, B. and Wapnish, P. 1985. *Animal Bone Archaeology: From Objective to Analysis*. Washington: Taraxacum.

Horwitz, L. K. 1987. Animal offerings from two Middle Bronze Age tombs. *Israel Exploration Journal* **37**, 251–5.

Horwitz, L. K. 2001a. Animal offerings in the Middle Bronze Age: food for the gods, food for thought. *Palestine Exploration Quarterly* **133**, 78–90.

Horwitz, L. K. 2001b. The contribution of archaeozoology to the identification of ritual sites, pp. 63–8 in Pike, S. and Gittin, S. (eds), *The Practical Impact of Science on Near Eastern and Aegean Archaeology* (Wiener Laboratory Monograph 3). London: Archetype Press.

Houlihan, P. F. 1996. *The Animal World of the Pharaohs*. London: Thames and Hudson.

Klein, R. G. and Cruz-Uribe, K. 1984. *The Analysis of Animal Bones from Archaeological Sites*. Chicago: University Chicago Press.

Lev-Tov, J. S. E. and Maher, E. F. 2001. Food in Late Bronze Age funerary offerings: faunal evidence from tomb 1 at Tell Dothan. *Palestine Exploration Quarterly* **133**, 91–110.

Llona, A. C. P. and Andrews, P. J. 1999. Amphibian taphonomy and its application to the fossil record of Dolina (middle Pleistocene, Atapuerca, Spain). *Palaeogeography, Palaeoclimatology, Palaeoecology* **149**, 411–29.

Payne, S. 1973. Kill-off patterns in sheep and goats: the mandibles from Asvan Kale. *Anatolian Studies* **23**, 281–303.

Shaffer, B. S. and Neely, J. A. 1992. Intrusive anuran remains in pit house features: a test of methods. *Kiva* **57**, 343–51.

Silver, I. A. 1969. The aging of domesticated animals, pp. 250–68 in Brothwell, D. and Higgs, E. S. (eds), *Science in Archaeology* (2nd edition). London: Thames and Hudson.

Wapnish, P. and Hesse, B. 2002. Mammal remains from the Early Bronze sacred compound, chapter 14 in Finkelstein, I., Ussishkin, D. and Halpern, B. (eds), *Megiddo III, The 1992-1996 Seasons*, Tel-Aviv: Emery and Claire Yass Publications in Archaeology.

Weissbrod, L. 2002. *Taphonomy of the Assemblage of Micromammal Remains from el-Wad Terrace*. Unpublished M. A. dissertation, University of Haifa, Haifa (in Hebrew).

Whisler, F. L. 1973. *Indian Cooking*. Nowega Press.

Whyte,T. R. 1991. Small-animal remains in archaeological pit features, pp. 163–76 in Purdue, J. R., Klippel, W. E. and Styles, B.W. (eds), *Beamers, Bobwhites, and Blue-Points: Tributes to the Career of Paul W. Parmalee* (Scientifc Papers 23). Springfield: Illinois State Museum.

Wissowa, G. 1894. 'Frosch', in Wissowa, G. (ed.), *Paulys Realencyclopädie der classischen Altertumvissenschaften*. Stuttgart: Alfred Druckenmüller Verlag.

Lior Weissbrod
Zinman Institute of Archaeology, University of Haifa, Haifa 31905, Israel and Department of Anthropology, Washington University in St. Louis, St. Louis, MO 63130, USA
E-mail: lweissbr@artsci.wustl.edu

Guy Bar-Oz
Zinman Institute of Archaeology,
University of Haifa,
Haifa 31905, Israel

9th ICAZ Conference, Durham 2002
Behaviour Behind Bones, (ed. Sharyn Jones O'Day et al.) pp. 25–33

4. The butchering patterns of Gamla and Yodefat: beginning the search for *kosher* practices

Carole Cope

The Roman Period sites of Gamla and Yodefat were destroyed by the Romans in AD 67 and were never resettled. Gamla is located in south-western Golan about six km east of the north shore of the Sea of Galilee. Yodefat was almost in the center of Lower Galilee, five kilometers north of the ancient Jewish capital of Sepphoris. Evidence presented by the principle investigators of the sites indicates that they were inhabited by observant Jews. The sites of Gamla and Yodefat offer a rare opportunity to reconstruct the animal treatment practices of the people of Israel on the eve of the Roman Conquest. The faunal material suggests a high regard for cleanliness as well as abhorrence of blood that might represent an early form of kosher.

Methods and materials

The faunal material from Gamla was studied by Horowitz in 1992 (*pers. comm.*). The present author had the opportunity to study the faunal materials (2,158 non-human bones from Gamla, and 2,278 from Yodefat) while focusing on the butchering patterns (Fig. 1, Fig. 2, and Fig. 3). This allowed sufficient time to produce the detail that was necessary for a precise anatomical analysis based on the species in question. This sort of anatomical description was used because the pattern may be followed or recreated by anyone using an atlas of the anatomy of domestic animals (Sisson 1910). I concluded that the butchering patterns at Gamla and Yodefat are extremely exact. The cuts are perfectly placed to sever origins and insertions of whole groups of muscles as well as arteries and veins. The only way this method may be described with any degree of accuracy is to be as precise as were the ancient butchers.

Gazella gazella	gazelle
Bos taurus	domestic cattle
Capra hircus	domestic goat
Ovis aries	domestic sheep
ovicaprines	sheep/goat
Sus scrofa	domestic pig

Fig. 1. Species list from Gamla and Yodefat.

Introduction

The laws of Kashrut have their origins in Biblical tradition. The Bible prohibits the consumption of animals that are "impure." This tradition divides the entire animal kingdom "pure" and "impure" creatures. The Bible also forbids consumption of animals that have cloven hooves (but do not chew the cud), as well as a long list of birds of prey, most insects, fish without fins and scales, and all shellfish or invertebrates except for some locusts and crickets (Cohen 1999). According to tradition, Moses received Torah from God at Sinai and transmitted it to Joshua who transmitted it to the Elders. The Elders transmitted Torah to the Prophets from the Prophets to the men of the Great Assembly (Council of Sages) (Farade 1991; Cohen 1999). These sages of the Assembly are those teachers with whom the first post-Torah scriptures are associated.

After the fall of the Second temple in AD 70, the rabbinic sages increasingly claimed not only the roles that were in former times the prerogative of the priesthood but Torah study as well (Farade 1991). Torah study not ritual sacrifice became the paramount religious act. Study and prayer became increasingly ritualized as acts of religious worship in their own right (Farade 1991). From the destruction of the Second Temple until the time of the earliest Rabbinic texts in the early third century, the rabbinic movement presumably went through major change. Unfortunately there is not a single datable text nor does archeological evidence fill the gap from the destruction of the Second Temple by the Romans to

Species	n	element	% cuts	Species	n	element	% cuts
Bos	5	atlas	100.0	ovicaprine	51	atlas	70.6
Bos	6	axis	83.3.	ovicaprine	21	axis	42.8
Bos	32	scapula	59.3	ovicaprine	44	scapula	50.0
Bos	25	humerus	45.0	ovicaprine	81	humerus	*
Bos	23	radius	52.0	ovicaprine	25	radius	44.0
Bos	0	ulna	*	ovicaprine	29	ulna	24.1
Bos	15	innominate	56.0	ovicaprine	76	innominate	36.8
Bos	16	femur	24.0	ovicaprine	64	femur	42.1
Bos	20	tibia	45.0	ovicaprine	59	tibia	57.6
Total	142				450		

*Fig. 2. Bones that most often show pre-*kosher *pattern cuts from Gamla. % cuts = percentage of cuts as described in the text; * = disarticulation cuts only.*

Species	n	element	% cuts	Species	n	element	% cuts
Bos	3	atlas	100.0	ovicaprine	6	atlas	66.7
Bos	3	axis	66.7	ovicaprine	1	axis	100.0
Bos	11	scapula	63.6	ovicaprine	28	scapula	50.0
Bos	14	humerus	42.9	ovicaprine	29	humerus	*
Bos	11	radius	54.5	ovicaprine	18	radius	44.4
Bos	10	ulna	*	ovicaprine	11	ulna	27.3
Bos	8	innominate	50.0	ovicaprine	25	innominate	36.0
Bos	16	femur	25.0	ovicaprine	29	femur	41.4
Bos	13	tibia	46.1	ovicaprine	34	tibia	55.9
Total	89				181		

*Fig. 3. Bones that most often show pre-*kosher *pattern cuts from Yodefat. % cuts = percentage of cuts as described in the text; * = disarticulation cuts only.*

approximately AD 220 (Sanders 1990; Farade 1991). It may be that after AD 70 the rabbis incorporated certain priestly traditions into daily ritual. What was sacred and separate became part of daily life. Before AD 70 the Bible recognized two types of slaughter, "sacred" and "profane"(Sanders 1990). Non-sacred slaughter was not a sanctified act and its purpose was simply to render an animal fit for food. After AD 70 the Tannanim understood the Torah as requiring that an animal could only be eaten if it was killed in the same way as sacrifice (Schiffman 1991). There is a time period of several hundred years from the Persian Period till AD 220 when text is rare or non-existent. However, it is probable that ideas concerning slaughter and butchering so precise and standardized as that of Gamla and Yodefat had been in the consciousness of the people for generations.

Slaughter

The method of slaughter used throughout the non-Jewish Mediterranean world was basically the same, the animal would have its head bent back and its throat would be cut straight across (Saunders 1990). The Jews considered that this "gentile" method of slaughter was impure because the victim strangled on its own blood which resulted in the "the bread of strangulation" (Saunders 1990). The Jewish method was much more precise and was designed to remove as much blood as quickly as possible. The animal's head was bent back and the knife passed downward on a slant from the carotid artery under the ear and across the ventral surface of the cervical vertebrae deeply enough to sever the windpipe. According to Talmud the first one third of the cut must be above the first ring of cartilage on the trachea. This is apparently to produce a pronounced slant to the cut (Saunders 1990). Dr. Mordachai Kizlev, Professor of Archeo-Botany at Hebrew University of Jerusalem, and an expert on *kosher* practices, informs me that any of the seven cervical vertebrae will do as long as the slant remains. However at Gamla and Yodefat most cuts were on the ventral surface of the first two cervicals– the atlas and axis vertebrae. Only one cervical vertebrae from farther down the neck was cut at Yodefat and none at Gamla.

The slant that was deemed a necessity for the animal to be rendered *kosher* may seem at first glance obsessive but

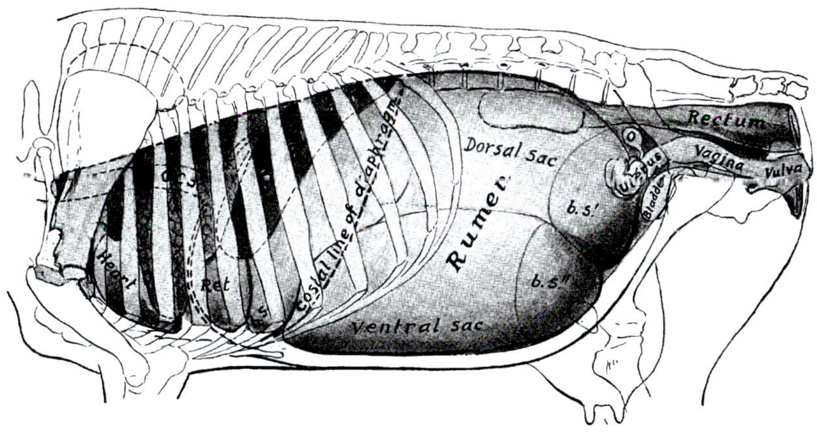

Fig. 4. Diaphragm muscle (modified from Sisson 1910).

is in fact highly practical. The slant that severed the carotids above the first cartilage of the windpipe would provide far wider surface areas and allow for heavier bleeding from the carotid artery. This would probably render the animal unconscious more quickly and provide it with a less traumatic death. In addition, more blood would have been lost during the period when the heart was still pumping. The slant would prevent the windpipe from closing on itself when the animal tried to breath thereby avoiding "the bread of strangulation" so repellent to observant Jews. At Gamla 57% of atlas and 40% of axis vertebrae had cuts across the ventral surface. At Yodefat most atlas and axis were too fragmented for analysis, however two atlases and one axis that were complete had cuts on the ventral surface.

The butchering pattern for Gamla

After slaughter the animal was skinned, hung, and gutted. At this point in most butchering patterns the animal would then be cut up and dismembered. At Gamla and Yodefat the order is somewhat different. The animal was skinned, hung, gutted and the muscles would be stripped from the bones along with the veins and arteries; only then would the carcass be disarticulated.

After gutting the pleural cavity was opened by the simple mechanism of removing the diaphragm muscle. This muscle divides the respiratory system from the digestive system and aids the animal in breathing (Fig. 4). For bovids, this muscles extends along the medial surface of the length and width of the rib cartilage, crosses the end of the 9th rib, and passes backwards and upwards in a curve so that its most distal part is at the level of the last rib. The medial surfaces of several of the last ribs of both *Bos* and ovicaprines show cuts that would correspond to the diaphragm. With the diaphragm removed the organs of the thoracic cavity, heart and lungs, may be easily stripped away along with the vena cava and vena portaea. These are the main trunks of arteries and veins that run down the body's interior along with the aorta (Fig. 5). At the level of the tuber ischii the aorta divides to become the branches of the iliac arteries while the vena cava runs along with it. Along the inner surface of the innominate they then blend with the psoas major and may be stripped along with the muscle which inserts on the trochanter minor of the femur. At Gamla 42% of *Bos* and 23.8% of ovicaprines exhibit a single cut on the trochanter minor.

At Gamla and Yodefat the form of butchery in use is known as "muscle stripping" (Cope 1999; Cope, *in press*). This technique is an ideal method for an urban situation without refrigeration (see section on muscle stripping). However the evidence suggests that while muscle stripping was the technique in use, the major objective was stripping the main arteries and veins.

The forequarter

We begin the discussion of the forequarter with a brief description of the brachial artery as it descends down the thoracic limb. The muscles and bones that relate to the arterial system can be more clearly understood in the context of muscle stripping if they are seen as a unit. The brachial artery emerges from the thoracic cavity at the first rib and passes downward across the origin of the coracobrachialis and the insertion of subscapularis on the medial scapula (Fig. 6). At the posterior border of the scapula it forms a major branch known as subscapularis artery which is bordered on one side by subscapularis and on the other by teres major and latissimus dorsi. The brachial artery itself proceeds down the medial surface of the humerus in association with latissimus dorsi, teres major, and coracobrachialis.

The scapula has four areas that exhibit cuts: 1) beneath the scapular spine, this cut is related to the infraspinatus and one origin of the deltoids on aponeurotic covering of

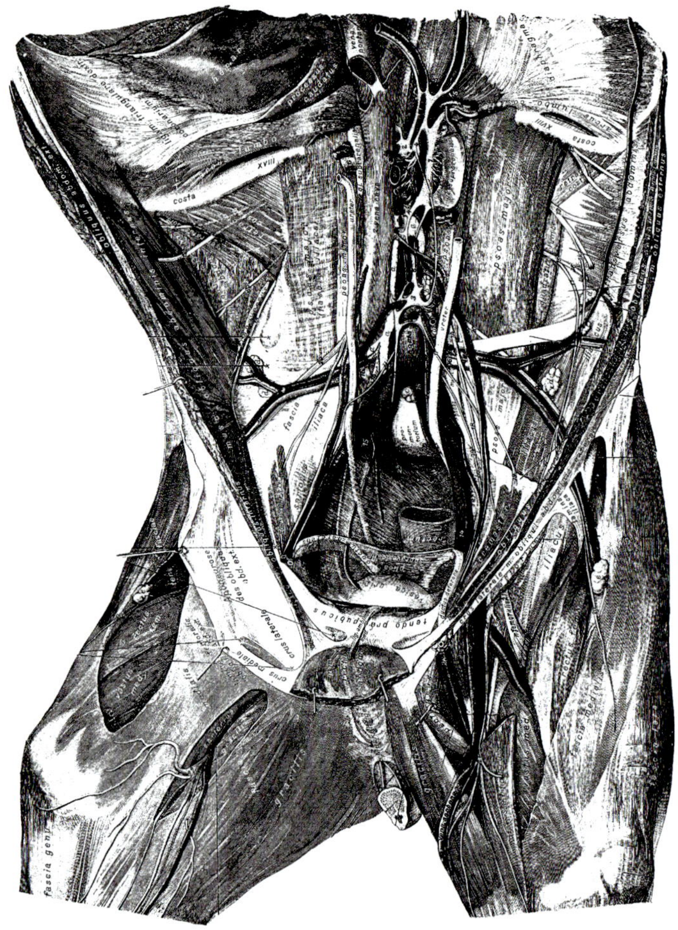

Fig. 5. Circulatory system (modified from Sisson 1910).

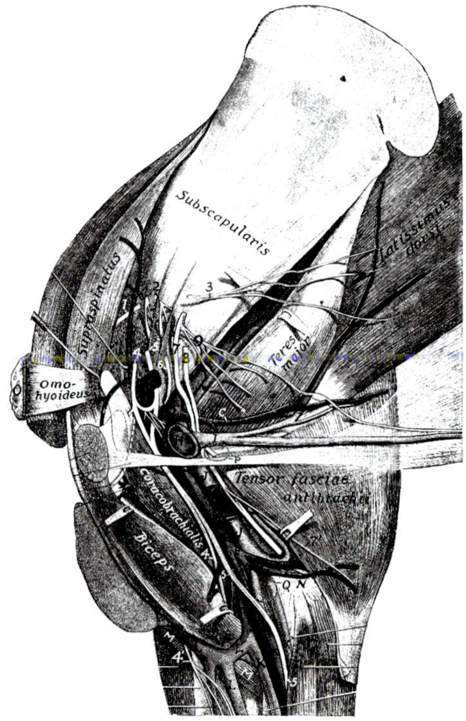

Fig. 9. Dissection of shoulder and arm (modified from Sisson 1910).

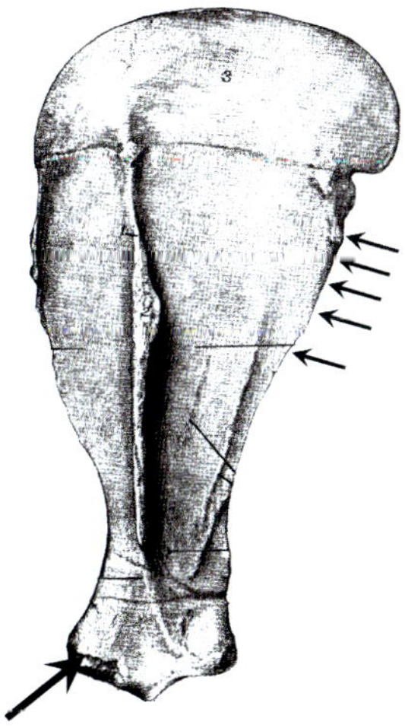

Fig. 7. Left scapula, medial and lateral views (modified from Sisson 1910). Arrows indicate areas that are consistently cut.

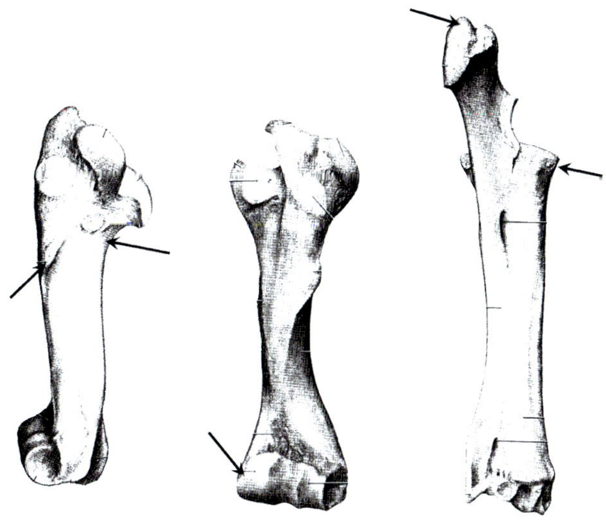

Fig. 8. Left humerus -lateral and anterior views; fused left radius and ulna-medial view (modified from Sisson 1910). Arrows indicate areas that are consistently cut.

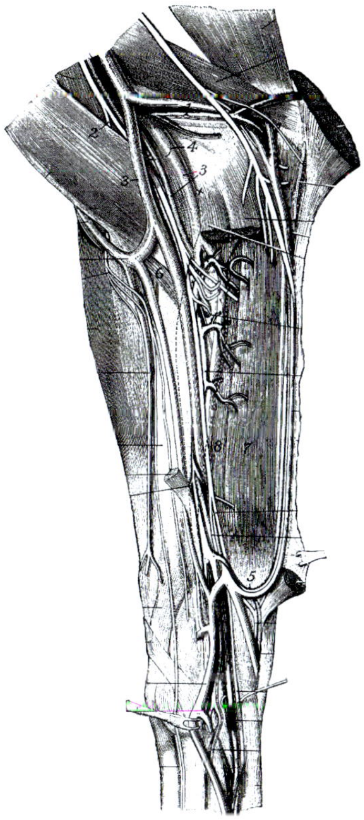

Fig. 9. Lower forelimb (modified from Sisson 1910).

infraspinatus; 2) on the tuber of the scapula for the origin of coracobrachialis, teres minor, and deep pectoral; 3) the posterior border of scapula. Cuts here would sever the origin of the long head of the triceps, teres major, tenser fasciae antebrachii, and one origin of the deltoid muscle as well as latissimus dorsi; 4) costal neck of scapula, these cuts are associated with the junction of the brachial artery, subscapular artery, and the brachial vein (Fig. 7). If these four areas are cut it is possible using the natural diversion between muscles (mesenteries) to strip muscles as well as arteries down to the radius and ulna without introducing blood and bacteria into the meat.

Cuts on the neck of the humerus correspond to the origin of medial head of triceps. In cattle and ovicaprines, there is often a cut on the upper deltoid tuberosity that would sever the biceps beneath its origin. On bovids the origin of this muscle is within the bicipital groove on the articulation of the humerus, this would make it very difficult to cut without disarticulating the forelimb. Cutting the biceps at this point may not be elegant but allows the muscle and associated brachial artery to be stripped without disturbing other structures which might complicate stripping the lower limb. Related to this cut is a single one just above the coronoid fossa on the anterior distal humerus, but only for ovicaprines (Fig. 8). For both cattle and sheep/goats, the origin of biceps is as just described. However the insertion for *Bos* is on the medial proximal radius where cuts as found 50% of the time at Gamla. The condition differs for ovicaprines in that the insertion bifurcates with one insertion on the proximal medial radius and another on the proximal ulna. Apparently the method for sheep/goats was a quick and dirty way, a short cut.

Stripping the lower forelimb seems to have been done the same way for cattle, sheep and goats. The tendons for flexor carpi radialis, and flexor carpi ulnaris are severed either on the medial distal humerus or on the proximal medial radius for the origins of these muscles as well as the insertion of biceps. These muscles could then be stripped along with common interosseous artery, and the median artery proceeds to become the common digital artery on the metacarpals (Fig. 9). Using only the cuts described for forequarters it is possible to strip a carcass

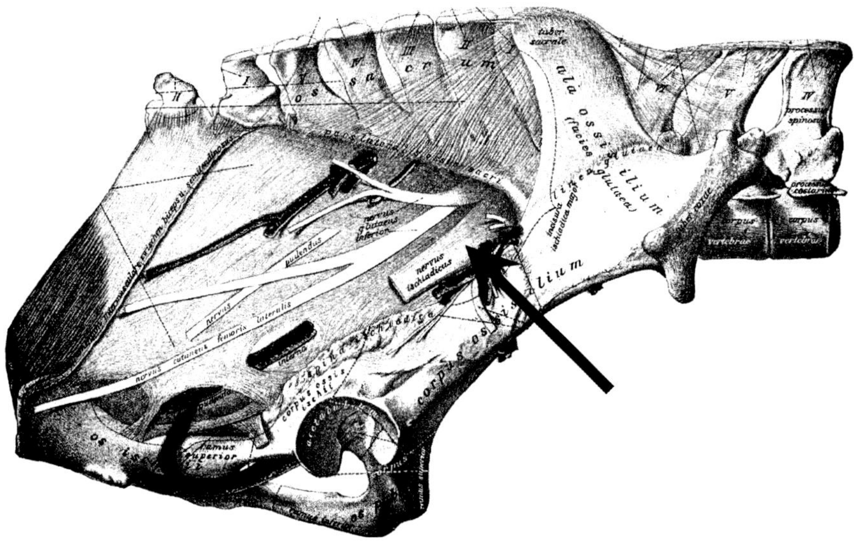

Fig. 10. Sciatic nerve (modified from Sisson 1910).

using only 10–12 cuts per limb. The anatomical knowledge of the ancient butchers of Gamla was detailed and the pattern seems to have been highly standardized. It now remains to be seen how extensive it was in time and space.

The hind quarter: innominate

Right and left sides of the pelvis are known individually as the innominates. Innominates of *kosher* animals at Gamla had a complex of three areas with cuts related to the origins of muscles that would allow stripping of the muscles of the femur related femoral arteries and sciatic nerve (Fig. 10). Cuts are so precisely placed that each one does double or even triple duty. Cuts along the ischium in front of the acetabulum notch sever an entire complex of structures. At Gamla 57% of sheep/goats and 40% of cattle exhibit one or more cuts in this area (Fig. 11).

The sciatic nerve pattern

The removal of the sciatic nerve is extremely important if meat is to be considered *kosher* by an observant Jew. The path traced by the sciatic nerve in the body is very similar for all mammals including human beings. It arises in bovids on the 5th and 6th lumbar vertebrae and passes through the greater sacral foramen and blends with the gluteal nerve and then passes downward on the lower section of the sacro-sciatic ligament and on the origins of the deep gluteal muscles. It then passes over the ischiatic spine of the innominate in front of the acetabular notch. This is one of the areas that is consistently cut on *Bos* and ovicaprines at Gamla. The area that is transected has a great deal going on with tendons for origins of the obturatorius internus muscle and quadratus femoris. It travels down the thigh laterally between adductor and semimembranosus and semitendinosus medially. It continues between the twin origins of gastrocnemius on the posterior distal femur where it becomes the tibial nerve. It then passes down the lower limb beneath the gastrocnemius until it finally terminates as the plantar nerve of the foot. All along the route traveled by the sciatic nerve down the hindquarter there are cuts on related bones with areas where origins of muscles or groups of muscles that would need to be stripped in order to effectively remove it easily. The area with cuts in front of the acetabular notch has already been described.

The second area on the innominate is along the ilium in front of the rectus depression, again this is an area with a lot going on anatomically. A few cuts here would sever origins of rectus femoris, sartorius gluteals, biceps femoris origin and branches of the gluteal arteries as well as iliac and femoral vessels. The third area is on the pubic tubercle. This is the site for attachment for the obliquus abdominis by the pre-pubic tendon along with the anterior of gracilis, origin of adductor and pectinus. With cuts on these three areas it is possible to strip the abdominals, the innominate of relevant muscles and veins, as well as most muscles of the femur along with the femoral arteries that travel down the medial surface of the thigh.

The femur

It is important to note here that most of the cuts that strip the femur were actually placed on the innominate. Cuts on the femur itself were often related to disarticulation (see section labeled skinning and disarticulation). Cuts that would strip muscles were found in four areas, the trochantor major, trochantor minor, beneath the femoral head and on the posterior distal just above the condyle (Fig. 12). The cuts on the trochantor major of the femur

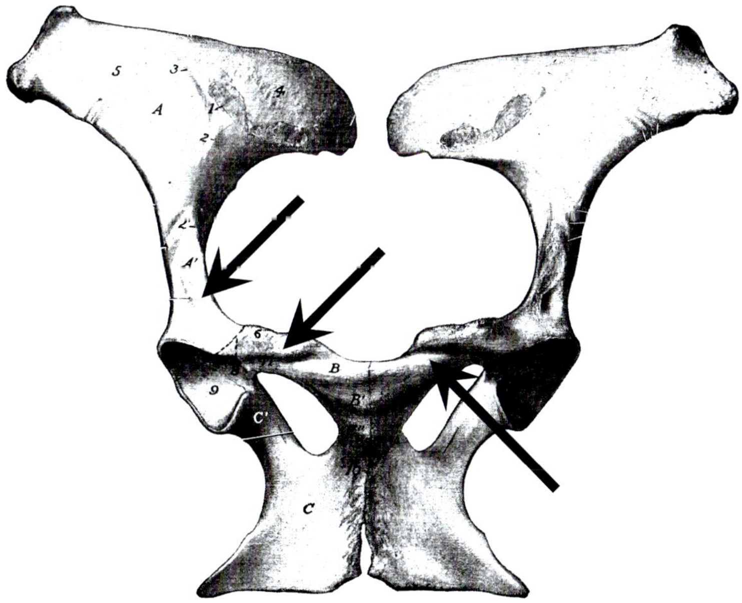

Fig. 11. Pelvis-ventral view (modified from Sisson 1910).

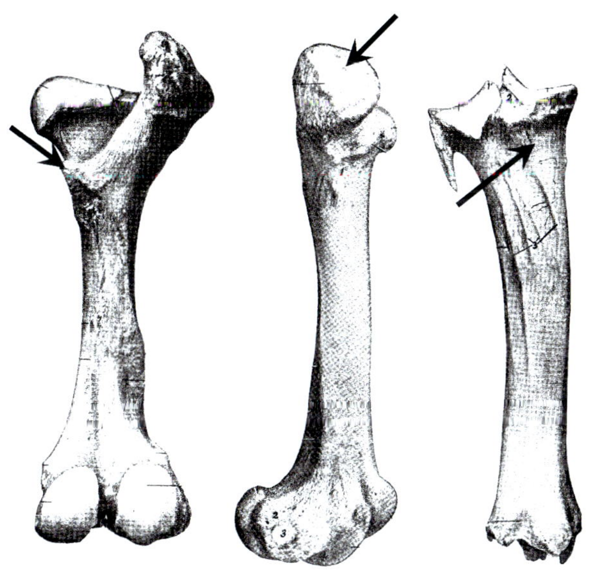

Fig. 12. Right femur, posterior and lateral view, left tibia posterior view (modified from Sisson 1910). Arrows indicate areas that are consistently cut.

correspond to the origin of vastus lateralis and the insertion of the gluteus medius. The gluteus medius needs to be stripped to expose the sciatic nerve and other important structures already described on the ischium. The single cut found beneath the femoral head would sever the origin of vastus medialis while the cut across the posterior distal would allow gastrocnemius and the sciatic nerve to be stripped at the point where it becomes the tibial nerve.

Tibia

Only two cuts are found with any consistency on the tibia. One on proximal posterior may be related to the muscle tibialis posterior or perhaps the cut that often is placed on the distal femur for the twin origins of gastrocnemius is simply placed a bit further down instead. The only other cut was placed on the medial malleolus of the distal tibia, which corresponds to the insertion of gastrocnemius (Fig. 12).

The differences in the patterns for *Bos* and ovicaprines

It would not be productive to include beasts of burden in this discussion as only two possible cuts were found on any bone identified as a 'beast of burden.' It should be noted here that bones of these horses and donkeys were at Gamla and Yodefat usually found in larger numbers, often articulated and frequently identified as the same individual. Under these circumstances it seems unlikely that such animals were seldom if ever eaten at Gamla or Yodefat.

For *Bos* and ovicaprines however, a clear pattern was apparent. The pattern was so similar for both categories that it differed only in small particulars. The differences in biceps removal found sometimes for ovicaprines has already been described. The only other difference observed was the tendency for *Bos* to have more and deeper disarticulation cuts than ovicaprines (see disarticulation section).

The "muscle stripping" method

The "muscle stripping" method has been previously described by the author (Cope 1999; Cope 2002), and was thought to be a technique unique to Palestinian butchers before the advent of refrigeration (Signora, *pers. comm.*). The carcass is suspended by its hind legs and the only cuts are those precisely placed on proximal and distal long bones. Between 25% and 75% of bones of ovicaprines and *Bos* displayed cuts that would correspond to origins or insertions of major muscle groups. Since most ungulate species carry nearly 75% of the weight of their muscle tissue high on the appendicular skeleton (Grand 1990), this method effectively removes it with little cutting. The advantages to the method include: 1) the carcass is left whole except for gutting and skinning and is suspended by the hind limbs; 2) the cuts are placed either medially or laterally dependent on origin or insertion of the muscle. This is in contrast to the Roman method in which most cuts are lateral and bones are often chopped or hacked through (Cope 1999); 3) this method also enables the butcher to use the mesenteries, the tissue that separates the individual muscles, to strip the muscle free. In this way the muscle underneath is not touched and, therefore, no bacteria is introduced. A carcass treated in this manner will keep up to a week without spoiling in a hot dry climate (O' Connor 1989).

"The Stripping Method" is characterized by few cut marks left on the bone, no or very few meat removal or periosteum removal marks, cuts on medial as well as lateral sides, and cuts precisely placed at the origins and insertions of major muscle groups. This pattern was likely mostly an urban one as much careful concern seems to have been placed in keeping the meat as fresh and clean as possible. The pattern is commonly seen in urban Palestinian areas before the advent of refrigeration (in the latter part of the twentieth century). When herders butcher their own herds the animal is left lying on its side and is quickly skinned, gutted, and bisected down the spine using an ax or heavy instrument. The carcass is then dismembered with whole limbs shared out among the woman folk for immediate preparation and family consumption. This is the method still in use by the Bedouin today (Horowitz, *pers. comm.*). Clearly, this method requires no particular sophistication or anatomical knowledge. The method in use in Palestinian areas before refrigeration as well as the one observed for Yod Fat however showed a remarkable degree of both sophistication and anatomical knowledge. An urban butcher butchering purchased stock for resale would need to protect his investment as well as protect and maintain his clientele. In such a situation the stripping method has obvious advantages, both hygienic and practical.

The observed method of "muscle stripping" has been recorded in Islamic sites that date from the 7th century through the present day, and was at first believed to be a native urban Palestinian pattern (Cope 1999; Cope 2002). However now that the pattern has been observed at Yodefat as well as Gamla some basic questions must be asked. Could the method observed in Islamic sites in Israel until the recent installation of refrigeration actually be far older than was first supposed? How far back in time could the method be traced? Is it possible that the pattern is actually a pre-Roman Conquest Jewish one? Could it be pre-diaspora *kosher*? Was the practice a religious one originally or a hygienic one meant to keep the meat fresh and clean as long as possible? Was touching the bone considered unclean? Was the original intention to avoid blood?

Professor Haskill Greenfield of the University of Ontario was trained as a *kosher* butcher by his father who was a *shoet*. According to Greenfield, touching the bone during slaughter in modern *kosher* practice renders the meat *unkosher*. Of course the bone must be touched sometimes, but a little as possible (Greenfield, *pers. comm.*). It may be that this method evolved along with urbanization out of necessity; if this is true, it has a long history in this part of the world.

Differences in the patterns for Bos and ovicaprines

Cuts on the long bones are consistently between 25% and 35% for ovicaprines. Cuts on *Bos* seem to be more common but again are consistently placed, that is the *Bos* femurs had the same pattern of cuts as the ovicaprines but instead of 23.7% for sheep/goats, 55% of cattle had the same cuts. Most of the extra cuts on *Bos* seem to be related to the dismemberment of an animal of such large size.

Future studies and future questions

Now that the pattern has been established for Yodefat and Gamla, the question must be asked how far does it extend in time and space? In what area or time period should we look for its origins? Could it have begun in the Bronze Age with the beginnings of urbanization? Could the Israelites have brought sophisticated ideas of food preparation back from the exile in Babylon? Why the abhorrence for blood? Can answers be found in ancient texts? Further studies of bones of other sites from earlier time periods in the area may help to narrow the search for the origins of a pattern that forms the very core of the tradition of an ancient people.

References

Cohen, J. O. 1999. *The Beginnings of Jewishness, Boundaries, Varieties, and Uncertainties*. Berkeley: University of California Press.

Cope, C. R. 1999. Faunal remains and butchery practices from Byzantine and Islamic contexts (1993-94 seasons), pp. 405–17, in Holum, K. G., Raban, A. and Patrich, J. (eds), *Caesarea Papers 2* (Journal of Roman Archaeology, Supplemental Series 35). Portsmouth.

Cope, C. R. 2002. Palestinian butchering patterns: their relation to traditional marketing of meat, pp. 316–9, in Buitenhuis, H., Choyke, A. M., Mashkour, M. and Al-Shiyab, A.H. (eds.), *Archaeozoology of the Near East V. Proceedings of the Fifth International Conference on the Archaeozoology of south-western Asia and Adjacent Areas* (ARC-Publicaties 62). Groningen: ARC.

Farade, S. D. 1991. *From Tradition to Commentary Torah and its Interpretation in the Midrash Sifire to Deuteronomy*. Binghamton, NY: State University New York Press.

Neiesner, A. 1991. Feeding cities: specialized subsistence economy in the Near East. *Journal of Anthropological Archeology* **7**, 1–55.

Neiesner, A. 1987. *Oral Tradition in Judaism: The Case of the Mishnah*. New York: Garland Publishing.

O'Connor, T. 1989. What shall we have for dinner? Food remains from urban sites, pp. 13–23 in Serjeantson, D. and Waldron, T. (eds), *Diet and Crafts in Towns* (BAR British Series 199). Oxford: British Archaeological Reports.

Sanders, E. P. 1990. *Jewish Law from Jesus to the Mishnah Fine Studies*. London: S.M. Press.

Schiffman, L. M. 1991. *From Text to Tradition: A History of Second Temple and Rabbinic Judaism*. Hoboken New Jersey: Ktov Publishing House, Inc.

Sisson, S. (1910). *A Text-Book of Veterinary Anatomy*. London: W. B. Saunders Company.

Carole Cope
430 5th Avenue, Redwood City,
CA 94063, USA
E-mail: cope@pacbell.net

9th ICAZ Conference, Durham 2002
Behaviour Behind Bones, (ed. Sharyn Jones O'Day et al.) pp. 34–40

5. Predynastic Egyptian bovid burial in the elite cemetery at Hierakonpolis

Sylvia Warman

At Hierakonpolis near Edfu a concentration of archaeological sites is undergoing investigation. One of the earliest of these sites dates to the early-mid Predynastic period – up to 3700 BC. The elite cemetery site at Hierakonpolis (HK-6) has provided a number of fascinating and unusual faunal remains. Several non-domestic and exotic species have been recovered over the long years of excavation, including elephant and baboon. The 1999 season produced a particularly interesting find: within grave number 19 a semi-articulated Bos *skeleton was found. Further examination revealed that the specimen was of a very large size, which coupled with the morphology of the skull and horncores, indicated that the individual might be a wild aurochs (*Bos primigenius*) rather than a domestic cow (*Bos taurus*). However, the suggestion that it might represent a specimen of lyre-horned cattle, a domestic type, is investigated. The specimen appeared to have been wrapped in matting and placed upon a bier. The wrapping of bones in matting treated with resin had been noted for human burials from the cemetery. The treatment of the cattle remains indicates a knowledge of body containment and wrapping which may be the first step towards the fully developed, formalised process of mummification, that is known to have been practised on both human and animal remains in the subsequent Dynastic period.*

Various aspects of this unusual burial are examined in this paper including: the skeletal elements present, the identification as an aurochs, the implications of earlier robbing/disturbance of the tomb, the treatment of the carcass prior to burial and associated archaeological and environmental finds from the grave. This find offers insights into the significance of this species within the death rituals of the elite population at Hierakonpolis. Additionally it allows the archaeologist to trace behaviours and practices better known in the later Dynastic periods back to this early urban centre of Egyptian civilisation.

Introduction

Hierakonpolis is in fact a collection of sites spanning from the Palaeolithic to Ptolemaic/Roman periods. The cemetery at locality 6 (HK-6) has been under archaeological investigation for over a century (Hoffman 1980). The nature of the burials, some being large stone built tombs, and the range and quality of the archaeological finds from them, have led to the interpretation of this as an elite cemetery for higher status groups. Another level of the social hierarchy is present at the nearby site of Adaima, which is believed to be a burial place for ordinary people (Adams, *pers. comm.*).

Although HK-6 is obviously a cemetery a great number of animal bones have been found in the graves. Their presence appears to be deliberate and a range of species, both wild and domestic, have been identified including: cow, dog, cat, baboon, and elephant. Both humans and animals are represented in the artefacts recovered from the graves. Several human figurines of pottery have been found; as have a number of delicate flints, worked into the shape of various animals including hippo and giraffe (Adams 2000a; Adams 2000b).

Fig. 1 illustrates locality 6 with the possible graves marked as ovals, all the excavated graves are numbered. Given the exceptional pedigree of the site based on previous discoveries, the find of a mostly complete skeleton of a very large specimen of *Bos* during the 1999 season was not unexpected. The preservation of the material and the manner of its burial was however of particular interest in terms of the investigation of the ritual and religious significance of this site to the Predynastic population who used it.

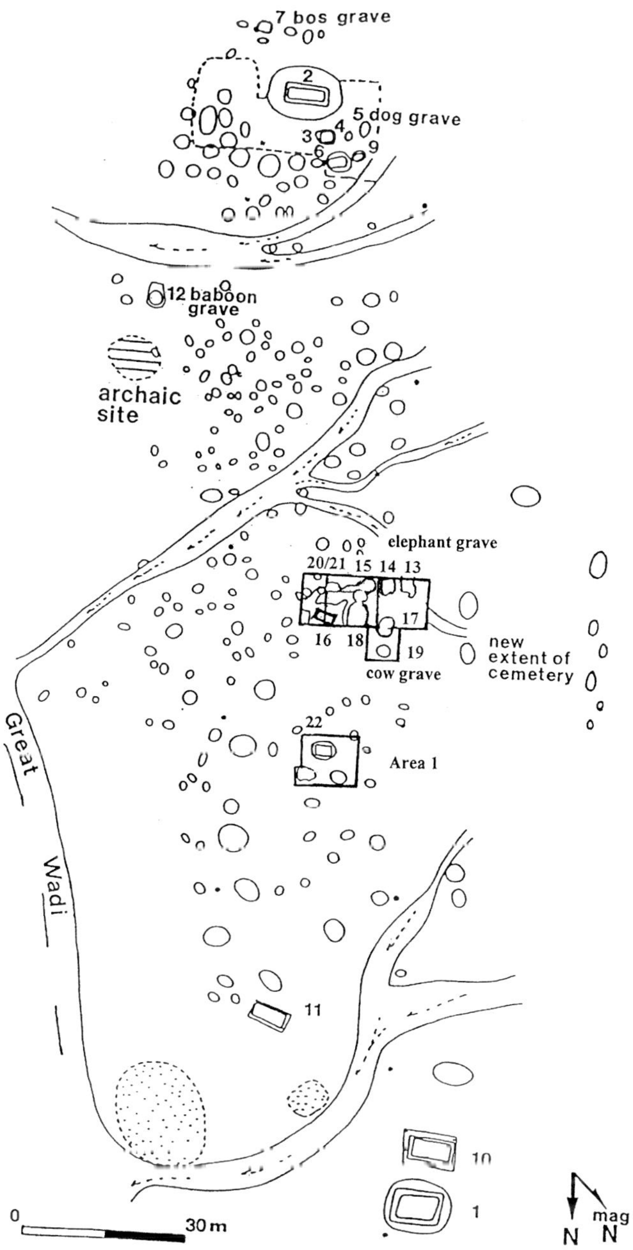

Fig. 1. Map of locality 6 (HK-6). All excavated graves are numbered. The grid squares under investigation during the 1997–2000 seasons are outlined. Ovals mark possible graves as yet unexcavated.

Grave 19

The bones, believed to be most of a single skeleton of an aurochs (*Bos primigenius*), were recovered from two contexts within grave 19. Fig. 2 is the excavator's sketch of this grave. Like the other graves excavated, this one had been robbed, creating a corona of fill (robbers' spoil) thrown up around the actual cut of the grave, this was context 3 which may overlap to some extent with the corona of grave 17. The less disturbed fill of grave 19 still *in situ* within the cut was context 30. The nature of the contexts in relation to the robbing event is discussed in more detail below. All the material from these two contexts examined in the 1999 season is shown in Fig. 4. The *Bos* bones form the main bulk of the sample although other species are present.

Zooarchaeological examination

On first examination two aspects of the material were very striking. Firstly size, all of the elements were much larger than any other *Bos* from HK-6, however as many of the long bones were not fully fused, or had missing epiphyses, length measurements could not be obtained. The second striking observation was that none of the bones showed clear evidence of butchery in the form of chop or cut marks. This finding, along with the fact that many elements were relatively complete suggested that this individual had not been intensively butchered. It would be possible for a skilled butcher with a good understanding of the anatomy of cattle, to remove the flesh without breaking the periosteal membrane and therefore without leaving cut marks on the bone surface (Luff 1994, 158–9). No evidence of burning is seen on the *Bos* remains, reinforcing the suggestion that this animal was not butchered or cooked.

The species present in contexts 3 and 30

Almost all the specimens identified to species level in context 30 are *Bos,* as are the majority of those in context 3. Within the material identified as *Bos* almost all elements are of a size and maturity to suggest that they may be derived from a single individual. The other taxa were dog, pig, sheep/goat, rodent, bird, and cow sized, sheep-sized, and cat-sized remains. Some of the more fragmented material was unidentifiable, but is included in the fragment count in Fig. 4.

The Bos skeleton – detailed description

The parts of the skeleton which were present will be described in detail. Of the skull, the frontal bone with the bottom third of both horncores was the largest single fragment. In addition parts of the maxilla, zygomatic, tympanic and occipital bones were recovered. All of one mandible and part of the other as well as numerous loose teeth both upper and lower were found. In terms of the axial skeleton, the atlas and axis were found although not complete, as well as numerous cervical, thoracic and lumbar vertebrae and a few caudal vertebrae. The epiphyses of the vertebral bodies were generally unfused and a large number of these unfused discs were found. Several ribs were found including the sternal ends and some very porous fragments of the sternum itself. The upper limbs are not particularly well preserved but most of the elements

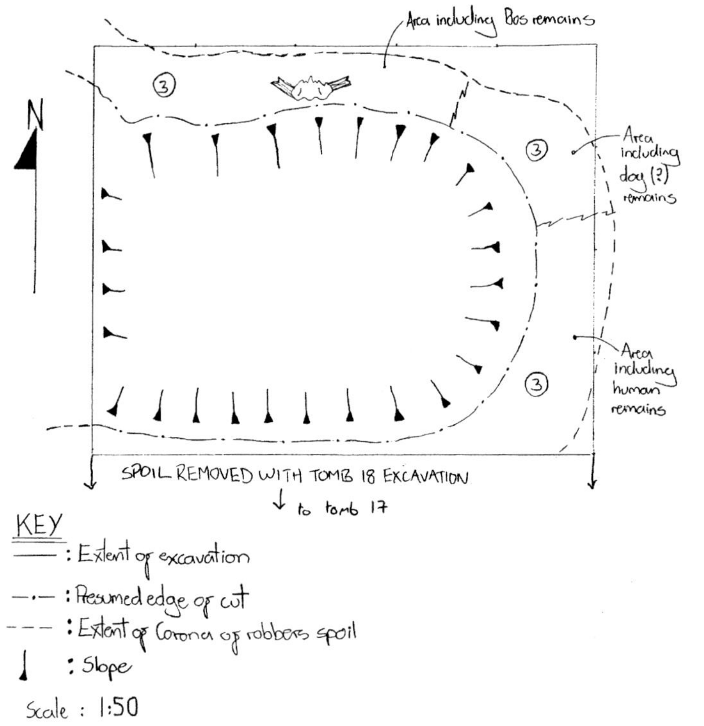

Fig. 2. Excavator's sketch of grave 19.

are represented in part. The articulation of the scapula and humerus are present but the distal part and shaft of the humerus were missing, as is the larger part of the blade of the scapula. The distal ends of both femora are present and most of the pelvis is missing apart from some fragments of the acetabulum. The lower limbs and feet were very well represented. In the fore limbs both radii are present as well as one ulna. The metacarpals are missing but most of the phalanges are present. In the hind limbs both tibia are present as well as both patellae, most of the tarsals and one of the metatarsals again, most of the foot bones are present.

Fig. 3 illustrates the distribution of the preserved elements throughout the skeleton. Although very few elements are complete, a great number of elements are represented in part. The only element types missing are the metacarpals and sacrum, however as adjacent elements are preserved, *e.g.* carpals and proximal phalanges, it is reasonable to assume these elements were deposited but have been destroyed or damaged beyond identification possibly in connection with the robbing event in the past.

The state of fusion of the long bones suggests that the age at death of the individual is likely to be between 2.5 and 3.5 years following Silver (1969). All of the permanent dentition is present; the lower third permanent molar is erupted but relatively unworn, confirming an estimated age at death of around three years (Hillson 1986, 206). The lack of a well preserved pelvis unfortunately means that the sex of the individual could not be determined.

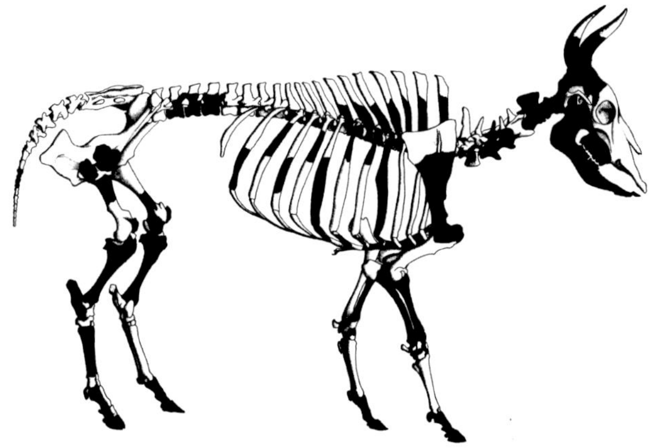

Fig. 3. Illustration of parts of the skeleton present (adapted from Dawson in Allen and Denslow 1968). Shaded areas indicate that a particular element or part of an element was present.

A note on the nature of the contexts

Almost all the graves had been robbed in antiquity and, as a result, a corona of more disturbed fill is seen in a ring around each cut, which in term is filled with less disturbed *in situ* material. In order to document this fact all graves are given two context numbers one for the true fill and one for the redeposited corona. In this grave it was noted that bone parts found in one context could be refitted with

Element	*Bos*	*Sus*	*Canis*	*Ovis/ Capra*	cow size	sheep size	cat size	rodent	bird	unid
Skull	49	–	–	–	82	1	–	–	–	–
Horncore	24	–	–	–	–	–	–	–	–	–
Maxilla	6	–	2	–	–	–	–	–	–	–
Upper teeth	14	–	–	–	–	–	–	–	–	–
Mandible	6	–	7	–	–	1	–	4	–	–
Lower teeth	10	–	1	–	11	1	–	–	–	–
Hyoid	1	–	1	–	–	–	–	–	–	–
Atlas	3	–	–	–	–	–	–	–	–	–
Axis	1	–	–	–	–	–	–	–	–	–
Scapula	4	–	–	–	–	–	1	–	–	–
Coracoid	–	–	–	–	–	–	–	–	1	–
Humerus	2	–	1	–	–	–	–	–	–	–
Radius	2	–	1	–	–	8	–	–	–	–
Ulna	1	–	1	–	–	1	5	–	–	–
Carpal	4	–	–	–	1	–	–	–	–	–
Metacarpal	–	–	–	–	–	–	–	–	–	–
Innominate	3	–	1	–	5	–	–	–	–	–
Sacrum	–	–	1	–	–	–	–	–	–	–
Femur	8		–	–	1	4	–	2	–	–
Patella	2	–	–	–	–	–	–	–	–	–
Tibia	4	–	–	–	–	–	–	–	–	–
Fibula	–	–	–	–	–	–	–	–	–	–
Talus	3	–	–	–	–	–	–	–	–	–
Calcaneus	3	–	–	–	–	1	–	–	–	–
Tarsal	1	–	–	–	5	–	–	–	–	–
Metatarsal	1	–	–	–	–	–	–	–	–	–
Metapodial	1	3	1	–	1	7	–	–	–	–
1st phalanx	4	–	2	5	–	–	–	–	–	–
2nd phalanx	6	–	–	–	–	–	–	–	–	–
3rd phalanx	3	–	–	1	1	–	–	–		–
Other vertebra	97	–	5	–	2	12	–	–	2	–
Sternum	2	–	–	–	–	–	–	–	1	–
Rib	46	–	5	–	34	9	11	–	–	–
Long bone	–	–	–	–	1	11	–	–	–	–
UNID	–	–	–	–	13	5	–	–	–	1400
No. of fragments	318	3	29	6	157	61	17	6	4	1400
MNI	1	1	2	1	–	–	–	–	–	–

Fig. 4. Animal bone by species and element from grave 19 (contexts 3 and 30). Note: MNI has not been calculated for material identified only to size group or broad taxonomic grouping.

those found in the other; for example, the unfused epiphyses and diaphyses of the long bones and the two pieces of the mandible shown in Fig. 5. The part from the undisturbed fill is darker due to the greater degree of moisture in this deposit, whilst the part retrieved from the corona is much lighter in colour, having been exposed to surface weathering or only buried at a very shallow depth. This provides insights into the process of robbing, and also validates the redeposited material by its direct association with the *in situ* material.

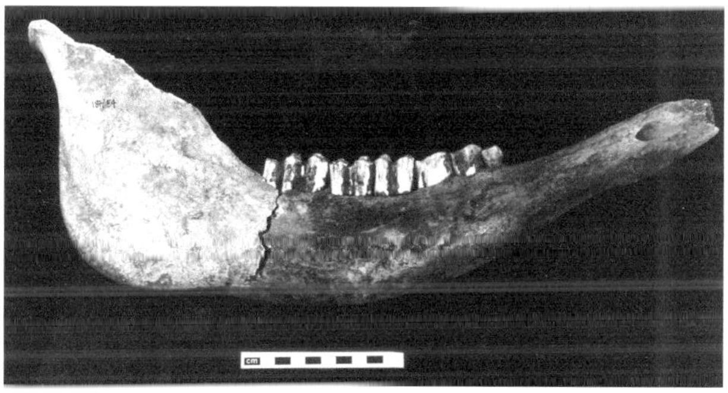

Fig. 5. Bos *mandible reconstructed from parts found in contexts 3 and 30.*

Metric data

It was obvious right from the start of the examination of the specimen that it was an individual of considerable size. The fact that the specimen was from a sub-adult individual meant that many standard dimensions could not be taken. In particular, none of the long bones were fully fused, thus it was not possible to take any greatest lengths which would enable a withers height estimation. Measurements of the widths of long bones in the parts that were fused and lengths and breadth of smaller elements clearly indicate that this was a very large individual (Fig. 6). Some measurements of the horncores were also taken in the hope of clarifying the identification as *Bos primigenius*. All dimensions are those defined by von den

Dimension	**BFcd**	**BFcr**	**GLP**	**LG**	**BG**	**SLC**	**SD**	**Bd**	**Bp**	**BFp**	**SDO**	**DPA**	**BPC**
Element													
Atlas	124.5												
Axis		114											
Scapula R			92.9	76.4	68.9	78.4							
Radius R							53.4		109.3	99.4			
Radius L							54.4		99.1	91.2			
Ulna R											67.1	84.3	58.3
Tibia R							49	77.9					
Tibia L							47.9	78.6					
Metatarsal R									63.5				

Dimension	**SD**	**Bd**	**Bp**	**GL**	**GB**	**GLPe**	**DLS**	**Ld**	**MBS**
Element									
Patella L				84.9	66.2				
Magnum R					47.6				
Scaphoid R					56.2				
1st Phalanx	33.8	38.8	41.9			73.5			
	34.9	38.1	41.4			78.6			
2nd Phalanx	35.8	35.3	41.4	55					
	32.8	33.9	38.9	51.7					
	33.5	34.5	39.7	51.4					
	33.2	33.6	40.6	57.5					
3rd Phalanx							82.4	67.2	25.7
							86.6	72	26.9
							92	71.3	31.5

Fig. 6. Measurements of post-cranial elements. All dimensions from von den Driesch (1976) using original abbreviations.

Driesch (1976). Most measurements were taken by the author in 1999; additional measurements supplied by Van Neer and Linseele were taken in 2001 and 2002.

My initial impression was that this was probably the remains of an aurochs, due to its large size and robustness, even though it was not yet skeletally mature. However, upon showing a slide of the skull and horncores, taken from a similar angle to that seen in Fig. 7, the possibility that this might in fact be an early domestic cow of the lyre-horned type (Grigson, *pers. comm.*) was suggested. Other authors have commented on the use of lyre-horned cattle for sacrificial and ritual purposes (Brewer *et al.* 1994).

The main methods for separating *primigenius* from *taurus* are a combination of skull and horncore measurements and non-metric features developed by Grigson (1976; 1978). Useful measurements include the circumference of the horncores at their base and the most useful non-metric feature is the sagittal profile. The circumferences of the horncores at the base (Fig. 8) are 280mm and 285mm for this specimen. Grigson's range for domestic *taurus* was 92–280mm (Grigson 1974, 365) and for *primigenius* it was 180–395mm (Grigson 1978, 138), thus this specimen is either right at the top of the range for *taurus* or low-mid range for *primigenius*. Obviously additional measurements are required to clarify this, but as the horncores do not survive to their full length and the skull is damaged, this was not possible.

As mentioned above, a number of non-metric features were also used by Grigson to separate the two taxa, these included the sagittal profile (as defined and used by Grigson 1976, Grigson 1978). Unfortunately, due to damage it was not possible to obtain a complete sagittal profile for this specimen. However, two other non-metric features can be observed in the skull: the shape of the ridge at the top of the frontal bone (frontal profile) when viewed from above and front on (Grigson 1976, 126). For the first feature the specimen shows a flat to slightly concave shape which is know as '*taurus* type 2' (*ibid.*), which is the shape seem most commonly in *primigenius* (Grigson 1978, 145). For the second feature this specimen showed a low single arch, '*taurus* type 1' which again is very common in *primigenius* (Grigson 1978, 145).

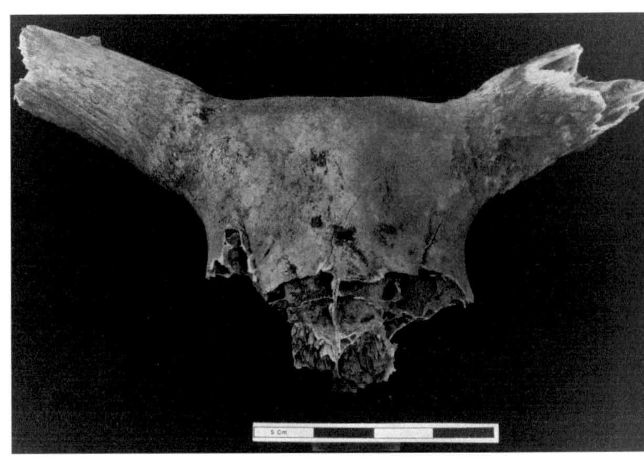

Fig. 7. Bos *skull and horncores.*

The final means for establishing the precise taxon of

	LCR	LMR	LPR	HM1	HP2	L	B	31	44 *	45 *	46 *
mandible right	165	101.9	58.3	65.1	52.7						
M1 lower right						26.2	17.2				
M2 lower right						31.7	15.8				
M3 lower right						44.6	15.6				
M2 lower left						31.7	16.4				
M3 lower left						44.5	15.1				
M1 upper left						27.8	20.9				
M2 upper left						33.2	23.2				
M2 upper right						32.9	23.5				
M3 upper						35.1	22.2				
skull								230			
horncore right									280	98	80
horncore left									285	95	84

*Fig. 8. Skull and tooth measurements. All dimensions from von den Driesch (1976) using original abbreviations. Notes: * Dimensions of horn cores taken by Van Neer and Linseele 2001/2; 31, least breadth between horncore bases; 44, horncore basal circumference; 45, greatest (oro-aboral) diameter of horncore base; 46, least (dorso-basal) diameter of horncore base.*

the specimen is to look at how the horncores curve as they leave the skull. The damage to the specimen means that only the bottom part can be examined. For *primigenius* there is a particular morphology called the '*primigenius* spiral' where the horns come out from the skull laterally then curve inwards and forwards, and finally, upwards and backwards (Grigson 1976, 132). Unfortunately, only the first stage is visible in this specimen and the horn cores appear splayed and angled upwards rather than just outwards. However, the morphology does not look quite right for a lyre-horned cow either; for example, the skull from Abydos illustrated in Clutton-Brock (1987, 68) shows horns which curve upwards much earlier so the tips point towards each other. The HK-6 specimen has horns, which are too splayed out to be of this type. With the material available, a firm identification as *primigenius* or *taurus* is not possible. However, when the range of other species found in the graves at HK-6 is considered, the interpretation of this as a wild or a domestic specimen is of less significance than the manner of the burial and the treatment of the body and associated finds.

The *Bos* skeleton in context – other finds in grave 19

Beyond the details already described (most parts of skeleton present, no butchery marks, no burning) other lines of evidence confirmed the unique nature of this burial. During the excavation of the undisturbed fill of the grave context 30, a rib was found with what looked like a vegetable deposit adhering to it. This was subsequently identified as a combination of resin and reed matting. Below this a more substantial deposit of plant material was found which was identified as a wood substrate with plaster and reed matting impressed into it. This was interpreted as a multilayered bier which lay beneath the body of the cow.

Other botanical remains included a linen bag and the well preserved flowers of a Nile herb *Ceruana pratensis* which would have been brought to the cemetery from the valley (Adams 2000b, 3). What is particularly interesting about these finds is that the resin and matting treatment of bones has been identified on human remains from HK-6 (Adams 2000b). Thus, a practice already identified in the treatment of the human dead appears to have been applied to an animal. It is intriguing that there were no human bones from the *in situ* fill of tomb 19; human remains were only found in the corona context 3, which could have been from the robbing of a neighbouring grave (#17). Ceramic finds from this grave comprised black-topped red and straw-tempered sherds which give a date of Naqada IC-IIA. This was subsequently confirmed by the radiocarbon date obtained from the plant material of 3720–3520 BC, (CAL BP 5670 to 5470) (Adams 2000b, 7). The other find was a red painted pottery human figurine. It is interesting to note that this grave contained the remains of an animal buried in the same fashion as humans and accompanied by an object, which apparently symbolised and/or represented the human form.

A previously excavated grave had also produced *Bos* remains in association with reed matting, this was tomb 7 excavated during Hoffman's project (1979–1985). The remains of three separate individuals were identified. Adams describes the find as the following,

> "Near the centre of the grave some bones were found in situ overlain by a single layer of grass matting, patches of which were also found near the edges. The bones consisted of part of a vertebral column, ribs and a scapula overlying the forelimb of another animal, which was also overlain by reed matting" (Adams 2000a, 33).

The bone was examined and it was concluded that it was not defleshed before burial. Unfortunately, a more detailed report on this material is not available as material from other localities at Hierakonpolis was analysed first (McArdle 1992).

Conclusion

This find is significant in that it has reinforced the interpretations drawn from the previous *Bos* grave (grave 7), particularly in terms of the treatment of the body and the use of a bier and the matting and resin. It is from such early forms of body containment and preservation that the rites of mummification of both humans and animals in the Dynastic period developed.

The fact that the skeleton does not show signs of butchery does, I believe, indicate that it was probably killed by slitting the throat, which is a preferred method for ritual sacrifice in many cultures, past and present. It was placed in the tomb on top of the bier, either whole or, at least, partly articulated. The lack of human remains juxtaposed with the presence of a human representation in the form of a figurine is striking.

It is unfortunate that a positive identification as either *Bos primigenius* or *Bos taurus* (lyre-horned) was not obtained. The application of other methods may clarify this in the future. Another avenue of evidence could be obtained from biomolecular examination of this specimen. A DNA study would also help positively classify the specimen. A sample submitted for isotopic analysis might reveal aspects of diet, which would presumably differ between domestic and wild populations. What is clear is that this specimen, whether wild or domestic, was treated in a very particular manner during burial. Given the previous discoveries in Tomb 7 (Adams 2000a, 34), this appears to be a repeated practice among the people who buried their dead at Hierakonpolis (HK-6).

Acknowledgements
Firstly, I would like to dedicate this paper to Barbara Adams who died in June 2002. I thank Barbara for inviting me to the site to study the animal bone material. The 1999 season at the site was sponsored by the National Geographic Society (grant number 6575–99). The excavation was directed by Ian Casey. I thank Jim Mower for his careful excavation of grave 19 and the sketch in Figure 2. Thanks to Richard Jaeschke, for on site conservation of the matting/resin covered rib and the bier materials. The photographs in Fig. 5 and Fig. 7 were taken by Laurent Bavay. Many thanks to Wim Van Neer and Veerle Linseele for taking additional measurements and photographs during the 2002 season. Fig. 3 is adapted from a drawing by Jan Dawson in Allen and Denslow 1968 (see references). Thanks also to Caroline Grigson for helpful discussion during and after ICAZ 2002. Many thanks to Sharyn Jones O'Day for organising the ICAZ session.

References

Allen, G. and Denslow, J. 1968. *The Clue Books: Bones*. Oxford: Oxford University Press.
Adams, B. 2000a. *Excavations in the Locality 6 Cemetery at Hierakonpolis 1979–1985* (Egyptian Studies Association Publication 4). London: Egyptian Studies Association.
Adams, B. 2000b. The 1999 season at the locality 6 Predynastic cemetery at Hierakonpolis, Upper Egypt: Protodynastic and faunal graves. Unpublished Report to the National Geographic Society.
Brewer, D. J., Redford, D. B. and Redford, S. 1994. *Domestic Plants and Animals: the Egyptian Origins*. Warminster: Aris and Phillips.
Clutton-Brock, J. 1987. *A Natural History of Domesticated Mammals*. London: British Museum (Natural History).
Grigson, C. 1974. The craniology and relationships of four species of *Bos* 1. Basic craniology: *Bos taurus* L. and its absolute size. *Journal of Archaeological Science* **1**, 353–79.
Grigson, C. 1976. The craniology and relationships of four species of *Bos* 3. Basic craniology: *Bos taurus* L. sagittal profiles and other non-measurable characters. *Journal of Archaeological Science* **3**, 115–36.
Grigson, C. 1978. The craniology and relationships of four species of *Bos* 4. The relationship between *Bos primigenius* Boj. and *B. taurus* L. and its implications for the phylogeny of the domestic breeds. *Journal of Archaeological Science* **5**, 123–52.
Hillson, S. 1986. *Teeth*. Cambridge: Cambridge University Press.
Hoffman, M. A. 1980. *Egypt before the Pharaohs*. London: Routledge and Kegan Paul.
Luff, R. 1994. Butchery in the Workmen's Village (VW), Tell-el-Amarna, Egypt, pp. 158–70 in Luff, R. and Rowley-Conwy, P. (eds), *Whither Environmental Archaeology* (Oxbow Monograph 38). Oxford: Oxbow Books.
McArdle, J. E. 1992. Preliminary observations on the mammalian fauna from Predynastic localities at Hierakonpolis, pp. 53–6 in Friedman, R. and Adams, B. (eds), *The Followers of Horus* (Oxbow Monograph 20). Oxford: Oxbow Books.
Silver, I. A. 1969. The ageing of domestic animals, pp 284–302 in Brothwell, D. and Higgs, E. S. (eds), *Science in Archaeology* (2nd edition). London: Thames and Hudson.
von den Driesch, A. 1976. *A Guide to the Measurement of Animal Bones from Archaeological Sites* (Peabody Museum Bulletin 1). Cambridge, MA: Harvard University, Peabody Museum of Archaeology and Ethnology.

Sylvia Warman
20A Moyser Road, Streatham,
London, U.K. SW16 6SA
E-mail: sylviawarman@hotmail.com

9th ICAZ Conference, Durham 2002
Behaviour Behind Bones, (ed. Sharyn Jones O'Day et al.) pp. 41–46

6. Typhonic bones: a ritual deposit from Saqqara?

Salima Ikram

This paper presents analysis of a ritual faunal deposit from Saqqara, Egypt dating to c. *2584–2565 BC. The remains of wild animals and fish not ordinarily found together in a cemetery might be the results of a royal funerary or hunting ritual associated with the establishment of divine order over the cosmos. This idea is further emphasized by the discovery of a harpoon in close association with the faunal remains.*

Introduction

It is an archaeological truth almost universally acknowledged that when one comes across a deposit that is incomprehensible it immediately enters the realm of "ritual." In most cases such deposits may eventually be explained quite prosaically as settlement or funerary detritus. This might explain the faunal assemblage found at West Saqqara, Egypt that is discussed below. For the present, however, this deposit remains firmly lodged in the domain of "ritual."

Polish-Egyptian Mission to West Saqqara

The West Saqqara Polish-Egyptian Mission's season in 2000 focused excavations in the area to the west of the Step Pyramid of King Djoser (*c.* 2584–2565 BC). A deep rock-cut dry moat surrounds the pyramid complex. During clearance of the moat area and above, the Mission discovered two rock-cut passages, one of which is clearly funerary, while the other, is not. This latter passage, Corridor 1, is horizontally cut into the eastern side of the dry moat and projects eastward, toward the pyramid (Fig. 1). It extends about 22 m into the rock, and, for the most part is 1.2 m to 1.8 m high, and 1.5 m wide (Mysliwiec 2000). The corridor abruptly terminates in a wall, with a small oblong chamber (4.15 m [EW] x 2.10 m [NS], with a maximum height of 1.8 m) leading off it to the northeast. The floor of the chamber is approximately 20 cm lower than that of the corridor, and the chamber's roof is correspondingly lower as well. The corridor was filled with debris almost to the roof, from 20 cm to 1 m below the ceiling. At least part of this debris appeared to be deliberately deposited (Kuraszkiewicz 2000). The stratified fill of the chamber was surmounted by a group of animal bones (Fig. 2 and Fig. 3), which was covered with a thin layer of dust, and in some instances, rock from roof-fall (Ikram 2000).

Beneath the faunal assemblage, to the depth of 70 cm, lay stratified layers of debris (sand, mud) with no finds save for one Sixth Dynasty (*c.* 2355–2195 BC) sherd. At about 70 cm, a wooden case, containing a wooden harpoon was found (Mysliwiec 2002). The case was constructed from a slender tree trunk that had been laterally sliced, and hollowed out (Fig. 4). The harpoon lay within. The weapon's long handle was faceted, with two of the faceted sides enhanced with serpents carved in bas relief and decorated with a criss-cross hatching. According to the ancient Egyptians, harpoons were supposed to strike their marks like snakes, which suggests that the serpent relief is sympathetic magic as well as ornamental. The harpoon's point was made of wood, and although sharp, could inflict little damage. To make it usable a metal or stone tip would have had to be lashed to the end; no evidence of this was found. Alternatively, this could have been a ritual or model weapon, similar to others found in Egypt (for example, the shield and spear models found in the First Intermediate Period (*c.* 2134–2040 BC) and Middle Kingdom (*c.* 2040–1640 BC) tombs). It was found lying along the southern wall of the chamber, about 20 cm above bedrock. Below the harpoon, placed on the floor of the chamber, was a group of pottery vessels dating to the second half of the

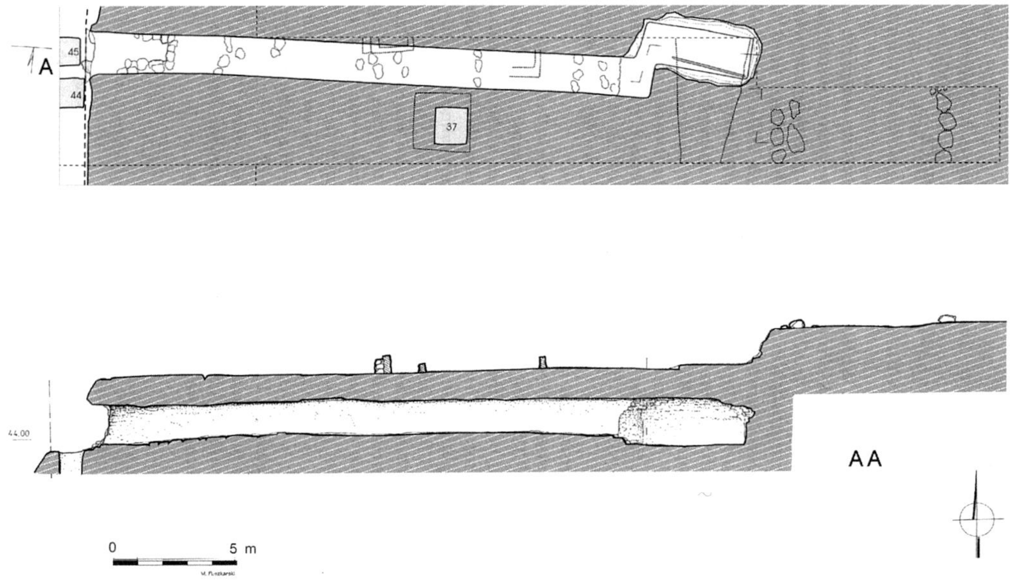

Fig. 1. Corridor 1. Plan (showing original and present outline of the chamber; position of harpoon, as well as structures on the rock surface) and E-W section (Drawing by M. Puszkarski).

Sixth Dynasty. The vessels included bowls, plates, miniature plates, a miniature beer jar, and four ceramic 'nails' (Rzeuska 2000).

The zooarchaeological assemblage

The faunal assemblage was arranged in a loose circle atop Chamber 1's stratified fill, with some disruption caused by bioturbation due to rodents, insects, other animals (dogs, jackals, foxes), and rock fall from the ceiling. For the most part, the bones were all on one level, with some bones being at a slightly lower level (varying between 0.5–2.0 cm) than others. Clearly, they are all part of a single depositional episode.

The evidence indicates that the whole assemblage: harpoon, pottery, and bones are a single related deposit. It is highly unlikely that the faunal assemblage was washed into the corridor and then neatly placed in a circle, as some of the excavators initially surmised. The arrangement of the bones and the taxa represented argue strongly against the hypothesis that the chamber was used as an animal's den and that the deposit is the result of several meals of a carnivorous creature. The bones are tidily arranged, and there is no soil covering them. Thus, it appears that after the harpoon was buried in the chamber, the chamber had been filled with dirt, and the bones were arranged on top.

Methodology

The bones were collected by hand, and the area was not sieved. For ageing, Grant (1982), Payne (1973), and Silver (1963) were used. A comparative collection for the fish

SPECIES	NISP
Clarias	73
Synodontis	7
E. asinus	8
S. scrofa	4
Alcelaphus buselaphus	2
Canis	39
Rodent	2
Artiodactyl	1

Fig. 2. Identified species list by NISP.

and all animals but the hartebeest were available in the field. The hartebeest was securely identified using the Cairo Agricultural Museum's collection.

The bones

The assemblage consisted primarily of cranial portions of animals including fish (catfish: *Clarias* sp., and *schall* or *gargoor* in Egyptian: *Synodontis* spp.), donkey (*Equus asinus*), pig (*Sus scrofa*), hartebeest (*Alcelaphus buselaphus*), and canid (*Canis familiaris/aureus*), and very few other body parts. Some of the body parts found here might be intrusive, belonging to a carnivore that probably came to feast off these remains. The species represented in this group are extremely unusual in Egyptian funerary or even ritual (*e.g.*, foundation deposits) contexts. The unifying theme amongst these diverse taxa is their association with wild, chaotic, and Typhonic forces, generally identified with the Egyptian god Seth, master of the desert and marginal areas.

Seth (or Typhon) balanced the gods Horus and Osiris,

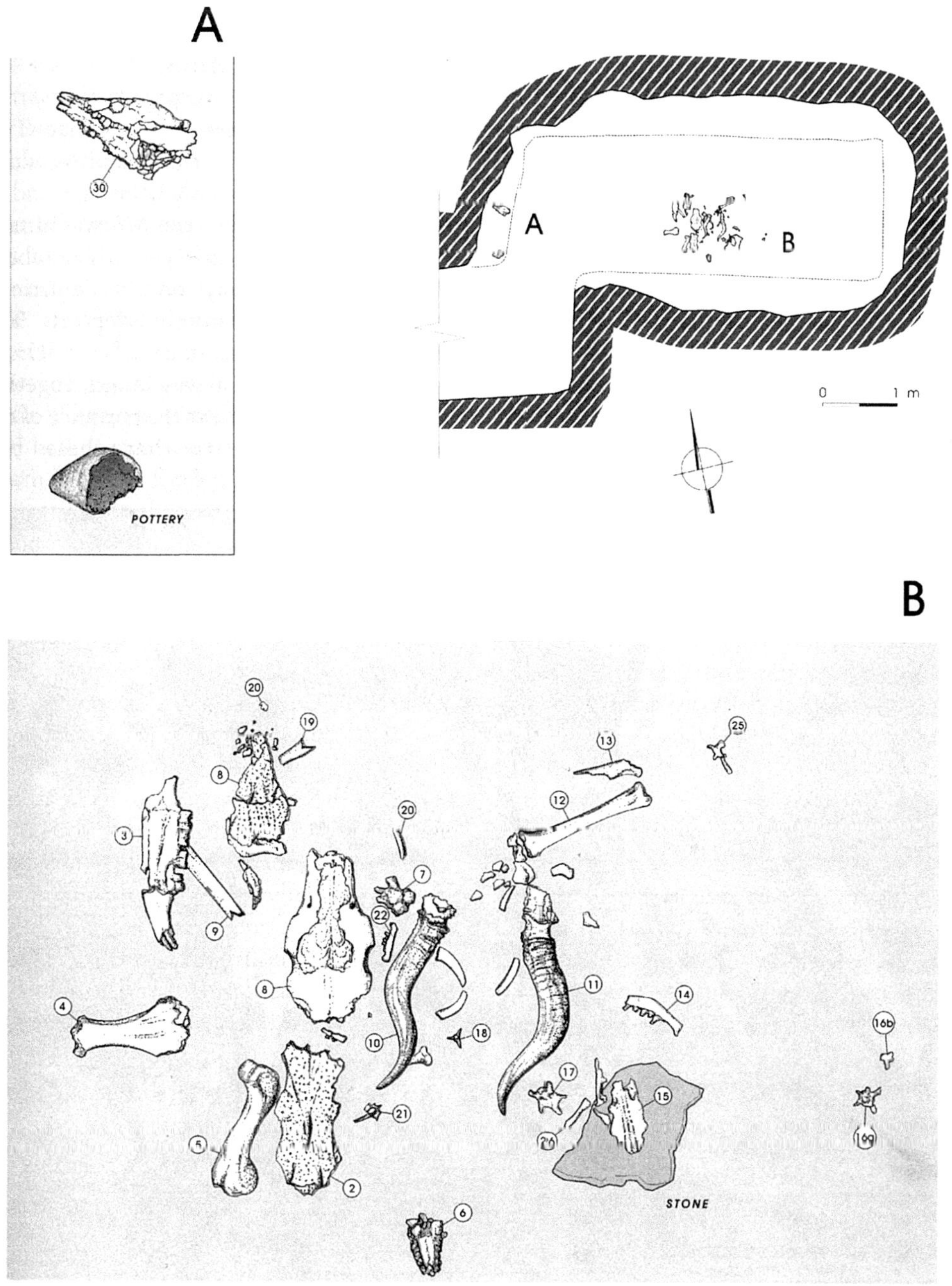

Fig. 3 The bone deposit in Chamber 1 (Drawing by M. Puszkarski).

rulers of Egypt and the Blessed Underworld, respectively, and proponents of *maat*, the ancient Egyptian idea of order, harmony, and balance. Seth, was god of chaos and untamed nature, and as such, certain dangerous and wild animals were under his domain, while other domesticated and benign creatures came under the dominion of Horus and Osiris. Thus, in Egyptian religion wild desert and savanna dwelling animals, as well as male hippopotami, come under Seth's domain, while domesticates and female hippopotami are under the control of Horus and Osiris.

The most dramatic find from the Chamber 1 faunal deposit was the horned skull of a hartebeest (*Alcelaphus buselaphus*). The preservation of the deposit was such that the keratin sheaths covering the horn cores were preserved, albeit damaged and splitting due to dryness. Only the horns and a part of the animal's skull were present. No portion of the maxilla or mandible were found as part of the assemblage, nor were other identifiable portions of the animal, although it is possible that some of the tiny long bone fragments of a large mammal found in the chamber (NISP = three) and in the corridor near the entrance (NISP = four) might be associated with this animal or the donkey (see below).

Hartebeests, like other antelope, were hunted and eaten

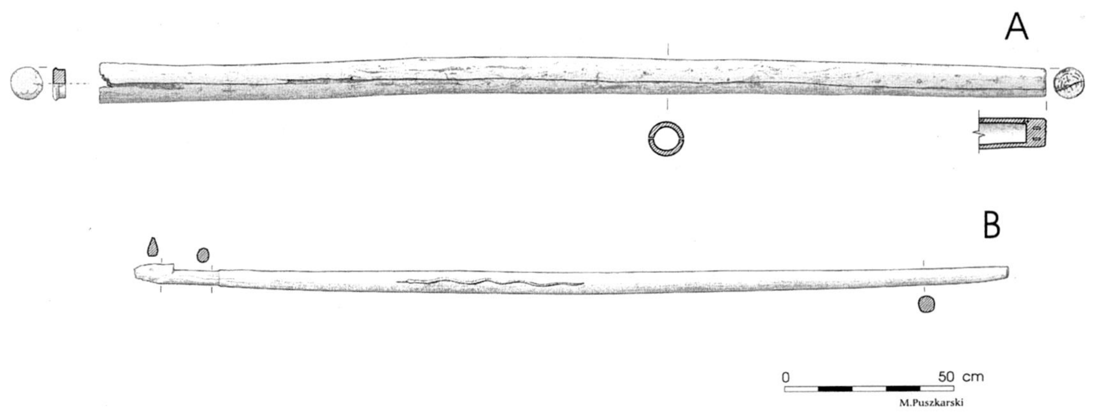

Fig. 4. The harpoon and its case (Drawing by M. Puszkarski).

by the Egyptians. However, they were also symbolic of the chaotic forces of nature, and hunt-scenes show the defeat of these animals, in order to establish order (*maat*) in the universe. For this reason hartebeests were offered to various divinities, thus demonstrating the establishment of a good, godly rulership over Egypt, and a defeat of all chaotic forces.

Another remarkable aspect of the assemblage was the large number of fish bones that it contained. Although it is difficult to be precise about the number of fish bones as many of the cranial plates separated during removal, while others remained fused and were not counted individually, eighty bones belonging to *Clarias* sp. and *Synodontis* spp. were identified, with the former dominating the assemblage. Because of the fragile nature of the bones, it was impossible to piece together all the parts of the fish. However, at least seven individual fish were identified in the deposit (calculations based on unique elements, such as pectoral spines and intact portions of crania), and perhaps as many as twelve. A minimum of two *Synodontis* were found in the assemblage. Cranial plates made up the majority of the piscian deposit; there were few vertebrae. This suggests that the heads of the fish were severed from their bodies and placed in the deposit. It is possible that a part was meant to signify the whole in this case.

A notable feature of the fish was their large size. By measuring the *Clarias'* supraoccipital, the estimated standard lengths of the deposited fish vary between 71 cm and 140 cm (the supraoccipitals ranged from 8–17 cm); the total lengths of the fish varied from 80 cm to approximately 150 cm. It should be remembered that these lengths are estimates and not precise. There is a plus/minus range of 10–20%, especially for the larger fish. The estimates of total length were based on Dr. O. Lernau's work (Fig. 5). These are indeed large specimens that might well be hunted with a harpoon. Generally, however, the harpoon or fish hunting spears that are shown in fishing scenes from Egyptian tombs tend to be smaller (Brewer and Friedman 1989, 21–5).

Length of supraoccipital (mm)	**Estimated Standard Length (cm)**	**Estimated Total Length (cm)**
8	71.5	80
10	87	97
12	101	114
15	130	139
17	140.5	156

Fig. 5. Estimates of standard and total lengths of fishes based on length of supraoccipital.

Catfish were consumed in ancient Egypt, and generally formed a large part of most people's diet. Fish were, however, at certain times in Egyptian history and in certain contexts, regarded as impure or taboo animals; thus, the Egyptians regarded fish with some ambivalence. However, the hieroglyph of a catfish was also part of the name of one of Egypt's earliest kings, Narmer. Furthermore, an ivory label dating to the reign of King Djer shows a catfish being brought as part of the offerings to the king on his *Sed* (rulership) Festival. *Synodontis* had a dangerous and slightly negative connotation for the Egyptians, due to the fact that adult fish eat decomposing organisms, and because these fish have large spines (Brewer and Friedman 1989, 67).

The other unusual animal that forms part of the assemblage is the pig (*Sus scrofa*; MNI = 2). The number of individuals was established by the two skulls found; a few fragments of limb bones may also belong to these animals, but due to their eroded and gnawed state the identification is unconfirmed. Although pork formed a substantial portion of consumed meat, especially in villages and non-royal contexts, there is very little evidence of pig in funerary deposits (Hecker 1982; Miller 1990; Ikram 1995, 29–33, 199–229; Redding 2000), save for some offerings found by the author at Saqqara and Abu Sir in mortuary contexts dating from Dynasty VI. Furthermore, Egyptian, Greek, and Roman texts of the fifth century BC onward stress the taboo nature of the pig in religious contexts as it is an animal associated with Seth, god of chaos. Like the catfish, the pig was a taboo animal in certain contexts.

The pig remains consisted of two crania, intact with the maxillae, one belonging to an old pig (the M3 and C were very worn down; the M3 was almost flat), and the other to a young pig (about 25 months old, based on the dentition). No mandibular elements from these animals were found. The younger animal was part of the circle, while the older beast was found, together with a catfish head, near the entrance of the chamber. It is possible that these two elements had been dragged away from the main deposit by some animal, or, if positioned at the periphery of the circle, had slid down toward the entrance. Note that Saqqara is near a military base and the blasting and artillery practice can cause tremors. Furthermore, the many earthquakes throughout Egypt's history might also be responsible for the disturbance of the deposit. A fragment of a scapula from a young pig and the very gnawed remains of a possible a juvenile pig's humerus, also formed part of the deposit.

Additionally, a donkey (*Equus asinus*) was identified from the Chamber 1 deposit. An entire skull, with the mandible separated from the rest of the head, and located at a slight distance away, was placed in a central position in the circle, oriented to the north. This was the only mandible of a large animal to be found in the deposit. According to the animal's dentition, the ass was well under two and half years of age, most probably between a year and half to two years old at the time of its death. The left foreleg of a donkey was also found, although it was not all articulated, as portions, especially the epiphyses, had been chewed by carnivores. The left scapula, humerus, radius, ulna, metapodia, and a carpal were all recovered from the area. The estimated age of these postcranial bones (humerus: proximal fusing, distal fused; radius: proximal fused, distal unfused, ulna: proximal unfused) suggests that these and the skull belonged to the same individual. Two vertebrae (centrum unfused) belonging to a large mammal, possibly the donkey, were also found. Forelimbs, especially from the left side, are an important part of sacred offerings (Ikram 1995, 50–1). In ancient Egypt, unlike in Rome, the left was the "better" or more sacred side as it was the side where the heart was located. Although donkeys were regarded as the ideal beast of burden, especially before the advent of the camel, they were also sometimes identified with Seth, especially if they came from a wild herd, which is not always easy to determine in the archaeological record.

Aside from the fish, the second largest number of identified bones (thirty, counting teeth in jaws as a unit) came from different canids: *Canis familiaris/C. aureus*. It is difficult to distinguish between dogs and jackals without more of the skeleton being present. From measurements of a mandible, one *C. aureus* is securely identified. There were at least three individuals present, perhaps four. Two of these were mature animals, represented by fragmentary cranial and dentary remains. Two almost complete mandibles with teeth, one left and one right (they did not fit together) were recovered. The left measured 61 mm from p1 to m3, while the right measured 67 mm from p1 to m3. A fragment of a third left mandible (i1-c) was also recovered, but due to the fragmentary nature of the bone, it is difficult to determine whether it is a separate animal, or part of the left mandible. Measurements suggest that it is part of a different individual, while the almost complete skeleton of a third was recovered. These remains include: mandible and maxilla fragments; atlas and axis; scapula (left); humerus (right and left) proximal unfused, distal fused; radius (left) proximal and distal unfused; and ulna (left) proximal fused; tibia (left) distal unfused, proximal unknown; femur (left) proximal and distal unfused; and five metapodia. Several vertebrae, some ribs, fragments of carpals and tarsals and a few fragmentary limb bones without epiphyses and evidence of gnawing as well as fragmentation were found and thought to be part of this animal. Based on dentition and epiphysial fusion, the age of death of the youngest animal can be calculated as being between eleven and fourteen months.

Portions of the skeleton belonging to the third and youngest canid were found beneath rock fall. It is possible that the two dog skulls formed part of the original deposit, and that the younger animal entered the chamber later and died as a result of a rock fall or natural causes. However, as it is not possible to date the rock fall, the skeleton of the young animal might be part of the original deposit. It is difficult to say why only one young complete animal was placed in the deposit; were it a mature animal one could argue that it was a hunting dog. Dogs and jackals were associated with Anubis and Wepwawet, two of the important funerary deities who led the deceased to the Afterworld. They were also associated with cemeteries.

A left mandible fragment of a sheep or goat, containing teeth (d2, d3, and d4) was also found in this deposit. Differences between sheep and goat are difficult to determine; generally for deciduous teeth, some distinctions can be made: generally d4s of goats have 'posts', while those of sheep do not (see Payne 1985). This particular d4 has what looks like the incipient beginnings of a post. The tooth is slightly worn, suggesting that the 'post' will not be forthcoming, but as its presence is uncertain, the identity for this element remains sheep/goat. Furthermore, the other teeth were somewhat damaged, and they could not be used for a definite identification. No other ovicaprid remains have been positively identified from the assemblage, although it is possible that a shaft fragment from a radius devoid of epiphyses might belong to an ovicaprid, and that some of the fragments of limb bones of medium sized mammals might also belong to these creatures. It is difficult to tell if these formed part of the original deposit, or if the remains were fragments that entered the deposit through some other means (washed in or dragged in by carnivores).

Conclusions

This mixed faunal assemblage of catfish, hartebeest, donkey, pig, dog, and perhaps an ovicaprid/small antelope is unique and its nature is unclear. All of these animals have Typhonic associations: the animals are identified with Sethian wild and chaotic aspects of nature, both on land and in the water (te Velde 1977). The placement of the bones in a circle roughly at the center of the chamber argues for a deliberate arrangement, as do the elements involved, mainly heads. The part might be symbolic of the whole, or the animals have been rendered harmless, as are hieroglyphs, by mutilation and the separation of the head from the torso. The donkey foreleg is an anomaly, as are the limb bones that have been positively identified to canids. The foreleg can be viewed as part of an offering, though the more complete canid remains are more difficult to explain. These could be from an animal that was attracted to the assemblage, who then got trapped by a rock fall. Given the degree of preservation with the keratin on the horns, one might expect some evidence of fur and cartilage to remain; however, the bones are all extremely clean.

It is possible that this deposit was related to a royal hunting ritual dedicated to the abolition of chaotic Typhonic beings and the establishment and maintenance of *maat*, or order in the cosmos. It certainly is unlike any other funerary or foundation deposit from ancient Egypt. Perhaps this ritual was associated with the Step Pyramid of Djoser, and was an event that was repeated throughout the Old Kingdom, thereby explaining the Sixth Dynasty pottery found at the base of the deposit.

The harpoon that forms a part of the deposit has yet to be dated using Carbon 14; however, the serpent carved on the case has tentatively been dated to the Djoser's reign on stylistic grounds. Harpoons were symbols of royal power in the early Old Kingdom until Dynasty Five. In the Pyramid Texts, royal funerary texts, Chapter 235 identifies the bones of the king with the point of a harpoon. Harpoons were also used to symbolize victory over enemies, as known from the text in the First Intermediate Period tomb of Ankhtifi at Moalla, after which time non-royal individuals were shown using these weapons, generally to hunt animals, especially hippopotami, and in some instances, pigs.

Certainly, a hippopotamus burial with the harpoon would be more in keeping with Egyptian beliefs than the animals that were found. The only animal forming part of this deposit that was likely to be hunted using a harpoon is the giant catfish. The fact that the harpoon had a blunt wooden point rather than a sharp wood or metal tip would also argue for a ritual rather than a real object. Thus, the ineffectualness of the harpoon, together with the fact that it would not be the first weapon of choice to bring down the animals that form the assemblage of trophy heads poses a problem in the interpretation of this deposit. Thus far, there is no clear indication of the purpose of this assemblage; however, until a more satisfactory answer can be found, it must remain, for the time being a ritual deposit.

Acknowledgements
I would like to thank the director of the West Saqqara Polish Mission, Dr. Karol Mysliwiec, for his kind invitation to participate in the excavation and for providing the opportunity of examining this unusual deposit. I am also grateful to Dr. Omri Lernau for his help with catfish measurements, and Dr. Richard Redding for the loan of his comparative collection of fish and jackal.

References

Brewer, D. and Friedman, R. 1989. *Fish and Fishing in Ancient Egypt*. Warminster: Aris and Phillips.

Grant, A. 1982. The use of tooth wear as a guide to the age of domestic ungulates, pp. 91–108 in Wilson, B., Grigson, C., and Payne, S. (eds), *Ageing and Sexing Animal Bones from Archaeological Sites* (BAR British Series 109). Oxford: British Archaeological Reports.

Hecker, H. 1982. A zooarchaeological inquiry into pork consumption in Egypt from prehistoric to New Kingdom Times. *Journal of the American Research Center in Egypt* **19**, 59–71.

Ikram, S. 1995. *Choice Cuts: Meat Production in Ancient Egypt*. Leuven: Peeters.

Ikram, S. 2000. Preliminary zooarchaeological report, 2000. *Polish Archaeology in the Mediterranean* **12**, 128–32.

Kuraszkiewicz, K. 2000. Remarks on corridor 1. *Polish Archaeology in the Mediterranean* **12**, 133–37.

Miller, R. 1990. Hogs and hygiene. *Journal of Egyptian Archaeology* **76**, 125–40.

Mysliwiec, K. 2000. West Saqqara Excavations, 2000. *Polish Archaeology in the Mediterranean* **12**, 115–6.

Mysliwiec, K. 2001. Le harpon de Sakkara, pp. 395–410 in Gyory, H. (ed.), *Mélanges Offerts à Edith Varga* (Bulletin du Musée Hongrois des Beaux-Arts, Supplément). Budapest: Musée Hongrois des Beaux-Arts.

Payne, S. 1973. Kill-off patterns in sheep and goats: the mandibles from Asvan Kale. *Anatolian Studies* **23**, 281–303.

Payne, S. 1985. Morphological distinctions between the mandibular teeth of young sheep *Ovis* and goats *Capra*. *Journal of Archaeological Science* **12**, 139–47.

Redding, R. 2000. The pig and the chicken: a parable on modeling human subsistence behavior in the archaeological record using historical and animal husbandry data. Unpublished conference paper presented at the Society for American Anthropology conference in Philadelphia 2000.

Rzeuska, T. 2000. The pottery. *Polish Archaeology in the Mediterranean* **12**, 138–40.

Silver, I. A. 1963. The ageing of domestic animals, pp. 283–302 in Brothwell, D. and Higgs, E. S. (eds), *Science in Archaeology*. London: Thames and Hudson.

te Velde, H. 1977. *Seth, God of Confusion*. Leiden: Brill.

Salima Ikram
American University in Cairo,
113 Sharia Qasr el-Aini, Cairo, Egypt
E-mail: salima@aucegypt.edu

9th ICAZ Conference, Durham 2002
Behaviour Behind Bones, (ed. Sharyn Jones O'Day et al.) pp. 47–53

7. Bones and bowls: a preliminary interpretation of the faunal remains from the Punic levels in Area B, at the temple of Tas-Silg, Malta

André Corrado, Anthony Bonanno and Nicholas C. Vella

The University of Malta has been excavating a section of the Phoenician/Punic sanctuary dedicated to Astarte at Tas-Silg (Malta) since 1996. The excavations are still going on, but a significant amount of faunal remains has already been retrieved, enabling preliminary analysis. These remains were recovered from well-documented levels and contexts, and are almost invariably in direct association with inscribed pottery and other artefacts indicative of ritual behaviour.

This paper seeks to give a preliminary interpretation of the mammal bone assemblage from only one of the three excavated areas (Area B). These results are of considerable importance, particularly in view of the fact that there are very few documented assemblages coming from Phoenician and Punic temples in the central and western Mediterranean. It is hoped that the full assemblage will help us to start understanding how this Levantine people adapted its religion and rituals as it migrated westwards, and how this adaptation incorporated the fauna then existing around the new settlements.

The Phoenicians themselves have left us in writing their system of offerings. The Marseilles tariff, listing formulae of offerings meant to regulate the sacrificial rituals at the temple of Baal Saphon at Carthage are written proof of what the western Phoenicians used to do. The faunal assemblage from Tas-Silg complements this textual evidence with zooarchaeological material for the first time.

Introduction

The Phoenician/Punic temple of Tas-Silg is situated on the southeastern coast of the island of Malta. The site lies on top of a hillock overlooking Marsaxlokk bay providing access to two other inlets to the east, making it an excellent location both strategically and for its scenic views (Fig. 1). The site at Tas-Silg spans almost 4000 years of human use and is the location of the ancient sanctuary of Phoenician Astarte (or Roman Juno), which was built over an earlier prehistoric megalithic construction. In Classical antiquity the sanctuary enjoyed international reputation as is attested by ancient authors such as Cicero who, in his orations against the corrupt Roman governor Verres, refers to the treasures of the sanctuary of Juno and the respect it received from pirates and Numidian princes (*Verr.* ii, 4, 103–4). Ptolemy, in his *Geography* (4.3.13), also gives the co-ordinates of the temple of Juno, which presumably was an important landmark, especially for mariners.

The site of Tas-Silg was extensively excavated in the 1960s by an Italian archaeological mission. The University of Malta opened its own dig in the southern sector of the site in 1996 with the intention, amongst its other aims, of gathering more evidence about the ritual activities that were taking place inside the sanctuary (Bonanno and Frendo 2000).

This paper gives an overview of the preliminary results of what is still, in all aspects, an ongoing archaeological project. Only a small part of the animal bone assemblage has been analysed to date, and this has been limited to the bones of macro-mammals. This is the first zooarchaeological study to be carried out on Malta by local scholars and the infrastructure for this kind of study was not available in the past. Our first task, therefore, has been that of setting up a zooarchaeological reference collection to enable the study. It is well known that it takes years to build a proper collection and while the one at the Department of Classics and Archaeology of the University of Malta is growing, it is still in its initial phases. Thus, it

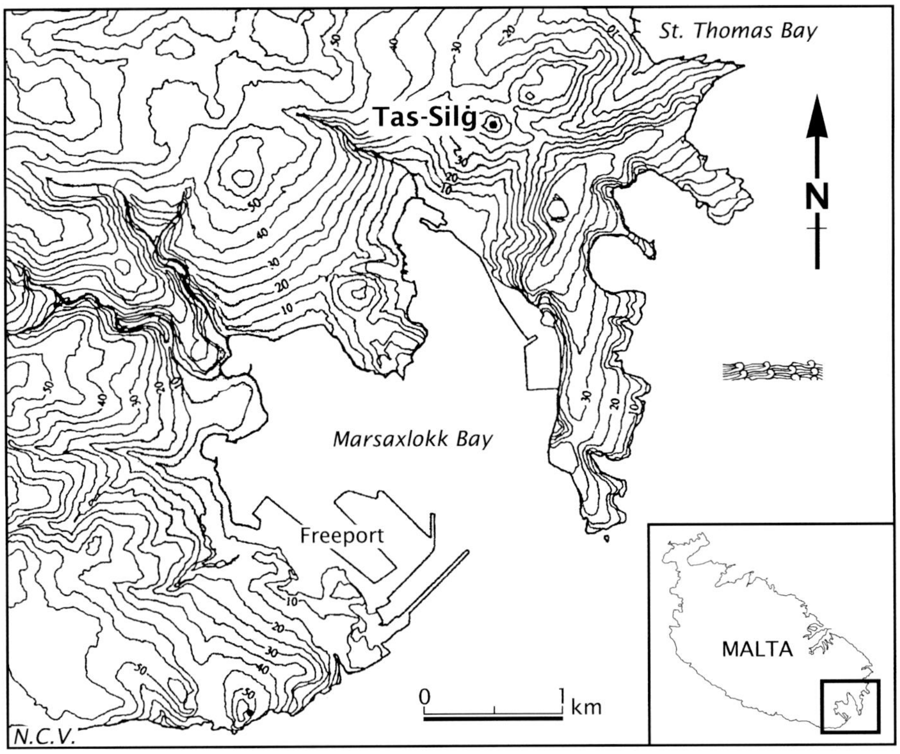

Fig. 1. Location of Tas-Silg, Marsaxlokk, Malta.

must be kept in mind that this study was conducted within the physical and practical constraints of our present reference material.

The zooarchaeological repertoire from Tas-Silg is our first glimpse at the material remains of activities about which, so far, we knew only from the ancient texts, if at all. The material studied, although partial, is enough to provide a first impression of ancient Punic ritual in the central Mediterranean, and allows initial discussion and comparisons with what has so far been deduced only from ancient textual sources. In this respect, perhaps the most important assemblages from the site are not the ceramic and artefactual ones as much as the environmental assemblages, in particular the animal bone repertoire that is the subject of this paper.

It is considered important that any interpretation of a faunal assemblage, particularly that resulting from ancient ritual behaviour should be contextualised. Our approach is not an exercise in pure zooarchaeological statistics. It allows data from other assemblages and textual sources to be juxtaposed with the faunal material, improving the possibility of interpreting evidence.

Available evidence

The discussion of ancient Phoenician and Punic ritual practices often takes place in the light of textual, iconographic, epigraphic and semantic evidence (Amadasi Guzzo 1988; Lipiński 1993). Indeed, this type of evidence has been privileged over the material remains of rituals, although there are now signs that call for a change of approach (Bonnet and Xella 1995, 319). The excavations at Tas-Silg offer us the opportunity to supplement the debate with non-textual evidence. The potential of such an approach should not be underestimated. To date, publications concerning environmental data retrieved from Phoenician and Punic non-funerary religious sites have been rather scant, with eleven sites having had their faunal material noted but no specialist reports published (Vella 1998, 285, Table 4.3).

The textual evidence concerning Phoenician and Punic rituals comes from two types of sources, namely epigraphic (the Marseilles and Carthage tariffs) and Biblical (the Book of Leviticus). The Marseilles Tariff (*Corpus Inscriptionum Semiticarum* I, 165) is an incomplete late fourth- or early third-century BC inscription that was affixed to the gate of the temple of Baal Saphon, the protector of seafarers, in the Phoenician metropolis, Carthage (Amadasi Guzzo 1967, 169–182). It is a list of formulae regulating the procedures to be followed during different types of sacrifices. The book of Leviticus, on the other hand, offers a collection of direct and indirect references to Canaanite and ancient Israelite sacrificial codes and traditions that have been found to be of relevance to Phoenician and Punic religious studies (Amadasi Guzzo 1988). If we follow common opinion that the book of Leviticus was compiled in the Post-Exilic period (6th century BC

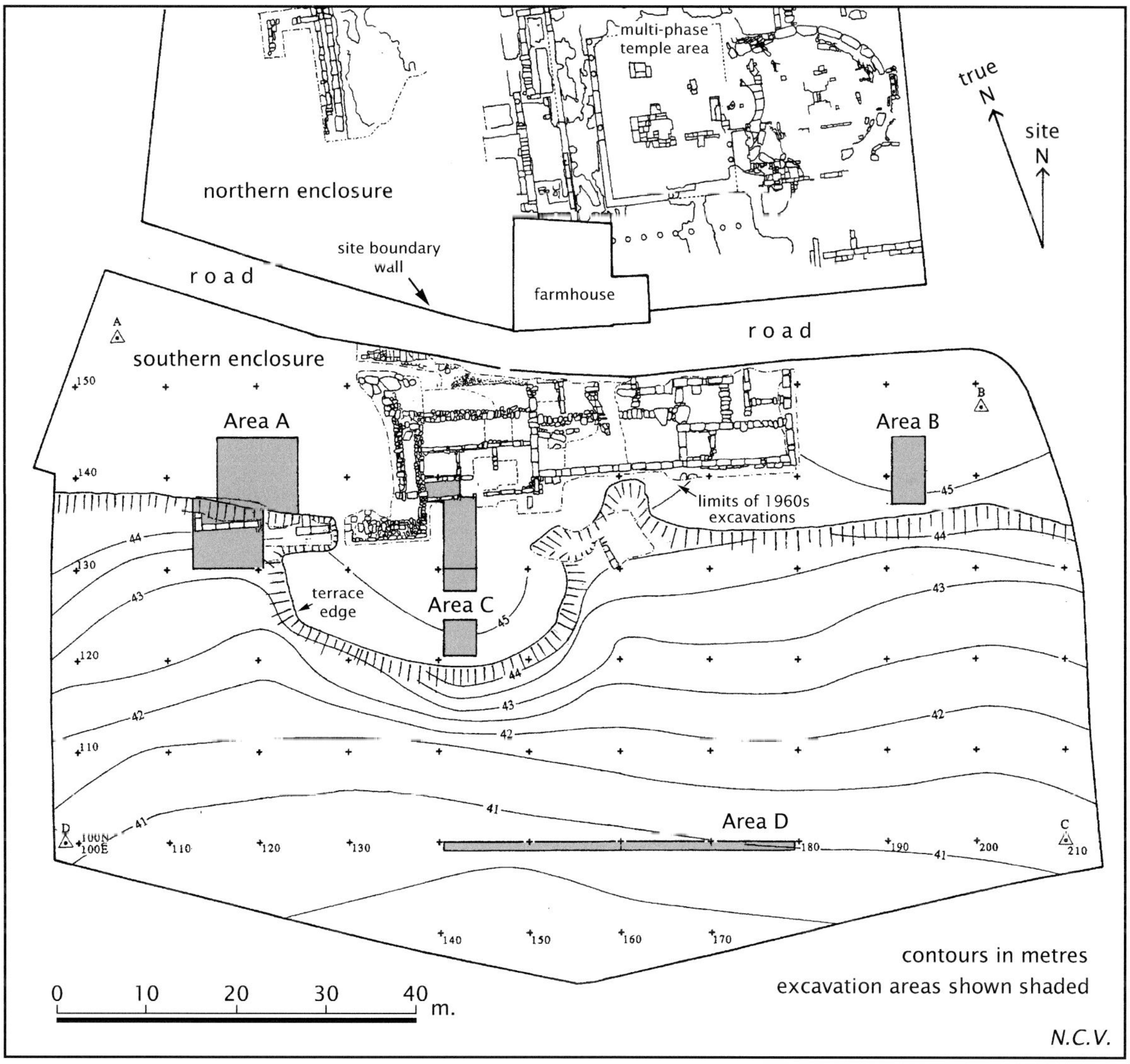

Fig. 2. Plan of Tas-Silg: archaeological site and locations of trenches.

onwards), these sources would stand about three hundred years apart and probably represent changing ritual practices that varied by region. Indeed, anomalies are detected within the biblical texts (for example between the Leviticus and Deuteronomy) as well as within the tariffs itself, as Amadasi Guzzo (1988) has pointed out, but have never been traced in the archaeological record.

The archaeological material under discussion comes from Area B, a 7 m x 4 m trench dug in the southern sector of the site at Tas-Silg (Fig. 2). The area has been meticulously excavated, with horizon plans being drawn at regular spit levels in order to aid in the reconstruction of the taphonomy and formation processes in play on site. Area B is on the edge of the sanctuary precincts and is characterised by a midden that contains material that was used in religious rituals (Mason and Vella 2000, 76–80) (Fig. 3). This is confirmed by the number of potsherds retrieved that contain short, stereotyped votive inscriptions (Frendo 2000, 98). Whether the actual formation of the midden is ritualistic in nature as suggested by earlier excavators has yet to be seen. In Leviticus (4:12) an area outside the sanctuary has been identified as a 'depository of ashes' utilised during atonement sacrifices (Amadasi Guzzo 1988, 100). This depository is mentioned in the context of sacrifices carried out in expiation of sins committed by anointed priests and the community. What we have in Area B may be a similar depository of ashes, especially since ash is present in varying intensity in the different layers.

While the ancient texts give the formula for the sacrifice, what happened between the sacrificial act and the final deposition of the material in the archaeological record is still largely unknown. This is ultimately our major point of focus. The Marseille Tariff provides an insight in that the inscription specifies which parts of the sacrificial victims, and which one of the three different types of sacrifices, were assigned to the priest and which parts were to be given to the members of the congregation

Fig. 3. Area B: detail of pottery sherds and bone fragments from the midden (SU1033).

making the offering. We cannot, as yet, translate the composition of our archaeological layers into a direct interpretation of ancient Punic ritual practices. Taphonomic evidence suggests that the deposition of many of the layers from Area B was of a secondary nature. These layers cannot therefore be taken to mirror the original spatial and contextual dimension of the material under study. The secondary nature of the deposits is confirmed by the fact that no animal bone articulations have been found in association and is further corroborated by the fact that only 16% of the ceramic material from one layer was found to be joining (Zammit 2001). These matching fragments pertained to vessels that were probably broken in situ when thrown on the heap. The secondary nature of the deposits, however, is no reason to forfeit valuable archaeological information. The midden itself is the result of conscious human actions closely related to ritual activity, and if correctly interpreted should still offer us a unique window onto past Punic ritual practices.

The composition of the assemblage

The assemblage on which this analysis is based is only part of the total number of bones retrieved from Area B, but as argued above, it is enough to highlight important trends in the archaeological record. 30% of the animal bones analysed have been identified to species, whereas a further 39% have been identified to a higher level including the 'small ungulate' group. The remaining 31% were unidentifiable to species. In spite of the fact that the assemblage is fairly fragmented, with the average fragment size being less than 1.5 cm in size, 64% of the fragments have been identified to the element. Most of the bones are covered with hard encrustations, typical of Maltese calcareous soils, limiting the study of surface striations, including butchery marks.

Ovicaprids (sheep and goat) are by far the most represented species from the assemblage analysed and they constitute 96% of the mammal bones identified. The remaining 4% is made up of cattle bones. The overwhelming presence of ovicaprids comes as no surprise and is to be expected in the context of a rocky, semi-arid Mediterranean island. Moreover, sheep and goat are frequently mentioned in the ancient texts referred to thus far as the proposed victims of sacrifices more than cattle or any other animals: lamb is the victim of the Passover sacrifice, a he-goat is specified for an expiation sacrifice for the sin of a priest while sheep and goat are also frequently mentioned in the Marseilles Tariff (Amadasi Guzzo 1988). Cattle are mentioned only in special instances such as the expiation of sins committed by an anointed priest or the Israelite community (Leviticus 4:3). This high percentage of ovicaprid remains allows more detailed study to concentrate on sheep and goat bones rather than any other species.

A preliminary analysis of the body-part representation gives an indication of which part of the carcass appears to be more represented on site. For this purpose the different elements identified have been grouped into body parts, loosely mirroring general butchery portions, that constitute an animal and the numbers have been corrected to allow for bias arising from the differential presence of elements in a single animal skeleton. The underlying principle behind this approach is that we are looking at an assemblage that is the result of butchery at some stage or other of its formation process and is therefore best analysed in terms of portions. The texts repeatedly speak of animals being dismembered and it is thus understandable that in the mind of the offering community the animals were also conceived as 'butchery portions' rather than as a 'whole'. Within the context of the sacrificial system large mammals must have functioned as 'parts' rather than 'complete' animals.

When the results are compared to the frequency of the same body-parts in a complete skeleton the resulting pattern seems to indicate an over-abundance of forequarter and hindquarter parts, which contain the meatiest parts of a carcass (Fig. 4 and Fig. 5). Against this background it is interesting that Leviticus (7:28–34) instructs the offerer to present a forequarter and reserve its fat to Yahweh and its meat to the priest. Notwithstanding this information, at

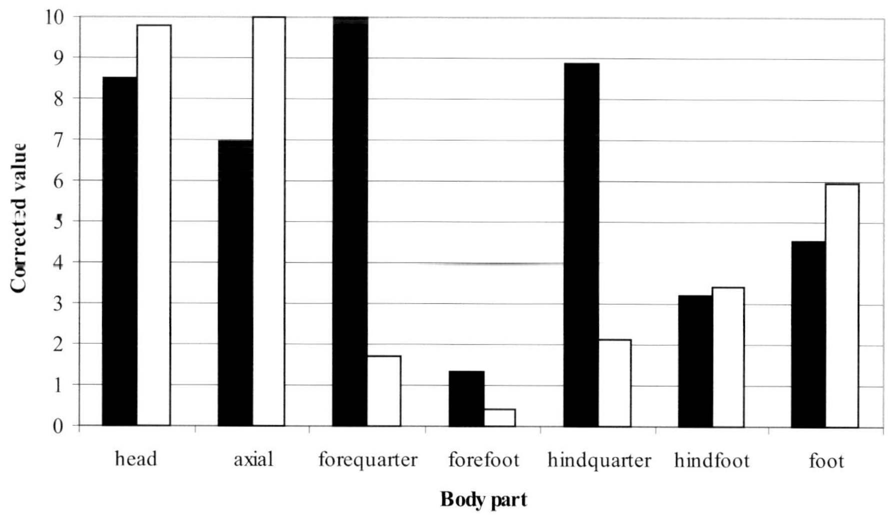

Fig. 4. Ovicaprid body-part representation. The white bars represent the values from a complete skeleton. The values from the Area B assemblage are black.

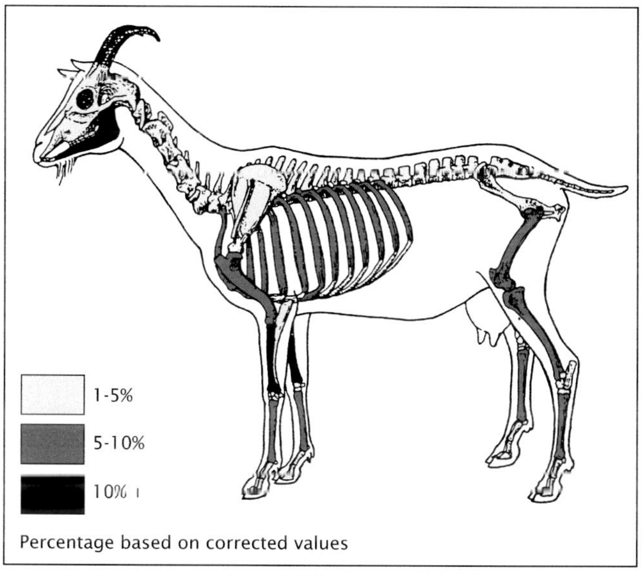

Fig. 5. Diagram showing the frequency of the different sheep/goat elements found in the Area B assemblage.

Tas Silg no part of the skeleton is completely missing, thus confirming that the whole skeleton was being processed and disposed of on site. This makes sense in a sacred context where the whole animal is the object of a ritual. What we know for sure from the ancient sources is that the different parts of the animal were not being subjected to the same treatment, depending on the kind of sacrifice being carried out (*e.g.*, Guzzo Amadasi 1988, 100). This must have further affected the variable survival of elements but it is difficult to identify between the remains of the different sacrifices in the archaeological record.

It is not far-fetched to imagine that some of the animals could have been purchased by the pilgrims on site rather than brought along from outside. The scene of the Jerusalem temple vendors (Luke 19:45) although much later in time, comes immediately to mind. Amadasi Guzzo explains how differences in the payments for each sacrifice between the Marseilles and the Carthage tariffs could in fact represent the amount paid for the animal (Amadasi Guzzo 1988, 117). As things stand however, there is no hard evidence to confirm this as yet.

The presence of the complete array of elements on site and, in particular, the apparent over-abundance of the meat-bearing parts point towards the consumption of the carcasses on site. This is also attested by the texts, with only few sacrifices, such as the holocaust (Leviticus 1) and the *šlem kll* (tariffs), involving the complete combustion of the carcass (Amadasi Guzzo 1988, 111). Leviticus (*e.g.*, 3:3–4, 3:9–10, 7:28–33) makes it clear that in the different sacrifices the animal was usually divided between Yahweh, the priest and the offerer. Yahweh usually got the greasiest interiors of the body including the stomach and the intestines and occasionally the tail, while the rest was shared between the priest and the offerer, to be consumed (7:32–3). As is stated in Leviticus (3.16). 'All fat belongs to Yahweh'. There are some differences in the biblical and epigraphic texts as to the parts owed to the priest, but the most common parts appear to have been the right hindquarter and the chest area together with the skin, which is altogether untraceable in the archaeological record at Tas-Silg.

Various layers from Area B are grey ashy deposits that show ample signs of burning. This may not be surprising in view of the fact that both the bible and the tariffs clearly indicate that most rituals involved the burning of at least part of the animal. Usually, it was those parts dedicated to the divinity that were incinerated with the exception of complete burning of the carcass as mentioned above. What is surprising in this context is the very low percentage of bones (2%) that exhibit any signs of burning, with only 7 fragments bearing signs of calcination and thus indicating exposure to high temperatures. In view of what the texts say, the more logical conclusion at this stage is that the ashes belong to, apart from the solid fuel, the remnants of those parts being offered to the deity, and therefore the non-bony organs. It is interesting that, with reference to the above mentioned depository of ashes, Leviticus (4:12) states that the bull offered for the atonement of the sins of a priest is to be carried outside the precincts to be burnt 'where the fatty ashes are thrown'.

Leviticus clearly affirms that in some cases animals were consumed on site (Amadasi Guzzo 1988, 104). Eating the animals on site must have involved cooking the animal. The lack of burnt bones seems to suggest boiling as the preferred cooking method. Boiling is mentioned in the Leviticus as the suggested way of cooking the animals with only a few cases of roasting (Amadasi Guzzo 1988, 105). The latter may also have been practised at Tas-Silg, in view of the small quantities of charred and blackened bones in the assemblage. Boiling seems to be also confirmed by the ceramic repertoire which includes large pots with everted rims to hold a lid, that are suitable for boiling (Sagona 2002, Cooking Pots Form IV:1; Skibo 1994). It is interesting to note that a good number of these vessels have been inscribed 'to Astarte' in Punic script prior to firing, indicating that they were intentionally produced for cultic use (Amadasi Guzzo 2000, 181; Quercia 2000; Sagona 2002, 222), thus further corroborating that boiling and eating the carcass on site were an intrinsic part of the ritual. It is also possible that some of the ash in the layers could also be the remnants of the combustible material that was used for cooking.

Another interesting aspect of the assemblage is the incidence of dog gnawing. Some of the bones show very clear evidence of gnawing, while many long bones are missing their articulations (Fig. 6). This might come as a surprise in a ritual environment and has yet to be studied more carefully. We are used to the western concept of sacred places that are out of bounds for animals as a sign of respect. Moreover, in our society animals would definitely not partake of any ritual meal. In Punic Malta, however, the practice may have been somewhat different. Dogs were present on site, their presence betrayed by the gnawing marks, and they were either being fed during the ritual meals we presume took place within, or in the environs of, the sanctuary or else the material was left lying around long enough for it to be scavenged by them.

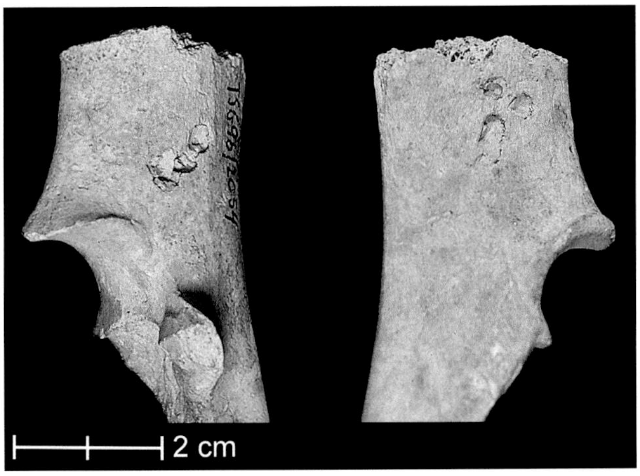

Fig. 6. A clear example of dog gnawing showing carnassial marks on the proximal-lateral (left figure) and proximal-medial (right figure) surfaces of the same ovicaprid ulna.

Further research

The animal bone assemblage from Area B, in its context, seems to derive from a midden area outside the temple precincts that was a final depository for ashes, bone and other non-organic artefacts closely associated with religious ritual. The pattern of animal remains so far examined fits into the ritual picture as drawn by ancient texts.

It is, however, still premature for complete explanations of past Punic ritual practices to be put forward and at this stage our research has perhaps created more questions than answers. Further work shall be based on a larger assemblage and shall also include the spatial and contextual analysis across the site. We already know, for example, that in other areas of the site different species are present, such as pig. We hope to provide a first glimpse at the remains of central Mediterranean Punic non-funerary religious rituals in the fourth and early third centuries BC and hopefully stimulate further research and discussion. The picture will become even more complete once the fish and the bird assemblages are also analysed. At this stage, however, it is already interesting to note that while the ancient texts mention birds as possible victims for sacrifices there is no mention of fish and other marine organisms either in Leviticus or in the Marseilles tariff. This trend has already been noticed elsewhere, even if in a completely different archaeological context and period (Lev-Tov and Maher 2001). Tas-Silg has already yielded a considerable number of bony fish remains, sea urchins and seashells (Schembri 2000). There is a possibility that variations, perhaps of a regional nature, existed in the rituals, and that the Maltese Punic community was making use of the largest commodity around it, the sea, especially since the land-based biota of an island would have been more limited than that of mainland areas.

References

Amadasi Guzzo, M. G. 1967. *Le iscrizioni fenicie e puniche delle colonie in occidente*. Roma: Istituto di Studi del Vicino Oriente, Università di Roma.

Amadasi Guzzo, M. G. 1988. Sacrifici e banchetti: Bibbia ebraica e iscrizioni puniche, pp. 97–122 in Grottanelli, C. and Parise, N. F. (eds), *Sacrificio e società nel mondo antico*. Bari: Edizione Laterza.

Amadasi Guzzo, M. G. 2000. Quelques tessons inscrits du sanctuaire d'Astarté à Tas Silg, pp. 181–96 in Aubet, M. E. and Barthélemy, M. (eds), *Actas del IV Congreso Internacional de Estudios Fenicios y Punicos*, vol. 1. Cádiz: Universidad de Cádiz.

Bonanno, A. and Frendo, A. J. (eds) 2000. Preliminary report of the University of Malta excavations at Tas-Silg (1996–1998). *Mediterranean Archaeology* **13** [2002], 67–114.

Bonnet, C. and Xella, P. 1995. La religion, pp. 646–62 in Krings, V. (ed.), *Civilisation Phénicienne et Punique: Manuel de Recherche*. Leiden: Brill.

Frendo, A. J. 2000. The inscriptions, pp. 97–9 in Bonanno, A. and Frendo, A. J. (eds), Preliminary report of the University of Malta excavations at Tas-Silg (1996–1998). *Mediterranean Archaeology* **13** [2002], 67–114.

Lev-Tov, J. S. E. and Maher, E. F. 2001. Food in Late Bronze funerary offerings: faunal evidence from Tomb 1 at Tell Dothan. *Palestine Exploration Quarterly* **133**, 91–100

Lipiński, E. 1993. Rites et sacrifices dans la tradition phénico-punique, pp. 257–81 in Quaegebeur J. (ed.), *Ritual and Sacrifice in the Ancient Near East*. Leuven: Peeters.

Mason, S. and Vella, N. C. 2000. The stratigraphy, pp. 71–83 in Bonanno, A. and Frendo, A. J. (eds), Preliminary report of the University of Malta excavations at Tas-Silg (1996–1998). *Mediterranean Archaeology* **13** [2002], 67–114.

Quercia, A. 2000. Pasti rituali nella Malta punica. *Malta Archaeological Review* **4**, 28–33.

Sagona, C. 2002. *The Archaeology of Punic Malta*. Leuven: Peeters.

Vella, N. C. 1998. *Ritual, Landscape, and Territory. Phoenician and Punic Non-Funerary Religious Sites in the Mediterranean: an Analysis of the Archaeological Evidence*. Unpublished Ph.D. dissertation, University of Bristol.

Schembri, P. J. 2000. The molluscan remains, pp. 102–9 in Bonanno, A. and Frendo, A. J. (eds), Preliminary report of the University of Malta excavations at Tas-Silg (1996–1998). *Mediterranean Archaeology* **13** [2002], 67–114.

Skibo, J. M. 1994. The Kalinga cooking pot: an ethnoarchaeological and experimental study of technological change, pp. 113–26 in Longacre, W. A. and Skibo, J. M. (eds), *Kalinga Ethnoarchaeology: Expanding Archaeological Method and Theory*. Washington: Smithsonian Institution Press.

Zammit, M. E. 2001. *Site Formation Processes at Tas-Silg: Analysis of the Ceramic Material from Stratigraphic Unit 1033*. Unpublished B.A. (Honours) dissertation, University of Malta.

André Corrado, Anthony Bonanno & Nicholas C. Vella
Department of Classics and Archaeology,
Archaeology Centre, Car Park 6, University of Malta,
Msida MSD 06, Malta
E-mails: acor1@um.edu.mt,
anthony.bonanno@um.edu.mt,
nicholas.vella@um.edu.mt

9th ICAZ Conference, Durham 2002
Behaviour Behind Bones, (ed. Sharyn Jones O'Day et al.) pp. 54–61

8. An Iron Age bone assemblage from Durezza Cave, Carinthia, Austria: detecting ritual behaviour through archaeozoological and taphonomical analyses

Alfred Galik

Throughout Central Europe's prehistory dwells, pits, postholes and natural cavities attracted humans to deposit people and animals. Therefore, these subsoil features sometimes yield special deposits consisting of bones associated with metal work and pottery. The assemblage from Durezza Cave is a unique Austrian Iron Age example discussed herein. Besides numerous human remains, pottery and bronze metal work, the vertical cave yielded complex parts of animal carcasses. The majority of these finds are homogeneously distributed and it is assumed that the cave was in use for 100 to 150 years, beginning in the late Hallstatt period (400/500 BC) up to the early La Tène period (300/400 BC).

The animal remains from this feature differ clearly from everyday disposals associated with settlements. Dog remains are most abundant in the Durezza assemblage, followed by horses. Other typical domesticates used as nutritional supply, such as ovicaprines, pigs and cattle are atypically less common in Durezza Cave. The large quantity of articulating animal bones, the inverse frequency of artificial modifications, and untypical distributions of breakage patterns in the animal bone assemblage reveal differences to domestic waste, as well. Besides these irregularities, a half a year seasonality can be detected in killing neonate and juvenile domesticates. I conclude that the animal assemblage at Durezza Cave should be interpreted as a result of distinct symbolically structured practices such as ritual worship.

Introduction

Durezza Cave is located at the Dobratsch massif near Villach in Carinthia, at Austria's most southern federal land. The vertically oriented cave developed as a result of tectonic and karst phenomenon in a small valley between the Dobratsch massif and the smaller hill Tscheltschnigkogel. A large block originally sealed up the cave's entrance and local cave explorers discovered this site, in the continuation of a doline. They started to dig a conically shaped sediment pile positioned at the bottom of the cave and below a thin and sterile loam stratum they recovered a large amount of human and animal remains. The cave explorers stopped their digging, and the remaining sediment was scientifically excavated under the direction of Dr. Gleirscher, Regional Museum of Carinthia, Klagenfurt (Galik 1997).

The majority of the remains date to the late Hallstatt period and the beginning of the La Tène period (Gleirscher 1997a). On one hand, the date is derived from traditional bronze ornaments, such as bronze fibulae (late Hallstatt culture, stages La Tène A – B1 300–405 BC) and bronze hair rings (400 – 550 BC, stages Hallstatt D2 – La Tène A). Two radiometric dates from human bones confirmed the archaeological results (Fuchs 1996, radiocarbon dating, Univ. Groningen, GrN-21825: 2450±50 BP; GrN-21826: 2490±50 BP).

An archaeological structure like Durezza Cave is unique in this regional area, but numerous other archaeological sites are reported in the vicinity of the Tscheltschnigkogel. Most of the sites in the region document activities from the Bronze Age, the Hallstatt period and the La Tène period. An example that does not reveal settling activities comes from Portal Cave, a small fissure in the limestone, south of Durezza Cave. Portal Cave yielded a large amount of ceramic potsherds from approximately 30 vessels. The remains are thought to be intentionally deposited and sacrificed (Dolenz 1972). The pottery remains document activities from the late Bronze Age (Bronze Age D to Hallstatt A, approx. 1350–1000 BC). Graveyards containing cremated people in urns, buried in small mounds are documented in the region, dating from the Hallstatt A (1000 BC) to the La Tène A/B period (300–400 BC, Gleirscher 1997b). The locality of the prehistoric settlement associated with Durezza Cave is unknown, although remnants of a building relating to the Hallstatt/La Tène period are reported from the Tscheltschnigkogel (Gleirscher 1997b). On top of the Tscheltschnigkogel are remains of a late Roman hill fort (5th–6th century AD, Piccottini 1989).

As I tried to find literature about comparable archaeological structures revealing similar bone assemblages, I soon realized that the crucial point was to determine, if such assemblages as that from Durezza Cave represent remains of sacrificial worship. Concerning human deposits, a debate is ongoing if people were sacrificed, or if there is another way to explain such assemblages. Similar subsoil structures appear to be surprisingly numerous in Central Europe, though comparable assemblages are quite rare (archaeozoological investigations, *e.g.*, Riedel 1977; Teichert 1977; Teichert 1978; Teichert 1985; Teichert 1985; Leja 1987). The cause of the data bias is often due to inefficiency of old excavation practices and investigation techniques. On the other hand, comparable assemblages often reveal a mix of remains from different cultural periods, including, for example, the Bronze Age, the Hallstatt, and the La Tène period.

Methods

As we started to excavate the cave, we were aware of this pivotal question in interpreting such a site, and we decided to employ fine excavation techniques (Galik 1997), although the campaign was restricted due to financial and time resources. All excavated sediments were washed through sieves with a mesh width of one millimetre. In order to reduce the loss of information, we re-excavated the already removed cave sediment.

To provide comparable data on animal remains, basic numerical counting techniques were used, such as NISP (number of identified specimens) and MNI (minimum number of individuals) following Grayson (1984). Some authors have pointed out justified critiques to the weakness of MNI procedure in the reconstruction of palaeopopulations (Fieller and Turner 1982, Ringrose 1993, Rogers 2000). However, the MNI procedure is used as a quantification method rather than as an attempt to reconstruct prehistoric animal populations in this investigation. Normally, MNIs include body side, estimated age, and defined types of bone fragments. Besides completely preserved long bones, fragments are divided in proximal, proxmid, mid, middistal and distal shaft fragments. Only overlapping shaft fragments must be counted to reduce overestimation of individuals. Unfortunately, this technique neglects paired elements, which occurred in large quantities in the Durezza Cave assemblage. Therefore, I used an extended MNI counting technique (MNIr) that takes results of re-articulation analyses, in this case paired elements, into consideration. For example, five dog humeri represent three right and two left-sided elements, including a pair of bones. MNI undoubtedly results in three individuals. MNIr consists of one individual derived from paired bones, two individuals from right and another one from left-sided humeri and results in four individuals.

Intensive ageing of animal remains was carried out to reconstruct potential kill seasons. Analysis of breakage patterns, modifications of natural and artificial nature on bones, and the aforementioned re-articulation analyses were employed to reconstruct the taphonomical history of the animal bone assemblage.

Material

The bones appeared disorderly scattered, and sometimes densely packed in the cave sediment. A remarkable quantity of complete anatomical elements passed on. They represent not only small and solid specimens, such as phalanges or tarsals, but also long bones. A few cut marks recognized on bones in all domestic species do not necessarily indicate tendencies in butchering behaviour (Galik 2002). They may also relate to removing hide, and disarticulating and de-fleshing carcasses. Analyses of breakage pattern revealed only a few bones that showed typical spiral-type, sharp-edged green bone fractures. Such pattern usually indicates fragmentation of fresh bones containing collagen (Lyman 1994). The majority of fragmented animal bones showed perpendicular fractures with fine-grained perpendicular edges. These bones are lacking in collagen and broke up after long-term deposit in the cave sediment. Digging produced most of these fragments. However, some perpendicular edges are covered with sinter. These bones were probably deposited at exposed positions and became fragmented by rock debris from the cave limestone.

Cut marks are sometimes found in combination with green bone fractures. Such pattern relates to carcass processing, like chopping and de-fleshing prior to disposing the bones into Durezza Cave. Besides artificial modifications, some animal remains show natural modifications such as carnivore gnawing marks (Fig. 1).

The faunal assemblage mainly consists of domestic animals, while wild animals are rare. Dog is most abundant followed by horse. Other domestic animals usually used for slaughter, like ovicaprines, pig and cattle, are atypically less abundant. The NISP and MNI values are impressive due to the abundance of dogs. However, the values do not indicate any age distribution of these animals as the result of age pattern analyses; a large part of the dog remains relate to adult individuals (Fig. 2 and Fig. 3). Other remains reflect different age distributions and reveal numerous

	NISP	Ec	MNI	MNIr	Cut marks	Carnivore gnawing	Pathology
Ungulates	118	–	–	–	4	1	–
Horse	1028	566	13	20	19	8	18
Cattle	520	92	5	5	12	–	5
Sheep/Goat	1026	320	24	36	7	16	7
Pig	361	142	7	9	7	4	4
Dog	2722	1484	45	55	14	69	132
Fox	59	25	3	4	–	–	–
Hare	41	12	4	5	4	7	–
Roe deer	39	20	1	1	–	–	–
Cat	35	19	3	3	–	–	–
Red deer	6	4	1	1	–	–	–
Birds/Chicken	20	3	3	4	–	–	1
Total	**5722**	**2626**	**109**	**143**	**67**	**105**	**167**

Fig. 1. Numerical distribution of elements (NISP), number of completely preserved elements (Ec), minimum number of individuals (MNI, MNIr), artificial and natural modifications (cut marks and carnivore punctures), and pathologies in animal species from Durezza Cave.

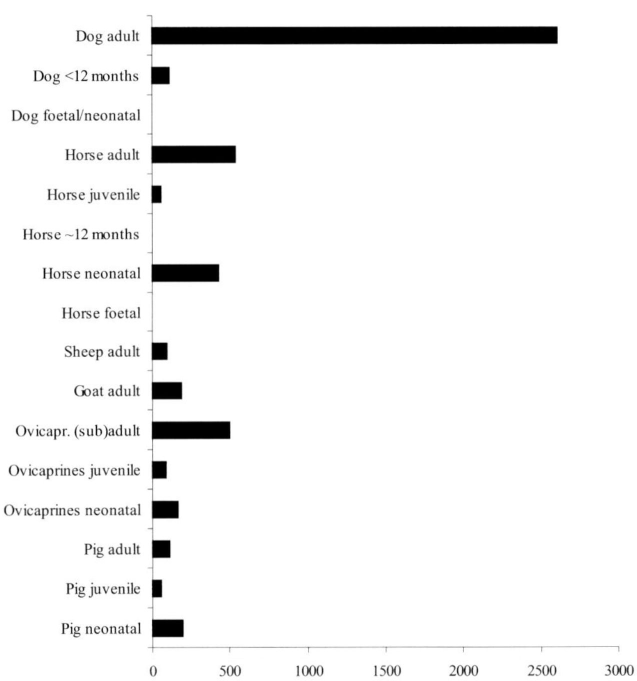

Fig. 2. Mortality patterns in domestic animals based on NISP values.

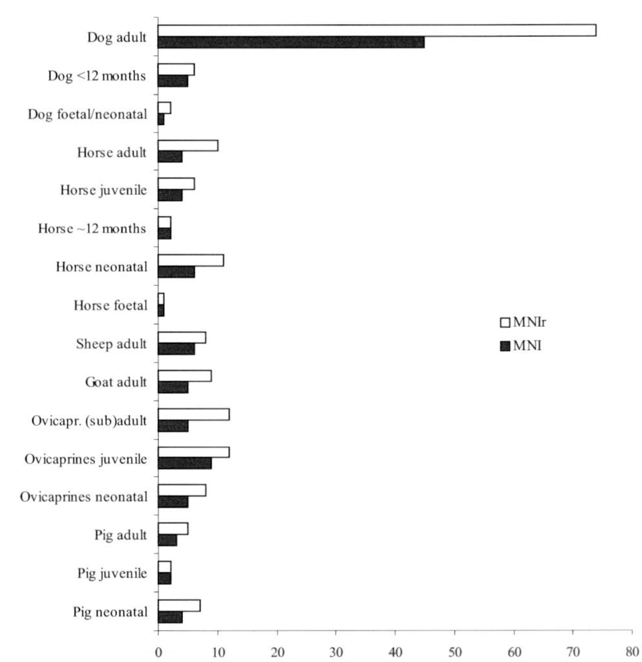

Fig. 3. Mortality patterns in domestic animals based on values derived from both MNI calculation methods.

neonatal and juvenile individuals, particularly in horses, ovicaprines and pigs.

Taphonomical history

It is expected that every being surviving the fall into the cave will certainly die inside. The chimney cave is impossible to leave because another exit does not exist. The limestone does not exhibit any artificially modified structures to enter or leave the cave. Snails and even large flies that entered the cave during the excavation did not find the way out.

An understanding of Durezza's taphonomical history is important in order to determine if this site is simply a natural trap, or if the remains reflect intentional deposits. Due to complicated sedimentary processes in chimney caves (Zapfe 1954), and loss of information caused by rough excavation methods used by the cave explorers, taphonomical history of archaeological finds from Durezza Cave is difficult to accurately reconstruct.

Archaeological finds were relatively homogeneously distributed inside the cave. Most remains correlate to the same time period. However, a few finds belong to the Roman period and to early Medieval times. These remains were uncovered by cave explorers in a niche in the upper part of the shaft and on top of the sediment pile. Only a single Roman potsherd was excavated within a middle stratigraphical layer. All other prehistoric sherds are intensively fragmented and do not give the impression that they were originally deposited as complete vessels (Gleirscher 1997a). However, sherds fracturing is indicative of secondary and/or natural fall in the cave and of already fragmented waste that was probably washed into the cave. Few Roman finds could have entered the cave before a large block sealed it up. On the other hand, they may have come into the cave after re opening the shaft.

An observation illustrates the problems in reconstructing taphonomic history. At the beginning of the excavation campaign, one cave explorer entered the cave and took photos. Then, he lost the camera's battery, and it simply disappeared. I found the shiny green battery the very last day of the excavations, located at the deepest place in a lateral cleft, associated with prehistoric human and animal bones.

In addition to fragmentary potsherds, highly fragmented animal remains should be expected, but the bone assem blage is in various states of preservation; the majority is well preserved. From an archaeozoological point of view, the animal bone assemblage might contain a few natural victims too, but they can hardly be recognised as such. An example is a nearly completely preserved cattle skeleton that was scientifically excavated. The remains are widely scattered through seven stratigraphical units, which illustrates complex taphonomical processes. The associated skeletal elements were confirmed by re-articulation analyses and morphological homogeneity (Galik 2002).

Wild animals, such as fox remains, roe deer elements, and wildcat bones probably represent naturally trapped victims. Few stag remains were washed into the cave, whereas artificially modified hare bones might relate to intentional input. Disposal of complete animals or dismembered carcasses is also a point worth pursuing. Gnawing marks that appeared on animal and human bones might help to answer this question. Some punctures relate to small teeth, like milk premolars of dogs or fox permanent dentition. Foxes may have fallen in the cave and gnawed exposed carcasses before they died. Such an explanation will hold true for some of the dogs as well. Numerous articulating elements appeared in the assemblage, but a reconstruction of complete dog skeletons remained impossible.

In addition to the aforementioned species, a large quantity of re-articulated elements appeared in other domestic animals. Only a minority of remains show artificial modifications, mainly cut marks. Articulating skeletal elements that bear such marks relate to the intentional input of animal carcasses.

Another clue to understand disposal of carcasses was obtained by comparing results between both MNI calculation methods. Fig. 3 depicts a heterogeneous distribution of left and right-sided elements. These results are confirmed by MNI analyses for each element in one species, revealing substantial differences in left and right-sided elements in all species (Galik 2002).

From a taphonomic point of view, a clear identification of naturally trapped victims versus intentionally deposited carcasses remains nearly impossible. But quantification results lead to the assumption that mainly dismembered animal parts were disposed into the cave. I argue that the majority of animal remains must be considered as part of artificial taphonomical processes. The same taphonomical history might be assumed for human remains, because they do not indicate substantial differences compared to animal remains. Anthropological investigations revealed similar results concerning paired human long bones (Fabrizii-Reuer and Reuer 1997).

Discussion

Soon after the discovery of the site, we were confronted with a typical, although compelling archaeological question: Why did people deposit humans and animals into the cave, and does this site serve an opportunity to elucidate the purpose of this behaviour? In order to interpret such sites, two models linked to catastrophic events may be employed. Sometimes arguments explain such bone accumulations as the result of war activities or epidemics. Durezza Cave is not large enough to hold a large number of carcasses, if they were disposed in one event. On the other hand, it seems unrealistic to argue that an epidemic killed humans and animals alike. However, Vergil (Georg. III, 478–566, cited in Graßl 1982) mentioned a pest that killed animals in Noricum (Roman Province in this region), but such an epidemic can hardly be proven on the basis of the skeletal elements composing the Durezza assemblage. Other reasons, as discussed in the following paragraphs, negate the aforementioned theories.

The contemporaneous prehistoric settlement, associated with Durezza Cave, is relatively unknown, but it was most likely over-built by the Late Roman hill fort at the top of

	Cattle	Ovicaprine	Pig	Horse	Dog	Chicken
Walpersdorf (Pucher 1998)	918	397	338	25	34	2
Inzersdorf (Pucher 1998)	166	53	20	2	5	–
Durezza	520	921	355	1026	2699	20

Fig. 4. Comparison of faunal compositions (NISP) from contemporaneous Iron Age settlements.

the Tscheltschnigkogel. If this was the case, the vertical cave is not far away and would offer a convenient place to discard domestic waste. If human remains are excluded from this discussion for a moment, another basic question arises: does the animal assemblage reflect patterns of structured behaviour and does it differ from every day domestic waste?

In comparison with contemporaneous animal assemblages from two settlements, Inzersdorf ob der Traisen and Walpersdorf in Lower Austria (Pucher 1998), the animal bones from Durezza Cave differ substantially. Pucher (1998) mentioned chopping marks on nearly every second bone and even long bones were longitudinally split. In contrast, only a few remains in the Durezza Cave assemblage show cut marks or indicate chopping of carcasses. Thus, the Durezza bone assemblage does not represent typical domestic waste, which usually consists of many heavily exploited and chopped bones.

In addition to differences in artificial modifications, faunal composition of domestic animals remarkably differs (Fig. 4). Dog remains, for example, usually appear in prehistoric domestic refuse, because they were sometimes consumed (for example, Kerth and Scherb 1998; Pucher 1998). In contrast to their representation in the Durezza Cave assemblage, dog bones are never predominant in domestic waste deposits from settlements in Austria's prehistory. Horse remains indicate the same discrepancy, because they are also very abundant in the Durezza assemblage, but infrequently distributed in domestic waste assemblages.

Another peculiarity of the Durezza assemblage makes an interpretation as domestic waste infeasible. A vast amount of articulating bones of domesticates was found, representing substantial meat bearing parts of animals. These remains were apparently deposited into the cave as they were still articulated. By considering usual exploitation of animals, such behaviour would indicate a tremendous waste of meat and marrow resources. This fact implies again that the input was a conscious occurrence and no simple disposal of waste.

However, Wilson (1996) mentioned that quantities of articulating elements could be deposited as a result of normal secular carcass processing, relating to medieval sites. For example, other archaeological structures, such as Roman ditch or pit fillings, where sometimes body parts of horses are identified (Kunst 1997; Kunst 2000; Czeika 2001). These bone assemblages obviously represent normal and secular carcass deposits.

Fig. 5. Bronze belt buckle from Stična (after Geupel 1972, cited in Teržan 1994).

Ritually structured disposals?

Ethnological and ethnographic approaches established frameworks to distinguish ritual deposits from secular archaeological finds. Ritual offerings assumed to appear as deposits produced in standardised, repetitive and stereotyped actions (Colpe 1968). Such rites and ceremonies are usually supposed to follow strict rules. If such structured patterns are traceable in bone assemblages, the material allows interpreting the nature of activities and to consider the purposes that produced these structures. People or religious representatives have to offer the sacred goods like their ancestors did, otherwise the sacrifices will not be accepted by the transcendent entities and the offering will fail (Thiel 1983). The sacred objects of symbolical value are offered with conscious purpose to revere or to communicate with supernatural, transcendent entities (Thiel 1983). Animals, especially dogs and horses, represent objects of such value. Although they are animals in daily use, they rarely occur as burial goods. If these animals were part of interments during the Iron Age, they were exclusively added to noble-people's burials in large mounds.

A belt buckle from the Hallstatt period found at Stična (Slovenia) likely displays a ritual, religious procession that involved a dog (Geupel 1972 cited in Teržan 1994; Fig. 5). The scene illustrated at the belt buckle might shed some light on the role dogs played in the activities at Durezza Cave. According to Teržan (1994) the scene alludes to the sacrifice of the woman. However, the dog depicted in supernatural size was probably a part of this worship and used as a sacred good.

The crucial question remains, what was the purpose of killing and disposing of the animals in the patterns observed in the Durezza assemblage? Literature concerning prehistoric vertical cave fillings often assumed ritual bloody slaughtering of animals and humans, sometimes in combination with anthropophagic activities (Erl 1953;

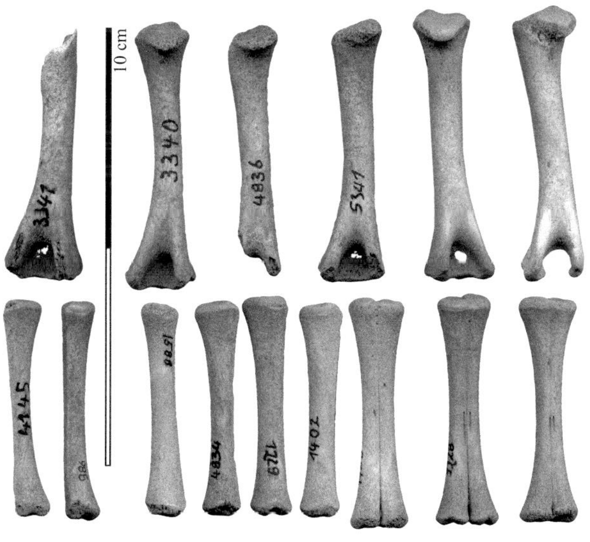

Fig. 6. Neonate ovicaprines humeri and metapodial remains.

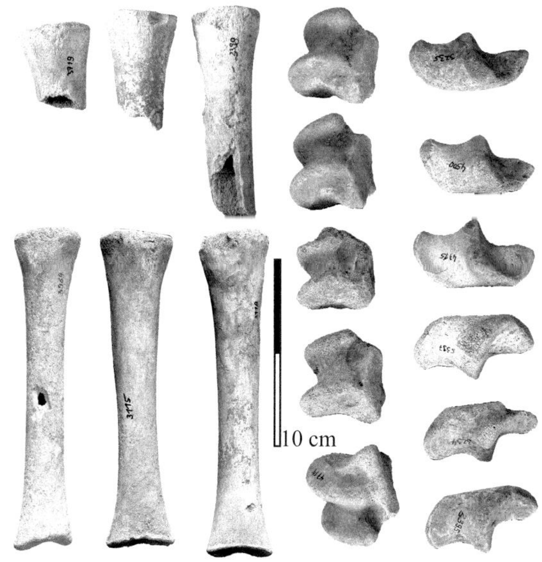

Fig. 7. Neonate foal metapodial and podial remains.

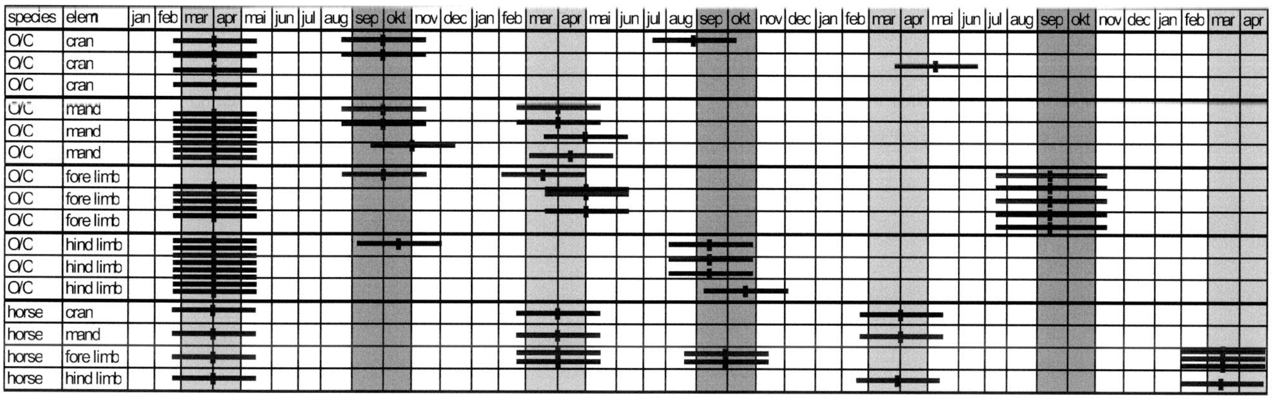

Fig. 8. Estimated kill seasons of ovicaprines and horses.

Behm-Blancke 1956; Behm-Blancke 1976; Maier 1984). However, the purpose of such ritual acts is usually related to vegetation and/or fertility cults, which are assumed to be typical of agrarian societies (Thiel 1983).

At Durezza Cave such ritual traditions, relating to a fertility cult, might be traced in the form of a periodic deposit of neonate ovicaprines and horses, assuming that the offspring was born in one season. The birth season for the old sheep race Heidschnucke, for example, is between March and April, more than 86% give birth at this period of time (Sambraus 1994, 84). The homogeneously sized neonate remains (Fig. 6 and Fig. 7) from Durezza Cave indicate individuals of nearly the same age that were deposited at the same time of year. MNIr of neonate ovicaprines indicates at least eight individuals. MNIr of neonate horses estimates at least eleven individuals. Input of these neonate individuals indicates a duration of at least 19 years of deposition. Besides neonate individuals, a vast amount of other animal remains is available that considerably extends the time of use.

Distribution of aged elements discerns a half a year seasonality in ovicaprines and horses (Fig. 8). This fact gives a hint for the reason of the killing and depositing. Neonates and juveniles were killed and offered at the beginning of spring and the beginning of autumn. This seasonally structured pattern alludes to the desire of Iron Age people for temperate climates and a good harvest or to get through a mild winter, for example. However, it is impossible to attest with this sample that prehistoric people did such a sacrifice every half year or that they did sacrifices after longer periods of time.

Consumption refuse?

In addition to well preserved body parts of dogs and other animals, a few animal remains appeared in the Durezza Cave assemblage that show typical patterns of consumption refuse in a wider and a narrower sense. These bones might indicate a different taphonomical history, and probably played a different role. They mainly represent cattle and pig, but are also evident among ovicaprine and horse remains. The bones often correspond to meat bearing parts around elbow or stifle joint and presumably indicate the input of bones after consumption. These bones are excellently preserved and most likely indicate a different taphonomical history than the finely fragmented pottery sherds. On one hand, meat consumption might have been part of a ritual activity. On the other hand, these modified bones also might be due to a different event that was excluded from the input of the animal carcasses.

Human burials?

Results revealed that the accumulation of human remains in Durezza Cave should be considered the outcome of regular interments (Fabrizii-Reuer and Reuer 1997). The MNI values estimate that 138 people consisting of 36 children, 41 men and 54 women, and seven unidentifiable individuals. However, in this regional area the dead were usually cremated at this time in prehistory. Evidences for completely buried bodies are rare, but sometimes available (Wedenig 1996). Dead people in regular interments have usually added a typical assemblage of traditional ornaments and other burial goods. Comparable goods are nearly missing in the assemblage from Durezza Cave (Gleirscher *et al.* 1997).

The scientific community offers two general opinions to explain the presence of human remains from subsoil features like Durezza Cave. One group argues that human remains have to be considered as secondary burials (Orschiedt 1997; Orschiedt 1999). In order to explain biased skeletal element representation or artificially modified human bones, the favoured interpretation usually is that macerated or dismembered human body parts were secondarily buried. For example, the human remains from Dietersberg Cave near Egloffstein (Kr. Forchheim, Germany) were originally considered to be typical of sacrificial worship behaviours (Erl 1953). The anthropologist Baum (1999) reinterpreted the accumulation of human remains as interments, because he did not trace any evidence for ritual behaviours associated with human bones, and he assumed these burials were interments of a social fringe group.

The other interpretation of such archaeological structures relates to human sacrifices, which are often reported from prehistoric sites in Central Europe (Hahnel 1994; Bockisch-Breuer 1996; Rind 1996; Gleirscher *et al.* 1997; Pertlwieser 1998; Green 2001). However, it is *probably* inappropriate to decide only from an archaeozoological point of view, if people were buried or sacrificed. Although I agree with Hill (1995) that human remains cannot be removed from the assemblage and interpreted separately, from an archaeozoological viewpoint, this site must be explained as a locality where persons deposited animals into the cave and correlated this activity with special conscious acts. The majority of the animal assemblage likely represents the remains of a ritual activity such as sacrificial worship.

The reconstruction of human behaviour from such a site requires, in addition to fine excavation techniques, an interdisciplinary approach of all involved archaeological sciences, which is sometimes still infeasible. In case of interpreting animal remains, prehistoric human cognitive faculties are often underestimated. The present analysis demonstrates the potential of archaeozoological investigation to contribute essential knowledge for the interpretation of ritual sites.

Acknowledgements
I am grateful to all my colleagues who supported me by discussing this topic, and especially Dr. Paul Gleirscher and Dr. Gerhard Forstenpointner. This paper is based upon my doctoral thesis, which was partly financed by the FWF (project number P11469 – PR).

References

Baum, N. 1999. Die Dietersberghöhle bei Egloffstein, Kr. Forchheim – von der Opferhöhle zum Bestattungsplatz. *Prähistorische Zeitschrift* **74**, 79–121.

Behm-Blancke, G. 1956. Bronze- und hallstattzeitliche Kulthöhlen im Gipsgebirge bei Bad Frankenhausen (Kyffhäuser). *Ausgrabungen und Funde* **1**, 276–7.

Behm-Blancke, G. 1976. Zur Funktion bronze- und eisenzeitlicher Kulthöhlen im Mittelgebirgsraum. *Ausgrabungen und Funde* **21**, 80–8.

Bockisch-Bräuer, Ch. and Zeitler, J. P. 1996. *Kulthöhlen. Funde, Deutungen und Fakten.* Nürnberg: Katalog zur Ausstellung der Naturhistorischen Gesellschaft e. V., Abteilung für Vorgeschichte.

Colpe, C. 1968. Theoretische Möglichkeiten zur Identifizierung von Heiligtümern und Interpretation von Opferplätzen in ur- und parahistorischen Epochen, pp. 18–39 in Jankuhn, J. (ed.), *Vorgeschichtliche Heiligtümer und Opferplätze in Mittel- und Nordeuropa* (Abhandlungen der Akademie der Wissenschaften Göttingen Philologisch-Historische Klasse 74). Göttingen.

Czeika, S. 2001. Entsorgung von Tierkadavern im römerzeitlichen Unterlaa, Wien. *Fundort Wien* **4**, 222–40.

Dolenz, H. 1972–1973. Archäologische Funde aus Stadt und Bezirk Villach. *Neues aus Altvillach* **9–10**, 9–56.

Erl, J. R. 1953. Die Dietersberghöhle bei Egloffstein. Abhandlungen der Naturhistorischen Gesellschaft zu Nürnberg **26, 5**, 219–302.

Fabrizii - Reuer, S. and Reuer E. 1997. Anthropologische Untersuchungen an 138 Skeletten der Durezza-Schachthöhle bei Villach, Kärnten. *Neues aus Alt-Villach* **34**, 125–212.

Fieller, N. R. J. and Turner, A. 1982. Number estimation in vertebrate samples. *Journal of Archaeological Science* **9**, 49–63.

Fuchs, M. 1996. Kärnten: KG Judendorf; SG Villach. *Fundberichte aus Österreich* **35**, 446.

Galik, A. 1997. Zur Grabungsmethodik und Fundsituation in der Durezza Schachthöhle. *Neues aus Alt-Villach* **34**, 17–29.

Galik, A. 2002. *The Late Hallstatt and Early La Tène Animal Bone Assemblage from the Vertical Durezza Cave near Villach, in Carinthia.* Unpublished doctoral thesis, University of Vienna.

Gleirscher, P. 1997a. Hallstattzeitliche Kleinfunde aus der Durezza-Schachthöhle. *Neues aus Alt-Villach* **34**, 31–53.

Gleirscher, P. 1997b. Spätbronze- und eisenzeitliche Fundstellen um Warmbad Villach. *Neues aus Alt-Villach* **34**, 55–86.

Gleirscher, P. Fabrizii-Reuer, S., Reuer, E. and Galik, A. 1997. Überlegungen zur Deutung der Durezza-Schachthöle. *Neues aus Alt-Villach* **34**, 213–38.

Graßl, H. 1982. Zur norischen Viehseuche bei Vergil. *Rheinisches Museum für Philologie N. F.* **125**, 66.

Grayson, D. 1984. *Quantitative Zooarchaeology: Topics in the Analysis of Archaeological Faunas.* Orlando: Academic Press.

Green, M. 2001. *Dying for the Gods.* Gloucestershire: Tempus Publishing Ltd.

Hahnel, B. 1994. Leichenzerstückelung, Kannibalismus, Amulett? *Fundberichte aus Österreich* **33**, 169–87.

Hill, J. D. 1995. *Ritual and Rubbish in the Iron Age of Wessex. A Study on the Formation of a specific archaeological Record* (BAR British Series 242). Oxford: British Archaeological Reports.

Kerth, K. and Scherb F. J. 1998. Die tierischen Ressourcen der frühlatènezeitlichen Siedlung auf der Ehrenbürg (Lkr. Forchheim). *Bayrische Vorgeschichtsblätter* **63**, 97–110.

Kunst, G. K. 1997. Equiden als Teile einer Abfallvergesellschaftung – Beobachtungen an einer Grabenverfüllung im Auxilliarkastell Carnuntum (Niederösterreich). *Beiträge zur Archäozoologie und Prähistorischen Anthropologie* **1**, 70–6.

Kunst, G. K. 2000. Archaeological evidence for equid use, sex structure and mortality in a Roman auxiliary fort (Carnuntum Petronell, Lower, Austria). *Anthropozoologica* **31**, 109–18.

Leja, F. 1987. Vorgeschichtliche Funde aus dem Kleebergschacht im Bärnhofer Wald, Ldkr. Amberg-Sulzbach (Oberpfalz). *Abhandlungen der Naturhistorischen Gesellschaft Nürnberg* **41**, 1–72.

Lyman, L. R., 1994. *Vertebrate Taphonomy.* Cambridge: Cambridge Manuals in Archaeology.

Maier, R. A. 1984. Schachthöhlen und Felstürme als urgeschichtliche Opferplätze. *Führer zu archäologischen Denkmälern in Deutschland* **5**, 204.

Orschiedt, J. 1997. Die Jungfernhöhle bei Tiefenellern. Eine Neuinterpretation. *Bericht des Historischen Vereines Bamberg* **133**, 185–98.

Orschiedt, J. 1999. Manipulationen an menschlichen Skelettresten. Taphonomische Prozesse, Sekundärbestattungen oder Kannibalismus. *Urgeschichtliche Materialhefte* **13**, 1–525.

Pertlwieser, M. 1998. 'Opfer für den Donnergott'. *Oberösterreichisches Museumsjournal* **1**, 3–4.

Piccottini, G. 1989. *Die Römer in Kärnten.* Klagenfurt: Universitäts Verlag Carinthia.

Pucher, E. 1998. Der Knochenabfall einer späthallstatt-/latènezeitlichen Siedlung bei Inzersdorf ob der Traisen (Niederösterreich), pp. 56–67 in Ramsl, P. C. (ed.), *Inzersdorf-Walpersdorf. Studien zur späthallstatt/latènezeitlichen Besiedlung im Traisental, Niederösterreich* (Fundberichte Österreich, Materialhefte A 6), Vienna: Bundesdenkmalamt.

Riedel, A. 1977. I resti animali della Grotta delle Ossa (Skocjan). *Estratto dagli Atti del Museo Civico di Storia Naturale – Trieste* **30**, 125–208.

Rind, M. 1996. *Menschenopfer vom Kult der Grausamkeit.* Regensburg: Universitätsverlag Regensburg.

Ringrose, J. T. 1993. Bone counts and statistics: a critique. *Journal of Archaeological Science* **20**, 121–57.

Rogers, A. R. 2000. Analysis of bone counts by maximum likelihood. *Journal of Archaeological Science* **27**, 111–25.

Sambraus, H. 1994. *Gefährdete Nutztierrassen.* Stuttgart: Ulmer Verlag.

Teichert, M. and Lepiksaar, J. 1977. Die Vogelknochen aus den urgeschichtlichen Kulthöhlen des Kyffhäusergebirges. *Alt-Thüringen* **14**, 108–44.

Teichert, M. 1978. Die Katzenknochen aus den urgeschichtlichen Kulthöhlen des Kyffhäusergebirges. *Alt-Thüringen* **15**, 32–67.

Teichert, M. 1985. Die Rinder-, Schweine- und Pferdeknochen aus den bronzezeitlichen Kulthöhlen des Kyffhäuser. *Beiträge zur Archäozoologie* **3**, 37–58.

Teichert, M. 1985. Die Schaf- und Ziegenknochen aus den bronzezeitlichen Kulthöhlen des Kyffhäuser. *Beiträge zur Archäozoologie* **3**, 3–36.

Teržan, B. 1994. Überlegungen zum sozialen Status des Handwerkers in der frühen Eisenzeit Südosteuropas. *Marburger Studien zur Vor- und Frühgeschichte* **16**, 664–6.

Thiel, J. F. 1983. *Grundbegriffe der Ethnologie* (Collectanea Instituti Anthropos 16). Berlin: Dietrich Reimer Verlag.

Wedenig, R. 2002. Das hallstattzeitliche Gräberfeld von Führholz in Unterkärnten. Ergebnisse der Grabungen seit 1989, pp. 18–23 in Ogris, A. and Wadl, W. (eds), *Bericht über den 22. Österreichischen Historikertag in Klagenfurt [4 bis 7. Mai 1999*] (Veröffentlichung des Verbands Österreichischer Historiker und Geschichtsvereine 31).Vienna: Österreichischer Historiker und Geschichtsvereine.

Wilson, B. 1996. Aspects of the literature on the theory and identification of ritual, pp. 11–6 in Anderson, S. and Boyle K. (eds), *Ritual Treatment of Human and Animal Remains* (Proceedings of the First meeting of the Osteoarchaeological Research Group). Oxford: Oxbow Books.

Zapfe, H. 1954. Beiträge zur Erklärung der Entstehung von Knochenlagerstätten in Karstspalten und Höhlen. *Beiheft zur Zeitschrift Geologie* **12**, 1–60.

Alfred Galik
Institute of Anatomy,
University of Veterinary Medicine Vienna,
Veterinärplatz 1,
A-1210 Vienna,
Austria
E-mail: alfred.galik@vu-wien.ac.at

9th ICAZ Conference, Durham 2002
Behaviour Behind Bones, (ed. Sharyn Jones O'Day et al.) pp. 62–65

9. Ritual feasting in the Irish Iron Age: re-examining the fauna from Dún Ailinne in light of contemporary archaeological theory

Pam Crabtree

Contemporary archaeological theory has emphasized the importance of the study of ritual in the analysis of prehistoric social life. This paper presents a re-analysis of the fauna from Iron Age site of Dún Ailinne, Co. Kildare, Ireland. Rather than emphasizing the paleoeconomic aspects of Iron Age subsistence, this contribution will emphasize the information that the Dún Ailinne assemblage can provide regarding Iron Age religion and ritual feasting.

Introduction

Archaeological theory has undergone a revolution in the past twenty years. The dominant processualist paradigm of the 1960s and 1970s has been criticized from many different sides. Some archaeologists have argued that the normative view of culture adopted by the processualists did not provide enough room for individual human actions and human agency. Others have been concerned that processual archaeologists did not pay enough attention to the issues of race, class, and gender (see, for example, Conkey and Spector 1984; Delle *et al.* 2000) that have dominated the theoretical debate in related disciplines including history, sociology, and social anthropology. Cognitive archaeologists and others have criticized the processual view of culture as the non-biological means by which humans adapt to their environments as leading to an overly functionalist view of past human societies in which many past behaviors are simply viewed as adaptive. One of the most powerful critiques of the New Archaeology, however, is that it neglected the symbolic dimensions of human behavior and human material culture. As Hodder (1991, 3) has noted, "If material culture, all of it, has a symbolic dimension such that the relationship between people and things is affected, then *all* of archaeology, economic and social, is implicated" (emphasis original). Although Binford (1962) called for the archaeological reconstruction of the technological, social, and ideological dimensions of past human societies in his seminal paper, "Archaeology as Anthropology," in practice, most processual archaeologists accepted Hawkes's (1954) "ladder of inference" which suggested that aspects of technology and economy were far easier to reconstruct from the archaeological record than were elements of social organization and religion.

Nowhere was this focus on technology and economy at the expense of other aspects of culture more apparent than in the nascent discipline of zooarchaeology. As the late Dexter Perkins (*pers. comm.*) explained to me when I was a student in his graduate faunal analysis course at the University of Pennsylvania in 1974, "If you have a choice between studying the bones from a domestic trash pit and those recovered from a temple, choose the bones from the trash pit, since those will be more informative about past economic behavior." At its most extreme, this viewpoint can be summarized as economy is everything, and culture is noise. We now recognize that the picture is far more complicated than we thought it was in 1974. Not only are archaeological animal bones subject to geological processes such as weathering and water transport, but decisions about refuse disposal have both an economic *and* an ideological dimension. The decision about whether to pile refuse in a midden or bury it in a pit may depend in part on culturally-specific notions of cleanliness, dirt, and hygiene. In addition, archaeological and faunal analytical research on ritual and ceremonial sites has shown that these sites have a great deal to tell us about past human belief systems and religious behavior.

In this paper I focus on the faunal remains from the

Fig. 1. Map of Ireland showing the location of Dún Ailinne.

Iron Age ceremonial site of Dún Ailinne in Ireland (Fig. 1). I was a member of the Dún Ailinne crew in the summer of 1972, and I co-ordinated the analysis of the extensive faunal collection from this site in the 1980s when I worked at MASCA in the University of Pennsylvania Museum. My initial interpretation of these faunal remains focused on paleoeconomy (Crabtree 1986; Crabtree 1990) at the expense of ritual and symbolic behavior. In this paper I will consider how these data might be re-interpreted in light of contemporary archaeological theory and interests.

Archaeological and historical background

Early medieval documentary sources (Grabowski 1990) indicate that Dún Ailinne was viewed as a symbol of the kings of Leinster. "It was depicted as a seat of the kingship and a site of important social or cultural activities (Grabowski 1990, 35). In 1837 Knockaulin Hill, Co. Kildare, Ireland was identified as Dún Ailinne (Wailes 1990, 10). A program of intensive archaeological research was conducted at the site between 1968 and 1975 under the direction of Prof. Bernard Wailes of the University of Pennsylvania. The excavation also served as an important training ground for a number of American archaeologists interested in later European prehistory.

The initial occupation of the site dates to the Neolithic period. Neolithic features discovered at the site include an irregularly circular ditch and a partial pit, disturbed by later excavation, which may have held a neolithic burial. In addition, the presence of an Early Bronze Age food vessel recovered from a pit indicates some limited Early Bronze Age activity at the site (Johnston 1990).

Most of the activity at Dún Ailinne, however, dates to the pre-Christian Iron Age period (see Wailes 1976 and Wailes 1990 for a complete description of the archaeology of Dún Ailinne.) Initially a circular trench supporting upright timber posts – forming a palisade or fence – was constructed at the site. The posts were removed from the trench, and a more elaborate series of structures was constructed on the site.

The next major structural phase includes three concentric circular trenches which would have held progressively larger upright posts. This structure has been reconstructed as a two-tiered standing or seating arrangement for people viewing the events that may have taken place in the center. The circular structure was flanked by fences which enclosed an avenue of posts forming an entrance to the main circular structure. During this phase, the site was also enclosed by an external bank and internal ditch (the reverse of a typical fortification). This feature links Dún Ailinne to the other henge monuments that are known from the British Isles.

These structures were dismantled and replaced by wooden constructions of a different design. The replacement structures included two concentric circles of posts enclosing an area of approximately 42 m in diameter. Within this structure were a free-standing circle of wooden posts and a central structure that has been tentatively reconstructed as a timber tower.

These final structures were also dismantled, but the site continued to be used. A rough paving of stones was laid over a portion of the site. Above this rough paving was a layer containing many animal bones, burnt stone, charcoal and ash, as well as thin lenses of humus. Wailes (1990, 15) has suggested that these deposits (termed Flame Phase by the excavator) represent a continuation of ritual feasting at the site after the demolition of the earlier structures. However, there is no evidence to suggest that the feasting continued into the Early Christian Period. The radiocarbon age determinations from the Iron Age features all lie between the 5th century BC and the 3rd century AD (Wailes 1990, 19)

Several lines of evidence indicate that Dún Ailinne was a ceremonial site. They include the inverted bank and ditch, the complete absence of residential structures, and the only minimal evidence for craft activities at the site.

The fauna from Flame Phase

Because of the periodic building and re-building at Dún Ailinne, the clearest evidence for ritual feasting at the site comes from the Flame Phase deposits. The Flame Phase deposits yielded a substantial proportion of the faunal remains that were recovered from the site. Of the 18,788 animal bones and fragments that were recovered during

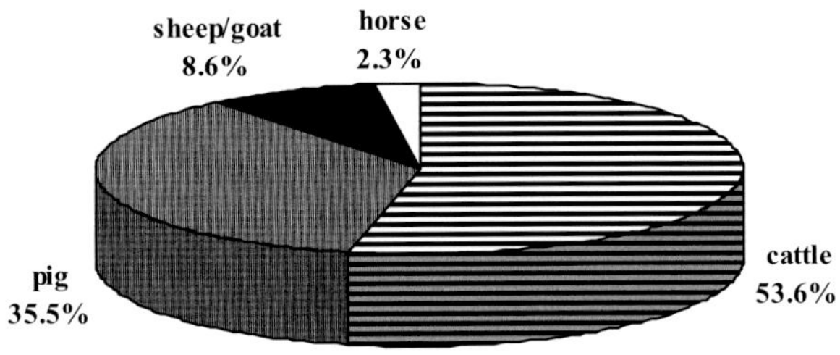

Fig. 2. Species ratios based on NISP for the main domestic mammals from Flame Phase contexts at Dún Ailinne.

the 1968–75 excavations, 5604 (30%) came from the Flame Phase deposits.

Virtually all the identified animal remains were fragments of cattle, pig, sheep/goat (overwhelmingly sheep), and horse bones. The species ratios, here based on NISP (Fig. 2), indicate that beef and pork were the preferred feasting foods and that smaller quantities of mutton and horseflesh were consumed.

The ageing data for the cattle teeth from Flame Phase are presented in the histogram (Fig. 3). Since the vast majority of the teeth were loose teeth, wear stages (following Grant 1975 and Grant 1982) were recorded for all deciduous 4th premolars and permanent 3rd molars. It is clear that the majority of the cattle were young animals, including a substantial number of suckling calves. These animals would have been costly to slaughter. As McCormick (1991) has correctly pointed out, if the calf is removed from its mother, the mother is likely to cease milk production. Therefore the cost of the feast includes both the meat value of the slaughtered calf and the additional value of the milk products that might have been produced if the calf had not been slaughtered. In short, the aging data indicate that the Dún Ailinne feasts would have been expensive to produce. The question we need to ask is what were the "kings of Leinster" (or the Lords of Dún Ailinne) gaining in return for these expensive feasts? Was it power, prestige, labor, or all of the above? Essentially, the archaeological problem is one of political economy rather than paleoeconomy.

Reinterpretation

In 1990 I argued that "the ceremonial feasts that took place at Dún Ailinne served both social and political functions and as a way to use up surplus animals produced by a dairy economy – young male calves and elderly females who were no longer milk producers"(Crabtree 1990, 24). McCormick (1991) subsequently published a powerful critique of the dairying model. However, recent biochemical research has suggested that dairying has a great antiquity in the British Isles (Copley *et al.* 2003). Nevertheless, my initial interpretation of the Dún Ailinne was based on an overly materialist perspective. Following the dominant theoretical paradigms of the 1960s and 1970s, I took it for granted that subsistence economy (including animal husbandry) lay at the core of ancient human societies, and that religious and ritual activities were, at best, epiphenomenal. I viewed ritual feasting as a functional outgrowth of early Irish subsistence practices.

Contemporary archaeological theory has cast ritual feasting, and ritual and ceremonial activity more generally, in a new light. For example, Dietler and Hayden (2001, 3) have argued that "it is crucial to recognize and understand feasting as a particular form of *ritual* activity" (emphasis original). As such, the foods consumed, their method of preparation and presentation, and the social and spatial contexts in which they are consumed are likely to differ significantly from day-to-day contexts of animal

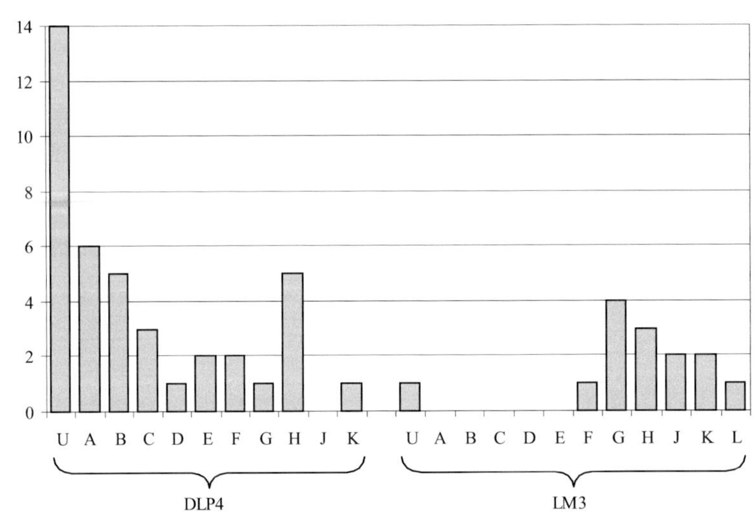

Fig. 3. Wear stages for the mandibular dp4 and M1 for cattle remains from Flame Phase at Dún Ailinne (u = unworn).

production and consumption. The faunal data from the Flame Phase contexts indicate that substantial quantities of veal, beef, and pork, and smaller quantities of mutton and horseflesh, were consumed periodically at Dún Ailinne during the pre-Christian Iron Age. The quantity and quality of the meat consumed there are unlikely to have been part of the regular diet of Irish Iron Age farmers. In particular, the consumption of a large number of veal calves can be seen as a form of conspicuous consumption, since the cost of slaughter includes not only the cost of the calf but also the loss of valuable dairy products. Moreover, these ritual meals were consumed in a spatial context that was far removed from the typical rural settlement of the pre-Christian period. The content and the context of these ritual feasts would have served to enhance the power and the prestige of the kings of Leinster/lords of Dún Ailinne.

Studying the Irish Iron Age presents a second interpretive challenge. Our knowledge of the history of this period is derived from early medieval textual material, which was written down after literacy was introduced to Ireland in the 5th century AD. In my initial analysis of the Iron Age animal bones from Dún Ailinne, I was perhaps too heavily influenced by historical accounts of dairy production in early medieval Ireland (see, for example Lucas 1958; Lucas 1960; Ó Sé 1948; Ó Sé 1949). While the degree of continuity that may have existed between pre-Christian Ireland and Early Christian Ireland is a subject of archaeological debate (*e.g.*, Mytum 1992), there is no question that the introduction of Christianity in the 5th century profoundly altered Irish ritual practices. It was inappropriate to view Iron Age ritual feasting through the lens of the historically-documented early medieval cattle economy. Instead the Dún Ailinne fauna reflect ritual practices that were replaced with the advent of Christianity in the 5th century.

The fauna from Dún Ailinne can provide an important lesson for zooarchaeologists. All faunal analysts recognize the importance of carefully identifying a faunal assemblage. If bones are mis-identified, all the conclusions that are drawn from those bones will be suspect. Careful identifications by themselves are not enough. The faunal data from Dún Ailinne were studied by no less than ten different zooarchaeologists, and I am quite certain that the basic identifications are correct. The data also must be interpreted within the appropriate theoretical framework. A paleoeconomic perspective was simply the wrong approach to the understanding of this ritual assemblage. However, since the basic data were carefully identified and archived on the computer (Campana and Crabtree 1987), it was possible to re-analyze these data in light of new theoretical perspectives on ritual feasting.

References

Binford, L. R. 1962. Archaeology as anthropology. *American Antiquity* **28**, 217–25.

Campana, D. V. and Crabtree, P. J. 1987. ANIMALS–A C language computer program for the analysis of faunal remains and its use in the study of Early Iron Age fauna from Dún Ailinne. *Archaeozoologia* **1**, 57–68.

Conkey, M. W. and Spector, J. 1984. Archaeology and the study of gender, pp. 1–38 in Schiffer, M. B. (ed.), *Archaeological Method and Theory, Vol. 7*. New York: Academic Press.

Copley, M. S., Berstan, R., Docherty, S., Mukherjee, A., Straker, V., Payne, S., and Evershed, R. P. 2003. The earliest direct evidence for widespread dairying in prehistoric Britain. *Proceedings of the National Academy of Science* **100 (4)**, 1524–9.

Crabtree, P. 1986. Dairying in Irish prehistory: the evidence from a ceremonial center. *Expedition* **28 (2)**, 59–62.

Crabtree, P. 1990. Subsistence and ritual: the faunal remains from Dún Ailinne, Co. Kildare, Ireland. *Emania* **7**, 2225.

Delle, J. A., Mrozowshi, S. A., and Paynter, R. (eds). 2000. *Lines that Divide: Historical Archaeologies of Race, Class, and Gender*. Knoxville: University of Tennessee Press.

Dietler, M. and Hayden, B. 2001. Digesting the feast: good to eat, good to drink, good to think, pp. 1–20 in Dietler, M. and Hayden, B. (eds), *Feasts: Archaeological and Ethnographic Perspectives on Food, Politics, and Power*. Washington, DC: Smithsonian Institution Press.

Grabowski, K. 1990. The historical overview of Dún Ailinne. *Emania* **8**, 32–6.

Grant, A. 1975. Appendix B: the use of tooth wear as a guide to ageing the domestic animals – a brief explanation, pp. 437–50 in Cunliffe, B. (ed.), *Excavations at Portchester Castle, Volume 1: Roman*. London: Society of Antiquaries.

Hawkes, C. 1954. Archaeological theory and method: some suggestions from the Old World. *American Anthropologist* **56**, 155–68.

Hodder, I. 1991. *Reading the Past: Current Approaches to Interpretation in Archaeology* (second edition). Cambridge: Cambridge University Press.

Johnston, S. A. 1990. The Neolithic and Bronze Age activity at Dún Ailinne, Co. Kildare. *Emania* **8**, 26–31.

Lucas, A. T. 1958. Cattle in ancient and Medieval Irish society. *O'Connell School Union Record* **1938–58**, 75–86.

Lucas, A. T. 1960. Irish food before the famine. *Gwerin* **3**, 1–36.

McCormick, F. 1991. Evidence of dairying at Dún Ailinne? *Emania* **8**, 57–9.

Mytum, H. 1992. *The Origins of Early Christian Ireland*. London: Routledge.

Ó Sé, M. 1948. Old Irish cheeses and other milk products. *Journal of the Cork Historical and Archaeological Society* **53**, 82–7.

Ó Sé, M. 1949. Old Irish buttermaking. *Journal of the Cork Historical and Archaeological Society* **54**, 61–7.

Wailes, B. 1986. Dún Ailinne: an interim report, pp. 319–38 in Harding, D. W. (ed.), *Hillforts: Later Prehistoric Earthworks in Britain and Ireland*. London: Academic Press.

Wailes, B. 1990. Dún Ailinne: a summary excavation report. *Emania* 7, 10–21.

Pam Crabtree
Anthropology Department,
New York University,
25 Waverly Place,
New York, NY 10003, USA
E-mail: PamDoug@compuserve.com

9th ICAZ Conference, Durham 2002
Behaviour Behind Bones, (ed. Sharyn Jones O'Day et al.) pp. 66–72

10. The economic and non-economic animal: Roman depositions and offerings

Roel C. G. M. Lauwerier

In archaeological contexts remains are often found of animals that were offered, purposefully deposited, or just dumped. Examples include offerings of animals at temples, building and site offerings, gravegifts, and the deposition of not-eaten species such as horses and dogs. In this paper I review the occurrence and possible meanings of offerings and depositions of animals from Roman times in the Netherlands. The consequences of non-economic uses of animals in relation to their potential use as food is also discussed. Finally, an attempt is made to quantify these phenomena and to examine the possible effects that a non-economic use of animals could have had on food economy and stockbreeding.

Introduction

Animals played many roles in people's lives in the past, just as they continue to do today. They were used for transport and traction, shorn for their wool, milked, slaughtered for their fat and meat, skinned for their hides, and so on. However, animals were not only used for economic reasons such as feeding people, working for them, and providing materials. Human beings, as thinking creatures, not only make practical use of their environment, they also interact mentally with the world around them, and animals are part of that world.

Examples of such non-economic use are manifold: dogs and cats have become our companions; in past and present horses are used as status symbols; the lamb is a symbol for the Christian god; the dog is an attribute of the regional native mother goddess Nehalennia in Roman Lower Germany; rabbits were used as prestigious gifts for medieval nobility; all kinds of animals have been sacrificed to gods; and so on.

Often these functions occur in combination. For example, a grave gift might be given for religious reasons, but many gifts also express status. Such non-economic functions often occur in combination with economic functions. Eating a Christmas turkey is an expression of that festivity and religious occasions, but it also fills the stomach. Another characteristic feature of *Homo sapiens* is that individuals and groups show habits and taste preferences for foods. This feature also affects a pure economic use of animals. In the past decade the interest in non-economic aspects of animals has increased (*e.g.*, Wilson 1992; Ryan and Crabtree 1995; Anderson and Boyle 1996; Bond 1996; Prummel and Bouma 1997; Bodson 2000; Lauwerier 2002).

The purpose of the present study is to focus on the dual function of animals. I aim to examine the possible effects that non-economic uses of animals had on food economy and stockbreeding in the Netherlands in Roman times. I proceed in three steps, answering three questions:

- what kinds of non-economic uses of animals are known from the Roman Netherlands and what was their meaning?
- what were the consequence of the non-economic uses on their economic uses, especially as food?
- did depositing and offering effect 'normal' economic patterns, such as stockbreeding strategies?

Material and method

A starting point for this study was the archaeozoological material found at excavations in the Netherlands from both sides of the Roman border (Fig. 1). However, as the examples in the introduction show, 'non-economic' is a broad concept. It ranges from rather concrete subjects like offerings and gravegifts to less tangible matters like status and symbols. Often these phenomena are difficult to identify in the archaeological record (*e.g.*, Hill 1996).

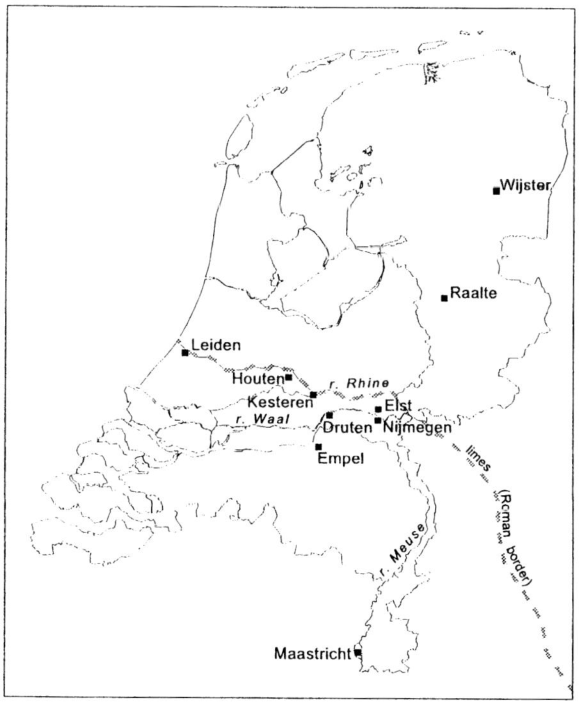

Fig. 1. Map of the Netherlands with the findspots mentioned in the text. The dotted line indicates the Roman imperial border.

In order to proceed by basing the current study on the archaeo(zoo)logical reality, my enquiry is restricted to phenomena that 'look like offerings and depositions'. According to their appearance, the archaeological reality, the archaeozoological material can be divided into 'normal offal', such as the waste from butchering and meals, and 'offerings and depositions' (Fig. 2). The animals and meat that contribute to the 'normal offal' had an economic meaning. On the other hand, the animals or animal products that caused the complexes that are interpreted as 'offerings and depositions' had a non-economic meaning. Unfortunately, what looks like normal waste may be the result of a religious act, for example, an offering may have an economic meaning at the same time. So, the meaning of these two categories 'normal offal' and 'offerings and depositions' can be both economic and non-economic. (An additional term, 'natural', should be added to the diagram to account for the animals that might have fallen into a well by accident).

This study focuses on a selection of the archaeozoological material that are obviously offerings and intentional deposits of animals or parts of them. Some remarkable examples from the archaeozoological literature were taken and grouped into five categories: offerings at temple sites, building and site offerings, grave gifts, depositions of horses in general, and a possible example of consumption of taboo foods. The arguments for the proposed meaning of these examples are given and the possible frequency or scale on

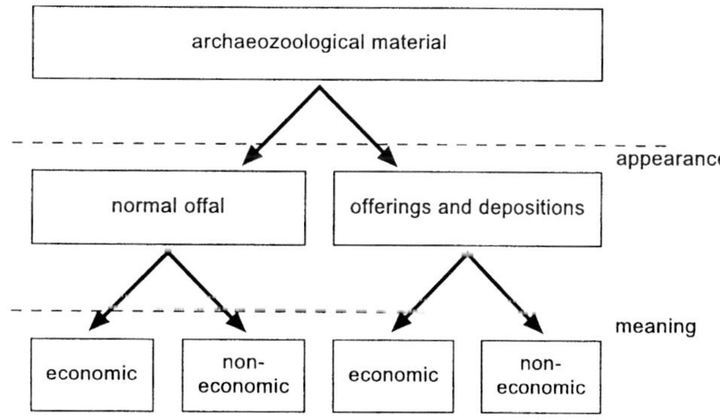

Fig. 2. Appearance and possible meaning of archaeozoological material at an excavation.

which the phenomena occurred is discussed. I then propose an answer to the question, if the non-economic use has any impact on food economy or breeding strategies.

Offerings and depositions

Offerings at temple sites

The first example originates from a known religious place: one of the temples at Elst. This temple was built by order of the Roman military authorities. Recent dating results show that this happened around AD 100 (personal communication T. Derks, Free University Amsterdam). The temple remained in use until into the 3rd century. The advantage with temples is that we know that they were places of ritual, although secular acts could have taken place at such holy spots. In a pit associated with the temple the skulls of a pig, a sheep, and a cow were found. They are considered to be the remains of a *suovetaurilia* sacrifice, a pig – sheep – cattle sacrifice, as is known from the classical literature and from several sculptures including Trajan's Column (Toynbee 1996, 134). It is the combination of archaeozoological information like species, sex and size, the frequency of occurrence of the extraordinary combination of skulls, and the context of the finds, which make the explanation as a *suovetaurilia* sacrifice plausible (Bogaers 1955; Lauwerier 1988, 118-9; Lauwerier 2002). This sacrifice was probably made on the occasion of the *lustratio* at the spot where the new temple of Elst was to be built after the destruction of the previous one.

However, because of the nature of this sacrifice (a consecration of the site of a new temple or military camp), this kind of sacrifice did not occur often. Moreover, the sacrificed animals are also normally consumed. Because of the low frequency of these offerings and the fact that normal consumption animals were used, I argue that this kind of offering did not have significant impacts on the food economy.

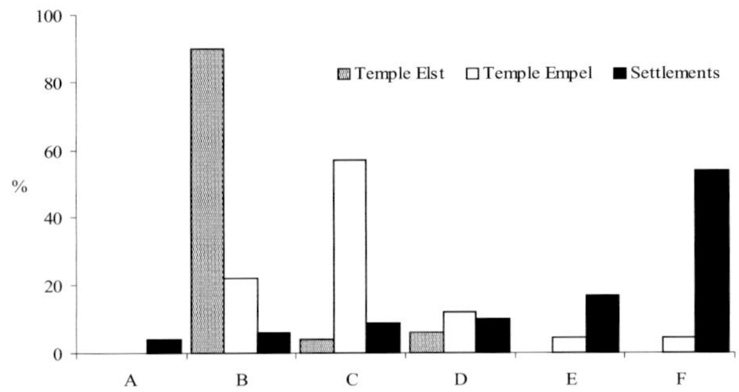

Fig. 3. Age distribution of slaughtered cattle (based on epiphysial fusion) of the temples at Elst and Empel compared to settlement sites from the eastern river area. Age (months): A 0–7/15; B 7/15–15/24; C 15/24–24/30; D 24/30–36/42; E 36/42–42/48; F >42/48

The remnants of the cattle and pigs give no indication for the selection of animals of a particular size or appearance. However, the skull of the ram shows a maximum diameter of the horncores of more than 6 cm, which is very large compared with measurements of other sheep from the area. Thus, I suggest that the animal had a striking external appearance in the time of sacrifice. This may be considered as evidence that animals were specially bred, imported, or selected for these occasions.

Most of the other bones found at Elst are considered to be the remains of sacrificed animals. In particular, the unusually young age of the slaughtered cattle indicates that these animals were sacrificed; a comparable phenomenon has been found at the contemporaneous temple at Empel (Lauwerier 1988; Seijnen 1994; Lauwerier 2002). Fig. 3 presents the age of the animals that were killed at the temple sites in comparison to other settlements. Clearly, young animals were selected. At Elst, ninety percent of the animals were killed during the second year of life. At Empel, most of them were killed around the age of two years. The temple sites show a pattern deviating very much from other settlement types, where much older cattle were usually slaughtered.

The question remains if this practice at temple sites had any economic consequences. The animals were offered, but the meat was probably normally eaten too. Although, until now no evidence for this has been found in the form of burned bone, it is presumed that only some small part of the animals were burned as a sacrifice to the god. In this sense, the offering probably had no effect on the balance of production and consumption of food. The deviation is the early age at which the animals were killed. The general pattern in the area is that most of the cattle were slaughtered at an adult age because they were kept primarily as a source of traction power and of manure for agricultural purposes (Lauwerier 1988). The young animals found at the temple sites have been withdrawn from this pool of animals that in the ordinary course of events would have been used for agricultural purposes. In other words, sacrificial animals were withdrawn from the pool of animals that could potentially be used for traction. It is not possible to quantify how many animals were involved, but looking at the archaeozoological material, this was a considerable number. New excavations on the temple site in 2002, confirm that a large number of animals were involved.

A selection of young animals could also indicate a preference for animals that proved to be less economically suitable, for example because they were too small or weak for use in agricultural labour. From the Roman agronomists we know that in Italy, in contrast to what was true for working animals, there were no quality requirements for sacrificial animals (Hauger 1921, 80). So, offerings could be a supplementary value for such animals. It was possible to add an extra value to animals that at first had only significance as food. In that case the offering of young animals, and the consumption of their meat afterwards provides a case for the fitting of ritual to the normal food economy.

Another sacrifice practice is known from a temple at Nijmegen dedicated to Fortuna (2nd century AD) (Zeiler 1996; Zeiler 1997). At this temple site mostly adult chickens were offered. Chickens were sacrificed as complete animals by putting them in the fire. Subsequently, most of the bones found were charred or calcinated. Other recovered remains include those of sheep, goats, cattle, and fish. Remains of a burned quail (*Coturnix coturnix*) were also found. Based on the bone assemblage characteristics, it was concluded that the chickens and the mammal legs were not defleshed before they were thrown into the fire (Zeiler 1997).

In contrast to the situation at Elst and Empel, remains from Nijmegen rituals indicate that the offerings produced a loss of meat that could have been eaten. As with the other sites, it is very difficult to estimate the quantity of the offerings. The studied samples give the impression that offerings of legs of mammals happened only occasionally. To get an idea of the impact of the more regular offerings of chicken, these sacrifices are compared with those of cattle. If the slaughter weight of a Roman cow is estimated between 200 and 300 kilograms, and the weight of a chicken as 1.5 kg, then 150 to 200 chickens would need to be slaughtered to generate the same amount of meat as one single cow. The finds, however, do not indicate large amounts of chickens. Considering that chicken meat has a higher economic value than that of cattle (Lauwerier 1988, 84-5), the influence of offering relatively small numbers of chickens in a temple like Fortuna at Nijmegen appears to have had little effect on the food economy.

Building and site offerings

A category of sacrifices specifically connected with the secular world is that of building and site offerings. At the

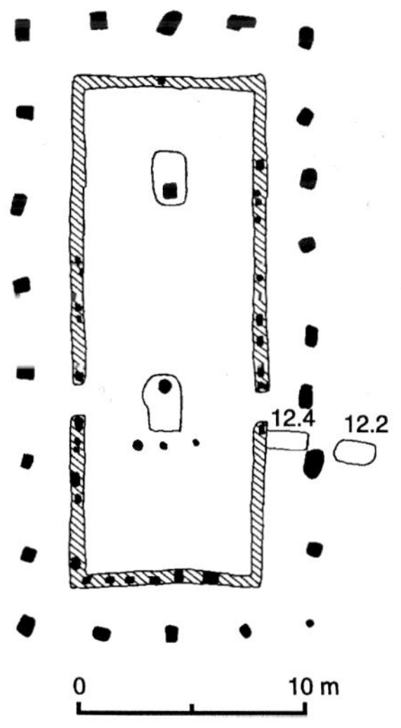

Fig. 4. Druten: one of the buildings with foundation deposits. The features 12.2 and 12.4 are the pits containing horse skeletons.

bottom of a sunken hut in Maastricht an almost complete skeleton of a quail (*Coturnix coturnix*) was found covered with an inverted intact plate of samian ware (Dijkman 1991). The find has been dated to around AD 400 (*ibid.*). It is unlikely that it was simply discarded as waste. Neither is it likely to have been a building offering because it was not buried under the sunken hut. Although the significance of this find is unclear, a ritual or magical explanation seems most appropriate. Importantly, this site and the Fortuna temple mentioned before are the only places in the Roman Netherlands where quail has been identified. This bird may have had a special status for ritual acts.

At the Druten farm, a *villa* with a fairly Batavian character, two horses were buried at the entrances to each of two buildings (dating: AD 70–2nd century) (Fig. 4) (Hulst 1978, 136; Lauwerier 1988, 104–11). One of these animals was 3.5 years at the time of death, the others were older. Although there is no real proof, (for example, in the form of a smashed cranium), it is presumed that these animals were killed as a sacrifice and buried (Lauwerier 1988). Excavations established that one of the burial pits was dug at the same time as one of the foundation trenches of the house, and therefore during the construction of the building. Based on this fact and the position of the horses, all four graves are considered to be building offerings.

In a site north of the Roman border, at the iron production settlement Raalte-Heeten, burials of horses, cattle, and also red deer were found associated with the fence and the entrance building. These phenomena date from AD 310/320 to 345. The position of the animal graves near the fence and the building's entrance, the regular forms of the graves (oval or rectangular, with a flat bottom and steep sides), plus the combination of cattle, horses, and especially red deer are indicative of site offerings (Lauwerier *et al.* 1999). Since preservation conditions for bones are extremely poor in the sandy area, possible traces of sacrifice, such as cut marks, could not be established. Less formal burials of cattle, from an earlier phase of the settlement (3rd century to AD 310/320), are associated with fragmentary traces of farmyards or houses. They are considered as possible building sacrifices.

At the native settlement of Wijster similar building and site offerings were found (Van Es 1967; Lauwerier and Robeerst 2001). In early phases of this settlement (AD 225–360) there appears to be a direct link between the distribution of horse and cattle graves and the location of houses and granaries. These animal graves are therefore regarded as building sacrifices. The animal graves dating to the following period (AD 360–395), and consisting of horses or combinations of horses and cattle were almost all situated against or near the enclosures, and are interpreted as site offerings. Again, real proof of active killing of the animals as a sacrifice could not be established on the very brittle and partly decayed bones.

For the inhabitants of the settlements mentioned, the killing of the animals must have been important for their spiritual life, but the killing and burying of these animals also was an economic loss. For the horses of the villa at Druten, this is especially true because these animals had a size that very well fitted the demands for big horses by the Roman army (Lauwerier and Robeerst 2001). The depositions of the normally eaten cattle and red deer in Heeten and Wijster were a direct loss of some hundreds of kilograms of usable meat per animal. On the other hand it is, of course, difficult to estimate if there was a profit in terms of status, how big this profit was, and if there was a possible indirect economic profit that was connected to this non-economic phenomena.

Grave gifts

A very common practice in the southern area was to bury animal food with the dead. This habit is reflected both in inhumation burials and in cremations (Fig. 5) (Lauwerier 2002). Chickens and pigs are the primary food items found in graves and cremations, but it is possible that boned beef was also occasionally given as a meal for the dead. Contrary to chicken and the meat of especially young pigs, beef normally leaves no traces on the plates, in the graves or in the cremations because it is not 'served' as meat on the bone.

Of course, the burying and/or burning of food withdraws a potential food source from the normal economy. Although the giving of food may have been a very common habit the economical implications were limited. In contrast to other phenomena, the impact this habit had is easier to estimate, simply because everybody only dies once. Even if everybody who died received, for example, a chicken and a part of a sucking pig, this 'loss' only occurred once. Thus, the impact of this ritual on the economy was limited.

Fig. 5. Nijmegen: dish containing the remains of a chicken and the head of a sucking-pig. Gravegift from the fourth century cemetery 'Margriet'.

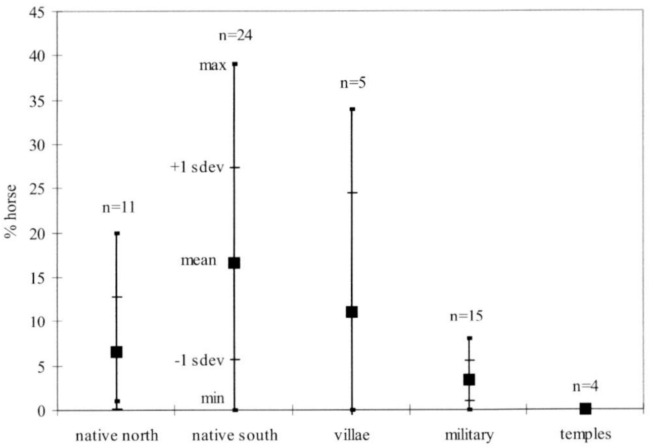

Fig. 6. Frequencies of horses from Roman period sites in the Netherlands (Relative compared to hand-collected material of cattle, sheep, goats and pigs) (Lauwerier 1999).

Depositions of horses

The last group of finds presented here is depositions of horses. In both the military settlements and most of the native settlements during Roman times there was a general avoidance of horsemeat. Likely this practice was related to the avoidance of eating 'comrades,' as is true in the modern western world for cats, dogs, and often also horses (Lauwerier 1999; Lauwerier and Robeerst 2001). Nevertheless, a consequence of the death of a horse was a loss of hundreds of kilograms of potential food.

For example, the cavalry horses from the army camp of Kesteren were dumped on a site outside the settlement (Lauwerier and Hessing 1992). At other forts, the dead dogs and horses were thrown into the river Rhine at a location not far from the settlement (van Wijngaarden-Bakker 1970; Lauwerier 1999). Together with a great deal of other refuse, the dead horses from the legionary fort of Nijmegen were thrown from a steep hill or buried near an exit of the army camp.

Based on archaeozoological material from a wide range of settlements in the Netherlands, it is concluded that horse meat was not eaten at military camps and was generally not eaten at native settlements south of the Roman border. Evidently, horse meat was only eaten at some settlements north of the Roman border. The reason for this food avoidance was probably based on the principle that one does not eat one's comrades. In general, the practice likely had no religious or magical background (Lauwerier 1999; Lauwerier and Robeerst 2001).

In most cases the death of an animal meant economic loss due to unused meat. Again, it is not easy to calculate the impact of this loss, but the contribution of the horse in the frequency distribution of mammals in settlements gives some indications.

Fig. 6 gives an overview of the proportion of horses in assemblages from various settlements. The percentages are relative to the other domesticated food mammals: cattle, sheep, goats and pigs. The data from the military settlements are not very informative. As has been discussed before, the carcasses of animals from these crowded settlements were not thrown away in the camps, but in places outside these settlements. What is shown is a reflection of deposition behaviour and not of the use of this animal. For the native settlements this is different. The offal found during excavation reflects the offal from the people that lived there. The average proportion of horse bones among the native sites north of the border is about 7%. The native sites inside the borders of the Roman Empire have a high score, most have more than 10% while the average is 16%.

The proportion of horse in the potentially available amount of meat is higher than the percentages based on NISP suggest. Counted bones of large animals such as horses and cattle represent far more meat than those of the smaller sheep and pigs. Another factor is that the bones of horses are less fragmented than those of the (in size) comparable cattle. This implies that one single counted fragment of a horse bone represents more meat than that of a cow. It can thus be stated that, for example, for the native sites south of the border, 16% of horse bones represents as a rough estimate more than a quarter of the potentially available amount of meat. This means that more than a quarter of the total available meat of domestic animals, the horsemeat, was not used as food. So, this behaviour, the avoidance of eating horsemeat must have had a considerable impact on food economy.

In addition, the avoidance of eating dogs and presumably also cats (although this last was very rare in this area in Roman times) meant economic loss because the meat was not used as food. Since the portion of these two species in the remains found is far less than those of horses, and these species have a rather low meat yield, the economic loss compared to horse meat is limited. (For those who like completeness it can also be stated that the non-existence of cannibalism implied economic loss in terms of lost proteins and calories. I leave this point out of further consideration.)

Eating 'taboo' food

The native settlement of Houten exemplifies a case in which the meat of an animal that was normally not eaten was consumed. In this settlement, south of the border, the bones of a five-year-old mare were found in one pit dated in the period from AD 50 to 300 (Laarman 1996; Lauwerier 1999; Lauwerier and Robeerst 2001). Remarkably, the bones were not anatomically connected and exhibit cutting and chopping marks. If this animal had been slaughtered for normal human consumption, not all the bones would have remained together since sections of the carcass would have been removed. It is suggested that the animal died from natural causes, the flesh was stripped off and may have been fed to the dogs (Laarman 1996). However, the butchering marks shown on the vertebra and ribs are characteristic of pieces of meat on the bone butchered for human consumption (see Laarman 1996, Figure 63a and 66).

Although the other assemblages from this site show that horsemeat was not eaten here, these remains suggest that a horse was consumed. This was not the normal pattern of consumption. About two hundred kilograms of horsemeat was eaten near the find pit within a short space. This extremely concentrated collection of bones reflect an exceptional consumption of a large quantity of meat from an animal that was not normally eaten. Therefore, I suspect that these were the remains of a ritual meal. Apart form some speculations its significance is unclear. But the effect of this supposed ritual was that meat that normally was not eaten, by the ritual became also part of food economy. However, if this single example is compared with all the other horses that were not eaten, the impact of such horse eating rituals on economy must be negligible.

Conclusion

Some examples have been presented of non-economic uses and/or non-economic habits or behaviours concerning animals in Roman times in the Netherlands. These examples include, offerings at temples, building and site offerings, gravegifts, and the avoidance of eating horse meat. Although these data do not form a complete data set of these phenomenon, they do give an introductory impression of them.

Based on the data presented, there are almost no archaeozoological indications of breeding animals used especially for offerings. The only exception is the impressive looking ram from the *suovetaurilia* offering at the temple of Elst. Occasionally some animals may have been bred for a specific appearance and ritual purpose. Another possibly ritual related activity was the catching of quails. While until now these birds only were found in a non-economic setting, the hunt for quails was possibly focussed on having animals for ritual purposes. Anyway, it seems to have been a small scale activity.

In general it is difficult to get an idea of the quantitative impact of non-economic use of animals on food economy. One example was found of a probable offering that provided an opportunity to break into normally not used food resources: the horse from Houten. Regarding the many not-eaten horses that are found elsewhere, the impact of this ritual on food-economy was small. Maybe the quails can also be considered in this perspective, because they are only known from ritual contexts. However, the small amount of sieved and studied samples from 'normal' refuse deposits from settlements can also explain why quail is not known from these contexts.

The effect of offerings at temples, as far as we know them, seems to be relatively narrow. In the case of Elst and Empel, where cattle were slaughtered, the slaughtering was probably only part of the ritual, but the meat was also used as food. The offering of these animals can be considered as an added value of these animals to their more common function in the food economy. At the Fortuna temple of Nijmegen, where animals were burned, it was the species, mainly chicken, that kept the effect of this ritual on food economy low.

The effect of the once-in-a-lifetime animal gravegifts at burials was small because of their frequency. Conversely, the building and site offerings-if they were real offerings-must have meant, locally and temporary, considerable economic loss or loss of food.

The non-rational avoidance of horsemeat-a generally observed habit south of the Roman border and somewhat less north of it-had the largest impact on food economy. A considerable percentage of the available animal food was not used.

In conclusion, the information about non-economic use of animals in the Netherlands in Roman times indicates that in general these phenomena had just a minor effect on food economy and stockbreeding, although locally or occasionally the consequences could be considerable. An exception, however, was the avoidance of eating horsemeat. This habit had a large and measurable influence on food-economy, due to the fact that roughly a quarter of the potentially available animal food was withdrawn from human consumption.

Acknowledgements

Thanks to my colleagues Menne C. Kosian for help with the illustrations, Herman A. Holleman for correcting the English and Sharyn Jones O'Day (University of Florida, USA) for comments and improving my text. I am grateful to Kristin Bosma and Froukje Veenman (University of

Groningen, the Netherlands) for organising the workshop 'Beyond bones' where we discussed this subject.

References

Anderson, S. and Boyle, K. (eds) 1996. *Ritual treatment of human and animal remains*. Oxford: Oxbow Books.

Bodson, L. (ed.) 2000. *Ces Animaux que l'Homme choisit d'inhumer. Contribution à l'Etude de la Place et du Rôle de l'Animal dans les Rites Funéraires* (Colloques d'Histoire des Connaissances Zoologiques 11). Liège: Université de Liège.

Bogaers, J. E. 1955. *De Gallo-Romeinse tempels te Elst in de Over-Betuwe:* (Nederlandse Oudheden 1). 's-Gravenhage: Staatsdrukkerij en uitgeverijbedrijf.

Bond, J. M. 1996. Burnt offerings: animal bone in Anglo-Saxon cremations. *World Archaeology* **28**, 76–88.

Dijkman, W. 1991. Een kwartel aan de kapoenstraat. *Archeologie in Limburg* **48**, 31.

Hauger, A. 1921. *Zur römischen Landwirtschaft und Haustierzucht*. Hannover: Verlag von M & H. Schaper.

Hill, J. D., 1996. The identification of ritual depositions of animals. A general perspective from a specific study of 'special animal deposits' from the Southern English Iron Age, pp. 17–32 in Anderson, S. and Boyle, K. (eds), *Ritual treatment of human and animal remains*. Oxford: Oxbow Books.

Hulst, R. S. 1978. Druten-Klepperhei. Vorbericht der Ausgrabungen einer Römischen Villa. *Berichten van de Rijksdienst voor het Oudheidkundig Bodemonderzoek* **28**, 133–51.

Laarman, F. J. 1996. The zoological remains, pp. 343–57 in Kooistra, L. I., *Borderland farming* (Thesis University of Leiden). Assen: Van Gorcum.

Lauwerier, R. C. G. M. 1988. *Animals in Roman Times in the Dutch Eastern River Area* (Nederlandse Oudheden 12). 's-Gravenhage: SDU Uitgeverij/Amersfoort: Rijksdienst voor het Oudheidkundig Bodemonderzoek.

Lauwerier, R. C. G. M. and Hessing, W. A. M. 1992. Men, horses and the Miss Blanche effect: Roman horse burials in a cemetery at Kesteren, the Netherlands. *Helinium* **32**, 78–109.

Lauwerier, R. C. G. M. 1999. Eating horsemeat: the evidence in the Roman Netherlands. *Archaeofauna* **8**, 101–13

Lauwerier, R. C. G. M., Groenewoudt, B. J., Brinkkemper, O. and Laarman, F. J. 1999. Between ritual and economics: animals and plants in a fourth-century native settlement at Heeten, the Netherlands. *Berichten van de Rijksdienst voor het Oudheidkundig Bodemonderzoek* **43**, 155–98.

Lauwerier, R. C. G. M. and Robeerst, J. M. M. 2001. Horses in Roman times in the Netherlands, pp. 275–90 in Buitenhuis, H. and Prummel, W. (eds), *Animals and Man in the Past* (ARC-Publicatie 41). Groningen: Archaeological Research and Consultancy.

Lauwerier, R. C. G. M. 2002. Animals as food for the soul, pp. 63–71 in Dobney, K. and O'Connor, T. (eds), *Bones and the Man; Studies in Honour of Don Brothwell*. Oxford: Oxbow Books.

Prummel, W. and Bouma, J. W. 1997. Animal offerings at Borgo Le Ferriere (Latium, Italy). *Anthropozoologica* **25–26**, 531–7.

Ryan, K. and Crabtree, P. J. 1995. *The Symbolic Role of Animals in Archaeology* (MASCA Research Papers in Science and Archaeology 12). Philadelphia: University of Pennsylvania.

Seijnen, M. 1994. Dierebotten en rituele maaltijden, pp. 162–73 in Roymans, N. and Derks, T. (eds), *De tempel van Empel*. 's-Hertogenbosch: Stichting Brabantse Regionale Geschiedbeoefening.

Toynbee, J. M. C., 1996. *Animals in Roman Life and Art*. London: Thames and Hudson.

van Es, W. A. 1967. *Wijster: a Native Village beyond the Imperial Frontier 150–425 A.D.* (Palaeohistoria 11). Groningen: J. B. Wolters.

van Wijngaarden-Bakker, L. H. 1970. Dierenresten uit het Castellum te Zwammerdam; voorbericht over de opgravingscampagnes 1988 en 1969. *Helinium* **10**, 274–8.

Wilson, B. 1992. Considerations for the identification of ritual deposits of animal bones in Iron Age pits. *International Journal of Osteoarchaeology* **2**, 341–9.

Zeiler, J. T. 1996. Geofferd aan de godin. Faunaresten uit de Fortunatempel te Nijmegen (2e eeuw n. Chr.). Unpublished report, City of Nijmegen.

Zeiler, J. T. 1997. Offers en slachtoffers. Faunaresten uit de Fortunatempel te Nijmegen (2e eeuw n. Chr.) *Paleo-Aktueel* **8**, 105–7.

Roel C. G. M. Lauwerier
Rijksdienst voor het Oudheidkundig Bodemonderzoek
(National Service for Archaeological Heritage),
P. O. Box 1600, NL-3800 BP Amersfoort,
The Netherlands
E-mail: R.Lauwerier@Archis.nl

9th ICAZ Conference, Durham 2002
Behaviour Behind Bones, (ed. Sharyn Jones O'Day et al.) pp. 73–76

11. Roman *suovitaurilia* and its predecessors

Barbara Wilkens

The term suovitaurilia *denotes a sacrifice of sheep, cattle and swine in a single ritual action. The Roman* suovitaurilia *as described by numerous Latin authors, finds direct antecedents in the pre-Roman ritual of the Italic populations of which some archaeological and literary evidence remains. This paper presents a recent study of faunal remains from the shrine of Monte d'Accoddi in Sardinia (Chalcolithic), which indicates that the Roman* suovitaurilia *derives from a long tradition of preceding rituals. In comparison with the classical Roman* suovitaurilia, *some differences can be noted concerning the sex and the age of the slaughtered animals. Faunal remains from the area of the archaic temples of Eraclea in southern Italy are noteworthy. However, in many cases it is not possible to recognize from the archaeological context the single ritual action and it is difficult to firmly establish that a given component of the assemblage is more than a simple sacrifice. Ritual remains of sheep, cattle and swine have also been recognized in more ancient periods of prehistory, but generally it is difficult to determine if the killing of the three species was contemporaneous.*

The literary evidence

Suovitaurilia, as it was practiced in the Classic Roman ritual, consisted of slaughtered equal numbers of bovines, ovines and pigs. This type of formal sacrifice was practiced both in official and private rituals. The latter practice was modest, performed on a limited scale. It may have been practiced as the purification for agricultural fields and for the commemoration of the dead [Caton (Cato) 1975].

As in all sacrifices, the age of the victims had to be within certain limits, although both young animals (*lactentes*) and adults (*hostiae maiores*) could be sacrificed. According to Varro (Varrone 1992), who also cites earlier references, piglets were retained ready for sacrifice and called "*sacres*", not before ten days after birth. According to Plinius [Plinio 1982 (*Naturalis Historia*, VIII, 77)] the piglet was pure and ready for sacrifice from the fifth day, and the lamb was ready from the seventh day. The calf was ready from the thirtieth day (Plinius, *ibid.*). In another section (*Nat.Hist.*, VIII, 70), Plinius states that the calf is ready for sacrifice when its tail reached the articulation of the hough. Plinius also writes that Coruncanius, a consul in 280 BC, issued a decree prohibiting ruminants from being declared pure victims before they were "*bidentes*" (literally: with two teeth). Since both bovines and caprovines have deciduous incisors at birth, this implies that sub-adult animals were being referred to at the time of the change of the incisors. This may thus be restrictive aimed at curtailing the sacrifice of young animals. The term "*bidens*" was later used only to refer to ovines of that particular age.

As stated by Cato [Caton 1975(*De Agricultura*, CL)], in the harvest sacrifice, suckling animals were used "for these things, to cleanse the property, the earth and the field and to carry out the cleansing, you are pleased by the sacrificing of these suckling suovitaurilia; Father Mars, for the same reason these suckling suovitaurilia please you". The animals were brought all around the property before being sacrificed. "The cleansing of the fields must be done in this way: give the order that the suovitaurilia be led around: With the will of the gods and that the result may be favourable, I entrust unto you, Manius, that you take care to cleanse, making these suovitaurilia go around my property, my fields and my earth for as much as you retain necessary that they may walk or be transported"(Cato, *De Agricultura*, CL).

In Book V of the Aeneid a sacrifice in honour of Anchises is described. "He slaughtered two "*bidentes*", according to the rites, and the same number of pigs and two black heifers, pouring out a cup of wine" [Virgilio 1963 (Vergilius, *Aeneis*, V, 141–143)]. Other offerings were made by the companions who "the cinders on the grass gather from beneath the skewers, roasting the meat" (Vergilius, *Aeneis*, V, 150–152). The colour black of the bovines is likely related to the fact that this type of sacrifice was for a dead person.

Public *suovitaurilia* that were practiced on important occasions involved adult and male animals: "when the Roman People act out the purifying rite of *suovitaurilia*, a boar, and ram and a bull are led around in a procession" [Varrone (Varro, *De Re Rustica*, II, 1)]. Sacrifice required that the victims were perfect. Bovines of certain breeds were preferred, above all the Epirotic and the Umbran White, which were impressive both for their size and their colour. According to Dumezil (1996, 559) the sacrifices made for sky gods were preferably white, while the colour black was usually used for sacrifices in honour of the gods of the night and the underworld and the colour red for Vulcan, the god of fire (Dumezil 1996). The gender of the animal may have also been of importance with reference to the divinity. Female victims may have been preferred for female divinities, castrates for Jupiter, and males for Mars (*ibid.*).

In exceptional circumstances special sacrifices were made, and in some cases the "*ver sacrum*" (sacred spring) was overridden. As reported by Titus Livius (Tito Livio 1997) during a crisis in which the republic was threatened by the Gauls and Carthaginians, the *Pontefex Maximus* consulted the people on the sacred spring.

"Do you want and do you order that these things come to pass? If the Republic of the Roman People of the Quirites in the next five years, as I want it to be saved, is to be saved in this struggle, in the Roman People's war against the Carthaginians, from the war with the Gauls who are on this side of the Alps, so the Roman People of the Quirites will give in spring from the swine, ovine, caprine and bovine herds and every thing which before was profane, that they be of Jupiter from the day that the people and the senate order" [Tito Livio 1997 (*Ab Urbe Condita Libri*, XXII)].

This case indicates that goats joined the three traditional species. All pigs, sheep, goat and cattle born during the sacred spring, were sacrificed to Jupiter. The solemn sacrifices enacted in Rome on this occasion led to the slaughter of three hundred white oxen, as well as other victims (*ibid.*).

In addition to animal victims, the sacrifices involved other offerings of non-preservable substances, such as wine and flat loafs of bread, which may not be directly observable in an archaeological assemblages. However, these items may be indicated by artefacts such as pottery or vessel fragments found in the sites (Dumezil 1996, 558).

A part of the meat and the internal organs from the sacrificed animals were cooked and eaten by those present at the ceremony. Generally cooking was done on skewers as is evidenced by burn marks that are frequently found on archaeological bones. However Varro notes that rams with horns, called *harriga*, received different treatment "These are the animals the sacrificial innards of which are cooked in a pot and not on skewers, as is written by Accius and as we read in the Pontifical Books" [Varrone 1992 (*De Lingua Latina*, V, 98)].

The archaeological data

One of the most ancient ritual assemblages in a Roman context are the remains of *suovitaurilia* that have been recognised in the votive deposit of *Lapis Niger* in the Roman Forum (De Grossi Mazzorin 1990). The oldest material from this deposit, from the 6th century BC includes the bones of ovines, pigs and bovines accompanied by remains of goats, dogs and vultures (*ibid.*). The remains from the ancient temple of S. Omobono at Rome (Tagliacozzo 1989) are from the same period. In this case caprovines and pigs dominate, while bones from bovines and dogs are present in lower numbers. This may indicate that more modest types of sacrifice were practiced, as well as the formal sacrifices. Informal sacrifices may have involved only one species amongst those of medium size. These practices increase the number of individuals of some species such as sheep or pig and make it difficult to recognise the contemporaneous sacrifice of an equal number of bovines, pigs and sheep, typical of *suovitaurilia*.

Other remains come from the sanctuary of *Satricum*, which was dedicated to a female divinity, probably *Mater Matuta* (Prummel 1996; Prummel and Bouma 1997). The ritual deposits span a long period, from the 8th to the 3rd century BC. These remains are in poor state of preservation, of which *c.* 10 % have been exposed to very high temperatures. The percentage of burnt remains is lower in the oldest levels and increases from the 5th century onwards. In this case, bovines, caprovines and pigs dominate the assemblage. This deposit is interpreted as an example of *suovitaurilia* (Prummel 1996; Prummel and Bouma 1997). This assemblage also contains remains of species including dog and deer, but these are rare.

The Italic context (*c.* 4th century BC) provides evidence for similar sacrifices. In the area of the ancient temples of Eraclea in Basilicata (Wilkens and Delussu 2002), several wells and graves containing the remains of single ritual actions have been uncovered. We recovered evidence for the contemporaneous presence of three species (sheep, cattle and swine); the treatment of the remains (anatomical selection, presence of burnt fragments and burial of the remains) clearly indicates that the assemblage represents a *suovitaurilia* (Fig. 1). Although, the site was within a Greek context (Eraclea was a Greek colony), the evidence of *suovitaurilia*, a sacrifice of Italic tradition, indicates that the indigenous substrate may have profoundly influenced the ritual practices, both with this form of *suovitaurilia* and with the sacrifice of dog. On the contrary,

	Bothros T89/64	Pit T89/64	Eschara T87/34	Eschara T87/18	Eschara T86/10	Eschara T92/103
Gallus gallus L.	–	–	1	1	1	1
Lepus capensis L.	–	–	–	–	–	1
Canis familiaris L.	1			–	1	–
Sus scrofa domesticus L.	4	1	2	2	4	6
Bos taurus L.	3	1	1	2	2	2
Ovis aries L.	1	–	–	–	–	2
Capra hircus L.	–	–	–	–	–	1
Ovis/Capra	–	1	2	2	3	–
Equus asinus L.	–	–	–	–	1	–

Fig. 1. Bird and mammal bones from Eraclea (number of individuals).

the presence of marine molluscs and chicken bones in some graves may be considered the result of Greek influence on traditions associated with this rite.

Of the swine, females have been identified in Bothros (T89/64), and new-borns in Eschara (T86/10 and T 92/103). Note that the term *eschara*, "hearth", was given by archaeologists to those pits particularly rich with burnt remains. The donkey bone from Eschara (T86/10) is likely intrusive and not related to the sacrifice.

Another example within the Italic context that can be associated with the *suovitaurilia* type ritual is described in *Tabulae Iguvinae (*Devoto 1975). Of the sacrifices codified in this 3rd/2nd century BC Umbran text, the expiatory sacrifice can be considered an Italic form of *suovitaurilia* that is very similar to the typical Roman type. The sacrifice took place close to the city gates and in surroundings woods. This involved the killing of certain animals in equal numbers: before the Trebulana gate three oxen; behind the Trebulana gate three pregnant sows; before the Tessenaca gate three oxen, and behind three suckling pigs; before the Veia gate three oxen, and behind three lambs; in the Giovio wood an unknown number of sheep and three mature calves; in the Coretio wood three mature calves. Evidence indicates that wine and pies were also included in the ritual (Devoto 1975, 28–43). Parts of the meat were distributed and eaten while other parts of the animals were buried or burnt. "Then the Mefa pie, the offerings, the meat, are taken from the sacrificial plate and are thrown into the fire" (Devoto 1975, 51). A part of the liquid and solid offerings and several vases were buried in graves.

From a zooarchaeological perspective, it may be difficult to distinguish between the remains from a normal sacrifice of domestic species repeated over time and that of *suovitaurilia*, typical of the Iron Age and the Roman world (in which there was a contemporary butchering of ovines, bovines and pigs in equal numbers). The problem is increased by a paucity of structures that can be clearly identified as having a ritual function. While the temple structures of the Roman, and pre-Roman cultures of the Iron age are easily identified, as they correspond to particular building canons that are distinct from those of habitations, in Italian pre- and proto-history well-defined and clearly distinct sacred structures are rare; the majority are composed of pits or cave deposits. These deposits have an unusual content including the presence of whole vases and animals, or articulated animal parts, evidence of particular butchering and cooking techniques, and in some cases, the presence of burials nearby. The nature of the deposits contributes to the assumption that these were indeed ritual areas. It is probable that many ritual areas are not recognised as such. Further, in the Roman period, most private sacrifices were carried out in the home or in the fields, and may be unobservable in domestic refuse.

The prevalence of bovines, pigs and caprovines in assemblages may be recognised in certain protohistoric cultural situations, even though it is difficult to arrive at clear distinctions of sex, due to the fragmented nature of the bones. In a Middle Bronze Age pit in Roca at Puglia (Wilkens 1995; Wilkens 1998), corresponding to a single ritual action, 31 % of the animal remains are pigs, 34 % are bovines and 29 % are non-distinguishable sheep and caprovines. Bones from other wild and domestic animals are also present in low numbers (other domestic mammals 1 %, wild mammals 3 %). In another pit at this site, also dating to the Middle Bronze Age, the assemblage was composed of 41 % pigs, 25 % bovines, and 28 % non-distinguishable sheep and caprovines. A small quantity of goat (0.5 %) and wild fauna was recovered in low percentages (3 %), in addition to a whole dog.

The author has recently begun studying the fauna from the Chalcholithic levels in the sanctuary of Monte d'Accoddi in North Sardinia (Fig. 2). This assemblage evidences the contemporaneous sacrifice of sheep, cattle and swine. It appears that this practice was widespread all over Italy and in the surrounding areas for a long time. The Iron Age populations may have simply borrowed from pre-existing cultures and rituals. This sanctuary's deposit

	NR
Pisces	12
Aves	1
Prolagus sardus Wagner	12
Vulpes vulpes L.	3
Canis familiaris L.	9
Sus scrofa meridionalis Forsyth Major	15
Sus scrofa domesticus L.	356
Cervus elaphus L.	5
Bos taurus L.	602
Capra hircus L.	4
Ovis cf. *musimon* Pallas	6
Ovis aries L.	214
Ovis/Capra	1559

Fig. 2. Faunal remains from Monte d'Accoddi, Sardinia (2000 excavations).

yielded bones of the three species in similar percentages, and supplied evidence indicating that certain sexes were chosen for the sacrifice. The ovines are mostly rams of large dimensions (at this point no females have been identified), while both sexes were represented in the pig bones. The determination of the sex of the bovines is difficult, however they are all quite large. Nevertheless, the examples that I have positively identified are male. A small amount of dog remains with butchering marks, and a few fragments of goat are also included in the assemblage; I interpret these as sacrificial victims. The presence of a fox and a *Prolagus* (lagomorph endemic to Sardinia) appear to have entered the deposit casually, as animals living in the site. A deer represented by a single tooth and some fragments of antler (one was worked) were also identified, however its role as sacrificial victim is as yet uncertain (Wilkens, *in prep.*).

Conclusion

In conclusion, I suggest that a type of solemn sacrifice including the three main domestic species present in the Mediterranean since the Neolithic developed early in prehistory, evolving in different cultures until it became the typical *suovitaurilia* of the Roman period. Interestingly, dog bones, a species that is not part of this ritual, is consistently present as an accompanying component in sites on the Italian peninsula and in Sardinia where *suovitaurilia* or similar forms of sacrifice were practiced. Dog is also part of the codified sacrifices in Tabulae Iguvinae. On the other hand, not all votive deposits that include ovine, pig and bovine bones can be reliably interpreted as *suovitaurilia*; the contemporaneous nature of the killings and the equal number of individuals of the three species must be supported in order to qualify as *suovitaurilia*. Assuming that a loss of some original remains may occur in the archaeologically recovered deposit, a collection of bones identified as pigs, bovines, and ovicaprines that are relatively close in number may indicate that the original deposited species were equal in number, and thus represent a *suovitaurilia*. The most convincing evidence is that recovered from pits which were originally dug with the intention of depositing the remains of a single sacrifice, as in the case of Eraclea and that of Monte d'Accoddi.

References

Caton, M. 1975. *De l'Agriculture*. Paris: Les Belles Lettres.

De Grossi Mazzorin, J. 1990. Ossa di avvoltoio, p. 274 in Cristofani, M. (ed.), *La grande Roma dei Tarquini*. Roma: L'Erma di Bretschneider.

Devoto, G. 1975. *Le Tavole di Gubbio*. Firenze: Sansoni.

Dumézil, G. 1996. *Archaic Roman Religion*. London: Johns Hopkins University Press.

Plinio, G. S. 1982. *Storia Naturale*. Torino: Einaudi.

Prummel, W. 1996. The sacrificial animals in deposit II of Borge Le Ferriere (Satricum), pp. 419–81 in Bouma, J. W. (ed.), *Religio Votiva: The Archaeology of Latial Votive Religion*. Drachten: Donkel and Donkel.

Prummel, W. and Bouma, J. W. 1997. Animal offerings at Borgo Le Ferriere (Latium, Italy). *Anthropozoologica* **25–6**, 531–7.

Tagliacozzo, A. 1989. Analisi dei resti faunistici dall'area sacra di S. Omobono, pp. 65–9 in Anonymous (ed.), *Il viver quotidiano in Roma arcaica* (exhibition catalogue). Edizioni Procom.

Tito Livio 1997. *Storia di Roma dalla Fondazione*. Roma: Newton.

Varrone, M. T. 1992. *Opere*. Torino: Unione Tipografica Editrice Torinese.

Virgilio, P. M. 1963. *Eneide*. Torino: Unione Tipografica Editrice Torinese.

Wilkens, B. 1995. Animali da contesti rituali nella preistoria dell'Italia centro-meridionale. *Padusa Quaderni* **1**, 201–7.

Wilkens, B. 1998. Le risorse animali, pp. 223–47 in Cinquepalmi, A. and Radina, F. (eds), *Documenti dell'Età del Bronzo. Ricerche Lungo il Versante Adriatico Pugliese*. Fasano: Schena editore.

Wilkens, B. and Delussu, F. 2002. I resti faunistici, pp. 299–342 in G. Pianu (ed.), *L'Agorà di Eraclea Lucana*. Roma: Carocci.

Barbara Wilkens
Dipartimento di Storia, University of Sassari,
viale Umberto I n°52, Sassari, Italy
E-mail: barbara.wilkens@tiscali.it

9th ICAZ Conference, Durham 2002
Behaviour Behind Bones, (ed. Sharyn Jones O'Day et al.) pp. 77–94

12. Gastronomy or religion? the animal remains from the *mithraeum* at Tienen (Belgium)

An Lentacker, Anton Ervynck and Wim Van Neer

The analysis of the animal consumption leftovers from a single feast held at the Roman temple of Mithras at Tienen (Belgium), and a comparison with the data available from other mithraea *in Europe, have made clear that the composition of the liturgical meals held within the context of the Mithras cult is not only determined by gastronomy. The choice of domestic fowl, the most frequent animal product prepared, is clearly guided by its symbolic meaning. The fact that the more than 250 domestic fowl served at this single banquet were most probably all cocks or cockerels indicates that the birds' crow, announcing the sun, was an important symbolic item within the cult. The mystery cult of Mithras indeed had a strong cosmological framework where the principle god was identified with the invincible Sun. That the age at slaughter of the domestic mammals, also served at the banquet, indicates that the festivity at Tienen was held on the occasion of the longest day of the year, again corroborates the relationship between the cock, as a sun-bird, and the Mithras cult. Our results call for increased attention to animal remains in archaeological investigations of ancient religions.*

Introduction: the cult of Mithras

Towards the end of the Roman period, many mystery cults existed in Europe, most of them having an Eastern origin. One of these religions was the cult of Mitras, that was especially popular from the second to the fourth century AD. Unfortunately, historical sources report little about this religion, primarily because Mithraism maintained strict secrecy about its teachings and practices, revealing them only to initiates and never in writing. At the other hand, an abundance of material evidence exists from the many Mithraic temples and artifacts that archaeologists have found scattered throughout the Roman empire. This forms our main information source about the religion. Until recently, mainly the monumental remains received most of the attention of the researchers, but there is at present a tendency to put more effort into the study of the small finds (see, *e.g.*, Martens and De Boe, *in press*). Animal remains, however, are still undervalued within the interpretations.

The central place within a Mithraic temple or *mithraeum* was traditionally decorated with a scene depicting the god Mithras killing a bull, the so-called *tauroctony* which is also the central theme around which the Mithras myth has been developed (Fig. 1). Traditionally, on this scene also the zodiac, a raven, a scorpion, a snake, a crater, a lion, and a dog were depicted. Following the theory of Ulansey (1989 and 1998), this symbolism can only be understood in a cosmologic framework. More precisely, the objects and animals depicted on this scene must be explained as representing the stellar constellations Corvus, Scorpio, Hydra, Crater, Leo and Canis Minor through which the celestial equator ran during the 'Age of Taurus (the Bull)'. During this period, which occurred roughly from 4000 to 2000 BC, the spring equinox of the sun was in the constellation of the Bull but it was believed to be slowly pushed towards the following constellations of the zodiac by Mitras. Each passage through a zodiac sign would take 2160 years. In that way, the spring equinox moved into the sign of Aries around 2000 BC, into Pisces around the beginning of our era, and now we are awaiting the 'dawning of the age of Aquarius' (Ulansey 1989; Ulansey 1998). Nowadays, we know that the movement of the equinox (the so called *processio*) is caused by a wobble in the rotation of the earth's axis and that the whole process is completed every 25,920 years. The confessors of the Mitras cult, however, believed that Mitras had powers above those of the cosmos and that he could move the whole universe, thus pushing the equinox and ending the Age of Taurus, here visualised as killing the

Fig. 1. The mithraic tauroctony *or bull-killing scene (after the frontispiece of Ulansey 1998). The central part depicts Mithras killing the bull, while a dog reaches for the blood shed. Also visible are a scorpion, a snake, a crater and a lion. On the sides of the representation the subdeities Cautes (with torch upright) and Cautopates (with torch held downwards) can be seen. Mithras' cape is decorated with stars, symbolising the night sky. On top of the cape sits a raven while a snake comes out of a tree in the background. Above this scene a zodiac is depicted. The upper part of the altar piece, above the zodiac, shows more mythical scenes from the life of Mithras.*

Bull itself. In later stages of the development of the religion, Mitras also became closely associated and even identified with the Invincible Sun (Merkelbach 1984; Ulansey 1989; Ulansey 1998).

The myth of Mithras is of course more elaborate than the bull-killing scene alone. Depictions on artefacts and monuments associated with the cult show the birth of Mithras from a rock, the god shooting an arrow to a rock, after which a spring arises, a dinner between Mithras and the Sun on the hide of the slain bull, etc. They also highlight the important role played by two subdeities, Cautes and Cautopates, representing the rising and declining sun. In contrast, iconographic material yielding information about the cult, and especially about the ritual acts involved in the worshipping of the god, is much rarer and often not well understood. We do know that we are dealing with an initiation cult, wherein the worshippers could climb through seven grades, and that some of the liturgical acts involved holding a communal meal or a procession (Merkelbach 1984). However, the details of these ritual acts remain obscure. In fact, the most solid information still comes from the architecture of the *mithraea*, the places where the rituals occurred, and from the small finds associated with them. More than hundred temples of Mitras have already been found within the boundaries of the former Roman Empire (Ulansey 1998, 11, Fig. 1.2). Recently, Flemish archaeology added a new site to the list when remains of a *mithraeum* were excavated at Tienen, in Northern Gaul.

A ritual pit near the *mithraeum* at Tienen

During the excavations at the Grijpenveld site at Tienen (Tirlemont, Belgium), the remains of a *mithraeum* were unearthed. A study of the structure of this temple, and of the cultural archaeological finds, has recently been published (Martens 2001; Vanderhoeven *et al.* 2001; Martens, *in press*). The analysis focused on a large pit found close to the temple building with straight walls and stairs leading to the bottom at one side. On the basis of the ceramics, this structure was filled during the second half of the 3rd century AD. It was hypothesised that the fill represents a single depositional event, a statement supported by the close chronological framework of the ceramics, and by the taphonomic characteristics of the fill. Generally, the artefacts from the fill can be subdivided in two groups: 1) a small number of finds clearly relating to the cult, and 2) remains from a large number of recipients representing what was needed to prepare and serve a banquet, possibly for more than hundred people. Artefacts included about a hundred each of boiling pots, cooking vessels, lids, plates and drinking goblets. Undoubtedly, the excavated context represents the remains of a large feast offered to the god Mithras (Martens 2001; Vanderhoeven *et al.* 2001; Martens, *in press*).

The present contribution deals with the animal remains from the *mithraeum* pit. These finds mainly come from the bottom layers of the fill and have been collected by dry sieving, using a mesh width of 5 mm. Additionally, a number of samples were taken, which were also wet-sieved, using meshes of 0.5 mm width. These samples contained small mammal and bird remains (smaller than 5 mm), unfortunately all fragmentary and unidentifiable to species level. The samples also contained a significant number of fish bones, a minority of which were identifiable to some extent.

A number of pits were found close to the *mithraeum*, which contained fill identical to that in the large pit. Some sherds from these pits articulated, illustrating that the whole set of pits was probably filled at the same time. The animal remains from these additional features, however, have not been studied into the same detail as those from the large pit, although it was immediately clear that they share the same characteristics. The animal remains from the smaller pits are discussed below, but only in relation to the estimations of minimum numbers of individuals (see further).

Inventory

Molluscs, amphibians, fish and wild birds

In total, nearly 14,000 animal remains were handcollected from the pit (Fig. 2). A full account of all identifications, measurements and other observations will be published as part of the final excavation report (Lentacker *et al.*, *in press*); only the data necessary for the interpretation within the ritual framework of the site is mentioned herein. The analysed material contained a small number of shell fragments from landsnails, and some amphibian remains. The handcollected fish bones included eel (*Anguilla anguilla*), cyprinids (Cyprinidae sp.), and the Spanish mackerel (*Scomber japonicus*). Amongst the cyprinids (members of the carp family), a single bone could be attributed to the genus *Leuciscus*; in our regions this genus includes the species orfe (*Leuciscus idus*), dace (*Leuciscus leuciscus*) and chub (*Leuciscus cephalus*) (Vandelannoote *et al.* 1998, 116–28). Another cyprinid bone probably belongs to the same genus but remained classified as 'cf. *Leuciscus* sp.'. The sieved samples (Fig. 2) revealed the presence of many more eel bones and some additional finds of the Spanish mackerel. New, however, was the find of skeletal remains belonging to small clupeids, individuals of the herring family (Clupeidae sp.). We suspect that these bones are from (small) herring (*Clupea harengus*) or sprat (*Clupea sprattus*).

The dimensions of the handcollected bones of eel show that at least two individuals are present, one of 70–80 cm, and one of 50–60 cm standard length. These large eels are not an import from the coast, but represent local catches. The same is true for the species belonging to the genus *Leuciscus,* which, at present, still occur in waters near Tienen (Vandelannoote *et al.* 1998, 116–28). The herring and sprat, however, are marine species (Poll 1947, 129–36; Vandelannoote *et al.* 1998, 231–5) which must have been brought to the site from the coast or from the estuary of the river Scheldt. Recent archaeological finds strongly suggest that this was associated with the production of fish sauce (*garum* or *allec*) on the North Sea coast (Van Neer and Lentacker 1994, Van Neer and Ervynck, *this volume*). Finally, the remains of the Spanish mackerel come from two individuals, both measuring between 30–40 cm. In Roman times, this fish was traded from South-European coasts to northern Gaul as a salted product (*salsamenta*) (Curtis 1991; Van Neer and Ervynck, *this volume*).

The bird bones from the *mithraeum* pit include following species: an unidentified duck (Anatidae sp.), the

species	recovery technique dry–sieved (5mm)	wet–sieved (0.5 mm)
unidentified landsnails (Helicidae sp.)	2	–
unidentified amphibians (Amphibia indet.)	26	–
eel (*Anguilla anguilla*)	18	74
herring or sprat (Clupeidae sp.)	–	35
orfe/dace/chub (*Leuciscus* sp.)	1	–
cf. *Leuciscus* sp.	1	–
unidentified cyprinid (Cyprinidae sp.)	3	–
Spanish mackerel (*Scomber japonicus*)	7	5
unidentified fish (Pisces indet.)	5	240
greylag or domestic goose (*Anser anser* f. domestica?)	15	–
mallard (*Anas platyrhynchos*)	1	–
unidentified duck (Anatidae sp.)	1	–
woodcock (*Scolopax rusticola*)	10	–
jackdaw (*Corvus monedula*)	1	–
unidentified songbird (Passeriformes sp.)	1	–
domestic fowl (*Gallus gallus* f. domestica)	7615	–
unidentified birds (Aves indet.)	1918	–
unidentified small mammals (Insectivora / Rodentia sp.)	41	–
common shrew (*Sorex araneus*)	1	–
woodmouse (*Apodemus sylvaticus*)	1	–
garden dormouse (*Eliomys quercinus*)	1	–
weasel (*Mustela nivalis*)	1	–
hare (*Lepus europaeus*)	4	–
dog (*Canis lupus* f. familiaris)	2	–
pig (*Sus scrofa* f. domestica)	278	–
cattle (*Bos primigenius* f. taurus)	77	–
sheep or goat (*Ovis ammon* f. aries/ *Capra aegagrus* f. hircus)	314	–
sheep (*Ovis ammon* f. aries)	1	–
vertebrae, large mammals	45	–
vertebrae, small to medium mammals	222	–
costae, large mammals	156	–
costae, small to medium mammals	710	–
unidentified mammals (Mammalia indet.)	2449	–

Fig. 2. Species composition of the animal remains from a large pit excavated near the mithraeum. *Represented are the finds numbers for both the dry-sieved (5 mm) specimens and the wet-sieved (0.5 mm) material. From the sieved sample only the fish remains have been counted.*

mallard (*Anas platyrhynchos*), the domestic goose (*Anser anser* f. domestica), the woodcock (*Scolopax rusticola*), the jackdaw (*Corvus monedula*) and an unidentified songbird (Passeriformes sp.). Most species are represented by a single bone, except in the case of the goose and the woodcock.

Domestic fowl

In sharp contrast to the low numbers of wild birds is the high frequency of domestic fowl (*Gallus gallus* f. domestica) (Fig. 2). More than half of the handcollected bones from the pit belong to this species. We suggest that the large number of unidentifiable bird remains may also be from domestic fowl, which would raise the species' frequency to more than two thirds of all handcollected bones! Fig. 3 lists the numbers of skeletal elements, demonstrating that virtually all bones of the body are present. Underrepresented bones include the easily fragmented mandibula, the cranium, vertebrae, and feet.

Our data indicates that the fill of the pit contained almost complete skeletons of a large number of domestic fowl. Taking into account the equal presence of bones from the left and right side of the body, the (subadult or adult) slaughtering age, and the frequency of diagnostic zones of each bone, this yields a MNI of 238 individuals (155 adults and 83 subadults: Fig. 4). When the domestic fowl remains from the other, smaller pits near the

skeletal element	subadult number	adult number	subadult %	adult %	total number
cranium	–	–	–	–	36
mandibula	–	–	–	–	107
scapula	117	253	31.6	68.4	370
coracoid	178	327	35.2	64.8	505
furcula	–				404
sternum	–	–	–	–	459
humerus	248	415	37.4	62.6	663
radius	150	453	24.9	75.1	603
ulna	201	456	30.6	69.4	657
scapholunar	–	–	–	–	6
cuneiform	–	–	–	–	2
carpometacarpus	–	–	–	–	395
phalanges of the wing	–	–	–	–	85
vertebrae	–	–	–	–	404
ribs	–	–	–	–	462
sacrum	–	–	–	–	129
pelvis	–	–	–	–	398
pygostyle	–	–	–	–	32
femur	139	336	29.3	70.7	475
tibiotarsus	268	504	34.7	65.3	772
fibula	31	129	19.4	80.6	160
tarsometatarsus	120	221	35.2	64.8	341
phalanges of the leg	–	–	–	–	150
total	–	–	–	–	7615

Fig. 3. Inventory of the remains of domestic fowl, per skeletal element. For some bones, a distinction was made between adult and subadult specimens.

mithraeum (see earlier), are taken into account, the total finds number roughly rises with another 20%. Assuming that the total MNI will increase proportionately, we estimate this number to be about 286. The frequency of subadult chickens within the population is estimated around 35% on the basis of the MNI calculations.

Measurements of the chicken bones were taken (following von den Driesch 1976); the most important results are summarised in Fig. 5. Remarkably, the greatest lengths show a unimodal distribution for all bones measured (data not represented here, see Lentacker *et al.*, *in press*). We expected that the values would show bimodal distributions (see, *e.g.*, Reichstein and Pieper 1986 for the large collection from Haithabu) because of the sexual dimorphism in domestic fowl (Benecke 1989). The patterns observed may thus indicate that only one sex is present within the adult domestic fowl population represented in this assemblage. In the absence of large assemblages from the Low Countries, we compare our osteometrical data with assemblages from the Roman period in northern France (Lepetz 1996, 71–6). The humerus length measurements gathered at Tienen correspond with the male part of the Roman fowl population from France (Fig. 6) (Lepetz 1996, 72, Fig. 88). That this conclusion is valid for all bones can be seen from a comparison of the data listed in Fig. 5 with the corresponding data from Roman northern France (Lepetz 1996, 72, Table LXXI). The subadult bones obviously could not be sexed. Therefore, the presence of females within this group cannot be excluded. We hypothesise that, because of the exclusively male composition of the adult population, this will also be the case for the subadult group. However, at this time there are no data to support this statement.

Mammals

A limited number of smaller mammal bones were collected by hand, including remains from the common shrew (*Sorex araneus*), the wood mouse (*Apodemus sylvaticus*), and the garden dormouse (*Eliomys quercinus*). These animals still occur in the area around the site (Lange *et al.* 1986). A single cranial fragment could be attributed to the weasel (*Mustela nivalis*), a carnivore that also lives in the area around Tienen. The same is true for the hare (*Lepus europaeus*), represented by four bones, of which one scapula shows evidence of butchery.

Bone fragments			humerus	ulna	tibiotarsus	tarsometatarsus
adult	complete	left	47	84	25	22
		right	54	82	24	18
	prox fr.	left	50	42	55	28
		right	47	59	56	33
	dist fr.	left	68	71	108	28
		right	67	62	104	26
	diaphysis		84	58	134	66
subadult	complete	left	42	33	31	17
		right	35	33	29	15
	prox fr.	left	25	29	25	2
		right	20	22	24	8
	dist fr.	left	40	32	52	19
		right	43	24	51	20
	diaphysis		41	26	54	39
Diagnostic zones						
adult	proximal	left	97	126	80	50
		right	101	141	80	51
	distal	left	115	155	133	50
		right	121	144	128	44
subadult	proximal	left	67	62	56	19
		right	55	55	53	23
	distal	left	82	65	83	36
		right	78	57	80	35
MNI estimation						
adult			121	155	133	51
subadult			82	65	83	36
all ages (sum)			203	220	216	87

Fig. 4. Summary of the estimation of the minimum number of individuals of domestic fowl.

skeletal element	measurement	min	max	mean	s dev	n
coracoid	GL	54.2	76.8	68.0	3.9	136
humerus	GL	68	82.8	75.4	2.9	95
	Bp	16.6	24	20.3	1.1	162
	SC	6.2	10	7.5	0.5	133
	Bd	14.1	17.8	16	0.8	188
ulna	GL	69	84.2	75.8	3.3	155
	Bp	7.7	10.8	9.5	0.5	244
	Dip	11.5	16	13.7	0.8	232
	SC	4.1	7	4.9	0.3	214
	Did	8.4	11.8	10.2	0.5	286
radius	GL	52.6	63.5	57.9	2.2	86
carpometacarpus	GL	36.1	46.7	41	1.9	203
	Bp	10.8	13.9	12.4	0.6	231
	Did	5.7	13.1	7.8	0.6	230
tibiotarsus	GL	106.6	134.2	120.9	6.7	39
	La	102.1	129.7	115.6	6.0	48
	Dip	17.3	25	22.1	1.3	131
	SC	5.7	12.9	7	0.8	168
	Bd	9.6	14.5	12.1	0.7	222
	Dd	6	15	12.9	0.8	214
tarsometatarsus	GL	72.8	91.8	82.4	5.0	38
	Bp	11.1	16.3	14.2	0.8	84
	SC	5.7	8.7	7.2	0.6	60
	Bd	11.8	15.8	14	0.9	69

Fig. 5. Summary of the measurements taken on the most important domestic fowl bones.

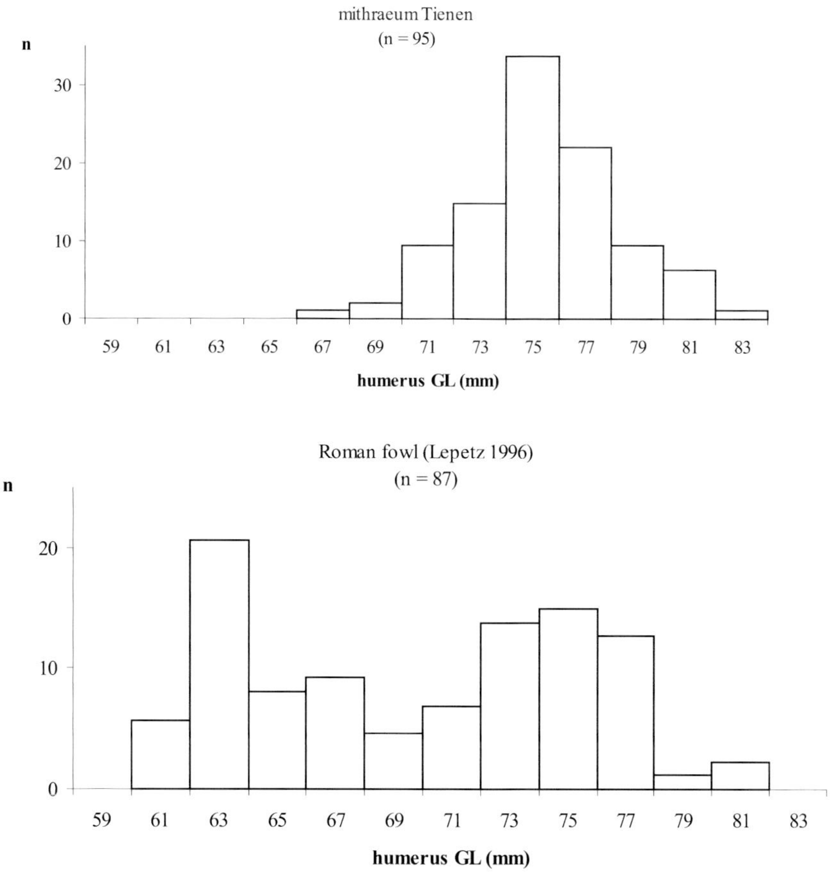

Fig. 6. Comparison of the distribution of the greatest length of the humerus of domestic fowl from the mithraeum *at Tienen, with that from Roman fowl from northern France (data compiled by Lepetz 1996) (n: finds numbers).*

The remains of domestic mammals from the fill include the bones of dog (*Canis lupus* f. familiaris), pig (*Sus scrofa* f. domestica), cattle (*Bos primigenius* f. taurus), and sheep or goat (*Ovis ammon* f. aries/ *Capra aegagrus* f. hircus). Only two dog bones were found. Although in Roman sites in Northern Gaul sheep almost always outnumber goat (see, *e.g.*, Lepetz 1996), in only one case (a sheep bone) it was possible to make the distinction between the two species (following Boessneck *et al.* 1964). This pattern is strongly related to the fact that all small ruminant bones come from young animals. The same is true for the pig remains. The cattle bones, however, come from older animals. Other differences exist between the remains of cattle and those of pig and sheep: the bones from pigs and sheep are often complete or articulating fragments, while the cattle remains are dominated by loose teeth and small fragments from large bones, often from parts of the skeleton that do not yield much meat. The only quality beef represented by the cattle bones is that from around the vertebrae and ribs. Indeed, the finds suggest that the pig and sheep remains represent complete skeletons, be it very partially preserved, while the cattle bones mainly comprise general habitation refuse, as found in all types of contexts from a variety of Roman sites.

The ages at death of the pigs and the sheep have been reconstructed by the observation of the molar wear stages (MWS) on the mandibles (data not represented here) (following Grant 1982). The mandibles also have been used to estimate the MNI for each species. The results show that at least 11 sheep have been slaughtered at the same, very young age (the age at death of a 12th individual could not be established). The MWS of 1 to 3 indicates that the age of death must have followed shortly after birth. The same is true for a MNI of 6 piglets, all slaughtered around the same age, shortly after birth (MWS 1 to 2). Additionally, a single pig mandible has been found, from an animal that reached an older age (MWS of 9). Another, rostral fragment of a pig mandible belonged to another individual (the elements do not fit together) but, on the basis of the eruption state of the canines, belongs to the same age group as the mandible with MWS 9. In total, the remains of at least 8 piglets must thus have been deposited in the pit. These MNI numbers become 10 for the pigs, and 14 for the sheep when the bones from the other, smaller pits near the *mithraeum* are taken into account. The age estimations from these additional animals are the same as for those from the large pit.

Taphonomy: from the preparation of the banquet to the removal of the waste

When the animal remains from the pit are subdivided into taphonomic groups (*sensu* Gautier 1987), they separate into several categories. First, the majority of the finds clearly represents consumption refuse, i.e; the fish, the birds (except perhaps for the jackdaw, a bird that is at present considered to be inedible), domestic fowl, hare, cattle, pigs and sheep. The landsnails and small mammal remains (shrew, woodmouse, garden dormouse, and weasel) represent intrusives.

Considering the consumption refuse, we now turn to the reconstruction of how the animal remains were deposited into the fill, and why they display the characteristics described above. Starting with the domestic fowl, the intraskeletal distribution can be explained by the deposition of the remains of complete individuals, be it that some parts of the body were often removed (heads, part of the neck vertebrae, feet). Taking into account the effects of unequal fragmentation, differential preservation, and unequal recovery, we suggest that the whole process started when complete chickens were beheaded and prepared for cooking. In many cases, this probably also involved cutting off the feet and the lower part of the legs, which could explain the low frequency within the collection of the tarsometatarsals and of the leg phalanges. Perhaps the neck of the animals was also removed, explaining the underrepresentation of vertebrae in the assemblage.

It is, of course, difficult to reconstruct the actual cooking process used to prepare the slaughtered chickens for consumption. However, burning traces on the bones suggest that the animals were fried, broiled or grilled rather than boiled. The rarity of the hare bones makes it impossible to say something about the cooking of this food product. The lambs and piglets seem to have been prepared as whole animals. Broiling on a spit was perhaps the practice used. The cattle bones do not reveal information that could illuminate the process of the preparation of beef. The fish bones point to the use of fish sauce, *salsamenta* and at least three freshwater fish (two large eels and a *Leuciscus* sp.). How the latter three food products were prepared remains unknown.

The observations made indicate that the fill of the pit mainly consists of table leftovers. Apparently, a small number of freshwater fish, salted fish products, some wild birds, around 286 chickens, 14 lambs, 10 piglets and a number of beef parts have been eaten and what was left was dropped in the large pit and in the other, smaller structures investigated. At least one of the dishes was spiced up with fish sauce. It is highly probable that the consumption took place near the pit, otherwise all the leftovers had to be transported and much of the small material would have been lost. Finally, the pit was filled completely, using sediment in which perhaps some residual bones were present. After the original deposition, the fill was invaded by more recent intrusives.

The timing of a single event

The hypothesis that the pit was filled during one event, put forward on the basis of the stratigraphy and the analysis of the ceramics (Martens, *in press*), is corroborated by the present study of the animal remains, more precisely by the ages of the slaughtered mammals from the large pit. The tooth rows of 11 of the 12 sheep all show the same wear stage and thus indicate a same age of death (taking into account the variation in the birth dates, and of tooth eruption and wear, see Ervynck 1997). Given the fact that, within a traditional system of sheep husbandry, lambs were born only once a year, this means that all sheep of which the bones were found within the *mithraeum* pit died in the same part of the year. Because the sedimentological characteristics of the fill indicate that the structure was not exposed for a long time, it must be concluded that all lambs died in the same part of the *same* year. This pattern is the same for the piglets from the pit; 6 out of 8 show the same molar wear stage (MWS). The other two show an older age, which, however, could refer to the same slaughtering season as that of the first six, taking into account that pig populations could have two litters a year (see further).

When the MWS stages are translated into estimations of real ages at death (Fig. 7), the interpretation about the timing of the deposition becomes clearer. On the basis of the eruption stages of the teeth (method explained in Ervynck 1997; eruption data following Habermehl 1975), the MWS of the tooth rows of 6 of the piglets can be translated into a slaughter age of two to three months. Assuming that, in a traditional breed, piglets were born around April (Williams 1977), this would point to June or July as the period of deposition. Two of the subadult pigs show ages that do not fit into this pattern (on the basis of the tooth eruption stages they were 8 months and somewhat more than 8 months old, respectively), but when we assume that they come from a litter that was born in September, this also points to June or July for the time of slaughtering. Within a traditional breeding system, pigs indeed can have two litters a year when there is enough food available, and late Summer or early Autumn is the period during which such a second farrowing could occur (Ervynck and Dobney 2002). Theoretically, the combination of pigs of 3 and 8 months old can also indicate a slaughtering moment in December (with two of them being born in Spring and six in early Autumn), but this possibility is contradicted by the presence of 11 lambs that were slaughtered around the age of three months (an interpretation again based upon the comparison of the known chronology of the eruption of teeth in sheep with the observations of eruption and wear made on the finds). Within a traditional animal husbandry system, sheep are only born once a year, in early Spring (Williams 1977), which again points to a slaughtering moment around June or July. All this would mean that the Mithras pit was filled at the beginning of Summer of a particular year during the second half of the third century AD.

	Oct.	Nov.	Dec.	Jan.	Feb.	Mar.	Apr.	May	Jun.	Jul.	Aug.	Sep.	Oct.	Nov.	Dec.	n
Pig																
1st farrowing	–	–	–	–	–	–	B	1	2	3	4	5	6	7	8	6
2nd farrowing	B	1	2	3	4	5	6	7	8	9	10	11	12	13	14	2
Sheep																
1st lambing	–	–	–	–	–	B	1	2	3	4	5	6	7	8	9	11

Fig. 7. Schematic representation of the reconstruction of the timing of the banquet on the basis of the slaughter ages of the lambs and piglets. Indicated are the ages (in months, B: birth) the animals reach within each month of the year. The combination of slaughtering ages of 2–3 and 8–9 months for the pigs and 3–4 for the sheep, in a single depositional event points to June-July for the timing of that event.

Reconstruction of the banquet

The foregoing observations and interpretations allow us to make a tentative reconstruction of the events that took place shortly before the filling of the investigated pit. Since the structure was filled in one event, this means that the consumption remains found could have been derived from one enormous banquet. The reconstruction of the minimum number of individuals tells us that, apart from the fish sauce and *salsamenta*, at least 3 fish, 286 chickens, a number of wild birds, (probably) a hare, 10 piglets, 14 lambs and a quantity of beef were served. On this basis, it may be proposed that at least 286 persons must have been present. If everybody was only offered half a chicken, this number can even be doubled! It should be noted that the estimation of the number of guests on the basis of food remains is significantly higher than that on the basis of the ceramics (somewhat higher than a hundred people). The guests thus likely shared plates and cooking pots. At the other hand, it is possible that we are dealing not with one meal but with a series of meals that were all served within a short period, for example during a festival of several days. In that case the number of attendants could have been considerably lower, perhaps the hundred estimated on the basis of the ceramics.

It is conceivable that a large group of people all brought with them one or more food items (as they did with their crockery: Martens, *in press*) to be consumed at the feast. Whether these food products came already prepared or were cooked and roasted at the feast, is difficult to establish. However, charcoal found in the pit suggests that a fire was built in the vicinity. Moreover, it has been demonstrated that in the case of the domestic fowl at least some slaughtering offal was present. Taking into account that most of the slaughtering refuse was probably deposited above ground, and thus not preserved, it is thus not excluded (but not proven) that the killing of the animals has taken place at the temple. For the cooking itself this may have been even more probably the case as suggested by the finds from other *mithraea*, where cooking places have been excavated (Schatzmann, *in press*). The excavation of the temple at Tienen has not revealed the presence of a kitchen but it must be stressed that within the temple domain there was hardly place for one (Martens, *in press*).

Gastronomy, or religion?

The evidence presented indicates that the large early summer event at the *mithraeum* at Tienen was a banquet. The question remains, however, if dining together was the only purpose of the gathering. Obtaining nourishment was probably not the primary goal of the festivity. On the contrary, we argue that the gathering at the temple had a higher meaning. Considering the slaughtering, it has been argued that it cannot be proven that this took place near or within the temple, although this possibility can also not be ruled out. The slaughtering could have been a purely practical matter, one that had nothing to do with the ritual sacrifice of living beings, but its possible symbolic or ideological nature remains hidden. We do not doubt, however, that the animals killed were consumed. It must indeed be stressed that we are not dealing with sacrificed animals that were left for the god, without being eaten by the worshippers. Of course, the act of consumption does not imply that the food served at the banquet could not represent offerings to the god; the deposition of the leftovers may have been the ritual act. That these leftovers were (most probably) first thrown on the ground and then buried in a pit (and not, for example, dumped on the compost heap) indicates that they were meant to be sent to the god. Within the Roman tradition, food leftovers became offerings when they were thrown on the floor, for the gods in the underworld, even when meals were taken at home.

It is known that a communal meal was part of the liturgy within the Mithras cult (Beck 2000, 145, note 3, and the references therein). The question must now be asked whether gastronomical quality was the only criterium upon which the ingredients for the banquet were selected. Without doubt, culinary value has played a role in the composition of the menu, since most of the animal food products consumed were of high quality. This is certainly indicated by the young slaughter ages of the piglets and lambs. Considering the overall frequency of species, the abundance of domestic fowl could also point in that direction. Chicken was a highly esteemed meat product within the Roman cuisine (Lepetz 1996). The Mithras context can also be compared with a bone assemblage consisting of general consumption refuse (which, at Tienen, was found on a location called 'Zijdelingse straat' during the excavation of a Roman street surface and its adjacent open space) (Fig. 8) (Verriest 1999). Even while taking into account possible taphonomic biases, it is clear that two characteristics discriminate the assemblages: the frequency of the chicken remains (high in the ritual deposit and negligible in the general refuse) and the frequency of the cattle bones (near absent in the ritual pit and extremely high in the general consumption refuse). It is commonly known that cattle, although being the most important meat supplier for large and small towns in Roman Gaul, was not regarded to deliver meat that was sought after in terms of gastronomy. The near avoidance of beef at the Mithras meal, together with the dominance of domestic fowl and the presence of young piglets and lambs, could suggest that people were bringing food items that were appreciated for their taste and, at the same time, for their market value. Their higher price would make them more meaningful as an offer. A parallel can be drawn with the food items presented as a meal for the deceased within funerary contexts. It has been shown that, in northern Gaul, pork is the most common funerary food gift in the earlier Roman contexts, while in the later contexts the meat of chicken is the most popular (Lepetz 1996). The choice of these animals is guided by their gastronomic status and not by their possible symbolic meaning within the beliefs around the passage to the hereafter.

Whether, except for the domestic mammals and the fowl, other animal products from the Mithras pit can corroborate the high gastronomic status of the assemblage is unclear. Game was extremely rare, and only represented by hare bones and a small number of the remains of wild birds. The appreciation (in Roman times) for the fish products found is also questionable. Fish sauce and *salsamenta*, being part of the Mediterranean cooking tradition, were probably highly valued in Gaul during the earlier period of Roman occupation, but whether they still had this status during the 3rd century AD is not at all sure (Van Neer and Ervynck, *this volume*). The culinary value of freshwaterfish is certainly hard to evaluate, given the scarce information about these food products from Gallo-Roman sites in Flanders, Belgium (Van Neer and Ervynck 1994).

A parallel to the finds at Tienen is provided by the archaeozoological study of the German site of Künzing (Fig. 8) (von den Driesch and Pöllath 2000). There, the excavation of the *mithraeum* also yielded a high frequency of chicken bones and a near absence of cattle remains. Comparing the finds with the general food refuse of the site, patterns similar to those found at Tienen can be observed. The generally higher frequencies of pig at the German site could be just a reflection of the presence of more woodland around that site, compared to the most probably deforested landscape of Roman Tienen. When the faunal material from the Crypta Balbi *mithraeum* at Rome is compared with that of a secular context from the same location, the only meaningful difference can be found in the significantly higher frequency of chicken bones at the temple (20% versus 7%, comparisons based upon NISP) (De Grossi Mazzorin, *in press*). However, caution must be taken when using these data because the contexts compared are not completely synchronous. At Martigny (Switzerland), the composition of the animal remains from the urban *mithraeum* (pig: 47%, chicken: 31%, sheep and goat: 17%, cattle: 5%) is quite different from that of other contexts from the same area (pig: 33%, chicken: 2%, sheep and goat: 16%, cattle: 44%), with chicken and cattle numbers again varying most significantly (Olive, *in press*, comparisons again based upon NISP). At Orbe-Boscéaz (Switzerland), the bone collection from a *mithraeum* associated with a large *villa* at the countryside (pig: 46%, chicken: 33%, sheep and goat: 4%, cattle: 16%) could be compared with faunal contexts from the *villa* itself (pig: 62%, chicken: 13%, sheep and goat: 6%, cattle: 18%). Here, the differences are less clear although a trend is again present in the chicken frequencies (Olive, *in press*). It must also be noted that both Swiss *mithraea* yielded highly comparable consumption remains, while town and *villa* differed considerably in this respect.

Additional support for a possible link between the Mithras cult and the dominance of chicken at the feasts can be found in the reports from other excavations of *mithraea*, although the lack of comparison with 'normal' contexts from the same period and area, and the occasional uncertainty about recovery methodology or taphonomical conditions, makes the evidence somewhat less substantial. Nevertheless, chicken bones were the most common animal remains recovered at the *mithraeum* of London (UK) (Macready and Sidell 1998, 213). This was also the case at the *mithraeum* at Septeuil (France), where the bones of domestic fowl reached 74% of the identified finds (n = 8612 identified remains) (Gaidon-Bunuel 2002). At the *mithraeum* at Zillis (Switzerland), domestic fowl were frequently found (Rageth, *in press*, cited by von den Driesch and Pöllath 2000) but its abundance is not as dominant as, for example, at Tienen. Domestic fowl was also abundant at the 'Kugelstein' site (Austria), a location also interpreted as a *mithraeum* (Adam *et al.* 1996, cited by von den Driesch and Pöllath 2000). Based on MNI estimations, domestic fowl was the most abundant species amongst the animal remains from the *mithraeum* in the army camp of Aquincum (Hungary) (Vörös 1991, cited by von den Driesch and Pöllath 2000). From a number of other *mithraea* only an occasional description of some animal remains is available (mostly because of the small number of finds), making it impossible to evaluate the numerical importance of domestic fowl remains (see, *e.g.*,

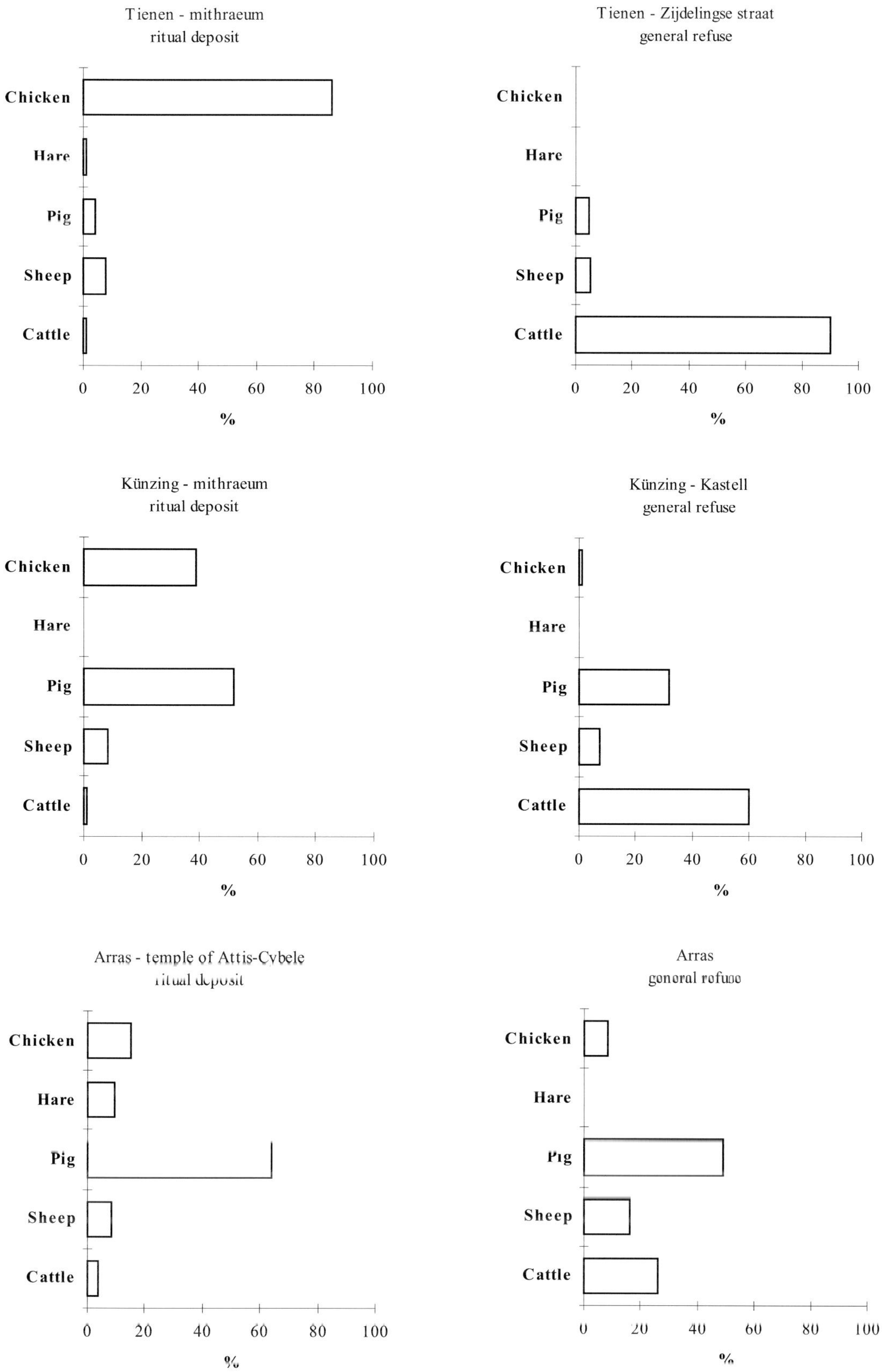

Fig. 8. Comparison between the species frequencies within the consumption refuse of a religious site and of a secular site, for the Roman settlements of Tienen, Künzing (data from von den Driesch and Pöllath 2000) and Arras (data from Lepetz 1996).

the 'second' *mithraeum* at Heidelberg, Germany (Hensen, *in press*), the *mithraeum* at Wiesloch, Germany (Hensen 1994), and other examples mentioned herein).

So far, nothing more has been suggested than that the high frequencies of domestic fowl remains within the context of *mithraea* are linked with the high quality of the meals served. However, it could be that the differences between the rich banquets and the general consumption refuse hide the symbolic meaning of some of the ingredients. A first indication for the latter statement comes from the comparison of the data for the Attis-Cybele temple at Arras in northern France, with a general context from the same town (Fig. 8) (Lepetz 1996, 15–28). It is clear that cattle are less frequent within the food refuse from the Cybele sanctuary while pig is more abundant. This again indicates the higher status of the food consumed in the temple. In the contexts associated with Cybele, however, chicken does not dominate as in the Mithras contexts, which could be seen as corroboration for a possible link between Mithras and domestic fowl. Recently, the frequencies of the species represented within the animal remains from the temple have been reviewed by the incorporation of faunal material from sieved samples (Jacques *et al.* 2002) but, given the fact that this adjustment has not been possible for the 'secular' contexts from this site, this does not help the interpretations.

It is of course beyond the scope of this paper to review all faunal assemblages from Roman temples but, in general, chicken bones are less frequent in Roman temples not devoted to Mithras compared to *mithraea* (Nickel 1999). Examples are provided by the excavations at Elst (The Netherlands) (Lauwerier 1988), Empel (The Netherlands) (Seijnen 1994), La Bauve à Meaux (France) (Magnan and Lepetz 2002), Jouars-Pontchartrain (France) (Blin and Lepetz 2002), Harlow (UK) (Legge *et al.* 2000) and Great Chesterford (UK) (Legge *et al.* 2000) although less detailed recovery methods could in some cases also have been responsible for a low frequency of bird remains. However, there are some examples of Roman temples not devoted to Mithras, where chicken remains were not rare. At the temple at Karden (Germany), possibly devoted to Vulcanus, there was a 'relatively high frequency of domestic fowl remains' (Benecke 1999). At the sanctuary of Mercurius at Uley (UK) chicken remains again formed significant numbers (Levitan 1993). Chicken bones sometimes even represent conspicuous finds at Roman temples not devoted to Mithras. An example is provided by a context of burned chicken bones found at the Fortuna temple at Nijmegen (The Netherlands) (n = 662, MNI = 16) (Zeiler 1996; Zeiler 1997). Another example is provided by the burned domestic fowl and songbird bones filling, almost exclusively, offering pits at the temple of Isis and Magna Mater at Mainz (Germany). At the same site, concentrations of unburned chicken and goose bones were also found. A third type of context of animal remains from the same site is formed by 'normal' consumption refuse, amongst which the skull, mandible and distal leg bones of chicken are frequent (is this the slaughtering offal connected with the other two types of contexts?) (Witteyer and Hochmuth 2002). In general, however, it can be concluded that, except for some examples, chicken bones are generally not dominating the finds assemblages of Roman (non-mithraic) temples.

Many of the studies of animal remains from *mithraea* mentioned above, only deal with the consumption leftovers as a whole, without detailed analysis per contextual unit. However, some reports mention special contexts in which domestic fowl certainly must have played a ritual role. At the Walbrook *mithraeum* at London, chicken bones were found in concentrations in the floor fillings (Macready and Sidell 1998, 208). There was also a collection of charred bones, including those of domestic fowl, found within a pot set beside an altar. Unfortunately, this assemblage was lost prior to detailed examination (Macready and Sidell 1998, 209). In the *mithraeum* at Carrawburgh (UK), chicken bones were found within the layers of heather covering the floor (Richmond and Gillam 1951, 16). Chicken and goose bones were also found in the wattle-revetments of the benches of the temple (Platt 1951a), and the skull of a chicken was excavated from a deposit below the altar (Platt 1951b). A recent review gives some more examples from Germany (Nickel 1999, 168). A cistern within the *mithraeum* of Nida-Heddernheim contained the remains of a cock and other birdbones (Huld-Zetsche 1986, 21). In a cavity in the walls of a later phase of the same temple bones of 'domestic birds' and 'ruminants' were found (Huld-Zetsche 1986, 30). In the *mithraeum* at Trier-Altbachtal a ritual pit was excavated, with a burnt layer, 'birdbones' and coins (Gose 1972, 110). At the *mithraeum* of Mundelsheim ceramic vessels were dug into the ground containing fragments of a cattle skull, and the bones of a piglet and a cock (Planck 1989, 180). Similar deposits were found at the *mithraea* of Dieburg (Behn 1928 cited by Wiblé 1995), Stockstadt (Schleiermacher 1928 cited by Wiblé 1995) and Martigny (Switzerland) (Wiblé 1995). Vermaseren mentions the presence of chicken remains at the *mithraea* of Pons Saravi (France), Köningshöffen (Germany), Schachadorf (Austria), Carnuntum (Austria) and Sarmizegetusa (Romania) (Gaidon-Bunuel 2002, 76, note 119, Table 7, based on Vermaseren 1956, 326; Vermaseren 1960, 138, 155, 213, 298) but it is unclear if these finds represent special deposits or consumption refuse. Nevertheless, the overall conclusion must be that a pattern of special deposits including the remains of domestic fowl does occur at *mithraea*. This, however, is not a completely exclusive trait of the Mithras temples. At Wavendon Gate, Milton Keynes (UK), for example, the remains of a cockerel have been found beneath a jar, close to a cult object (a so-called solar or Taranis wheel), clearly in a non-mithraic context (Williams *et al.* 1996).

The fact that only males were present in the slaughtered population is the most convincing argument that the abundance of domestic fowl at the *mithraeum* at Tienen

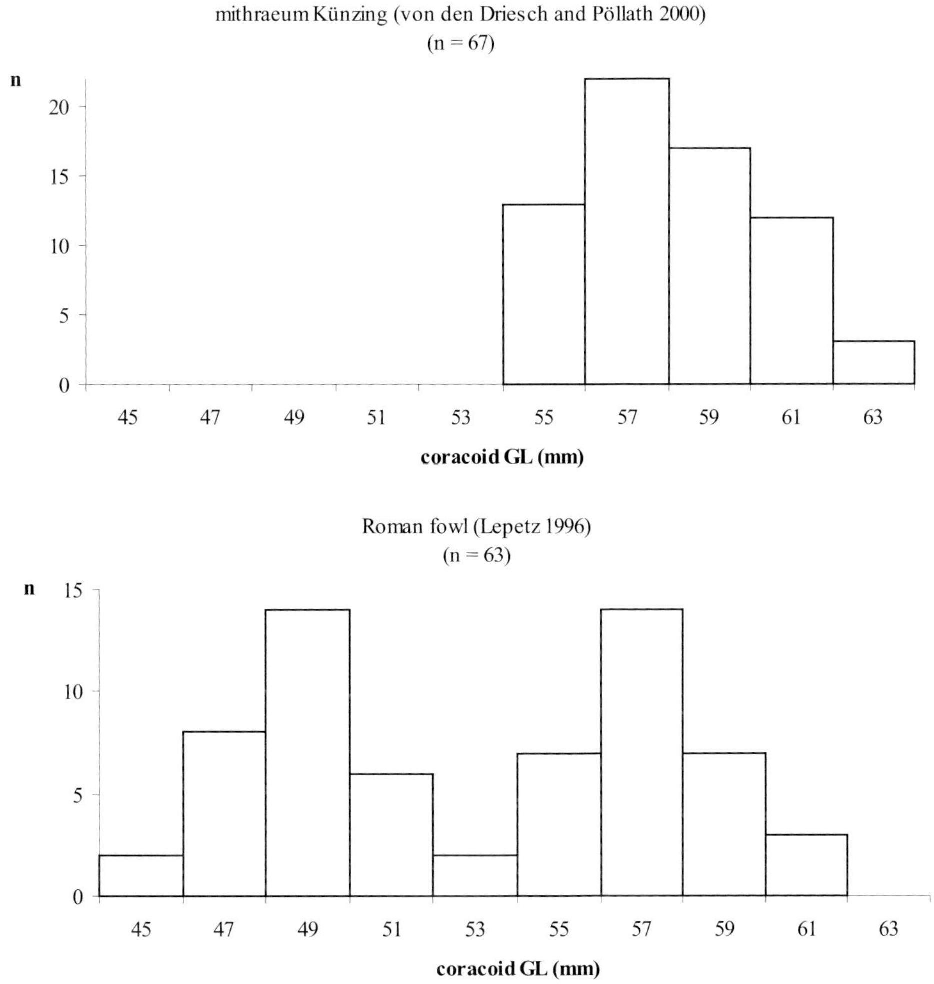

Fig. 9. Comparison of the distribution of the greatest length of the coracoid of domestic fowl from the mithraeum *at Künzing (data from von den Driesch & Pöllath 2000), with that from Roman fowl from northern France (data compiled by Lepetz 1996) (n: finds numbers).*

cannot be explained by gastronomy alone. Whether this was also the case in other *mithraea* is difficult to assess since information about sex ratios is often not available from the archaeozoological reports. An example can be found in the bone report from Künzing, although for that site sex estimations could be made afterwards because the measurements of the domestic fowl remains were fully published; after comparing the measurements from Künzing (von den Driesch and Pöllath 2000) with those from Tienen and from northern France, it becomes clear that all chickens slaughtered at the former site were male (Fig. 9)! At the Crypta Balbi *mithraeum* at Rome, two chickens from layers associated with the temple were male while the sex of a third, labeled 'female?', could not be established beyond doubt (De Grossi Mazzorin, *in press*, Table 4, US 72 and 59). However, these sex estimations are based upon the morphology of the tarsometatarsus, and perhaps the spur-less individual labeled 'female?' may have been a young male? When the measurements of the greatest lengths of the chicken bones from these layers (De Grossi Mazzorin, *pers. comm.*) are compared with the material from Tienen, it can be demonstrated that 5 out of 7 completely preserved bones are from males. One bone shows a greatest length with a value falling between the distributions for males and females, and another was most probably female. Quite remarkably, the domestic fowl remains from a layer deposited after or during the destruction of the *mithraeum* came from both male and female individuals, in almost equal numbers (De Grossi Mazzorin, *in press*, Table 4, US 54). However, these sex estimations are again based upon the morphology of the tarsometatarsal and must thus be taken with caution.

Of the 13 tarsometatarsals recovered at the *mithraeum* of London, 11 could be sexed and they all proved to be male (Macready and Sidell 1998, 213). Amongst the chicken bones found in the wattle-revetments of the benches of the *mithraeum* at Carrawburgh (UK), tarso-metatarsals were identified bearing large spurs. However, the possible presence of female birds was not discussed (Platt 1951a). The chicken bones from the *mithraeum* of Septeuil (France) represent *'principalement des sujets mâles'* (Gaidon-Bunuel 2002). From the data published on the tarsometatarsals (Gaidon-Bunuel 2002, Table 5), it can be calculated that at least 83% of the individuals were male, while 12% was of unknown sex (MNI = 116). Not fitting within this pattern of male dominance, however, are the data from the Swiss *mithraea* of Martigny and Orbe-Boscéaz, where the ratios of male to female birds are 7:4 and 4:6 (Olive, *in press*). However, it is not clear upon which criteria these sex estimations are based.

It must now be questioned whether the selection for male domestic fowl is an exclusive characteristic of the animal remains from *mithraea*. Therefore, the data must be taken into account from temples not devoted to Mithras, but with a high frequency of chicken amongst their consumption waste. At the temple of Karden (Germany), probably devoted to Vulcanus, young cockerels and hens were found amongst the domestic fowl remains, an observation being based upon osteometrical analysis (Benecke 1999). At the Gallo-Roman sanctuary of Halatte (France) (not a *mithraeum*) 90% of the chicken remains were from females (Lepetz 2000, 200). At the sanctuary of La Bauve à Meaux (France) chickens were rarely found (7% of the identified remains, n = 35,000) and the remains consisted mainly of chicks and young hens (Magnan and Lepetz 2002). However, based on the presence or absence of a spur on the tarsometatarsal, the domestic fowl remains from the Fortuna temple at Nijmegen (The Netherlands) were estimated to be derived from at least 8 cocks and 1 hen, while it remained possible that all individuals were male (see the discussion about the absence of a spur in young males earlier) (Zeiler 1997, 105). At Beaumont-sur-Oise (France), within the context of a Roman small town, a 4th century bone deposit was found, of which 48% represented domestic fowl (total number of identified finds = 2397). Within this chicken population virtually all animals were male! Most probably, this context represents a deposit with a 'special', perhaps ritual nature, since it was excavated from a small room within a house, and comprises the leftovers of a single meal (Lepetz 1996, 21, 130–1). The finds from Beaumont should be considered with caution due to our insufficient knowledge of their ideological context.

In conclusion, a dominance of male fowl is the typical pattern (although not exclusive) for *mithraea*. Clearly, gastronomy cannot account for the relationship between Mithras and the cock. Otherwise female domestic fowl would be found at *mithraea* and chickens would have been abundant in all Roman temples. The link between the cock and the cult of Mithras must therefore be a symbolic one.

The symbolic meaning of the cock, and other animals

The question remains if archaeological data can shed light upon the symbolic meaning of the cock within the Mithras cult. The main problem associated with such an interpretation is the lack of direct historical data on the cult and the conflicting theories about the Mithras myth. As has been explained, following Ulansey (1989 and 1998), the Mithras religion was established within a cosmological framework, thus explaining why the animals and figures represented in the *tauroctony* are regarded as symbols for astronomical constellations (Fig.1). These animals, as part of the myth, likely also played a role in the cult. The cock, however, is uncommonly represented in the main bull-killing scene. However, important within this context are the two sub-deities of the Mithras myth depicted at the sides of the *tauroctony*: Cautes, holding a torch upright, and his counterpart, Cautopates, holding its torch downwards. The first figure represents the rising sun while the second symbolises sunset (Vermaseren 1956, 196, 202, Fig. 135; Merkelbach 1984, 285, both cited by Gaidon-Bunuel 2002, 77, note 131) (on the iconography of these torchbearers: Beck 2000, 157, note 54 and the references cited therein). Alternatively, they could represent the growth season (spring), and autumn (Ulansey 1998, 57-60). Importantly, in some representations the cock is associated with Cautes (see a fresco in the *mithraeum* at Santa Maria Capua Vetere: Merkelbach 1984, 122, 285, Fig. 26). It is even said that on a famous brooch found in Ostia, and now in the Ashmolean Museum (Oxford), the cock replaces the figure of Cautes (Merkelbach 1984, 122, 295, Fig. 39; Weiss, *in press*). A connection between the cock and the cult can thus be found in the fact that the cock is the animal that announces the rise of the sun with its crow. There also exists a representation of Cautes and Cautopates where the torches of both figures are replaced by birds, possibly cocks (Beck 2000, 157, note 55).

The importance of the sun within the cosmological framework of the Mithras doctrine is clear. Indeed, within the myth there is a close association between Mithras and the Sun god, reflected in the cult by the two highest initiation grades, *i.e.* the Father (substituting Mithras) and the *Heliodromus* or Sun-runner (representing the Sun), and by the mythical story of the dinner between both figures. Mithras is the superior of the Sun, becoming the *Sol Invictus*. According to Merkelbach (1984, 85), the cock is one of the symbols of the sixth initiation grade, which is the *Heliodromus*.

A further corroboration for the association between the cock and the rising sun can be found in the timing of the single event of which the remains were studied, that is in June or July. We suggest that the banquet was held at the moment of the summer solstice, which in the northern-hemisphere occurs on June 21 or 22 and is the moment within the year characterised by the longest day. According to recent studies, both the winter and summer solstices (shortest and longest day) were very important within the liturgical year of the religion, being connected with the concepts of mortality and immortality, with the descent and ascent of the souls (Beck 2000, 145, note 2). One text even describes the gathering of the Mithraists at Virunum, on the 26th of June (Beck 1998). Our evidence supports the possibility that the announcement of the longest day made the cock an important animal for a feast at that time of the year.

Rare historical evidence underlining the special place of the cock within the religion of Mithras can be found in eastern sources, although it must be stressed that the eastern religion involving Mithras differs significantly from the west-Roman religion discussed here. Nevertheless, the Zend-Avesta prescribes offerings of fowl to Mithras and the accounts of the temple at Dura Europos mention the purchase of fowl for the preparation of the ritual banquets (Richmond and Gillam 1951, 16, notes 14–16). The classical author Ambrosiaster reports that, during one of the initiation rites, "... their hands bound with chicken guts, are propelled over trenches filled with water; then comes someone with a sword and severs the guts" (Beck 2000, 146, note 10). Why precisely chicken guts were chosen (and whether these came from male or female individuals) is not clear. However, Ambrosiaster's description points to the possibility that completely unknown ritual acts were performed in which domestic fowl played a significant role. In fact, the only known iconographical reference to the use of cocks within the cult comes from the *mithraeum* under the Santa Prisca in Rome where a fresco depicts a procession of men, one holding a cock (Merkelbach 1984, 310–1, Fig. 57).

The confrontation between the special status of the cock in the Mithras cult and the symbolic meaning of this animal in other Roman religions remains unresolved. It is known that the cock and the male goat were companions of Mercurius (Schindler 1965). This is illustrated by the excavation of the temple devoted to a Romano-Celtic representation of Mercurius at Uley (UK), where goats where more numerous than sheep, and where domestic fowl was abundant (Woodward 1992, 79, cited by King 2002). Furthermore, cocks and hens were animals typically sacrificed for Aesculapus (Filtzinger 1980). On the basis of the abundance of chicken remains at Karden (Germany) (Benecke 1999), a link may be made between domestic fowl and Vulcanus. The same is true for a possible link between the cock and Fortuna, based upon the finds at Nijmegen (The Netherlands) (Zeiler 1997, 105). Generally, the cock seems to have been regarded as one of the supreme animals figuring into religious practices (see for example, Plinius the Elder's *Naturalis Historiae*, book X, cited by Gaidon-Bunuel 2002, 77, note 129). The bird also had a special status in Celtic cults surviving into Roman times. This is illustrated by the cockerel found in connection with a solar or Taranis wheel at Wavendon Gate (UK) (Williams *et al.* 1996). These observations may appear to contradict the hypothesis of a special place for the cock within the Mithras cult. However, the evidence from the special deposits within the *mithraea* and the complete absence of hens in contexts such as the one excavated at Tienen still support the notion that the cock was special within the context of Mithras.

The proposed symbolic meaning of the cock necessitates a reconsideration of the other species represented within the pit excavated near the Tienen *mithraeum*. The dog, referring to the constellation of Canis Minor (Ulansey 1998, 47), is also typically depicted on the tauroctony but the remains from this species found at Tienen are probably reworked bones buried in the ground where the pit had been dug. At the 'second' *mithraeum* at Heidelberg (Germany), dog bones were found amongst a small bone assemblage (Hensen, *in press*), but the taphonomic status of these finds is uncertain.

The jackdaw, or 'black birds' in general, represent another case. The raven, referring to the stellar constellation of Corvus (Ulansey 1998, 47), figures on the Mithraic bull-killing scene. It is the symbol for one of the initiation stages and plays a part in a catasterism myth about drought and water, in which Apollo (the sun god), Corvus, Crater, and Hydra play a role (Beck 2000, 153, note 36). We suggest that the jackdaw from Tienen's *mithraeum* served as an *ersatz* for the raven. Comparable finds from other *mithraea* also exist. Within the *mithraeum* at Wiesloch the bones of a raven (*Corvus corax*) were found under the sherds of an oil amphora (Hensen 1992 and 1997, cited by von den Driesch and Pöllath 2000, 153). A raven bone was also found at the *mithraeum* of London (Macready and Sidell 1998, 214), while remains of a carrion crow (*Corvus corone*) were found at Martigny (Olive, *in press*). However, crow (*Corvus corone*) and jackdaw (*Corvus monedula*) bones were also found within the Hercules temple at Empel (The Netherlands) (Seijnen 1994), and bones from the crow or the rook (*Corvus frugilegus*, another 'black bird') were part of the animal remains from the Gallo-Roman sanctuary of Halatte (France) (Lepetz 2000). A Romano-British ritual deposit at Garforth (UK) included the remains of a small dog, a piglet, a goat, and a partial raven skeleton (Jaques 2000). Seijnen (1994) mentions that bones from the crow are regularly found at Roman cult places but gives no further references to support this statement.

The remains of two large eels also comprise an intriguing find from the *mithraeum* at Tienen. On the *tauroctony* a snake is depicted, referring to the constellation of Hydra. Hydra, however, is the watersnake, an animal easily substituted by an eel, especially in a part of the world where watersnakes are not present, and where taxonomical biological knowledge had not yet reached its present standards. The symbolic meaning of eels within the Tienen context is further corroborated by the fact that large freshwater fish are rarely documented from Gallo-Roman sites in Flanders, Belgium (Van Neer and Ervynck 1994). This suggests that freshwater fish were probably gastronomically unimportant for the people of that time and must thus have been present at the banquet for other reasons. It should be noted that Hydra plays an important role in the myth in which Apollo, Corvus and Crater also figure (see above) (Beck 2000, 153, note 36).

The slaughter ages of the animals eaten in the temple at Tienen were very young. This is also the case at Carrawburgh (UK) (Fraser 1951; Richmond and Gillam 1951, 12, 21, 24), Künzing (Germany) (von den Driesch and Pöllath 2000), Zillis (Switzerland) (Rageth, *in press*, cited by von den Driesch and Pöllath 2000), and at the *mithraeum* in the army camp of Aquincum (Hungary) (Vörös 1991, cited by Olive, *in press*). At the two Swiss sites of Martigny and Orbe-Boscéaz, it was demonstrated that slaughter ages were considerably younger in the *mithraea* than elsewhere (Olive, *in press*). However, these young slaughter ages do not seem to be an exclusive trait of the temples for Mithras. The remains of young animals are generally found in all Roman temples (Nickel 1999, 168–9).

Finally, the *mithraeum* at Tienen did not yield any evidence for the real killing of a bull near or within the temple complex. This observation matches with what has been found in all other *mithraea* from which the animal remains have been studied.

Conclusion

Our analysis of the animal remains from the ritual deposit near the *mithraeum* at Tienen has revealed much information about the Mithras cult. The food leftovers indicate that ingredients of high gastronomic quality were served at a banquet. However, the selection of the animals killed for the festivities also reflects the symbolic meaning of certain species within the Mithras cult. This is certainly the case for the cock, announcer of the Sun, and probably for the jackdaw (representing the constellation of Corvus) and the eel (Hydra). On the basis of the slaughtering ages of the mammals consumed, the banquet was held at the summer solstice, the time of the year characterised by the longest day. This time frame further underlines the symbolic role of the cock.

The comparison of the animal remains from the *mithraeum* at Tienen with that of other archaeozoological assemblages from *mithraea* in different parts of the western Roman empire, has shown that the characteristics of Tienen's *mithraeum* fit into a larger pattern. These characteristics appear to be an essential part of the Mithras religion. However, the analysis of animal remains from excavated *mithraea* has certainly not yet reached its full potential. Such finds were frequently overlooked and/or understudied in the past, and animal remains only figured as anecdotal evidence in many publications. Archaeozoologists now realise that such faunal assemblages may reveal much about the Mithras cult (von den Driesch and Pöllath 2000, 153) but an in-depth analysis of the place of animals within the religion can only be made when historians fully consider data from archaeozoological analyses. We believe that this will lead to an improved understanding of the ritual role of certain animals within the Mithras cult and myth.

Acknowledgements
The authors wish to thank their colleague Marleen Martens for the cooperation within this project. She organised the excavations for the Institute for the Archaeological Heritage of the Flemish Community and the city of Tienen, in collaboration with the Gewestelijke Ontwikkelingsmaatschappij (Regional Development Society) of the province of Vlaams-Brabant. Jacopo De Grossi Mazzorin must also be thanked, for information about the Crypta Balbi material, and Jorn Zeiler for information about the Fortuna temple at Nijmegen. Wim Van Neer's contribution to this paper represents research results of the Interuniversity Poles of Attraction Programme – Belgian State, Prime Minister's Office – Federal Office for Scientific, Technical and Cultural Affairs.

References

Adam, A., Czeika, S. and Fladerer, F. A. 1996. Römerzeitliche Tierknochenfunde aus zwei Höhlen am Kugelstein bei Deutschfeistritz, Steiermark – Hinweise auf den Mithraskult? *Mitteilungen Anthropologische Geselschaft Wien* **125–126**, 279–89.

Beck, R. 1998. *Qui mortalitatis causa convenerunt.* The meeting of the Virunum Mithraists on June 26, A.D. 184. *Phoenix* **52**, 335–44.

Beck, R. 2000. Ritual, myth, doctrine and initiation in the mysteries of Mithras. New evidence from a cult vessel. *Journal of Roman Studies* **90**, 145–80.

Behn, F. 1928. *Das Mithrasheiligtum zu Dieburg* (Römisch-Germanische Forschungen 1). Berlin, Leipzig: W. de Gruyter & Co.

Benecke, N. 1989. Zum Geslechtsdimorphismus am Skelett mittelalterlicher Haushühner (*Gallus gallus* f. domestica). *Zoologische Abhandlungen Staatliches Museum für Tierkunde Dresden* **45 (3)**, 23–33.

Benecke, N. 1999. Die Tierreste, pp. 158–67 in Nickel, C. (ed.), *Gaben an die Götter. Der Gallo-Römische Tempelbezirk von Karden (Kr. Cochem-Zell, D)* (Archéologie et Histoire Romaine 3). Montagnac: Editions Monique Mergoil.

Blin, O. and Lepetz, S. 2002. Un sanctuaire du vicus de Jouars-Pontchartrain, cité des Carnutes: rituels et vestiges matériels, pp. 40–2 in Lepetz, S. and Van Andringa, W. (eds), *Archéologie du Sacrifice Animal en Gaule Romaine. Rituels et Pratiques Alimentaires. Volume des Pré-Actes.* Paris: CNRS.

Boessneck, J., Müller, H.-H. and Teichert, M. 1964. Osteologische Unterscheidungsmerkmale zwischen Schaf (*Ovis aries* Linné) und Ziege (*Capra hircus* Linné). *Kühn-Archiv* **78 (1–2)**, 1–129.

Curtis, R. I. 1991. *Garum and Salsamenta. Production and Commerce in Materia Medica.* Leiden: Brill.

De Grossi Mazzorin, J. in press. I resti animali del mitreo della Crypta Balbi. testimonianze di pratiche cultuali, in Martens, M. and De Boe, G. (eds), *Roman Mithraism. The Evidence of the Small Finds* (Archeologie in Vlaanderen Monografie 5). Zellik and Tienen: Institute for the Archaeological Heritage of the Flemish Community.

Ervynck, A. 1997. Detailed recording of tooth wear (Grant, 1982) as an evaluation of the seasonal slaughtering of pigs? Examples from Medieval sites in Belgium. *Archaeofauna* **6**, 67–79.

Ervynck, A. and Dobney, K. 2002. A pig for all seasons? Possible approaches to the assessment of second farrowing in archaeological pig populations. *Archaeofauna* **11**, 7–22.

Filtzinger, Ph. 1980. *Hic saxa loquuntur – Hier reden die Steine* (Kleine Schriften zur Kenntnis Römische Besetzungsgeschichte Südwestdeutschland 25). Stuttgart: Gesellschaft fur Vor und Frühgeschichte in Wurttemberg und Hohenzollern.

Fraser, F. C. 1951. Report upon mammalian bones from the mithraeum at Carrawburgh, pp. 89-90 in Richmond, I. A. and Gillam, J.P. (eds), *The Temple of Mithras at Carrawburgh* (Archaeologia Aeliana fourth series XXIX). Newcastle: The Society of Antiquaries of Newcastle upon Tyne.

Gaidon-Bunuel, M.-A. 2002. La cuisine du mithraeum de Septeuil, pp. 70–83 in Lepetz, S. and Van Andringa, W. (eds), *Archéologie du Sacrifice Animal en Gaule Romaine. Rituels et Pratiques Alimentaires. Volume des Pré-Actes.* Paris: CNRS.

Gautier, A. 1987. Taphonomic groups: how and why? *ArchaeoZoologia* **I (2)**, 47–52.

Gose, E. 1972. *Der Gallo-Römische Tempelbezirk im Altbachtal zu Trier* (Trierer Grabungen und Forschungen VII). Mainz: von Zabern.

Grant, A. 1982. The use of tooth wear as a guide to the age of domestic animals, pp. 91–108 in Wilson, B., Grigson, C. and Payne, S. (eds), *Ageing and Sexing Animal Bones from Archaeological Sites* (BAR British Series 109). Oxford: British Archaeological Reports.

Habermehl, K.-H. 1975. *Die Altersbestimmung bei Haus- und Labortieren* (second edition). Berlin and Hamburg: Paul Parey.

Hensen, A. 1992. *Das Mithräum im Römischen Vicus von Wiesloch (Rhein-Neckar-Kreis).* Unpublished MsA thesis, University of München.

Hensen, A. 1994. Das Mithräum im Vicus von Wiesloch. *Archäologische Nachrichten aus Baden* **51/52**, 30–7.

Hensen, A. in press. Das zweite Mithräum von Heidelberg, in Martens, M. and De Boe, G. (eds), *Roman Mithraism. The Evidence of the Small Finds* (Archeologie in Vlaanderen Monografie 5). Zellik and Tienen: Institute for the Archaeological Heritage of the Flemish Community.

Huld-Zetsche, I. 1986. *Mithras in Nida-Heddernheim* (Museum für Vor- und Frühgeschichte Frankfurt am Main Archäologische Reihe 6). Frankfurt am Main: Museum für Vor- und Frühgeschichte.

Jacques, A., Lepetz, S. and Van Andringa, W. 2002. Vestiges de repas et identification d'un siège de collège à Arras-Nemetacum. pp. 22–30 in Lepetz, S. and Van Andringa, W. (eds), *Archéologie du Sacrifice Animal en Gaule Romaine. Rituels et Pratiques Alimentaires. Volume des Pré-Actes* Paris: CNRS.

Jaques, S. D. 2000. Technical report. Vertebrate remains from Garforth, West Yorkshire (site code B1530A). Unpublished technical report from the Environmental Archaeology Unit, York 2000 (10).

King, A. 2002. Animal remains from temples in Roman Britain a review, pp 20–1 in Lepetz, S. and Van Andringa, W. (eds), *Archéologie du Sacrifice Animal en Gaule Romaine. Rituels et Pratiques Alimentaires. Volume des Pré-Actes* Paris: CNRS.

Lange, R., Van Winden, A., Twisk, P., De Laender, J. and Speer, C. 1986. *Zoogdieren van de Benelux. Herkenning en Onderzoek.* Amsterdam: Jeugdbondsuitgeverij.

Lauwerier, R. C. G. M. 1988. *Animals in Roman Times in the Dutch Eastern River Area* (Nederlandse Oudheden 12). Amersfoort: Rijksdienst voor het Oudheidkundig Bodemonderzoek.

Legge, A. J., Williams, J. and Williams, P. 2000. Lambs to the slaughter: sacrifice at two Roman temples in southern England, pp. 152–7 in Rowley-Conwy, P. (ed.), *Animal Bones, Human Societies.* Oxford: Oxbow Books.

Lentacker, A., Ervynck, A. and Van Neer, W. in press. The symbolic meaning of the cock. The animal remains from the mithraeum at Tienen, in Martens, M. and De Boe, G. (eds), *Roman Mithraism. The Evidence of the Small Finds* (Archeologie in Vlaanderen Monografie 5). Zellik and Tienen: Institute for the Archaeological Heritage of the Flemish Community.

Lepetz, S. 1996. *L'Animal dans la Société Gallo-Romaine de la France du Nord* (Revue archéologique de Picardie numéro spécial 12). Amiens: Société de Préhistoire du Nord et de Picardie.

Lepetz, S. 2000. Les restes osseux animaux du sanctuaire gallo-romain de la forêt d'Halatte (commune d'Ognon, Oise). Vestiges sacrificiels et reliefs de repas? *Revue archéologique de Picardie* **n° spécial 18**, 197–200.

Levitan, B. 1993. Vertebrate remains, pp. 257–301 in Woodward, A. and Leach, P. (eds), *The Uley Shrines. Excavation of a ritual complex on West Hill, Uley, Gloucestershire. 1977–9* (Historic Buildings and Monuments Commission Archaeological Report 17). London: Historic Buildings and Monuments Commission.

Macready, S. and Sidell, J. 1998. The animal bones, pp. 208–15 in Shepherd J. D. (ed.), *The Temple of Mithras.* London: English Heritage.

Magnan, D. and Lepetz, S. 2002. La Bauve à Meaux, sanctuaire suburbain de la cité des Meldes, pp. 34–7 in Lepetz, S. and Van Andringa, W. (eds), *Archéologie du Sacrifice Animal en Gaule Romaine. Rituels et Pratiques Alimentaires. Volume des Pré-Actes* Paris: CNRS.

Martens, M. 2001. Een tempel voor Mithras in de vicus van Tienen. *Hermeneus* **73(3)**, 244–51.

Martens, M. in press. The mithraeum in Tienen, small finds and what they can tell us, in Martens, M. and De Boe, G. (eds), *Roman Mithraism. The Evidence of the Small Finds* (Archeologie in Vlaanderen Monografie 5). Zellik and Tienen: Institute for the Archaeological Heritage of the Flemish Community.

Martens, M. and De Boe, G. (eds) in press. *Roman Mithraism. the Evidence of the Small Finds* (Archeologie in Vlaanderen Monografie 5). Zellik and Tienen: Institute for the Archaeological Heritage of the Flemish Community.

Merkelbach, R. 1984. *Mithras.* Königstein: Hain.

Nickel, C. 1999. *Gaben an die Götter. Der Gallo-Römische Tempelbezirk von Karden (Kr. Cochem-Zell, D)* (Archéologie et Histoire Romaine 3). Montagnac: Editions Monique Mergoil.

Olive, C. in press. La faune exhumée des mithraea de Martigny (Valais) et d'Orbe-Boscéaz (Vaud) en Suisse, in Martens, M. and De Boe, G. (eds), *Roman Mithraism. The Evidence of the Small Finds* (Archeologie in Vlaanderen Monografie 5). Zellik and Tienen: Institute for the Archaeological Heritage of the Flemish Community.

Planck, D. 1989. Ein römisches Mithräum bei Mundelsheim, Kreis Ludwigsburg. *Archäologische Ausgrabungen in Baden-Württemberg* **4**, 177–83.

Platt, M. I. 1951a. Report upon bird bones from the wattle-revetments of benches in mithraea II and III at Carrawburgh, p. 91 in Richmond, I. A. and Gillam, J. P. (eds), *The Temple of Mithras at Carrawburgh* (Archaeologia Aeliana, fourth series XXIX). Newcastle upon Tyne: the Society of Antiquaries of Newcastle upon Tyne.

Platt, M. I. 1951b. Report upon the skull of a common fowl from the ritual deposit below the altars of Mithras, Carrawburgh, pp. 91–2 in Richmond, I. A. and Gillam, J. P. (eds), *The Temple of Mithras at Carrawburgh* (Archaeologia Aeliana, fourth series XXIX). Newcastle upon Tyne: the Society of Antiquaries of Newcastle upon Tyne.

Poll, M. 1947. *Faune de Belgique. Poissons Marins.* Bruxelles: Musée royal d'Histoire naturelle de Belgique.

Rageth, J. in press. Neue Beiträge zur spätrömischen Kulthöhle in

Zillis. *Zeitschrift Schweizerische Archäologie und Kunstgeschichte.*

Reichstein, H. and Pieper, H. 1986. *Untersuchungen an Skelettresten von Vögeln aus Haithabu (Ausgrabung 1966–1969)* (Berichte über die Ausgrabungen in Haithabu 22). Neumünster: Karl Wachholtz.

Richmond, I. A. and Gillam, J. P. 1951. *The Temple of Mithras at Carrawburgh* (Archaeologia Aeliana fourth series 29). Newcastle upon Tyne: the Society of Antiquaries of Newcastle upon Tyne.

Schatzmann, A. in press. Möglichkeiten und Grenzen einer funktionellen Topographie von Mithrasheiligtümern, in Martens, M. and De Boe, G. (eds), *Roman Mithraism.The Evidence of the Small Finds* (Archeologie in Vlaanderen Monografie 5). Zellik and Tienen: Institute for the Archaeological Heritage of the Flemish Community.

Schindler, R. 1965. Gallorömische Götter, Kulte und Heiligtümer im Saarland. *Berichte der Staatliche Denkmalpflege im Saarland* **12**, 79–106.

Schleiermacher, L. 1928. Das zweite Mithräum in Stockstadt a. M.. *Germania* **12**, 46–56.

Seijnen, M. 1994. Dierebotten en rituele maaltijden, pp. 162–173 in Roymans, N. and Derks, T. (eds), *De Tempel van Empel. Een Hercules-Heiligdom in het Woongebied van de Bataven* (Graven naar het Brabantse Verleden 2). 's-Hertogenbosch: Stichting Brabantse Regionale Geschiedbeoefening.

Ulansey, D. 1989. *The Origins of Mithraic Mysteries.* Oxford: Oxford University Press.

Ulansey, D. 1998. *Die Ursprünge des Mithraskults. Kosmologie und Erlösung in der Antike.* Stuttgart: Theiss.

Vandelannoote, A., Yseboodt, R., Bruylants, B., Verheyen, R., Coeck, J., Maes, J., Belpaire, C., Van Thuyne, G., Denayer, B., Beyens, J., De Charleroy, D. and Vandenabeele, P. 1998. *Atlas van de Vlaamse Beek- en Riviervissen.* Wijnegem: WEL.

Vanderhoeven, A., Martens, M., Ervynck, A., Cooremans, B. and Van Neer, W. 2001. Interdisziplinäre Untersuchungen im römischen Vicus von Tienen (Belgien). Die Integration von ökologischen und archäologischen Daten, pp. 13–33 in Frey, M. and Hanel, N. (eds), *Archäologie Naturwissenschaften Umwelt* (BAR International Series 929). Oxford: British Archaeological Reports.

Van Neer, W. and Ervynck, A. 1994. New data on fish remains from Belgian archaeological sites, pp. 217–29 in Van Neer, W. (ed.), *Fish Exploitation in the Past. Proceedings of the 7th Meeting of the ICAZ Fish Remains Working Group* (Annalen van het Koninklijk Museum voor Midden-Afrika. Zoologische Wetenschappen 274). Tervuren: Koninklijk Museum voor Midden-Afrika.

Van Neer, W. and Ervynck, A. this volume. Remains of traded fish in archaeological sites: indicators of status or bulk food? pp. 203–214 in O'Day, S. J., Van Neer, W. and Ervynck, A. (eds), *Behaviour Behind Bones. The Zooarchaeology of Ritual, Religion, Status and Identity.* Oxford: Oxbow Books.

Van Neer, W. and Lentacker, A. 1994. New archaeozoological evidence for the consumption of locally-produced fish sauce in the northern provinces of the Roman Empire. *Archaeofauna* **3**, 53–62.

Vermaseren, M. J. 1956. *Corpus Inscriptionum et Monumentorum Religionis Mithriacae I.* Den Haag: M. Nijhoff.

Vermaseren, M. J. 1960. *Corpus Inscriptionum et Monumentorum Religionis Mithriacae II.* Den Haag: M. Nijhoff.

Verriest, S. 1999. *Archeozoölogisch Onderzoek van de Romeinse Vindplaats "Zijdelingsestraat" te Tienen.* Unpublished MSc. thesis, University of Gent.

von den Driesch, A. 1976. *A Guide to the Measurement of Animal Bones from Archaeological Sites* (Peabody Museum Bulletin 1). Harvard: Peabody Museum.

von den Driesch, A. and Pöllath, N. 2000. Tierknochen aus dem Mithrastempel von Künzing, Lkr. Deggendorf, pp. 145–62 in Schmotz, K. (ed.), *Vorträge des 18. Niederbayerischen Archäologentages.* Rahden, Westfalen: Verlag Marie Leidorf.

Vörös, I. 1991. Das Haus des Tribuni Laticlavii aus dem Legionslager vom 2.–3. Jh. in Aquincum. Das Tierknochenmaterial des Mithräums. *Budapest Régiségei* **28**, 155.

Weiss, M. in press. Die Mithras-Brosche im Ashmolean Museum, Oxford, in Martens, M. and De Boe, G. (eds), *Roman Mithraism. The Evidence of the Small Finds* (Archeologie in Vlaanderen Monografie 5). Zellik and Tienen: Institute for the Archaeological Heritage of the Flemish Community.

Giblé, F. 1995. Le mithraeum de Forum Claudii Vallensium/ Martigny (Valais). *Archäologie der Schweiz* **18**, 2–15.

Williams, A. 1977. *Backyard Pig and Sheepbook.* Dorchester: Prism Press.

Williams, R. J., Hart, P. and Williams, A. T. L. 1996. *Wavendon Gate, a Late Iron Age and Roman Site in Milton Keynes* (Buckinghamshire Archaeological Society Monograph Series 10). Aylesbury: Buckinghamshire Archaeological Society.

Witteyer, M. and Hochmuth, M. 2002. Sacrifice and food offerings to Isis and Magna Mater in Mainz, pp. 57–8 in Lepetz, S. and Van Andringa, W. (eds), *Archéologie du Sacrifice Animal en Gaule Romaine. Rituels et Pratiques Alimentaires. Volume des Pré-Actes* Paris: CNRS.

Woodward, A. 1992. *Shrines and Sacrifice.* London: Batsford.

Zeiler, J. T. 1996. Geofferd aan de godin. Faunaresten uit de Fortunatempel te Nijmegen (2e eeuw n.Chr.). Unpublished internal report for the Community of Nijmegen (the Netherlands).

Zeiler, J. T. 1997. Offers en slachtoffers. Faunaresten uit de Fortunatempel te Nijmegen (2e eeuw n.Chr.). *Paleo-aktueel* **8**, 105–7.

An Lentacker and Anton Ervynck
Institute for the Archaeological Heritage
of the Flemish Community,
Koning Albert II-laan 19, Box 5,
B-1210 Brussel, Belgium
E-mail: lentacker.ervynck2@yucom.be

Wim Van Neer
Royal Museum of Central Africa, IUAP 05/09,
Leuvensesteenweg 13, B-3080 Tervuren, Belgium
and Katholieke Universiteit Leuven,
Laboratory of Comparative Anatomy and Biodiversity,
Ch. de Bériotstraat 32, B-3000 Leuven, Belgium
E-mail: vanneer@africamuseum.be

9th ICAZ Conference, Durham 2002
Behaviour Behind Bones, (ed. Sharyn Jones O'Day et al.) pp. 95–100

13. Prehispanic guinea pig sacrifices in southern Perú, the case of El Yaral

Juan Rofes

The study of 112 naturally mummified guinea pigs found buried underneath the floor of four residences at El Yaral, a prehispanic village located in the Osmore drainage to the extreme south of Perú, has documented ritual practices involving the sacrifice of these animals. Evidence concerning pre and post-Conquest guinea pig use is presented. These data indicate continuity through time in ritual practices and beliefs related to this animal up to the present day.

Introduction

Domestic guinea pigs (*Cavia porcellus*) are medium-sized rodents (Rodentia: Caviidae) whose importance may be easily stated, both economically and ritually, in the present Andean society. Ethnohistorical and ethnographic data record their use as food, diagnostic medical devices, divinatory agents, and healing properties; they were also used for sacrifices and as welfare propitiatory offerings (*e.g.*, Polo de Ondegardo [1559] 1906; Acosta [1590] 1962; Guamán Poma de Ayala [1613] 1980; Arriaga [1621] 1968; Cobo [1653] 1964; Gilmore 1948, Gade 1967; Andrews 1972–74; Escobar and Escobar 1979; Bolton 1979; Bolton and Calvin 1981; Cobo 1990; Salomon and Urioste 1991; Archetti 1992; Morales 1995). Most ethnohistorians, ethnographers and archaeologists have assumed that these animals fulfilled the same functions in prehispanic times.

One way to test the validity of these assumptions is through analysis of the archaeological record. In this respect, the El Yaral site offers remarkable evidence of how the Chiribaya people from the Osmore drainage in Moquegua (Fig. 1) (Ghersi 1956; Buikstra 1988; García 1988; Jessup 1990; Jessup 1991a; Jessup 1991b; Owen 1992; Rice 1993; Owen 1995) utilized these animals in ritual during the first centuries of the second millennium AD.

In this paper I present a brief review of the Chiribaya culture and the main characteristics of the location. I then contrast a series of archaeological and ethnographic data

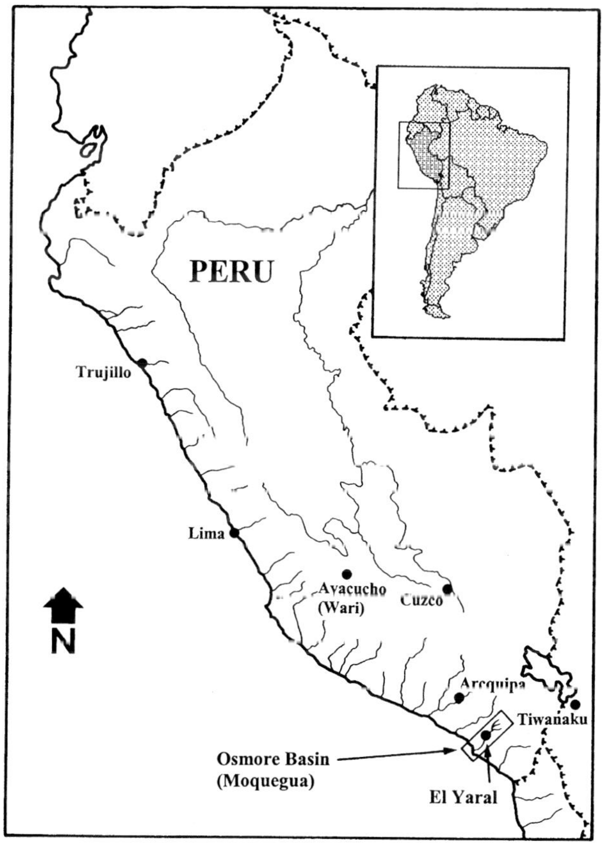

Fig. 1. The El Yaral site in the Osmore River's middle valley, southern Perú.

that help to understand the complex meaning of the guinea pig sacrifice for the inhabitants of El Yaral and for the Andean people in general.

The Chiribaya culture

The origin and initial development of the Chiribaya culture is uncertain. Stanish (1990) and Jessup (1991a and 1991b) suggested that Chiribaya could have developed from the Tiwanaku colonization of the southwestern valleys, around AD 1000. Tiwanaku is one of the main pre-Inca cultures whose core area includes the highland region (Altiplano), south of Lake Titicaca. With the decadence of Tiwanaku around AD 1100 (Ortloff and Kolata 1993; Binford *et al.* 1997), gradual changes in the archaeological record reflect the weakening of the altiplanic influence in Moquegua. This period has been called Tumilaca (*c.* AD 900/950–1050) in the highlands area (Bawden 1989; Goldstein 1990a; Goldstein 1990b), and Ilo-Tumilaca/Cabuza (*c.* AD 1000–1250) in the coast (Owen 1992). According to Bruce Owen (1992 and 1995), Chiribaya appeared as an autonomous coastal entity shortly after the beginning of the Tumilaca phase. This culture is either the result of a southern immigration event or it is a local development of the Ilo-Tumilaca/Cabuza tradition group on the coast.

Whatever the true origins of the Chiribaya culture, we know that its influence in Moquegua was concentrated on the coast and lower valley of the Osmore river; it also reached the middle valley. This last area was shared with the Tumilaca groups during the first part of the Late Intermediate period (*c.* AD 1000-1050), with Chiribaya persisting until *c.* AD 1375 (Jessup 1991a; Jessup 1991b; Owen 1992; Owen 1995).

The archaeological site

El Yaral is situated in what is now known as the Moquegua Department, in the southern Peruvian subtropical desert zone. It is located at the lowest part of the Osmore river middle valley (3300 feet above sea level), 31 miles from the coast and 15.5 miles from the modern city of Moquegua (Fig. 1). The site extends 547 yards, occupying 32 acres of land on the right bank of the river. It is positioned on a stone and sand hill where more than 330 inhabitable terraces are widely dispersed, and reaches 396 feet above ground level (Rice 1993).

El Yaral consists of structures made of internal and perimetrical reed walls and wood posts (willow and carob tree trunks). The building techniques vary slightly from one structure to the other, however, the size of the units and the number of their internal subdivisions differs significantly.

The room floors were levelled with thin layers of sand, gravel, clay and boulders. Most of these floors contain offerings of domestic guinea pigs, as well as sacrificed llamas and alpacas. The extreme dryness of the sediment, together with its minimal acidity and the lack of saline sediments, have created an excellent environment for the conservation of the camelid and guinea pig remains buried under these buildings. Due to this aridity, a natural mummification process occurred and its results prove to be informative about ritual in the past.

Guinea pig sacrifices at el Yaral

Archaeological and ethnographic data

The oldest published evidence documenting guinea pig burials in residential areas describes deposits under a terrace floor at the Chavín de Huantar site (Burger 1992) dating to the late Janabarriu period (*c.* 400–200 BC). Evidence also comes from the Nasca city of Cahuachi, belonging to the Early Intermediate period (*c.* 200 BC–AD 500), where an offering of 23 young guinea pigs was buried under the floor of structure number 19 (Silverman 1988; Silverman 1993). All of the Cahuachi animals had their bellies cut open, just as Andrews (1972–74) and other ethnographers have described in present day Andean communities. For the Inca period (*c.* AD 1480–1540), Sandweiss and Wing (1997) documented naturally mummified guinea pig offerings in the site of Lo Demás, from the floors of communal residences. All of the rodents at this site were young animals; one had its belly cut open and another had its throat slit.

Some 700 years after Cahuachi and about 300 years before Lo Demás, sacrificing and burying guinea pigs was a widespread practice at the Chiribaya community of El Yaral, Moquegua. The 112 guinea pig mummies discovered in a single layer under the floor of the four structures studied at El Yaral were recovered in very good condition (Rofes 1998). Both young and infant guinea pigs were preferred for these sacrifices (90.2%), and infant animals were most frequently chosen (61.6%).

Following information recorded by Bolton (1979) in Santa Barbara (Cusco), I assume that the preference for infant guinea pigs in El Yaral, Cahuachi and Lo Demás was a response to economical circumstances. By using the smallest animals for sacrifices the amount of invested meat is minimal. The term "invested" rather than "lost" is used since every offering to the gods was meant to beg for favours, appease their rage, or ensure the guinea pigs' owner's welfare. The Huarochirí myths, transcribed by father Francisco de Avila, repeatedly references these beliefs (Salomon and Urioste 1991).

Adult animals were also sacrificed, but with much less frequency (9.8%). The finding of infant, young, and adult guinea pigs suggests that these rodents were bred *in situ*. This hypothesis is also supported by the large accumulations of excrements found on the floor of the households, mostly in their largest rooms.

Within the biggest subdivisions of the four studied structures, evidence of cooking, including hearths, mills, and vegetable refuse was found. I argue that those guinea

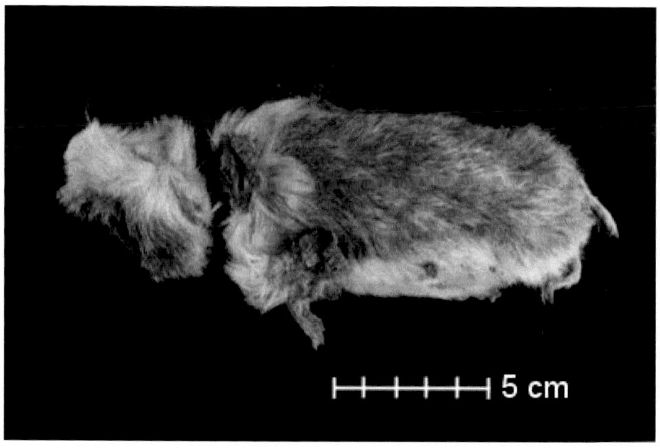

Fig. 2. Beheaded guinea pig.

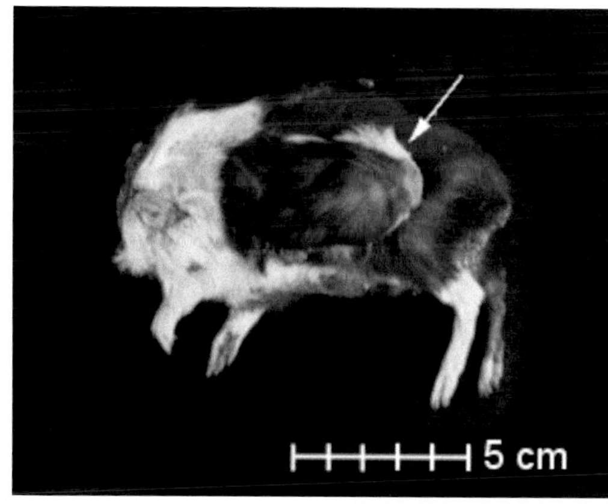

Fig. 3. Beheaded guinea pig with the head over the body.

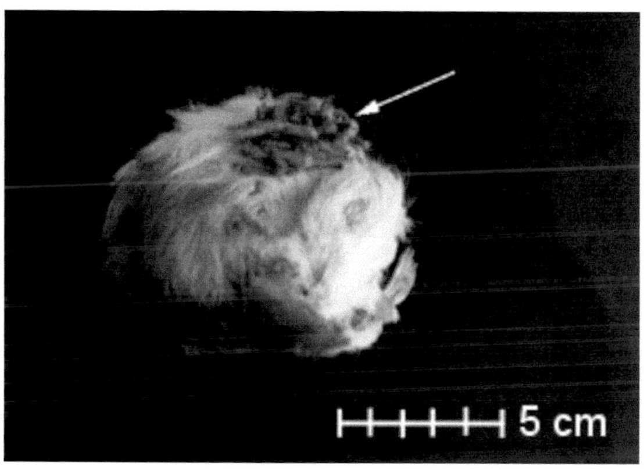

Fig. 4. Guinea pig with an offering made of molle *sticks.*

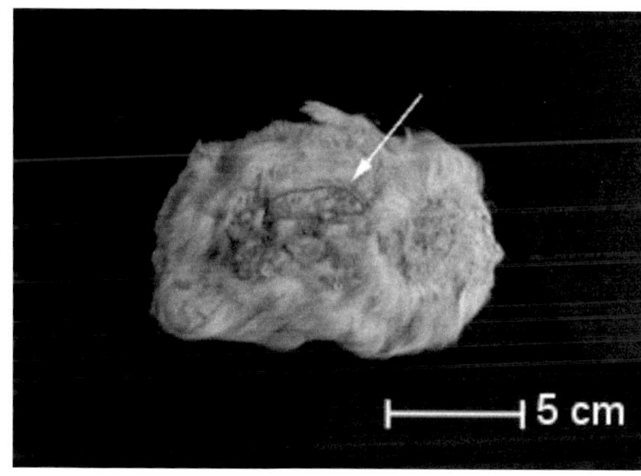

Fig. 5. Beheaded guinea pig with a little basalt knife over its body.

pigs were easily bred inside these rooms, since the heat of the stoves attracted them and they could feed on the falling crumbs of human consumption. Furthermore, in every house there was a room especially devoted to the preparation of *molle* (*Schinus molle*) or corn beer (Rice 1993). Gade (1967) and Escobar and Escobar (1976) state that the best food for guinea pigs was the leaves surrounding the corn grain and other waste resulting from its brewing.

In present day Andean communities 5 to 15 of these animals are usually bred in any given house (Gilmore 1948). Nevertheless, this number may occasionally rise up to 40 or 50 guinea pigs (Gade 1967). The largest quantity of sacrificed guinea pigs (50) was found at the 70th structure, yet it is unlikely that its inhabitants killed such a large amount at a single time. Instead of this, gradual seasonal offerings likely followed the fluctuations of the litters' pattern. Bolton and Calvin's postulate (1981), suggests a perfect adaptation of the sustaining ritual cycles to the vital cycles of the domestic animals in the prehispanic rural communities. I believe that this must have become a fact in El Yaral. Particularly true in the cases of llamas and alpacas, which served as basic meat and ritual sources (Wheeler *et al.* 1995); and also to the guinea pigs, being a complement to both aspects (Rofes 1998).

The guinea pigs sacrifice: ritual and consumption

The frequent sacrifice and burial of camelids and guinea pigs underneath homes may reflect how worried the inhabitants of El Yaral were about their welfare as well as that of the community's. The sacrifice of a guinea pig was performed by breaking its neck, cutting its throat, or beheading the animal (Fig. 2). In the last case, the head was placed either over the body or under it (Fig. 3). However, we do not rule out the natural death of some guinea pigs without clear evidence of cultural modification.

Many times guinea pigs were buried together with little offerings, such as *coas* (small *molle* sticks) (Fig. 4), lumps of animal fat (most probably of camelids), coloured threads, silver and turquoise beads, copper plates, coca and *molle* leaves, cotton, and feathers. On one occasion a small basalt knife, which was probably used to behead the guinea pig, was found over the body of the animal (Fig. 5), the tool had dry bloodstains on its cutting edge. The

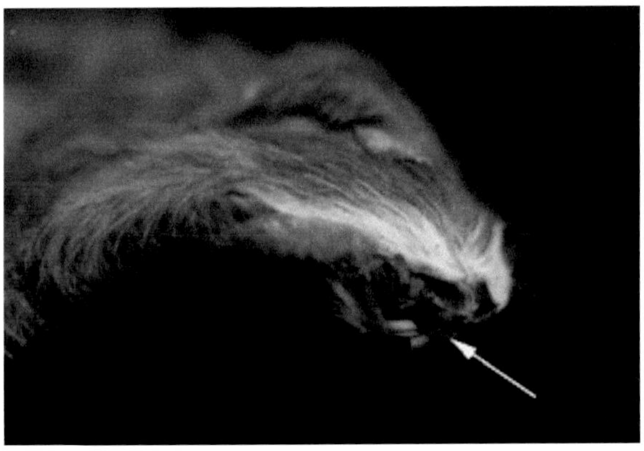

Fig. 6. Guinea pig with coca *leaves inside its mouth.*

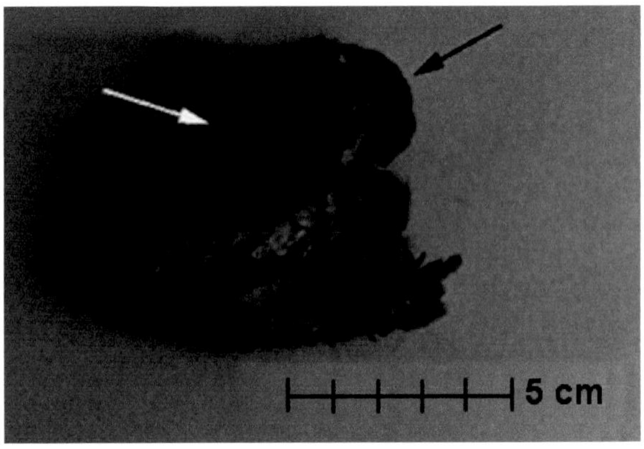

Fig. 7. Guinea pig with two little cloth parcels tied to its sides.

finding of coca leaves (*Erythroxylon coca truxcillensis*) in one of the sacrificed guinea pigs' mouth (Fig. 6) and over another's body coincides with Gade's ethnographic studies (1967). This finding establishes, with documentary evidence, the antiquity of the Andean custom of offering guinea pigs and coca leaves to the spirits and gods, just as it is practiced today.

When the inhabitants of El Yaral chose the guinea pigs, which would be buried next to the camelids, it appears that they looked for colour coordination among the animals. This fact is demonstrated in many cases where the guinea pigs' hair colour and that of the alpacas and llamas' was exactly the same. Furthermore, in some other cases these guinea pigs were buried at the bases of house columns. This was meant to be a "ritual reinforcement" to guarantee that the house was stable and durable. Burying large camelids, due to space and stability reasons, was not the best option to achieve this target. It is worth mentioning the case of a guinea pig which had two little brown cloth parcels tied to its sides similar to the ones llamas used to carry in commercial caravans (Fig. 7).

One of the most important features of domestic guinea pigs, in comparison to their close relative and wild ancestor (*Cavia aperea*), is the colour of the hair. *Cavia porcellus* have a wide range of tones. In El Yaral, there are a wide variety of colours (brown, orange, grey, cream, white and beige) which, in most cases, have different patterns; although mono-coloured guinea pigs may also be found. The total lack of completely black animals in these burials is not surprising. This fact might be explained in relation to the belief that black guinea pig meat had healing properties, especially in breathing illnesses, which is reported by Bolton (1979). As soon as a completely black guinea pig was born, it was immediately destined for medical use.

As a complementary hypothesis to explain the high number of domestic animal sacrifices at the site, it has been postulated that a drastic reduction of net precipitation may have affected the entire central Andean region and possibly the western portion of South America between AD 1000–1100. Rainfall remained low for the next 300-400 years, being one of the main reasons for the collapse of the Tiwanaku civilization (Ortloff and Kolata 1993; Stine 1994; Binford *et al.* 1997; Klein *et al.* 1999; Seltzer *et al.* 2000). This situation should have hit the Moquegua valley to some extent during the occupation of El Yaral. Rainfall would have decreased severely and reduced ground water recharge and spring and river flows at the Osmore drainage. Hence, the magnitude of guinea pig and camelid sacrifices may also be a situation partially brought on by environmental stress.

Ethnographical comparison may also explain the low frequency of guinea pig consumption remains recovered from the household floors. Almost all accounts (Gade 1967; Andrews 1972–74; Escobar and Escobar 1979; Bolton 1979; Bolton and Calvin 1981; Archetti 1992; Morales 1995) agree that the region's present day rural inhabitants prefer eating guinea pig, generally, at specific religious and pagan festivities. Gade (1967) also reports that guinea pigs are usually eaten after collective community works such as sowing or harvesting. In both cases, the remains were left at the same communal spot where the guinea pigs were eaten. The atmosphere being probably that of a big party or "game time" as described by Andrews at the modern Andean community of Paucartambo during the Epiphany day celebration:

> "A crowd of over 150 people gathered in front of the steward's house (the sponsor) and formed a circle of 30 foot in diameter. Two female teenagers and an old woman got into the circle, taking a guinea pig tied to the other side of a 6 foot 4 inch rope. The women danced following the rhythm of a tune played by a small band. As they were dancing, they were dragging the guinea pigs. After a while, eight couples joined them and started dancing as well. The women raised the guinea pigs and

> threw them to either the dancing men or those who were watching. Moreover, the women tried to rub the guinea pigs over the men's faces and necks as the men tried to stop them by using their arms. This celebration lasted for an hour. It all made the participants and the crowds laugh. Once it was over, everybody went to a square which was at the lowest part of the town, where the show was repeated .... The number of spectators rose to 275 and there was so much joy and laughter that it caused a convulsion in some participants and spectators" (Andrews 1972–74, 132–3).

According to this account, guinea pigs would rarely be eaten inside the houses, which helps explain infrequent findings of consumption remains on floors. On the other hand, many taphonomical agents, such as digging and scavenging animal activities or the cleaning of living floors, may also explain the low number of these remains.

There is no evidence to suggest that any guinea pig was cut open for medical or foretelling reasons in El Yaral. Nevertheless, such evidence was found in one guinea pig of Lo Demás (Sandweiss and Wing 1997) and in all the animals of Cahuachi (Silverman 1988; Silverman 1993).

Guinea pigs do not seem to have played an important role as a component of the human burial offerings in El Yaral. At the two excavated graveyards, just one case was reported (Lozada and Torres 1991). However, at the Chiribaya coastal sites, guinea pig offerings in human cemeteries were common in Chiribaya Alta (Wheeler 1990) and Chiribaya Baja (Ghersi 1956; Sonia Guillén, *pers. comm.*, 1998).

Conclusions

The extraordinary ritual evidence, together with the large accumulation of excrements found on the house floors indicates that guinea pigs were bred at the site of El Yaral. While camelids may have been the main source of meat and rituals in El Yaral, guinea pigs were a great and effective supplement in both instances. In the ritual terrain, 112 mummified guinea pigs have been recovered from under the houses, most of them extremely well conserved. These offerings were probably intended to calm the gods' rage, beg their support, and ensure the owner's and the community's welfare (Rofes 1998).

In spite of the significant time difference, the guinea pig mummies from El Yaral and those from Cahuachi and Lo Demás are very similar. These three sites become a link between the prehispanic past and the ethnographic record of the ritual sacrifice, the healing properties, and elaborate beliefs systems concerning these animals.

The fact that so few records of guinea pig consumption have been found may be partly due to conservation problems. Nevertheless, we must bear in mind that most of the guinea pig population was cooked and eaten by the whole community, generally during religious festivities or after common working tasks such as harvesting and sowing, as it is still done today in traditional Andean communities.

Acknowledgements
The archaeofaunistic material required for our research was kindly provided by Antonio Oquiche, director of the Contisuyo museum in Moquegua, with the permission of Dr. Don Rice, responsible for the excavation of the four houses in El Yaral where the samples come from. Special thanks to Dr. Jane Wheeler and Dr. Philippe Béarez for their constructive criticism, advice and support, as well as for having corrected the original manuscript. Many thanks to my friend Xavi Plaza for helping me with the translation of the paper.

References

Acosta, J. de 1962. *Historia Natural y Moral de las Indias* (originally published 1590). Edmundo O'Gorman (ed.). México D.F.: Fondo de Cultura Económica.

Andrews, D. H. 1972–74. On the ethnozoology of the Guinea Pig. *Ñawpa Paccha* **10–12**, 129–34.

Archetti, E. 1992. *El Mundo Social y Simbólico del Cuy*. Quito: CEPLAES.

Arriaga, P. J. de 1968. *Extirpación de la Idolatría del Piru* (originally published 1621), pp. 191–277 (Biblioteca de Autores Españoles 209). Madrid: Ediciones Atlas.

Bawden, G. 1989. The Tumilaca site and Post-Tiwanaku occupational stratigraphy in the Moquegua Drainage, pp 287–302 in Rice, D., Stanish, C. and Scarr, P. (eds), *Ecology, Settlement and History in the Osmore Drainage, Peru, II* (BAR International Series 545). Oxford: British Archaeological Reports.

Binford, M., Kolata, A., Brender, M., Janusek, J., Seddon, M., Abbot, M. and Curtis, J. 1997. Climate variation and the rise and fall of an Andean civilization. *Quaternary Research* **47**, 235–48.

Bolton, R. 1979. Guinea pigs, protein, and ritual. *Ethnology* **18**, 229–52.

Bolton, R. and Calvin, L. 1981. El cuy en la cultura peruana contemporánea, pp 261–326 in Lechtman, H. and Soldi, A. M. (eds), *Runakunap Kawasayninkupaq Runaskankunapa.La Tecnología en el Mundo Andino*. Mexico D.F.: Universidad Nacional Autónoma de México.

Bulkstra, J. 1988. Chiribaya: an integrated approach to the biological and cultural development of an Andean society. Proposal for research support to the Archaeology Program of the National Science Foundation.

Burger, R. L. 1992. *Chavín and the Origins of Andean Civilization*. London: Thames and Hudson.

Cobo, B. 1964. *Historia del Nuevo Mundo* (originally published 1653), Libros I–X (Biblioteca de Autores Españoles 91). Madrid: Ediciones Atlas.

Cobo, B. 1990. *Inca Religion & Customs* (originally published 1653; translated and edited by Roland Hamilton). Austin: University of Texas Press.

Escobar, G. and Escobar, G. 1976. Observaciones etnográficas sobre la crianza y los usos del cuy en la región del Cuzco. *Antropología Andina* **1–2**, 34–9.

Gade, D. W. 1967. The guinea pig in Andean Culture. *The Geographical Review* **57**, 213–24.

García, M. 1988. *Excavaciones de Dos Viviendas Chiribaya en el Yaral, Valle de Moquegua*. Tesis de Bachillerato, Universidad Católica Santa María (Arequipa, Perú).

Ghersi, H. 1956. Informe sobre las excavaciones en Chiribaya. *Revista del Museo Nacional* **25**, 89–119.

Gilmore, R. M. 1948. Fauna and ethnozoology of South America; in Steward, J.H. (ed.), *Handbook of South American Indians*, 6 (Smithsonian Institution Bureau of American Ethnology Bulletin 143). Washington D.C.: Smithsonian Institution.

Goldstein, P. 1990a. La Ocupación Tiwanaku en Moquegua. *Gaceta Arqueológica Andina* **18–19**, 75–104.

Goldstein, P. 1990b. La cultura Tiwanaku y la relación de sus fases cerámicas en Moquegua, pp. 31–58 in Watanabe, L., Moseley, M. and Cabieses, F. (eds), *Trabajos Arqueológicos en Moquegua, Peru*, 2. Lima: Programa Contisuyo del Museo Peruano de Ciencias de la Salud Southern Peru Copper Corporation.

Guamán Poma de Ayala, F. 1980. *Nueva Corónica y Buen Gobierno* (originally published 1613; transcribed by Franklin Pease) (Biblioteca Ayacucho 75). Caracas.

Jessup, D. 1990. Rescate arqueológico en el Museo de Sitio de San Gerónimo, pp. 151–66 in Watanabe, L., Moseley, M. and Cabieses, F. (eds), *Trabajos Arqueológicos en Moquegua, Peru*, 3. Lima: Programa Contisuyo del Museo Peruano de Ciencias de la Salud Southern Peru Copper Corporation.

Jessup, D. 1991a. General trends in the development of the Chiribaya culture, south-coastal Peru. Paper presented at the 56th Annual Meeting of the Society for American Archaeology, New Orleans.

Jessup, D. 1991b. Chiribaya: an ethnic group. Paper presented at the 90th Annual Meeting of the American Anthropological Association, Chicago.

Klein, A., Seltzer, G. and Isacks, B. 1999. Modern and last local glacial maximum snowlines in the Central Andes of Peru, Bolivia, and Northern Chile. *Quaternary Science Reviews* **18**, 63–84.

Lozada, M. C. and Torres, E. 1991. Mortuary excavations at La Yaral, Southern Perú. Department of Anthropology Unpublished report, University of Chicago.

Morales, E. 1995. *The Guinea Pig: Healing, Food and Ritual in the Andes*. Tucson: University of Arizona Press.

Ortloff, C. and Kolata, A. 1993. Climate and collapse: agro-ecological perspectives on the decline of the Tiwanaku state. *Journal of Archaeological Science* **20**, 195–221.

Owen, B. 1992. Coastal colonies and the collapse of Tiwanaku: the coastal Osmore Valley, Peru. Paper presented at the 57th Annual Meeting of the Society for American Archaeology.

Owen, B. 1995. Warfare and engineering, ostentation and social status in the Late Intermediate Period Osmore Drainage. Paper presented at the 60th Annual Meeting of the Society for American Archaeology.

Polo de Ondegardo, J. 1906. Los errores y supersticiones de los indios, sacadas del tratado y averiguaciones que hizo el licenciado Polo (originally published 1559). *Revista Histórica* **1**, 207–31.

Rice, D. 1993. Late Intermediate Period domestic architecture and residential organization at La Yaral, Perú, pp. 66–82 in Aldenderfer, M. S. (ed.), *Andean Domestic Architecture*. Iowa City: University of Iowa Press.

Rofes, J. 1998. *La Utilización de Recursos Faunísticos en el Yaral, Asentamiento Chiribaya en la Cuenca del Osmore, Sierra Baja de Moquegua*. Tesis de Licenciatura, Pontificia Universidad Católica del Perú.

Salomon, F. and Urioste, G. 1991. *The Huarochirí Manuscript: A Testament of Ancient and Colonial Andean Religion* (manuscript by Francisco de Avila [*ca.* 1598–1608]; translated by Salomon, F. and Urioste, G.L.; transcribed by Urioste, G.L.). Austin: University of Texas Press.

Sandweiss, D. and Wing, E. 1997. Ritual rodents: the guinea pigs of Chincha, Peru. *Journal of Field Archaeology* **24 (1)**, 47–58.

Seltzer, G., Rodbell, D. and Burns, S. 2000. Isotopic evidence for late Quaternary climatic change in tropical South America. *Geology* **28 (1)**, 35–8.

Silverman, H. 1988. Cahuachi: Non-urban cultural complexity on the south coast of Peru. *Journal of Field Archaeology* **15**, 403–30.

Silverman, H. 1993. *Cahuachi in the Ancient Nasca World*. Iowa City: University of Iowa Press.

Stanish, Ch. 1990. Economías agrarias Post Tiwanaku en la cuenca del río Moquegua, pp. 115–60 in Watanabe, L., Moseley, M. and Cabieses, F. (eds), *Trabajos Arqueológicos en Moquegua, Peru, 2*. Lima: Programa Contisuyo del Museo Peruano de Ciencias de la Salud Southern Peru Copper Corporation.

Stine, S. 1994. Extreme and persistent drought in California and Patagonia during mediaeval time. *Nature* **369**, 546–9.

Wheeler, J. C. 1990. Faunal remains from Chiribaya Alta. Unpublished Report.

Wheeler, J. C., Russel, A. and Redden, H. 1995. Llamas and alpacas: pre-conquest breeds and post-conquest hybrids. *Journal of Archaeological Science* **22**, 833–40.

Juan Rofes
Laboratorio de Arqueozoología,
Departamento de Biología,
Universidad Autónoma de Madrid,
E-28049 Madrid, Spain
E-mail: juan.rofes@uam.es

9th ICAZ Conference, Durham 2002
Behaviour Behind Bones, (ed. Sharyn Jones O'Day et al.) pp. 101–113

14. Animals from the Maya underworld: reconstructing elite Maya ritual at the Cueva de los Quetzales, Guatemala

Kitty F. Emery

A well-preserved zooarchaeological assemblage has been recovered from the Cueva de los Quetzales at the Classic Maya site of Las Pacayas, Guatemala. The cave location and artifactual remains both suggest that the animal remains found in the cave were ritually derived or were the result of elite activities. Zooarchaeological analysis reveals a diverse faunal population including cave residents, exotic species, and dietary favorites of the ancient Maya. These remains do not, however, provide clear evidence of the nature of this deposit as resulting from elite cave-associated rituals, as opposed to debitage from the royal dining tables, or deposition of remains from natural cave fauna. The distinction is particularly important for current debates concerning the role of cave rituals and the politics of landscape control in the ancient Maya world. A detailed analysis of the zooarchaeological correlates of different scales, types, and functions of cave-associated rituals tests the hypothesis that the faunal materials recovered from the Cueva de los Quetzales are the result of cave rituals. These correlations are then used to define the specific nature of those rituals as private or public, and inclusionary or exclusionary. Although based on preliminary data, this analysis explores the potential of structural modeling in the analysis of zooarchaeological assemblages whose identification as ritual is compromised.

Introduction

As the physical manifestations of otherwise invisible religious and social beliefs, rituals are particularly attractive to archaeologists. Their formalized, symbolic, and repetitive nature makes them likely to appear in the archaeological record, and likely to reflect a standard set of behaviors as opposed to a diversity of singular events (Flannery 1976, 132; Rappaport 1999, 24). Rituals are also intriguing to zooarchaeologists primarily because they are often associated with animal sacrifice, feasting, and animal related symbolism. But how do we determine that a faunal assemblage actually derives from ritual activity?

Ritual animal assemblages can often be so defined on the basis of archaeological context and taxonomic content – because in ritual situations both the scene and its players should be notably unusual (Rappaport 1999; Sponsel 2001, 178). But in some cases even seemingly ritual contexts and taxa cannot be considered sufficient evidence of ritually derived animal assemblages. In these cases, before we can designate an assemblage as "ritual," we must find clear zooarchaeological correlates of ritualized behavior in the faunal remains themselves. These zooarchaeological correlates must be based on culturally specific definitions of type and scale of ritual because symbolic behavior and its zooarchaeological correlates are highly variable within and between cultures. Animal remains resulting from feast and fast, for example, will be very different. And the strictures associated with fasting will differ considerably on the basis of ethnicity, religious beliefs, and resource availabilities.

Here we present an example of one faunal assemblage whose ritual status cannot be easily defined by context or taxonomy: remains from the Cueva de los Quetzales, or Cave of the Quetzal Bird, located in the Petén jungles of Guatemala (Fig. 1). This cave system runs beneath the site of Las Pacayas, an elite political center occupied during the periods of earliest societal complexity in the Maya world (300 BC – AD 500). The sole entrance to this cave system is a vertical shaft opening into the center of the main politico-religious complex at the site (Brady and Rodas 1994; Brady and Rodas 1995; Brady 1997). Other such cave systems are recognized as ritual loci by several

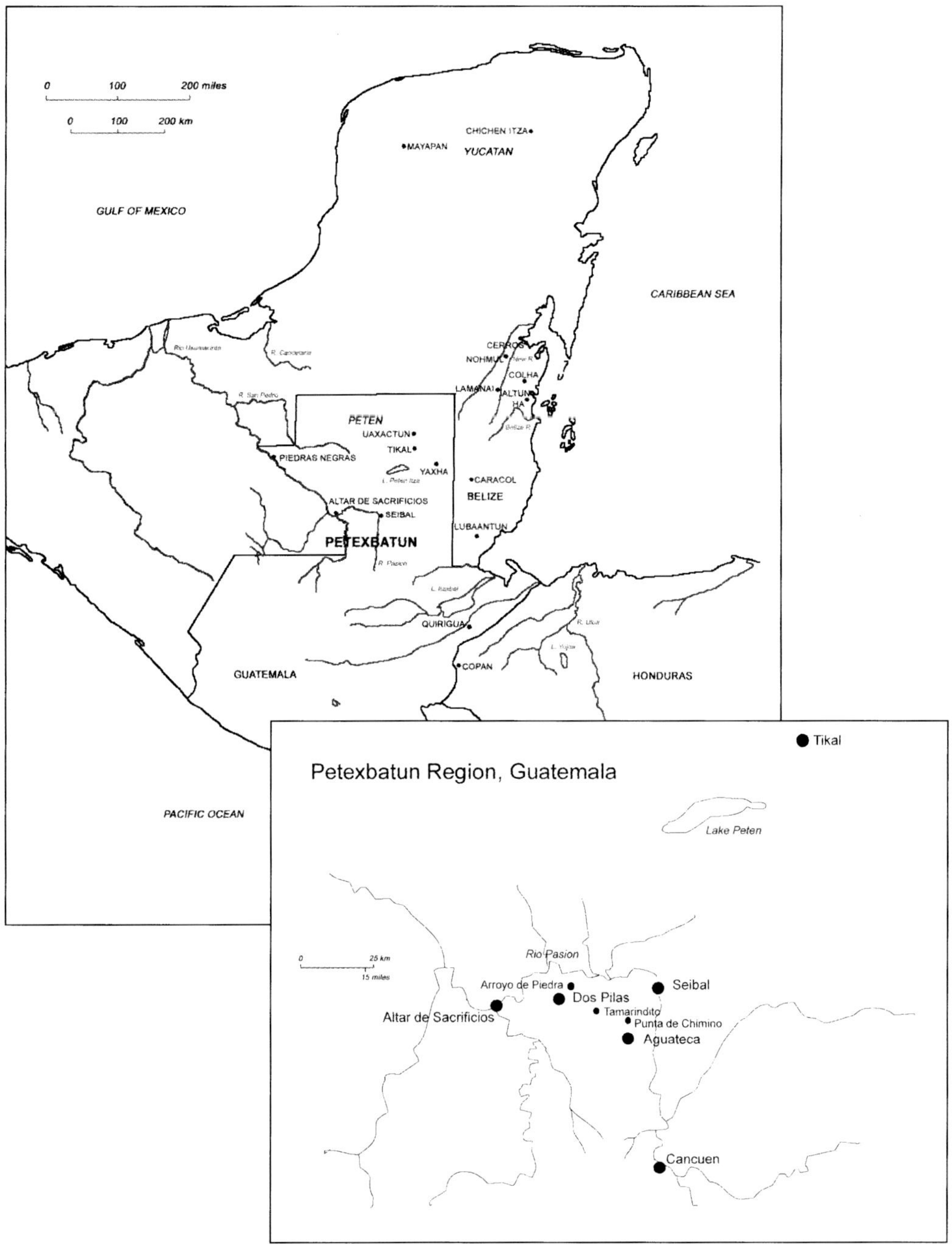

Fig. 1. Map of the Maya world and the Guatemalan Petén with the site of Las Pacayas and other sites mentioned in the text. The Cueva de los Quetzales lies beneath the main ceremonial plaza of Las Pacayas.

authors (Brady and Rodas 1994, Reeder *et al.* 1998, Thompson 1975). Brady has suggested that the artifactual remains recovered beneath this opening resulted from cave-associated rituals (Brady and Rodas 1995; Brady 1997). The artifact assemblage is clearly ritual in nature, including large collections of musical instruments (both portable ceramic drums and ocarinas) as well as other ceremonial goods (Brady and Rodas 1994; Helton 1997; Reeder *et al.* 1998)

In the ancient Maya world, caves held enormous mystical and religious importance. They were the portals to the underworld and the link to the ancestors and gods who controlled time, life, and the vagaries of nature (Thompson 1970, 267–76; Bassie-Sweet 1991; Stone 1995). But caves were also politically important as the legendary places of human origin, the birthplaces of the divine Maya rulers, and the ultimate homes to the spirits of these rulers (Pohl 1983, 99; Brady 1989, 55–64; Bassie-Sweet 1991, 77;

Tedlock 1996, 91–102; Brady 1997; Christenson 2000, 70–86). Correlations between political centers and cave entrances suggest that the political elite made a conscious effort to claim ownership over these underworld entrances, and presumably control access to them (Brady 1997, 604, 611). In view of the role caves played in ancient Maya beliefs, it seems reasonable that the faunal remains collected from the Cueva de los Quetzales would represent cave-related ceremonies of the elite Maya residents of the site of Las Pacayas.

However, in this case both context and taxonomy are complicated by the fact that the species symbolically associated with cave rituals included both those that normally inhabit caves and those that were favorite elite Maya foods. As a metaphor for the passage to the underworld, there is no better messenger than the bat (a cave resident). The deer was the ultimate symbol of fertility and rebirth, but was also the favorite food of the Maya elite. Clearly then, to prove that the Cueva de los Quetzales assemblage is one derived from ritual activity, specific zooarchaeological correlates go beyond the taxonomic characteristics of the zooarchaeological assemblage.

In addition, because today, and likely in the past, Maya cave rituals range from solitary petitions for fertility to community-wide royal accession rites (Vogt 1969, 387; Thompson 1970, 268; Heyden 1981; Bassie-Sweet 1991, 77–80), even cave-related zooarchaeological correlates may vary enormously between rituals. So the structure of ancient Maya cave ritual must be defined before patterned animal use behaviors can be linked to these rituals. The following discussion explores the potential of structural modeling in the analysis of zooarchaeological assemblages whose identification as ritual is compromised.

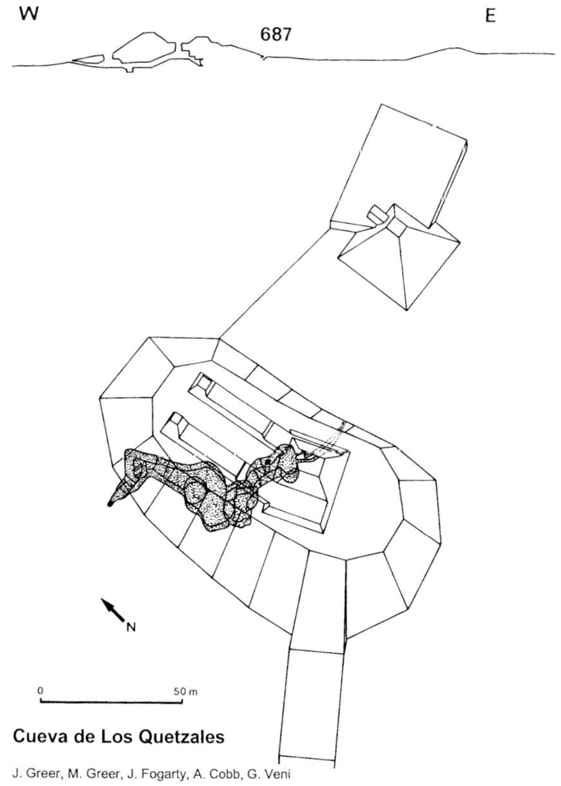

Fig. 2. Profile and plan of the Cueva de los Quetzales and associated ceremonial architecture of the site of Las Pacayas. Note the location of the 'sky-light' opening centrally located between the two main temple-mounds, and the conical dump of matrix (where the faunal assemblage was recovered).

The faunal assemblage

The Cueva de los Quetzales stretches under one of two politico-ceremonial complexes centrally located in the site of Las Pacayas (Fig. 2). During occupation of the site, the sole entrance to the cave was a "sky-light" opening into the main ceremonial complex. All other cave entrances were intentionally blocked when Las Pacayas was constructed, although today a second entrance has been opened by recent looting (Brady and Rodas 1994; Brady and Rodas 1995; Brady 1997, 608). Directly below this entrance lies a conical matrix "dump" some 3 meters deep, containing artifacts and ecofacts tossed in from above, presumably from the ceremonial plaza (Brady and Rodas 1994; Brady and Rodas 1995). The site of Las Pacayas, located approximately 12 km east of the Petexbatún region, was constructed atop a natural hill that was later transformed into a stepped pyramidal base for the political core of the site (Escobedo *et al.* 1994; Brady 1997, 608). Residential structures surrounded this political center, and the residents of the site would have been supported by extensive agriculture and trade with neighboring communities.

In 1990, both the cave and Las Pacayas were surveyed and test pitted by members of the Atlas Arqueológico de Guatemala, directed by Dr. Juan Pedro Laporte. Excavations were directed by Lic. Hector Escobedo and Dr. James Brady (Brady and Rodas 1994; Escobedo *et al.* 1994; Brady and Rodas 1995). The cave deposit was trowel excavated and matrices were water-screened using a 4 mm gauge mesh (Brady and Schwegman 1994). The excavators considered all materials from this deposit to be part of a single unstratified provenience unit because they noted no cultural or natural subdivisions (Brady, *pers. comm.*). The identifications presented here represent approximately a third of the total faunal assemblage. Identifications were begun in 1999 and 2000 by Emery, was continued by Emery and Stevens in 2001 at SUNY Potsdam, and was completed by Emery in 2002. All identifications are based on comparison with specimens from SUNY Potsdam and the Florida Museum of Natural History. All calculations are presented as number of identified specimens (NISP) although elements have been refitted where possible to create an approximation of the minimum number of elements (MNE).

Remains from the neighboring site of Arroyo de Piedra (approximately 15 km away) are used as a comparative sample. The assemblage is similar in size, and comes from chronologically contemporaneous elite deposits, in this case residential. These remains were excavated by the Vanderbilt Petexbatún project of which Brady and Escobedo were both a part, and identified by the author, so excavation and analytical methods are comparable (Escobedo 1994; Escobedo 1997; Emery 1997). Again, NISP has been used as the quantitative unit.

Taphonomic conditions require separate mention here since cave assemblages present special circumstances of preservation and intrusion. Cave deposits, protected from the elements and in direct association with limestone bedrock, are generally well preserved in comparison to other tropical assemblages. There is no evidence of water activity in the Cueva de los Quetzales during or after the occupation of Las Pacayas (Brady and Rodas 1994). Non-aquatic intrusive animals however, should be considered in a discussion of taphonomic agents for this cave deposit. Two factors argue against a significant role for intrusive animals either as inclusions in the deposit or as taphonomic agents. First, all other cave entrances had been blocked by the ancient Maya, leaving only this entrance, which, in the center of a bustling administrative center, would have been inaccessible to wild animals. Second, the remains come from a single conical deposit, which is admittedly an unlikely spot for intrusive animals conveniently to have died.

Defining ritual

A structural model of ancient Maya cave rituals must begin with a set of definitions. Ritual is generally described as the set of formalized behaviors through which religious constructs are generated and reproduced (Rappaport 1999, 23, 27). In zooarchaeological terms these behaviors include animals in their symbolic roles as performative or communication devices (see Douglas 1994). Animals act as metaphors (for conditions of nature including seasons, for social conditions like rulership or age, and for special circumstances like those of war), as totems (of the soul, of lineages, of social groups or communities), as offerings (as sacrifices or as foods both in redistributive and competitive feasts or as tithes), and as measures for quantitative comparison between individuals and groups (of, for example, sacredness, or power). Current interest in feasting as one facet of politically motivated ceremonial activity in hierarchical societies has generated a robust literature on ritual types and their archaeological correlates. The general discussion of ritual types presented here is based on a range of comparative literature (Drennan 1976; Flannery 1976; Goody 1982; Hayden 1995; Dietler 1996; Hayden 1996; Mennell 1996).

As in most complex hierarchical agricultural societies, Maya rituals can be separated by scale as public (occurring at the community or regional level) or private (performed by individuals or households) (Goody 1982, 99; Mennell 1996, 32). They can also be divided on the basis of group membership as exclusionary (used to sanctify divisions between groups) or inclusionary (promoting wide scale solidarity within communities, regions, or trade networks) (Brumfiel and Earle 1987; Clark and Blake 1994; Hayden 1995, 27; Dietler 1996, 92–7). Ancient Maya rituals varied as well by type (Drennan 1976; Flannery 1976, 332), celebrating a range of ritual cycles including life transition cycles (birth, maturity, death), time cycles (calendrical, annual, or longer cycles), and political cycles (transfers of power from one ruler to the next, cycles of conflict in stylized warfare). Combining these, a continuum of rituals can be envisioned that included private rituals, public exclusionary rituals, and public inclusionary rituals. Each of these ritual categories was associated with specific formalized behaviors and material correlates depending on whether the ritual being celebrated was a life transition, time, or political cycle.

Literature on the specific details of ancient Maya rituals include ethnographic analogies, iconographic analyses, and ethnohistoric documents (Redfield and Rojas 1934; Tozzer 1941; Bunzel 1952; Vogt 1993) as well as several excellent discussions of ritual in general and feasting in particular (*e.g.*, LeCount 2001). Private rituals in the Maya world are, and likely were in the past, celebrated at the individual or family level as a communication with the gods or family ancestors (Vogt 1993, 192; LeCount 2001). These emphasized individual sacrifices and the use of the sacred, and were usually celebrations of individual transitions and seasonal calendrical cycles (Tozzer 1941). Archaeological remains suggest that contexts were private or household locations, and trash was probably considered ceremonial (following Walker 1995). Zooarchaeological markers of private rituals would include the symbolic use of sacred species or rare individuals, sacrifices of individual animals, and the use of individual performative or transformational paraphernalia.

Public Maya rituals were used to legitimate social hierarchy, sanctify interregional exchange, or manage extremes in resources through redistribution or tithing (Tozzer 1941; McAnany 1995, 8). Today these vary from rituals that include groups within a community to those that include the entire community or polity. Public exclusionary rituals were likely celebrated within extended families, class groups, or occupational groups, and they emphasized solidarity within the group and exclusion of outsiders, so their communication was between and among group members. The contexts for these celebrations were ceremonial or uncommon communal locations. These rituals celebrate life cycles or annual time cycles (Tozzer 1941, 163; LeCount 2001), but as in most hierarchical societies, they also emphasized competition between groups as competitive feasting or displays of material wealth (Hayden 1995; Dietler 1996). As exclusionary rituals they likely emphasized the rare and inaccessible.

Zooarchaeological markers at this level might include the use of lineage or group totem species, the use of exotics or high status goods in competitive displays, a show of control of managed animals through sacrifice of certain special individuals, small scale performance or dance, and an emphasis on high quality and exotic, expensive goods.

Ancient public inclusionary rituals were community- and polity-based, and likely celebrated annual or period-ending calendrical cycles or political cycles to emphasize the solidarity of the community or polity and legitimate social order and hierarchy (LeCount 2001). Ethnographic evidence and ethnohistoric literature suggests that this level of ritual was associated with large scale feasting as redistribution from the elite core, with tithing back into the core as offerings, with sacrifices emphasizing quantity not quality of goods, and with theatrical performances involving elaborate costuming and music (Tozzer 1941; Farriss 1984; Vogt 1993). Ritual deposition of trash is unlikely to have occurred since most trash was probably not directly ceremonial.

At each level then, the symbolic use of animals reflects the ritual type. Contexts, totemic representation, offering type, and measures all directly reflect the level of ritual from individual-private to broadest public ritual. Metaphors and offerings may reflect ritual scale or the cycle being celebrated. The next step in the analysis of these remains is the recognition and definition of the zooarchaeological correlates of these structurally defined ritual types in the Cueva de los Quetzales faunal remains.

Searching for marker of ceremony in the assemblage

The following analysis defines the patterns in taxonomy, element distribution, and population characteristics found in the Cueva de los Quetzales faunal assemblage. However, it is through the comparison of these patterns with those found in another, non-ritual assemblage from the site of Arroyo de Piedra, that special patterning might be distinguished. Data is presented here as comparative analyses, and always as relative frequencies to avoid sample size biases in the comparisons.

Taxonomy: sacred animals

Several of the species found more often at the Cueva de los Quetzales than at Arroyo de Piedra were either sacred or ritually important (Fig. 3). The felines are underworld deities associated with dynastic ritual (Roys 1965; Peterson 1980; Saunders 1994) and specifically with caves (Tedlock 1986, 128; Stone 1995, 23, 43), crocodiles represent the earth itself (Puleston 1977; Pohl 1983, 65, 80), bats are the messengers of the gods (Benson 1988), and it is out of the maw of the snake that the ancestors are reborn (Peterson 1980; Aguilera 1985; Reifler-Bricker 1990). Other species hold roles in certain cave-related rituals. The opossum was the year-bearer and the iguana a required sacrifice in new-year renewal rituals (Tozzer 1941, 137–41; Pohl 1983, 79, 103), while fish and deer were offerings demanded by the gods during the transfer of rulership and at period endings (Tozzer 1941, 134, 155 56; Pohl 1981; Pohl 1983, 63, 74). The deer was particularly associated with fertility and life-renewal ceremonies carried out in caves (Pohl 1981; Pohl 1983) and may have been sacrificed there, although most of the cave-related deer remains appear to be crania, antlers, and teeth (Pohl 1983, 89; Pohl and Pohl 1983). The dog was also important in the renewal of the new-year and during political transitions (Danien 1997; White *et al.* 2001, 92), but was often buried with the royalty as a guide during the dangerous journey to the afterlife (Tozzer and Allen 1910, 360; Pohl 1983, 70).

As a reminder though, some of these species are natural cave residents (bats for example) or are found near or inside cave entrances (snakes and jaguars), or associated with nearby exposed rocks (iguanas, etc.). Other species are favored foods (deer and dog), and some may simply be better preserved in this environment than at surface sites (fish for example). Because of these complications, it would be naïve to consider the Quetzales animals in terms of generalized ritual metaphors. It is important instead to search for patterns that would indicate these species do reflect specific ritual activities. Here we consider them in terms of three relevant categories: 1) sacred species, those that are used only in ritual situations (in the Cueva de los Quetzales assemblage these include such species as jaguars; the cane toad, whose poisons were used as hallucinogens; and the stingrays, whose spines were used in blood-letting sacrifices); 2) exotics, those that would signify use in exclusionary rituals that define the power of the elite (here mainly marine species); and 3) artifacts representing ritual paraphernalia, including adornments, musical instruments, and the like (Fig. 4).

While the Cueva de los Quetzales and Arroyo de Piedra assemblages are equivalent in terms of the use of exotics (an elite prerogative) and artifactual adornments, the use of primarily sacred species is clearly greater at the Cueva de los Quetzales. This is a good indication of rituals carried out at the private and public exclusionary levels. However this interpretation continues to rest on the complicated taxonomic issues. Here our 'sacred' species include bats, marine toads, and other potentially intrusive species, so this cannot act as our only evidence.

Taxonomy: managed animals

An analysis of certain "managed" species provides data that are consistent with the conclusion of ritually specific use of taxa in the Cueva de los Quetzales. Managed ritual species include those that, although not always domesticated, were husbanded, raised, or penned in anticipation of their ritual use. In the ancient Maya world, only the dog was fully domesticated during the period of occupation of the Cueva de los Quetzales (Schwartz 1997). However

	Cueva de los Quetzales			Arroyo de Piedra
Taxonomic Category	**NISP**	**Cumulative NISP**	**% NISP**	**Comparative %NISP**
Freshwater Molluscs	–	29	2.58	0.31
Jute (*Pachychilus* spp.)	1			
Apple Snail (*Pomacea flagellata*)	9			
River Clam (*Psoronaias* spp.)	16			
Winged River Clam (*Nephronaias* spp.)	3			
Marine Molluscs	–	6	0.53	0.61
Olive Snail (*Oliva* spp.)	5			
Marine Bivalve (Pelecypoda. marine)	1			
Stingrays (Dasyatidae)	1	1	0.09	0.00
Bony Fishes (Osteichthyes)	196	211	18.77	0.31
Freshwater Gars (Lepisosteiformes)	5			
Freshwater Catfishes (Ictaluridae)	10			
Marine Toad (*Bufo marinus*)	12	12	1.07	0.00
Reptiles	4			
Crocodiles (*Crocodylus* spp.)	3	3	0.27	0.00
Turtles (Testudines)	153	173	15.39	12.27
Giant Musk Turtle (*Staurotypus triporcatus*)	2			
Mud and Musk Turtles (*Kinosternon* spp.)	2			
Slider (*Trachemys scripta*)	8			
Central American River Turtle (*Dermatemys mawii*)	8			
Lizards (Lacertilia)	24	33	2.94	0.31
Iguanas (Iguanidae)	9			
Snakes (Serpentes)	63	63	5.60	0.00
Birds (Aves)	9	12	1.07	0.15
Galliform Birds (Galliformes)	3			
Bats (Chiroptera)	13	13	1.16	0.00
Armadillo (*Dasypus novemcinctus*)	178	16	1.42	0.00
Opossums (Didelphidae)	33	33	2.94	0.15
Carnivores. small	–	6	0.53	0.00
Procyonids (Procyonidae)	3			
Weasels (Mustelidae)	1			
Coatis (*Nasua narica*)	2			
Grey Fox (*Urcyon cineoargenteus*)	5	5	0.44	0.00
Domestic Dog (*Canis familiaris*)	67	71	6.32	0.77
Wild Cats (Felidae)	17	21	1.87	0.61
Indeterminate Dogs/Cats (Canidae/Felidae)	8			
Deers and Peccaries (Artiodactyla)	8			
Tapir (*Tapirus bairdii*)	0	0		0.31
Peccaries (Tayassuidae)	10	12	1.07	2.76
Brocket Deer (*Mazama americana*)	24	26	2.31	2.61
White–tailed Deer (*Odocoileus virginianus*)	107	114	10.14	13.19
Agoutis and Pacas (Dasyproctidae)	–	72	6.41	1.23
Paca (*Agouti paca*)	47			
Agouti (*Dasyprocta punctata*)	25			
Rodentia. small	12	12	1.07	0.31
Rabbits (*Sylvilagus* spp.)	6	6	0.53	0.15
Other Mammals	170			
TOTAL NISP	1283	1124		652

Fig. 3. List of all taxa recovered in the Cueva de los Quetzales faunal assemblage. Taxa are presented by common name, but are listed in taxonomic order.

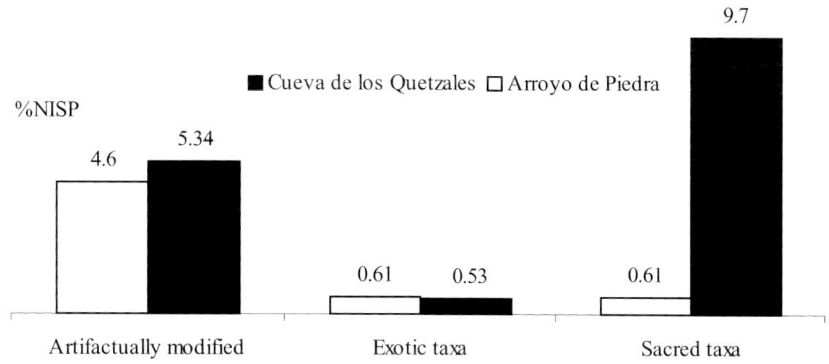

Fig. 4. Bar chart of relative frequencies of sacred and exotic taxa as well as artifactually modified remains. Frequencies are relative NISPs charted as a percent of the total faunal assemblage for each site.

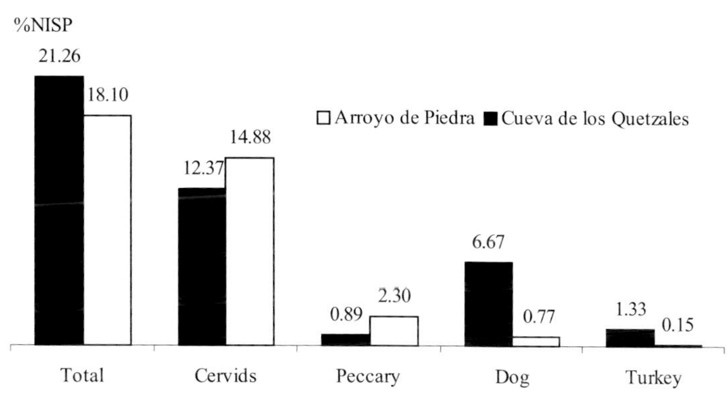

Fig. 5. Relative frequency of "managed" taxa including domestic dog, deer, peccary, and galliform birds.

ethnographic and ethnohistoric literature suggest that in addition to the domestic dog, other species including the deer, turkey, and peccary were captured and penned for fattening prior to a ritual event during the Postclassic period (Pohl and Feldman 1982; Pohl 1983; Carr 1996; White *et al.* 2001). In 16th century communities, select deer and dogs were raised by the priestly elite as a stable resource specifically for elite feasts and for ritual use (Tozzer 1941; Pohl 1985; White *et al.* 2001, 92). Pohl and Feldman (1982) documented that women raise, care for, and even breast-feed young animals in modern communities, a practice also documented in the ethnohistoric literature (Tozzer and Allen 1910; Tozzer 1941). Classic period Maya artwork often depicts women carrying or feeding dogs, and deer tethered and even being ridden (Tozzer and Allen 1910; Pendergast 1974; Danien 1997; White *et al.* 2001). Dog remains recovered from caches by White (2001) provide chemical evidence that these individuals have been specially fed apart from other dogs in the communities. Similar chemical studies of deer remains suggest that some individuals of this species were also intentionally fed (White *et al.* 1993; White *et al.* 2001; White *et al., in press*). It is possible that this management of certain species for ritual uses has very ancient antecedents.

In fact, both Arroyo de Piedra and the Cueva de los Quetzales have approximately the same proportion of deer and peccary in their assemblages suggesting their use was an elite prerogative but not limited to ritual activity (Fig. 5). However, the Cueva de los Quetzales assemblage has significantly more dog (the only real domesticate of the Early Classic Maya world), and turkey (a later domesticate introduced from Mexico and an important sacrificial offering during Postclassic and Historic periods). These observations are not in themselves sufficient evidence of ritual management of dogs, and certainly do not provide any argument for the ritual management of deer populations. However it is intriguing that higher frequencies of domesticates occur in the cave assemblage.

Population characters

In view then of the greater frequency of dog bones in the Cueva de los Quetzales assemblage, it is useful to look beyond simple taxonomy to the specific characteristics of the various animal populations – in particular at the age

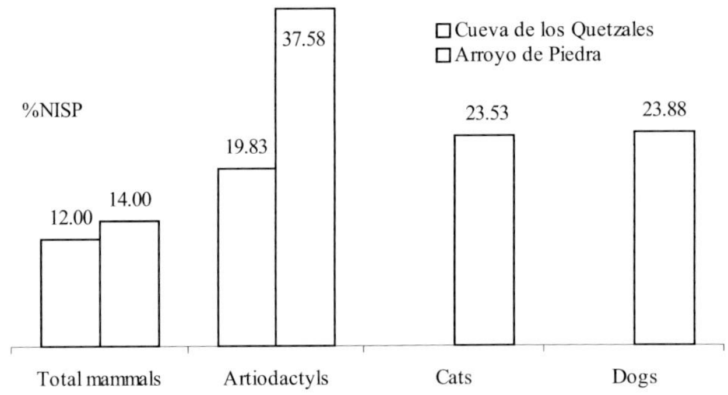

Fig. 6. Relative distributions of immature and mature individuals as represented by elemental epiphyseal fusion and tooth eruption. These are presented here as frequency of immature elements as a percent of total aged mammals for each site.

distributions. Age is often used as a marker of animal management, whether in hunted or domesticated populations. It is important to recognize specific age groups culled from populations that are otherwise represented in the zooarchaeological assemblages by natural or hunted distributions. Most individuals in Maya faunal assemblages are adult, and very few young individuals are found as dietary debitage. However, young individuals are common in Maya ritual deposits (Pohl 1981; Pohl 1983, 62; Wing and Scudder 1991; Carr 1996) and are more frequent in elite deposits, suggesting a status mediated control over the culling of young individuals in both wild and domestic populations.

The age analysis discussed here is preliminary and classifies animals only as immature vs. mature based on epiphyseal fusions and tooth eruption. Overall, the age of the Cueva de los Quetzales population is fairly similar to the age of the Arroyo de Piedra animals but, in fact, the proportion of immature individuals of the "managed" and sacred species is much higher in the cave record (Fig. 6). In particular, the complete absence of immature dogs or cats at Arroyo de Piedra and the high proportion of immature artiodactyls at the Cueva de los Quetzales are both notable divergences.

These results point to the use of specific individuals of managed populations. An emphasis on immaturity might be a symbolic metaphor for rebirth and renewal, and immature animals often play a role in period-ending rituals (Pohl 1981; Pohl 1983). However, it is important that the emphasis is on immature individuals only among potentially managed species. This suggests that the pattern represents public exclusionary rituals because elites with preferential access to these animals would be more likely to present them as quality sacrifices or feast foods in an exclusionary arena than in an inclusionary redistribution festival.

Element distribution: body portions

Element distributions can provide further tests of the ritual model. Modern Maya rituals often include a ceremonial use of the cranium and its ritual deposition (Pohl 1981; Pohl 1983). As well, elite deposits at some sites have a greater frequency of meat-bearing elements (Pohl 1994) suggesting preferential access to these portions. Unfortunately, there is no overt evidence of decapitated crania or other clear indication of sacrifice or ritual activity in the Cueva de los Quetzales remains, so once again it is the more subtle patterning that must provide clues to ritual behavior. Overall, the intermediate mammal group (dominated by dogs and cats) and large mammals (mainly artiodactyls) are represented by far greater proportions of cranial and axial elements at the Cueva de los Quetzales than at Arroyo de Piedra (Fig. 7 and Fig. 8). To test the possibility that this reflects a greater overall skeletal completeness, Stiner's "index of completeness" (1994) can be used to calculate a ratio of limb to non-limb elements (Fig. 9). Completeness is indeed higher for both intermediate and large mammals at the Cueva de los Quetzales. The separate comparison of dog and deer remains at the Cueva de los Quetzales indicates that this trend holds true for these specific taxa as well and in these cases can be used as a marker for sacrifice or burial deposition (Reitz and Wing 1999, 114).

Element distribution: sides

Another element distribution pattern that has been suggested for the Maya realm, and particularly for caves, is the symbolic emphasis of the "left" as representative of the underworld, the direction of the setting sun, and the holy (Gossen 1974). Mary Pohl (Pohl 1983, 89–90; Pohl and Pohl 1983; Pohl 1985) has found a statistically higher proportion of left elements of birds and deer in a few other ritual deposits, suggesting a differential use of body side in ceremonial contexts.

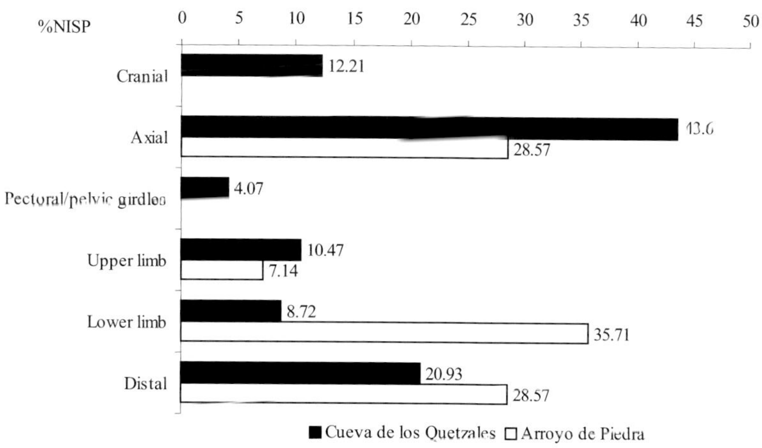

Fig. 7. Distribution of elements by body portion for intermediate mammals.

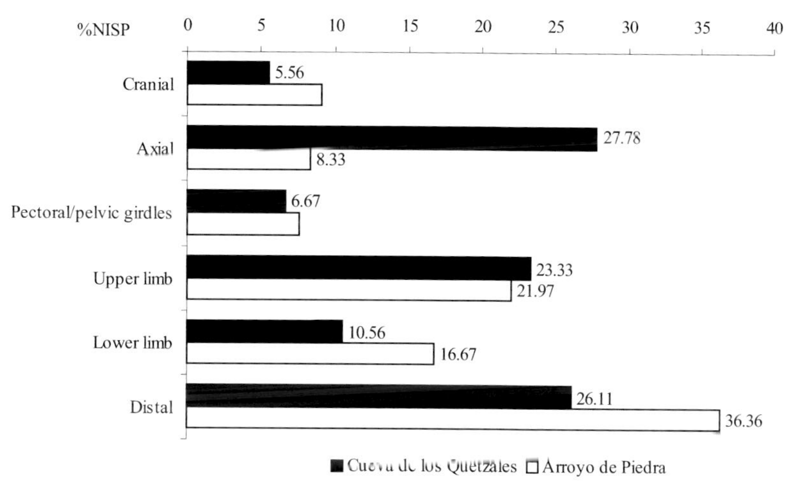

Fig. 8. Distribution of elements by body portion for large mammals.

At both the Cueva de los Quetzales and Arroyo de Piedra the majority of the mammals (including dogs), are represented by a standard proportion of left to right elements (Fig. 10). However, the artiodactyls, and particularly the deer remains, are clearly more frequently left elements than right in the Cueva de los Quetzales assemblage. Intriguingly the majority of the Cueva de los Quetzales galliform bird elements are also lefts (there are admittedly very few of these). Ethnohistoric documents record the fact that during certain ceremonies the priest is given a cut of deer (Landa 1978, 62–3; Taube 1988, 244), and perhaps that was from the left side. Possibly the ritually imbued left side was habitually sacrificed to the gods. This metaphorical use of animal carcass portions as sacrifices or offerings likely represents private or public exclusionary rituals and not large-scale festivities or feasting during which whole animals would have been used.

Discussion

While many ritual faunal assemblages can be so defined on the basis of the ceremonial contexts in which they are found and the sacred taxa that they contain, this is not the case for animal remains found in caves in the Maya world. At first glance, the Cueva de los Quetzales appears to be a ceremonial context (beneath the main ritual complex of a Maya site), and the remains recovered from within the cave contain species with definite sacred meaning for the ancient Maya. However, the same species that symbolically represent caves are often the natural residents of caves. As well, it is possible that the remains found within the cave were not associated directly with cave ritual, but were derived instead from non-cave related surface rituals or feasting by the residents of the elite center of Las Pacayas. Since ceremonial trash is often

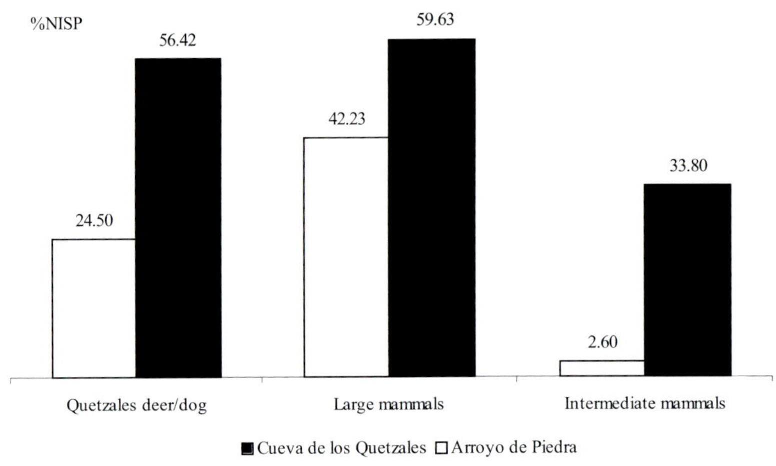

Fig. 9. Skeletal completeness measured as a ratio of limb to non-limb elements.

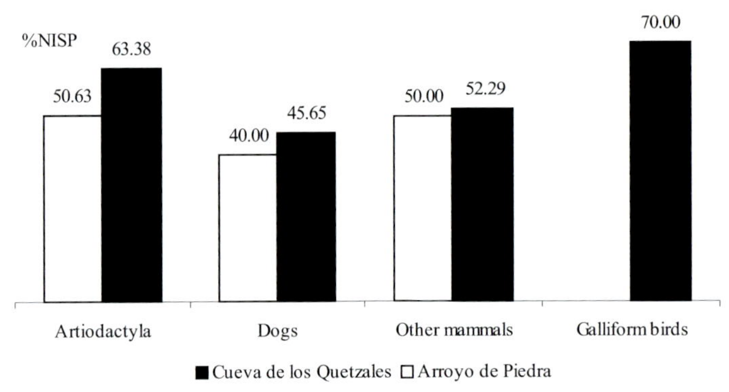

Fig. 10. Proportion of left sided elements for selected taxa (note the 50% starting point for these frequency distributions).

deposited in a sanctified location, it is quite possible that the material was the trash from surface rituals or even from secular but elite feasting. It is important therefore to prove both that the remains recovered in the Cueva de los Quetzales were derived from ritual activity, and that those rituals were in fact associated with the cave itself.

The correlation between zooarchaeological markers and cave-related ritual behavior provides excellent evidence that this was indeed a ritual deposit. The comparison between the Cueva de los Quetzales remains and another elite assemblage from the same period at the site of Arroyo de Piedra shows that secular elite deposits do not contain the same animal types (including non-intrusives), nor do they contain the same distributions of managed species, age groups, or body portions. The differences observed between the assemblages from the two elite centers vary specifically in patterns that are directly associated with cave rituals. The emphasis on symbolic underworld species, on deer and dog, on young individuals and on left elements, are all key to cave rituals.

Nevertheless, it is clear that a simple designation of "ritual" or even "cave ritual" is insufficient to understand the zooarchaeological correlates of ritual behavior, or to properly describe either emic symbolic roles or etic functional or structural interpretations. The results of this attempt to correlate different scales and types of Maya cave rituals with the zooarchaeological distributions that identify these in the archaeological recordare interesting despite the preliminary nature of this analysis. In many ways the Cueva de los Quetzales assemblage resembles other Early Classic elite deposits. High quantities of exotic materials and managed species including deer, dogs, and galliform birds suggest that in both situations the elite used these markers as expressions of status, power, and wealth. These markers would have been used competitively as symbolic metaphors as well as quantitative measures.

However, in many subtle ways, the Cueva de los Quetzales assemblage is different from the secular Arroyo de Piedra remains. The Cueva de los Quetzales fauna are more often sacred animals with metaphorical links to the underworld, to royalty, and to rites of death, renewal, and fertility. The dog, a symbolic link to the underworld, and the only Maya managed domesticate at this time, is significantly more frequent at the Cueva de los Quetzales. More importantly, at the Cueva de los Quetzales, this managed species as well as deer and cats are much more frequently represented by juvenile elements and higher skeletal completeness ratios. In combination this suggests sacrifice or ritual deposition of managed species as whole-body carcasses or, in the case of deer, body portions. The deer body portions intriguingly emphasize the left side of the carcass, suggesting that when deer body portions were offered to the underworld gods they were from the left side of the body – the side associated with the underworld and the heart.

Although the results presented here are too preliminary to completely substantiate this proposal, it is possible to suggest that the zooarchaeological remains from the Cueva de los Quetzales indicate public exclusionary rituals that focus on competitive elite rites of renewal (of annual and political cycles) and include sacrifice of juvenile dogs, cats, and deer, and offerings of deer body portions. The fact that the body portions do not overemphasize the meaty haunches suggests that the ritual significance of side was more important in these offerings than the quality of the food it represented.

This conclusion has broader implications for the elite control of landscape and of the symbolic center of the Maya universe. The Maya elite clearly positioned themselves to control entrance to this and other caves or portals to the underworld, but they may also have used ritual to sanctify the exclusivity of elite access to these portals. Their control of these underworld portals may have been celebrated in performative rituals displayed as expressions of power to visiting elites from other polities or to other individuals of the non-ruling nobility.

Regardless of the conclusions that are raised by this analysis, it may be useful to emphasize the fact that in certain situations, we must ask how to define ritual assemblages in a zooarchaeological sense before using presumed ceremonial deposits as descriptors for ceremonial activities involving animals. Zooarchaeology can link faunal remains to specific scales and types of rituals and therefore to the cultural meaning behind the ritualized behavior.

Acknowledgements

Funding for the Las Pacayas/Los Quetzales excavations were generously provided by a Wenner-Gren grant to Juan Pedro Laporte, and permission for zooarchaeological analysis was granted by the Instituto de Antropologia e Historia de Guatemala. Bevin Stevens did some of the Quetzales faunal identifications and her enthusiastic assistance is gratefully acknowledged. Comments by Jim Brady have significantly improved this paper, and the author takes full responsibility for all remaining errors. My thanks to Sharyn Jones O'Day for including me in the original ICAZ session from which this paper was derived.

References

Aguilera, C. A. 1985. *Flora y Fauna Mexicana: Mitologia y Tradiciones*. Mexico City: Editorial Everest Mexicana.

Bassie-Sweet, K. 1991. *From the Mouth of the Dark Cave: Commemorative sculpture of the Late Classic Maya*. Norman, OK: University of Oklahoma Press.

Benson, E. P. 1988. The Maya and the bat. *Latin American Indian Literatures Journal* **4(2)**, 99–124.

Brady, J. E. 1989. *An Investigation of Maya Ritual Cave Use with Special Reference to Naj Tunich, Peten, Guatemala*. Ph.D. Dissertation, UCLA.

Brady, J. E. 1997. Settlement configuration and cosmology: the role of caves at Dos Pilas. *American Anthropologist* **99(3)**, 602–18.

Brady, J. E. and Rodas, I. 1994. Analisis de la naturaleza de los depositos ceremoniales en cuevas Mayas: investigaciones recientes en la Cueva de los Quetzales, pp. 540–43, in Laporte, J. P. and Escobedo, H. L. (ed.), *VII Simposio de Investigaciones Arqueologicas en Guatemala*. Guatemala City: Museo de Antropologia e Historia y Asociacion Tikal.

Brady, J. E. and Rodas, I. 1995. Maya ritual cave deposits: recent insights from the Cueva de los Quetzales. *Institute of Maya Studies Journal* **1(1)**, 17–25.

Brady, J. E. and Schwegman, A. S. 1994. Nuevos metodos para la investigacion de cuevas Mayas, in Laporte, J. P. and Escobedo, H. L. (eds), *VII Simposio de Investigaciones Arqueologicas en Guatemala*. Guatemala City: Museo de Antropologia e Historia y Asociacion Tikal.

Brumfiel, E. M. and Earle, T. K. (eds). 1987. *Specialization, Exchange and Complex Societies*. Cambridge: Cambridge University Press.

Bunzel, R. 1952. *Chichicastenango, a Guatemalan village* (American Ethnological Society Publication 22). Locust Valley, NY: American Ethnological Society.

Carr, H. S. 1996. Precolumbian Maya exploitation and management of deer populations, pp. 251–61 in Fedick, S. L. (ed.), *The Managed Mosaic: Ancient Maya Agriculture and Resource Use*. Salt Lake City: University of Utah Press.

Christenson, A. J. 2000. *Popol Vuh: The Mythic Sections*. Provo, UT: Brigham Young University Press.

Clark, J. E. and Blake, M. 1994. The power of prestige: competitive generosity and the emergence of rank societies in lowland Mesoamerica, pp. 17–30 in Brumfiel, E. M. and Fox, J. W. (eds), *Factional Competition and Political Development in the New World*. Cambridge: Cambridge University Press.

Danien, E. 1997. The ritual on the Ratinlixul vase: pots and politics in highland Guatemala. *Expedition* **39**, 37–48.

Dietler, M. 1996. Feasts and commensal politics in the political economy: food, power and status in prehistoric Europe, pp. 86–126 in Wiessner, P. and Schiefenhovel, W. (eds), *Food and the Status Quest: An Interdisciplinary Perspective*. Providence, RI: Berghahn Books.

Douglas, M. 1994. The pangolin revisited: a new approach to animal symbolism, pp. 25–36 in Willis, R. (ed.), *Signifying Animals: Human Meaning in the Natural World*. London: Routledge Press.

Drennan, R. D. 1976. Religion and social evolution in formative mesoamerica, pp. 345–68 in Flannery, K. V. (ed.), *The Early Mesoamerican Village*. New York: Academic Press.

Emery, K. F. 1997. *The Maya collapse: a Zooarchaeological Inquiry*. Ph.D. Dissertation, Department of Anthropology, Cornell University.

Escobedo, H. L. 1994. Investigaciones arqueologicas y epigraficas en Arroyo de Piedra: un centro secundario en la region Petexbatún pp. 429–37 in Laporte, J. P. and Escobedo, H. (eds), *VII Simposio de Arqueologia Guatemalteca, 1993*. Guatemala: Ministerio de Cultura y Deportes, IDAEH, Asociacion Tikal.

Escobedo, H. L. 1997. Arroyo de Piedra: sociopolitical dynamics of a secondary center in the Petexbatún region. *Ancient Mesoamerica* **8(2)**, 307–20.

Escobedo, H. L., Samayoa, J. M., and Gomez, O. 1994. Las Pacayas: Un nuevo sitio arqueologico en la region Petexbatún, pp. 515–22 in Laporte, J. P. and Escobedo, H. L. (eds), *VII Simposio de Investigaciones Arqueologicas en Guatemala*. Guatemala City: Museo de Antropologia e Historia y Asociacion Tikal.

Farriss, N. M. 1984. *Maya Society Under Colonial Rule: Collective Enterprise of Survival*. Princeton, NJ: Princeton University Press.

Flannery, K. V. (ed.). 1976. *The Early Mesoamerican Village*. New York: Academic Press.

Goody, J. 1982. *Cooking, Cuisine and Class: A Study in Comparative Sociology*. Cambridge: Cambridge University Press.

Gossen, G. 1974. *Chamulas in the World of the Sun*. Cambridge, MA: Harvard University Press.

Hayden, B. 1995. Pathways to power: principles for creating socioeconomic inequalities, pp. 15–85 in Price, T. D. and Feinman, G. M. (eds), *Foundations of Social Inequality*. New York: Plenum Press.

Hayden, B. 1996. Feasting in prehistoric and traditional societies, pp. 127–47 in Wiessner, P. and Schiefenhovel, W. (eds), *Food and the Status Quest*. Oxford: Berghahn Books.

Helton, C. 1997. *An Analysis of Polychrome Ceramics from Los Quetzales Cave at the Site of Las Pacayas, Guatemala: typology, structure and design*. M.A. Thesis, Department of Anthropology, Brigham Young University Provo UT.

Heyden, D. 1981. Caves, gods and myths: world-view and planning in Teotihuacan, pp. 143–211 in Benson, E. P. (ed.), *Mesoamerican Sites and World-Views*. Washington, DC: Dumbarton Oaks Research Library.

Landa, D. de 1978. *Yucatan Before and After the Conquest* (translated by W. Gates). New York: Dover Publications.

LeCount, L. J. 2001. Like water for chocolate: feasting and political ritual among the Late Classic Maya at Xunantunich, Belize. *American Anthropologist* **103(4)**, 935–53.

McAnany, P. A. 1995. *Living with the Ancestors: Kinship and Kingship in Ancient Maya Society*. Austin: University of Texas Press,.

Mennell, S. 1996. *All Manners of Food: Eating and Taste in England and France from the Middle Ages to the Present*. Urbana: University of Illinois Press.

Pendergast, D. M. 1974. *Excavations at Actun Polbiche*. Toronto, ON: Royal Ontario Museum.

Peterson, J. F. 1980. *Precolumbian Flora and Fauna: Continuity of Plant and Animal Themes in Mesoamerican Art*. San Diego: Mingei International Museum.

Pohl, M. D. 1981. Ritual continuity and transformation in Mesoamerica: reconstructing the Ancient Maya Cuch Ritual. *American Antiquity* **46(3)**, 513–29.

Pohl, M. D. 1983. Maya ritual faunas: vertebrate remains from burials, caches, caves, and cenotes in the Maya Lowlands, pp. 55–103 in Leventhal, R. and Kolata, A. (eds), *Civilization in the Ancient Americas*. Albuquerque: University of New Mexico Press.

Pohl, M. D. 1985. The privileges of Maya elites: prehistoric vertebrate fauna from Seibal, pp. 133–45 in Pohl, M. D. (ed.), *Prehistoric Lowland Maya Environment and Subsistence Economy* (Papers of the Peabody Museum of Archaeology and Ethnology 77). Cambridge, MA: Harvard University.

Pohl, M. D. 1994. The economics and politics of Maya meat eating, pp. 121–149 in Brumfiel, E. M. (ed.), *The Economic Anthropology of the State* (Monographs in Economic Anthropology 11). New York: University Press of America.

Pohl, M. D. and Feldman, L. H. 1982. The traditional role of women and animals in the Lowland Maya economy, pp. 295–311 in Flannery, K. (ed.), *Maya Subsistence*. New York: Academic Press.

Pohl, M. D. and Pohl, J. M. 1983. Ancient Maya Cave Rituals. *Archaeology* **36**, 28–32, 50–1.

Puleston, D. E. 1977. The art and archaeology of hydraulic agriculture in the Maya lowlands, pp. 449–67 in Hammond, N. (ed.), *Social Process in Maya Prehistory: Studies in Memory of Sir Eric Thompson*. London: Academic Press.

Rappaport, R. A. 1999. *Ritual and Religion in the Making of Humanity*. Cambridge: Cambridge University Press.

Redfield, R. and Rojas, A. V. 1934. *Chan Kom: A Maya Village* (Carnegie Institution of Washington Publication 448). Washington DC: Carnegie Institute.

Reeder, P., Brady, J.E. and Webster, J. 1998. Geoarchaeological investigations on the Northern Vaca Plateau, Belize. *Mexicon* **20**, 37–41.

Reifler-Bricker, V. 1990. Faunal offerings in the Dresden Codex, pp. 285–93 in Robertson, M. G. and Field, V. M. (eds), *The Sixth Palenque Round Table, 1986*. Norman, OK: University of Oklahoma Press.

Reitz, E. J. and Wing, E. S. 1999. *Zooarchaeology*. Cambridge Manuals in Archaeology. New York: Cambridge University Press.

Roys, R. L. 1965. *Ritual of the Bacabs: A Book of Maya Incantations*. Norman, OK: University of Oklahoma Press.

Saunders, N. J. 1994. Tezcatlipoca: jaguar metaphors and the Aztec mirror of nature, pp. 159–77 in Willis, R. (ed.), *Signifying Animals: Human Meaning in the Natural World* (One World Archaeology 16). London: Routledge.

Schwartz, M. 1997. *A History of Dogs in the Early Americas*. New Haven, CT: Yale University Press.

Sponsel, L. E. 2001. Do anthropologists need religion, and vice versa? adventures and dangers in spiritual ecology, pp. 177–202 in Crumley, C. L. (ed.), *New Directions in Anthropology and Environment: Intersections*. Walnut Creek, CA: Altamira Press.

Stiner, M. C. 1994. *Honor Among Thieves: A Zooarchaeological Study of Neandertal Ecology*. New Jersey: Princeton University Press.

Stone, A. J. 1995. *Images from the Underworld: Naj Tunich and the Tradition of Maya Cave Painting*. Austin, TX: University of Texas Press.

Taube, K. A. 1988. *The Ancient Yucatec New Year Festival: the Liminal Period in Maya Ritual and Cosmology*. Ph.D. Dissertation, Yale University.

Tedlock, B. 1986. On a mountain road in the dark, pp. 125–38 in Gossen, G. H. (ed.), *Symbol and Meaning Beyond the Closed Community: Essays in Mesoamerican Ideas*. Albany, NY: Institute for Mesoamerican Studies, University of Albany.

Tedlock, D. 1996. *Popol Vuh* (revised edition). New York: Touchstone.

Thompson, J. E. S. 1970. *Maya History and Religion*. Norman, OK: University of Oklahoma Press.

Tozzer, A. M. 1941. *Landa's Relacion de las Cosas de Yucatan* (Papers of the Peabody Museum of American Archaeology and Ethnology 18). Cambridge MA: Harvard University.

Tozzer, A. M. and Allen, G. M. 1910. *Animal Figures in the Maya Codices* (Papers of the Peabody Museum vol. 4, no. 3). Cambridge, MA: Harvard University.

Vogt, E. 1969. *Zinacantan*. Cambridge, MA: Harvard University Press.

Vogt, E. 1993. *Tortillas for the Gods: A Symbolic Analysis of Zinacanteco Rituals*. University of Oklahoma Press, Norman, OK.

Walker, W. H. 1995. Ceremonial trash?, pp. 67–79 in Skibo, J., Walker, W. H. and Nielson, A. (eds), *Expanding Archaeology*. Salt Lake City: University of Utah Press.

White, C. D., Healy, P. F. and Schwarcz, H. P. 1993. Intensive agriculture, social status, and Maya diet at Pacbitun, Belize. *Journal of Anthropological Research* **49**, 347–75.

White, C. D., Pohl, M. D., Schwarcz, H. P. and Longstaffe, F. J. 2001. Isotopic evidence for Maya patterns of deer and dog use at Preclassic Colha. *Journal of Archaeological Science* **28**, 89–107.

White, C. D., Pohl, M. D., Schwarcz, H. P. and Longstaffe, F. J. in press. Feast, field and forest: deer and dog diets at Lagartcro, Tikal, and Copan, in Emery, K. F. (ed.), *Maya Zooarchaeology: New Directions in Method and Theory*. Los Angeles: Institute of Archaeology, UCLA Press.

Wing, E. S. and Scudder, S. J. 1991. The exploitation of animals, pp. 84–97 in Hammond, N. (ed.), *Cuello: An Early Maya Community*. Cambridge: Cambridge University Press.

Kitty F. Emery
Curator of Environmental Archaeology,
Florida Museum of Natural History,
University of Florida Box 117800,
Gainesville, Florida 32611–7800, USA.
E-mail: kemery@flmnh.ufl.edu

9th ICAZ Conference, Durham 2002
Behaviour Behind Bones, (ed. Sharyn Jones O'Day et al.) pp. 114–127

15. Observations on the religious content of the animal imagery of the 'Gran Coclé' semiotic tradition of pre-Columbian Panama

Richard Cooke

The explicit and flamboyant pre-Columbian art of the 'Gran Coclé' culture area of central Panama is replete with animal images that have invited many interpretations by art historians, anthropologists and archaeologists. Zoomorphic images represent different grades of realism, which permit the application of modern taxonomic principles to the identification of specific taxa, ranging from Class to genus and, in a few cases, even species. The distribution of these taxa on art objects can be compared with that of animal remains in archaeofaunal samples from dietary and ritual contexts in order to reconstruct the way the people of 'Gran Coclé' made use of the regional life assemblage of organisms for food and materials and how they may have related individual taxa to cognitive behaviour and supernatural environments. A wide variety of invertebrates and vertebrates from coastal, littoral and terrestrial habitats are depicted. There are also some striking absences. Although there are many non-natural representations of animals, especially composite creatures that incorporate features of more than one taxon, it is possible to identify their component parts by focussing on anatomical details. Taxa that appear to have a special cognitive significance include marine worms, crabs and other inshore crustaceans, sting-rays, sawfish and sharks, crocodilians, marine and freshwater turtles, certain kinds of birds, bats, dogs, and deer. Interpreting the relationship of animal images to religiosity is fraught with difficulties since pre-Columbian isthmian peoples were pre-literate. In this paper the opinions of several scholars are discussed and the author's personal views expressed vis-à-vis *zoogeography, ethology, taxonomy, aesthetics, dualism, shamanism, and social affinities between humans and animals.*

Introduction

I discuss how certain animal images and geometric symbols used by the pre-Columbian inhabitants of the 'Gran Coclé' culture area of central Panama to decorate artefacts may have been related to a supernatural environment and, in this sense, be imbued with a 'religious' significance (Cooke, *this volume*, Fig. 1). By focussing on taxonomic and behavioural information exhibited by these images, I consider the degree to which modern non-native observers can, or cannot, identify specific categories of animals – ranging from Class to species – in a flamboyant semiotic system that has invited many commentaries from archaeologists and art historians (Lothrop 1937; Lothrop 1942; Linares 1976; Helms 1977; Linares 1977; Helms 1979; Helms 1981; Cooke 1984; Briggs 1989; Benson 1992; Cooke 1992b; Cooke 1993; Helms 1995; Labbé 1995; Helms 1997; Cooke 1998; Helms 1998; Helms 2000). I emphasize four themes: 1) real and imaginary animals in the context of the 'life assemblage' (Klein and Cruz-Uribe 1984, 3) that would hypothetically have existed in pre-Columbian times in 'Gran Coclé' (located between 79°–82° West and 7°–9° North in the Neotropical realm), 2) groups of recognisable animal taxa whose depiction *together* on artefacts can be assumed to have some cognitive 'meaning', 3) anthropomorphic (or, at least, bipedal) forms, which appear to refer to shamanism (or similar ceremonial and ritual activities), and 4) pictorial relationships between geometric symbols, animals and anatomy that help us identify the biological prototypes of these symbols.

'Real' animals and regional faunas

I envision iconographic studies as an integral part of regional archaeozoological research, one that is not necessarily as biologically objective as osteology, but should, nonetheless, be taken seriously. In my opinion, all animal classes depicted on 'Gran Coclé' art, whose affinity with a living taxon can be established sensibly by reference to painted or modelled anatomical details, would have been present within the 'Gran Coclé' culture area during the time period under consideration (~cal BC 200 and cal AD 1550). The term 'sensible' subsumes zoogeographic likelihood as well as observational objectivity: for example, widespread species such as the Virginia opossum (*Didelphis marsupialis*), raccoon (*Procyon*), and coati (*Nasua nasua*) are more likely models for mammals with clearly depicted black masks (Lothrop 1942, figs. 317, 421b; Helms 1995, fig. 75), than the spectacled bear (*Tremarctos ornatus*) (Helms 1992; Helms 1995, 59–71), a species which may be present in mountains near the Colombian border, but has not been recorded elsewhere in Central America in Holocene times (Cooke 1998, 114, note 6).

Once identified, the presence and abundance in art of a specific taxon can be compared with those of taxa whose body parts have been recorded in regional archaeofaunal assemblages in order to acquire as broad a view as possible, not only about how the pre-Columbian peoples of 'Gran Coclé' utilised available animals for food, raw materials, tools, ornaments, pets and trade goods, but also how they perceived their participation of these animals in their cognitive universe, i.e., as symbols, totems and emblems, and as actors with humans in a supernatural environment (Cooke 1992b; Cooke 1998). The fact that many animal taxa have uneven distributions in 'Gran Coclé' art and in dietary, artesanal and ritual archaeofaunal samples suggests that regional pre-Columbian peoples indeed viewed the components of the regional fauna in many different ways (Cooke, *this volume*). The white-tailed deer (*Odocoileus virginianus*) was important pictorially, industrially *and* nutritionally (Cooke 1992b). Sea turtles, crocodiles, bats, dogs and monkeys were seemingly *not* used for food although they were painted and modelled, and their body parts were used for making tools and ornaments (monkeys less often than the other three) (Cooke 1992b; Sánchez and Cooke 1998). Molluscs, catfish (Ariidae, Parauchenipteridae and Pimelodidae), puffer fish (Tetraodontidae: *Guentheridia formosa, Sphoeroides* spp.), and some rodents (*e.g. Agouti paca*), are regularly found in middens, but were never illustrated (Linares 1977). Some abundant groups of Neotropical organisms, which are visually and behaviourally striking – *e.g.*, butterflies, sloths (Bradypodidae), and tapirs (*Tapirus*) – are absent in identifiable forms both on the art, and (in the case of vertebrates) in sampled archaeological bone samples (Linares 1976; Linares 1977, 67; contra Helms 1995, figs. 20, 22, and Labbé 1995, fig. 23). Notwithstanding sampling biases (extant archaeofaunal samples represent a much more restricted area of 'Gran Coclé' than do artefact samples) such selectivity likely reflects human behavioural traits that comply with the term 'religiosity', such as taboos and social or genealogical affinities between humans and animals (Helms 1979, 70–108). (For invertebrate distributions in kitchen middens, see Cooke *et al.* 1996; Hansell 1979; Carvajal 1999. For vertebrate distributions, see Cooke and Ranere 1989; Cooke 1992a; Cooke and Ranere 1992a; Cooke and Ranere 1999; Jiménez 1999; Jiménez and Cooke 2001; Cooke and Ranere, *in press*).

Lothrop (1942, 28), who first contextualised and described the 'Gran Coclé' art style, contended that 'in no case does the....artist portray the animals of nature (but) instead the beasts from a world of never recorded and long forgotten myths.' His opinion was based, understandably, on the fact that many icons are depicted in an *anatomically* unrealistic manner, i.e., two-legged 'birds' with 'saurian' claws. If we accept this imprecision as an intrinsic element of *style* or *aesthetics*, however, many such zoomorphs clearly depict or incorporate *real*, rather than just fantastic or mythical animals (Linares 1977). What is more, they comprise a surprisingly large number of invertebrate and vertebrate animal classes, all of which are noticeable and/or widespread denizens of the Neotropical realm, being found in estuaries, mangroves and rivers, and, in the case of terrestrial habitats, not necessarily in 'jungles' (Helms 2000, 8), but rather in grasslands, wooded savannas, and gallery and hilltop forests. The landscape had been impacted by agricultural activities for several millennia before the 'Gran Coclé' semiotic system became manifest (Cooke and Ranere, 1992; Piperno and Pearsall 1998, 209–27, 286–97; Cooke and Ranere, *In press*; Cooke, *this volume*).

Sometimes, 'Gran Coclé' animal icons are sufficiently realistic to allow unequivocal identifications to *species*. A well-known example is the king vulture (*Sarcoramphus papa*) (Ladd 1964, plate 4 a and b; Labbé 1995, figs. 64, 145; Cooke 1998, fig. 4.2a). Large white spots on a dark ground (Fig. 1a) identify the spotted eagle ray (*Aeteobatus narinari*) (cf. Lothrop 1942, figs. 211b and 212c; Helms 2000, 102). The white-tailed deer's 'badge' (Fig. 1c) is branching antlers, emphasized even when the icon is devoid of a body (the only other deer genus in Panama [*Mazama*] has erect straight or bifurcate antlers) (Lothrop 1942, figs. 85, 423; Benson 1992; Cooke 1992b; Labbé 1995, figs. 63 and 114; Helms 2000, 43). It appears that the need to obey stylistic and aesthetic principles gave the artisan a licence to manipulate chromatics while honouring biological observation. For example, the bird in Fig. 3b has a *black* body. The only bird species in Panama with a curved red bill and red legs is the ubiquitous white ibis (*Eudocimus albus*). Adult ibises have a white body. Since the background of the ceramic vessel is pale, the artist was obliged to paint a black body (cf. Labbé 1995, figs. 14 and 15).

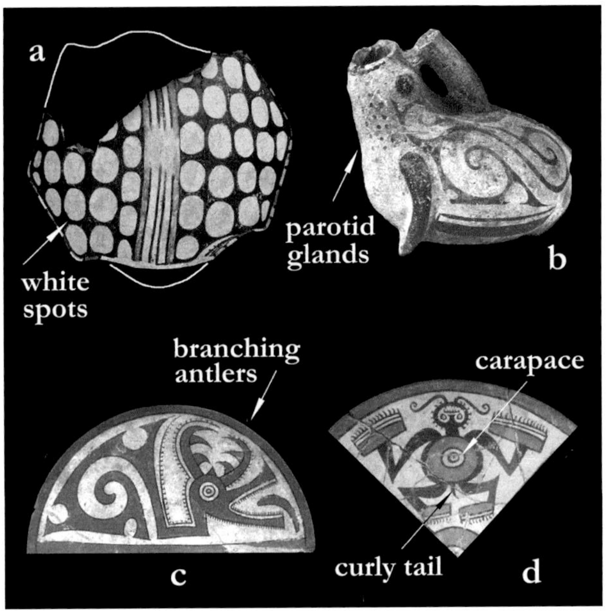

*Fig. 1. a: polychrome effigy plate painted like an eagle ray (*Aeteobatus narinari*), Cerro Juan Díaz, Los Santos, Panama (photo: R. Cooke), b: effigy vessel, probably depicting the marine toad (*Bufo marinus*), Sitio Conte, Panamá (photo: P.S. Briggs), c: half a polychrome plate decorated with the head of a white-tailed deer (*Odocoileus virginianus*), Museo de Antropología, Panamá (photo: R. Cooke), d: icon of a tailed (freshwater) turtle from a polychrome plate, Museo de Antropología, Panamá (photo: C. Hansen).*

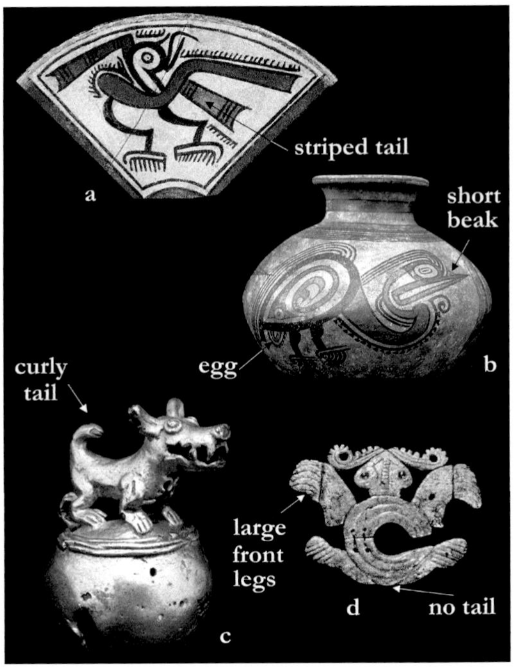

*Fig. 2. a: a backward-looking bird with striped tail, referred by many scholars to the great curassow (*Crax rubra*). One of four icons on a polychrome plate, Museo de Antropología, Panama City (photo: R. Cooke) b: Crested bird, which apparently is laying an egg, from a polychrome jar, Museo de Antropología, Panama City (photo: R. Cooke), c: cast gold-copper bell that depicts a dog, Museo de Antropología, Panamá (photo: J. Maiquez), d:* Spondylus *shell pendant shaped like a sea turtle from Playa Venado, Panama, Dumbarton Oaks Museum, Washington D.C. (photo: L. A. Sánchez).*

Helms (1977 and 1979, 97–108 and 2000, 55–75) insists that subtle anatomical details, i.e., the shape and position of pupils, crests, 'beards', teeth and colouration, signal particular reptile species (*e.g.*, the green iguana [*Iguana iguana*], the baslisk [*Basiliscus basiliscus*] and boa [*Boa constrictor*]), which, she believes, are imbued with great religious significance (as I discuss in later pages). She has also perceptively suggested (Helms 2000, 120, figs. 6.15 and 6.16) that crustacean icons, which exhibit a pointed protuberance in front of their heads and have thin pincers, may be penaeid shrimp whose rostral spikes prick unwary handlers (Lothrop 1942, figs. 374 and 375). The hirsuteness and lack of antlers of a male ungulate zoomorph (Fig. 3c) suggest that it represents a peccary (*Tayassu*). The raised area painted with black dots on an anuran effigy (Fig. 1b) orients one's glance to the swollen parotid glands of a bufonid toad (probably the marine toad, *Bufo marinus*) (Briggs 1989, fig. 21; Cooke 1989). The clearly demarcated round spots of other anurans modelled in clay allude to small dendrobatid ('poison dart') frogs (Fig. 4f). The large front legs and tailless condition of realistic *Spondylus* and cast gold-copper turtles signal a marine species (Chelonia) (Fig. 2d; Lothrop 1937, fig. 118 e,f; Sánchez and Cooke 1998, fig. 7 e,f), and a short, curly tail (Fig. 1d), a terrestrial species. In other modelled and painted images a striped head is the badge of the ubiquitous painted terrapin (*Trachemys scripta*) (Lothrop 1942, figs. 208, 280, 413a; Labbé 1995, fig. 137). A fish effigy from Sitio Conte (Fig. 3a) presents anatomical details that appear to highlight the venomous pre-opercular and opercular spines of a toadfish (*e.g.*, *Batrachoides, Daector*) in addition to capturing impressionistically the large head, wide mouth and prominent eyes that characterise these common inshore genera. Linares (1977, 65–7) proposed that the iconographic pre-eminence of pricking, stinging and biting animals, such as crabs, sawfish, sharks, sting-rays and toadfish, suggests that they are metaphors for qualities of human behaviour that would have been respected in competitive rank societies. The fact that the skeletal parts of these taxa, however, are frequently found in kitchen middens indicates that they were regularly used for food, as well as for making tools and ritual objects (Cooke and Ranere 1999; Jiménez and Cooke 2001).

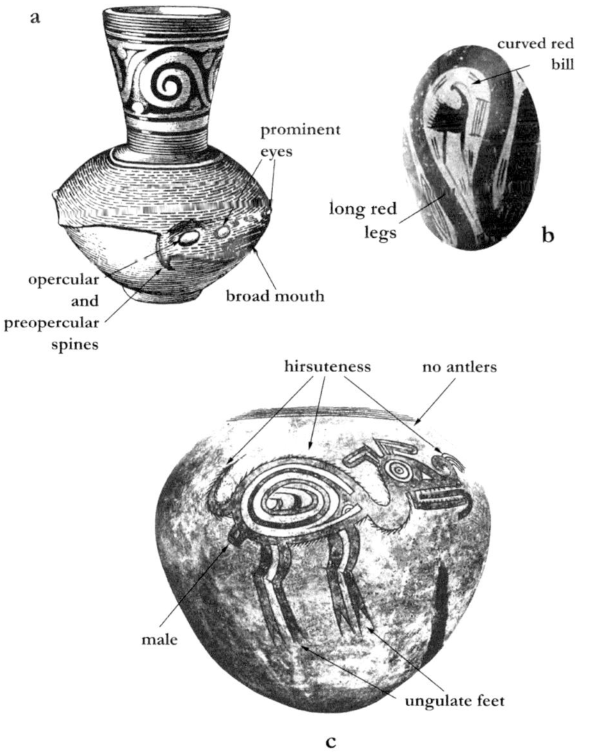

*Fig. 3. a: polychrome effigy vessel of a fish with features typical of a toadfish (*Batrachoides *or* Daector*), Sitio Conte, Coclé, Panamá (after Lothrop 1942, fig. 212a), b: detail of a Tonosí style polychrome vessel which may depict a white ibis (*Eudocimus albus*) with inverted colouration for stylistic reasons, Museo de Antropología, Panamá (photo: R. Cooke), c: polychrome jar decorated with a mammal with characteristics of a peccary (*Tayassu*), Coclé, Panamá (after Lothrop 1942, fig. 466b).*

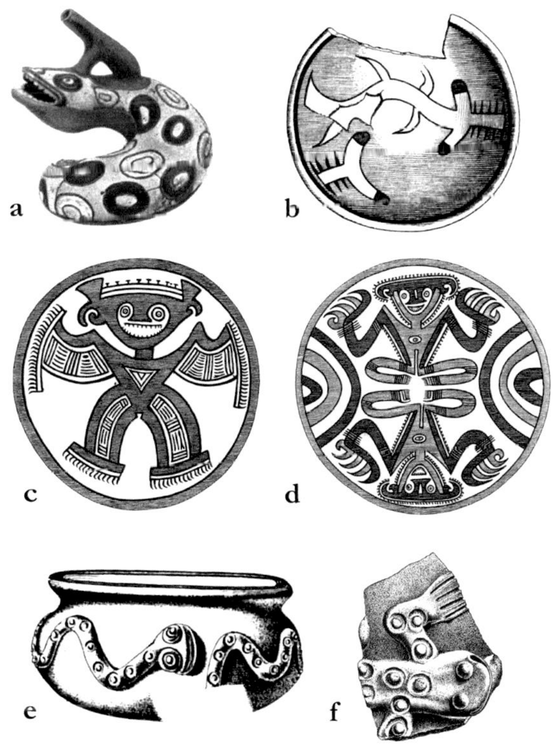

*Fig. 4. a: polychrome effigy jar shaped like a marine eel (perhaps Muraenidae), Sitio Conte, Panama (after Lothrop 1942, Plate 1d), b: polychrome plate depicting a fish with characteristics of sawfish (*Pristis*) and hammerhead shark (Sphyrnidae), Pearl Islands, Panama (after Lothrop 1942, fig. 442b), c: human figure with psittacid ears, and possible vulvar notch. The arms and legs are decorated with the double-direction line motif, probably simulating tattoos or body paint, from a polychrome plate, Sitio Conte, Coclé, Panama (after Lothrop 1942, fig. 91a), d: paired and seated human figures with pstittacid ears and vulvar notches, from a polychrome plate, Sitio Conte, Panama (after Lothrop 1942, fig. 57b), e: modelled serpentiform creature on a monochrome vessel, perhaps a muraenid eel (cf.* Myrichthys tigrinus*), Sitio Conte, Coclé, Panamá (after Lothrop 1942, fig. 347b), f: modelled frog, perhaps Dendrobatidae, Sitio Conte, Coclé Panama (Lothrop 1942, fig. 348).*

It has also been apparent to commentators that 'Gran Coclé' artisans recognised the importance of certain details of physiology and behaviour for signalling specific taxa for their audience. Gill slits and male copulatory organs are emphasized on some shark icons. Rays are depicted with well-marked olfactory pits (Lothrop 1942, figs. 31,150, 212 c,d; Linares 1977, 63; Helms 2000, fig. 16.14). It seems likely that the artist responsible for the crab icon in Fig. 5a wants the observer to know it is a female because of the visual prominence given to the semicircular appendage at the posterior edge of its carapace (Linares 1977). This is a dorsal image. When crabs are viewed thus the abdomen is not generally visible unless the animal is an incubating female (in fact, some female rock crabs press the extended abdomen and egg clutch against the substrate, an action that is, presumably, a form of maternal care [Christy, *pers. comm.*, 2003]) (Fig. 5b). The only other visual clue to the sex of a crab would be claws of greatly unequal size, which is a characteristic of males of some taxa, including *Cardisoma crassum* – the most abundant crab species in archaeofaunal samples (Cooke *et al.* 1996). However, it is likely that the principles of bilateral symmetry on Coclé art would be inimical to a realistic depiction of differential claw size.

In my opinion, it is the *behavioural* characteristics of domestic dogs that stand out on many curly-tailed, boisterous icons, which I believe represent this taxon (Fig. 2c; Cooke 1992, fig. 4). A polychrome plate has captured exquisitely a group of birds that are surely humming birds

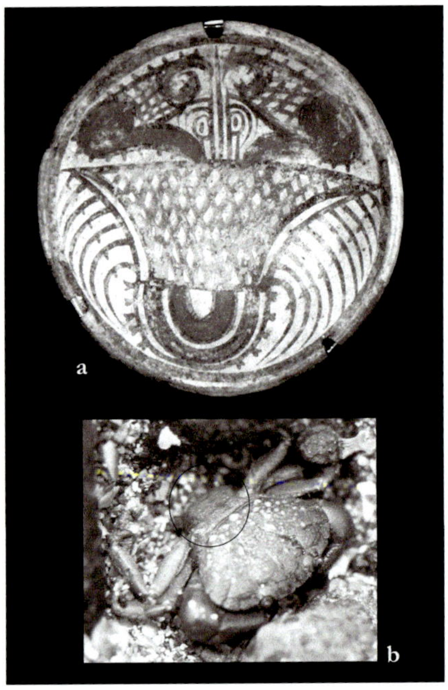

Fig. 5. a: image of a crab with a semicircular motif appended to the posterior end of its carapace. This appears to signal that the abdominal flap is extended. The shape of the carpus suggests a large xanthid crab, such as Ozius *or* Eucides *(photo: R. Cooke), b: An incubating female crab (Xanthidae:*Eurypanopeus*). The crab has extended its abdominal flap and is pressing it against the substrate (photo: J. Christy).*

Fig. 6. Polychrome plate that depicts a group of birds, probably humming birds (Trochilidae), Museo de Antropología, Panamá City (photo: R. Cooke).

(Trochilidae) fussily feeding at flowers (Fig. 6; also Cooke 1984; Labbé 1995, fig. 42). Linares (1977) and Helms (1995, 1997, 2000) suggest that the avian model for a crested bird with a striped tail, which often looks backward (Fig. 2a; cf. Lothrop 1942, 29–30) is the great curassow (*Crax rubra*) whose chestnut-bodied female sports a striped tail. Linares (1977, 61) proposed that the curassow's confident strutting behaviour and crest were metaphors for the comportment of important people, like warriors, who in real life would have worn plumed head-dresses. Lothrop (1942, 30) likened these birds' poses on polychrome plates to those of 'dancers'. Helms (1995, 40–1) averred that the backward glance reflected this large species' 'habit of passing the head back over the shoulder and wings.' Some bird images (Fig. 2b; Lothrop 1942, figs. 104 and 225b) are apparently laying an egg. Some reptile images appear to depict embryos inside eggs (Labbé 1995, fig. 139; Lothrop 1942, fig. 192 d). The insect image (perhaps a dragon-fly) on Fig. 7e may be symbolizing metamorphosis from a larval form into an adult.

The identifications that commentators have proposed for some of these ethologically informative icons are debatable. For example, Helms (1995, 44ff) favoured a tinamou (*Tinamus major)* as the model for one of the egg-laying bird icons (Lothrop 1942, fig. 192d) on account of this ground-dwelling species' prodigious egg-laying abilities. As Helms points out, however, tinamous are *not* crested and they have shortish necks; so if the crest and neck are taxonomic 'badges', the bird being signalled is *not* a tinamou. The only bird species present in the region that has a long thin crest is the boat-billed heron (*Cochlearius cochlearius*; Ridgely and Gwynne 1989, plate 2,1), but one would expect this species' enormous beak to be signalled if the artisan wished its identity to be clear. Many 'Gran Coclé' bird icons stress crests, forked tails and hooked beaks; but these traits are shared by so many Neotropical taxa that one can only guess the identity of most of them (Cooke 1984; Cooke 1986; Cooke 1998, fig. 4.3).

It can be inferred from the above examples that pre-Columbian artisans of 'Gran Coclé' *wanted* and *expected* knowledgeable observers to refer certain animal icons correctly to real animals in a familiar environment, and, when necessary, devised methods of highlighting salient anatomical and behavioural features to ensure an unambiguous taxonomic identification notwithstanding repetitive decorative and stylistic formalities. Even when we are unable to assign a particular image to genus or species on the basis of the visual information it contains, this does not necessarily mean that it was taxonomically irrelevant to its pre-Columbian beholders whose own classification of animals is, of course, unfathomable and unrelatable to modern taxonomy. This 'signalling' of diagnostic features of an animal's anatomy applies to biologically coherent taxa and fantastic creatures alike – Lothrop's (1942, 28) 'monsters' or 'dragons'. Composite creatures come in many guises, but they are not necessarily garbled: the badges of their component animals are carefully highlighted. An especially frequent composite animal is bipedal – presumably the signal for a bird – but it has pert mammalian ears and/or the branched antlers of a white-

Fig. 7. a: polychome plate which represents two crocodilian heads connected by a serpentiform shape marked, according to some scholars, like a boa (photo: Bowers Museum of Cultural Art, Los Angeles), b: above: schematic representation of a boa body markings, below: hypothetical derivation from this (the double-direction line motif) (after Helms 2000, fig. 2.5), c: serpentiform image, perhaps a marine worm (Polychaeta) (after Lothrop 1942, fig 94d), d: polychrome effigy jar of a mammal which has the double-direction line pattern along its arms and legs, Museo de Antropología, Panama City (photo: R. Cooke), e: polychrome plate with an image of an insect, perhaps a dragon-fly, Museo de Antropología, Panama City (photo: R. Cooke), f: polychrome plate that depicts a composite creature (bird + deer), Museo de Antropología, Panama City (photo: R. Cooke).

tailed deer (Fig. 7f; Lothrop 1942, figs. 51, 61). Several elongate fish-like icons have bilateral projections on the head, which commentators usually identify as hammerhead sharks (*Sphyrna*). There are, in fact, two fish-like icons with laterally expanded heads. One has the eyes placed in the *correct* position for *Sphyrna* – at the extremities of an appropriately expanded 'hammer' (Linares 1977, fig. 36). On a more frequent icon (Lothrop 1942, fig. 485; Labbé 1995, 14; Helms 2000, fig. 6.8b), the eyes are painted at the *corners* of the *thin* projections, while the head sports a *rostral spike*, which is sometimes depicted as a sawfish rostrum (Helms 2000, fig. 6.8). I propose that the icon with the correct eyes and head proportions is indeed signalling a hammerhead shark, while the latter is a rendering of a composite mythical animal (shark + saw-

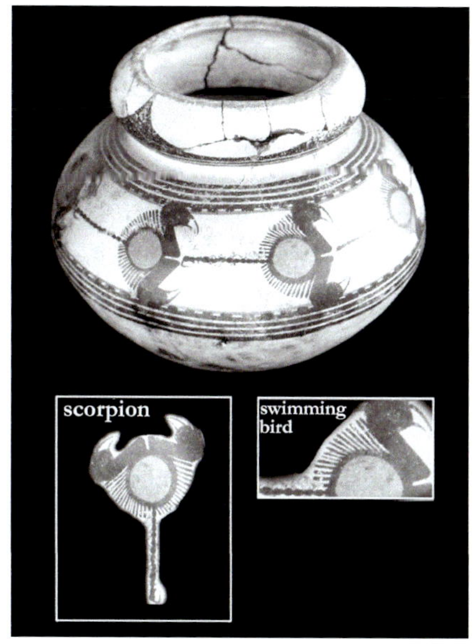

Fig. 8. Polychrome jar of the Tonosí style: a fine example of 'split representation', by which two animals are represented in a single icon, private collection, Panama City (photo: R. Cooke).

fish). This association is explicit on a polychrome plate found on the Pearl islands (Fig. 4b).

It is formally reasonable to attribute a *uniquely* cognitive significance to these combinations of different animals on a single image, i.e. that they describe transmutations of one animal into another, cosmological pairs or mythical beings – ideas which can be expected to have been an important component of 'Gran Coclé' thinking, as I shall discuss later. Nevertheless, the prevalence of 'split representation', in which a single icon represents two or more realistic animals *depending on the way it is observed*, makes more straightforward explanations viable, i.e., that artists enjoyed aesthetic or stylistic ruses – visual tricks, if you like! A typical example (Fig. 8) depicts a quite realistic scorpion viewed from above and a swimming grebe-like bird in profile (see also Linares 1977, fig. 30 a).

'Explaining' combinations of animals and humans

Observers of 'Gran Coclé' art are invariably impressed by its balance, which is expressed geometrically, chromatically and iconologically (Lothrop 1942; Linares 1977; Helms 1993; Helms 1995). Frequently the 'canvas' of an object is divided into sectors separated by lines and scrolls. Colours, zoomorphs and motifs are repeated, alternated, opposed and juxtaposed. Since dualism is a pervasive concept in cognitive systems everywhere one would expect it to have influenced the thoughts behind this kind of decoration, i.e., lower and upper worlds; life and death; male and female; human and non-human; day and night; real and unreal; 'deer' and 'curassow.' Bicephaly, bifurcation, mirror images and the depiction of animals in pairs - sometimes of the same taxon, sometimes not - also allude to the cognitive importance of twinning. Mary Helms has produced detailed and interesting analyses of such relationships making extensive use of ethnographic analogy. In the following section I offer a few ideas of my own *vis-à-vis* the biological identity of animal icons on 'Gran Coclé' art whose depiction in conjunction, not only with other identifiable animal taxa, but also with humans, suggests a specific religious significance, i.e. in the context of myth, legend, cosmology and 'shamanic' transformations.

Big lizards, crocodiles and boas

In her book *The Curassow's Crest* Helms argued that the sinuous bodies of many images – whatever the identify of the animals depicted as their heads and/or tails – represent *snakes,* specifically the boa (*Boa constrictor*), in her opinion 'the isthmian version of the formidable great snake, widely found in the cosmologies and iconographies of Mesoamerica and lowland South America' whose body represents 'pure, undifferentiated life energy, energy that lies at the heart of all other life-forms as *fons et origo* and that also connotes, in its pure, featureless, un-shaped serpentine form, the life force that existed at the dawn of time before articulated life-forms had been created or organized' (Helms 2000, 12–3). Onto this ubiquitous serpentiform body form Helms grafts details gleaned from the myths and cosmologies of other Neotropical peoples, such as the body parts of humans and animals in the process of being devoured and an iguana-boa crasis with crest and beard, which, she proposes, symbolises a terrestrial/arboreal environment in contraposition to the aquatic environment of the crocodile (Helms 2000, 73–5). Labbé (1995, 37–9) also grants a primary symbolic role to serpents 'as....the concept of vital force, essentially binary in nature, expressing both a 'male' and 'female' aspect'.

Another idea that Helms (1977, 1979, 2000) has championed is that the green iguana (*Iguana iguana*) and the basilisk lizard (*Basiliscus basiliscus*) were more important participants in the 'Gran Coclé' supernatural environment than the two isthmian crocodile species – the cayman *(Caiman fuscus*) and American crocodile (*Cocodrilus acutus*) - to which most other commentators have attributed a widespread *saurian* image, often depicted in a bipedal stance (Fig. 9). In view of the importance of reptiles in the contemporary myths of isthmian peoples like the Bribri, Buglé (Bokotá), Kuna and Emberá (Helms 1977; Helms 1979,102-8), it is reasonable that iguanas, basilisks, boas *and* crocodiles should have been prominent actors in the pre-Columbian myths and religion of 'Gran Coclé'. But since I argue that animals in art comprise but one part of a many-tiered perception and expression of an

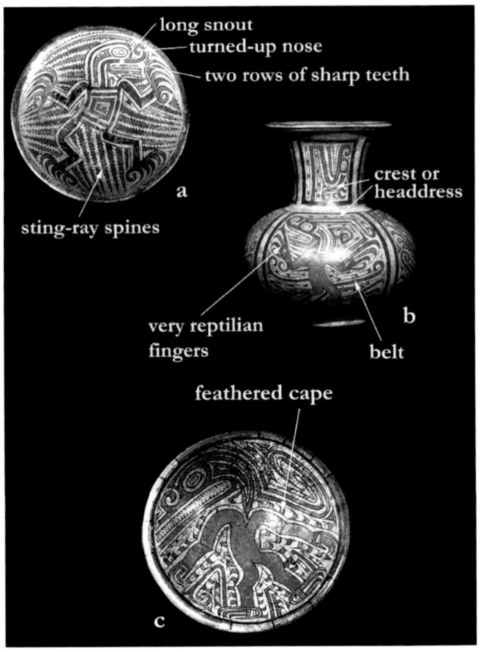

Fig. 9. Images of an image of a bipedal crocodilian. a: Sitio Conte, Coclé, Panama (Photo: R. Cooke), b and c: private collections, unprovenienced (photos: Bowers Museum of Cultural Art) (c: see Labbé 1995, fig. 58).

entire regional 'life assemblage', it is not idle to scrutinise critically the anatomical details, which commentators have accepted as signals of particular species.

Minimally, not all sinuous-bodied animals can be accepted *a priori* to represent 'snakes'. A life-like toothed eel with prominent round spots recalls *Echidna nocturna* and *Muraena lentiginosa*, in-shore muraenid eels, which can deliver a nasty bite (Fig. 4a; Allen and Robertson, 1994, 43–8). Since large round spots are not a characteristic of any common snake in 'Gran Coclé' another modelled form (Fig. 4e; Ladd 1964: fig. 52b) may be pointing towards another spotted eel, such as *Myrichthys tigrinus* (Allen and Robertson 1994, 53), whose bones occur in kitchen middens (Jiménez and Cooke 2001).

Another ubiquitous image is that of a sinuous animal delineated by a multi-flecked stripe that parallels the body outline and appears to represent its 'legs.' Alternating horizontal and vertical lines are depicted along the body (Fig. 7c; Lothrop 1942, figs. 70, 94d, 95). Some of these images have a smaller number of thick 'legs' (Lothrop 1942, figs 71, 94c) and others have one or two pairs of typical (conventionalised) clawed legs behind the head (Lothrop 1942, fig. 94 a,b). Since the head often sports a 'tongue' that ends in double scrolls commentators relate this being to a serpent (Labbé 1995; Lothrop 1942). This is not an exclusive relationship, however, because a double scroll motif generally protrudes from the snout of realistic *turtle* images (Figs. 1d, 2d). In my opinion, marine worms (Polychaeta), and sea scorpions (*Squilla*) (Lothrop 1942, fig. 71) are more likely to be the models for the referred serpentiform images with lateral flecks, than snakes (see also Linares 1977, 65). Admittedly, polychaete worms do not have well-defined heads. But the hairiness or prickliness implied by the image in Fig. 7c seems antithetical to a basic characteristic of snakes – their slipperiness or smoothness.

Helms has constructed an elaborate argument to explain the double-direction line pattern used to fill the bodies of the above-mentioned images as a derivation from boa body markings and, thence, its function as a 'kenning' (replacement symbol) for a mythical iguana-boa (Helms 2000, figs. 2.5 and 4.15). This observation underlines an analytical dilemma that all commentators on 'Gran Coclé' art have confronted: are the several geometric motifs regularly employed on 'Gran Coclé' art – frequently as proxies for tattooing or painting on human and animal arms and legs (Fig. 7d) – 'kennings' of a particular animal taxon and, if so, which analytical procedures demonstrate this relationship? Or are these geometric motifs divorced from life forms and merely decorative? Theoretically, specific geometric motifs can be related back to a particular zoomorph by the standard stylistic procedure of tracing stages along a sequence from real to abstract to geometric, for example, with respect to a bird image with outstretched wings and tail (Sánchez and Cooke 1998, fig. 3) and to a laterally depicted crocodilian head, whose diachronic geometricisation accompanies a general stylistic shift in 'Gran Coclé' painted ceramics from curvilinear to rectilinear designs (Cooke 1985; Labbé 1995; Cooke 1998). It is conceivable that the double-direction line pattern is indeed an extreme simplification of the oval and rectangular markings attributed by Helms to a bicephalous boa (Fig. 7a,b; Helms 2000, fig. 4.2). In this case, boas are 'kenned' with humans and a host of other creatures whose bodies are decorated with the double-direction line pattern (*e.g.*, Figs. 4c and 7e). This relationship, however, would gain validity if there were clear intermediate stages between this ultimate degree of abstraction and *realistic* depictions of boas, which, as far as I can tell, are absent on 'Gran Coclé' art – even in the earliest phases of the development of polychrome pottery when animal images are particularly life-like. *Boa constrictor* is a widespread, frequently encountered animal on today's landscape and its remains are ubiquitous in archaeofaunal bone samples around Parita Bay (Cooke and Ranere 1989; Cooke and Ranere 1992). Therefore, snakes may not have been cognitively primordial at all, but, rather, items worthy of the cooking pot, like pacas (*Agouti paca*), but unworthy of being advertised.

The 'saurian' / 'crocodile god' (Lothrop 1942, 83) / 'dancing crocodile' motif to which I referred earlier is one of the most striking and prevalent in 'Gran Coclé' art. It comes in two guises, one naturalistic (often horizontally depicted), and the other humanised (usually vertical, either frontal or in profile). It incorporates a number of conventions: a) a long snout, with an upper and lower row of strong pointed and isomorphic teeth, b) a crest-like emanation in front of the eyes and/or on top of the head, c) a backward-facing scroll or curlicue at the end of the snout, and d) noticeably reptilian feet sometimes with quite realistically depicted long phalanges and claws (particularly Helms 2000, fig. 5.9). Frequently the bipedal mode is painted against a background of alternating red and bluish serrate spikes, interpreted universally as 'sting-ray spines' (Fig. 9a). When the image is humanised it is often frontally depicted wearing belts with heads at each extremity – these often (but not always) mirrors of the principal icon's head –, necklaces, ligatures, and ear rods, and brandishing offensive weapons like *macanas* or sword-clubs (Lothrop 1942, figs. 149, 193, 223, Plates I a, II a,b; Ladd 1964, Plate 7a; Cooke and Bray 1985, fig. 9; Labbé 1995, fig. 56, 18, table 4; Cooke 1998; Cooke *et al.* 2000, fig. 8.9 b; Cooke *et al.*2003).

Helms has highlighted details, which, she believes, confirm the identity of this image as a 'crested *iguana*', namely, small ticks, curls and lines painted or (on metalwork) embossed between the snout and the eyes and underneath the mandible. She argues that these elements reproduce iguanas' facial filaments and dewlap (Helms 2000, figs. 4.8, 4.10). Apparently she no longer thinks that the humanised variant represents the spectacled bear (Helms 1995, 69, fig. 80). She also notes that many of these icons are depicted with a *horizontal* eye, which, she proposes, mirrors the shape of a closed iguana eye and not an open, vertical crocodile eye (Helms 2000, 57–9; Ladd 1964, plate 7a, fig. 42b). Her argument that the crests that bedeck the icon's head are derived from iguanid lizard crests is neatly woven around careful observation of these reptiles' anatomy and behaviour (Helms 2000, 63–5). There are, however, some incongruencies: for example, the animals depicted on the belts of a Sitio Conte embossed gold plaque (Helms and Sharer 1992, plate 2), which were originally interpreted by Helms (1977; 1979, 105–6 and figs. 17, 79, 199) as basilisk lizards on account of their tricuspid teeth, actually have *mammalian* ears and tongues, so it is non-intuitive to interpret them as *canid* or *felid*, in which case the big tricuspid teeth would be carnassials. This interpretation fits in with the proposal that one guise of the bipedal saurian is that of 'hero-hunter' (Helms 1979, fig. 17; Helms 2000, 90-2). When 'beards' and 'whiskers' are depicted they may be a badge for *humanness*, rather than of *'iguana-ness'*. The Spanish marvelled at the bearded warriors in this region of Panama (Gaspar de Espinosa, in Jopling 1994, 54).

Realistic crocodiles are widespread in many media (Lothrop 1942, fig. 229; Bray 1992, fig. 3.8; Labbé 1995, figs. 52 and 117; Cooke 1998, 97–100 and fig. 4.5b; Sánchez and Cooke 1998, fig. 6). The long-toothed and noticeably upturned and incurved snout is preserved in extremely abstract painted and modelled icons, which belong to the last few centuries of the pre-Columbian era (Ladd 1964, fig. 13; Cooke 1998, fig. 4.5 c, e and f). Realistic iguanas are much rarer. Two are quite explicit

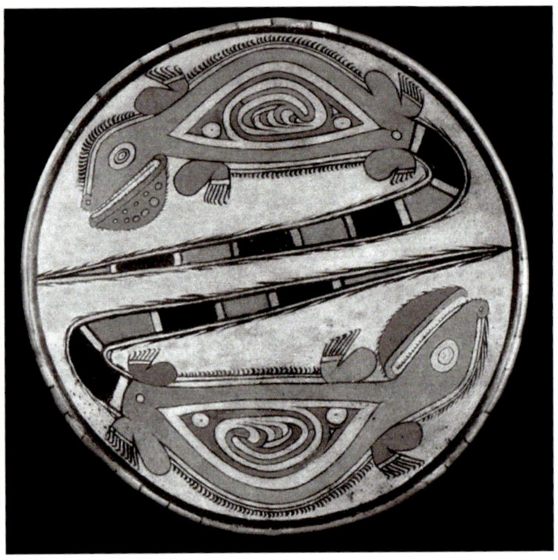

*Fig. 10. Realistic representation of a green iguana (*Iguana iguana*) on a polychrome plate,* n.b., *the long, clearly striped tail and the sub-mandibular skin flap (after Labbé 1995, fig. 43).*

and highlight the striking long striped tail and sub-mandibular skin flap of the green iguana (*Iguana iguana*), which is very prominent in males in breeding condition (Fig. 10). Helms (1995, fig. 76) attributes the realistic iguana image illustrated by Lothrop (1942, fig. 147b [also Cooke 1992b, fig. 5a]) to a coati. Cooke (1992b, fig. 5b) illustrates another iguana in cast gold.

To sum up, I propose that Lothrop's 'crested dragon' motif, epitomised by the illustrations to which I have referred, is based on a *crocodilian* and not a basilisk or iguana. When 'Gran Coclé' artisans depict iguanas they highlight anatomical details that are typical of this species and are different from those that signal crocodilians. The iconographic pre-eminence accorded to crocodilians, especially after cal AD 750, contrasts with the total absence of their bones in 'Gran Coclé' kitchen middens that date to the time period under consideration although crocodiles provided teeth for ornaments, were food items in the same region before cal BC 200 (Cooke and Ranere 1989; Cooke and Ranere 1992b), and were regularly hunted in another area of Panama (Bocas del Toro, Caribbean) (Wing 1980). Bearing in mind the abundance of crocodilians around Parita Bay today, where they are a serious menace to fishers and bathers, I doubt whether pre-Columbian peoples hunting without firearms could have severely impacted local cayman and American crocodile populations. Pascual de Andagoya, a soldier of Pedrarias' hordes, mentions that crocodilians were abundant and a threat to human life in the eastern half of Panama Bay in the decade AD 1510–1520 (in Jopling 1994:34).

Turtles and other animals

Turtles are another widespread animal form on 'Gran Coclé' art. I have already given my reasons for believing that both marine and freshwater taxa are depicted. Although Lothrop (1942) accepted their iconographic pre-eminence in the Sitio Conte ceramic sample, in two- and three-dimensional representations, Helms (2000) does not. She mistakenly attributes the only turtle icon she accepts (a striped-headed and tailed one) to a marine species (1995, 98, fig. 19). In five graves at Sitio Conte, whole turtle carapaces were buried with the dead (Lothrop 1937, figs. 205, 212, 215, 222 and 245). Lothrop (1942, 35) calls them 'large sea turtles' although, to judge from field drawings, they are more likely to be emydid, probably *Trachemys scripta,* and they are not particularly large even for this species (<40 cm carapace length). All these graves were occupied by high rank or high status people (Briggs 1989); Lothrop proposes that the turtles are food offerings for them on their journey to the other world. Linares (1977, 61) suggested that the whole shells may have been ritual paraphernalia, such as sounding boards. Modern Ngobére-speaking Guaymí rub the pitch-covered neck of the carapace to produce a humming noise – an instrument much-used at the *balsería* ceremony (Cooke, *this volume*). Some small vessels depict a modelled human head above an effigy of a striped-headed terrapin (Lothrop 1942, fig. 208 c and e) – an association that alludes to an emblematic or mythological significance (i.e., clan membership or a mythical human-turtle personage). I refer later to another instance of the possible cosmological role of turtles.

Parrot beaks and ears

Seated and standing human figures with ears shaped like psittacid (parrot, parakeet or macaw) beaks are frequently depicted on polychrome plates (Fig. 4 c and d; Lothrop 1942, 37 and figs. 57–9, 91). Lothrop equates them with 'turtles' and a 'turtle god'. Equally baffling is Helms' (1995, figs. 63, 64) proposal that they are spectacled bears! Helms (1995, 83–9, 2000 *passim*) summarises the metaphorical and metonymic significance of detached parts of human and animal bodies among Neotropical Native Americans, including extant isthmian peoples. She argues, for example, that a particular kind of crest, which is related to the icon she interprets as a curassow, is a severed human leg (Helms 2000, 33–53). Helms' interpretation of a pair of seated personages with parrot-ears as 'a section of a boa body…and a boa rectangle' is, however, unrealistically syllogistic (Helms 2000, fig. 5.6). On the other hand, her observation that standing and sitting human figures with parrot ears are likely to be *female* is perspicacious: many are depicted with a vulvar notch and with ligatures on arms and legs – a sartorial item that is a characteristic of modern Kuna *women* (Lothrop 1942, fig. 90; Helms 2000, 34–5) (see also Labbé, 1995, for a discussion of vulva symbolism and Helms 2000, 76–96, for a particularly detailed and interesting discussion of gendering).

Parrot-like birds among some South American peoples are associated, on one plane, with acute hearing and clever speech, and, on another, with seeing into the future and eavesdropping – activities that are typical of shamans and healers (Cooke 1984). Some cast gold pieces – not necessarily made in 'Gran Coclé', but part of the metallurgical tradition to which this region belonged (Bray 1992) – depict parrot-like birds seemingly whispering in a human's ear. Some anthropomorphic icons sport long tongues, which terminate in a parrot head (Lothrop 1942, plate 1h). Although I have suggested (Cooke 1984) that some kind of shaman-partner or human-avatar relationship is one possible *raison d'être* of the parrot beak metonym, the fact that the seated and standing figures with parrot ears are female, can be considered contradictory. A burial at Sitio Sierra contained the partial skeleton of a macaw buried alongside an unusually small adult male who owned a pelican bone flute and a greenstone necklace (Cooke, *this volume*).

Shamanic transformations

One of the most influential figures of South American anthropology in the last century was Gerardo Reichel-Dolmatoff who gifted the discipline several monographs, which describe the life-ways and cognitive systems of Native American groups who live in tropical forests. His analyses of Tukano, Desana and Kogi myths, cosmology, folk taxonomy, curing and shamanism, including hallucinogen use, upon which he relied in a classic evaluation of the meaning of Colombian gold-work, are bibliographic *sine qua non* (*e.g.*, Reichel-Dolmatoff 1971, 1975, 1976, 1977, 1978a,b, 1987, 1990). It would be most unusual if the principal behavioural correlates of the kind of transformational shamanism that Reichel-Dolmatoff described were not present in the pre-Columbian society of 'Gran Coclé'. Nevertheless it is intemperate to attribute a primary role to psychotropic substances in the light of current ethnographic knowledge for the isthmus (barring a surfeit of tobacco smoke and hot *Capsicum* peppers). As far as I know there is not a shred of evidence that pre- or post-Columbian peoples in central and western Panama (outside areas settled by people of Mesoamerican descent) ever used 'hallucinogenic' mushrooms (Cooke 1989; Cooke 1998; *contra* Helms 1995, 78–81; Helms 2000, 149–57).

Paraphernalia derived from large spotted cats and marine shells, which were associated in graves with fine gold- and stone-work and incense burners, suggest that these are likely to have been the property of people who engaged in ceremonial activities, such as those of shamans, curers and chanters (Cooke, *this volume*). Other animal products found in certain graves in large numbers, such as stingray spines, may have had some kind of ceremonial function although the possibility that they were weapons used in war or for hunting and fishing (as giggs) cannot be ruled out. The transformation of the shaman into other beings, sometimes real animals, sometimes imagined ones, and his voyages to other worlds in their company, are integral episodes of the shamanic experience. Labbé (1995) stressed the role of shamanism in his interpretation

Fig. 11. a: polychrome plate, which exhibits a standing figure clothed like a shaman (after Labbé 1995, fig. 30), b: polychrome plate with a central bipedal figure that suggests a human dancing with an animal mask, Museo de Antropología, Panamá (photo, R. Cooke).

of 'Gran Coclé' art. Dancing with musical instruments, wearing animal masks, plumed crowns and feathered capes, singing in a loud voice, bloodletting, imbibing unusual substances, and sexual posturing are commonplace props for and preludes to knowledge-acquiring flights of fantasy in the ethnographic present. Although it is a hyperbole to assume that every icon of a spread-eagled bird – an ancient and ubiquitous motif on 'Gran Coclé' art (Cooke *et al.* 2003) – represents a shaman transformed into an 'eagle,' 'swallow-tailed kite' or 'king vulture', or that every composite animal has shamanic overtones, several bipedal zoomorphs do give the impression that they represent real humans transformed into animals. This transformation may be symbolised by animal masks: a bipedal figure (Fig. 11b) seems to wear an antlered and snouted 'mask', which recalls that of the *cucuá* dancers of the mountains of Coclé who use bark cloth capes and a snouted and antlered deer mask (see also Cooke and Bray 1985, fig. 10). The standing figure illustrated in Fig. 11a exhibits sting-ray spines; bells on wrists and legs; an object which may be a rattle; a baton or wand; and a suggestion of a feathered or animal skin cape and belt. In this case the human-animal association is not explicit although the details strongly suggest shamanic activities. Some bipedal crocodile figures (Fig. 9c; Labbé 1995, fig. 58) appear to wear feathered capes. Although Labbé refers to these figures as 'shamans-in-combat' (1995, figs. 56, 58, 111 and 142) or 'shamanic transformation themes' (1995, figs.

107 and 109), this explanation vies with Helms' well-researched analyses that give a stronger weight to *myth*, i.e., culture heroes, hunter-heroes, and founder figures. Some icons of the bipedal crocodilian that are embossed on gold and were found in very rich graves at Sitio Conte ape the dress of humans of high rank or status (Cooke, *this volume*; *see*, particularly, Hearne and Sharer 1992, plate 1). This fact, in addition to the very ubiquity of crocodile icons in cemeteries bordering Parita Bay, suggests strongly that social affiliation is implied, *i.e*, that crocodilians were genealogically or politically important for a sector of the 'Gran Coclé' population.

Cosmological twosomes and threesomes

Although I do not share Helms' and Labbé's optimism that it is possible to identify the role of particular celestial bodies in 'Gran Coclé' cognition, such as the sun, some associations of recognisable zoomorphs do point towards a cosmological structure of some kind. A modelled clay vessel (Fig. 12a), found just outside the currently accepted eastern boundary of 'Gran Coclé' (Lake Madden or Alajuela), depicts a turtle suspended by a raptorial bird and resting on the head of a crocodile – a trilogy that alludes to a cosmological scheme typical of some Amazonian groups, of the sky as bird, the earth as turtle and the underworld as crocodile. The size of the front legs suggest that the turtle is a marine taxon and the carbuncled beak that the bird is a king vulture. A modelled trichrome effigy vessel, which depicts a crocodile with a stylised turtle on its back (Fig. 12c), may be thematically related. This vessel, wrought in the 'El Hatillo' style, represents the latest period in the stylistic development of 'Gran Coclé' polychrome pottery and was probably made soon before or just after Spanish contact. Fig. 12b depicts an icon, whose body form is avian (outstretched wings, forked tail); the head is painted like that of curly-tailed turtle images; the wings end in a square-mouthed creature and heads at each end of the tail have bifid tongues. In this case, the possible quadripartite combination (bird + turtle + crocodilian + marine worm?) is suggestive of a tiered cosmological scheme.

Helms (2000, 33–53) has presented an interesting analysis of an animal pair: the great curassow and the white-tailed deer. She proposes that they are related chromatically, thematically and behaviourally as species that 'hold anomalous positions betwixt and between the wild and the human' because they are often penned up and the young are cared for by women. She also remarks (Helms 2000, 47) that both have a habit of looking backwards, the deer when on the alert and the curassow because it constantly rearranges its back feathers (as an owner of a male and female curassow, I can vouch that they almost perpetually engage in this behaviour!). The curassow is a forest bird that requires extensive tracts of mature trees. Curassow bones have not been reported from 'Gran Coclé' archaeofaunal samples (Cooke and Ranere, *in press*). The image in question is a backward-looking crested bird (Fig. 2a; Lothrop 1942: figs. 43–5), which sports a striped tail and, sometimes, what appear to be ligatured legs. These details point to female curassows; males, which have all black tails, would surely be signalled by the yellow knob on their bills and by their striking pure white bellies. This hypothetical female curassow is also paired on clay vessels with a forward-looking bird with crests and a thickened lower mandible – as though the artist were signalling a pelican (*Pelecanus*) (Lothrop 1942, fig. 48b) – and with turtles (Lothrop 1942, fig. 56c).

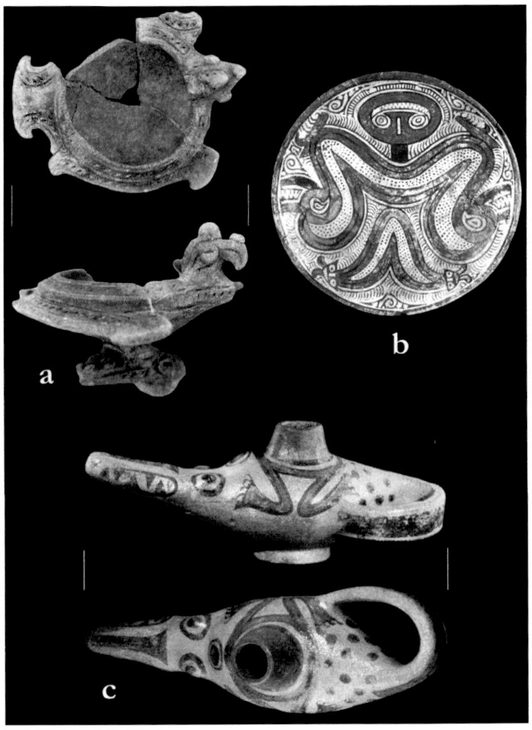

Fig. 12. a: modelled and incised vessel that appears to depict a king vulture (top), a marine turtle (centre) and a crocodilian (bottom), Lake Alajuela, Panama (photo: R. Cooke), b: polychrome plate with a creature consisting of four parts (perhaps a turtle head, bird wings with crocodilian heads, and a marine worm tail, Museo de Antropología, Panama City (photo: R. Cooke), c: effigy vessel of a crocodilian with an animal on its back (perhaps a turtle) (photo: Bowers Museum of Cultural Art, Los Angeles).

The crested caracara (*Polyborus plancus*) is a plausible alternative model for the bird with the backward glance. This visibly abundant raptor in the Pacific plains of 'Gran Coclé' has a confident, strutting gait. It has a prominent crest, striped tail and a big *hooked* beak (as *all* the relevant stripe-tailed icons do; the curassow's beak is short and unhooked). Its bones are present in refuse dumps at Cerro Juan Díaz (Cooke and Ranere, *in press*).

There is documentary evidence that, in spite of the

white-tailed deer's being by far the most abundant mammal species in dietary archaeofaunas around Parita Bay during and before the time period under consideration, its hunting was governed by some sort of social and dietary control: one soldier remarks that unlike ordinary people warriors were not allowed to eat any mammal meat (Cooke and Ranere 1989; Cooke 1992b; Cooke and Ranere 1992; Jopling 1994, 34; Cooke and Ranere, *in press*). Another soldier comments that 'the chief and principal Indians .... are like Dominican and Carthusian monks, because they do not eat any meat of any kind and under any condition whatsoever, except fish and iguanas, although deer and other game are all over the place' (Jopling 1994, 65). (The fact that *all* mammal meat was tabooed by warriors and important people suggest that there was a symbolic and structural relationship between hunting and war; whether it was deer meat or not may, therefore, be unimportant).

Geometric symbols and animals: a brief comment

Ever since Lothrop (1937, 1942) first described the 'Gran Coclé' semiotic system, commentators have been intrigued by, and have attempted to explain, the elaborate and often repetitive bands, blocks and body part terminations constructed with a set of standard geometric motifs, such as Ys, Vs, Cs and Ss. As Helms (2000, 127) succinctly comments, these are 'deceptively simple but very sophisticated design motifs,' which, in some cases, probably do symbolise life-forms, but, as I have argued, are difficult to relate by normal stylistic processes to specific animal classes. That these designs were painted on the human body is apparent from contact-period ethnographic literature, from human effigy vessels with decorated arms and legs, and from finds of clay roller stamps, which would have been used to apply the designs – probably made with the juice of the plant species that are still used by isthmian peoples, i.e., 'jagua' (*Genipa americana*) and annatto ('achiote') (*Bixa orellana*) (Lothrop 1942, fig. 368). Spanish captain Pascual de Andagoya (in Jopling 1994, 34) remarked that in the territory of Escoria (modern Santa Maria river; Cooke *this volume*, Fig. 1) 'there was a generation of Indians older than the others and of better quality: among them were knights who were very proud of being brave: all over their breasts and arms they were decorated with chains and knots' (my translation).

To judge from the ethnographic record for extant Native American peoples in tropical lands, it is not unlikely that geometric designs also reflect cosmology, social organisation, and gendering. Helms has offered some intriguingly complex explanations for the zoological origin of some of these motifs. In addition to stressing, predictably, but in my view undemonstrably, the undulating line–snake connection, she proposes more bizarre interpretations: for example, that the Y- or line-gap motif is connected to 'the haunting call of the often cosmologically significant tinamou (Tinamidae), "possibly the egg-laying bird to which I referred earlier", because the call of the tinamou is described as a single long-drawn-out note sometimes with a dip in the middle' (Helms 2000, 128, fig. 7.3). If I were reducing the call of the commonest tinamou over most of 'Gran Coclé' *Crypturellus soui*) to a geometric form, I may well depict it as lines interrupted by notches although I would not draw the lines straight, but pointing upward!

We are on firmer ground when we assess the *anatomical position* of geometric motifs. On a very large proportion of zoomorphs, YC scrolls are depicted within the body cavity of an organism, as if they were its entrails, or, more poetically, its 'inner essence' or 'life force'. This is very clear, for example, on the icons of the bipedal crocodilians, deer and birds, and on the purported shaman figures I have just described (Fig. 11). In Helms' (1995, 77) words, 'indigenous Panamanian art forms, both verbal and plastic, have utilized expository styles which emphasize the revelation of 'that which is within', indicating that the real value or something (and someone) is found in qualities located inside the being or thing.'

Conclusion

The animal icons depicted on the 'Gran Coclé' semiotic tradition of central Panama are so descriptive, colourful, and repetitive that it is logical that anthropologists, archaeologists and art historians should have dedicated so much time to attempting to identify the taxa represented and relate them to natural and supernatural environments. Although it is presumptuous to ignore merely decorative and aesthetic criteria for the selection and depiction of images by the pre-Columbian artists, the use of animals in this semiotic system as a whole can be construed as relating to religiosity in its broadest sense. The only sensible methodology for its interpretation is social anthropology. The details, of course, will remain more contentious than the intent: specialists will carry on arguing whether a particular zoomorph can be best explained in terms of myth and legend, 'ethnic' and social affiliation, shamansim, cosmology or dualism and will continue to search the literature for analogies with recently disappeared or surviving societies in order to justify their opinions. Come what may, any interpretation of an extinct, pre-literate and largely unfathomable semiotic system benefits from, firstly, doing our level best to be as accurate as possible with taxonomy and biology in a sound zoo geographic and cultural historical context; secondly, comparing the iconographic prevalence of an animal with its presence, absence and abundance in other contexts (such as kitchen refuse, graves and workshops); thirdly, assuming that both the artist and the audience knew how to identify animals, were aware of their diagnostic external characteristics and behavioural quirks, and were sometimes concerned about displaying them realistically and sometimes not; and, finally, refraining from assuming that all pre-Columbian peoples in the Neotropics lived in unbroken rainforests and were awed only by 'jaguars', 'anacondas' and 'harpy eagles'.

References

Allen, G. R. and Robertson, D. R. 1986. *The Complete Divers' and Fishermens' Guide to Fishes of the Tropical Eastern Pacific*. Bathurst: Crawford House Press.

Benson, E. 1992. Motifs, meanings and myths: the iconograpjy of Sitio Contegold plaques, pp. 22–31 in Hearne, P. and Sharer, R. J. (eds) 1992. *River of Gold: Precolumbian Treasures from the Sitio Conte*. Philadelphia: University of Pennsylvania Museum of Archaeology and Anthropology.

Bray, W. M. 1992. Sitio Conte metalwork in its pan-American context, pp. 33–6 in Hearne, P. and Sharer, R. J. (eds), *River of Gold: Precolumbian Treasures from the Sitio Conte*. Philadelphia: University of Pennsylvania Museum of Archaeology and Anthropology.

Briggs, P. S. 1989. *Art, Death and Social Order: the Mortuary Arts of Pre-Conquest Central Panama*. (BAR International Series 550). Oxford: British Archaeological Reports.

Carvajal, D. R. 1998. *Análisis de Cuatro Componentes en el Rasgo CH Excavado Mediante la Microestratigrafía: el Caso de Cerro Juan Díaz*. Graduation thesis, Universidad Nacional de Colombia, Facultad de Ciencias Humanas, Santa Fé de Bogotá (available on CD-ROM through the Smithsonian Tropical Research Institute, Panama).

Cooke, R. G. 1984. Birds and men in prehistoric central Panama, pp. 243–81 in Lange, F. W. (ed.), *Recent Developments in Isthmian Archaeology*. (BAR International Series 212). Oxford: British Archaeological Reports.

Cooke, R. G. 1985. Ancient painted pottery from central Panama. *Archeology* **July/August**, 33–9.

Cooke, R. G. 1986. El motivo del ave de las alas desplegadas en la metalurgia de Panamá y Costa Rica, pp. 139–53 in Plazas, C. (ed.), *Metalurgia Precolombina*. Banco de la República, Bogotá. (Also published without illustrations, pp. 307–18 in Reichel, E. (ed.) *Arqueología de las Américas* [Memorias del 45 Congreso Internacional de Americanistas, Bogota, 1985]). Santa Fé de Bogotá: Fondo de la Promoción de la Cultura.

Cooke, R. G. 1989. The use of anurans for food by tropical Amerindians: archaeological and ethnographic evidence. *ArchaeZoologia* **3**, 133–42.

Cooke, R. G. 1992a. Prehistoric nearshore and littoral fishing in the eastern tropical Pacific: an ichthyological evaluation. *World Archaeology* **6**, 1–49

Cooke, R. G. 1992b. Preliminary observations on vertebrate food avoidance by the Precolombian Amerinds of Panama, with comments on the relevance of this behaviour to archaeozoology and palaeoenvironmental reconstruction, pp. 59–107 in Ortiz-Troncoso, O., and T. van der Hammen, T. (eds), *Archaeology and Environment in Latin America*. Amsterdam: Instituut voor Pre- en Protohistorische Archeologie Albert Egges van Giffen, Universiteit van Amsterdam.

Cooke, R. G. 1993. Animal icons and Precolumbian society: the Felidae, with special reference to Panama pp. 169–208 in Graham, M. M. (ed.), *Reinterpreting Prehistory of Lower Central America*. Niwot CO: University of Colorado Press.

Cooke, R. G. 1998. The Felidae in Pre-Columbian Panama: a thematic approach to their imagery and symbolism, pp. 77–121 in Saunders, N. J. (ed.), *Icons of Power: Felid Symbolism in the Americas*. London: Routledge.

Cooke, R. G. this volume. Rich, poor, shaman, child: animals, rank, and status in the 'Gran Coclé' culture area of pre-Columbian Panama, pp. 271–84 in O'Day, S. J., Van Neer, W. and Ervynck, A. (eds), *Behaviour behind Bones. The Zooarchaeology of Ritual, Religion, Status and Identity*. Oxford: Oxbow Books.

Cooke, R. G. and Bray, W. M. 1985. The goldwork of Panama: an iconographic and chronological perspective, pp. 35–49 in Jones, J. (ed.), *The Art of Precolumbian Gold: the Jan Mitchell Collection*. London: Weidenfield and Nicholson, London.

Cooke, R. G., Isaza, I., Griggs, J., Desjardins, B. and Sánchez, L. A. 2003. Who crafted, exchanged and displayed gold in pre-Columbian Panama? pp. 91–118 in Quilter, J. and Hoopes, (eds), *Gold and Power in the Intermediate Area*. Washington D.C.: Dumbarton Oaks.

Cooke, R. G., Norr, L., and Piperno, D. R. 1996. Native Americans and the Panamanian landscape: harmony and discord between data sets appropriate for environmental history, pp. 103–26 in Reitz, E. A., Newsom, L. A., and Scudder, S. J. (eds), *Case Studies in Environmental Archaeology*. New York: Plenum.

Cooke, R. G. and Ranere, A. J. 1989. Hunting in prehistoric Panama: a diachronic perspective, pp. 295–315 in Clutton-Brock, J. (ed.), *The Walking Larder: Patterns of Domestication, Pastoralism and Predation*. London: Unwin Hyman.

Cooke, R. G. and Ranere, A. J. 1992. Human influences on the zoogeography of Panama: an update based on archaeological and ethnohistorical evidence, pp. 21–58 in Darwin, S. P. and Welden, A. P. (eds), *Biogeography of Mesoamerica. Proceedings of a Symposium (Mérida, Yucatán, México, October 26–30, 1984)*. Tulane: Special Publication of the Mesoamerican Ecology Institute, Tulane University.

Cooke, R. G. and Ranere, A. J. 1999. Precolumbian fishing on the Pacific coast of Panama, pp. 103–22 in Blake, M. (ed.), *Pacific Latin America in Prehistory: the Evolution of Archaic and Formative Cultures*. Pullman: Washington State University Press.

Cooke. R. G. and Ranere, A. J. in press. Influencias humanas sobre la vegetacion y fauna de vertebrados de Panamá: actualización de datos arqueozoológicos y su relación con el paisaje antrópico, in Leigh, E. G (ed.), *Ecología y Conservación en Panamá*. Panama: Smithsonian Tropical Research Institute.

Cooke, R. G. and Sánchez, L. A. 1998. Coetaneidad de metalurgia, artesanías de concha y cerámica pintada en cerro Juan Díaz, Panamá. *Boletín Museo del Oro* **42**, 57–85.

Cooke, R. G., Sánchez, L. A. and Udagawa, K. 2000. Contextualized goldwork from 'Gran Cocle', Panama: an update based on recent excavations and new radiocarbon dates for associated pottery styles, pp. 154–76 in McEwan, C. (ed.), *Precolumbian Gold: Technology, Style and Iconography*. London: British Museum Press.

Hansell, P. 1979. *Shell Analysis. A Case Study from Central Pacific Panama*. Unpublished Master's Thesis, Department of Anthropology, Temple University, Philadelphia.

Hearne, P. and Sharer, R. J. (eds) 1992. *River of Gold: Precolumbian Treasures from the Sitio Conte*. Philadelphia: University of Pennsylvania Museum of Archaeology and Anthropology.

Helms, M. W. 1977. Iguanas and crocodilians in tropical American mythology and iconography with special reference to Panama. *Journal of Latin American Lore* **3**, 51–132.

Helms, M. W. 1979. *Ancient Panama: Chiefs in Search of Power*. Austin: University of Texas Press.

Helms, M. W. 1981. *Cuna Molas and Coclé Art Forms*. Philadelphia: Institute for the Study of Human Issues.

Helms, M. W. 1993. Cosmological chromatics: color-related symbolism in the ceramic art of ancient Panama, pp. 209–52 in Graham, M. M. (ed.), *Reinterpreting Prehistory of Central*

America. Niwot: University of Colorado Press.

Helms, M. W. 1995. *Creations of the Rainbow Serpent: Polychrome Ceramic Designs from Ancient Panama*. Albuquerque: University of New Mexico Press.

Helms, M. W. 1997. The curassow and the severed leg: cross-cultural implications of ancient Panamanian ceramic motifs. *Journal of Latin merican Lore* **20**, 275–316.

Helms, M. W. 1998. The spectacled bear in iconic imagery of ancient Panama and Colombia, pp. 129–41 in Oyuela-Caicedo, A., and Raymond, J. S. (eds), *Recent Advances in the Archaeology of the Northern Andes* (Institute of Archaeology Monograph 39). Los Angeles: UCLA.

Helms, M. W. 2000. *The Curassow's Crest. Myths and Symbols in the Ceramics of Ancient Panama,* University of Florida Press, Gainesville.

Ichon, A. 1980. *L' Archéologie du Sud de la Péninsule d' Azuero, Panama*. Études Mésoamericaines - Serie II. México D.F.: Mission Archéologique et Ethnologique Francaise au Méxique.

Jiménez, M. 1999. *Explotación de Vertebrados Acuáticos y Terrestres por los Indígenas Precolombinos en Cerro Juan Díaz, Los Santos, durante el Periodo 300–700 d.C.* Graduation thesis, Escuela de Biología, Universidad de Panamá, Panama City (available through the Smithsonian Tropical Research Institute, Panama).

Jiménez, M. and Cooke, R. G. 2001. La pesca en el borde de un estuario neotropical: el caso de Cerro Juan Díaz (Bahía de Parita, costa del Pacífico de Panamá). *Noticias de Arqueología y Antropología*. Buenos Aires: Grupo NaYa, CD-ROM.

Jopling, C. (ed.). 1994. *Indios y Negros en Panamá en los Siglos XVI y XVII: Selecciones de los Documentos del Archivo General de Indias*. Woodstock, VT: C. Lutz.

Klein, R. G. and Cruz-Uribe, K. 1984. *The Analysis of Animal Bones from Archaeological Sites*. London: University of Chicago Press.

Labbé, A. J. 1995. *Guardians of the Life Stream: Shamans, Art and Power in Prehispanic Central Panamá*. Los Angeles: Bowers Museum of Cultural Art.

Ladd, J. 1964. Archaeological investigations in the Parita and Santa María zones of Panama. *Smithsonian Institution Bureau of the American Ethnology, Bulletin* **193**. Washington DC.

Linares, O. F. 1976. Animals that were bad to eat were good to compete with: an analysis of the Conte style from ancient Panama, pp. 1–20 in Young, P. D. and Howe, J. D. (eds), *Ritual and Symbol in Native South America* (University of Oregon Anthropological Papers 9). Eugene: University of Oregon Press.

Linares, O. F. 1977. *Ecology and the Arts in Ancient Panama: on the Development of Rank and Symbolism in the Central Provinces* (Studies in Precolumbian Art and Archaeology 17). Washington D.C.: Dumbarton Oaks.

Lothrop, S. K. 1937. *Coclé: an Archaeological study of Central Panama, Part 1* (Memoirs of the Peabody Museum of Archaeology and Ethnology 7). Cambridge: Harvard University Press.

Lothrop, S. K. 1942. *Coclé: an Archaeological Study of Central Panama, Part 2* (Memoirs of the Peabody Museum of Archaeology and Ethnology 8). Cambridge: Harvard University Press.

Piperno, D. R. and Pearsall, D. M. 1998. *The Origins of Agriculture in the Lowland Tropics*. San Diego: Academic Press.

Reichel-Dolmatoff, G. 1971. *Amazonian Cosmos: the Sexual and Religious Symbolism of the Tukano Indians*. Chicago: University of Chicago Press.

Reichel-Dolmatoff, G. 1975. *The Shaman and the Jaguar: a Study of Narcotic Drugs among the Indians of Colombia*. Philadelphia: University of Temple Press.

Reichel-Dolmatoff, G. 1976. Desana curing spells: an analysis of some shamanistic metaphors. *Journal of Latin American Lore* **2**, 157–219.

Reichel-Dolmatoff, G. 1977. Training for priesthood among the Kogi of Colombia, pp. 265–88 in Wilbert, J. (ed.), *Enculturation in Latin America: an Anthology*. Los Angeles: UCLA Latin American Center Publications.

Reichel-Dolmatoff, G. 1978a. Desana animal categories, food restriction, and the concept of colour energies. *Journal of Latin American Lore* **4**, 243–91.

Reichel-Dolmatoff, G. 1978b. The loom of life: a Kogi principle of integration. *Journal of Latin American Lore* **4**, 5–27.

Reichel-Dolmatoff, G. 1987. The Great Mother and the Kogi universe: a concise overview. *Journal of Latin American Lore* **13**, 73–113.

Reichel-Dolmatoff, G. 1990. *Orfebrería y Chamanismo: un Estudio Etnográfico del Museo del Oro*. Medellín: Editorial Colina.

Ridgely, R. S. and Gwynne, J. A. Jr. 1989. *A Guide to the Birds of Panama with Costa Rica, Nicaragua and Honduras* (second edition). Princeton: Princeton University Press.

Sánchez, L. A. and Cooke, R. G. 1998. ¿Quién presta y quién imita?: orfebrería e iconografía en 'Gran Coclé', Panamá. *Boletín del Museo del Oro* (Colombia) **42**, 87–111.

Wing, E. S. 1980. Aquatic fauna and reptiles from the Atlantic and Pacific sites, pp. 194–215 in Linares, O. F. and Ranere, A. J. (eds), *Adaptive Radiations in Prehistoric Panama* (Peabody Museum Monographs 5). Cambridge: Harvard University Press.

Richard Cooke
Smithsonian Tropical Research Institute,
Apdo. 2072, Balboa, Ancón, Panama,
República de Panamá
E-mail: cooker@naos.si.edu or cominata@hotmail.com

9th ICAZ Conference, Durham 2002
Behaviour Behind Bones, (ed. Sharyn Jones O'Day et al.) pp. 128–143

16. Identifying ritual use of animals in the northern American Southwest

Robert J. Muir and Jonathan C. Driver

Ethnographic data suggest that numerous species of vertebrates were incorporated into the ritual and ceremonial life of the Puebloan peoples of the American Southwest. We use spatial and contextual analysis of faunal remains from prehistoric sites to examine this long-standing tradition. Through comparison to ethnographic accounts of Puebloan ritual, we argue that non-random distributions of faunal remains, coupled with associations of certain species with certain contexts, suggest that deposition of some species was influenced by ideological considerations. Specifically, some faunal deposits represent intentional 'ritual interments' of animals while others are accumulations of 'ritual refuse' resulting from repeated ritual activities associated with communal deer hunting and other activities. Examination of temporal changes in the location and frequency of such deposits suggest that the scale and nature of ritual activities are strongly correlated with changes in settlement patterns. We conclude with a discussion of the implications of our findings with respect to the role of ritual in the development and organization of large aggregated Pueblo communities.

Introduction

The role of ritual in the development of complex societies has become the focus of considerable recent archaeological discourse and research (*e.g.*, Aldenderfer 1993; Kolb 1994; Hayden 1995; Grier 1996; Kuijt 1996). In the American Southwest archaeologists have turned to examining ritual as a behavior that can be developed, controlled, and manipulated for social, political, and economic purposes and was instrumental in the formation of large aggregated settlements and integration of multi-settlement communities (*e.g.*, Potter 1997; Muir 1999a; Muir 1999b; Muir 2000; Potter and Perry 2000; Walker and Lucero 2000; Walker 2001). These investigations have incorporated examination of settlement patterns, architecture, ceramics, faunal remains, and other datasets. Given the ample ethnographic data indicating that numerous species of animals were incorporated into the ritual and ceremonial life of the Puebloan peoples of the American Southwest, we believe that zooarchaeological analyses have great potential to contribute to this field of inquiry. In this paper we examine zooarchaeological evidence for ritual use of animals at prehistoric Pueblo sites in the Northern San Juan region of Colorado and discuss the role of ritual in the development and organization of large aggregated communities.

While the term 'ritual' has been defined by others to encompass a broad range of human and animal behaviors (Rappaport 1999, 24), we restrict it to the physical expression of symbolic relationships and spiritual beliefs through formal behavioral routines. This usage excludes instinctive 'display' behaviors of animals and pathological neurotic/compulsive behaviors of humans from the realm of ritual. While many aspects of Pueblo life could be described as 'ritualized', we specifically focus on those that we believe should be most apparent in the archaeological record. These include highly ritualized special events and ceremonies that would have involved elaborate paraphernalia, sacrificial offerings, or consistently repeated formalized rites. These may include both large scale communal events or smaller scale 'private' rituals.

The complexity and abstract nature of ritual activities present obvious difficulties in determining their nature, context, function, and meaning. As physical manifestations of ideology and symbolic relationships, ritual activities may vary significantly cross-culturally and be difficult or impossible to understand precisely. However as argued and illustrated by Walker (1998) archaeologists do not need to diagnose symbols in order to be able to identify sacred spaces or learn about ritual site formation processes. Cultural deposits resulting from ritual activities, whatever their purpose and meaning, can be distinguished from those resulting from other types of human behavior by nature of their physical, rather than symbolic, variability (Walker 1995; Walker 1998, 251; Walker and Lucero 2000). We discuss patterning among deposits of faunal remains which we believe reflect ritual activity. These deposits are primarily distinguished by an abundance of species not commonly found among more ubiquitous refuse midden deposits. While this characteristic is not *by itself* necessarily indicative of ritual behavior, the physical context and predominant taxa present among these assemblages are in all cases consistent with ethnographic accounts of ritualized procurement and use of animals in dances, ceremonies, and other ritual society activities among historic Pueblo communities.

Background: study area and prehistory

The Northern San Juan region is situated in the northern portion of the Colorado Plateau, which extends throughout southern Utah, southwestern Colorado, and northern Arizona and New Mexico. The region is defined by the San Juan River to the south, Cedar Mesa to the west, the La Plata and Abajo Mountains to the north and the Animas River to the east. The local landscapes feature broad plateaus, mesas, and deeply entrenched canyons. Modern environmental zones range from desert grasslands, to ponderosa pine-Douglas fir forests.

Prehistorically the region was occupied by semi-sedentary subsistence horticulturalists known as the Anasazi. The Anasazi are one of the three major prehistoric cultural groups of the American Southwest ancestral to modern Pueblo people. Syntheses, review, and discussion of the culture history of the Anasazi in the northern San Juan region are abundant (*e.g.*, Cordell 1984; Cordell and Gumerman 1989; Rohn 1989; Minnis and Redman 1990; Varien 1999; Lipe *et al.* 1999). The later phases of Anasazi prehistory in the northern San Juan consist of three periods, Pueblo I, II and III, dating from about 750 to 1275 AD. After 1275 the region was completely depopulated, probably through mass migrations to the south. For all periods the Anasazi were subsistence farmers, relying heavily on maize supplemented by beans, squash, domestic turkey and wild plants and game.

During Pueblo I through III times the basic settlement unit was the "unit pueblo", consisting of a few contiguous masonry rooms and a subterranean structure, usually to the south of the rooms. In Pueblo I times this structure was a fairly shallow pithouse, which seems to have had a domestic function. Over the centuries this was modified to become a deep, circular masonry-lined structure. For many years archaeologists have referred to these as "kivas", a term that describes similar structures used primarily for ritual purposes in modern Puebloan communities of the Southwest. It is likely that modern kivas are derived from prehistoric kivas, because they share similar traits and because there is evidence that ritual activity took place in prehistoric kivas. However, there are some important differences. In prehistoric sites there seems to have been one kiva per family, whereas in modern pueblos there are only a few kivas in villages with populations in the hundreds. In prehistoric Anasazi sites there is evidence for other kinds of structures that may have served community-wide ritual needs, including over-sized "great kivas", masonry towers, and a variety of multi-walled structures (including the D-shaped structures referred to later).

Unit pueblos were typically scattered across the Southwestern landscape as small hamlets. It is assumed that these were components of larger communities, often centred around a community structure such as a great kiva. At certain times and locations the scattered unit pueblos tended to aggregate into more densely packed settlements, although individual unit pueblos can usually be discerned.

Modern puebloan people occupy tightly clustered villages (to which the colonizing Spanish applied the term "pueblo"). Perhaps more than any other native American groups they have successfully maintained their cultural independence from Spanish and American influences through secrecy, an elaborate system of religion and ritual, and passive resistance. Social organization of the historic and recent pueblos does not fit easily into the evolutionary schemes favored by archaeologists of the 1970's and 1980's. Pueblos were not controlled by chiefs, but neither were they egalitarian. Influential individuals seem to have exerted control of different aspects of village life, and these individuals were often the heads of sodalities concerned with dancing, healing, religion, hunting and warfare. In spite of this evidence for what have been termed heterarchical systems of organization, there is ongoing debate regarding the complexity of sociopolitical, ideological and economic systems in ancestral Pueblo communities. While some scholars believe that southwestern cultures were largely egalitarian throughout the prehistoric period (Graves *et al.* 1982; Graves and Reid 1984; Reid 1985; Johnson 1989; Reid and Whittlesey 1990; Brandt 1994), others see evidence of social stratification at several times and places (Upham 1982; Upham and Plog 1986; Lightfoot and Upham 1989, Wilcox 1991; Lipe 1995; Larson *et al.* 1996; Lekson 1999). Much of this debate has focused on the "Chacoan phenonmenon" of the Pueblo II period (*c.* AD 900–1150) which, to some researchers, represents the high point of social complexity in the northern American Southwest (Irwin-Williams 1972; Vivian 1990; Lekson 1999). Recently attention has turned to large aggregated communities common to the later Pueblo periods (Pueblo III and IV) and the social and economic systems that produced them. Our research has focused primarily on the Pueblo III period.

At the beginning of the Pueblo III period in the northern San Juan there is a shift from a settlement system consisting of many small, highly dispersed habitation sites to increasingly clustered settlements centered around larger central pueblos (Rohn 1989, 158; Adler 1992). There is also evidence for an overall population increase from Pueblo I through Pueblo III times (Dean *et al.* 1994; Varien 1999; Duff and Wilshusen 2000). Collectively these changes resulted in significant increases in local human population density. Our previous research suggests that these settlement changes were accompanied by changes in the scale and organization of hunting practices and ritual activities (Muir 1999a; Muir 1999b; Muir and Driver 2002).

Animals and ritual in the American Southwest

Animals feature prominently in the oral histories, cosmogony, and symbolism of historical Pueblo societies and the extensive use of animals in ritual and ceremony is well documented ethnographically (Tyler 1964; Tyler 1975; Tyler 1979; Gnabasik 1981). That such ritual use of animals was also a part of prehistoric Pueblo culture is readily evident from iconography, artifacts, and mortuary remains (E. Hill 2000, 362). The ritual use of animals by Pueblo peoples includes their use as raw materials for costumes and other ritual paraphernalia, as food for specific ritualized events, and for ritual sacrifice.

The ethnographic record of the American Southwest is rich with references to ceremonial activities that involved the use of costumes, masks, ornaments, musical instruments and other paraphernalia made primarily from skins, furs, feathers and bones of a variety of animals (Gnabasik 1981). The hides, horns/antlers, bones and hoofs of deer and other artiodactyls are all noted as having been used as dance 'equipment' at various Pueblos. Costumes were made from hides, masks were frequently adorned with antlers or horns, and musical instruments were commonly made from 'deer leg bones', scapulae, and hoofs. The pelts of many carnivores were also used as ceremonial costumes, robes and masks. In some cases cranial bones, phalanges and caudal vertebrae may have remained attached to the skins (Gnabasik 1981, 41–2). The wings and/or feathers of a wide variety of birds are repeatedly noted as having been essential components of costumes, masks, prayer-bundles, prayer-sticks and other ritual items (Ladd 1963; Gnabasik 1981; Akins 1985). Rattlesnakes were used for specific ritual purposes, including as altar paraphernalia (Stevenson 1894, 77) and as live participants during 'snake handling' ceremonies (Gnabasik 1981, 231). Other live animals were also used similarly, as indicated by Lange's (1959, 360) account of the Santiago's Day ceremony at Cochiti where squirrels were captured, brought to the center of the village and then released.

Many animals are also documented as having been processed and consumed in ritual contexts. These predominantly include mammals such as deer, pronghorn, cottontails and jackrabbits (Gnabasik 1981). Ceremonies during which these animals were consumed included community wide activities, such as the all-night summer solstice ceremony at Laguna (Parsons 1920, 59), as well as more exclusive rituals held by one or two select members of specific societies, such as the tending of the warrior society scalps at San Felipe (White 1932, 13; White 1974a, 37).

The procurement of animals used in ceremonial contexts was usually a highly ritualized activity, conducted by selected members of the community (hunting or war society members) under the supervision of ritual specialists (Anell 1969; Parsons 1918; Parsons 1920; Parsons 1921; White 1932; Underhill 1946; White 1974a; Parsons 1977). This is well documented by numerous accounts of communal rabbit, deer and antelope hunts. Communal rabbit hunts were commonly held to provide a supply of meat for ritual specialists for ceremonial purposes (*e.g.*, Parsons 1918, 173; Parsons 1921, 162; White 1932, 52; White 1974a, 31–40; Anell 1969, 61; Parsons 1977, 70). For example, Whitman (1947, 137–8) observes that at San Ildefonso the war captain and his assistants would formally organize and conduct a rabbit hunt in order to obtain a supply of meat for dancers during the spring *Tede Share* ceremony. Similarly at Zia, deer hunts in which only the men could participate, were held at the request of the war captain to supply meat to the *cacique* (ritual specialist) for ceremonial purposes (White 1974b, 301–2), and at Jemez the mountain lion society was principally responsible for providing the *cacique* with deer and rabbit meat (Parsons 1977, 70).

Deer hunting ritual

As an example of the pervasive presence of ritual activity in the economic life of the puebloans, we can consider ritual aspects of deer hunting, as well as the role of deer in ceremonial and ritual activities. Unless otherwise noted, the following data are from summaries by Tyler (1975) and Gnabasik (1981).

Preparation for deer hunting involved songs, sexual abstinence, the manufacture and deposition of offerings, prayers and other special behaviors. Songs were a way of communicating with the spirit world to induce the appropriate beings to supply game; they were also a way of attracting deer to the village. At Isleta the hunt chief was secluded for four days before a communal hunt and attempted to draw deer to the village by blowing smoke towards the mountains. Many puebloan hunters made prayer sticks that were deposited at shrines inside or outside the village. At Santa Ana a special mixture of minerals was prepared to place in deer tracks. At Taos hunters addressed prayers to a minor deity, "dirt boy" who inhabited a pit near the house door; dirt boy was related to deer and helped hunters find game (Parsons 1970, 103).

Deer carcasses were treated with care after death. At the kill site the animal was oriented with its head towards the pueblo (Parsons 1918, 186; Parsons 1920, 66). Hunters carried small fetishes representing the mountain lion, and these were "fed" on deer meat or blood (Parsons 1970, 103; White 1974b, 303). (Much larger mountain lion images were incorporated into shrines or altars visited before the hunt). Hunters from Zia spread corn meal over the freshly killed deer (White 1974b, 303).

Once the deer had been returned to the pueblo it was usually laid out in the hunter's house and dressed or decorated with various items including corn meal or pollen (most communities), beads and bells (Laguna), dresses and belts (Santa Ana), feathers (Taos) or a blanket (Zuni). After some communal hunts all the prey might be taken to the *cacique's* house for butchering and distribution, but ceremonial treatment of the hunter's portions (including the head and hide) would still occur. Deer skulls were often placed on the roof of the house. In some cases deer bones were kept away from dogs, or deposited at shrines (Parsons 1929; Cordell 1977, 458).

As well as being the focus of ritual, deer also supplied meat, hides, antler and hooves for consumption or use at ceremonies. Communal hunts supplied meat for ceremonial purposes. Deer were associated with rain, and lived in the mountains – the place where clouds gather. Killing deer and bringing them to the village was thought to encourage rain to come, and hunting songs sometimes incorporated a plea for rainfall. Tyler has outlined the relationship between creation myths, deer, people, agriculture and the spirit world. Given this intimate and complex relationship, it is not surprising that artifacts made from deer featured in many ceremonies, especially the kachina dances.

Kachinas are spirit beings who used to visit the pueblos to bring rain, food and entertainment. A long time ago they ceased their visits, but left their costumes for people to wear (White 1935). The costumes included elaborate masks, which disguised participants from other members of the village. There were a great many kachina spirits, some of whose masks include deer antlers or whose costumes incorporate deer hide or hooves. At Zuni some kachinas were thought to die and be transformed into deer. It is reported that at the end of the very sacred Shalako ceremony the kachinas are pursued by young men; when a Shalako kachina was caught it indicated that the catcher would have successful hunting.

These abbreviated examples show that deer were closely related to the spiritual life of pueblo communities. Similar examples can be found for other artiodactyls, such as pronghorn antelope and bighorn sheep. We should therefore be prepared to find evidence for ritual treatment of artiodactyls in ancestral puebloan sites.

Identifying ritual through zooarchaeology

As illustrated by the examples above, the ethnographic record provides ample evidence of ritual use of animals by Puebloan peoples and it is reasonable to assume that such activities are represented among the archaeological remains of ancient Pueblos. The identification of these activities requires that we distinguish deposits that are the result of ritual activities from those that are the result of more mundane and utilitarian human behavior. In this paper we discuss three general types of deposits: 1) common refuse, 2) ritual interments, and 3) ritual refuse. These are idealized constructs and it is important to note that in reality these deposits may be distorted, overlapping, or mixed in the archaeological record. We also recognize that these three types do not encompass all animal bone deposits potentially present at a site, nor do we expect that *all* animal remains associated with ritual activities will be deposited in a manner which is distinctly different from that of common refuse.

The first type, 'common refuse', refers to deposits of animal remains that are the product of common daily subsistence activities. These are most often mixed with other 'domestic' refuse (broken pots, exhausted tools, etc.) and represented among high density midden deposits distributed within abandoned structures and throughout non-structural areas within a given site. They may also, however, include lower density refuse accumulations in activity areas such as animal processing, tool making or cooking areas. The second type, 'ritual interment' includes deposits of animal remains that are the direct result of intentional ritualized human activities, involving deliberate disposal of animal remains. This includes animals that are interred in prepared pits or placed in specific locations, and are occasionally accompanied by grave goods (E. Hill 2000, 364). Such deposits are often readily identified during excavation and can be interpreted in detail with reference to stratigraphic, architectural and other contextual information (*e.g.*, J. D. Hill 1995, J. D. Hill 1996; Walker 1998; E. Hill 2000). The term 'ritual refuse' is used here to refer to deposits that include abandoned, lost, or informally discarded remnants of animals that were procured in ritual contexts or used for ritual purposes. This type of deposit is generally more difficult to identify during excavation and is less readily interpreted in terms of specific associations and stratigraphic context.

Ritual interments

Deposits consisting of virtually complete skeletons or unprocessed articulated elements are the most readily apparent indications of ritual interment of animals. Unusual (improbable or odd) groups of elements or isolated crania are also often cited as evidence of ritual activity. While such remains are not inherently diagnostic of ritual, when other contextual information is taken into consideration a strong argument for ritual activity can often be made. The occurrence of such anomalous remains among archaeological deposits has been interpreted as evidence of ritual interment by several researchers. So-called 'special animal deposits' have been interpreted as evidence of ritual interment on the basis of formal patterning (or 'structured deposition') through space and time (J. D. Hill 1995; J. D. Hill 1996). For example, Stallibrass (1996) argues that odd groups of cattle remains (crania, hides and feet) found in an odd location (the middle of a peat bog) is evidence of ritual interment, based on their similarity to contemporaneous human burial practices. Hayden (1997, 98) cites numerous examples of dog crania or complete skeletons situated in the center of ancient pithouses as evidence of abandonment ritual. Driver (1999) has suggested that two raven skeletons found among the Palaeoindian component of the Charlie Lake Cave site in northeastern British Columbia may be evidence of ritual activities, on the basis of their location, condition, and association with artifacts.

In the American Southwest animal interments are commonly found at archaeological sites. Erica Hill (2000) has documented numerous incidents of virtually complete animal skeletons from Pueblo sites, suggestive of ritual interment based on the location and condition of the remains. These include single or multiple interments of domestic dogs, raptors, turkey, and psittacids (macaws and parrots). The remains are distinguished as 'ritual interments' due to the consistent formal structure of the deposits. While some may be the result of systematic disposal of animals that have served a ceremonial purpose, Hill (2000, 389) argues that the context and condition of many of the interments are evidence of ritual sacrifice of animals as dedicatory offerings, as indicated by decapitation, placement in the center of abandoned structures, and/or burial within house floors.

Ritual refuse

The discovery of animal species that are unlikely to have been hunted for food often raise suspicions about the social context of the remains. Much like non-utilitarian artifacts (*e.g.*, figurines and beads), such animals are often assumed to be symbols of status, deities, or identity, and possibly the material remains of ritual activity. In the American Southwest 'wild birds', particularly raptors, and wild mammalian carnivores fall into this category.

There are few ethnographic references that indicate that wild birds were eaten and there is general consensus that falcons, hawks, eagles were not dietary items (Henderson and Harrington 1914; Beaglehole 1936; Vivian and Mathews 1965). However, wild birds are commonly mentioned in ethnographic literature in association with spiritual beliefs and their body parts, particularly their feathers, were often used in ceremonies. As argued by Akins (1985, 381):

> "Probably the most important use of wild birds was to provide feathers for ceremonies... Prayer feathers were important parts of masks and ceremonial costumes, used in altar arrangements, and placed in fields at planting time. Prayer sticks were specially made and expended within a few days of manufacture (Judd 1954), producing a perpetual need."

Though it is improbable that the use of feathers will be directly represented archaeologically, the skeletal remains of birds procured for the purpose of obtaining feathers may be. In some cases entire bird wings were used in ceremonies (Gnabasik 1981, 179). Presumably this would include articulated skeletal elements as well as bird skins and feathers. It has been observed that birds such as kestrels, macaws, and parrots, were sometimes kept as pets or as a supply of feathers (Tyler 1979, 198; Gnabasik 1981, 263).

Similarly, wild mammalian carnivores were not common sources of meat but rather were primarily sought for their pelts and/or ritual use. In particular, the crania, claws, and pelts of these animals were used as raw materials for ceremonial costumes and altar paraphernalia (Gnabasik 1981).

While deliberate interment of these animals may be the most obvious indication of their ritual significance, more subtle evidence may also be present. Processing, storage, handling, and use of these animals is likely to result in their remains being deposited under less deliberate circumstances. A 'residue' of ritual refuse may gradually accumulate in areas, which are common places for storage, manufacturing, or use of ritual paraphernalia. Such deposits may consist of concentrations of relatively rare taxa, bone fragments from broken ritual paraphernalia, and/or discarded bone waste from artifact manufacturing.

A second type of ritual refuse involves animals which may have been a staple component of the diet, but which were treated with reverence and ritual due to their economic, social, or ideological value. In the American Southwest such treatment was especially common for artiodactyls, particularly deer, as discussed above. For example, among the Papago, Underhill (1946, 105) observed ritualized 'quarantine' of deer remains:

> "After the meat was eaten, no bones might be thrown on the ground lest a dog should carry them to a house of a menstruating woman. If this should happen, no more deer could be caught. All bones must be placed on the ceremonial house, where they remained for a year. Even those who took meat home to other members of the family must first remove the bones."

Similarly, at Zia the postcranial bones of deer obtained during communal hunts were discarded outside of the pueblo and protected from ravaging by dogs, while the head and horns were ceremonially 'dressed' and displayed on the roofs of the houses of the hunters (White 1974b, 303–4). As noted by Gnabasik (1980, 47–8) such behavior would influence bone deposition and preservation:

> "...deer bones and eventually even the skull are deposited or buried outside the pueblo proper. With the separate treatment of the skull, deer skulls should not be found with any other deer bones, especially since the post-cranial skeleton seems to have at least been tossed on the midden or possibly in a special place of it or near it, while the skulls (and antlers?) are buried further away from the pueblo, probably at or near a shrine. The presence of the skulls with their antlers on the house tops is to be noted, as is the claim that they are only taken out and buried when they are very old and probably deteriorating."

Such treatment of remains should influence the distribution of faunal remains, and may produce distinctive patterning if ritual activities were persistent. Concentrations of large game remains in locations other than cooking or processing locations may be indications of ritualized quarantine or display of animals obtained during communal hunts and/or for specific ceremonies. While the majority of remains may ultimately be discarded amongst other refuse, some specimens may be overlooked during clean-up and a 'residue' of ritual refuse may remain.

Ritualized treatment of the bones of smaller game such as cottontail and jackrabbit is *not* documented ethnographically. Most accounts suggest that the majority or all of the meat obtained during communal rabbit hunts was taken to the office of the *cacique* or a particular society, where it was stored for ceremonial use, while the bones were commonly tossed out with other domestic trash (Gnabasik 1981, 108-24). In at least three communities (Cochiti, Laguna, and Zuni) rabbit meat was not eaten at all by the leaders of the hunting society as it was considered poisonous to them (Lange 1959, 130, 271–2).

Zooarchaeological evidence for ritual use of animals

While other researchers have used detailed stratigraphic/contextual analyses to identify ritual deposits on the basis of structured depositional patterning (*e.g.*, J. D. Hill 1995; J. D. Hill 1996; Walker 1998), we have taken a different approach. In our analyses we have focused on identification of *compositional patterning* (Ferring 1984) through comparisons of depositional contexts. This has included examination of the composition of faunal remains

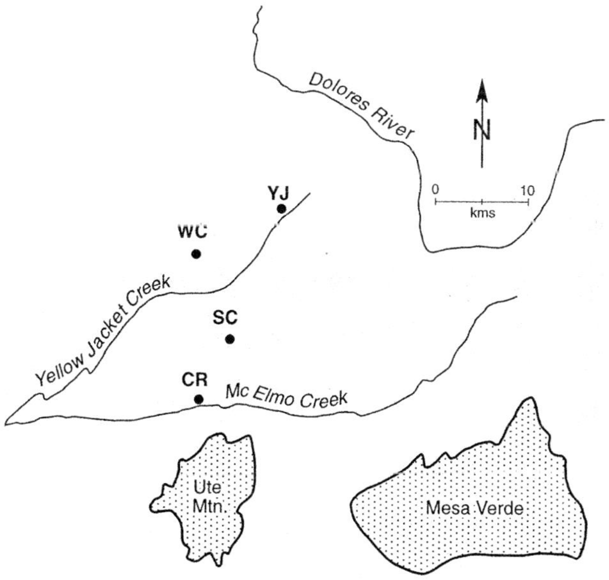

Fig. 1. Map of the central portion of the northern San Juan Region, indicating the location of Castle Rock (CR), Sand Canyon (SC), Wood Canyon (WC), and Yellow Jacket (YJ) pueblos. The small sites discussed in the text are all located within a few kilometers of Sand Canyon and Castle Rock pueblos.

deposits according to architectural block, structure type, and deposit type (*e.g.*, midden, floor, room fill, roof collapse, etc.). Though we believe that detailed stratigraphic analyses may have also been valuable in the identification of ritual deposits, for several reasons we have not included this approach in our analyses. First, many of the sites we are dealing with were occupied for relatively short periods of time (i.e., 30 to 50 years), consequently there is little opportunity to look for consistent and repeated structured depositional patterning through time. Second, while excavators have meticulously described and interpreted each deposit that they encountered, in many cases the specific disposition or stratigraphic context of individual faunal remains has not been documented in detail. Finally, examination of compositional patterning is readily applicable to comparisons between site assemblages; as it focuses on identification of very general, 'robust' patterns it is unlikely to be affected by minor variations in excavation methods and stratigraphic interpretation.

Our analyses have included examination of faunal assemblages from seventeen 12th and 13th century AD Pueblo III sites in the northern San Juan Region (Fig. 1). These include four large sites: Sand Canyon, Yellow Jacket, Woods Canyon, and Castle Rock pueblos, and thirteen smaller sites (Fig. 2). The large sites each consist of an aggregated community of multiple households ranging in momentary population from about 100 to 700 individuals (Adler 1992, Bradley 1992; Lipe 1992, 124;

Site Name* (site number)	Estimated Number of Structures			Great Kivas	Plaza	Biwalled structures	Faunal Remains: Total NISP
	Rooms	Kivas	Towers				
Large Sites (pueblos)							
Yellow Jacket [1] (5MT5)	600–800	192	18	1	yes	Biwalled tower	9132
Sand Canyon [2] (5MT765)	421	93	12 to 14	1	yes	D–shaped structure	17628
Woods Canyon [3] (5MT11842)	unknown	50	9		yes	D–shaped structure	2184
Castle Rock [4] (5MT1825)	40–75	13–16	3		yes	D–shaped structure	2485
Small Sites (homesteads and hamlets)							
Green Lizard [5] (5MT3901)	20	2	0				3580
Lillian's Site [6] (5MT3936)	7–10	1	1				435
Kenzie Dawn Hamlet [6] (5MT5152)	6–10	2	0				1475
Lookout House [6] (5MT10459)	5–10	2	1				517
Catherine's Site [6] (5MT3967)	5–7	2	1				844
Roy's Ruin [6] (5MT3930)	5–7	1	1				184
Shorlene's Site [6] (5MT3918)	5–7	1	1				206
G and G Hamlet [6] (5MT11338)	3–5	1	0				152
Lester's Site [6] (5MT10246)	3–5	2	0				379
Stanton's House [6] (5MT10508)	3–5	1	1				2252
Saddlehorn Hamlet [6] (5MT262)	3–5	1	1				621
Mad Dog Tower [6] (5MT181)	1	1	1				37
Troy's Tower [6] (5MT3951)	0	1	1				159

*Fig. 2. List of sites discussed in this article, indicating estimated numbers of architectural structures present and faunal assemblage sample sizes. * References for architectural data: 1) Ortman* et al. *2000; Kuckelman 1997, 2) Lipe and Ortman 2000; Bradley 1992, 3) Lipe and Ortman 2000, 4) Lipe and Ortman 2000; Varien and Kuckelman 1999, 5) Huber and Lipe 1992, 6)Varien and Kuckelman 1999.*

Lipe 1995). Each of these sites is believed to have been the center of a larger residential/economic community, encompassing many smaller pueblo homesteads and hamlets located nearby. The small sites that we have examined are all located within a few kilometers of Sand Canyon and Castle Rock Pueblos.

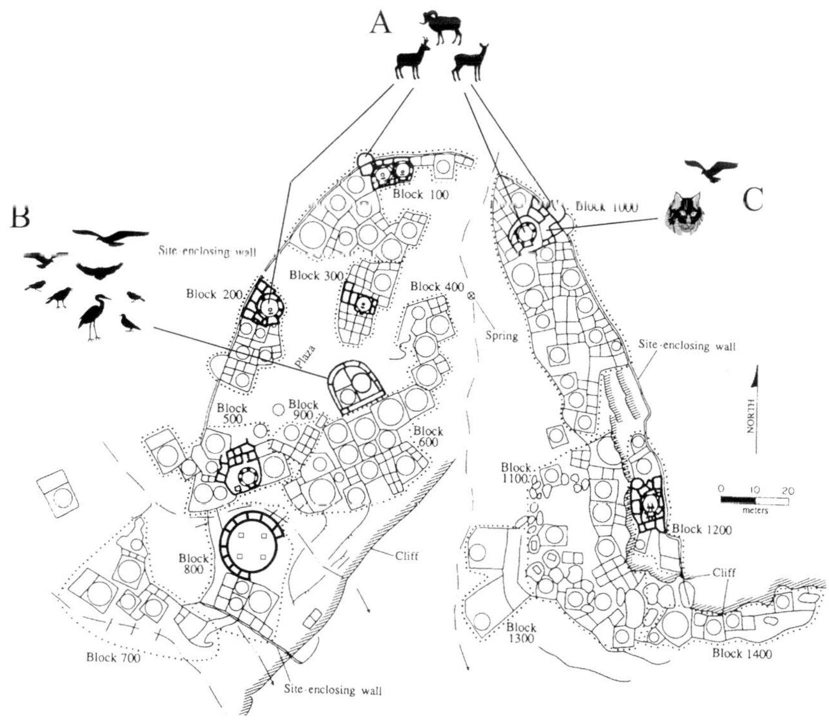

Fig. 3. Plan map of Sand Canyon Pueblo (after Bradley 1992, 80) indicating locations of 'ritual refuse' and 'ritual interment' deposits. Bolding indicates extensively excavated structures.

Sand Canyon Pueblo

Sand Canyon Pueblo (Fig. 3) is a very large late Pueblo III site (occupied *c.* AD 1250–1280) located in the central portion of the northern San Juan Region. It is situated at the head of a small canyon that divides the pueblo into roughly equal eastern and western portions. The site consists of an estimated 420 rooms; 90 kivas; 14 towers; an enclosed plaza; a D-shaped, biwalled structure; and a great kiva (Bradley 1992, 79). Most of the structures are enclosed within a masonry wall that arcs across the west, north, and eastern margins of the site.

Detailed spatial analyses of the faunal remains from Sand Canyon Pueblo have been conducted by Muir (1999a). This involved examination of patterning of faunal remains deposits through contingency (two-way cross-tabulation and Pearson's chi square analyses), K-means cluster, and diversity analyses. Collectively, these analyses revealed that the composition of faunal remains deposits was largely uniform throughout the Pueblo. In particular midden deposits display remarkable consistency. All of the midden deposits are clearly and consistently dominated by turkey (*Meleagris gallopavo* Linnaeus) remains, but also contain significant quantities of lagomorphs. Both turkey and lagomorphs are documented as having been primary sources of meat among many historic Pueblos (Gnabasik 1981). Small quantities of other taxa, including artiodactyls are also consistently present among the midden deposits, but those taxa that were probably not commonly consumed as food (i.e., wild birds and carnivores) are rare and unevenly distributed. Many instances of roof collapse, room-fill, and floor deposits also conform to this general pattern, though the overall density of remains tends to be lower among these latter contexts.

The middens and other similar deposits presumably represent 'common refuse' accumulations from multiple activity areas throughout the site. They include deposits located within the central plaza, adjacent to room blocks, outside the site-enclosing wall and along the lower slopes of the central drainage. Despite these disparate origins the composition of the deposits are all remarkably similar (Muir 1999a, 90). These distributions are consistent with what would be expected of animal remains resulting from domestic activities on a daily basis, while a more discrete distribution would be expected of those resulting from special events or ceremonies (Muir 1999a, 110).

However, the spatial analyses also revealed some anomalous compositional patterns. Among these are three that we believe are consistent with ritual interment and/or ritual refuse accumulations. They include.

1. concentrations of artiodactyl remains within architectural blocks containing D-shaped towers (blocks 100, 200, and 1000) and particularly high concentrations of artiodactyl remains among roof fall deposits of the towers themselves (Fig. 3, A);
2. an abundance of wild bird remains associated with

the D-shaped, biwalled structure located near the center of the site (Fig. 3, B); and
3. a localized concentration of *Lynx* and birds of prey remains within the courtyard of block 1000 (Fig. 3, C).

We believe that the first two patterns are examples of 'ritual refuse' accumulations while the latter may be the result of 'ritual interment'. Pearson's chi square analyses indicate that all of these patterns are statistically significant (i.e., $p < 0.05$, using Yates' correction for sparse samples where appropriate).

The high frequency of artiodactyl remains associated with the D-shaped towers was apparent in the contingency analysis; however, the cluster analysis further clarified the nature of this distinct pattern. Relatively high concentrations of artiodactyl remains are clearly associated with the roof fall deposits of both D-shaped towers which were investigated. In contrast the interiors (deposits found below wall and roof fall) of these structures display relatively few such remains. It is notable that at least three species of artiodactyl: deer (*Odocoileus* sp. Rafinesque), pronghorn antelope (*Antilocapra americana* Ord), and bighorn sheep (*Ovis canadensis* Shaw) are evident among the remains from both structures. Artiodactyl concentrations are also found within (i.e., among deposits below wall and roof fall) the kivas associated with the 'tower blocks' (blocks 100 and 1000). Collectively this information indicates that the pattern is not simply due to the chance deposition of a single animal on the surface of the site after abandonment. Clearly the concentration of remains is the result of consistent and repeated human activity. These concentrations are by far the most prominent and intriguing of the identified spatial patterns.

The distribution of artiodactyl remains is suggestive that the 'tower blocks' were associated with communal hunting and feasting activities, perhaps acting as hunting or war society houses (or offices). As indicated in ethnographic accounts (e.g., Parsons 1977, 70; White 1974b, 303), apportionment of game obtained during communal hunts was usually controlled by specific individuals or societies. Ritual specialists, and society heads received specific portions or a prescribed number of animals. Also, society 'offices' were used for storage of meat obtained during these hunts. It seems likely that a disproportionate quantity of game, such as that identified in association with the 'tower blocks', would be found in or near the 'offices' or homes of such societies or individuals. The concentrations of artiodactyl remains on the roofs of the towers within blocks 100 and 1000 are particularly consistent with ethnographic references to the ritualized display of animals on roof tops after a communal hunt (cited above).

Our analyses also indicated that an abundance of wild bird remains clearly distinguish the D-shaped, biwalled structure from all other architectural units at the site. The remains from the D-shaped structure were found in association with all major context types (Fig. 4), and account for 64% of the wild bird remains recovered. At least 8 species of wild bird are present and potentially more are represented by the general categories Corvidae; small passeriform; small bird; and falconid. Only 12 species of wild bird were identified among the entire excavated faunal assemblage from Sand Canyon Pueblo.

Given the architecturally unique nature of the biwalled structure it would not be surprising if it served as a focal point for community ceremonies. The abundance of wild bird remains may reflect the use of this structure for preparation and storage of costumes, masks, prayer sticks, prayer bundles and other similar items. Although it is recognized that the ethnographic data suggest that feathers were the primary bird elements incorporated into such items, entire wings were sometimes used and the skeletal remains of birds obtained for their feathers are likely to be found within or near locations where ritual paraphernalia is prepared and stored.

Though it is probably impossible to determine the true ritual significance of the bird remains, the ethnographic record provides a basis for speculation. All avian taxa recovered from the D-shaped structure are documented as having had specific spiritual associations among historic Pueblos (Ladd 1963; Tyler 1979, Gnabasik 1984). Though a wide array of associations are represented by these taxa, some common themes are apparent. A number of the taxa are documented as having been associated with water, specifically rain, including: mourning dove (*Zenaida macroura* Linnaeus), sandhill crane (*Grus canadensis* Linnaeus), raven (*Corvus* sp. Linnaeus), and owl (Strigiformes). Of these the relationship between the mourning dove and rain is most firmly documented. Tyler cites various ethnographic accounts of traditional Puebloan stories in which doves, particularly the mourning dove, invoke rain or are indicators of rain pools and springs (Tyler 1979, 105–12). Similarly Tyler notes that the sandhill crane is associated with the "Rain Bird" kachina (*kwapiye ohuwah*) described as "a bird that comes only with continuous rain" (Henderson and Harrington 1914, 46, cited in Tyler 1979, 129). Flocks of ravens and crows are also associated with rain clouds, which are in turn associated with kachina spirits "passing over the villages occasionally to bring rain" (Tyler 1979, 173). Finally, owls have a fairly tenuous association with rain indicated only by a rain making game involving the use of owl feathers documented by Parsons (1939, 774, cited in Tyler 1979, 164).

It is notable that these four taxa also have spiritual links to agriculture. According to Tyler (1979, 106) doves are associated with winnowing of grain due to their seed eating habits. The sandhill crane (*Grus canadensis* Linnaeus) is associated with the harvest, specifically as a guardian of harvested corn, but also as a bringer of seeds (Tyler 1979, 128–9). Crows and ravens can also be thought of as birds of agriculture due to their habit of 'joining in on the harvest' (Tyler 1979, 173), and owls, (particularly

Taxon: Common Name	Context						Total NISP
	Kiva Floor	Kiva Roof	Room Floor	Room Roof	Midden	Other*	
Poor–will	–	–	–	1	–	–	1
Mourning Dove	–	–	–	1	–	2	3
Sandhill Crane	–	2	–	–	–	–	2
Raven	–	–	–	–	4	6	10
Jay, Crow, Raven	–	–	3	3	8	16	30
Small Passeriform	–	2	–	1	6	–	9
Small Bird	–	4	–	1	–	10	15
Great Horned Owl	1	–	–	–	–	–	1
Owl		–	–	1	–	–	1
Hawk	–	–	–	–	7	–	7
Falconid	–	1	–	–	1	2	4
Turkey Vulture	–	–	8	–	–	1	9
Totals	1	9	11	8	26	37	92

*Fig. 4. Frequency of wild bird remains found in association with the D-shaped, biwalled structure at Sand Canyon Pueblo (after Muir 1999a, 120, Table 21). * 'Other' contexts include disturbed and indeterminate deposits.*

the great horned owl) are associated with the bringing of hot weather for the ripening of corn and other crops (Tyler 1979, 165).

Most of the birds of prey are associated with war and hunting to various degrees. The hawk has a particularly strong association, though the smaller falconids (specifically the kestrel) also have lesser 'spiritual roles' in such activities (Tyler 1979). Crows and ravens are associated with war and death, likely because they eat carrion (Tyler 1979, 181–2), as does the turkey vulture which is specifically associated with recovery of war-dead and purification, after battle (Tyler 1979, 225–9). This theme of death extends to the sandhill crane and owls. As well as being guardians of corn, sandhill crane are considered guardians of clowns, *kachina* dancers and the dead (Tyler 1979, 129). Owls, particularly the burrowing owl, are associated with night, the underworld and the god of death (Tyler 1979, 164).

These are the major spiritual themes represented by the birds identified among the D-shaped structure deposits. Common associations have been emphasized here, but it is important to recognize that the taxa also have many other spiritual associations which have not been discussed, some of which are even contradictory to those presented above. For example, while crows and ravens are associated with rain, they are also an omen of drought (Tyler 1979, 180). Such contradictions are not uncommon and reflect the variety of roles that birds played among the different Pueblo communities and the limitations and potential problems of trying to synthesize this diversity. However, perhaps more important than the consistencies, is the range of associations that are represented. Clearly the birds are not all linked by a single theme, such as agriculture, hunting, or war, but represent at least two or three different activities. This suggests that the D-shaped structure may have acted as a multi-purpose ceremonial facility for the Pueblo as whole, rather than as the 'house' or 'office' of a single (*e.g.*, war or hunting) society. Admittedly this interpretation is quite speculative; however, it is consistent with the unique architectural nature of the D-shaped structure and its central location within Sand Canyon Pueblo.

Much like the concentration of wild bird remains associated with the biwalled structure, the concentration of bobcat or lynx (*Lynx* sp. Kerr) and bird of prey remains within the 1000 block courtyard is not likely the result of refuse disposal or storage of meat. However, the significance of this pattern is questionable. The concentration includes only a single deposit and likely consists of a small number of individual skeletons. A closer examination of the data reveals that only one bird of prey species is actually represented among the deposit: the American Kestrel (*Falco sparverius* Linnaeus). Furthermore it appears that only one individual kestrel is represented based on the frequency of elements (MNI = 1). Similarly the *Lynx* remains recovered from the deposit collectively produce a MNI value of one. Some of the lynx bones have been culturally modified (cut and polished) and it is clear that an entire skeleton is not present. The deposit was interpreted by the excavator as "construction fill", and is situated in the corner of the courtyard. The excavator did not document any evidence for a pit feature or other form of 'structured' deposit.

Though direct evidence is sparse, it is possible that the remains represent the ritual interment of a complete American kestrel along with some *Lynx* sp. bone artifacts. Both bobcat and kestrel are spiritually associated with hunting and war (Tyler 1975; Tyler 1979). That the concentration occurs within Block 1000 (a tower block) is particularly intriguing given the evidence for communal hunting of artiodactyls associated with this block discussed above. Ethnographically both war and hunting societies are documented as having been principally responsible for the organization of communal hunts of deer and other large game. The presence of bobcat and kestrel interments would be consistent with the interpretation that the 1000 Block functioned as a war or hunting society house or office.

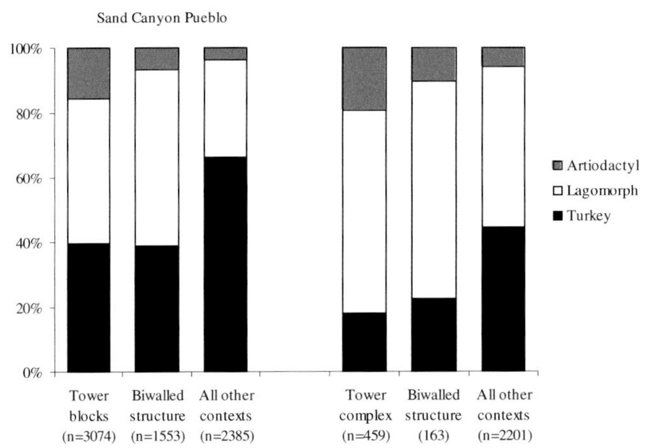

Fig. 5. Relative abundance of lagomorphs, artiodactyl, and turkey remains according to context from Sand Canyon Pueblo compared to those from similar contexts at Yellow Jacket Pueblo.

Other large Pueblo III sites

Faunal assemblages from three other large Pueblo III sites have been examined in an attempt to find compositional patterning similar to that observed at Sand Canyon Pueblo. The assemblages from these sites are not as large as that excavated from Sand Canyon Pueblo and in some cases, stratigraphic and contextual control is less precise due to disturbance of the site deposits (particularly at Yellow Jacket and Woods Canyon Pueblos). However our analyses have revealed several interesting similarities and differences.

Yellow Jacket Pueblo is a very large site occupying nearly 100 acres and containing approximately 192 kivas, 600–800 rooms, 18 towers, a great kiva and a biwalled structure (Kuckelman 1997; Lipe and Ortman 2000). The site is situated at the head of Yellow Jacket Canyon, approximately 15 km north/northeast of Sand Canyon Pueblo. Based on dendrochronological data and seriation of ceramic assemblages the site appears to have grown gradually during the Pueblo II and Pueblo III periods.

Yellow Jacket is the only site we have examined that displays deposits similar to the 'ritual refuse' accumulations identified at Sand Canyon Pueblo; specifically, marked (and statistically significant, $p < 0.05$) concentrations of artiodactyl remains in association with tower structures. The concentrations are found within a unique architectural complex consisting of multiple towers, including a biwalled, tower/kiva structure. Compared to other portions of the site the tower complex contains more than three times the average relative frequency of artiodactyl remains. In this respect the similarities in compositional patterning at Sand Canyon and Yellow Jacket Pueblo are remarkable (Fig. 5).

Deposits containing concentrations of wild birds or wild carnivores were not observed anywhere at Yellow Jacket Pueblo. In fact, these taxa are exceedingly rare among the site assemblage, each group representing less than 1% of the identified remains.

Castle Rock Pueblo is located at the lower end of Sand Canyon, about 7.5 kms south/southwest of Sand Canyon Pueblo, with which it is contemporaneous (Kuckelman 2000). An aggregated settlement on a much smaller scale than Sand Canyon Pueblo, it includes 19 kivas, most associated with room blocks (Fig. 6). Whereas Sand Canyon Pueblo was a planned community, Castle Rock seems to have grown through accretion as unit pueblos clustered around a prominent butte that gives the site its name. In the northwest of the site there is a D-shaped structure, but unlike the elaborate building complex at Sand Canyon Pueblo, this was simply an enclosing masonry wall. There was at least one tower, a slightly oversized kiva and a square kiva, all of which might have had special functions.

The fauna from Castle Rock contains a variety of wild birds, mainly birds of prey (Driver 2000). These include a large hawk (*Buteo* sp.), unidentified Falconiformes (also consistent with *Buteo*), *Falco* sp. (probably *F. peregrinus* Tunstall), American kestrel (*F. sparverius*), a small owl (*Aegolius* sp. Linnaeus) and a large owl. Other birds include raven (*Corvus corax* Linnaeus), a large duck (*Anas* sp. Linnaeus), dove or pigeon (Columbiformes), and quail (Phasianidae). One individual *Buteo* was deposited as a complete skeleton of a large juvenile bird. It had been placed above the fill of a hearth in a circular structure, probably a kiva. This seems to be an example of deliberate ritual interment, probably sacrificial and probably associated with the final use of the structure. Assuming that the large Falconiformes are also *Buteo*, clustering of these specimens can be seen around the

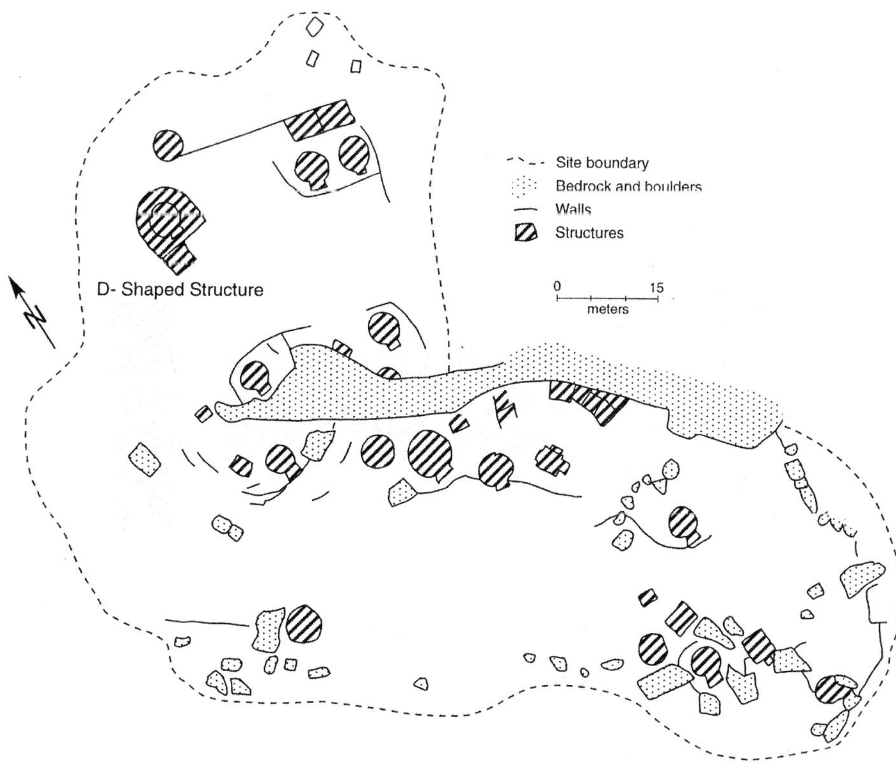

Fig. 6. Plan map of Castle Rock Pueblo.

rectangular kiva (that also produced the kestrel remains), and near the tower (an area that also produced all the owl bones). This rather patchy distribution suggests that ritual interment of complete birds in certain locations may have been practiced, although it is also possible that the incomplete skeletons are simply accumulations of ritual refuse, as postulated for Sand Canyon Pueblo. There was no concentration of wild birds detectable near the D shaped structure, but this is a much less substantial building than the D-shaped structure at Sand Canyon Pueblo.

Deliberate ritual interment of hawks resembles practices recorded for eagles captured and sacrificed by the Hopi (Voth 1912; Beaglehole 1936). The Hopi took young eagles from nests and raised them for a few months. They were then ritually killed, plucked and skinned for their feathers and down. The bodies were buried carefully in a special cemetery outside the Hopi villages.

Unlike Sand Canyon Pueblo, there was no statistically significant difference in the distribution of artiodactyls remains between structures at Castle Rock Pueblo.

Woods Canyon Pueblo is located 10 km north/northwest of Sand Canyon Pueblo. The site consists of at least 50 kivas, associated room blocks, a plaza, 15 towers, a D-shaped building (possibly biwalled), and a water reservoir (Lipe and Ortman 2000; Ortman *et al.* 2000). It is believed that the site grew gradually during Pueblo III times.

Overall, the Woods Canyon faunal assemblage is very similar to that of the other Pueblo III sites. However there is no significant spatial patterning among artiodactyls, birds of prey, or carnivores. Though Woods Canyon Pueblo contains multiple towers and a D-shaped structure similar to the biwalled structure at Sand Canyon Pueblo, there do not appear to be any correlations between structure types and faunal distributions. Instead, artiodactyls, birds of prey and carnivores are extremely rare and not associated with any particular structure type.

Small sites in the Sand Canyon locality

The small sites examined in this study are all associated with the Sand Canyon community and probably each represent settlements of one to three households. While the faunal assemblages from the small sites in the Sand Canyon locality are generally too small to allow detailed analysis of compositional patterning, the samples can be pooled in order to look at general chronological trends. In order to examine temporal patterns the small sites have been grouped according to their primary Pueblo III occupations as either 'early', 'middle' or 'late'. Early sites include those which were occupied and abandoned prior to the establishment of Sand Canyon Pueblo (*c.* AD 1250). The 'middle' sites include all those which appear to have been established prior to Sand Canyon Pueblo, but which continued to be occupied for sometime after AD 1250. The final group – the 'late' sites – includes all those which do not appear to have been occupied until after AD 1250.

When the assemblages are pooled it is apparent that wild birds, particularly birds of prey are more abundant among the 'early' sites compared to the small 'late' sites, while the middle period sites display an intermediate value. Though birds of prey represent only a little over 1% (1.2%) of the 'early' site assemblages (excluding rodents, reptiles and amphibians), this is much higher than the average frequency of these remains among the small middle (f = 0.2%) or late (f = 0.1%) period sites. Chi square analysis confirmed that this pattern is statistically significant (i.e., $p < 0.05$, using Yates' correction for sparse samples) despite relatively small samples. It is also notable that the bird of prey remains were recovered from three of the five 'early' sites, so the relatively high frequency is not simply the result of an isolated cluster encountered at one site.

Discussion and conclusions

We have argued elsewhere that the patterns evident among the faunal assemblages in the Sand Canyon Locality reflect community organization with respect to the location, scale, and perhaps control, of ritual activities, and in particular ritualized communal hunting (Muir 1999a; Muir 1999b, Muir and Driver 2002). We believe that by the late Pueblo III period, Sand Canyon Pueblo was a focus of ritual activity within the locality, while the frequency or intensity of ritual activities at the small sites had diminished. This argument is supported by the relative abundance of wild birds and artiodactyls at these sites, as well as the distribution of remains within Sand Canyon Pueblo. Specifically, the faunal data indicate that wild birds and artiodactyls are initially fairly common among most of the small sites, however, by the 'late' Pueblo III period they are restricted almost exclusively to Sand Canyon Pueblo. If we accept that the frequency of wild birds is an indicator of the occurrence of ritual activities, this pattern suggests that rituals which were initially conducted within the hamlets, presumably by and for the benefit of the residents of each individual site, changed in terms of scale and location by the late Pueblo III period. The D-Shaped, biwalled structure at Sand Canyon Pueblo appears to have been the location of much of this later activity, though other structures may have also played a role. It seems probable that these structures functioned as specialized communal ritual facilities for the inhabitants of Sand Canyon Pueblo as well as for the residents of the small contemporaneous sites located nearby.

Based on apparent similarities in ritual refuse accumulations we argue that similar developments occurred at Yellow Jacket Pueblo. The 'tower complex' at Yellow Jacket Pueblo appears to have functioned in a similar capacity to the 'tower blocks' at Sand Canyon Pueblo – as the focus of ritualized communal hunting activity. Although data from smaller sites in the Yellow Jacket locality are not available, we anticipate that they would display diminished abundance of artiodactyls and wild birds through time, similar to the small sites of Sand Canyon, reflecting the emergence of Yellow Jacket Pueblo as the center of local ritual activities during late Pueblo III times.

Correlations between structure types and faunal distributions were not observed at Woods Canyon Pueblo and were of minor significance at Castle Rock Pueblos for birds of prey only. It is important to note that although these are considered to be large sites, they are substantially smaller than either Sand Canyon or Yellow Jacket Pueblo. It appears that whatever activities were creating such concentrations at the larger sites do not appear to have occurred at these smaller pueblos. This may indicate fundamental differences in the nature and organization of ritual activities such as communal hunting, corresponding to community size (Muir and Driver 2002).

Ritual and the development of large aggregated communities

Changes in frequency and distribution of faunal remains through the Pueblo III period indicate that significant social and economic reorganization accompanied the development of the large aggregated communities in the northern San Juan region (Muir 1999a; Muir and Driver 2002). In particular, there is evidence of increased centralization, scale, and perhaps control of ritualized subsistence events, such as communal hunts. These activities appear to have been organized by specific individuals, households, or societies at the very largest sites, such as Sand Canyon and Yellow Jacket Pueblos.

We believe that community aggregation during the late Pueblo III period may have been largely a response to increased competition for limited wild resources, due to regional population increases, environmental degradation, and regional economic competition. These factors prompted the need to pool labor and knowledge in an integrated communal effort. The development of large community centers for the coordination of communal ritual events, would have encouraged cooperation, group unity, and community identity and facilitated activities requiring large groups, such as communal hunts and other large scale subsistence activities (*e.g.*, planting and harvest). The apparent function of the biwalled, D-Shaped structure at Sand Canyon, and the Tower Complex at Yellow Jacket Pueblo, as ritual facilities certainly lends support to the existence of societal or community based organization. That evidence for such activities is only evident at the largest sites in the region suggests that community population size may have been a critical variable with respect to economic success. There is considerable evidence for an increased level of inter-community hostilities during the late Pueblo III period (Lipe 1995; LeBlanc 1999; Kuckelman *et al.* 2000) and it is possible that a number of men would have been required to remain on-hand to protect each settlement. Consequently only very large communities could spare enough men to conduct successful long range hunting trips. It appears that the inhabitants of very large sites such as Sand Canyon and Yellow Jacket Pueblos had the most success. Smaller communities such as Castle Rock and Woods Canyon Pueblos either had little success in such endeavors, or did not attempt them, and instead primarily obtained meat through garden trapping and turkey husbandry.

Acknowledgements
This research was conducted with cooperation, assistance, and support from the staff of Crow Canyon Archaeological Center and Simon Fraser University. We thank all of the individuals who have been involved in the Sand Canyon, Castle Rock, Woods Canyon, and Yellow Jacket archaeological projects, particularly, Karen Adams, Bruce Bradley, Melissa Churchill, Donna Glowacki, Kristin Kuckelman, Ricky Lightfoot, Bill Lipe, Scott Ortman, Melita Romasco, and Mark Varien. We also thank Bob Birtch, Ginnie Dunlop, and Mary Etzkorn for their assistance with graphics. Sharyn Jones O'Day and two anonymous reviewers provided valuable comments on an earlier draft of this paper. This research was funded in part by grants from the Social Sciences and Humanities Research Council of Canada and Simon Fraser University.

References

Adler, M. A. 1992. The upland survey, pp. 11–23 in Lipe, W. D. (ed.), *The Sand Canyon Archaeological Project: a Progress Report* (CCAC Occasional Paper 2). Cortez: Crow Canyon Archaeological Center.

Akins, N. J. 1985. Prehistoric faunal utilization in Chaco Canyon: Basketmaker III through Pueblo III, pp. 305–445 in Mathien, F. J. (ed.), *Environment and Subsistence of Chaco Canyon* (Chaco Canyon Studies, Publications in Archaeology 18E). Santa Fe: National Park Service, U.S. Department of the Interior.

Aldenderfer, M. 1993. Ritual, hierarchy, and change in foraging societies. *Journal of Anthropological Archaeology* **12**, 1–40.

Anell, B. 1969. *Running Down and Driving Game in North America* (Studia Ethnographica Upsaliencia 30). Sweden: Berlingska Boktrijckeriel.

Beaglehole, E. 1936. *Hopi Hunting and Hunting Ritual* (Yale University Publications in Anthropology 4). New Haven: Yale University Press.

Bell, C. 1992. *Ritual Theory, Ritual Practice*. Oxford: Oxford University Press.

Bradley, B. A. 1992. Excavations at Sand Canyon Pueblo, pp. 79–97 in Lipe, W. D. (ed.), *The Sand Canyon Archaeological Project: a Progress Report* (CCAC Occasional Paper 2). Cortez: Crow Canyon Archaeological Center.

Brandt, E. A. 1994. Egalitarianism, hierarchy, and centralization in the Pueblos, pp. 9–23 in Wills, W.H. and Leonard, R.D. (eds), *The Ancient Southwestern Community: Models and Methods for the Study of Prehistoric Social Organization*. Albuquerque: University of New Mexico Press.

Cordell, L. S. 1977. Late Anasazi farming and hunting strategies: one example of a problem in congruence. *American Antiquity* **42**, 449–61.

Cordell, L. S. 1984. *Prehistory of the Southwest*. Orlando: Academic Press.

Cordell, L. S., and Gumerman, G. J. 1989. Cultural interaction in the prehistoric southwest, pp. 1–18 in Cordell, L. S. and Gumerman, G. J. (eds), *Dynamics of Southwest Prehistory*. Washington: Smithsonian Institution Press.

Dean, J. S., Doelle, W. H. and Orcutt, J. D. 1994. Adaptive stress: environment and demography, pp. 53–86 in Gumerman, G. J. (ed.), *Themes in Southwest Prehistory*. Santa Fe: School of American Research Press.

Driver, J. C. 1999. Raven skeletons from Paleoindian contexts, Charlie Lake Cave, British Columbia. *American Antiquity* **64**, 289–98.

Driver, J. C. 2000. Faunal remains from Castle Rock Pueblo, in Kuckelman, K. (ed.), *The Archaeology of Castle Rock Pueblo: A Thirteenth-Century Village in Southwestern Colorado*, (CCAC Web Publication, www.crowcanyon.org/researchreports/castlerock). Cortez: Crow Canyon Archaeological Center.

Duff, A. I. and Wilshusen, R. H. 2000. Prehistoric population dynamics in the northern San Juan Region, A.D. 950–1300. *Kiva* **66**, 167–90.

Ferring, C. R. 1984. Intrasite spatial patterning: its role in settlement-subsistence systems analysis, pp. 116–26 in Hietala, H. J. (ed.), *Intrasite Spatial Analysis in Archaeology*. Cambridge: Cambridge University Press.

Graves, M., Holbrook, S. and Longacre, W. 1982. Aggregation and abandonment of Grasshopper Pueblo, Arizona. *Journal of Field Archaeology* **9**, 193–206.

Graves, M. and Reid, J. J. 1984. Social complexity in the American Southwest: a view from east-central Arizona, pp. 266–75 in Upham, S., Plog, F., Batcho, D. G. and Kauffman B. (eds), *Recent Research in Mogollon Archaeology* (The University Museum Occasional Papers No. 10). Las Cruces: New Mexico State University.

Grier, C. 1996. *Whaling and Social Inequality in Classic Thule Eskimo Societies: a Central Canadian Arctic Study*. Unpublished M. A. thesis. Arizona State University, Tempe.

Gnabasik, V. R. 1981. *Faunal Utilization by the Pueblo Indians*. Unpublished M. A. thesis, Eastern New Mexico University.

Hayden, B. 1995. Pathways to power: principles for creating socioeconomic inequalities, pp. 15–86 in Price, T. and Feinman, G. (eds), *Foundations of Social Inequality*. New York: Plenum Press.

Hayden, B. 1997. *The Pithouses of Keatley Creek: Complex Hunter-gatherers of the Northwest Plateau*. Fort Worth: Harcourt Brace College Publishers.

Henderson, J. and Harrington, J. P. 1914. *Ethnozoology of the Tewa Indians* (Bureau of American Ethnology, Bulletin 56). Washington D.C.: Smithsonian Institution.

Hill, E. 2000. The contextual analysis of animal interments and ritual practice in southwestern North America. *Kiva* **65**, 361–98.

Hill, J. D. 1995. *Ritual and Rubbish in the Iron Age of Wessex: A Study on the Formation of a Specific Archaeological Record* (BAR British Series 242). Oxford: British Archaeological Reports.

Hill, J. D. 1996. The identification of ritual deposits of animals: a general perspective from a specific study of 'special animal deposits' from the southern English Iron Age, pp. 17–32 in Anderson, S. and Boyle, K. (eds), *Ritual Treatment of Human and Animal Remains: Proceedings of the First Meeting of the Osteoarchaeological Research Group held in Cambridge on 8th October 1994*. Oxford: Oxbow Books.

Huber, E. K. and Lipe, W. D. 1992. Excavations at the Green Lizard Site, pp. 69–78 in Lipe, W. D. (ed.), *The Sand Canyon Archaeological Project: a Progress Report* (CCAC Occasional Paper 2). Cortez: Crow Canyon Archaeological Center.

Irwin-Williams, C. (ed.) 1972. *The Structure of Chacoan Society in the Northern Southwest, Investigations at the Salmon Site – 1972* (Contributions in Anthropology 4/3). Portales: Eastern New Mexico University.

Johnson, G. A. 1989. Dynamics of Southwestern prehistory: far outside – looking in, pp. 371–90 in Cordell, L. S. and Gumerman, G. J. (eds), *Dynamics of Southwestern Prehistory*. Washington: Smithsonian Institution Press.

Kolb, M. 1994. Ritual activity and chiefly economy at an upland religious site on Maui, Hawaii. *Journal of Field Archaeology* **21**, 417–36.

Kuckelman, K. A. 1997. Preliminary results of testing and report of 1997 excavations at Yellow Jacket Pueblo (site 5MT5), Montezuma County, Colorado. Unpublished report submitted to the Archaeological Conservancy, Albuquerque, and the Colorado Historical Society, Denver.

Kuckelman, K. A., Lightfoot, R. R. and Martin, D. L. 2000. Changing patterns of violence in the Northern San Juan Region. *Kiva* **66**, 147–66.

Kuijt, I. 1996. Negotiating equality through ritual: a consideration of Late Natufian and Prepottery Neolithic A Period mortuary practices. *Journal of Anthropological Archaeology* **15**, 313–36.

Ladd, E. J., 1963. *Zuni Ethno-Ornithology*. Unpublished MSc Thesis, Department of Anthropology, University of New Mexico.

Lange, C. H. 1959. *Cochiti: a New Mexico Pueblo, Past and Present*. Austin: University of Texas Press.

Larson, D. O., Neff, H., Graybill, D. A., Michaelsen, J. and Ambos, E. 1996. Risk, climatic variability, and the study of Southwestern prehistory: an evolutionary perspective. *American Antiquity* **61**, 217–41.

LeBlanc, S. A. 1999. *Prehistoric Warfare in the American Southwest*. Salt Lake City: University of Utah Press.

Lekson, S. H. 1999. *The Chaco Meridian: Centers of Political Power in the Ancient Southwest*. Walnut Creek: Altamira Press.

Lightfoot, K. G. and Upham, S. 1989. Complex societies in the prehistoric American Southwest: a consideration of the controversy, pp. 3–30 in Upham, S., Kent, K. G. and Jewett, R. A. (eds), *The Sociopolitical Structure of Southwestern Prehistoric Societies*. Boulder: Westview Press.

Lipe, W. D. 1992. Summary and concluding comments, pp. 121–33 in Lipe, W. D. (ed.), *The Sand Canyon Archaeological Project: a Progress Report* (Occasional Paper 2). Cortez: Crow Canyon Archaeological Center.

Lipe, W. D. 1995. The depopulation of the Northern San Juan: conditions in the turbulent 1200s. *Journal of Anthropological Archaeology* **14**, 143–69.

Lipe, W. D. and Ortman, S. G. 2000. Spatial patterning in northern San Juan villages, AD 1050–1300. *Kiva* **66**, 91–122.

Lipe, W. D., Varien, M. D. and Wilshusen R. H. (eds) 1999. *Colorado Prehistory: A Context for the Southern Colorado River Basin*. Boulder: Colorado Council of Professional Archaeologists.

Minnis, P. E. and Redman, C. L. 1990. *Perspectives on Southwestern Prehistory*. Boulder: Westview Press.

Muir, R. J. 1999a. *Zooarchaeology of Sand Canyon Pueblo, Colorado*. Unpublished Ph.D. thesis, Simon Fraser University.

Muir, R. J. 1999b. Intensification of communal ritual: zooarchaeology of the Sand Canyon locality. Paper presented at the 64th Annual Meeting of the Society for American Archaeology, Chicago, Illinois.

Muir, R. J. and Driver, J. C. 2002. Scale of analysis and zooarchaeological interpretation: Pueblo III faunal variation in the northern San Juan region. *Journal of Anthropological Archaeology* **21**, 165–99.

Ortman, S. G., Glowacki, D. M., Churchill, M. J. and Kuckelman, K. A. 2000. Pattern and variation in Northern San Juan Village histories. *Kiva* **66**, 123–46.

Parsons, E. C. 1918. Notes on Acoma and Laguna. *American Anthropologist* **20**, 162–86.

Parsons, E. C. 1920. Notes on Isleta, Santa Ana, and Acoma. *American Anthropologist* **22**, 56–69.

Parsons, E. C. 1921. Further notes on Isleta. *American Anthropologist* **23**, 149–69.

Parsons, E. C. 1929. *The Social Organization of the Tewa of New Mexico* (American Anthropological Association Memoir 36). Menasha, WI: The American Anthropological Association.

Parsons, E. C. 1970. *Taos Pueblo*. New York: Johnson Reprint.

Parsons, E. C. 1977. *The Pueblo of Jemez*. Ann Arbor: University Microfilms.

Potter, J. 1997. *Communal Ritual, Feasting, and Social Differentiation in Late Prehistoric Communities*. Unpublished Ph.D. Dissertation, Arizona State University.

Potter, J. 2000. Ritual, power, and social differentiation in small-scale societies, pp. 295–316 in Diehl, M. (ed.), *Hierarchies in Action: Cui Bono?* Carbondale: Center for Archaeological Investigations, Southern Illinois University.

Potter, J. and Perry, E. M. 2000. Ritual as a power resource in the American Southwest, pp. 60–78 in Mills, B. J. (ed.), *Alternative Leadership Strategies in the Prehispanic Southwest*. Tucson: University of Arizona Press.

Rappaport, R. 1999. *Ritual and Religion in the Making of Humanity* (Cambridge Studies in Social and Cultural Anthropology). Cambridge: Cambridge University Press.

Reid, J. J. 1985. Measuring social complexity in the Southwest, pp. 167–74 in Thompson, M., Garcia, M. and Kense, F. (eds), *Status, Structure and Stratification: Current Archaeological Reconstruction. Proceedings of the 16th Chacmool Conference*. Calgary: University of Calgary.

Reid, J. J. and Whittlesey, S. M. 1990. The complicated and the complex: observations on the archaeological record of large Pueblos, pp. 184–95 in Minnis, P. and Reman, C. (eds), *Perspectives on Southwestern Prehistory*. Boulder: Westview Press.

Rohn, A. H. 1989. Northern San Juan prehistory, pp. 149–78 in Cordell, L. S. and Gumerman, G. J. (eds), *Dynamics of Southwest Prehistory*. Washington: Smithsonian Institution Press.

Stallibrass, S. 1996. Sacred cows from Solway: bog bodies on the Anglo-Scottish border, pp. 35–44 in Anderson, S. and Boyle, K. (eds), *Ritual Treatment of Human and Animal Remains: Proceedings of the First Meeting of the Osteoarchaeological Research Group held in Cambridge on 8th October 1994*. Oxford: Oxbow Books.

Stevenson, M. C. 1894. The Sia, pp. 3–157 in Powell, J. W. (ed.), *Eleventh Annual Report of the Bureau of American Ethnology, 1899–90*. Washington D.C: Smithsonian Institution.

Tyler, H. A. 1964. *Pueblo Gods and Myths*. Norman: University of Oklahoma Press.

Tyler, H.A. 1975. *Pueblo Animals and Myths*. Norman: University of Oklahoma Press.

Tyler, H.A. 1979. *Pueblo Birds and Myths*. Norman: University of Oklahoma Press.

Underhill, R. M. 1946. *Papago Indian Religion*. New York: Columbia University Press.

Upham, S. 1982. *Polities and Power: An Economic and Political History of the Western Pueblo*. New York: Academic Press.

Upham, S. and Plog, F. 1986. The interpretation of prehistoric political complexity in the central and northern Southwest: toward a mending of the models. *Journal of Field Archaeology* **13**, 223–38.

Varien, M. D. 1999. *Sedentism and Mobility in a Social Landscape: Mesa Verde and Beyond*. Tucson: University of Arizona Press.

Varien, M. D. and Kuckelman, K. A. 1999. Introduction, in Varien, M. D. (ed.), *The Sand Canyon Archaeological Project: Site testing* (CCAC Web Publication, www.crowcanyon.org/researchreports/sitetesting). Cortez: Crow Canyon Archaeological Center.

Vivian, R. G. 1990. *The Chacoan Prehistory of the San Juan Basin.* San Diego: Academic Press.
Vivian, R. G. and Matthews, T. W. 1965. *Kin Kletso: A Pueblo III Community in Chaco Canyon, New Mexico* (Southwestern Monuments Association, Technical Series 6/1). Arizona: Globe.
Voth, H. R. 1912. Notes on the Eagle Cult among the Hopi Indians, pp. 107 9 in Voth, H. R. (ed.), *Brief Miscellaneous Hopi Papers* (Field Museum of Natural History Publication 157, Anthropological Series XI, No. 2). Chicago: Field Museum of Natural History.
Walker, W. H. 1995. Ceremonial trash? pp. 67–79 in Skibo, J. M., Walker, W. H. and Nielsen, A. E. (eds), *Expanding Archaeology.* Salt Lake City: University of Utah Press.
Walker, W. H. 1998. Where are the witches of prehistory? *Journal of Archaeological Method and Theo*ry **5**, 245–308.
Walker, W. H. 2001. Ritual technology in an extranatural world, pp. 87–106 in Schiffer, B. (ed.), *Anthropological Perspectives on Technology.* Albuquerque: University of New Mexico Press.
Walker, W. H. and Lucero, L. J. 2000. The depositional history of ritual and power, pp. 130–47 in Dobres, M. and Robb, J. (eds), *Agency in Archaeology.* London: Routledge.
White, L. 1932. *The Pueblo of San Felipe* (Memoirs of the American Anthropological Association 38). Washington.
White, L. 1935. *The Pueblo of Santo Domingo, New Mexico* (Memoirs of the American Anthropological Association 43). Washington.
White, L. 1974a. *The Pueblo of Santo Domingo, New Mexico.* Millwood, New York: Kraus Reprint.
White, L. 1974b. *Zia - the Sun Symbol Pueblo.* Albuquerque: University of Albuquerque Press.
Whitman, W. 1947. *The Pueblo Indians of San Ildefonso: A Changing Culture* (Columbia University Contributions to Anthropology 34.). New York: Columbia University Press.
Wilcox, D. 1991. Changing contexts of Pueblo adaptations, A.D. 1250–1600, pp. 128–54 in Spielmann, K. (ed.), *Farmers, Hunters, and Colonists: Interaction Between the Southwest and the Southern Plains.* Tucson: University of Arizona Press.

Robert J. Muir and *Jonathan C. Driver*
Department of Archaeology,
Simon Fraser University,
Burnaby BC V5A 1S6, Canada
E-mail: bmuir@sfu.ca

9th ICAZ Conference, Durham 2002
Behaviour Behind Bones, (ed. Sharyn Jones O'Day et al.) pp. 144–152

17. Facts and fantasies: the archaeology of the Marquesan dog

Sidsel N. Millerstrom

Dog burials and dog bones have been uncovered at several archaeological sites since the advent of modern archaeology in the Marquesas Islands (French Polynesia) in the mid 1950s. In addition, a large number of dog petroglyphs have been documented on the islands of Nuku Hiva, Hiva Oa, and Ua Huka. However, information on the Marquesan Canis familiaris *is fragmented and the data has never been synthesized. It appears that the dog held a privileged position among some of the island tribes. Spatial distributional information on dog petroglyphs shows that the images where both restricted to specific regions and valleys. In this paper I discuss archaeological evidence for the presence of dogs in the Marquesas Islands and relate this data to the spatial distribution of dog images.*

Introduction

Dogs, along with pigs and jungle fowl or chicken, were part of the main triage that accompanied the early human colonizers into most of the Polynesian islands (Titcomb 1969; Kirch 1984, 2000). Easter Island is an exception: it appears as if the dog never made it to that isolated island. In New Zealand the dog occurred but pigs and chickens were not present (Davidson 1984, 129). Prehistorically the Polynesian domestic dogs (they never became feral) were used as pets (Oliver 974, 276–7; Valeri 1985, 47), they were venerated and they served as religious offerings (Luomala 1960, 190–240; Valeri 1985, 46). Dog teeth were also fashioned into ornaments, dog bones were made into tools, and dog furs were used in clothing (Titcomb 1969; Oliver 1974).

Throughout the Pacific dogs were eaten (Titcomb 1969) although most were apparently consumed in a ritual context (Handy 1923, 1927; Oliver 1974; Valeri 1985). In Tahiti for instance, dog was considered a luxury food item (Oliver 1974, 276; Titcomb 1969, 24–30). According to Titcomb (1969, 1), "...it was highly regarded as an article of food, more so than the pig." However, in 1774 when George Forster, the naturalist on Captain James Cook's second Pacific voyage (1772–1795), visited the southern Marquesan island group he did not list dogs as part of the foods consumed by the indigenous people (Forster 1977). Neither is the dog mentioned in Marquesan legends collected in the 1890s (Terell 1988) and 1920s (Handy 1930). This led some ethnologists to believe that dogs did not exist in prehistoric Marquesas.

However, since the advent of modern archaeological excavation in the Marquesas Islands (French Polynesia), in the mid 1950s, dog burials and disarticulated dog bones were uncovered at several sites. In addition, a large number of dog petroglyphs have been recently documented on the islands of Nuku Hiva, Hiva Oa, and Ua Huka (Millerstrom 1985a; Millerstrom 1985b; Millerstrom 1990; Millerstrom 1992; Millerstrom 1997; Millerstrom 2001). Nonetheless, information on the Marquesan *Canis familiaris*, is fragmented and the data has never been synthesized. Except for *Dog and Man in Polynesia* (1969) by Margaret Titcomb, and a few isolated articles written about the Polynesian dog (*e.g.*, Anderson 1981, 15–20; Bay-Petersen 1983), surprisingly limited information has been published that deals specifically with the Marquesan prehistoric dog.

It appears that the domestic dog was never numerous in the islands, but as I will demonstrate, among some of the Marquesan tribes, the dog held a privileged position in the late prehistoric–early-historic sociopolitical sphere. My research in the archipelago shows that spatial distributional information on dog images found in the rock art repertoire are restricted to specific regions as well as valleys. Why then was the dog present only on some of the islands and why was it restricted to only certain localities? What symbolic and ritual functions did the Marquesan dog serve in prehistory? What did the

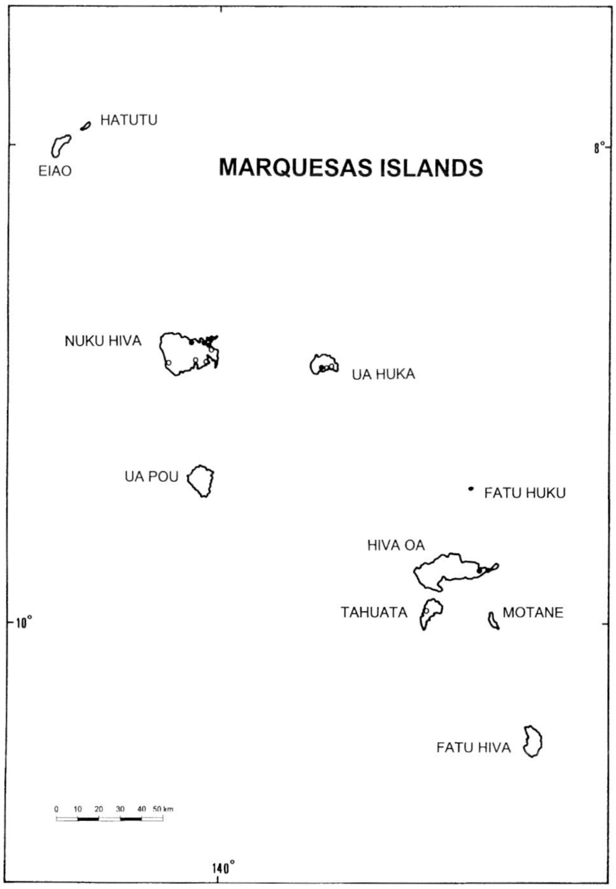

Fig. 1. Marquesas Islands. Black spots indicate the locations of dog petroglyphs; circles show location of archaeological sites where dog bones were excavated.

prehistoric dog look like? In this essay I address these issues and then I relate the data to the spatial distribution of dog images on rocks.

The Marquesan Archipelago

Situated in the eastern part of the Pacific Ocean, the Marquesas Archipelago is located approximately 1300 km northeast of Tahiti, *c.* 300 km from Easter Island, and some 3500 km from the islands of Hawaii. The land surface covers approximately 1048.3 square km (Millerstrom 2001). The string of islands extends approximately 370 km from the northwest to the southeast and is situated between 138° 20′ and 140° 30′ west longitude and between 7° 50′ and 10° 35′ south latitude (Brousse *et al.* 1978, 9–74). The Marquesas Islands consist of eight volcanic islands that were previously inhabited (some for a short period), a number of islets and a few surf-beaten rocks. Today only six islands are permanently settled.

The archipelago is geographically separated into two distinct groups (Fig. 1). To the northwest are the islands of Nuku Hiva, Ua Pou, Ua Huka, and the three presently uninhabited islets of Eiao, Hatu Iti, Hatutu, and the four banks of Hinakura, Motu One, Clark and Lawson. To the southeast are the islands of Hiva Oa, Tahuata, Fatu Hiva, and the four presently uninhabited islets of Fatu Huku, Motane, Fatu Huku, and Thomasset (Brousse *et al.* 1978, 9–74)

The archipelago is characterized by dramatic ridges towering up to 1200 m above sea level, breathtaking jagged peaks, deep and narrow valleys, and high cliffs that fall abruptly into the ocean. Perennial rivers, streams, waterfalls, and intermittent seasonal streams cut through the dense vegetation on the windward side of the islands. The lush tropical forest region stands in stark contrast to the arid desert area on the leeward part of the islands. Naturally, the wet-dry dichotomy influenced human settlement patterns (Kirch 1984; Kirch 2000; Kirch and Hunt 1997). Thus, early communities settled in the well-watered windward parts of the islands, while the marginal leeward areas began to be occupied when population increased, and when available arable land and resources diminished. The archipelago is today a part of French Polynesia.

Archaeology of the dog

In many respects the archeology of the Marquesan dog starts with the work of Karl von den Steinen [1928(II), 77–86]. Von den Steinen, a German physician with an interest in ethnography, worked in the islands in the 1880s. At *me'ae* I'ipona (also called Oipona), a large temple site with complex architecture and numerous megalitic stone statues in Puama'u, Hiva Oa (Millerstrom 1985a; Millerstrom 1985b; Millerstrom 1986; Millerstrom 1990; Ottino 1992) he collected a megalithic sculptured stone-head with two quadrupeds each placed at each corner of the mouth. The stone sculpture is presently located in the Museum für Völkerkunde, Berlin. At the same site, situated on a cube formed base of another sculpture von den Steinen found two additional quadrupeds (Fig. 2). Von den Steinen's informant told him that the quadrupeds represented pigs, rats, or dogs. Since it was assumed that dogs did not existed in the islands prehistorically but were a European introduction, von den Steinen concluded that the images represented rats.

Archaeologist Ralph Linton of the Bayard Dominick Expedition, Honolulu, worked in the archipelago in 1920–21 with E. S. Craighill Handy and his wife Willowdean Chatterson Handy (Linton 1923; Linton 1925). Linton concentrated on ceremonial sites. While he mapped and described some 170 structures of which seventy two were tribal ceremonial sites (*tohua*) and fifty-eight shrines (*me'ae*), he observed several large petroglyphs boulders depicting anthropomorphs and zoomorphs. He did not, however, mention seeing dog images. Although Linton also recorded the sculptures at *me'ae* I'ipona, it appears that he failed to notice the quadrupeds situated on the

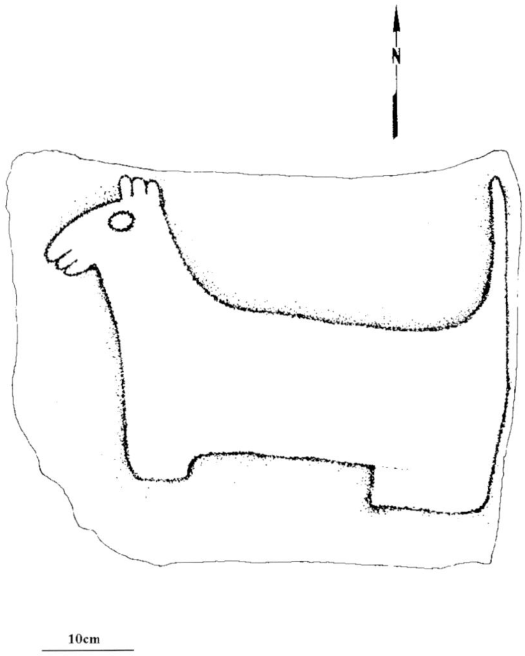

Fig. 2. Dog petroglyph situated on the base of a sculpture at me'ae *Iopona, Puamau, Hiva Oa.*

base of a statue as had already been noted by von den Steinen. Although von den Steinen worked in the islands some 30 years before Linton and the Handys', because of World War I, von den Steinen's data was not published until a few years after Linton's work. In fact, Linton (1925) felt that he found so few petroglyphs that he declined to speculate on the images' symbolic and ritual functions.

Pioneering stratigraphic excavation by Robert C. Suggs (1961), a member of the American Museum of Natural History Expedition to the archipelago in 1956 uncovered dog bones at Ha'atuatua Beach Dunes (Fig. 3). Ha'atuatua is located on the north coast of Nuku Hiva. Radiocarbon age determination placed the Ha'atuatua site (Nhaa 1) to *c.* 120 BC or what Suggs has coined the Settlement period (150 BC to AD 100). Although the early dates are contested (Kirch 1986; Spriggs and Anderson 1993; Kirch and Ellison 1994) the site represents, to some archaeologists, the earliest evidence of human colonization in East Polynesia. A gallery of pecked images, some representing dogs, was discovered under several inches of coquina in a nearby dry riverbed (south of Nhaa1, area A; see Suggs 1961, Plate 11A). Suggs (1961, 60) concluded that the images were contemporary with the dune site.

At the same time the Norwegian Expedition, organized and financed by Thor Heyerdahl, was engaged in archaeological research on Marquesas (Heyerdahl and Ferdon 1965). His team mapped and excavated *me'ae* I'ipona (Heyerdahl 1965, 123–50), the same temple site that von den Steinen had examined nearly 80 years earlier. In his interpretation of the site Heyerdahl relied mostly on information provided by von den Steinen and argued that the figures did not signify rats at all but represented llamas (see Fig. 2). This, of course, bolstered Heyerdahl's theory that people from South America migrated to East Polynesia prehistory, a theory that is rejected by most Oceanic archaeologists. Excavation at *me'ae* I'ipona yielded radiocarbon age determinations for the architecture that, perhaps also, provided relative dates for the stone sculptures. Heyerdahl (1965, 123–51) suggested that the sculptures were placed at *me'ae* I'ipona approximately AD 1400–1500 (uncalibrated dates). Von den Steinen [1928(II), 84–86] calculated the age of the site based on genealogical information, arguing that the temple site was constructed approximately AD 1700–1750.

These projects were followed by archaeological research in Hane Valley, Ua Huka (Sinoto and Kellum 1965; Sinoto 1966; Sinoto 1968). At the Hane dune site (MUH 1), Yosihiko Sinoto (1969, 107; 1979, 119) with the Bishop Museum, Honolulu, uncovered a drilled dog canine in Phase I (Initial settlement from AD 300–600), and one premolar and two dog burials in Layer V (Area B). Analysis on the Hane faunal assemblage by Patrick V. Kirch (1973) showed that the dog was not common in the Hane site. Based on the limited dog bones uncovered at the Hane site Kirch (1973, 29–38) believed that the dog was not a major food source in the Marquesas and that it became extinct sometime between AD 1600 and 1800.

Sinoto (1979; 1983) also worked at the Ha'atuatua dune site to see if excavation would yield early age dates similar to that which would support Suggs analysis. However, Sinoto's excavation yielded much later dates, thus he proposed that the archipelago was first settled approximately AD 300–600. Consistency, the results of both Suggs' and Sinoto's carbon analysis has since troubled archaeologists and has led to spirited debates (Kirch 1986; Spriggs and Anderson 1993; Kirch and Ellison 1994). To settle the debate, a multi-year archaeological research project was undertaken by University of Hawaii, Département Archéologie du Centre Polynésien des Sciences Humaines, and Université Française du Pacifique, Tahiti. None of the ten radiocarbon dates this project uncovered supported Suggs' hypotheses of early colonization, nor did the data show evidence of occupation in the early periods of Marquesan prehistory as proposed by Sinoto (Rolett and Conte 1995).

Pascal Sellier (2000), a French physical anthropologist that recently excavated at Manihina, a sand dune located in a valley west of Hane (Ua Huka), uncovered three dog skeletons among several human burials. One dog skeleton appeared to have been placed in a coffin. The site dates to the late prehistoric-early-historic period.

Finally, at Hanamiai, Tahuata, Rolett (1998) found dog bones in all the excavated levels from the time of settlement AD 1025–1300 (Hanamiai Phase I) to the early historic period AD 1800–1850 (Hanamiai Phase V). Some of the dog bones that were excavated by Rolett had cut marks which suggested that they were eaten, skinned or perhaps both. Rolett's (1998, 92) conclusion concurred with Kirch

Island, Valley	Site	Site type	Material evidence for dog	Reference
Nuku Hiva				
Ha'atuatua	NHaa 1	Sand dune	Tooth (canine) graver and bones. Settlement Period	Suggs (1961, 133, 144, 181, 195)
Uea Valley	NBM 1	Rock shelter	Bones. Expansion Period–Classic Period	Suggs (1961, 22, 195)
Taiohae	*Me'ae* Kakahoautea	Tribal ceremonial complex	Bones. Historic Period (?)	Suggs (1961, 26, 195)
Hapa'a	House–platform		Burial	Suggs (1961, 27, 195)
Ua Huka				
Hane Valley		Beach dune	Pendant (canine) Phase I, Initial Settlement Burials. Phase II. Developmental Stage	Sinoto (1969, 107–10)
Manihina		Beach dune	Burials. Late prehistoric–early historic period	Sellier (2000)
Tahuata				
Hanamiai	*Tohua* Aimaha	Ceremonial site	Bones, some with cut–marks. AD 1025–1850	Rolett (1998, 92)

Fig. 3. Distribution of archaeological sites where dog bones have been excavated.

analysis that dog was never an important food source in prehistoric Marqueses and that they seemed to have almost disappeared, a least at Hanamiai and Hane, by the early historic period.

Additional and complimentary archaeological evidence is found in the substantial rock art inventory. I will now turn to the domestic dog depicted in the petroglyph repertoire that has been documented in the archipelago in the last sixteen years (Millerstrom 1985a; Millerstrom 1985b; Millerstrom 1990; Millerstrom 1992; Millerstrom 1997; Millerstrom 2001).

The Marquesan dog petroglyph distribution – a case study

A multi-year rock art project conducted in the Marquesas between 1984 and 1989 recorded and collected information on over 6000 individual petroglyphs, 110 wall paintings, and 81 sculptures (Edwards 1985a; Edwards 1985b; Millerstrom 1985a; Millerstrom 1985b; Edwards 1989; Millerstrom 1990; Millerstrom 1992; Edwards and Millerstrom 1995; Millerstrom 1997; Millerstrom 2001; Millerstrom and Edwards 1998). This long-term project was initiated in 1984 by Maeva Navarro, the former director of *Département d'Archéologie du Centre Polynésien des Sciences Humaines, "Te Anavaharau,"* Tahiti (French Polynesia) and headed by Edmundo Edwards, the former chief archaeologist with the Department. The center is now under the name *Service de la Culture et du Patrimoine*. This project was the first archaeological survey in French Polynesia with a specific focus on rock art and, in many respects, covered previously uncharted territories. Island-wide the majority or 66.3 percent of the images represented geometric figures; anthropomorphs constituted 22.7 percent, while animals only represented 5.4 percent (Millerstrom 1990, 92). Of all the animals depicted in the rock art for instance, fish, turtle, lizard and birds, the dog is the most prevalent zoomorphic image found in Marquesas.

Dog figures are restricted to three islands Nuku Hiva, Hiva Oa and Ua Huka. Even on these islands the dog images are centered in at *me'ae* Oipona in Puamaou, and Eiaone Valley (Hiva Oa), and Vaikivi (Ua Huka). However, dog images occur in great numbers only on the north cost of Nuku Hiva. They are particularly numerous in the valleys of A'akapa, Ha'atuatua, and Hatiheu (Fig. 4). As Fig. 5 shows, 162 dog figures, or 79.4 percent of all the dogs documented in the archipelago have been recorded in Hatiheu Valley. Elsewhere, a few stylistically different dog figures have been documented in Vaipaee (Ua Huka), Eiaone and Puamau (Hiva Oa).

In order to place the images in a cultural context, between 1991 and 1996 I conducted settlement pattern studies in the western section of Hatiheu Valley. One of the reasons for selecting the valley was because a great number of rock art had already been located and recorded. In addition, the western side was uninhabited and the sites appeared to be less disturbed. The research area was centered on the area between Vaiu'ua and Puhi'oho, two of the four major rivers in the valley. All visible structures and features were mapped and described. Age determination of excavated charcoal fragments suggested that the area was well occupied from AD 1400 and onwards. Nonetheless, the majority of the surface architecture dates to the late-prehistoric-early-historic period.

The research area was clearly divided into an agri-

Fig. 4. Dog petroglyphs from A'akapa, Nuku Hiva.

Island	Valley	Frequency	Percent
Nuku Hiva			
	Hatiheu	162	79.4
	Ha'atuatua	7	3.4
	A'kapa	21	10.3
	Ma'atea	1	0.5
Subtotal		191	93.6
Ua Huka			
	Vaikivi *me'ae*	5	2.5
Subtotal		5	2.5
Hiva Oa	Puamau Oipona *me'ae*	4	1.96
	Eiaone	4	1.96
Subtotal		8	3.9
Total		204	100

Fig. 5. Distribution and frequency of dog petroglyphs in the archipelago.

cultural segment with taro fields and agricultural terraces to the west and what I call a chiefly settlement with elite architecture and tribal ceremonial complexes to the east. A total of 1282 images, including 54 figures depicting dogs, were found (Fig. 6). Some of the images were situated within a cluster of four tribal ceremonial complexes where they were linked to the temple structure, the chief's habitation area, and perhaps in one case the warriors platform, within the complexes. However, the majority of the images, including the dogs, were found on elaborate and complex architecture which were distributed in the chiefly occupational zone called Ototemoui, located on a small ridge above the tribal ceremonial plazas (Fig. 7). Only a couple of boulders containing geometric motifs and anthropomorphs were discovered in the agricultural section of the research area (Millerstrom 2001).

The cultural context of dog petroglyphs elsewhere in Hatiheu Valley has yet to be fully investigated, however, at *tohua* Maikuku, a tribal ceremonial site located in the upper part of Hatiheu Valley, dogs seem to have taken on a cult activity. Several dog images, two superimposed, along with numerous anthropomorphic faces, are found on boulders that form an alleyway into the tribal ceremonial plaza. Above the *tohua*, situated on a narrow ridge, the surface of a large boulder facing the *tohua* is filled with dog petroglyphs.

Most often the Marquesan dog images are peculiar looking (Fig. 8): they do not correspond to descriptions by the early navigators nor do they look like the drawings of dogs made by artists from the early post-contact era. The Marquesan dogs images show that the necks and the bodies are exaggerated in length. The tails are long and curved over the back while the ears and muzzle may be pointed, square, or rounded. The legs are short and in one case from Hatiheu Valley the paws were pointed in the wrong direction. They are often associated with anthropomorphs.

Probably the earliest drawings of the Polynesian dog are pen and wash drawings by Sydney Parkinson, the artist on Captain James Cook's first voyage to the Pacific between 1768–1771 (Joppien and Smith 1985, 134–36, Plates 1.74 and 1.78). Two different drawings dated from 1769 (with perhaps the same dog?) show harbor scenes in Raiatea, Society Islands, each with a small dog on board a double canoe. Other early versions of dogs are depicted, for instance, in two paintings from Hawaii dated to 1818 and 1849 (Forbes 1992, 69, 132). While it is possible that these dogs were introduced from Europe, they appear similar to the dog depicted by Sydney Parkinson in 1769 (Joppien and Smith 1985). Although the historic evidence is scarce, the typical early post-contact dog appears different than what the Marquesan images depict. The early post-contact dog is white or spotted, small to medium size, with pointed snout and ears, and a long tail. Could the Marquesans of the past have forgotten what the dog looked like or did it matter how they depicted the dog?

Image classification	Number of image types	Percentage of image types
I. ANTHROPOMORPHS		
1. Anthropomorphic stick figure	266	20.8
2. Square-bodied anthropomorphic figure	14	1.1
3. Double outline anthropomorphic figure	14	1.1
4. Open body anthropomorphic figure	18	1.4
5. Naturalistic human figure	36	2.8
6. Profile squatting human figure	1	0.08
7. Human and dog combination		
8. Human and bird combination		
9. Human and fish combination		
10. Human-lizard	2	0.2
11. Humanoid	8	0.6
12. *Tiki*	4	0.3
13. Anthropomorphic body part	8	0.6
14. Anthropomorphic face/eyes	106	8.3
Subtotal	477	**37.3**
II. ZOOMORPHS		
1. Dog	54	4.2
2. Lizard	5	0.4
3. Turtle	29	2.3
4. Fish	13	1.0
5. Octopus	1	0.08
6. Bird	8	0.6
7. Unidentified creature	4	0.3
Subtotal	114	**8.9**
III. MATERIAL OBJECT		
IV. GEOMETRIC MOTIF		
1. Curvilinear motif	478	37.3
2. Linear motif	119	9.3
3. Cupule	3	0.2
4. Polishing grove		
5. Exotic geometric figure	91	7.1
Subtotal	691	**53.9**
Total	1282	**100**

Fig. 6. Petroglyphs from the western section of Hatiheu Valley showing the various groups of images documented in the study area.

Site name	Site Type	Frequency	Percentage
Tohua Kamuihei I	Tribal ceremonial complex	2	3.7
Tohua Kamuihei II	Tribal ceremonial complex	0	
Tohua Tahakia	Tribal ceremonial complex	7	13.0
Tohua Hikokoa	Tribal ceremonial complex	0	
Ototemuoui	Chiefly area	45	83.3
Total		54	100

Fig. 7. Architectural association of dog petroglyphs in the study area.

Why are the dog images associated with the chiefly class?

Based on the paucity of dog bones uncovered at Marquesan archaeological sites, I suggest that the scarcity of dogs in the islands is precisely the reason why dogs were elevated as an ideological object that took on an enormous cultural significance among the power structure of the Marquesan tribes of Nuku Hiva in the late prehistoric early historic period. It is likely that the dog images served as a visual emblematic symbol expressing solidarity between the Marquesan polities who inhabited the north coast of the island. Those images associated with the warriors and the temple platforms within the *tohua* proper perhaps served to legitimate the dog as a ritual object as well as confirmed their high status within the tribes.

Regional tribal identity and Hatiheu politics

In the past, two political divisions were recognized on Nuku Hiva: the western portion of the island was occupied by the Tei'i; and the Taipi constituted the eastern division (Crook 1790s; Handy 1923, 31; Pechberty 1996, 165). Traditionally the Tei'i and the Taipi were bitter enemies (Handy 1923, 31). The matter of political divisions is complicated, however. Marquesan society was fluid and in the event of war, alliances frequently shifted (Garcia 1843; Handy 1923, 31). Some of the same clan names documented in the protohistoric period by Crook (1790s) appeared, some eighty years later, to have occupied different valleys. For instance, the Tu'uoho, one of the clans inhabiting Hatiheu, also occurred in Hakaui; likewise the Naiki tribe

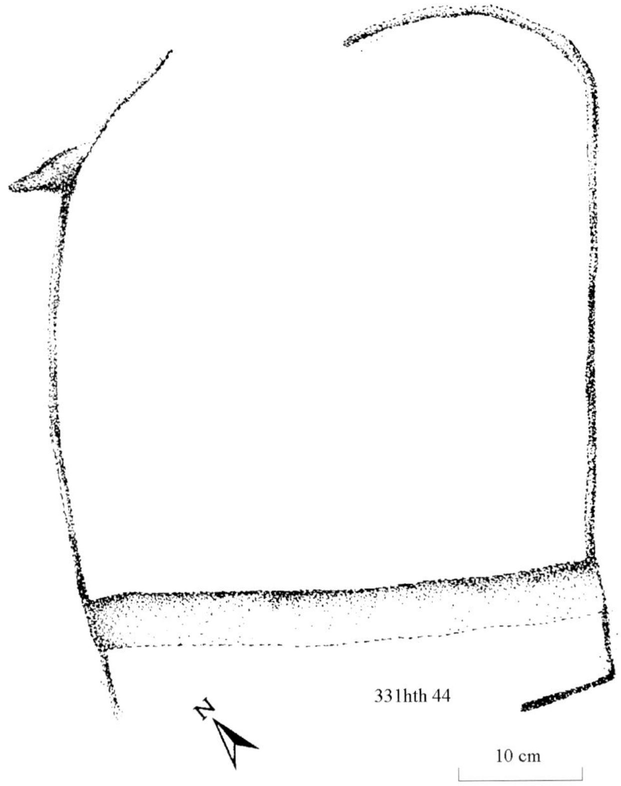

Fig. 8. Dog petroglyph linked to the temple structure at tohua *Kamuihei I, Hatiheu Valley, Nuku Hiva.*

of Pua (north coast) were also present in Ua Huka (probably Vaipaee), Hiva Oa and Ua Pou (von den Steinen 1925, 17; Kellum-Ottino 1971, 43). This connection may explain the presence of dog images in Ua Huka.

To summarize, archaeological evidence shows that the Polynesian domestic dogs not rats or llamas were present in the Marquesas Islands since initial settlement and into the early part of the historic period. However, excavation suggests that dogs were scarce, they were eaten, and they were, in some cases, venerated to the point where they were intentionally buried with humans. Rock art research clearly demonstrates that the dog images were exotic looking and on Nuku Hiva Island, where the images are only found on the northeast coast, dog images were associated with chiefly architecture and sometimes on religious and chiefly structures within the tribal ceremonial complexes. This suggests that they served as a regional identity and tribal solidarity.

More research on the Marquesan dog is certainly needed. However, studies on dog mtDNA, similar to those done on the Pacific rat (Matiso-Smith *et al.* 1998), is underway at the University of Auckland (Matiso-Smith, *pers. comm.*, 2002). Nevertheless, to my knowledge, little has been written on the morphology of the bone structure, or a study of the dog burials found at Hane and Manihina, Ua Huka: nor has the movement of dogs from Near to Remote Oceania been traced linguistically. In addition, a study of the socio-economic role of the dog is sorely needed to place the dog in a broader Marquesan and Oceanic context.

Acknowledgements

I wish to thank Sharyn Jones O'Day for inviting me to participate in the session "Beyond Calories: The Zooarchaeology of Ritual and Religion" at the ICAZ Conference, Durham, UK, 2002. It is always a great pleasure to be associated with Sharyn. I greatfully acknowledge my two daughters Jessica Gyps Millerstrom and Nikolett Fouassier, my greatest supporters, for reading and commenting on earlier drafts of this paper. Field research is never done in isolation. Edmundo Edwards, various volunteers, field assistants, and I worked together in the Marquesas from 1984 to 1989. I thank them for friendship and support. While a considerable amount of work on rock art has taken place in the archipelago in the 1990s, especially on Nuku Hiva, much of my work is based on those first field seasons. I especially wish to acknowledge Heidy Baumgartner Lesage for assistance in the field as well as in helping in many other capacities. As always, I am grateful to the Government of French Polynesia for permitting me to do field research in the archipelago.

References

Anderson, A. 1981. Pre-European hunting dogs in the South Island, New Zealand. *New Zealand Journal of Archaeology* **3**, 15–20.

Bay-Peterson, J. A. 1983. Competition for resources: The role of pig and dog in the Polynesian agricultural economy. *Société des Océanistes* **77(39)**, 121–9.

Brousse, R., Denizot, M. and Salvat, B. 1978. Etude géomorphologique des Iles Marquises. *Cahiers du Pacifique* **21**, 9–74.

Crook, W. P. 1790s. Account of the Marquesas Islands (in Geogre M. Sheahan's typescript *Marquesas Source Materials* 1952). Unpublished manuscript. Sidney: Mitchell Library.

Davidson, J. M. 1984. *The Prehistory of New Zealand.* Auckland: Longman Paul.

Edwards, E. 1985a. *Eléments de description du tohua Pehe Kua, vallée de Puamau, Hiva Oa.* Tahiti: Département d'Archéologie, Centre Polynésien des Sciences Humaines Te Anavaharau.

Edwards, E. 1985b. *Archaeological survey of Hiva Oa and Ua Huka.* Tahiti: Département d'Archéologie, Centre Polynésien des Sciences Humaines Te Anavaharau.

Edwards, E. 1989. Peintures rupestres et pétroglyphs des Iles Marquises. *Te Ana* **1**, 65–71.

Edwards, E. and Millerstrom, S. 1995. Peintures rupestres de la Vallée de Eiaone à Hiva Oa. *Bulletin de la Société des Etudes Océaniennes* **267**, 5–17.

Forbes, D. W. 1992. *Encounter with Paradise; Views of Hawaii and its People, 1778–1941.* Honolulu: Academy of Arts.

Forster, G. 1777. *A Voyage Round the World ...Resolution, Commanded by Captain James Cook. Vol. 2.* London: Whik.

Forster, J. R. 1778. *Observations made during a Voyage round the World on Physical Geography, Natural History, and Ethic Philosophy.* London: G. Robison.

Gracia, M. 1843. *Lettres sur les Iles Marquises ou Mémoires pour servir a l'Etude Religieuse, Morale, Politique, et Statistique des Iles Marquises et de l'Océanie Orientale.* Paris: Gaume Frères Libraires.

Handy, E. S. C. 1923. *The Native Culture in the Marquesas* (Bernice P. Bishop Museum Bulletin 9). Honolulu: B. P. Bishop Museum.

Handy, E. S. C. 1927. *Polynesian Religion* (Bernice P. Bishop Museum Bulletin 34). Honolulu: B. P. Bishop Museum.

Handy, E. S. C. 1930a. *Marquesan Legends* (Bernice P. Bishop Museum Bulletin 69). Honolulu: B. P. Bishop Museum.

Heyerdahl, T. and Ferdon, E. N. (eds). 1965. *Reports of the Norwegian Archaeological Expedition to Easter Island and the East Pacific* (Monographs of the School of American Research and the Kon-Tiki Museum 24/2). Stockholm: Forum Publishing House.

Heyerdahl, T. 1965. The statues of the Oipona *me'ae*, with comparative analysis of possibly related stone monuments, pp. 123–51 in Heyerdahl, T. and Fendon, J. E. N. (eds), *Reports of the Norwegian Archaeological Expedition to Easter Island and the East Pacific* (Monographs of the School of American Research and the Kon-Tiki Museum 24/2). Stockholm: Forum Publishing House.

Joppien, R. and Smith, B. 1985. *The Art of Captain Cook's Voyages, The Voyages of the Endeavour 1768–1771. Vol. 1.* London: Yale University Press.

Kellum-Ottino, M. 1971. *Archéologie d'une Vallée des Iles Marquises, Evolution des Structures de l'Habitat à Hane, Ua Huka* (Société des Océanistes 26). Paris: Musée de l'Homme.

Kirch, P. V. 1973. Prehistoric subsistence pattern in northern Marquesas Islands, French Polynesia. *Archaeology and Physical Anthropology in Oceania* **8**, 24–40.

Kirch, P. V. 1982. Advances in Polynesian prehistory: three decades in review. *Advances in World Archaeology* **1**, 51–97.

Kirch, P. V. 1984. *The Evolution of the Polynesian Chiefdoms* (New Studies in Archaeology). Cambridge: Cambridge University Press.

Kirch, P. V. 1986. Rethinking east Polynesian prehistory. *Journal of the Polynesian Society* **95**, 9–40.

Kirch, P. V. 2000. *On the Road of the Winds; An Archaeological History of the Pacific Islands before European Contact.* Berkeley: University of California Press.

Kirch, P. V. and Ellison, J. 1994. Paleoenvironmental evidence for human colonization of remote Oceanic islands. *Antiquity* **68**, 310–21.

Kirch, P. V., and Hunt, T. L. (eds). 1997. *Historical Ecology in the Pacific Islands; Prehistoric Environmental and Landscape Change.* London: Yale University Press.

Linton, R. 1923. *The Material Culture of the Marquesas Islands.* (Memoirs of the Bernice P. Bishop Museum 8/5). Honolulu: B. P. Bishop Museum.

Linton, R. 1925. *Archaeology of the Marquesas Islands* (Bernice P. Bishop Museum Bulletin 23). Honolulu: B. P. Bishop Museum.

Luomala, K. 1960. The native dog in the Polynesian system of value, pp. 190–240 in Diamond, S. (ed.), *Cultural In History: Essays in Honor of Paul Radin*, New York: Columbia University Press

Matiso-Smith, E., Roberts, R. M., Irwin G. J., Allen, J. S., Penny, D. and Lambert, D. M. 1998. Patterns of prehistoric human mobility in Polynesia indicated by mtDNA from the Pacific rat. *Proceedings of the National Academy of Science Online* **95(25)**, 15145–50 (http:/www.pnas.org/cgi/content/full/95/25/15145).

Millerstrom, S. 1985a. Rock art in the Marquesas Islands; A preliminary report. *La Pintura, American Rock Art Research Association, Newsletter* **12**, 7.

Millerstrom, S. 1985b. Up-date on Marquesan rock art. *La Pintura, American Rock Art Research Association Newsletter* **12**, 5–6.

Millerstrom, S. 1986. *Field Reports from the Marquesas Islands.* Tahiti: Département d'Archéologie, Centre Polynésien des Sciences Humaines Te Anavaharau.

Millerstrom, S. 1990. *Rock Art of Marquesas Islands, French Polynesia: a Case Study of Hatiheu Valley, Nuku Hiva.* Master of Arts in Anthropology, San Francisco State University.

Millerstrom, S. 1992. Report on the Marquesas Islands rock art project. *Pacific Arts* **6**, 19–25.

Millerstrom, S. 1997. Carved and painted rock images in the Marquesas Islands, French Polynesia. *Archaeology in Oceania* **32**, 181–96.

Millerstrom, S. 2001. *Images Carved in Stones and Settlement Patterns: Archaeology in Haitheu Valley, Nuku Hiva, The Marquesas Islands, French Polynesia.* Ph.D. Dissertation, University of California at Berkeley.

Millerstrom, S. and Edwards, E. 1998. Stone sculptures of the Marquesas Islands (French Polynesia), pp. 5–10 in Stevenson, C. M., Lee, G. and Morin. F. J. (eds), *Easter Island in Pacific Context, South Seas Symposium. Proceedings of the Fourth International Conference on Easter Island and East Polynesia, August 1997.* Albuquerque: Easter Island Foundation, University of New Mexico.

Oliver, D. L. 1974. *Ancient Tahitian Society. Vol I.* Honolulu: The University Press of Hawaii.

Ottino, P. 1992. Archéologie et festival des arts, le cas de Iiopona à Puamau, Hiva Oa. *Bulletin de la Société des Etudes Océaniennes* **21(7/8)**, 256–57.

Pechberty, D. 1996. *La vie aux îles Marquises; Récits de missionnaires (1797–1842). Vol. 6.* Uvea, Wallis: Te Fenua Foou.

Rolett, B. V. 1998. *Hanamiai: Changing Subsistence and Ecology in the Prehistory of Tahuata (Marquesas Islands, French Polynesia)* (Yale University Publications in Anthropology 84). New Haven, Conn.: Department of Anthropology, Yale University.

Rolett, B. V. and E. Conte. 1995. Renewed investigation of the Ha'atuatua Dune (Nuku Hiva, Marquesas Islands): a key site in Polynesian prehistory. *Journal of the Polynesian Society* **104**, 195–228.

Sellier, P. 2000. Burial practices of ancient Marquesas Islanders: excavation at Manihina, Ua Huka Island. Paper presented at Pacific 2000, Marquesan History and Prehistory, Hawai'i Preparatory Academy, Kamuela, Hawai'i Island.

Sinoto, Y. H. 1966. A tentative prehistoric cultural sequence in the Northern Marquesas Islands, French Polynesia. *Journal of the Polynesian Society* **75**, 287–303.

Sinoto, Y. H. 1968. Position of the Marquesas Islands in East Polynesian prehistory, pp. 111–8 in Yawata, I. and Sinoto, Y. H. (eds), *Prehistoric Culture in Oceania.* Honolulu: B. P. Bishop Museum Press.

Sinoto, Y. H. 1969. An archaeologically based assessment of the Marquesas as a dispersal center in East Polynesia, pp. 105–32 in Green, R. C. and Kelly, M. (eds), *Studies in Oceanic Culture History* (Pacific Anthropological Records 11). Honolulu: B. P. Bishop Museum.

Sinoto, Y. H. 1979. The Marquesas, pp. 110–34 in Jennings, J. D. (ed.), *The Prehistory of Polynesia.* Cambridge, MA: Harvard University Press.

Sinoto, Y. H. 1983. An analysis of Polynesian migrations based on the archaeological assessments. *Journal de la Société des Océanistes* **39**, 57–67.

Sinoto, Y. H. and Kellum, M. 1965. Preliminary report on excavation in the Marquesas Islands, French Polynesia. Mimeographed report to the National Science Foundation. Honolulu: B. P. Bishop Museum.

Spriggs, M. and Anderson, A. 1993. Late colonization of East Polynesia. *Antiquity* **67**, 200–17.

Suggs, R. C. 1961. Archaeology of Nuku Hiva, Marquesas Islands, French Polynesia. *Anthropological Papers of the American Museum of Natural History* **49(1)**, 1–205.

Terrell, J. (ed.). 1988. Von den Steinen's Marquesan Myths. *Pacific Studies* **14(4)**, 159-73.
Titcomb, M. 1969. *Dog and Man in the Ancient Pacific with Special Attention to Hawaii* (Bernice P. Bishop Museum Special Publication 59). Honolulu: B. P. Bishop Museum.
Valeri, V. 1985. *Kingship and Sacrifice: Ritual and Society in Ancient Hawaii*. Chicago: The University of Chicago Press.
von den Steinen, K. 1925 [1969]. *Die Marquesaner und ihre Kunst. Vol. I. Tatauierung*. Berlin.
von den Steinen, K. 1928 [1969]. *Die Marquesaner und ihre Kunst. Vol. II. Plastik.* Berlin.

Sidsel N. Millerstrom
Postdoctoral Fellow,
Archaeological Research Facility,
Oceanic Archaeology Laboratory,
University of California at Berkeley,
Berkeley, CA 94720, USA
E-mail: miller@sscl.berkeley.edu

9th ICAZ Conference, Durham 2002
Behaviour Behind Bones, (ed. Sharyn Jones O'Day et al.) pp. 153–161

18. Past and present perspectives on secular ritual: food and the fisherwomen of the Lau Islands, Fiji

Sharyn Jones O'Day

Secular rituals, everyday routines of social life, have often been a secondary consideration for anthropologists who have typically focused on religious rituals. Nevertheless, secular ritual practices are formally patterned and of symbolic and archaeological importance. Ethnographic studies in Fiji and Polynesia illustrate that religion, economics, and politics are tightly associated. These institutions hinge on social organizational principles of hierarchy, which are reproduced in practice (through behavior). In Fiji and Polynesia secular and religious rituals can be heuristically divided, but are related within this social system. In this paper, I discuss my recent archaeological and ethnographic work that focuses on secular rituals involving subsistence and food, in the Lau Islands of Fiji. Topics including food production, food distribution, consumption patterns, and refuse disposal are addressed. My preliminary research indicates that animal resource exploitation and consumption patterns are affected by accessibility that is both environmental and cultural. In this case, cultural accessibility refers to patterns of food distribution and consumption that are regulated by an ideology of hierarchy, and played out in ritual actions. I argue that food-associated rituals in the past may be better understood through ethnoarchaeological and historical approaches.

Introduction

This paper focuses on secular rituals associated with subsistence and food, both past and present. I draw on my dissertation research, which centers on a group of islands at the eastern extent of what is politically Fiji, and culturally a mix of Polynesia and Melanesia (Fig. 1). Specifically, my research is in the Lau Archipelago, on the island of Nayau (Fig. 2). Geographically, Lau is situated between the large main islands of Fiji and Tonga. The Lau Group is a cluster of 80 islands, 29 of which are inhabited today, extending north to south across 450 km of ocean. The main Fijian islands of Viti Levu and Vanua Levu are located about 200 km west and 100 km northwest of Lau, respectively. Lau lies about 320 km west of Tonga.

Culturally as well as biogeographically, the Lau Islands are the meeting point between Polynesia and Melanesia. While the precise timing and origins of the cultural relationship between Tonga and Lau remains unknown, the two areas regularly interacted by the 17th century and likely well before 1000 BP (Hocart 1929; Reid 1990; Best 2002). The 17th century Lauan people were intermediaries between chiefly lineages of Tonga and Fiji, and a "House of Fiji" (*Ha'a Fale Fisi*) was firmly established in Tonga (Hocart 1929; Reid 1990). At European contact the Lauans were engaged in exchange networks that spanned much or all of Fiji, Tonga, and Samoa. Anthropologists have suggested that Lau was a central political node between Polynesia and the main Fijian Islands (Hocart 1929; Roth 1953; Best 1984).

On Nayau I use ethnographic and archaeological research to explore the role of food in culture and the material correlates of subsistence oriented social interactions. In particular, I study contemporary food associated behaviors in order to compare archaeological materials from precolonial household contexts. For the purposes of this paper, I focus on gender and age specific food production (especially inshore fishing and collecting), distribution, consumption, and refuse disposal among the inhabitants of Nayau's three villages (Salia, Narocivo, and Liku). I argue that secular rituals may be understand through the material remains left by the people of the past and by comparison with the social world of living peoples. In

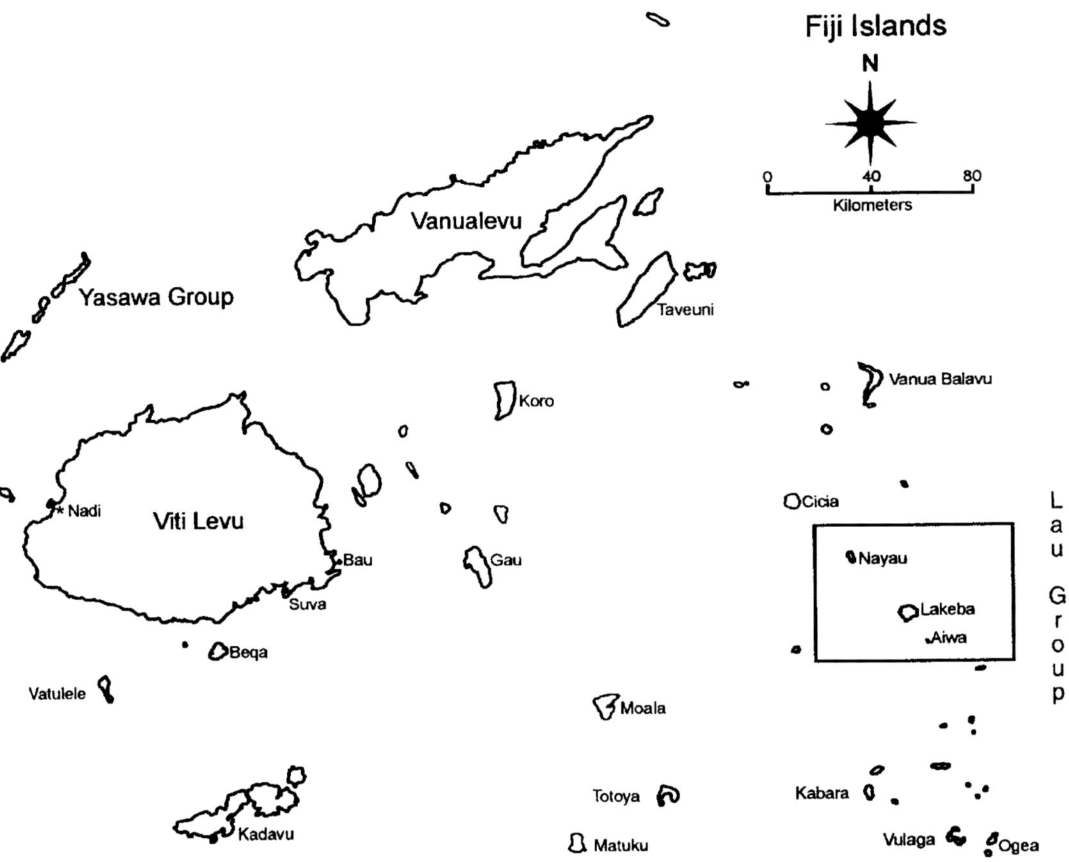

Fig. 1. The Fiji Islands denoting island locations that are the focus of the research described herein (Aiwa, Lakeba, and Nayau indicated by the square).

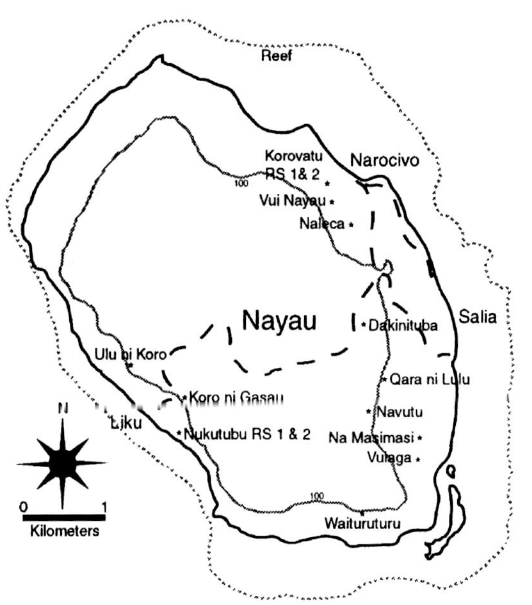

Fig. 2. Nayau, showing three villages (Salia, Narocivo, Liku) and the known archaeological sites by name. Contour line is 100 m and — = trails.*

Fiji, a traditional ideology of hierarchy is reinforced and perpetuated through social norms and rituals surrounding food and food-associated activities.

Ritual and hierarchy

Over the years anthropologists have generated a variety of definitions for ritual and even more perspectives from which the phenomena can be interpreted (*e.g.*, Frazer 1955; Levi-Strauss 1963; Firth 1967; Turner 1969; Geertz 1973; Leach 1965; Bourdieu 1977; Tambiah 1985; Rappaport 1999). I recognize two principle divisions or types of ritual. The first is *religious* ritual, for example, human sacrifice and other activities that take place on sacred sites or temples. The second type is classified as *secular*. Initially, secular ritual may be more difficult to imagine, especially from an archaeological perspective. Yet, secular rituals are found in everyday practice and as such are more common than religious rituals. Examples of secular rituals include the way people eat (Douglas 1975) and the way they orient their houses (Kirch 1996; Toren 1999). Indeed, I posit that formal patterned behavior, the same behavior that produces the archaeological record (Shennan 1993), is the true essence of secular ritual. I will demonstrate this point in the following discussion. In my usage here, specific to the culture areas of Polynesia and Fiji, *secular ritual* refers to formal patterned behavior that is a medium for the expression of hierarchy.

Rituals are repetitive patterned behaviors that have social meaning, which may be subconscious or conscious. Broadly defined, ritual is, "a complex sociocultural medium, variously constructed of tradition, exigency, and self-expression; it is understood to play a wide variety of roles and to communicate a rich density of over determined messages and attitudes" (Bell 1997, xi). Following Bell, I believe that, "the efficacy of ritual activity lies in its ability to have people embody assumptions about their place in the larger order of things" (*ibid.*). Ritual's vocabulary consists of gestures, words, and actions. Rituals are also a means of expression. Leach (1965, xiv, 12–16) argued that ritual was a communicative aspect inherent in *all behaviors*. He did not make a distinction between sacred and secular, believing that this is an arbitrary division. For Leach, ritual symbolizes "the system of socially approved 'proper' relations between individuals and groups" (1965, 15). Undoubtedly, ritual actions facilitate the expression of a person's identity and self to the community (Leach 1965; Leach 1976; Deverell 1986). Turner (1992), Becker (1995), and Toren (1990 and 1999) have clearly and definitively demonstrated this phenomena in Fiji from an ethnographic perspective.

What are the implications of secular rituals for archaeology? Importantly, secular rituals as evidenced by patterned everyday practices produce material residues that compose the archaeological record. All human interactions in Fiji are guided by socially correct and expected behaviors. These behaviors are determined by social morals and standards that are learned at childhood, strictly enforced throughout one's life cycle, and reinforced though continued action; 'proper behaviors' are patterned and repeated in all social situations. It should be noted that virtually all situations are 'social' because village life is communal, individuals are virtually never alone and, unlike Western norms, the community rather than the individual is the focus (see Toren 1990; Becker 1995; Lawson 1997; Toren 1999). Although the view of secular ritual I employ is broad, it recognizes that ritual – as patterned behavior – is accessible to archaeologists through material correlates. This approach provides an avenue to examine thought and the ideational realm, which are often overlooked or simply assumed to be inaccessible to archaeologists.

Worldview

In Fiji and Polynesia secular and religious rituals are related within a holistic social system. Unlike the Western tradition of dividing a given society into more easily approachable categories, in this cultural reality the parts form a conflated totality, often making categories difficult to delineate when examining ethnographically recorded behavior (*e.g.*, Leach 1965; Weismantel 1988; Weiner 1992). Hocart (1970, 256) argued that in Fiji there is no religion, "...only a system that in Europe has been split up into religion and business." Economic and food related activities were at the time of European colonization, and are today, intricately related to politics and cosmology. These rituals characterize the social and cosmic order of society by repeatedly establishing hierarchies based on birth order, age, and gender.

In Lau and Polynesia rituals associated with food production and the harnessing of raw materials from the natural world, the realm of the gods, is described as the 'work of the gods,' a fundamentally religious process (Handy 1927; Firth 1967). Humans reaffirm and recreate their hierarchical relation to the gods, or ancestors, through ritualized patterned acts of production. Few anthropologists have included or addressed women's work (or labor other than planting, which is a male activity) in this formulation. The assumption that only male-associated food production holds religious significance is unlikely (Ortner 1981; Linnekin 1990; Weiner 1992; Ortner 1996). The remarkably widespread ideology surrounding food production (as sacred and ritualized) suggests an institution that is anchored in the deep values of Fijian and Polynesian society. This ideology has endured Colonization and Christianization, suggesting that it existed long before Western contact. Neo-traditional secular rituals as they are practiced today – including subsistence production and food consumption and distribution patterns – provide points of connection between social and material phenomena (*e.g.*, food, spatial organization of the house and village, cooking technologies).

Fijian domestic life is the structural parallel to more grandiose rituals, both of which renew connections between cosmological creation, the ancestors, and humans (see Eliade 1959). The practices of eating food, and drinking *yaqona* (also known as *kava*, a mild narcotic made into a traditional drink), are ritualized forms of sacrifice and acts of *mana* (supernatural efficacy) (Sahlins 1983; Toren 1998; Toren 1999). These rituals are strictly patterned by proscribed actions. Despite colonialism and Christianization of the Fijian Islands, "... production, exchange, tribute, and consumption... are still distinctly Fijian... [and] ...eating together still defines the household..." (Toren 1998, 113). Every meal is structured as a feast (when the primary food is prepared cooked or raw) at a much smaller scale and less formal, just as every man is treated as a chief in his home. Hocart notes that a "small private feast is often called an oven (*lovo*)" (Hocart 1929, 75), a *lovo*, or earth oven, is used at least once a week to prepare food, but often more regularly depending on the household and occasion. Normal patterns of food consumption and what constitutes a 'meal' at a basic level may be viewed as small-scale feast (but, with less quantity overall, no pudding, and lacking an abundance of meat of domestic animals – chicken, pig, dog, cow) (Becker 1995,

58; Kirch 2001). A proper 'meal,' according to traditional Fijian and Polynesian customs, is composed of 'true food,' starchy crops (*e.g.*, taro, yams, sweet potatoes, cassava, breadfruit) and flesh food or 'relish' (most often the products of inshore fishing – shellfish and reef fish, less often meat of domestic animals) (Hocart 1929, 137, 140; see Kirch and O'Day 2003). A prayer is said before each meal, just as it was *before* the islands were Christianized (Handy 1927; Hocart 1929; Thompson 1940; Toren 1999). Hocart recorded that in Lau a woman sometimes prepared for her husband, "...a special evening meal called the awaiting feast (*i wawa*)" (Hocart 1929, 137).

Ethnographers have documented fundamental ideological divisions between male and female social space in Fiji. This ideology is played out in division of labor, in food production activities, and in household and village spatial orientation and organization (Fig. 3). Fijians refer to natural and human constructed landscapes in terms connotating rank, which literally translate into 'high' and 'low' (Hocart 1929; Sahlins 1962; Toren 1990; Toren 1999). Women are associated with the 'low' and/or 'inside' areas in space (within the household, village, and beyond) while men are associated with the 'high' and 'outside' space (*ibid.*). For example, the offshore area of the island/reef is referred to as the *waitui*, or noble sea. Likewise, the most sacred door of a house is often the 'honored door,' facing the sea (Sahlins 1962; Toren 1990; Toren 1999). The eastern, male side of the house is the 'noble side.' In the household women position themselves 'below' (*i ra*), by the 'common entrance,' or, literally, the 'door of the cooking irons' (*darava i sue*) (Toren 1999, 52, Fig. 2.2). This data conforms well with ethnographic work elsewhere in Polynesia (*e.g.*, Hocart 1929; Thompson 1940; Gailey 1987; Malm 1999). In this worldview, women assume the lower, inside position within the household and village, both ideologically and physically. Women are associated with the inshore area and the village generally; the 'inside' is the axis of women's primary social domain and the focus of their labor. A corollary of this ideology is evident in the fact that most men – including those holding public power – consider women's work to be a non-prestigious activity (Sahlins 1962; Kirch and Dye 1979; Malm 1999). Worldview is mapped into Fijian villages and individual house plans. This point is critical from an archaeological perspective as evidence of a direct link between ideological phenomena and material phenomena.

Fig. 3. Liku village, Nayau. Inshore-female-inside areas and offshore-male-outside areas indicated.

Food production in Lau

On Nayau women are the primary producers of inshore marine resources (*e.g.*, reef fish, shellfish, and seaweed), which constitute the majority of protein consumed each day by the islands' inhabitants. Of course, there are exceptions to the age and sex make-up of inshore fishing groups, but most often these are comprised of women, adolescents, and children. Importantly, in Fijian society men are generally considered 'adults' only after marriage. Thus unmarried men may accompany and assist women and children with inshore fishing tasks. Women also prepare and serve most of the food in both domestic and feasting contexts. This situation offers an opportunity to investigate gender specific subsistence production and associated social relations, in addition to the material correlates of this economic system.

Until recently, women's, adolescent's, and children's roles in food production have largely been ignored in the ethnographic literature of the Pacific Islands (*e.g.*, Meehan 1982; Kahn 1986; Gailey 1987; Malm 1999; Bird and Bliege Bird 2000; Thomas 2002). This statement is not a negative assessment of earlier works. A simple reason for the male focus of the past is access. In Fiji cross-gender interviews are difficult if not impossible, due to the fact that traditionally, respect is shown to the opposite sex by avoidance. For example, men on Nayau generally avoided me, and when necessary would answer questions only through my husband (who served as a sort of gender-specific interpreter). Clearly, male researchers are permitted access to certain parts of social life, while female anthropologists have access to other interactions. Naturally, during the course of my research I was readily permitted to observe women and their daily routines and to participate in female-associated activities.

Theoretical concerns

With a recent increased ethnographic and archaeological interest in activities associated with marine food production, we can expect to become more familiar with these aspects of society and their relation to ideology and ritual. A focus on the social realms of production and food generally (both tightly tied to women in most societies geographically and temporally; see Bruch 1973; Meigs 1983; Scheper-Hughes 1992; Sobo 1993; Becker 1995; Bordo 1997; Counihan and Van Esterik 1997; Counihan 1999) is especially relevant to zooarchaeology as it promises to inform about the social relations behind food refuse and food technologies (*e.g.*, pottery and food processing artifacts). Faunal remains and pottery make up a large portion of archaeological materials that are commonly recovered in the Pacific Islands and beyond. The aforementioned studies are therefore relevant to archaeologists and zooarchaeologists interested in social issues including ritual and religion.

Many recent 'ethnoarchaeological' studies on Pacific Island food production are entrenched in functionalist and evolutionary type approaches (*e.g.*, Bird and Bliege Bird 2000; Bird *et al.* 2002; Thomas 2002). Such theories are unable to discern aspects of food production, distribution, and consumption that are not strictly economic (although, it is noted that these studies contribute data to a purely economic view of human/animal behaviors). Evolutionary functionalist approaches typically focus on counting calories, issues of maximizing return rates, efficiency, determining costs and fitness, managing risk, predator-prey relations, etc. (see Noss and Hewlett 2001). Meanwhile, ethnographers who study living humans and consider themselves social and cultural anthropologists have long recognized that the role of food in society amounts to much more than calories (Meigs 1983; Sahlins 1983; Weismantel 1988; Scheper-Hughes 1992; Sobo 1993, Becker 1995; Keating 1998; Toren 1998; Counihan 1999; Toren 1999; Keating 2000; Holtzman 2001). Formal evolutionary theory will never account for history, ideology, emic conceptions, the complexity of social environments, and the impacts of these issues on material remains (nor do practitioners of the approach generally attempt this understanding). These aspects of human society are left to anthropologists employing cultural and/or historical approaches (*e.g.*, Marshall 1985; Sahlins 1985; Kirch and Green 1987; Weiner and Schneider 1989; Linnekin 1990; Thomas 1991; Weiner 1992; Nelson 1997; Toren 1999; Kirch and Green 2001; Best 2002; Kirch and O'Day 2003).

While a cultural approach to the past lacks a formal body of scientific theory, it should be viewed as a complement to evolutionary, formal approaches. At a minimum, it is obvious that each view is relevant to examining different sets of questions. The two approaches taken together might even be considered to result in a balanced interpretation of the past, one that accounts for economics, ideology, and long-term history.

Inshore fishing

Women, teenagers, and children on Nayau and throughout Lau, perform the vast majority of the inshore fishing. This includes collecting invertebrates and seaweed by hand and netting, trapping, and spearing fish. Women are not strictly prohibited from other forms of subsistence production, but females ranging from young to late middle ages do focus on this form of production, and it is central to everyday subsistence. Conversely, trolling outside of the reef, which is almost exclusively practiced by men, forms a much smaller portion of everyday food intake. Fishes collected offshore are generally marked for special occasions, particular exchanges, or may be given as tribute to high-ranking individuals.

After interviewing and working with Fijian women, I suggest that inshore marine fishing strategies are comparable to land based hunting in terms of skill and knowledge requirements. As researchers in Polynesia have stated, fishing is by no means a passive activity (Kirch and Dye 1979; Malm 1999). The same is true of what has been termed "collecting and gathering;" this activity is often considered less important than other forms of food production, despite its significant protein contribution and the degree of skill and mastery of technology involved. Inshore fishing and collecting is demanding work, requiring a great degree of skill, speed, and knowledge of the environment. I can personally vouch for the arduousness of the activity, having participated in numerous expeditions during the course of my research. Fijians highly value an individual's ability to work hard and spend much time discussing and reflecting on this. Indeed, strength and the ability to work hard are equated with health and social connectedness (Becker 1995, 50).

The practical knowledge of fisherwomen, adolescents, and children is expressed in their daily routines of food production and participation in communal village life. Certain fish are highly prized, while others are simply noted for their "sweetness." The fish each have specific names and specific taste and value characteristics that all who participate in fishing expeditions know and discuss.

Food distribution, preparation and consumption

In Fiji, a person's personal achievements are indexed by their connectedness with and performance of care in the social matrix of the family, clan, and village (Becker 1995, 57). Food distribution and exchange, especially in situations where food comes from women, is conditioned by an ethos of nurturing and care. During and after fishing, women and their helpers (adolescents and children) are congratulated for certain choice catches and the participants are generally jovial. *Choice fish are generously given away to either the owner of the net that was used, often a high-ranking male related within the clan, or divided up among the participants along lines of rank, to be taken home by each person.* In this way, daily routines form and contribute to the personality of individuals and social institutions/ideology. Becker (1995, 66-7) states that feeding and food sharing in the household is the primary means of conceiving and maintaining social relations and organizing domestic activity.

When a net haul is brought on shore the women are quick to prepare the fish by scaling and gutting it. While preparing the fish the group often stops on the beach and consumes some of the smaller catch and invertebrates raw. In addition, the fishing party eats the fish gonads while cleaning the animals, before taking the catch back to the village. These behaviors certainly add to the diet, but may be totally invisible archaeologically. Another notable pattern related of food consumption occurs when women are fishing and commonly collect and consume invertebrates as snacks. This material never makes it to the village, nor does it enter the archaeological record.

Women take great pride in their fishing skills and their inshore haul. When a girl or women enters into a village with a long string of fish over one shoulder, everyone looks on enviously and comments on the fish, while the woman moves with an erect pride that fails to conceal her excitement. She will take her fish to her house where it will be prepared (most often by boiling or frying and mixing it with coconut milk), and distributed along lines of rank to all the household members. Men provide 'true food' (starch and root crops) through their role as gardeners, while women provide 'relish' or sea-foods. Together, these items create the proper meal, as discussed above (Hocart 1929, 137; Toren 1999, 75). The woman serves the fish to her husband, father, and/or older brother before eating.

In Lau men eat before women and children, leaving what are generally thought to be non-choice portions of fish and other foods. In my experience in Lau, men almost always eat the fish heads and larger fish, while women and children eat smaller fish, the tails, and suck the meat from what the men have leftover. The same principle applies to the heads and bodies of pigs and other terrestrial animals. Here hierarchies of age and gender are expressed, created, and perpetuated each time people eat. Rather than looking at these practices as a negation of female power, I suggest that the way Fijian women use food is as an agent of empowerment (also see Becker 1995, 73). Through patterns of food distribution women display both their generosity and competence as caregivers, at the same time reinforcing hierarchical social relations that are a hallmark of Fijian and Polynesian social systems.

Discovering secular ritual in the past

In conclusion, I briefly discuss the archaeological data I collected thus far, and my future research expectations and strategies to discover secular rituals in the past. I am currently analyzing the material remains from test excavations in seven households on Nayau. I also focus on marine resource use to better understand how ethnographically recorded behaviors are reflected by zooarchaeological evidence. That is, in modern refuse middens and in archaeological refuse associated with households. This work is still in progress, so my interpretations at this point are preliminary.

My analysis of zooarchaeological remains from Nayau indicates a dependence on inshore marine resources in the past, just as research has documented in the present. Archaeologically, I aim to determine if food remains simply reflect patterns influenced by the local availability of resources, or if there is a discernable cultural aspect to this patterning. By comparing assemblages from various households within each of Nayau's three villages, I expect that prehistoric animal resource exploitation patterns were, just as they are now, affected by accessibility that is environmental and cultural. By cultural accessibility, I mean practices that are regulated by an ideology of hierarchy. If high-ranking persons within the village eat more choice foods and more abundant amounts of luxury foods (see Kirch and O'Day 2003), then this pattern should be reflected in whatever remains constitute the deposited record that can potentially be excavated archaeologically.

Refuse disposal, household spatial organization and history

An understanding of food disposal patterns holds the most potential for tracing archaeological remains to the ideology of the past. Unfortunately, as I have witnessed on Nayau, much of food refuse is fed to the pigs or is consumed by dogs. Recognizing and documenting this unfortunate taphonomic reality begs the question: What behaviors result in a household's archaeological record? An ethnoarchaeological approach will assist in answering this question.

The Fijian *lovo*, or earth oven, provides a promising avenue of research into ritual. The *lovo* is the most commonly encountered sub-surface archaeological feature in my experience excavating on three islands in Lau. This feature, commonly used in contemporary cooking practices on Nayau and throughout Fiji and Polynesia, is constructed by digging a hole in the ground and filling it with fuel for fire (usually wood chips and small branches, then larger fuel/wood to burn). Then, the wood is lit and topped with stones. While the stones are heating, the food (root crops, fish, pig, or any other item) is wrapped in leaves (usually banana, palm, or breadfruit leaves). When the stones are hot they are spread out and covered with leaves, on which the wrapped food is placed. After the oven is full of food, it is covered with old palm leaf baskets, sacks, rubbish, and anything that will burn, and is buried under more rocks and dirt (see Hocart 1929, 138). *Lovo* leave an unmistakable archaeological signature of burned and fire-cracked rock, ash, charcoal lenses, and loose sediment. They typically contain numerous faunal remains (fragments of mammal bone, fish bones, and shell) and often artifacts (rubbish deposited to fill the hole). The *lovo* is also a key indicator of secular and more formal ritual cooking preparations. Today, food is prepared in a *lovo* at least once a week, often on Sunday, and sometimes more frequently depending on household activities (*e.g.*, if visitors are present or during certain important occasions). Thus, *lovo* are material representations of secular rituals; as Hocart stated, a *lovo* constitutes a mini-feast (1929, 75). A micro-scale analysis of their contents should prove informative, as I have demonstrated elsewhere (O'Day 2001). I will continue to explore the contents of *lovo* features, in order to uncover archaeological patterns and variability.

Household spatial organization may also provide salient clues to finding and deciphering ritualized behavior in prehistory. Initially, I am targeting late prehistoric and contact period households for faunal, spatial, architectural, and other material data that might yield insights into these patterns. Because there appears to be strong continuity between past and present spatial organization, I expect that these material correlates of social behaviors may be indicative of similar behaviors in the past. I also expect that the extent of inshore fishing activities, the degree of the marine contribution to the overall diet of people of various rank, and the social value of this activity may have all varied through time in ways perceptible in the zooarchaeological remains. I have started by carefully mapping archaeological households and villages and will move on to map modern households and villages.

Eating together defines the household (Toren 1990; Becker 1995; Toren 1998; Toren 1999), while domestic groups (often extended families related through brothers who live in close proximity) are defined by a shared kitchen or cookhouse (Sahlins 1962). Based on surveys of Nayau, these elements are visible archaeologically and are virtually identical to modern patterns of domestic group organization within villages. Clusters of house mounds and a shared kitchen structure have been documented archaeologically and mapped on previous visits to Nayau. These will be excavated to examine patterns and/or variations in material remains (faunal, lithic, and pottery).

In Fiji, chiefly houses are traditionally, "...built on an earth foundation (*yavu*), whose height denotes the rank of its owners" (Toren 1999, 53). The pattern persists today in Lau the house of the village and/or island chief is always built on a *yavu* elevated above that of that of the surrounding commoners (in addition, chiefly houses are often built in a prominent or central position in the village). Likewise, the relative status of a given prehistoric house's inhabitants should be clearly evidenced in the relative height of the *yavu*. During surveys of Nayau numerous house mounds of various sizes were documented and mapped in detail. I expect that excavations of these structures will reveal patterns in zooarchaeological remains that are also indicative of rank. To bolster this *in situ* data, oral traditions (*Tukutuku Raraba*), recorded in 1938 by the Native Land Commissions, refer to some *yavu* by name. The *yavu* name also refers to a senior line of a clan (*matagali*) or lineage (*yavusa*), which still exist on Nayau; the clans themselves maintain certain oral traditions and can add to an understanding of the people that inhabited houses built on particular *yavu*.

One way to detect differences in patterns of food consumption between archaeological households is by calculating and comparing diet breadth. Inter-assemblage variability can be tested using the Shannon-Wiener Measure of niche breadth (Krebs 1989) or diversity when applied to zooarchaeological data. This will measure the degree of specialization found in, for example, commoner *vs.* elite households. The measure quantitatively characterizes the variety of animals used within an assemblage and the relative importance of the species present. Biomass estimates also may provide valuable data for determining the relative importance of faunal groups within and between assemblages (Reitz and Wing 1999). Additionally, simple qualitative comparisons (*e.g.*, absence and/or presence of certain food items, evidenced by faunal remains) may be informative. When used in conjunction with other zooarchaeological measures and archaeological data a clearer picture of prehistoric subsistence may emerge.

Conclusions

Female organized fishing rituals and their products are clearly an expression of personal identity. Becker (1995) and Toren (1990; 1998; 1999) have thoroughly documented the relationship between the procurement and sharing of food resources and the recreation and maintenance of correct nurturing behavior. On Nayau, inshore fishing expeditions are vested community efforts that build, regulate, and negotiate social relations symbolically through food production, and subsequent distribution, exchange, and consumption. Food that is given away or exchanged is a show of proper behavior that 'makes sense' within the Fijian worldview. The simultaneous ideological core structures of *care* and *hierarchy* are reinforced through living, giving, and eating.

Everyday living patterns, including working, sharing, eating, and the construction of social space are secular rituals. Through these rituals Fijian quotidian domestic life is the structural parallel to grandiose rituals, both of which renew connections of cosmological creation to ancestors and humans. I do not assume, *a priori*, that there is an idealized ancient form of social and economic organization that remains fixed and structurally predictable – rather, I will look to the archaeology and to the ethnography to determine if parallels exist. Only archaeology can answer questions of history and long term social stasis and/or change. Everyday rituals surrounding the production, distribution, and consumption of food, which are regulated by hierarchy, ultimately produce the archaeological record. Through these rituals, I hope to find the articulation between ideology, history, and the material phenomena that archaeologists and zooarchaeologists deal with every day.

Acknowledgements
This research was conducted over two trips to the Lau Islands, Fiji, totaling five months of data collection in 2001 and 2002. I gratefully acknowledge Na Gone Turaga Na Tui Lau, Tui Nayau Ka Sau ni Vanua ko Lau, The Right Honourable Ratu Sir Kamisese Mara for allowing me to conduct research on Nayau. I thank the Fiji Museum and Sepeti Matararaba for their important assistance. I am especially grateful to the people of Nayau and Lakeba for welcoming and facilitating my research, in addition to providing me with a wealth of local knowledge. *Vinaka vakalevu* Jack, Sera, Bola, Akanise, Colati, and Rusila. I also thank David Steadman and the Florida Museum of Natural History for support and use of multiple laboratories and comparative collections. This research was funded by an NSF grant to D. W. Steadman (EAR 9714819) and a NSF Graduate Student Research Fellowship to S. J. O'Day.

References

Becker, A. E. 1995. *Body, Self and Society: A View From Fiji*. Philadelphia: University of Pennsylvania Press.

Bell, C. 1997. *Ritual: Perspectives and Dimensions*. Oxford: Oxford University Press.

Best, S. L. 1984. *Lakeba: the Prehistory of a Fijian Island*. Ph.D. Thesis, University of Auckland. Ann Arbor, MI: University Microfilms.

Best, S. L. 2002. *Lapita: A View From the East* (New Zealand Archaeological Monograph 24). Auckland: New Zealand Archaeological Association Publications, Auckland Museum.

Bird, D. W. and Bliege Bird, R. 2000. The ethnoarchaeology of juvenile foragers: shellfishing strategies among Meriam Children. *Journal of Anthropological Archaeology* **19**, 461–76.

Bird, D. W., Richardson, J. L., Veth, P. M. and Barham, A. J. 2002. Explaining shellfish variability in middens on the Meriam Islands, Torres Strait, Australia. *Journal of Archaeological Science* **29**, 457–69.

Bordo, S. 1997. Anorexia nervosa: psychopathology as the crystallization of culture, pp. 226–50 in Counihan, C. and Van Estrik, P. (eds), *Food and Culture: A Reader*. London: Routledge.

Bourdieu, P. 1977. *Outline of a Theory of Practice* (Cambridge Studies in Social Anthropology). Cambridge: Cambridge University Press.

Bruch, H. 1973. *Eating Disorders*. New York: Basic Books.

Counihan, C. M. 1999. *The Anthropology of Food and Body: Gender, Meaning and Power*. London: Routledge.

Counihan, C. and Van Estrik, P. (eds) 1997. *Food and Culture: A Reader*. London: Routledge.

Deverell, B. 1986. Introduction, pp. iv-ix in Deverell, G. and Deverell B. (eds), *Pacific Rituals: Living or Dying?* Suva, Fiji: Institute of Pacific Studies of the University of the South Pacific.

Douglas, M. 1975. Deciphering a meal, pp. 249–75 in Douglas, M. (ed.), *Implicit Meanings: Essays in Anthropology*. London: Routledge & Kegan Paul.

Eliade, M. 1959. *The Sacred and the Profane: The Nature of Religion*. New York: Harcort Brrace Jovanovich.

Firth, R. 1965. *Primitive Polynesian Economy*. London: Routledge.

Firth, R. 1967. *The Work of the Gods in Tikopia*. New York: Athlone Press.

Frazer, J. G. 1955 [1911]. *The Golden Bough: A Study in Magic and Religion* (3rd edition). London: Macmillan.

Gailey, C. 1987. *Kinship to Kingship: Gender, Hierarchy, and State Formation in the Tongan Islands*. Austin: University of Texas Press.

Geertz, C. 1973. *The Interpretation of Cultures*. New York: Basic Books.

Handy, E. S. C. 1927. *Polynesian Religion* (B. P. Bishop Museum Bulletin 34). Honolulu: Bishop Museum Press.

Hocart, A. M. 1929. *Lau Islands, Fiji* (B. P. Bishop Museum Bulletin 62). Honolulu: Bishop Museum Press.

Hocart, A. M. 1970 [1936]. *Kings and Councillors*. Chicago: Chicago University Press.

Holtzman, J. 2001. The food of elders, the "ration" of women: brewing, gender, and domestic processes among the Samburu of Northern Kenya. *American Anthropologist* **103**, 1041–58.

Kahn, M. 1986. *Always Hungry, Never Greedy: Food and the Expression of Gender in a Melanesian Society*. Cambridge: Cambridge University Press.

Keating, E. 1998. *Power Sharing: Language, Rank, Gender and Social Space in Pohnpei Micronesia*. Oxford: Oxford University Press.

Keating, E. 2000. Moments of hierarchy: constructing social stratification by means of language, food, space, and the body in Pohnpei, Micronesia. *American Anthropologist* **102**, 303–20.

Kirch, P. V. 1996. Tikopian social space revisited, pp. 257–74 in Davidson, J. Irwin, G., Leach, F., Pawley, A. and Brown, D. (eds), *Oceanic Culture History: Essays in Honour of Roger Green* (New Zealand Journal of Archaeology Special Publication). Auckland: New Zealand Archaeological Association.

Kirch, P. V. 2001. Polynesian feasting in ethnohistoric, ethnographic, and archaeological contexts: a comparison of three societies, pp. 168–84 in Dietler, M. and Hayden, B. (eds), *Feasts: Archaeological and Ethnographic Perspectives on Food, Politics, and Power*. Washington DC: Smithsonian Institution Press.

Kirch, P. V. and Dye, T. 1979. Ethnoarchaeology and the development of Polynesian fishing strategies. *Journal of the Polynesian Society* **88**, 53–76.

Kirch, P. V. and Green, R. C. 1987. History, phylogeny, and evolution in Polynesia. *Current Anthropology* **28**, 431–56.

Kirch, P. V. and Green, R. C. 2001. *Hawaiki, Ancestral Polynesia: An Essay in Historical Anthropology*. Cambridge: Cambridge University Press.

Kirch, P. V. and O'Day, S. J. 2003. New archaeological insights into food and status: a case study from pre-contact Hawaii. *World Archaeology* **34**, 484–97.

Krebs, C. J. 1989. *Ecological Methodology*. New York: Benjamin Cummings.

Lawson, S. 1997. Chiefs, politics, and the power of tradition in contemporary Fiji, pp. 108–18 in White, G. M. and Lindstrom L. (eds), *Chiefs Today: Traditional Pacific Leadership and the Post-Colonial State*. Stanford, CA: Stanford University Press.

Leach, E. R. 1965. *Political Systems of Highland Burma: A Study of Kachin Social Structure*. Boston: Beacon Press.

Leach, E. R. 1976. *Culture and Communication*. London: Cambridge University Press.

Levi-Strauss, C. 1963. *Structural Anthropology*. New York: Basic Books.

Linnekin, J. 1990. *Sacred Queens and Women of Consequence: Rank, Gender, and Colonization in the Hawaiian Islands*. Ann Arbor: University of Michigan Press.

Malm, T. 1999. *Shell Age Economics: Marine Gathering in the*

Kingdom of Tonga, Polynesia. Lund, Sweden: Department of Sociology, Lund University.
Marshall, Y. 1985. Who made Lapita pots? A case study in gender archaeology. *Journal of the Polynesian Society* **94**, 205–34.
Meehan. B. 1982. *Shell Bed to Shell Midden*. Canberra City, Australia: Australian Institute of Aboriginal Studies.
Meigs, A. S. 1983. *Food, Sex, and Pollution: A New Guinea Religion*. New Brunswick, NJ: Rutgers University Press.
Nelson, S. M. 1997. *Gender in Archaeology: Analyzing Power and Prestige*. Walnut Creek, CA: Alta Mira Press.
Noss, A. J. and Hewlett, B. S. 2001. The contexts of female hunting in Central Africa. *American Anthropologist* **103**, 1024–40.
O'Day, S. J. 2001. Excavations at the Kipapa Rockshelter, Kahikinui, Maui, Hawai'i. *Asian Perspectives* **40**, 279–304.
Ortner, S. B. 1981. Gender and sexuality in hierarchical societies: the case of Polynesia and some comparative implications, pp. 359–409 in Ortner, S. B. and Whitehead, H. (eds), *Sexual Meanings: The Cultural Construction of Gender and Sexuality*. Cambridge: Cambridge University Press.
Ortner, S. B. 1996. *Making Gender: The Politics and Erotics of Culture*. Boston: Beacon Press.
Rappaport, R. A. 1999. *Ritual and Religion in the Making of Humanity* (Cambridge Studies in Social and Cultural Anthropology 110). Cambridge: Cambridge University Press.
Reid, A. C. 1990. *Tovata I & II*. Suva, Fiji: Fiji Museum.
Reitz, E. J. and Wing, E. S. 1999. *Zooarchaeology*. Cambridge: Cambridge University Press.
Roth, G. K. 1953. *The Fijian Way of Life*. Oxford: Oxford University Press.
Sahlins, M. D. 1962. *Moala: Culture and Nature on a Fijian Island*. Ann Arbor, MI: University of Michigan Press.
Sahlins, M. D. 1983. Raw women, cooked men, and other "great things" of the Fiji Islands, pp. 72–93 in Brown, P. and Tuzin, D. (eds), *The Ethnography of Cannibalism*. Washington, DC: Society for Psychological Anthropology.
Sahlins, M. 1985. *Islands of History*. Chicago: University of Chicago Press.
Scheper-Hughes, N. 1992. *Death Without Weeping: The Violence of Everyday Life in Brazil*. Los Angeles: University of California Press.
Shennan, S. 1993. After social evolution: a new archaeological agenda? pp. 53–9, in Yoffee, N. and Sherratt, A. (eds), *Archaeological Theory: Who Sets the Agenda?* Cambridge: University Press.
Sobo, E. J. 1993. *One Blood: The Jamaican Body*. Albany, NY: State University of New York Press.
Tambiah, S. J. 1985. *Culture, Thought, and Social Action: An Anthropological Perspective*. Cambridge, MA. Harvard University Press.
Thomas, F. R. 2002. An evaluation of central-place foraging among mollusk gatherers in Western Kiribati, Micronesia: linking behavioral ecology with ethnoarchaeology. *World Archaeology* **34**, 182–208.
Thomas, N. 1991. *Entangled Objects: Exchange, Material Culture, and Colonization in the Pacific*. Cambridge, MA: Harvard University Press.
Thompson, L. M. 1940. *Southern Lau, Fiji: An Ethnography* (B. P. Bishop Museum Bulletin 162). Honolulu: Bishop Museum Press.
Toren, C. 1990. *The Social Construction of Hierarchy in Fiji*. London: Athlone Press.
Toren, C. 1998. Cannibalism and compassion: transformations in Fijian concepts of the person, pp. 95–115 in Keck, V. (ed.), *Common Worlds and Single Lives: Constituting Knowledge in Pacific Societies*. Oxford: Berg.
Toren, C. 1999. *Mind, Materiality and History: Explorations in Fijian Ethnography*. London: Routledge.
Turner, J. W. 1992. Ritual, habitus, and hierarchy in Fiji. *Ethnology* **31**, 291–302.
Turner, V. 1969. *The Ritual Process: Structure and Anti-Structure*. Ithaca, NY: Cornell University Press.
Weiner, A. B. 1992. *Inalienable Possessions: The Paradox of Keeping-While-Giving*. Berkeley: University of California Press.
Weiner, A. B. and Schneider, J. (eds). 1989. *Cloth and the Human Experience*. Washington DC: Smithsonian Institution Press.
Weismantel, M. J. 1988. *Food, Gender, and Poverty in the Ecuadorian Andes*. Philadelphia, PA: University of Pennsylvania Press.

Sharyn Jones O'Day
Department of Anthropology,
University of Florida, Turlington Hall,
P.O. Box 117305,
Gainesville, FL 32611-7305, USA
E-mail: srjo@ufl.edu

Part 2

Equations for inequality: the zooarchaeology of identity, status and other forms of social differentiation in former human societies

edited by Wim Van Neer and Anton Ervynck

9th ICAZ Conference, Durham 2002
Behaviour Behind Bones, (ed. Sharyn Jones O'Day et al.) pp. 164–176

19. Early evidence of economic specialization or social differentiation: a case study from the Neolithic lake shore settlement 'Arbon-Bleiche 3' (Switzerland)

Elisabeth Marti-Grädel, Sabine Deschler-Erb, Heide Hüster-Plogmann and Jörg Schibler

Dating by dendrochronology shows that the Neolithic lakeshore settlement 'Arbon Bleiche 3' was founded in 3384 BC and abandoned 3370 BC, after a conflagration. In the excavated area (1100 square metres) – probably half of the former village – 27 buildings were identified. The site has yielded over 32,000 identifiable bones. Domesticated animals (cattle, pig, sheep, goat and dog) and hunted animals (mainly red deer) were equally important for the subsistence. In terms of bone weight, cattle are most frequently represented among the domesticated animals. The distribution of the bones shows that there are considerably higher amounts of pig bones in the southern part of the village whereas cattle are dominant in the northern part. The distribution of the fish bones indicates a similar subdivision within the village by showing higher amounts of coregonid fishes in the southern than in the northern part. These distribution patterns remain the same throughout the duration of the settlement and can neither be related to the chronology of the buildings nor to taphonomic factors. However, the finds do not give evidence for the existence of hierarchical differences among the inhabitants of the village either. The presence of locally produced vessels in foreign styles leads us to the hypothesis that a part of the village population of Arbon Bleiche 3 were immigrated people, probably with different food habits or different economic traditions.

Introduction

In this paper we present the distribution of the faunal remains from the lakeshore settlement of 'Arbon-Bleiche 3' with the aim of elucidating the economic and social organization, as well as the consumption habits of its inhabitants. Studies of bone distributions within Neolithic sites from Central Europe are typically characterised by relatively small bone quantities, bad preservation conditions, long settlement periods, or small excavation sectors within the settlements (*e.g.*, Stampfli 1976a; Stampfli 1976b; Johansson 1981; Chaix 1989; Stampfli 1992; Arbogast *et al.* 1995). At Arbon-Bleiche 3, however, the conditions for the analysis of the distributions of the archaeological finds are much more favourable. The quantity and quality of preservation of the archaeological remains, the size of the excavation area, and the clear settlement structures, are extraordinary. In this paper we can present only summaries of the most relevant results of the bone identifications and of the taphonomic analyses, and only a selection of the bone distributions will be discussed. A detailed presentation of the results will be published in 2004 in the third volume of the series *Die jungsteinzeitliche Seeufersiedlung Arbon-Bleiche 3, Archäologie im Thurgau.*

The site of Arbon-Bleiche 3

The Neolithic site Arbon-Bleiche (47.30°N, 9.25°E; cultural layers *c.* 394 and 396 m a.s.l.) is situated on the south-western, Swiss, shore of Lake Constance (Fig. 1). Originally the settlement was situated directly on the shore of a large, shallow, bay. At present it lies about 600–700 m inland, because the bay gradually silted with sediment deposits from influent rivers. Rescue excavations at the site were carried out between 1993 and 1995, during which an area of about 1100 square metres was investigated. A slight inclination could be observed in the northern area of the excavation (Fig. 2). South of this, the settlement area sloped slightly towards the original lakeshore.

The original extent of the village is not known, but it is estimated that about half of its surface has been excavated

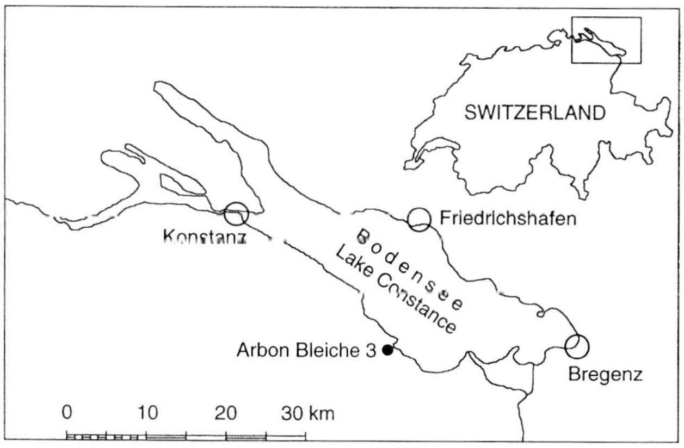

Fig. 1. Location of the site Arbon Bleiche 3.

(Leuzinger 2000). In the humid soil, the conditions of preservation for organic remains such as wood, botanical macroremains and fish scales were outstanding. The remains of piles preserved *in situ* permitted the reconstruction of a total of 27 more or less complete building plans (Fig. 2). By dendrochronology it was possible to establish the construction year of each building. The construction period is dated to between 3384 and 3376 BC. Eight construction "stages" (or periods) could be discerned, as parts of the gradual enlargement of the village. Almost all of the buildings had a ground surface of 4 to 8 metres. Two buildings (17 and 25), that were clearly smaller and which were probably not inhabited, are the exceptions. Based on the building remains an elevated construction can be assumed for all of the buildings. The village was abandoned subsequent to a fire shortly after 3370 BC. It therefore existed for only 15 years. The site was not resettled in prehistoric times. Shortly after the village was abandoned it was covered by a thick layer of sand due to a rise of the water level of Lake Constance. This was caused by a climatic deterioration ascertained in central Europe to between 3370 and 3000 BC (Gross-Klee and Maise 1997).

The cultural layers, which were 5 to 40 cm thick, encompass only one phase of occupation, meaning that after the gradual enlargement of the settlement (finished in 3376 BC) all the houses were inhabited at the same time. The archaeological finds can be typologically attributed to the transition between the Pfyn and Horgen cultures (de Capitani and Leuzinger 1998; de Capitani *et al.* 2002). Aside from the typical local pottery, which is in the Pfyn tradition with Horgen elements, there are vessels with parallels in the early phase of the Baden Culture (Boleraz group) of south-eastern Europe. The importation of flint provides evidence of cultural and economic contact with northern France and northern Italy.

The greater part of the archaeological level has remained below the ground water table after the settlement was abandoned. This explains why the conditions of

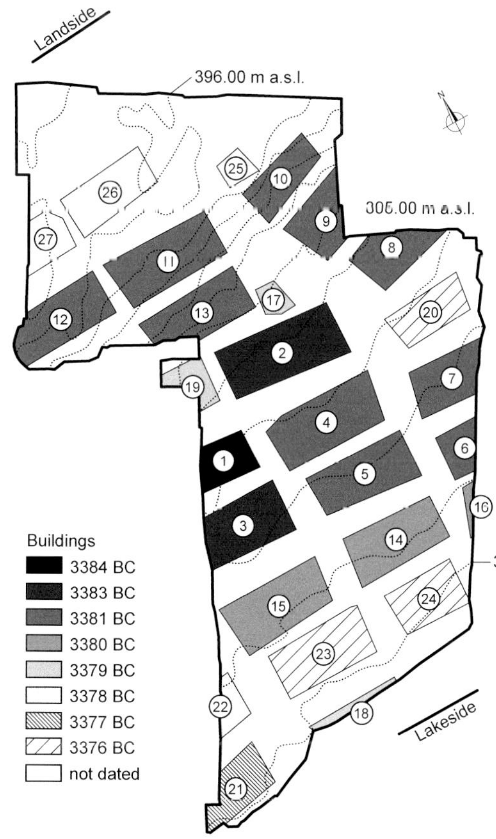

Fig. 2. Settlement plan with altitudes, building plans, and construction phases (after Leuzinger 2000, 12, Fig. 4. and Leuzinger 2002, 21, Fig. 10). The buildings are numbered according to the construction chronology.

preservation for organic remains were so good. However, there was still some variation in the preservation of the archaeological levels within the surface excavated. The most northern area was drained in AD 1944 in order to gain arable land. This led to the decomposition of the organic finds and to the erosion of the archaeological levels within this zone (buildings 12, 26, 27, northern parts of buildings 10 and 11). In the most southerly zone the preservation of the finds is good, although the sediments seemed to be affected by occasional flooding. In the area of excavation in between, the preservation was best, which is reflected in the quantity of finds in all find categories.

Materials and methods

The bone material from Arbon-Bleiche 3 consists of over 70,000 hand-collected remains, gathered per square metre. Furthermore, it comprises about 20,000 remains from small animals from sieved samples. Most of the bone finds represent consumption refuse. All animal bones that could be identified were recorded individually. The exact location of the find (square, layer), the osteological identification (species, skeletal element, the individual's age, fragmentation, dimensions), the weight and the state of preservation (surface preservation, presence and state of broken edges, traces of burning etc.) were documented. For each animal species the individual bones were plotted on a map, as were the fragment proportion and weight proportion using the square metre grid. The aim of the representation by proportions per square metre was to render find concentrations visible. In order to compare the proportions of animal species for the individual buildings directly, we also grouped the bone material by building. Only buildings that were at least half excavated were selected for this comparison. The bones from all square metre plots that were enclosed by a building or intersected by its walls were considered to belong to that particular building. A total of 19 buildings could be analysed by this method. Sixteen of these provided bone quantities large enough to calculate meaningful proportions.

Thanks to dendrochronology the duration of occupation of each of the buildings is known to the year. Therefore we can calculate find concentrations representing the total number or weight of bones of individual species deposited in the ground per square metre and per year. In contrast to the proportions expressed in percentages this method of calculation has the advantage that the values of the individual species do not influence one another.

Results

Taphonomy

The differences in preservation conditions within the cultural layer throughout the settlement are not only visible in terms of find distributions but also visible in the surface preservation of the hand-collected bones. In almost all parts of the excavation the preservation of the bones is very good. However, very few or no bones were found in the most northern part of the excavation (buildings 12, 26, 27, the northern part of buildings 10 and 11) and the few bones recovered were in a poor state of preservation. The proportion of loose teeth, which preserve better than bone, is especially high in this area. This pattern is of course explained by the drainage of this part of the site in 1944 AD, aimed at gaining arable land. In contrast, the preservation of the bone remains from the analysed sieved samples that stem from the middle and south of the excavated area is excellent.

In general, the rounding of the broken edges of the bones is slight, indicating rapid deposition of the finds. The preliminary results of the micromorphological analysis of the cultural layer point towards the same conclusion (Meyer and Rentzel, *pers. comm.*). The location of bones that fit together and that were probably separated only after deposition (epiphyses-diaphyses) indicate a relatively slight displacement of the bone refuse after deposition. Most of the fitting fragments were found in the same, or in the neighbouring, square metres and were therefore within 2 to 3 metres distance at the most. Most of the fitting bone fragments were scattered in all directions, as was also the case with the pottery (de Capitani 2002, 176). This shows that the distribution of the finds does not appear to have been significantly influenced by the inclination of the area nor by flooding during rain or by the transgression of the lake. The investigation of the average bone weight per species and square metre leads to the same conclusion: the bone fragments were not sorted by size or weight as it would be the case if the finds' deposition had been strongly disturbed by flooding. The distribution of bones with traces of carnivore teeth also provides evidence for little disturbance of the original deposition. Bones with bite marks are often found along the passages between the buildings. Had the bones been subject to great disturbance by flooding we would expect a more even distribution covering the whole settlement area. A mixing or redeposition of the bone remains from the sieved samples can also be deemed minimal. The fish scales recovered in the area excavated show concentrations created by man. Strong floods would have distributed the scales more homogeneously.

Unidentifiable bone fragments among the hand-collected bones were not recorded individually, but the quantities of bones that could be attributed only to 'animal groups' are about the same all over the settlement area. It is therefore probable that the proportion of identifiable bones is also more or less the same within the excavated area.

No systematic differences in the age distribution of the individual species within the settlement were found. There is no relationship with possible differences in preservation and deposition (for example of the more fragile bones of young animals).

The skeletal composition of the bones showed some variations within the settlement area for some of the most frequently represented species. In the southern part of the village vertebrae (cattle, red deer, wild boar) and ribs (cattle, red deer, sheep and goat) tend to be more frequent than in the north, while skull fragments (cattle; sheep and goat) are better represented in the buildings of the northern area than in those close to the shore. In the skeletal composition of pig there were no significant differences between settlement areas or buildings. The described differences among the skeletal distributions cannot be explained by differential preservation. It is important to note that they have no influence on the proportions of the species per square metre. Differences in proportions could also be ascertained by considering individual skeletal parts alone (skulls, vertebrae, ribs, extremities).

The estimation of bone loss (for literature see Gross *et al.* 1990) indicates that a maximum of 10 to 20% of the original bone quantity has been preserved and that all species were equally affected. In this the material record in Arbon-Bleiche 3 is exceptional. As a rule a maximum

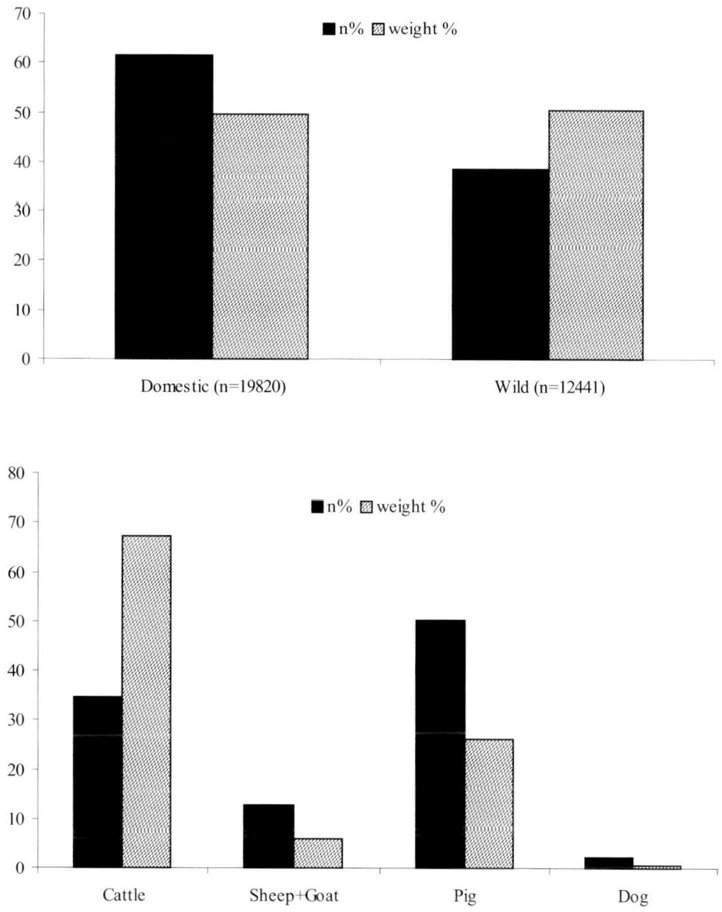

Fig. 3. Proportions of the large animal remains based on number of fragments and bone weight for domestic versus wild animals, and for the individual domestic species.

of only 1 to 10% of the original refuse is found in lakeshore settlements (Gross *et al.* 1990).

Summary of the most important identification results

Over 32,000 hand collected animal bones could be identified to species level. Over 60% of these bones are from domesticated animals (Fig. 3 top). Based on bone weight, the proportion of domestic animals versus wild animals is well balanced. In terms of both fragment numbers and bone weight, about two thirds of the bones of wild animals are from red deer (*Cervus elaphus*).

According to fragment number, pig is the best represented among the domesticated animals (Fig. 3 bottom). Based on bone weight, however, cattle together with red deer were the chief suppliers of meat during the entire settlement period. Because of the large number of individuals (MNI) and the broad age spectrum of the species recovered, we can conclude that the bone remains represent the entire duration of the settlement and that the village was inhabited throughout the year.

Among the 20,214 identifiable bone fragments from the sieved samples, 44% are from fish (Fig. 4). 43% are very small fragments, usually weighing less than 0.01g, from larger mammals as well as a few remains from small mammals. 11% of the bones are from amphibians, mainly represented by frogs (*Rana temporaria*). Only 2% of the bones are from birds.

Overall distribution of the bone material and proportional distribution per square metre

The quantity of hand-collected bones varies with the quality of the preservation of the archaeological level within the settlement (Fig. 5). As stated before, hardly any bone remains were recovered in the most northern area of the site. The largest number of finds comes from the middle area (around building 3).

The distribution of the bone fragments of wild and domesticated animals separately is similar to the overall distribution of bones. The distribution of the proportion of wild animals per square metre shows their importance in the diet within the settlement. A high proportion of wild animals in one isolated square metre may be a random

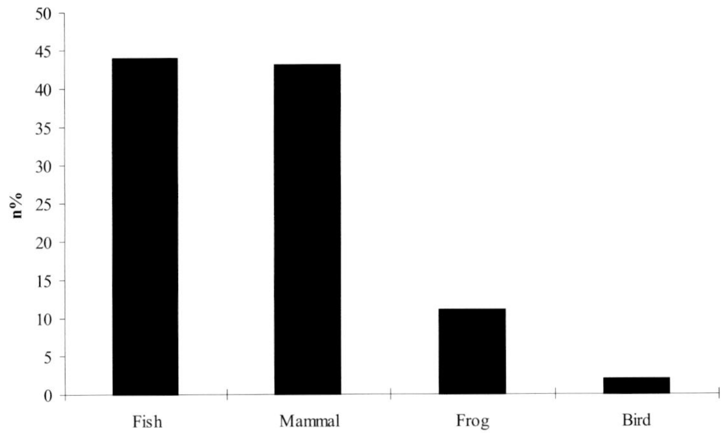

Fig. 4. Identification results of the animal bones recovered from the sieved samples (n=20214).

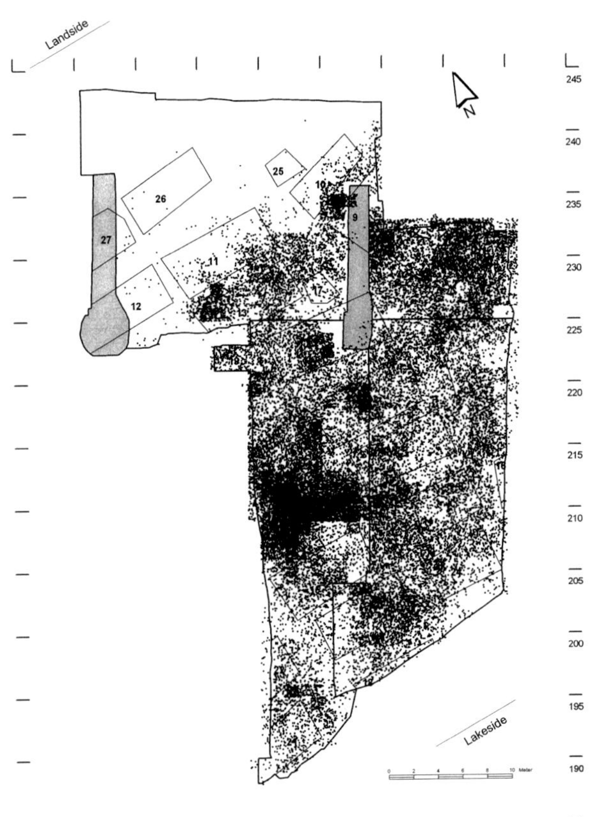

Fig. 5. Overall distribution of the animal bones in the settlement. As the bones were collected by square metre, the find position was randomized within a square metre for individual mapping. The two greyish coloured sections in the north of the site represent older excavations.

effect due to the small quantity of bone material in this specific square metre. However, entire settlement zones with high proportions per square metre can be discerned, for example in the areas with buildings 1 and 2, 8 and 24 (Fig. 6 left). Based on proportional weight, the wild animal remains sometimes exceed 75% of the identifiable animal bones per square metre (Fig. 6 right). In these settlement areas the consumption of game was apparently above average. In other regions of the settlement, for example in the buildings 5, 7, 14 and 23, the proportion of wild animals per square metre was low.

In most of the square metres the proportion of red deer bone fragments (the dominant game animal found in all the areas of the settlement) represents at least 60%, according to the bone weight over 80% of the wild animal bones. High bone concentrations of individual wild animal species, for example *of Meles meles* (badger) and *Mustela putorius* (polecat) could be established (Fig. 7). These concentrations may be an indication of specialized hunting within individual households.

There are clear differences in the distribution of the different domestic animal species. While the proportion of wild animals shows strikingly high values in certain areas of the village, domesticated cattle and pig show different representations in well-separated zones. While high proportions of cattle per square metre characterize the northern part of the settlement, pigs are best represented in the area closer to the shore (Fig. 8 and 9). In the north-eastern part the proportions of both species are similar, based on fragment numbers. The separation between the northern and southern parts of the village is well marked by the proportional bone weight. Cattle bones dominate in the entire settlement area and were thus the most important species for the subsistence of the site (Fig. 10). Domestic pig, however, have much higher values in the southern part of the settlement than in the north according to proportional weight (Fig. 11). A sort of 'dividing line' between these two areas can be drawn from

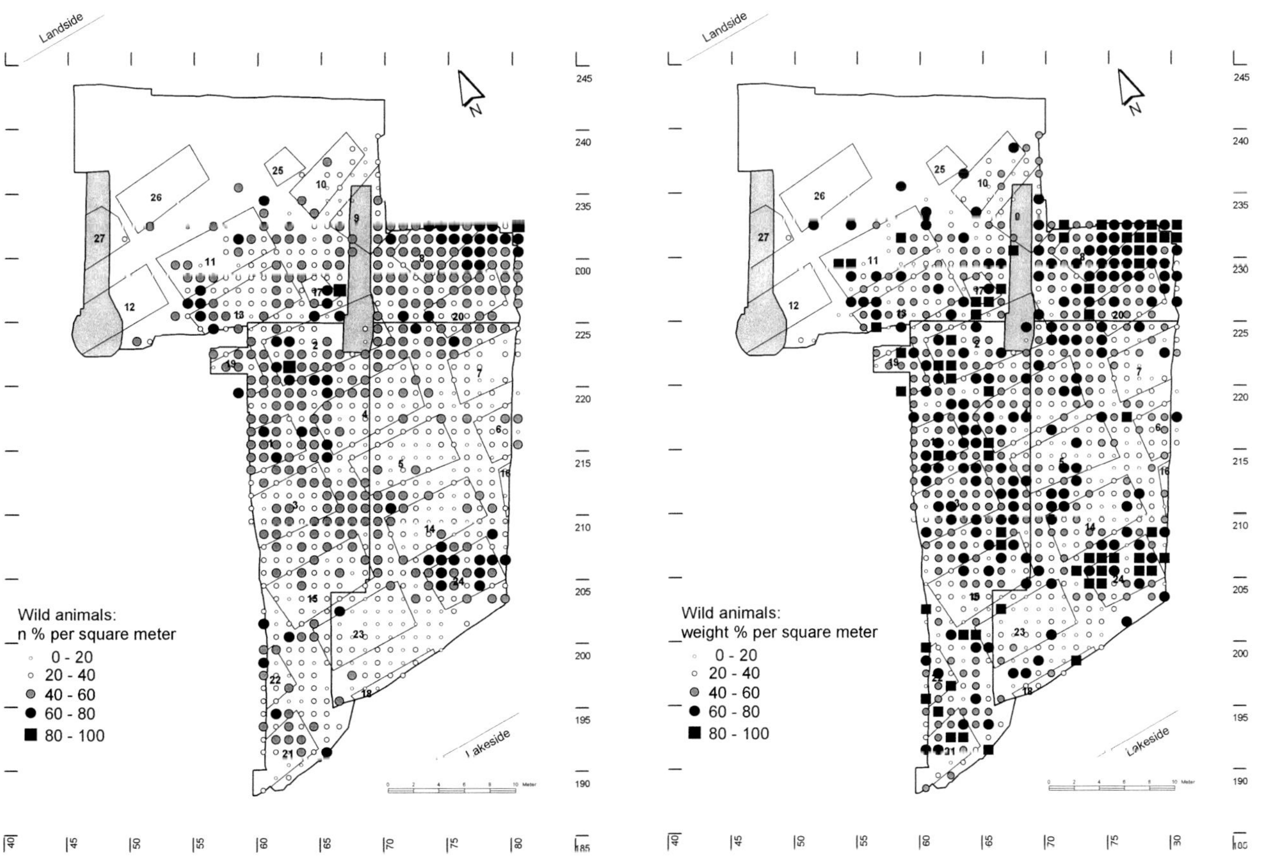

Fig. 6. Proportion of the wild animals within the bones of domestic and wild animals per square metre, based on fragment numbers (left), and on bone weight (right).

west to east, through the coordinates 60/215 and 80/225 by the buildings 1, 4 and 7.

The proportion of ovicaprid bones does not reflect such a division within the village (Fig. 12). Find accumulations can be observed in areas far apart from one another, in the north-western settlement area as well as in the area close to the lake shore. Remains of dog bones are found mostly in the southern part of the settlement (Fig. 13). The concentration north of building 21 consists of an almost entire skeleton recovered there. A further concentration was found in the north-eastern settlement area whereas dog bones are conspicuously rare in the north-western area.

The composition of the bones within the individual buildings

The analysis of the bones by building showed that buildings 8 and 24 had the highest proportion of wild animals (Fig. 14). In these buildings 50 to 60% (fragment number) or around 70% (bone weight) of the remains were from wild animals. The proportion of wild animals is also high in the buildings 1, 2 (only in terms of fragment number) and 20.

Among the domestic animals, cattle are best represented in the buildings of the north-eastern settlement area with fragment proportions between 50 and 60% (Fig. 15 left). Domesticated pig attains a proportion of about 20 or 25% (buildings 11, 13, 2) to almost 40% (building 1). In all the buildings in the village half closest to the shore this relationship is inverted: here domestic pig is best represented with proportions of about 50 to 70%, while cattle attains smaller proportions of only 20 to almost 40%. In the buildings in the north-eastern area of the settlement (buildings 10, 8, 20), and in building 4, the difference in terms of fragment proportion is less than 5%. Both species are therefore equally represented there. Above average proportions of ovicaprids (>20%) were found in the buildings 2, 11 and 13 (northern settlement area) and in buildings 23 and 24 (southern settlement area). Dog is rare (<5%) in all the buildings.

Based on proportional bone weight, cattle were the most important domestic animal for the diet in all the buildings (Fig. 15 right). The proportional weight of domestic pig is, however, significantly higher in the lakeside buildings when compared with the buildings in the northern settlement area. Pork apparently played a larger nutritional role for the inhabitants of this area than it did in the rest of the settlement.

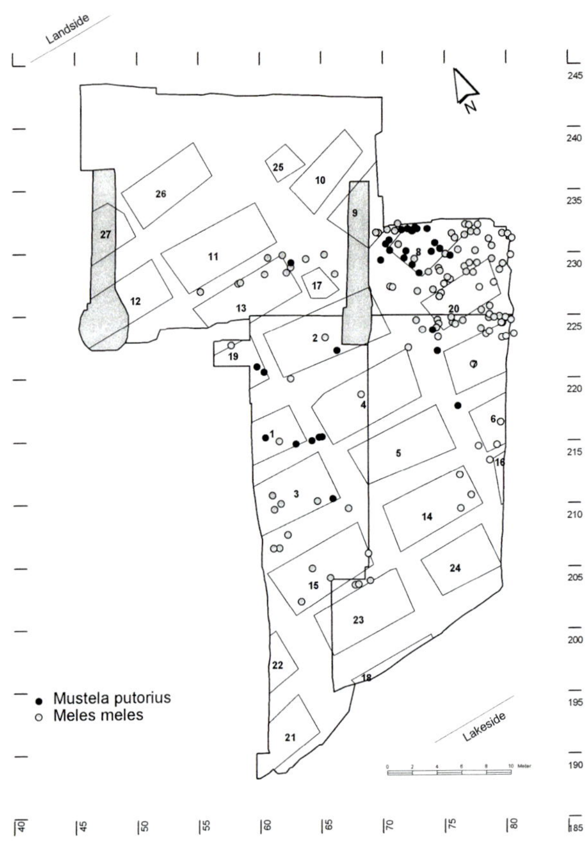

*Fig. 7. Distribution of the bones of badger (*Meles meles*) and polecat (*Mustela putorius*). As the bones were collected by square metre, the find position was randomized within a square metre for individual mapping.*

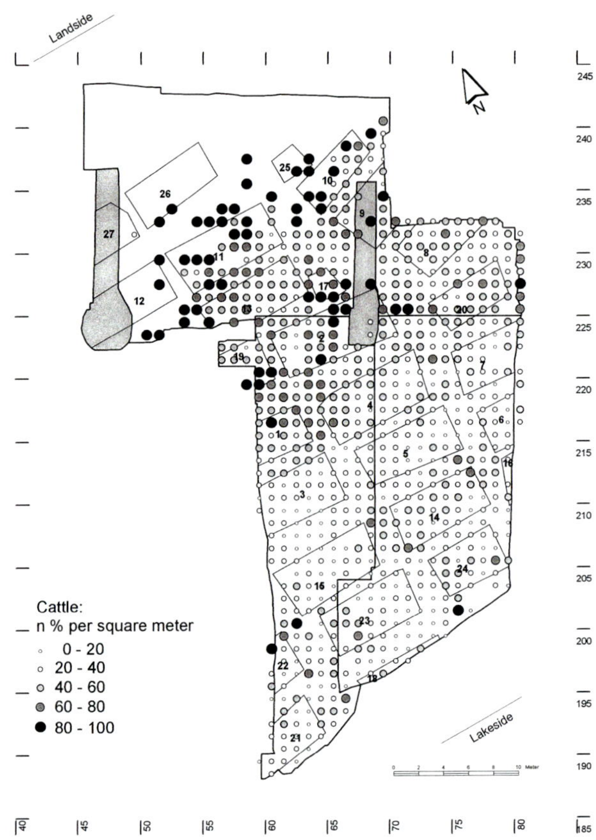

Fig. 8. Proportion (n%) of cattle among the bones of domestic animals per square metre.

There is no relationship between the differences in the distributions revealed by the proportions either per square metre or by households and the house construction chronology. The distributions described apply to both the bone remains from the cultural layer and to those from the layer of burnt debris, which can be considered to be a 'snapshot' of the situation shortly before the destruction of the settlement. In buildings 4 and 20 the layer of burnt debris contained more bone fragments of domestic pig than cattle. In this last settlement phase the boundary based on the proportions of domestic pig and cattle may have been situated somewhat more to the north than was indicated when considering the entirety of the bone material. In the buildings in the north-west of the excavation (buildings 11, 13, 1 and 2), the burnt layer contains a clear increase in the proportion of ovicaprids.

Distribution of small animal remains

The proportions of domestic animal species are not the only ones found unequally distributed over the excavated area. The same can be said for fish remains, which consisted mainly of coregonid fishes (*Coregonus* sp.), perch (*Perca fluviatilis*), pike (*Esox lucius*) and different cyprinids (Cyprinidae).

No data are available yet for the buildings in the very far northern area. However, the apparent differences are analogous to the results obtained from the analysis of the mammal bones. A sort of boundary appears south of buildings 1, 4, and 20 (Fig. 16), as was the case with the proportions of cattle and pig. In the area farther from the lake, fishing in shallow waters seems to have dominated. Cyprinids, perch and pike comprise half to three quarters of all the fish consumed. The fish remains from the buildings and passages close to the water show a high proportion of coregonid fish, which made up as much as 50% of the fish consumed. Possibly more elaborate fishing in open water was practised here.

Find concentrations per building on the basis of bone weight per square metre and year

In order to compare the find concentrations, the buildings are grouped in Fig. 17 according to the proportional percentages of cattle and pig and to their position from north to south: on the left the buildings with a high

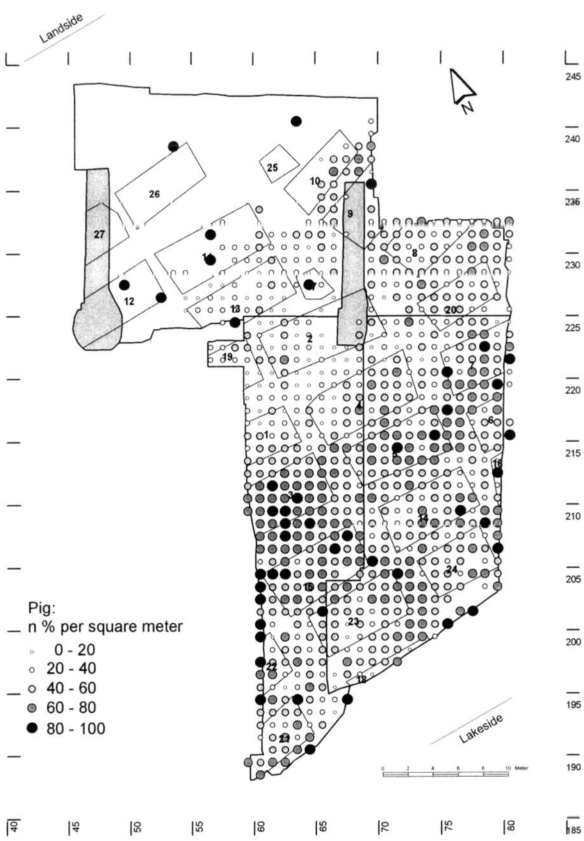

Fig. 9. Proportion (n%) of domestic pig among the bones of domestic animals per square metre.

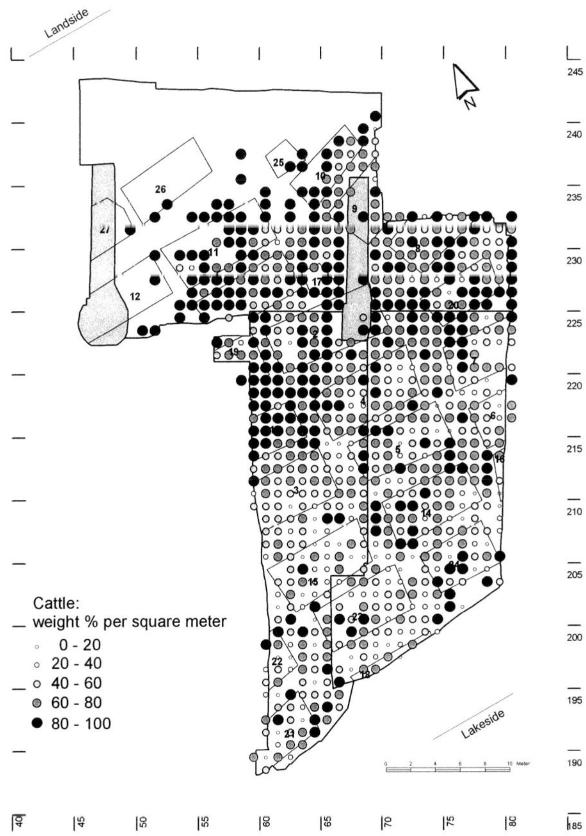

Fig. 10. Proportion (weight%) of cattle among the bones of domestic animals per square metre.

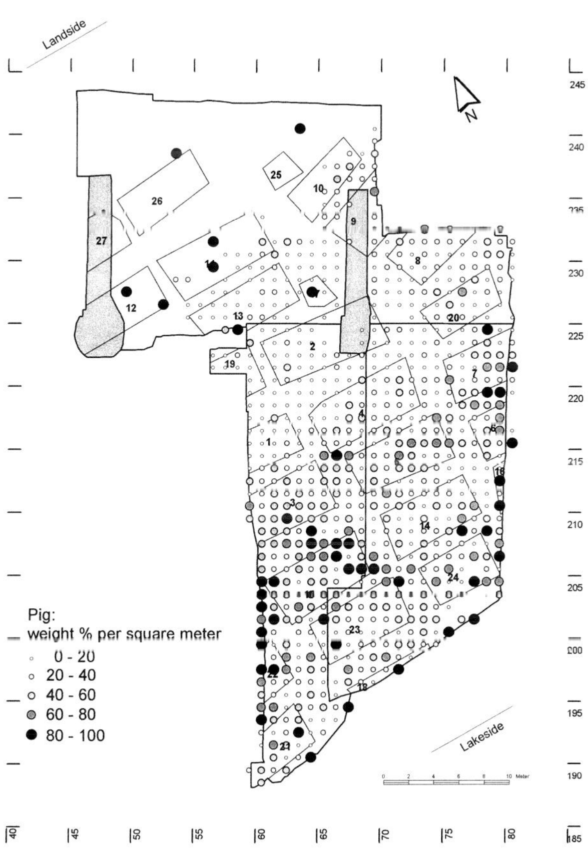

Fig. 11. Proportion (weight%) of domestic pig among the bones of domestic animals per square metre.

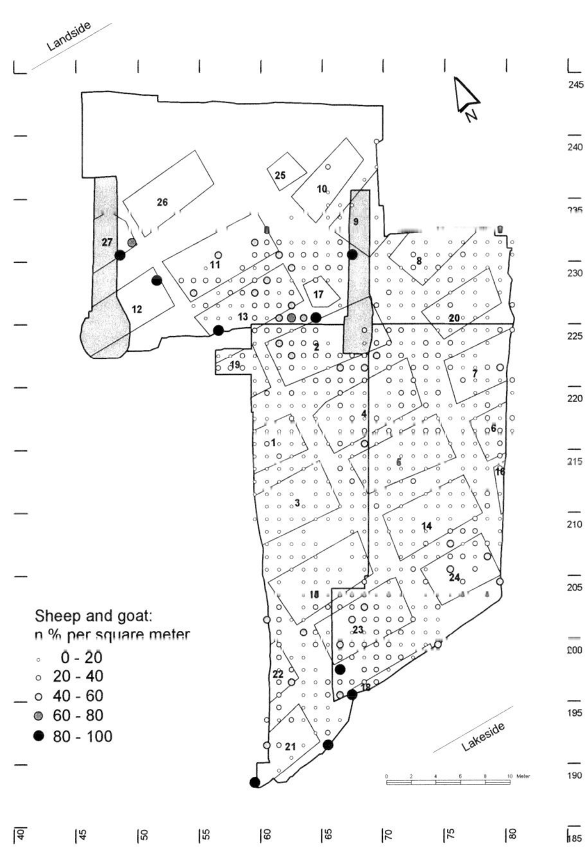

Fig. 12. Proportion (n%) of sheep and goat among the bones of domestic animals per square metre.

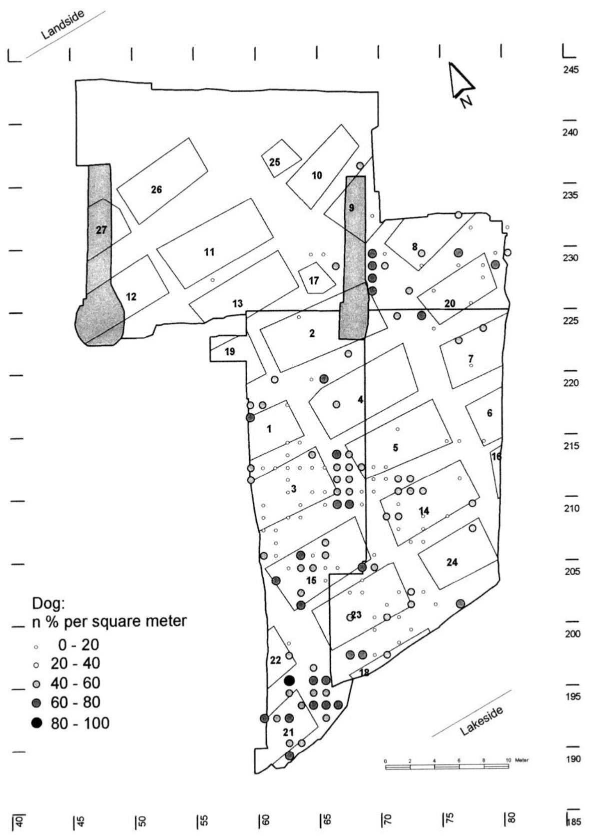

Fig. 13. Proportion (n%) of dog among the bones of domestic animals per square metre.

proportion of cattle, and on the right the buildings with a high proportion of pig. The group in the middle comprises the buildings that have approximately the same proportions of cattle and pig based on fragment numbers.

The find concentration varies tremendously between the buildings (Fig. 17 top). This is in part a result of the differential preservation of the cultural level. However, it also raises the question whether vegetables, fruits or fish may have played a more important role in buildings with a lower find concentration of mammal bones. In addition, different numbers of inhabitants per building, and the relationship or economic unity between buildings may also have influenced the find density.

There are large differences in the find concentrations of domestic and wild animal bones between the buildings. About half the buildings have a higher concentration of wild or domestic bones respectively (Fig. 17 center). Very high concentrations of wild animal bones can be found in buildings 20, 8 and 3. The high number of wild animal remains is evident in buildings 20 and 8 based on the relative proportions as well as the find concentrations. The concentrations show that large quantities of wild animal bones were also deposited in building 3, However, because an almost equal amount of domestic animal bones was deposited there, the relative proportion of wild animals is less conspicuous.

The find concentrations of bones of domestic animals show that the quantity of pig bones is more important for

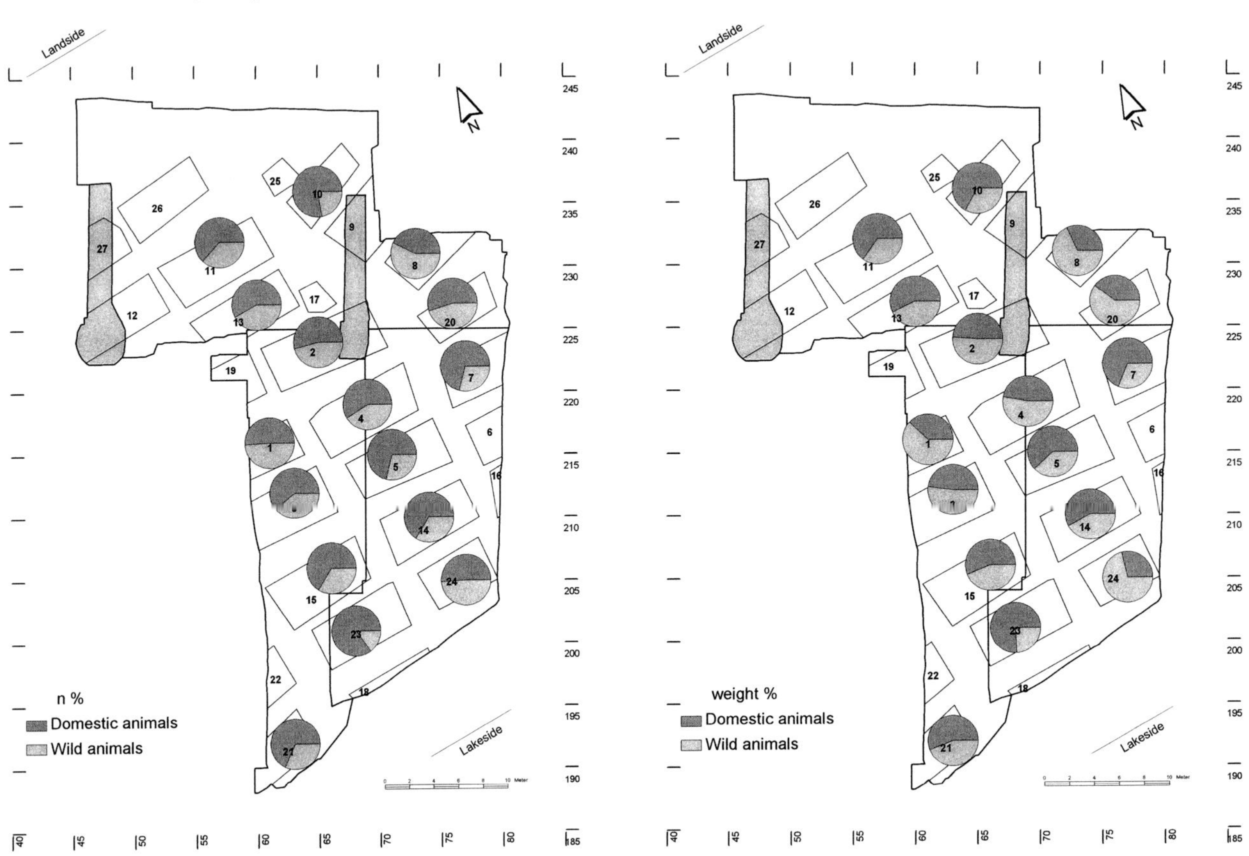

Fig. 14. Proportion of domestic and wild animals in the individual buildings by fragment numbers (left), and bone weight (right).

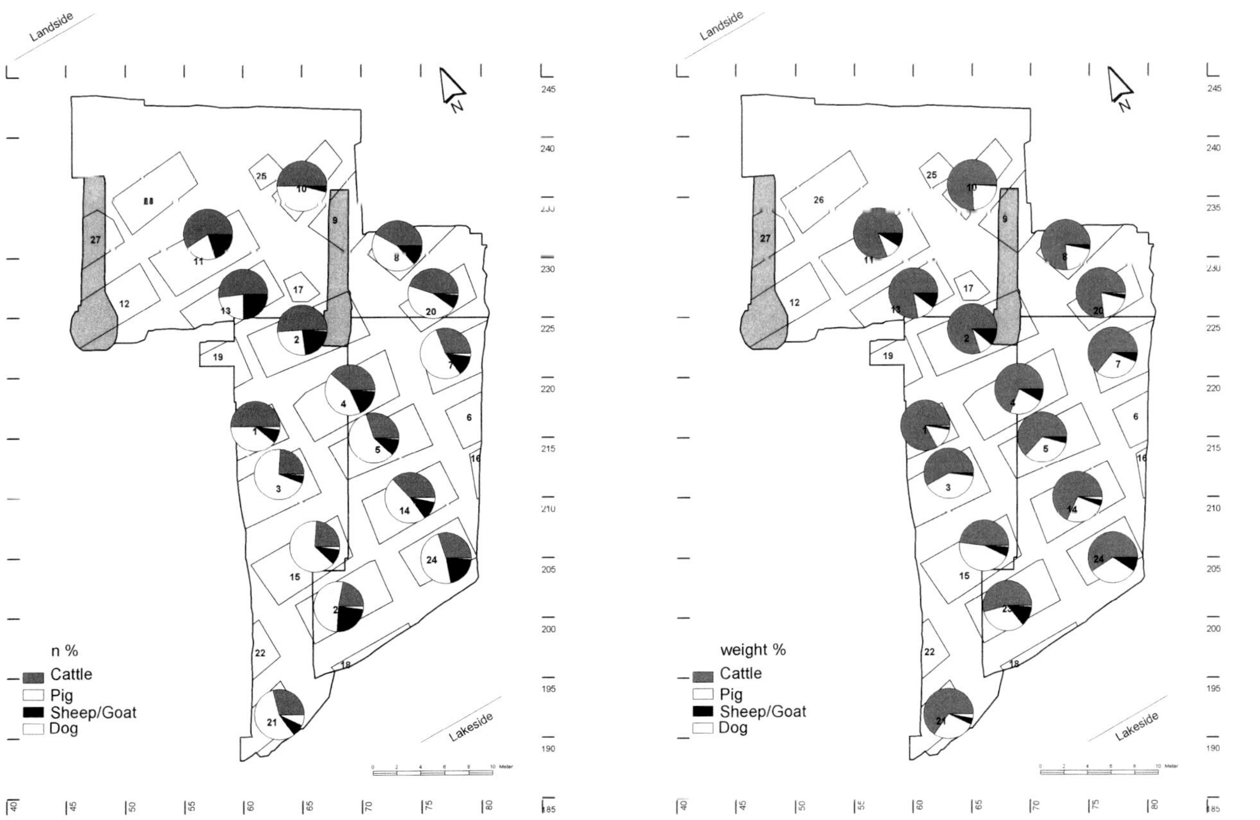

Fig. 15. Proportion of the individual domestic animals in the individual buildings by fragment numbers (left), and bone weight (right).

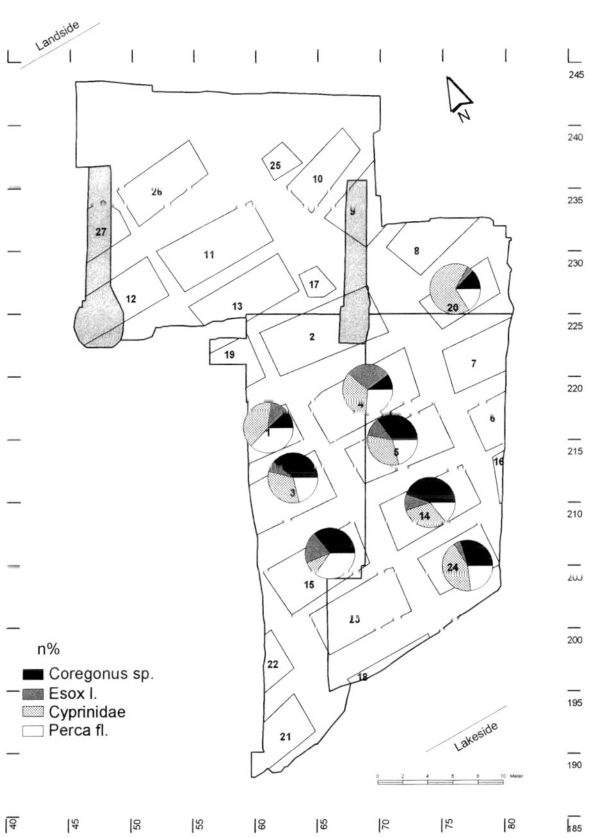

Fig. 16. Proportions (n%) of the different fish taxa in the individual buildings

the characterization of the different areas of the settlement than the cattle bones (Fig. 17 bottom). This can be illustrated with a comparison between buildings with similar cattle bone find concentrations, i.e. buildings 11 and 7, or buildings 13, 8, 14 and 23, or buildings 20 and 3. This implies that in the southern part of the village an extra share of pork was consumed.

Discussion

The taphonomical analyses of the animal bone remains from Arbon Bleiche indicate rapid deposition and only slight displacement of the bone refuse in all settlement areas. The investigation of the ceramics (DeCapitani 2002) and also the preliminary results of the micromorphological analysis of the cultural layer (Meyer and Rentzel, *pers. comm.*) have led to the same result. The different distributions that are found for the bones can therefore not be attributed to natural phenomena but must be considered as the consequence of the inhabitants' behaviour with regard to consumption and refuse disposal. The subdivision of the village on the basis of the distributions of the bones of domestic cattle, pig and different species of fish suggests social differences within the village population, as it can also be observed in bone refuse from more recent time periods. However, there are no finds at Arbon Bleiche that give evidence for hierarchical differences!

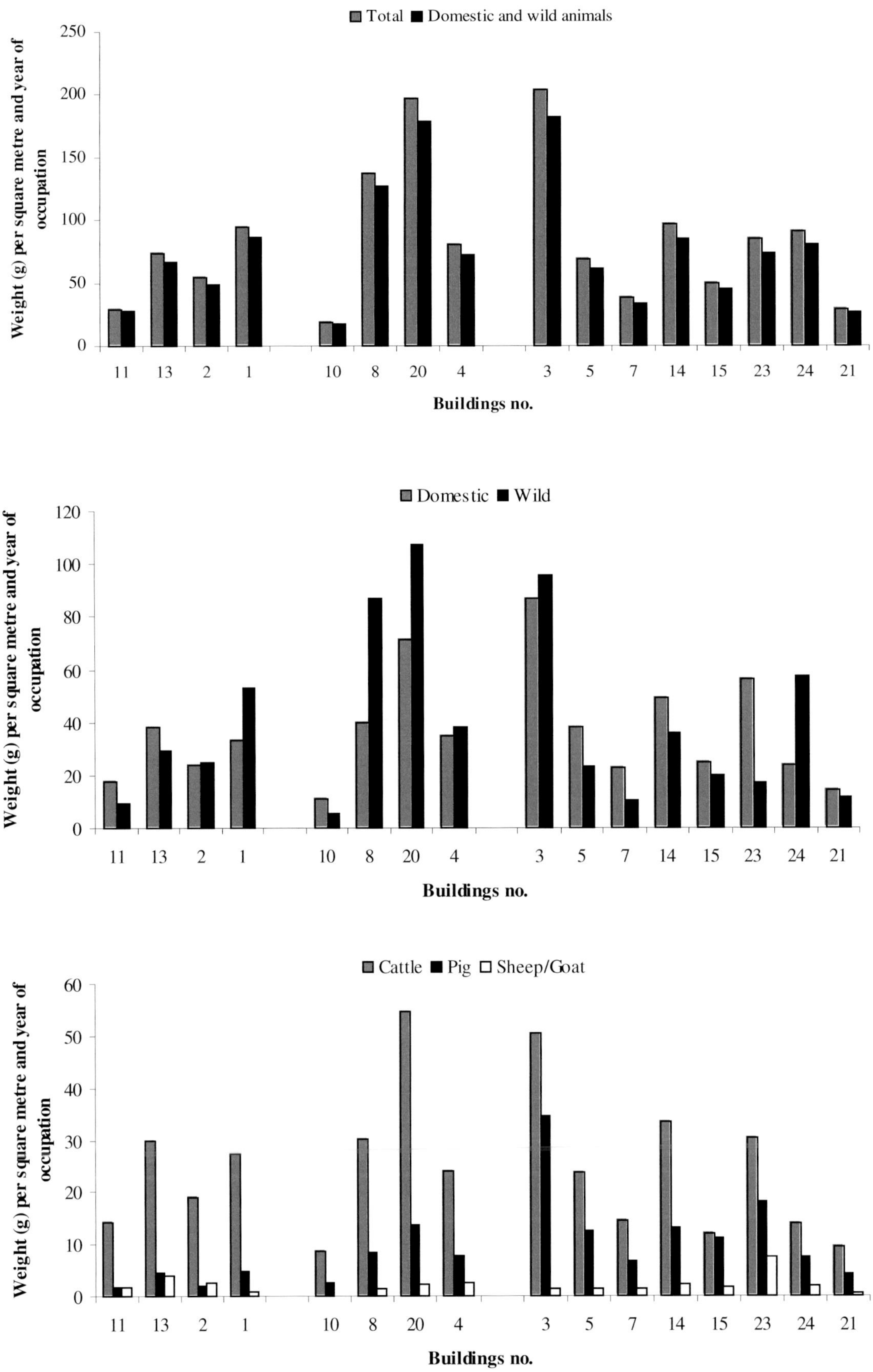

Fig. 17. Find concentration of bones per building (weight in grams/square metre/year) for total bone (top), domestic and wild animal species (center), and domestic cattle, pig, sheep/goat (bottom).

The size and construction of the buildings (Leuzinger 2000) are more or less uniform as in all other Neolithic lake dwellings in Switzerland (Hasenfratz and Gross-Klee 1995). The distribution and quality of the archaeological finds (DeCapitani *et al.* 2002), as well as the age structure and skeletal composition of the animal bones per building, give no indication of the existence of a socially privileged person or group within the village population. The described variation in the skeletal composition of some species within the settlement area cannot be connected with possible differences in the quality of the meat consumed and again, therefore, gives no evidence for the existence of a hierarchical structure of the village population .

What else could then be the reason for the different bone distributions of cattle, pig and fish within the settlement? Within ethnography, the dichotomy of a society or population is known as moiety (Vivelo 1981). An individual's membership to a group is dependent on parentage or descent and implicates the participation in ritual practices and the commitment to observe certain dietary rules. In the case of Arbon Bleiche 3 it is possible that the different distributions of bones of cattle, pig and fish could be caused by two groups of inhabitants with different food habits or different economic and fishing traditions. The presence of Boleraz vessels in Arbon Bleiche 3, which according to present knowledge were produced locally at Arbon, together with the traditionally autochthonous ceramic types (Pfyn/ Horgen cultures), could be an indication for this.

Aside from the ceramic finds there is possibly a further reference to the Baden culture in Arbon-Bleiche 3. The finding of a wooden yoke, as well as metrical and pathological changes on cattle foot bones and vertebrae that can be attributed to burdens, indicate that in Arbon-Bleiche 3 cattle were used as draught animals, most probably to plough the fields. This represents the earliest indication of this new achievement in our region. The south-eastern region of Europe was important for the diffusion of cattle as draught animals for ploughs and wagons (Benecke 1994). The cultural association ascertained in the archaeological finds in Arbon-Bleiche 3 may thus be an indication for an immigration of people from the east. Assuming that the indigenous and the immigrated inhabitants of Arbon Bleiche 3 lived in different parts of the village, this could perhaps explain the heterogeneous horizontal distribution of the animal bone refuse in Arbon Bleiche 3. However, the distribution of foreign ceramic forms within the site does not allow further interpretationsl: for the time being.

Conclusion

The study of the sedimentation and taphonomy of the animal bones leads to the conclusion that the horizontal bone distributions cannot be attributed to natural phenomena. Neither do they show any relationship to the chronology of the buildings' construction. The distribution of the bone refuse must therefore be explained by the inhabitants' behaviour with regard to consumption and refuse disposal. Neither the detailed study of the bone assemblage composition by building, nor the distributions of the archaeological finds give evidence for the existence of hierarchical differences within the village population. The differences in the bone assemblage compositions therefore primarily reflect economic specialisation and/or different consumption patterns. The presence of locally produced vessels in the tradition of the allochthonous Baden Culture leads us to the hypothesis that a part of the village population of Arbon Bleiche 3 were perhaps immigrated people with different food habits or different economic traditions.

References

Arbogast, R.-M., Pétrequin, A.-M. and Pétrequin, P. 1995. Le fonctionnement de la cellule domestique d'après l'étude des restes osseux d'animaux: le cas d'un village néolithique du lac de Chalain (Jura, France). *Anthropozoologica* **21**, 131–46.

Benecke, N. 1994. *Archäozoologische Studien zur Entwicklung der Haustierhaltung in Mitteleuropa und Südskandinavien von den Anfängen bis zum Ausgehenden Mittelalter* (Schriften zur Ur-und Frühgeschichte 46). Berlin: Akademie Verlag..

Chaix, L. 1989. Die Tierknochenfunde der spätneolithischen Siedlung Zug-Schützenmatt, pp. 133–9 in Elbiali, N. (ed.), *Zug, Schützenmatt- eine Seeufersiedlung der Alteren Horgener Kultur* (Tugium 5). Zug: Regierungsrat des Kantons Zug.

de Capitani, A. and Leuzinger, U. 1998. Arbon-Bleiche 3, Siedlungsgeschichte, einheimische Traditionen und Fremdeinflüsse im Übergangsfeld zwischen Pfyner und Horgener Kultur. *Jahrbuch der Schweizerischen Gesellschaft für Ur-und Frühgeschichte* **81**, 237–49.

de Capitani, A. 2002. Die Gefässkeramik, pp. 135–276 in de Capitani, A., Deschler-Erb, S., Leuzinger, U., Marti-Grädel, E. and Schibler, J. (eds), *Die Jungsteinzeitliche Seeufersiedlung Arbon-Bleiche 3. Funde* (Archäologie im Thurgau 11). Frauenfeld: Veröffentlichungen des Amtes fur Archäologie des Kantons Thurgau.

de Capitani, A., Deschler-Erb, S., Leuzinger, U., Marti-Grädel, E. and Schibler, J. 2002. *Die Jungsteinzeitliche Seeufersiedlung Arbon-Bleiche 3. Funde* (Archäologie im Thurgau 11). Frauenfeld: Veröffentlichungen des Amtes für Archäologie des Kantons Thurgau.

Gross, E., Jacomet, S. and Schibler, J. 1990. Stand und Ziele der wirtschaftsarchäologischen Forschung an neolithischen Ufer- und Inselsiedlungen im unteren Zürichseeraum (Kt. Zürich, Schweiz), pp. 77–100 in Schibler, J., Sedlmeier, J. and Spycher, H. P. (eds), *Festschrift für Hans R. Stampfli, Beiträge zur Archäozoologie, Archäologie, Anthropologie, Geologie und Paläontologie*. Basel: Helbing and Lichtenhahn.

Gross-Klee, E. and Maise, C. 1997. Sonne, Vulkane und Seeufersiedlungen. *Jahrbuch der Schweizerischen Gesellschaft für Ur- und Frühgeschichte* **80**, 85–94.

Hasenfratz, A. and Gross-Klee E., 1995. Siedlungswesen und Hausbau, pp. 195–229 in *Die Schweiz vom Paläolithikum bis zum Frühen Mittelalter*, SPM II- Neolithikum. Basel: Verlag Schweizerische Gesellschaft für Ur-und Frühgeschichte.

Johansson, F. 1981. Horizontale Fundverteilung, pp. 106–11 in Becker, C. and Johansson, F. (eds) *Tierknochenfunde, Zweiter Bericht* (Die neolithischen Ufersiedlungen von Twann 11). Bern: Staatlicher Lehrmittelverlag Bern.

Leuzinger, U. 2000. *Die Jungsteinzeitliche Seeufersiedlung Arbon-Bleiche 3. Befunde* (Archäologie im Thurgau 6). Frauenfeld: Veröffentlichungen des Amtes für Archäologie des Kantons Thurgau.

Leuzinger, U. 2002. Befund, pp. 18–21 in de Capitani, A., Deschler-Erb, S., Leuzinger, U., Marti-Grädel, E. and Schibler, J. (eds), *Die Jungsteinzeitliche Seeufersiedlung Arbon-Bleiche 3. Funde* (Archäologie im Thurgau 11). Frauenfeld: Veröffentlichungen des Amtes für Archäologie des Kantons Thurgau.

Stampfli, H. R. 1976a. *Osteo-archäologische Untersuchung der Tierknochenmaterials der Spätneolithischen Ufersiedlung Auvernier-La Saunerie*. Solothurn: self-published.

Stampfli, H. R. 1976b. Die Tierknochen von Egolzwil 5. Osteo-archäologische Untersuchungen, pp. 125–39 in Wyss, R. (ed.), *Das Jungsteinzeitliche Jäger-Bauerndorf von Egolzwil 5 im Wauwilermoss* (Archäologische Forschungen 1976). Zürich: Archäologische Abteilung des Schweizerischen Landesmuseum.

Stampfli, H. R. 1992. *Die Tierknochen aus den Jungsteinzeitlichen Siedlungen Egolzwil 3 und Egolzwil 4*. (Archäologische Schriften Luzern 1). Luzern: Kantonsarchäologie Luzern.

Vivelo, F.R. 1981. *Handbuch der Kulturanthropologie: eine Grundlegende Einführung*. Stuttgart: Klett-Cotta.

Elisabeth Marti-Grädel, Sabine Deschler-Erb, Heide Hüster-Plogmann and Jörg Schibler
Institut für prähistorische
und naturwissenschaftliche Archäologie (IPNA),
Spalenring 145,
CH-4055 Basel, Switzerland
E-mail: clisabcth.marti@unibas.ch

9th ICAZ Conference, Durham 2002
Behaviour Behind Bones, (ed. Sharyn Jones O'Day et al.) pp. 177–189

20. Levels of social identity expressed in the refuse and worked bone from Middle Bronze Age Százhalombatta-Földvár, Vatya culture, Hungary

Alice M. Choyke, Maria Vretemark and Sabine Sten

The composition of refuse bone and worked bone assemblages is clearly affected by embedded traditions linked to social identity. This identity can be seen to exist on both a regional and a very local level in the stable period of the Middle Bronze Age in Hungary. Some aspects of the way animals were butchered and bones worked operated on an unconscious level related to conservative traditions passed on within the family. Other aspects have a deliberate symbolic content related to how people or groups of people saw themselves. The Vatya culture is known for a line of hill-forts running through Transdanubia in Hungary. Work has been carried out at a number of these hill-forts since the 1950's. Thus, refuse bones and worked bones from eight of these sites became available for archaeozoological study. In the beginning of the 90's more intensive, fine-grain excavations began at Százhalombatta–Földvár which continue to the present day. The possibility emerged of comparing the mass of bones from earlier excavations and looking in detail at animal processing and bone working at one particular locality. This provides a foundation with which to assess multi-level social cultural identity as reflected in human animal relationships in this period. The paper will particularly focus on differences in the treatment of horse and dog at these Vatya hill-forts. Ritual slaughter of animals with their burial in pits within settlements, as a regional phenomenon linking settlements will be discussed.

Introduction: material culture and social identity

Prehistoric assemblages are inherently more difficult to interpret from the point of view of social reconstruction than historical collections. This difficulty extends to understanding whether similar and different patterns observed in faunal and worked bone assemblages between regions and within regions are the result of environmental limitations or of cultural preferences related to trade and tradition, the latter reflecting aspects of the social identities of groups and even individuals. Prehistoric researchers must work through analogy, suggestion and interpretation of perceived patterns rather than from the more solid, "objective" textual base available from later proto-historical and historical periods (the degree to which such texts actually represent reality on the ground or themselves offer a distorted picture of reality being a question for another paper altogether!).

The fortified site of Százhalombatta–Földvár was chosen as a case study for the interpretation of social identity through refuse and worked bones because it is an outstandingly well excavated site for Hungary and it is part of a well defined regional unit called the Vatya culture in the literature (Bóna 1975). The sites assigned to this so-called Vatya culture are characterized by a hierarchy of settlements from high status fortified, stratified settlements, to stratified settlements, to single-layer hamlets, and urn cemeteries. These sites contain basically similar pottery, metal objects, lithic and bone tool assemblages. These sites are, thus, clearly more closely related to each other than to other contemporary settlement agglomerations to the east within the Carpathian Basin. The exact nature of the relationship between the settlements over this proscribed region as a whole will probably never be known. However, recognizing patterns related to social imperatives marks an advance in knowledge (away from typo-chronological problems) concerning different levels of social integration as reflected in animal keeping practices and bone manufacturing tradition. These will be examined from the points of view of (1) the Bronze Age in the Carpathian Basin, (2) regional differences between the Vatya culture and other Bronze Age polities, and (3) local social identities as exemplified by the Százhalombatta assemblage. The method used to highlight differences and similarities between these cultural units will simply be comparative. More rigorous statistical analyses will have to wait until more sites have been excavated at the same high level of resolution as Százhalombatta–Földvár.

Social identity and style

It has been noted that a great deal of behavior has a communicative function without communicative intent, with three interactor roles of sender, emitter, and receiver (Schiffer 1999, 67). There are several levels of "sender" and "receiver" operating at different scales in this particular prehistoric system: throughout the Carpathian Basin, regional – between larger polities (Vatya culture and other), within individual polities (within the Vatya culture), within settlements (such as Százhalombatta) and on the individual level (assertion of individual identity). One not only signals to outsiders to mark community or define ethnic boundaries; internal reinforcement is at least as important (Hodder 1982, 52–6). Within the latter process, individual identity is the most difficult to assess. The "emitter" in this case study would be forms of animal exploitation and types of tools.

Social and stylistic comparisons and the corresponding choices can be made at very different levels of consciousness or intent (Wiessner 1985, 161). People keep animals and view them in culturally idiosyncratic ways. Special emphasis on certain kinds of animal keeping may come to be part of a people's identity and ultimately have an effect on ritual treatment of animals. Edmund Leach (1976, 45) has noted that humans engage in ritual in order to transmit collective messages to themselves. Here we compare the ways animals were exploited and suggest that the choice of how animals are used has a strong social as well as functional component.

Another important way social identity is maintained, is through the style of objects. Style is rooted in cultural choice (David and Kramer 2001) whether conscious or not. Wobst (1977 and 1999, 323), one of the first archaeologists to deal with this subject, noted that messages emitted in artifacts (bone and antler tools in this case) tend to be "simple, invariant and recurrent". He sees style as a way to help maintain and further differentiate between socially differentiated groups at little cost (Wobst 1999, 327–8). However, others would argue that this implies that the message is always a deliberate one related only to decoration and form. While these purposeful emblematic (Wiessner 1984, 191) aspects are very important and seen across social boundaries in the bone and antler tools discussed here, there is also a passive, latent aspect to style inherent to what Sackett (1990, 56–7) has called the *isochrestic* choices that lie behind artifact manufacture. This really, is dependent on how people learn to make bone tools. Most of the bone and antler tools from Bronze Age Hungary were certainly produced by non-specialists, with techniques passed on from parent to child. David and Kramer (2001, 215) note that people who regularly assist each other in making and doing things and have frequent opportunities for observing each other using these tools produces similarities in practice. Social expectations, thus, kept form and manufacturing techniques of Bronze Age bone and antler tools relatively stable within communities and even over larger areas and long periods of time. To put it another way, certain kinds of tools were used because the tasks they were used in (function) demanded them but the choice of which bones they were made from and the details of their manufacture were culturally determined.

Finally, a great variety of style was available to Bronze Age craftspeople which had common sources: a shared symbolic reservoir of motifs and techniques. This reservoir accounts for similarities and many different levels of social complexity.

The Bronze Age and Százhalombatta–Földvár

The Middle Bronze Age, multi-layer site of Százhalombatta–Földvár lies on a high loess plateau on the right bank of the Danube some 30 km south of Budapest in Western Hungary (Poroszlai 1992; Poroszlai 2000). The settlement was initially unfortified until its second period of occupation. At that time a ditch and rampart system was established. Over two-thirds of the site has been destroyed by the clay pits of a modern brickyard and some erosion by the Danube. Today, the site is 200m long and 100m wide. Traces of the defensive system can be seen only in its northeastern part.

The first excavations (Kovács 1969, 161–8) were carried out in the southern part of the site which was most endangered by clay extraction, although the narrow trenches opened then were really aimed at establishing a chronological sequence for the site rather than understanding intra-site relationships. Already at that time (Kovács 1969, 161–8; Kovács 1984) it became clear that the chronological sequence begins with the Nagyrév culture (*c.* 2400–2100 calBC) continuing without break through the Middle Bronze Age Vatya culture (*c.* 2100–1600 calBC) and Vatya-Koszider (1630–1508 and 1480–1458 calBC) cultures and finally, from the end of the Middle Bronze age, the Koszider culture (*c.*1400–1200 calBC; radiocarbon dates: Raczky *et al.* 1992, 42–7). The uppermost level is mixed with Celtic, Iron Age and Koszider finds.

Geographically, the Vatya culture is defined by an oval ring of 30 fortified settlements in Transdanubia, along the Danube, and on into the Danube-Tisza interfluve (Fig. 1). This area enclosed a total of 300 sites including, as has been stated, fortified and stratified sites, unfortified stratified sites, simple hamlets and cemeteries. It has been

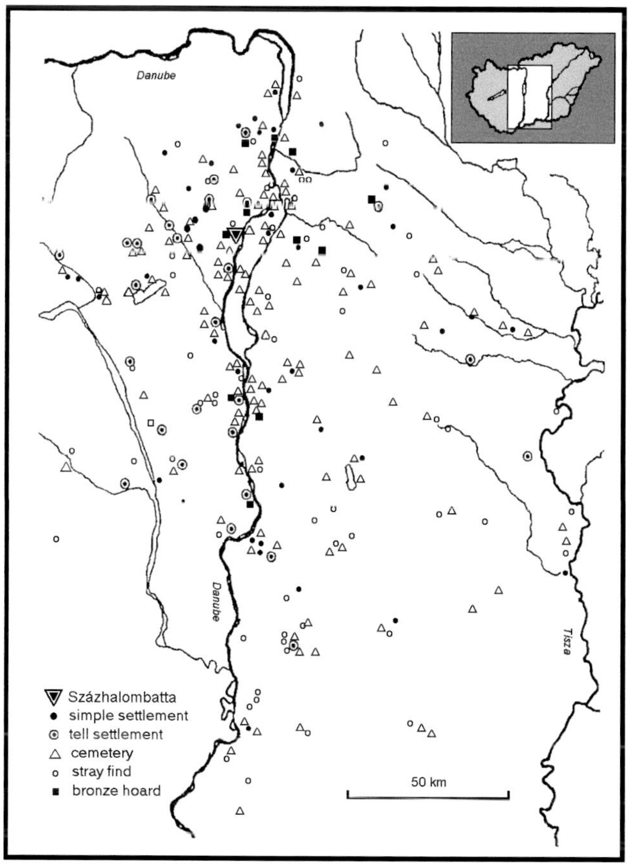

Fig. 1. Map of Vatya culture territory in Transdanubia, along the Danube and in the Danube-Tisza interfluve (Kovács 1982).

proposed that the fortified sites were established to control passage and communication at various points across this region. Many Vatya culture hill-forts are located at the mouth of broad valleys (Nováki 1952). A total of eight fortifications were located on the high western bank of the Danube including the site that is the focus of this study (Vicze 2000, 120). It has been proposed that one might expect to find one fortified stratified settlement, one possible stratified unfortified settlement, three single-layer settlements and three cemeteries in individual micro-regions with a diameter of 8–10 km (Tóth 1990). To date, two high-status fortified settlements in addition to Százhalombatta, one or two stratified sites, and 9 single-layer sites were discovered during field-walks within the 20 km long valley of the Benta stream (Vicze 2000, 120).

The Vatya site distribution is in sharp contrast to the territorial organization of archaeological cultures in eastern Hungary. Although the geographical situation of the two areas was largely the same as a continuation of the Great Hungarian Plain, the landscape towards the east was divided into smaller polities with an even distribution of fortified tell sites.

Description of the material

The consumption refuse studied here comes largely from the excavations conducted by Ildikó Poroszlai from 1989 to 1991 (Poroszlai 2000, 14). Although not sieved and covering a relatively small area in the southwestern part of the site, the careful hand collection resulted in the recovery of even quite small bones. In addition, the special finds of dog crania from the new excavations were also included in the discussion.

The bone tool material in contrast comes from two sources: Ildikó Poroszlai's excavations (Poroszlai 2000) as well as the current fine-grained joint excavations with a team from Sweden (Göteborg University and the Swedish National Heritage Board). Material from the first excavations conducted by Tibor Kovács in 1963 and 1969 (Kovács 1969, 161–8) was not included here. He dug in trenches so small that house contexts could not really be observed. Also, in keeping with the fashion of the times the site layers were taken down in artificial levels of about a shovel blade's length (20 cm). Most of the soil was removed with shovels so that small finds were probably missed for the most part. Consequently, the assemblage from the 1963–1969 excavations is not really comparable to that of either Poroszlai's carefully hand-collected excavations or of the field campaigns currently going on. The recent excavations also include sieving campaigns but this does not seem to produce more small bone tools from the Vatya levels.

The layers excavated at the Százhalombatta–Földvár site are numbered starting from the latest levels down.

- Level I: Mixed Celtic and Iron age finds together with a few Koszider finds from the end of the Middle Bronze age
- Level II: Koszider phase of the Vatya culture (*c.* 1630–1508 and 1480–1458 BC)
- Level III: Vatya culture (*c.* 2100–1600 BC)
- Level IV: Vatya culture (*c.* 2100–1600 BC)
- Level V: Nagyrév culture – an intermediate phase
- Level VI: Nagyrév culture (*c.* 2400–2100 BC)

The analysis presented here concentrated mostly on the refuse and worked bones from the Vatya and Koszider layers.

The refuse bone

Following years of Bronze Age research in Hungary, the faunal material from quite a fair number of major settlements has been studied (Choyke and Bartosiewicz 1999; Choyke and Bartosiewicz 1999–2000). For the most part, the assemblages are marked by a very high percentage of domesticates based on the number of identifiable specimens (NISP) (Fig. 2). At some of these sites, in particular settlements associated with the Vatya culture, up to 98% of the bone refuse material from the assemblages chosen

Species	Level						Sum
	VI	V	IV	III	II	I	
Cattle (*Bos taurus* L. 1758)	254	204	455	353	253	79	**1598**
Horse (*Equus caballus* L. 1758)	6	8	35	9	14	7	**79**
Sheep/Goat (Caprinae)	90	132	302	241	179	58	**1002**
Sheep (*Ovis aries* L. 1758)	8	8	10	10	8	2	**46**
Goat (*Capra hircus* L. 1758)	–	1	1	1	2	–	**5**
Pig (*Sus domesticus* Erxl. 1777)	55	50	132	100	64	40	**441**
Dog (*Canis familiaris* L. 1758)	8	9	28	22	27	1	**95**
Lynx (*Lynx lynx* L. 1758)	–	–	–	–	1	–	**1**
Aurochs (*Bos primigenius* Boj. 1827)	–	1	–	–	–	–	**1**
Wild boar (*Sus scrofa* L. 1758)	–	–	1	–	1	1	**3**
Red deer (*Cervus elaphus* L. 1758)	2	12	8	–	2	–	**24**
Brown hare (*Lepus europaeus* Pall. 1778)	1	–	2	–	2	–	**5**
Hamster (*Cricetus cricetus* L. 1758)	–	2	–	–	–	–	**2**
Bird (Aves)	1	–	1	1	–	–	**3**
Human (*Homo sapiens* L. 1758)	1	–	2	2	–	–	**5**
Total NISP	**426**	**427**	**977**	**739**	**553**	**188**	**3310**
Non–identifiable	139	173	174	173	127	62	847

Fig. 2. The number of fragments (NISP) per species in the different chronological layers of Százhalombatta–Földvár. Levels VI-V originate from the Early Bronze Age Nagyrév culture during (2000–1700 BC). Levels IV–II represent the Middle Bronze Age Vatya culture (1700–1400 BC). The material in level I is mixed, originating from both the Late Bronze Age and Iron Age period. The hamster remains are intrusive.

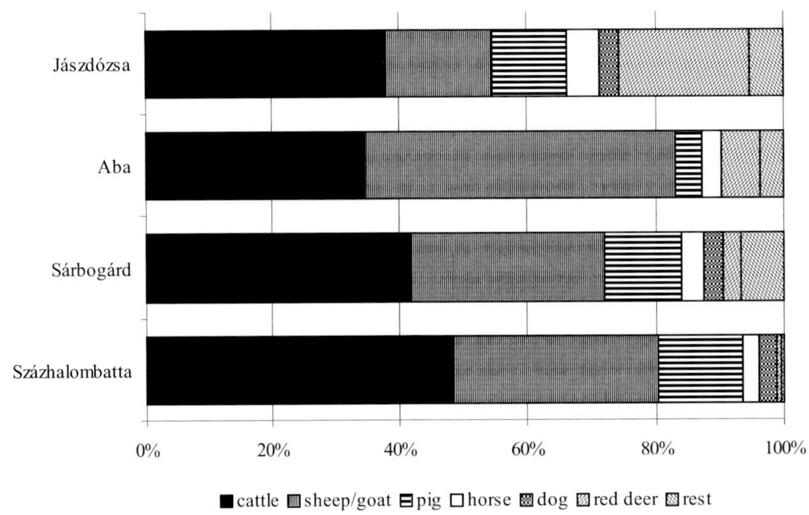

Fig. 3. Comparison of the frequency of species found at Vatya sites and eastern tell sites as exemplified by Jászdózsa–Kápolnahalom on the northern edge of the Great Hungarian Plain.

for study may originate from domestic animals as exemplified also by the faunal material from Százhalombatta (Vretemark and Sten, *in press*). The percentages of pig and horse were markedly lower at all these sites (Fig. 3) compared to the highs of up to 60% at Early Bronze Age Bell beaker sites. Generally, within the Middle Bronze Age Vatya culture, horse was no longer seen as a meat animal and seems to have been transformed into a high-status species. A dependence on domesticates is, however, generally characteristic of this period in the Carpathian Basin. Meat provisions were secured by animal husbandry at Vatya settlements, mainly by keeping cattle, sheep and, to some extent, goat.

One can see that there is a clear, general trend for more game being eaten on the eastern sites as opposed to the general case at Vatya sites (Choyke 1984; Choyke 1987; Choyke 1998; Choyke and Bartosiewicz 1999; Choyke and Bartosiewicz 1999-2000). Antler tools were used everywhere at this time in approximately equal numbers, showing that red deer would have been available everywhere. However, for some reason it was more intensively exploited towards the east (Choyke 1987; Choyke 1998). It has been suggested that some settlements, such as Jászdózsa-Kápolnahalom in the northeastern part of the Hungarian Plain (end of the Early Bronze Age [*c.* 2450–2430 BC], Middle Bronze Age [1936–1870 BC]

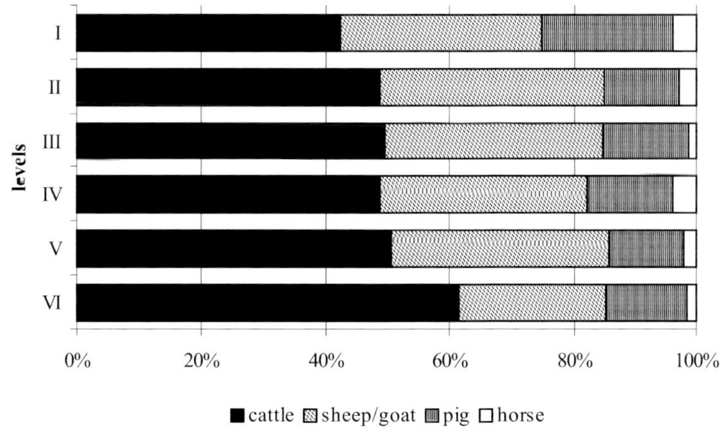

Fig. 4. The relative distribution of species from the earliest (VI) to the latest (I) levels at Százhalombatta–Földvár.

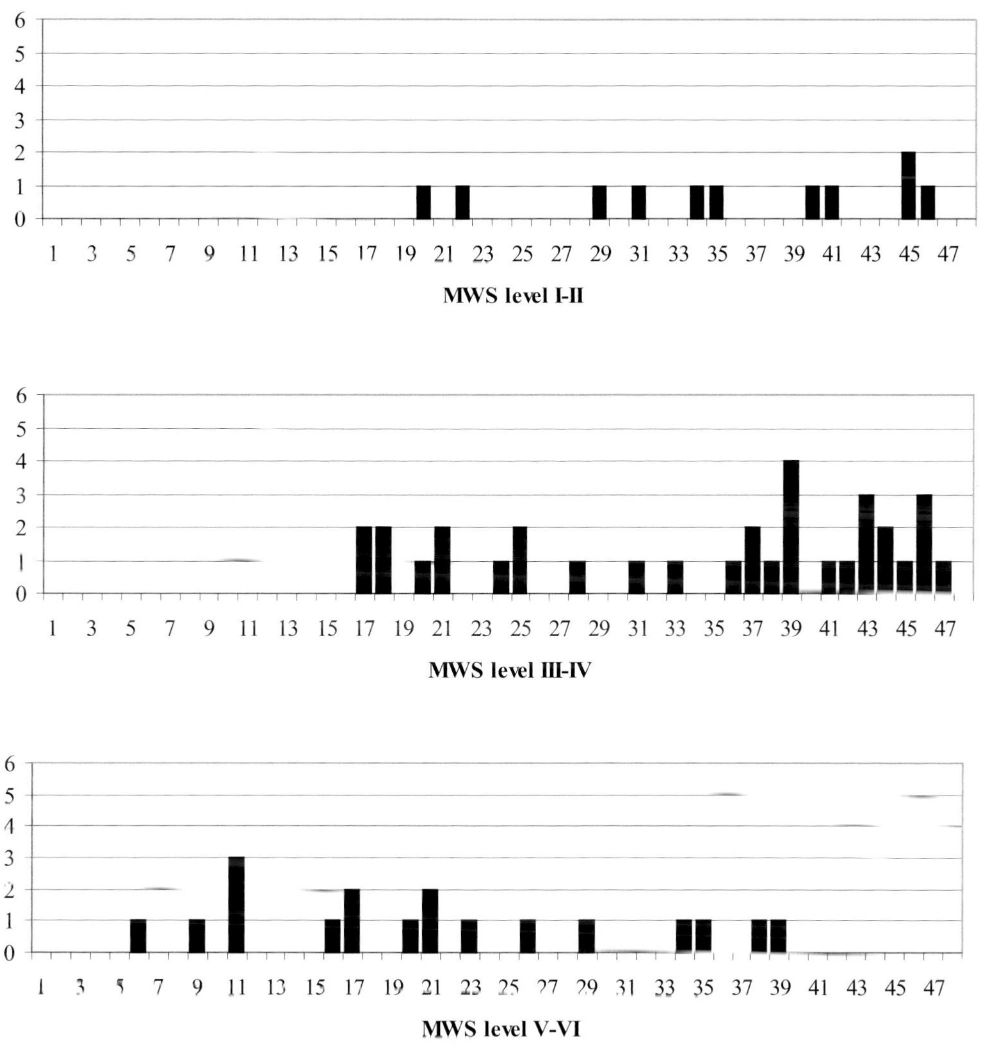

Fig. 5. Absolute distribution of the age at death evaluated by mandible wear stages (MWS) on lower jaws from sheep/ goat (in accordance with Grant 1982). The higher the value, the older the age of slaughter of the individual. The diagram in the bottom represents the jaws from levels V–VI or the Early Bronze Age. Diagrams in the middle and top represent levels III–IV and I–II which represents the Middle Bronze Age and Late Middle Bronze age respectively.

and Late Bronze Age [1566–1530 BC]; Raczky *et al.* 1992, 43), might have specialized in trading in deer hides (Choyke 1987; Choyke 1998). The red deer skulls placed on poles behind the palisade and found within one of the defensive ditches surrounding the tell suggest that this species had a totemic/symbolic meaning here.

The dominant species at all levels at Százhalombatta being cattle, it is clear that the percentage of the remains of this species declined through time (Fig. 4). The species that simultaneously increased in importance was sheep. The proportion of sheep NISP reached between 30–40 % in the refuse from the meat producing domesticates, which is a high figure compared to other coeval sites, including those of the Vatya culture. This increase occurred in level V, representing the transition from the Early Bronze Age Nagyrév culture to the Middle Bronze Age Vatya culture. The increase in the percentage of sheep indicates that the landscape was opening up and was more intensively exploited. An open, drier pastureland on its way towards deforestation spread around the settlement of Százhalombatta–Földvár. Whether this is an entirely cultural phenomenon or also related to environmental limitations is unclear at present. It will be interesting to see whether this trend continues at other sites in the Benta stream valley.

The possibility of utilizing sheep, not only for meat, but also for wool production, clearly increased the importance of this species, which is probably the reason for the high relative frequency of sheep bones during phase V at Százhalombatta–Földvár. The age distribution of the species also supports this assumption. There is a clear trend showing that the proportion of sheep slaughtered at higher ages (Fig. 5), increased from the oldest to the later chronological layers In level VI–V, the percentage of younger animals is higher than that of the older animals, which indicates that the primary aim of sheep husbandry was meat production.. However, by levels IV and III, which represent the Vatya culture at this site, the percentage of older sheep slaughtered, clearly dominates over younger animals. This is very typical for a wool producing population. It cannot be excluded that there was even surplus production of wool, that could, via trade and barter, have been used to acquire other important materials such as metal ingots from the west, in what is now Austria. This development seems to be connected with the Vatya culture in this limited area (Vretemark and Sten, *in press*).

Apart from cattle, sheep, goats and pigs, horse was also an important factor in the animal economy even if the percentage in this settlement assemblage only reached about 5 % of NISP. The horse bones represent different parts of the animal, from the cranium to the trunk and the extremities. Bones from the less meaty body parts, such as distal limb segments are often complete, while bones from meatier parts, such as the shoulder blades and the proximal extremity bones, have been butchered. Butchering marks noted on horse bone from meaty body regions, such as the shoulder blade and thigh bone, clearly

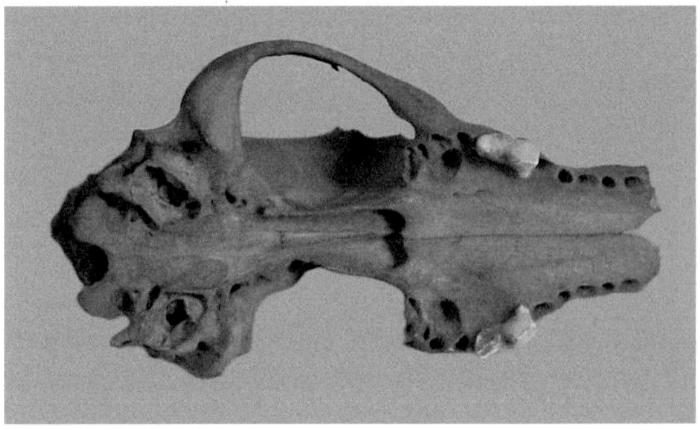

*Fig. 6. Dog cranium found in a Vatya house foundation at Százhalombatta–Földvár (*norma basilaris*). Photo: M. Vretemark.*

show that horse meat was eaten even if this was perhaps not the main purpose of horse breeding. Elsewhere, the great number of intact bones from this species has been interpreted as meaning that horse meat was not consumed in this period (Choyke 1984). However, it is also possible that butchering marks were less consistently recognized by Choyke, earlier in her career. Thus, the question of whether horses were exploited differently at his site cannot be resolved for the moment. This Early Bronze Age Bell-Beaker sites in Hungary contain the bones of horse which have clearly been broken up for their marrow, in fact, butchered in the same way as cattle (Bökönyi 1980; Choyke 1984; Kyra Lyublyanovics, *pers. comm.*). By the Nagyrév period at the end of the Early Bronze Age horse was no longer a regular part of the diet and its bones become much less common on sites (Bökönyi 1980; Choyke 1984).

The number of bones from dogs in the Százhalombatta material is striking, particularly among the bones from the Middle Bronze Age Vatya culture layer. The dog bones display no butchering marks so there is no evidence to suggest that dog meat was eaten. If one examines the body parts represented, cranial fragments clearly dominate compared to skeletal elements from the rest of the body. There are many jaws and complete or fragmented dog skulls (Fig. 6). It seems probable that this is not a random phenomenon and reflects ritual treatment of this species at this particular site. One interpretation is that some dog bones found in house foundations originate from building sacrifices, where certain body parts were intentionally chosen. Skulls and jawbones can in this context, represent the whole animal as does the paw found in another foundation (perhaps symbolizing the speed of this animal; Fig. 7). As a form of clearly defined ritual behavior this use of dogs may be related to hunting, herding and guarding, activities clearly related to adult male identity in many societies across the world (Morris 1998, 69–71). The reasons given for keeping dogs in Malawi, for example, include (1) a guard dog for protection from thieves and wild animals, (2) beautifying the household, as a

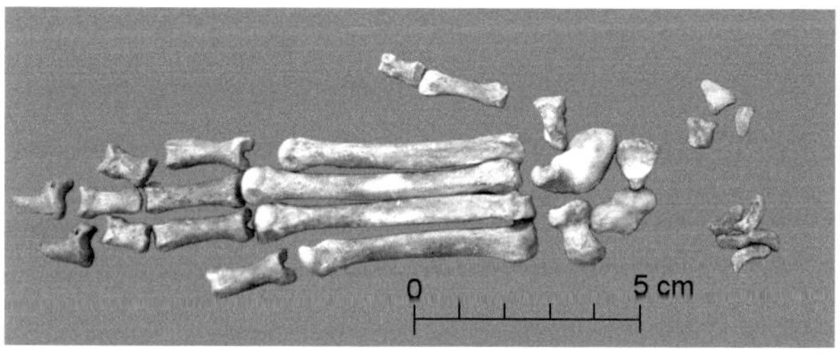

Fig. 7. Complete dog paw found in small pit associated with the foundation of a Vatya house at Százhalombatta–Földvár. Photo: M. Vretemark.

flower, and (3) help in hunting (Morris 1998, 86).

The age structure of the dogs shows a dominance of fully grown animals at Százhalombatta.The importance of dogs was probably to a high degree related to animal husbandry, where they were presumably used primarily for herding, protection and guarding the herds, sheep in particular. These functions are of vital importance in a culture where identity and wealth was comprised of domestic animal stock and where the productive yield from husbandry represented survival. It is surely this role, which in its transferred form, is the basis of the custom of using dogs in particular, as building sacrifices. Even a dead dog was regarded as providing protection for houses, people, animals and possessions. Pendants made of dog teeth and metapodia found at Százhallombatta-Földvár may also be seen in this context.

The percentage of overall game NISP consists only of between 1–2 % in the different layers. It is clear that the relative frequency of wild animal bones is highest in the material from the earliest levels, VI–IV (c. 2000–1600 BC). After that, the percentage of bone from game is reduced. Of the species represented, only red deer occurs regularly while lynx, aurochs, wild boar and hare are represented only by a few bone fragments. Hunting had therefore no direct importance in meat provision, but may have satisfied other important needs, for example the supply of antlers (they can have been found as shed antlers as well), skin and leather of special quality and raw material like wild boar tusk for tools and ornaments. Apart from this, hunting, presumably with dogs, surely had social significance as a highly ritualized male pursuit. Thus, even if hunting did not have a great significance in Vatya society as a whole and at this settlement in particular, the symbolic importance of hunting must have been a phenomenon characteristic of all coeval settlements in the Carpathian Basin. The ritual power of various animals as described in Morris (1998, 68) was also expressed through a variety of decorative objects made from tusk, antler and bone from all sorts of wild animals.

	large rib scraper	caprine tibia scraper	large mandible smoother	skates	hare metapodium pendants	small double points	flat butted heavy duty	oblique butt heavy duty	hafted beam or crown	horse bridle ornaments	Class II awls	Class II scrapers	Class I awls	Class I scrapers	**Total**
Százhalombatta-Földvár EBA	5	–	–	–	2	–	2	–	–	–	4	3	–	5	**21**
Százhalombatta-Földvár MBA	28	–	–	3	9	18	11	2	–	5	10	18	23	20	**133**
Százhalombatta-Földvár LBA	1	–	–	–	–	–	1	–	–	–	–	–	4	7	**13**
Western Hungary															
Lovasberény-Mihályvár MBA	41	–	–	–	–	–	–	–	–	–	–	3	2	42	**88**
Pákozd-Várhegy MBA	27	–	–	–	2	–	3	–	1	2	8	21	13	22	**99**
Sárbogárd-Cifrabolondvár MBA	3	1	–	–	–	–	5	–	–	–	–	1	–	4	**14**
Aba-Belsőbárándvár MBA	1	–	–	–	–	1	16	–	–	–	–	–	–	2	**20**
Vál-Pogányvár MBA	–	–	–	–	–	1	5	–	–	–	–	–	–	–	**6**
Igar-Vámpuszta MBA	–	–				–	2	–	–	–	1	–	3	2	**8**
Danube-Tisza interfluve															
Mende-Leányvár MBA	1	–	–				5	–	–	2	–	–	–	7	**15**
Csongrád-Vidresziget MBA	6	–	–	–	–	–	–	–	–	–	–	–	–	3	**9**
Jászdózsa-Kápolnahalom E-MBA	–	15	4	–	–			10	–	–	–	–	–	–	**29**
Jászdózsa-Kápolnahalom MBA	–	12	15	–	–	–	–	24	22	4	3	–	10	27	**117**
Jászdózsa-Kápolnahalom LBA	–	6	2				–	19	13	–	–	–	4	23	**67**
Jászdózsa-Kápolnahalom stray	3	3	3	–	–	–	–	6	4	–	–	–	4	12	**35**
Eastern Hungary															
Törökszentmiklós-Terehalom MBA	–	7	6	–	–	–	–	–	–	–	2	1	3	3	**22**
Tiszaug-Kéménytető stray	3	16	–	–	–	–	–	1	–	–	–	–	15	21	**56**
Tiszaug-Kéménytető MBA	–	9	2	–	–	1	–	3	–	–	–	–	10	10	**35**
Tiszaug-Kéménytető EBA	2	2	2	–	–	1	–	–	1	–	–	–	3	5	**16**
Gyulavarsánd-Laposhalom MBA	1	–	–	–	–	2	8	45	25	–	–	–	2	2	**85**
Total	**122**	**71**	**34**	**3**	**13**	**24**	**58**	**110**	**66**	**13**	**28**	**47**	**96**	**225**	**910**

Fig. 8. The representation of various tool types at Százhalombatta–Földvár and other sites mentioned in the text. For the typology, see Choyke (1984).

Worked bone and antler

Bronze Age worked bone, antler and tooth share a number of characteristics (Fig. 8). The bone tools at most sites from the Bronze age in Hungary are overwhelmingly made from domestic animals, something directly related to availability. The only real exception to this is, again, the site of Jászdózsa where the red deer bone is chosen for manufacturing in parallel with the increase of its bones in the refuse bone assemblage (Choyke 1984). The distribution of types found at Vatya sites compared to worked bones from eastern Middle Bronze Age sites can be seen in Fig. 9.

As can be seen, the only artifact type which is shared universally across the Carpathian Basin in this period is the ornamented horse bridle (Choyke 1979; Choyke 1983). As mentioned earlier, horse itself seems to increase in status during this period. This may be related to the movement of people in well-established exchange systems. Such constant movement served to make these displayed objects increasingly uniform as discussed in David and Kramer (2001, 215). The curvilinear meander design of this antler bridle ornament (Fig. 10) was particularly favored on a variety of decorative objects. Such antler ornaments are emphatically Class I, that is, planned objects (Choyke 1997; Choyke 2001) of a high enough quality to suggest that they might have been produced by specialized, possibly itinerant craftspeople, thus symbolizing an emblematic style which crossed political boundaries. Displaying such easily recognizable objects would have served to reinforce both the high status of the users and their membership in a global economic system in much the same way that a crucifix represents the wearer's Christianity, cross-cutting specific political affiliation.

Another important feature shared by Middle Bronze

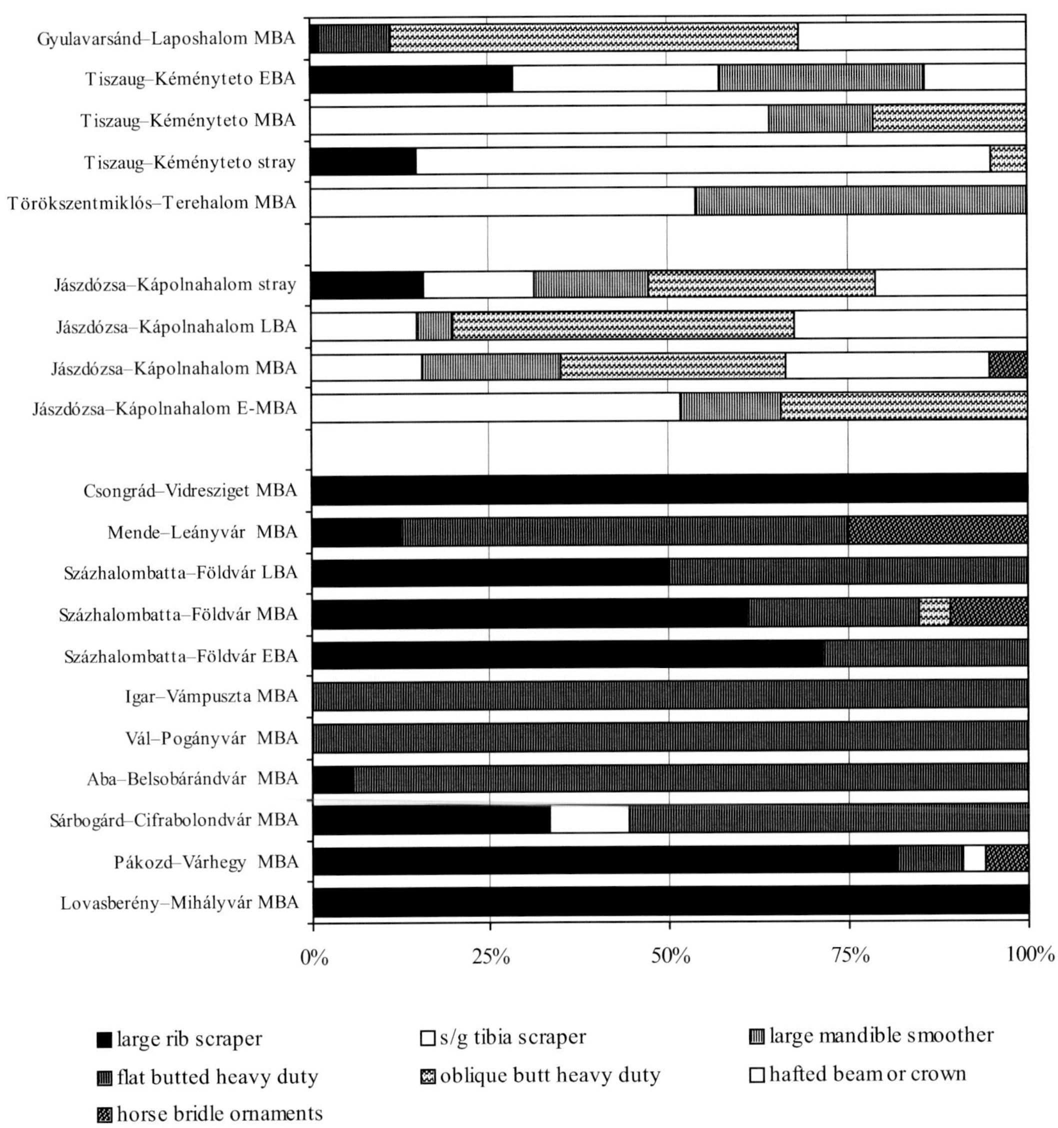

Fig. 9. Comparison of main tool types between Vatya culture and eastern Bronze Age sites.

Fig. 10. Horse bridle strap ornament made from antler, the curvilinear meander design was carved with metal tools. Such fine objects may have been produced by itinerant specialists. Photo: J. Rasson.

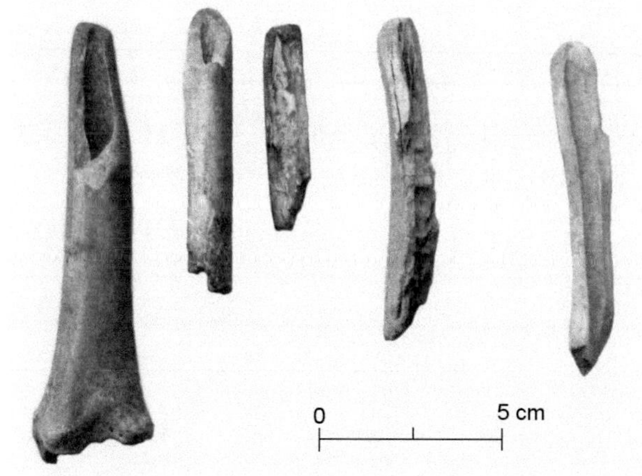

Fig. 11. Crudely made scrapers typical of the Middle Bronze Age in general which fall towards the Class II end of the manufacturing continuum. Photo: J. Rasson.

Age worked bone assemblages in Hungary is the relatively poor quality of manufacturing (Choyke 1997; Choyke 2001) of the majority of these artifacts. Only Class I tools are carefully planned, made from selected raw materials and with increasing amounts of work invested in their manufacture. Such ornaments or tools were intended for specific long-term, repeated tasks such as hide preparation and were often repaired (curated) as they broke during use. At the other end of the continuum are Class II tools which are made from bones which broke in a lucky way. They tend to be utilized rather than specifically worked and are intended for expedient use. A typical example of a tool type of a Class II character are bone scrapers from Százhalombatta which are based on the spiral fractures of long bones broken during food preparation (Fig. 11). Like many of the other tools from this site, not a great deal of energy was expended on their manufacture.

A comparison of the worked bone and antler objects from Vatya sites in general and Százhalombatta in particular (Fig. 12) helps us assess the degree to which this site shares bone manufacturing traditions with some other sites of the Vatya culture (Choyke 2000). The greater variability of types at Százhalombatta is certainly related to larger sample size.

Heavy duty antler tools (Fig. 13) made from the burr and beam of red deer antler were a continuous part of the standard tool inventory in the Carpathian Basin since the late Neolithic (Bakalov 1979; Choyke 1984; Periši 1984; Russell 1990). However, people living in Vatya settlements used a special kind with a flat hollowed out butt-end such as the top artifact in Fig. 13. Presumably, this end could be fitted with a separate groundstone or metal blade. Some of these tools are finely polished with designs on them. In keeping with their "Vatya" identity the people living at our settlement also used this type. In the east of the country heavy duty burr and beam tools tend to have oblique butts (Choyke 1984) (see Fig. 9). Making everyday tools in this particular way represents a variety of Sackett's passive, latent style (Sackett 1990, 56–7).

Another important general Vatya culture type, also present at Százhalombatta, are cattle rib scrapers (Fig. 14). They can be highly polished by heavy use but are not heavily modified. Such tools, of course, can appear anywhere in the region but are consistently found on sites of the Vatya culture (Choyke 1984). They represent a specific cultural marker for Vatya worked bone. Functionally the same tool is made from sheep/goat tibiae in the east (Choyke 1984; Choyke and Bartosiewicz 1999–2000) (see Fig. 9).

Within the Vatya sphere two worked bone types appear exclusively at Százhalombatta. They are both tools rather than decorative objects meant for display, so they also represent a different function and a latent style aimed at other households at the settlement. The first type are skates made from cattle radii (Fig. 15). This represents a unique type in the Middle Bronze Age, however, similar objects can be found at (Early Bronze Age) Bell Beaker sites around Budapest. There may be a regional tradition in the manufacture of such tools which does not extend much beyond this part of the Vatya territory. The second type

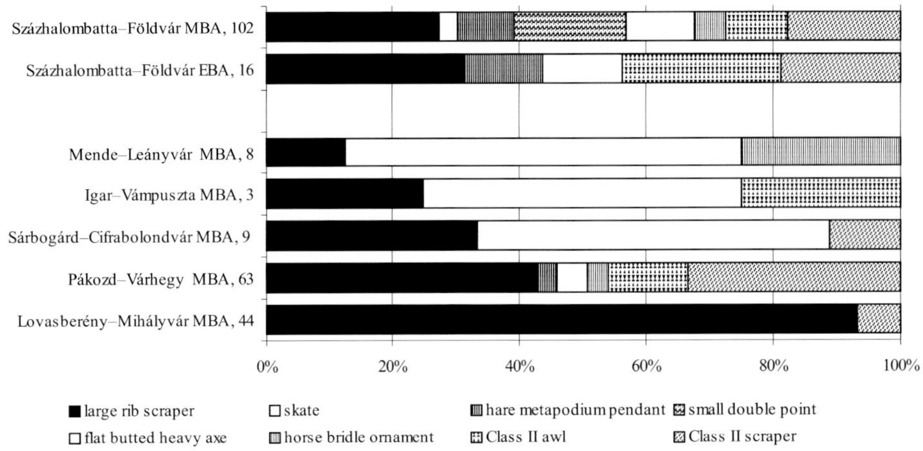

Fig. 12. Comparison of the main tool types found at Vatya period sites and Százhalombatta–Földvár.

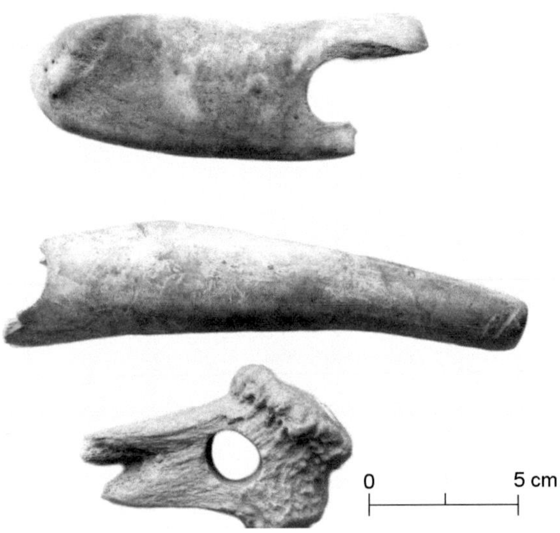

Fig. 13. Tools with flat, hollowed out butt-ends (top) are typical of Vatya period sites. Antler tine "pick" fragment (center). Heavy duty burr and beam antler tools (bottom) are characteristic of the late Neolithic in the Carpathian Basin. Photo: J. Rasson.

Fig. 14. Polished scrapers made with cattle ribs are particularly characteristic of Vatya period sites. Photo: J. Rasson.

are small double points from both rib and long bone diaphysis splinters (Fig. 16). Some of these objects are heavily polished and used. They range in size from 5 to 2.5 cm. Although such tools were not found on Vatya sites (Choyke 1983; Choyke 1984; Choyke and Bartosiewicz 1999–2000), larger specimens are reasonably common at major tell sites near rivers (Choyke and Bartosiewicz 1999-2000). The tiny double points are unique to Százhalombatta and were obviously used in a special task. It is suggested here that they may have been used as body decoration in which case they are intended as active symbols of a group's social identity.

Finally, we come to two special decorative sets of objects related to the ritual symbolism encountered first with the use of dog crania and paws as house sacrifices. Pendants made from dog canines with drilled roots are also present although it is a standard practice at many prehistoric sites to modify dog canines in this fashion. This phenomenon would not be worth of further comment if there was not an example of a second upper dog molar which is polished and also drilled through the root (Fig. 17). The second set of objects, which are also almost unique to this site (the single exception comes from Pákozd–Várhegy: Fig. 12), are drilled metapodia from

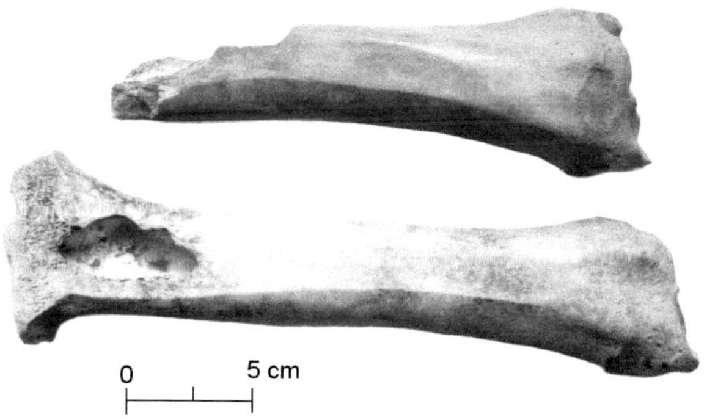

Fig. 15. Skates made from cattle radii appear locally on sites of the Early Bronze Age, Bell Beaker culture however, they are only found at Százhalombatta–Földvár in the Middle Bronze Age. Photo: J. Rasson.

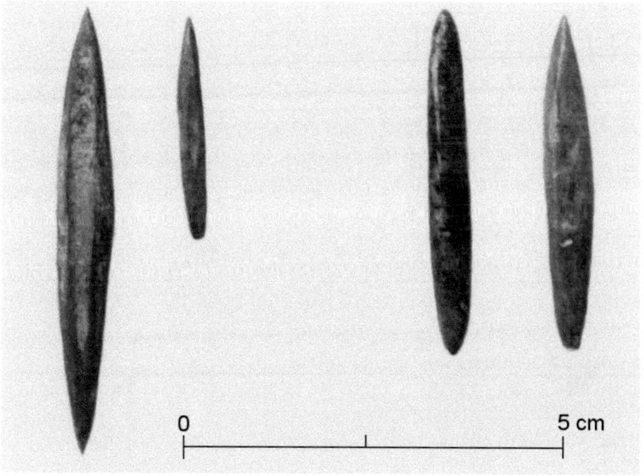

Fig. 16. Small double points are typical of the Százhalombatta–Földvár material but otherwise uncommon on Vatya period sites. The smallest points may be body decoration characteristic of this settlement alone. Photo: J. Rasson.

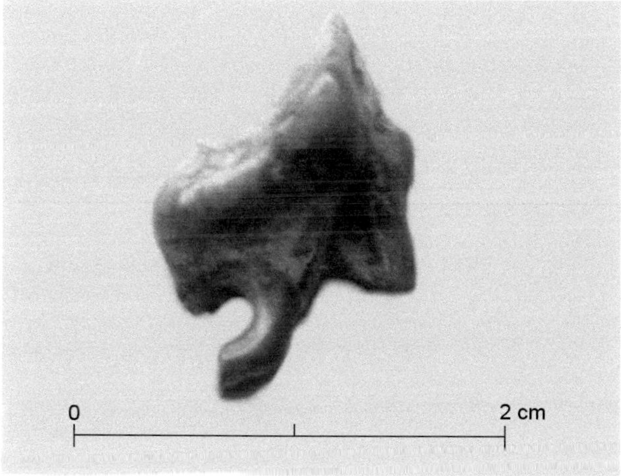

Fig. 17. Drilled and polished dog second molar. Photo: J. Rasson.

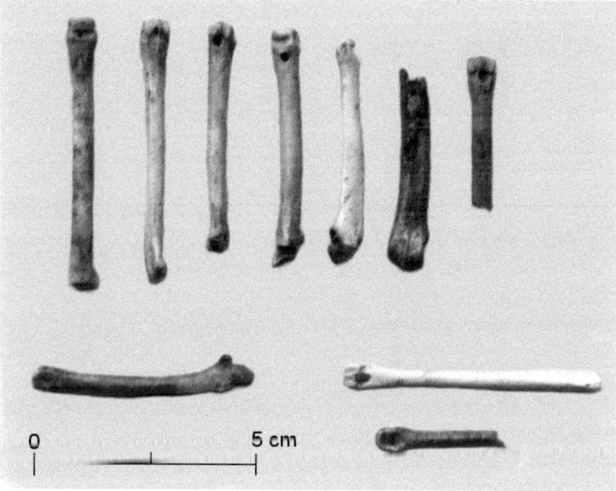

Fig. 18. Drilled hare and one dog metapodial. Photo: J. Rasson.

hare, with an example from dog, probably not accidentally (Fig. 18). These small game animals are not otherwise common in the bone refuse material (Fig. 2) although they must have thrived in the grassland surrounding the settlement. What seems important here, is their possible symbolic meaning: speed connected with wilderness and hunting as opposed to the village, the domestic sphere. Male social identity may be important in this case (Morris 1998, 69). It is also possible that the use of these metapodia pendants was meant as an identifying feature of people from this locality. It will be interesting to see whether other sites in the Benta valley contain such ornaments.

Conclusions

The composition of refuse bone and worked bone assemblages is clearly affected by embedded traditions linked to social identity in the Middle Bronze Age in the Carpathian Basin. This identity can be seen to exist on several levels. Similar traits from the region at large come from a shared pool of symbols and traditions. They also exist at the level of smaller defined regions such as that represented by the area encompassed by the Vatya culture. Finally a local level of social identity can also be detected in the ways dogs are treated and in special bone tools such as skates, tiny double points, drilled dog teeth and metapodials. The drilled dog teeth as well as hare and dog metapodial pendants may perhaps relate to hunting and possibly male identity.

Other aspects of identity relate to the way animals were used and bone working operated on an unconscious level related to conservative traditions passed on within the regions, settlements and households where people had multiple opportunities to observe and compare each other. The ritual use of dogs as house sacrifices and decorative elements such as horse bridle ornaments, small double points, drilled dog teeth and drilled hare metapodia deliberately communicate information to other communities or within settlements.

The case for identifying patterns in refuse bone vs. worked bone and antler would be greatly strengthened if one could compare degrees of similarity on a statistical basis. Unfortunately, although new excavations at the site of Százhalombatta–Földvár are of fine resolution (with consistent sampling, screening and flotation), this was not the case for the other Middle Bronze Age sites discussed here. Here, a simple comparison is made of tendencies based on counts of individual specimens and of the presence or absence of types. It may very well be that, as the number of properly excavated sites from this period increases, the picture of social identity in the Middle Bronze Age will also be sharpened and even modified.

References

Bačkalov, A. 1979. *Predmeti od kostii roga u preneolitu I neolitu Srbije* (Stone and Bone Tools in Preneolithic and Neolithic Serbia). Beograd: Archaeologia Jugoslavica.

Bóna, I. 1975. *Die Mittlere Bronzezeit Ungarns und ihre Südöstlichen Beziehungen*. Budapest: Akadémia Kiadó.

Bökönyi, S. 1980. The importance of horse domestication in economy and transport, pp. 15–21 in Sörbom, P. (ed.), *Transport Technology and Social Change* (Tekniska Museet Symposia 2). Stockholm: Daedalus.

Choyke, A. M. 1979. A classification of the bone and antler tools from the Bronze Age hill-fortress of Pákozdvár. *Alba Regia* **17**, 9–21.

Choyke, A. M. 1983. Előzetes jelentés a tiszaug-kéménytetői ásatás csontszerszámairól (Preliminary report on the bone tools from Tiszaug-Kéménytető). *Archeológiai Értesítő* **109/1**, 35–41.

Choyke, A. M. 1984. An analysis of bone, antler and tooth tools from Bronze Age Hungary. *Mitteilungen des Archäologischen Instituts der Ungarischen Akademie der Wissenschaften* **12/13**, 13–57.

Choyke, A. M. 1987. The exploitation of red deer in the Hungarian Bronze Age. *Archaeozoologia* **I (1)**, 109–16.

Choyke, A. M. 1997. The bone manufacturing continuum. *Anthropozoologica* **25–26**, 65–72.

Choyke, A. M. 1998. Bronze Age red deer: case studies from the Great Hungarian Plain, pp. 157–78 in Anreiter, P., Bartosiewicz, L., Jerem, E. and Meid, W. (eds), *Man and the Animal World. Studies in Memoriam Sándor Bökönyi*. Budapest: Archaeolingua Kiadó.

Choyke, A. M. 2000. Refuse and modified bone from Százhalombatta–Földvár. Some preliminary observations, pp. 97–102 in Poroszlai, I. and Vicze, M. (eds), *Százhalombatta Archaeological Expedition Annual report 1*. Százhalombatta: Matrica Museum.

Choyke, A. M. 2001. A quantitative approach to the concept of quality in prehistoric bone manufacturing, pp. 59–66 in Buitenhuis, H. and Prummel, W. (eds), *Animals and Man in the Past* (ARC-Publicatie 41). Groningen: ARC.

Choyke A. M. and Bartosiewicz, L. 1999. Bronze Age animal keeping in Western Hungary, pp. 239–49 in Jerem E. and Poroszlai, I. (eds), *Archaeology of the Bronze Age and Iron Age: Experimental Archaeology, Environmental Archaeology, Archaeological Parks*. Budapest: Archaeolingua Series Major.

Choyke A. M. and Bartosiewicz, L. 1999–2000. Bronze Age animal exploitation in the Central Great Hungarian Plain. *Acta Archaeologica Academiae Scientiarum Hungaricae* **51**, 43–70.

David, N. and Kramer, C. 2001. *Ethnoarchaeology in Action*. Cambridge: Cambridge University Press.

Grant, A. 1982. The use of tooth wear as a guide to the age of domestic animals, pp. 91–108 in Wilson, B., Grigson, C. and Payne, S. (eds), *Ageing and Sexing Animal Bones from Archaeological Sites* (BAR British Series 109). Oxford: British Archaeological Reports.

Hodder, I. 1982. *Symbols in Action: Ethnoarchaeological Studies of Material Culture*. New York: Cambridge University Press.

Kovács, T. 1969. A százhalombatta bronzkori telep (The Bronze Age settlement at Százhalombatta). *Archeológiai Értesítő* **96**, 161–9.

Kovács, T. 1982. Befestigungsanlagen um die Mitte 2. Jahrtausends v. u. Z. in Mittelungarn, pp. 279–91 in Chropovsky, B. and Herrmann, J. (eds), *Beiträge zum Bronzezeitlichen Burgenbau in Mitteleuropa*. Berlin: Zentralinstitut für Alte Geschichte und Archäologie and Nitra: Archeologický Ústav Slovenskej Akadémia Vied.

Kovács, T. 1984. Die Vatya-Kulture, pp 219–33 in Tasi, N. (ed.), *Kulturen der Frühbronzezeit des Karpatenbeckens und Nordbalkans*. Beograd.

Leach, E. 1976. *Culture and Communication*. Cambridge: Cambridge University Press.

Morris, B. 1998. *The Power of Animals*. Oxford: Berg.

Nováki, Gy. 1952. Fejér megye őskori földvárai (Prehistoric hill-fortresses in Fejér country). *Archeológiai Értesítő* **79**, 3–17.

Periši, S. 1984. *Predmeti od Kosti, Roga I Kamena* (Stone and Bone Tools in Kamena). Katalog XIII. Beograd: Muzej Grada Beograda.

Poroszlai, I. 1992 Százhalombatta-Földvár, pp. 153–5 in Fodor, I., Meier-Arendt, W. and Raczky, P. (eds), *Bronzezeit in Ungarn. Forschungen in Tell-siedlungen an Donau und Theiss*. Frankfurt am Main: Dezernat Kultur und Freizeit der Stadt Frankfurt a.M.

Poroszlai, I. 2000. Excavation campaigns at the Bronze Age tell site at Százhalombatta–Földvár I: 1989–1991; II. 1991–1993, pp. 13–74 in Poroszlai, I. and Vicze, M. (eds), *Százhalombatta Archaeological Expedition, Annual Report 1*. Százhalombatta: Matrica Museum.

Raczky, P., Hertelendi, E. and Horváth, F. 1992. Zur absolute Datierung der bronzezeitlichen Tell-Kulturen in Ungarn, pp. 42–7 in Fodor, I., Meier-Arendt, W. and Raczky, P. (eds), *Bronzezeit in Ungarn. Forschungen in Tell-Siedlungen an Donau und Theiss*. Frankfurt am Main: Dezernat Kultur und Freizeit der Stadt Frankfurt a.M.

Russell, R. 1990. The bone tools, p. 521–48 in Tringham, R. and Krstiç, D. (eds), *Selevac: A Neolithic Village in Yugoslavia* (Monumenta Archaeologica 15). Los Angeles: University of California Press.

Sackett, J. 1990. Style and ethnicity in archaeology: the case for isochrestism. pp. 32–43 in Conkey, M. and Hastorf, C. (eds), *The Uses of Style in Archaeology*. Cambridge: Cambridge University Press.

Schiffer, M. 1999. *The Material Life of Human Beings: Artifacts, Behavior and Communication*. London: Routledge.

Tóth, E. 1990. Négy évtized régészeti kutatési Bács-Kiskun megyében (1949–1989) [Four decades of archaeological research in Bács-Kiskun county (1949–1989)]. *Cumania* **12**, 81–116.

Vicze, M. 2000. Background to the field-survey, pp. 119–130 in Poroszlai, I. and Vicze, M. (eds), *Százhalombatta Archaeological Expedition, Annual Report 1*. Százhalombatta: Matrica Museum.

Vretemark, M. and Sten, S. *in press*. Diet and animal husbandry during the Bronze Age. An analysis of animal bones from Százhalombatta-Földvár, Hungary. In Poroszlai, I. and Vicze, M. (eds), *Százhalombatta Archaeological Expedition, Annual Report 2*. Százhalombatta: Matrica Museum.

Wiessner, P. 1984. Reconsidering the behavioral basis for style: a case study among the Kalahari San. *Journal of Anthropological Archaeology* **3**, 190–234.

Wobst, H. M. 1977. Stylistic behavior and information exchange, pp. 317–42 in Cleland, C. (ed.), *Papers for the Director: Research Essays in Honor of James B. Griffin* (Anthropological Papers 61). Michigan: University of Michigan, Museum of Anthropology.

Wobst, H. M. 1999. Style in archaeology or archaeologists in style, pp. 118–32 in Chilton, E. S. (ed.), *Material Meanings: Critical Approaches to the Interpretation of Material Culture*. Salt Lake City: University of Utah Press.

Alice M. Choyke
Aquincum Museum,
Zahony u. 4,
H-1031 Budapest, Hungary
E-mail: H13017cho@ella.hu

Maria Vretemark
The County Museum of Västergötlands,
Box 253, S-532 23 Skara, Sweden
E-mail: kansliet@skaramus.se

Sabine Sten
The National Historical Museum,
Box 5428,
S-11284 Stockholm, Sweden
E-mail: sabine.sten@historiska.se

9th ICAZ Conference, Durham 2002
Behaviour Behind Bones, (ed. Sharyn Jones O'Day et al.) pp. 190–197

21. Animal husbandry and centralized cultures. How social and political factors can influence rural lifestyle

Giovanni Siracusano

This paper examines the changes that took place in the faunal assemblages at Zeytinli Bahçe Höyük (Urfa, Turkey) from the end of the 4th millennium BC onwards, and these changes are compared with what is known from the archaeozoological record of neighbouring sites and regions. In the Urfa/Adiyaman zone, along the Middle and Upper Euphrates River, and in Northern Mesopotamia, pig breeding seems to have been a deeply-rooted tradition since Neolithic times. North of the Taurus Mountains, however, cattle seem to have played a major part in animal rearing. Both species markedly declined in favour of sheep and goats, which became by far the most common type of livestock everywhere during the last centuries of the 4th millennium, in connection with the influence of centralised organisational systems. With the changes that occurred after the so-called Uruk culture "collapse", pig again became the dominant species reared in the Upper Mesopotamian regions, while in the northern Upper Euphrates region cattle keeping gained in importance.

The archaeological setting

Work on the archaeological excavations at Zeytinli Bahçe Höyük (Fig. 1) began in 1999 under the direction of M. Frangipane and was conducted by the Italian Archaeological Mission to Eastern Anatolia from Rome's "La Sapienza" University, in conjunction with the Urfa Museum. This project forms part of the wider archaeological salvage campaign organised by the Middle East Technical University of Ankara, in connection with the construction of the Carchemish and Ilisu dams on the Euphrates and the Tigris, respectively (Tuna *et al.* 2001; Tuna *et al.* 2002). The excavations have brought to light a fairly small site (about 2.6 hectares) with a long settlement sequence. This has yielded archaeological remains dating back to the period running from the second half of the 4th millennium BC, corresponding to the Middle and Late Uruk periods, to the end of the first millennium AD. The youngest occupation phase of the site is characterised by a type of fortified Late Byzantine "farm", with a cemetery built on top of it. The series of settlements created with juxtaposed terracing have often re-used or demolished earlier structures, making the stratigraphic sequence rather difficult to establish. Some sectors, for example the northern, which was occupied by a medieval necropolis and by Roman and Byzantine structures, appear to have been considerably disturbed. However, the soundings taken on the western side of the tell, seriously eroded by the river, have brought to light a long and detailed stratigraphic sequence of levels from the Late Chalcolithic period (six building levels) and Early Bronze Age I (EBA I; 12 main buildings) in a deposit of about ten metres depth. The sequence clearly documents the changes that took place between the Late Uruk period and the Early Bronze Age (Frangipane and Buçak 2001; Frangipane *et al.* 2002; Frangipane *et al., in press*).

This paper addresses only the faunal remains discovered by these soundings in the western sector of the tell at Zeytinli Bahçe Höyük, with the material being subdivided into the three main phases identified in the stratigraphic sequence: Late Uruk (3300–3000 BC), Early Bronze IA and Early Bronze IB (*c.* 3000–2600 BC). The Middle Uruk levels were only brought to light in the most recent 2002 excavation campaign, and therefore those data have not been included here.

As early as the second half of the 4th millennium, the Uruk culture spread throughout Greater Mesopotamia. Major socio-cultural changes took place in the final centuries of the 4th millennium BC in particular,

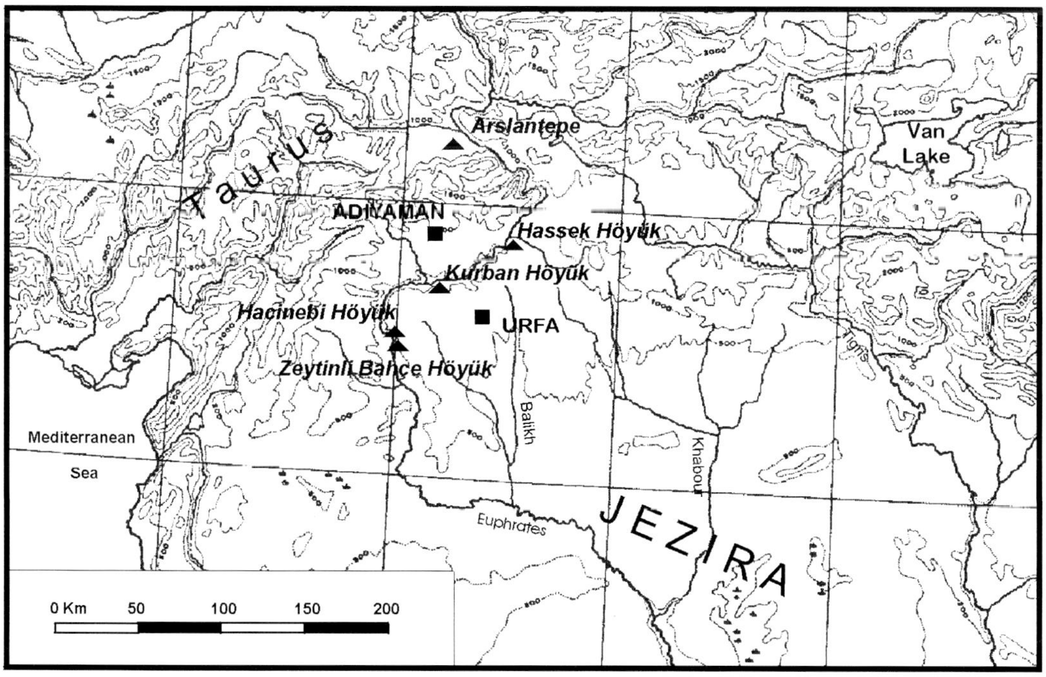

Fig. 1. Location of the sites and regions mentioned in the text.

influencing most of the Mesopotamian world with the emergence of centralised political and economic structures, and the strengthening of the elites. During this period a cultural and organisational model began to establish itself, gradually accompanied by more or less explicit forms of settlement by southern groups, and then spread throughout the whole of Greater Mesopotamia. This transformed the earlier social, cultural, economic and political system, standardising a vast area running from the lower Euphrates and the Tigris to the Taurus spurs (Frangipane 1996). At the turn of the 3rd millennium this dominant model seems to have gone into crisis in the northern areas, and the regional system inside Upper Mesopotamia and its northern border was changed once again. A divide occurred, opening up and separating the sites along the northern Upper Euphrates Valley (north of the Taurus Mountains), and those in the southern sectors of the valley (south of Taurus) and central-eastern Jezira, from the two sectors of the Euphrates Valley to the north and the south of the Taurus, with different developments in the mountainous zone and the more specifically Syro-Mesopotamian zone (Frangipane 1998).

Zeytinli Bahçe Höyük is a settlement on the left bank of the Euphrates, to the south of the Taurus chain, in the region called Urfa/Adiyaman, which is the southernmost of the three areas into which the upper course of the river divides the region in both environmental and cultural terms. The long stratigraphic sequence makes it possible to detail the development of the settlement from the mid 4th millennium and the early centuries of the 3rd millennium. The settlement exhibits an extraordinary cultural continuity running throughout the Uruk period, the transition to Early Bronze Age, and finally the EBA IA and IB cultures. The archaeological remains provide evidence for the establishment of a well-entrenched culture that would appear to have maintained its traditions without any major upheavals, despite the events of history. Characteristically, the dwellings, which appear to have been reconstructed several times, do not show any significant changes, and even though there were modifications in terms of the size, shape and layouts of the houses in the Early Bronze IB period, the new model of the houses and the more highly planned urban layout that can be found in this late phase are repeated once again with extraordinary regularity in the highly numerous construction levels belonging to this period. The fact that the architectural continuity, and ceramic production, persisted over such long timeframes suggests a strong cultural continuity through the final phases of Late Chalcolithic and the early period of the Bronze Age.

Analysis of the fauna

During the first three years of the excavations of the Zeytinli Bahçe site a fairly large number of bone remains were discovered (Siracusano 2002). By the end of the 2001 campaign, over 3000 of them had been identified (Siracusano, *in press*). They come from every level that was excavated: from Late Chalcolithic and Early Bronze IA and IB in the western sectors, Middle Bronze Age and Iron Age in the higher conical mound, and from Roman

	Late Uruk						Early Bronze Age					
	NISP	% of domestic	% of total	MNI	% of domestic	% of total	NISP	% of domestic	% of total	MNI	% of domestic	% of total
Equus caballus	0	0.0	0.0	0	0.0	0.0	1	0.1	0.1	1	1.1	1.0
Equus asinus	7	0.9	0.9	2	5.0	4.3	2	0.1	0.1	1	1.1	1.0
Equus sp.	3	0.4	0.4	0	0.0	0.0	2	0.1	0.1	0	0.0	0.0
Bos taurus	88	11.6	11.4	6	15.0	12.8	142	8.7	8.6	11	12.2	11.5
Ovis aries	23			4			52			12		
Capra hircus	22			4			35			8		
O/C	426						785			23		
O/C total	471	62.3	60.9	17	42.5	36.2	872	53.3	52.6	43	47.8	44.8
Sus domesticus	180	23.8	23.3	12	30.0	25.5	612	37.4	36.9	30	33.3	31.3
Canis familiaris	7	0.9	0.9	3	7.5	6.4	5	0.3	0.3	4	4.4	4.2
Total domestic animals	**756**	**100.0**	**97.8**	**40**	**100.0**	**85.1**	**1636**	**100.0**	**98.6**	**90**	**100.0**	**93.8**
Cervus elaphus	3		0.4	1		2.1	6		0.4	2		2.1
Dama dama	5		0.6	2		4.3	0		0.0	0		0.0
Cervidae	1		0.1	0		0.0	2		0.1	0		0.0
Gazella sp.	6		0.8	3		6.4	4		0.2	1		1.0
Sus scrofa	0		0.0	0		0.0	1		0.1	1		1.0
Lepus capensis	2		0.3	1		2.1	10		0.6	2		2.1
Total game	**17**		**2.2**	**7**		**14.9**	**23**		**1.4**	**6**		**6.3**
Total mammals	**773**			**47**			**1659**			**96**		
Aves	9			1			11			4		
Anser sp.	1			1			2			2		
Anas sp.	0			0			2			1		
Pisces	3			1			8			2		
Total	**13**			**3**			**23**			**9**		
Vulpes vulpes	1			1			5			1		
Mustelidae	0			1			3			1		
Testudo graeca	45			2			0			0		
Micromammalia	34			1			32			4		
Brachyura	1			1			1			1		
Total	**81**			**6**			**41**			**7**		
Grand total	**867**			**56**			**1723**			**112**		

Fig. 2. Animal remains from Zeytinli Bahçe Höyük: Late Uruk levels compared to Early Bronze Age levels (1999/2001).

and medieval-Byzantine age in the upper levels. This article will deal only with the end of the Late Chalcolithic period and the Early Bronze Age.

Seven species of domestic animals were identified: ass, horse, cattle, goat, sheep, pig and dog (Figs. 2–4). A number of wild species were also identified. Even though there were not very many of them, the most representative species of semi-desert and open woodland respectively, are certainly gazelle (presumably *Gazella subgutturosa*) and deer (represented by fallow deer *Dama dama* and red deer *Cervus elaphus*). Goat and sheep account for over half of the bone remains identified in all the stratigraphic sequences, and there are always more sheep than goats, in a ratio of 1.5 to 1. The histograms (Fig. 5), based on the NISP (Number of Identified Specimens), indicate a fairly stable situation from Late Uruk to EBA IA, with sheep and goats by far the most common animals there, followed by pigs accounting for some 24 percent, and cattle accounting for about 15 percent. Further analysis of the NISP shows a slight reduction in the frequency of sheep, goats and cattle in the transition between EBA IA and EBA IB. The variation in the frequency of pigs is much more marked. They account for about 40 percent of the identified remains in EBA IB. While the architectural structures and archaeological remains show no particular breaks in continuity, the sharp increase in the domestic pig population in EBA IB seems to act as an indicator of major changes occurring during the course of the early centuries of the 3rd millennium.

As far as animal husbandry strategies are concerned, as one would have expected, there was an improvement with a general increase in sizes between the earlier periods and the Roman/Byzantine Age. However, we still know very little about this aspect because the number of finds to be used as evidence is low. For the moment the most we can say is that while the sheep and goats stood about 60 cm high at the withers in the Late Uruk levels, this had risen to 63 cm in the Early Bronze Age. Judging from the only two remains from Late Uruk, the estimated size of the pigs is about 65 cm. The only withers height that could be

	NISP	% of domestic	% of total	MNI	% of domestic	% of total
Equus caballus	1	0.3	0.3	1	2.8	2.6
Equus asinus	2	0.6	0.5	1	2.8	2.6
Equus sp.	1	0.3	0.3	0	0.0	0.0
Bos taurus	54	15.1	14.7	5	13.9	12.8
Ovis aries	12			5		
Capra hircus	7			2		
O/C	193			9		
O/C total	212	59.2	57.8	16	44.4	41.0
Sus domesticus	86	24.0	23.4	11	30.6	28.2
Canis familiaris	2	0.6	0.5	2	5.6	5.1
Total domestic animals	**358**	**100.0**	**97.5**	**36**	**100.0**	**92.3**
Cervus elaphus	3		0.8	1		2.6
Gazella sp.	4		1.1	1		2.6
Sus scrofa	0		0.0	0		0.0
Lepus capensis	2		0.5	1		2.6
Total game	**9**		**2.5**	**3**		**7.7**
Total mammals	**367**			**39**		
Aves	2			2		
Anser sp.	1			1		
Total	**3**			**3**		
Grand total	**370**			**42**		

Fig. 3. Animal remains from EBA IA levels at Zeytinli Bahçe Höyük (1998/2001)

	NISP	% of domestic	% of total	MNI	% of domestic	% of total
Equus caballus	0	0.0	0.0	0	0.0	0.0
Equus asinus	0	0.0	0.0	0	0.0	0.0
Equus sp.	1	0.1	0.1	0	0.0	0.0
Bos taurus	88	6.9	6.8	6	11.1	10.5
Ovis aries	40			7		
Capra hircus	28			6		
O/C	592			14		
O/C total	660	51.6	51.1	27	50.0	47.4
Sus domesticus	526	41.2	40.7	19	35.2	33.3
Canis familiaris	3	0.2	0.2	2	3.7	3.5
Total domestic animals	**1278**	**100.0**	**98.9**	**54**	**100.0**	**94.7**
Cervus elaphus	3		0.2	1		1.8
Cervidae	2		0.2	0		0.0
Sus scrofa	1		0.1	1		1.8
Lepus capensis	8		0.6	1		1.8
Total game	**14**		**1.1**	**3**		**5.3**
Total mammals	**1292**			**57**		
Anas sp.	2			1		
Anser sp.	1			1		
Aves	9			2		
Pisces	8			2		
Total	**20**			**6**		
Vulpes vulpes	5			1		
Mustelidae	3			1		
Micromammalia	32			4		
Brachyura	1			1		
Total	**41**			**7**		
Grand total	**1353**			**70**		

Fig. 4. Animal remains from EBA IB levels at Zeytinli Bahçe Höyük (1998/2001)

evaluated for domestic cattle from the Early Bronze Age levels was around 122 cm.

A pastoral economy always depends on two main components: the environment and the culture. Because of the site's location, close to a wide river in the Upper Euphrates region, strong seasonality in rainfall and semi-arid conditions and environmental changes due to climatic shifts, as seen at the end of 3rd millennium (Weiss *et al.* 1993), may have been limited. It is unclear for the moment if the apparent disappearance of gazelle from the beginning of the 3rd millennium onwards is due to environmental changes, or to effects of over-hunting or competition with livestock.

The site must have been influenced, even to a lesser degree, by the sequence of events of the age throughout which Uruk culture flourished and decayed. Dwelling structures have been brought to light by the excavations, but no palaces or places of worship, and there are no signs of any particular changes in the purposes for which the structures were used. However, in the course of the first part of the 3rd millennium sequence of residential settlement levels that followed one another without interruption, very interesting changes occurred in animal husbandry practices, as already indicated.

The major change in the frequency of the remains of the domestic pigs (from 24 percent in the final phase of Uruk to over 40 percent in Early Bronze IB) must certainly have been due to cultural and organisational changes. The return to similar conditions to those that had existed in the region in the earlier Neolithic and Chalcolithic periods may suggest a resumption of earlier traditions. That is, if one considers that the regions of south-eastern Anatolia south of the Taurus Mountains are an area where domestication of pig was very early, and rearing was widespread until the end of Chalcolithic, before the so-called Uruk expansion (Wattenmaker and Stein 1986; Schäffer and Boessneck 1988; van Wijngaarden-Bakker 1989; Cavallo 1996; Zeder 1998a), whereas the sites north of the Taurus Mountains seem to have had a greater emphasis on cattle during the Chalcolithic (Bökönyi 1983; Schäffer and Boessneck 1988). The widespread presence of pigs among the domestic animals is typical of a stable sedentary population. Greater consumption of pigs, which were reared as the main source of animal protein at the beginning of the 3rd millennium in many sites of the lower part of the Upper Euphrates region, might be construed as a choice to raise animals in a way that was closer to their own traditions, both because the environment was favourable to it, and because the rural economy had been designed in this way for a long time. It is possible that the organisation of this husbandry in the 3rd millennium was not the same as in the pre-urban periods, fitting in with new and more sophisticated needs (Zeder 1998a), but it is very likely that this choice was connected with the customs, knowledge, and ability acquired during many centuries of a "specialised" rearing practice. And it is not by chance, in my opinion, that the species components - in terms of percentages - shown by the faunal remains in both regions, north and south of Taurus, reproduce, in the Early Bronze Age, the same patterns seen in the same areas in earlier periods.

Culture, traditions and livestock

A comparison of the data from Zeytinli Bahçe with other sites in Eastern Anatolia for which the fauna has been studied and published reveals a similar tendency to a return to local Chalcolithic traditions during the course of the Early Bronze Age, even though this varied in time and took different forms in different sites, after a period in the so-called Late Uruk phase where a split occurred, evidenced by a sharp increase everywhere in the numbers of sheep and goats.

In the first half of the 4th millennium (Pre-Uruk, Late Chalcolithic), despite the fact that sheep and goats were very numerous and ovicaprine keeping was much more common throughout the Near East, one finds that in the north of the Taurus it was cattle, together with sheep and goats, that constituted the most commonly raised livestock (see Arslantepe VII/LC3–4 in Fig. 5). In the settlements along the middle course of the Euphrates, in the region dominated by Adiyaman-Urfa and more generally south of the Anti-Taurus, the archaeozoological finds show a certain preference for more widespread raising of pigs (see Hacinebi B1/LC3, Hacinebi B2 LLC/LC4 and Hassek LU/LC5 in Fig. 5) (Behm-Blancke 1984; Frangipane and Siracusano 1998; Bigelow 1999). The changes that took place in the societies in the Upper Euphrates in the latter half of the 4th millennium, connected with the arrival of the Uruk culture, caused changes to take place in the subsistence strategies of the local communities, and everywhere there was an increased specialisation in sheep and goats (see Arslantepe VIA/LC5, Hacinebi B2 U/LC4, Zeytinli Bahçe LU/LC5, in Fig. 5). The livestock raised, with a particular emphasis on one domestic species, the sheep, nevertheless continued to exhibit regional "preferences". Once again, apart from the fact that the sheep and goats were the most numerous domestic animals by far, there was still a distinctive local character among the livestock, with cattle as the second most numerous species in the North (see Arslantepe VIA /LC5 and Arslantepe VIB/ EBA I in Fig. 5) and pigs remaining a "distinctive" species in the herd structure south of the Taurus (see Hacinebi B2 U/LC4, Zeytinli LU/ LC5, Zeytinli EBA IA, Hassek LU/ LC5 in Fig. 5). The very high percentage of pig at Hassek Höyük even in the Late Uruk period, when all the other sites show a radical change in favour of ovicaprines, may be related to its very small size and suggests that small hamlets were less involved in the major changes that took place at the main sites. This would stress the continuity in local husbandry traditions in the rural environment.

The establishment of the Uruk culture at the end of the

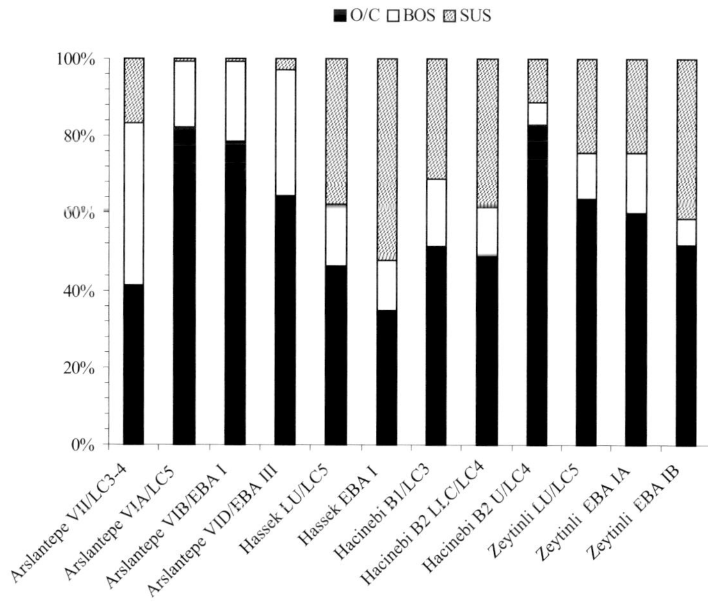

Fig. 5. Proportions of cattle, sheep/goat and pig, based on NISP, for the various sites and periods discussed in the text. The chronology used for the 4th millennium was recently proposed by Rothman (2001, 7). Arslantepe VII/LC3–4 (Local Late Chalcolithic 3–4), Arslantepe VIA/LC5 (Late Uruk = Late Chalcolithic 5), Arslantepe VIB (Early Bronze Age I), Arslantepe VID (Early Bronze Age III), Hassek LU/LC5 (Late Uruk = Late Chalcolithic 5), Hassek EBA I (Early Bronze Age I), Hacinebi B1/LC3 (Local Late Chalcolithic 3), Hacinebi B2 LLC/LC4 (Local Late Chalcolithic 4), Hacinebi B2 U/LC4 (Middle Uruk = Late Chalcolithic), Zeytinli LU/LC5 (Late Uruk = Late Chalcolithic 5), Zeytinli EBA IA (first phase of Early Bronze I), and Zeytinli EBA IB (second phase of Early Bronze I).

4th millennium, based on a centralised organisational pattern, to a certain extent standardised the herds, with a sharp preference for sheep and goats at all the sites that have been analysed. This is also the case in those settlements in which there were what one may presume to have been "colonial" intrusions (Hassek Höyük, Hacinebi) (Boessneck 1992; Bigelow 1999), as well as in those which developed a local Uruk culture (Arslantepe) and in one way or another where the fundamental features were derived from the Uruk culture (Zeytinli Bahçe) (Frangipane 1996; Frangipane 1998).

Of all the species raised, sheep and goats are known to be the most easily adaptable, because they can adapt to a wide variety of diets and have a greater tolerance for the semi-desert and scrubland environments on the margins of the floodplain of the Euphrates that were not being cropped. As far as the circulation of goods was concerned, contact with a centralised state certainly offered more possibilities for trade, which meant that there would be a marked specialisation in the choice of livestock raised. It was also likely that the sheep and goats had a higher intrinsic value than cattle and pigs. They would require less investment for their maintenance and would be easier to dispose of than cattle (each head of which must have represented a large capital investment). The same number of herders could look after a larger number of sheep, which could also be moved in large numbers much more easily than would be possible with pigs.

The environment was certainly of primary importance in the pastoral traditions of the villages. But considering the short timeframe we are dealing with here, in which no drastic climate changes occurred, it is the other, cultural, component that in addition to affecting the social and political order of the territory could also cause even major changes in the subsistence economy, leading to substantial changes in strategy. In other words, the data on the fauna in the sites that we have examined suggest that the cultural influence of a Late Uruk organised state, with new approaches and new demands, had very powerful repercussions on the economic and social fabric of the communities in Upper Mesopotamia (which includes the mountain regions of the Upper Euphrates Valley in this case) and also influenced the animal husbandry strategies in the zone (Zeder 1998b).

Once the influence of the Uruk culture had been weakened or had disappeared altogether, and the trend towards a centralised territorial organisation had been weakened or had disappeared in the course of the Early Bronze Age period (2900–2000 BC), the herds were reconfigured at different times, depending on the zone or site, along the lines of their local traditions (see Zeytinli EBA IB, Arslantepe VID EBA III, and Hassek Höyük

EBI in Fig. 5). In other words, while the herding of cattle increased again in the north, in the south pig-raising became more common again.

At Zeytinli Bahçe the full return to ancient traditions does not appear to have taken place immediately, but only in the second phase of the Early Bronze Age I, and took place even later – in the second half of the 3rd millennium – at Arslantepe (Bökönyi 1983) and Kurban Höyük (Wattenmaker and Stein 1986; Frangipane and Siracusano 1998, Fig. 4). Not only in the north, where a far-reaching cultural and socio-political change took place in which the centralised Mesopotamian-type society disappeared for ever, but also in the south where urbanisation seemed to lose momentum in the Euphrates Valley, the sites, particularly the small ones which were perhaps no longer organised along state territorial lines, probably moved towards a more closed and to a certain extent "autarchic" subsistence economy, in which pigs traditionally represented one of the main food sources in the meat diet. The rural economy of the villages and the small centres must have been characterised by domestic animal husbandry, and pig rearing would have met these more limited requirements. The fact that there are no particular variations in the size of the domestic animals throughout the prehistoric period at Zeytinli Bahçe would seem to confirm the fact that the traditions which the rural village populations preserved for millennia continued to persist despite the political and social changes.

References

Behm-Blancke, M. R. 1984. Hassek Höyük. *Istanbuler Mitteilungen* **34**, 31–149

Bigelow, L. 1999. Zooarchaeological investigations of economic organization and ethnicity at Late Chalcolithic Hacinebi: a preliminary report. *Paleorient* **25**, 83–9.

Boessneck, J. 1992. Besprechung der Tierknochen und Molluskenreste von Hassek Höyük. *Istanbuler Forschungen* **38**, 58–74.

Bökönyi, S. 1983. Late Chalcolithic and Early Bronze Age I animal remains from Arslantepe (Malatya) Turkey: a preliminary report, pp. 581–98 in Frangipane, M. and Palmieri, A. (eds), Perspectives on Protourbanization in Eastern Anatolia: Arslantepe (Malatya). An interim report on the 1975–1983 campaigns. *Origini* **12**, 287–668.

Cavallo, C. 1996. The animal remains – a preliminary account, pp. 475–566 in Akkermans, P. M. M. G. (ed.), *Tell Sabi Abyad – The Late Neolithic Settlement. Report on the Excavations of the University of Amsterdam (1988) and the National Museum of Antiquities Leiden (1991–1993) in Syria*. Istanbul: Nederlands Historisch-Archeologisch Instituut.

Frangipane, M. 1996. *La nascita dello Stato nel Vicino Oriente*. Roma-Bari: Laterza.

Frangipane, M. 1998. Changes in Upper Mesopotamian/Anatolian relations at the beginning of 3rd millennium BC. *Subartu* **6**, 195–218

Frangipane, M. and Siracusano, G. 1998. Changes in subsistence strategies in East Anatolia during the 4th and 3rd millennium BC, pp. 237–46, in Anreiter, P., Bartosiewicz, L., Jerem, E. and Meid, W. (eds), *Man and the Animal World*. Budapest: Archaeolingua.

Frangipane, M. and Buçak, E. 2001. Excavations and research at Zeytinli Bahçe Höyük, 1999, pp. 84–131 in Tuna, N., Özturk, J. and Velibeyoglu, J. (eds), *Salvage Project of the Archaeological Heritage of the Ilisu and Carchemish Dam Reservoirs. Activities in 1999*. Ankara: Middle East Technical University (METU) Keban Project and Carchemish Projects Publications.

Frangipane, M., Alvaro, C., Balossi, F. and Siracusano, G. 2002. The 2000 campaign at Zeytinli Bahçe Höyük., pp. 57–99 in Tuna, N., Özturk, J. and Velibeyoglu, J. (eds), *Salvage Project of the Archaeological Heritage of the Ilisu and Carchemish Dam Reservoirs. Activities in 2000*. Ankara: Middle East Technical University (METU) Keban Project and Carchemish Projects Publications.

Frangipane, M., Balossi, F., Di Nocera, G. M., Palmieri, A. M. and Siracusano, G. in press. The 2001 excavation campaign at Zeytinli Bahçe Höyük (Birecik – Urfa): preliminary results, in Tuna, N., Özturk, J. and Velibeyoglu, J. (eds), *Salvage Project of the Archaeological Heritage of the Ilisu and Carchemish Dam Reservoirs. Activities in 2001*. Ankara: Middle East Technical University (METU) Keban Project and Carchemish Projects Publications.

Rothman, M. 2001. *Uruk Mesopotamia and its Neighbors. Cross-cultural Interactions in the Era of State Formation*. Santa Fe: SAR Press.

Schäffer, J. and Boessneck, J. 1988. Bericht über die Tierreste aus der halafzeitlichen Siedlung Çavi Tarlasi (Nisibin/Osttürkei). *Istanbuler Mitteilungen* **38**, 37–62.

Siracusano, G. 2002. Preliminary study of the fauna at Zeytinli Bahçe Hoyük, pp. 96–99 in Tuna, N., Özturk, J. and Velibeyoglu, J. (eds), *Salvage Project of the Archaeological Heritage of the Ilisu and Carchemish Dam Reservoirs. Activities in 2000*. Ankara: Middle East Technical University (METU) Keban Project and Carchemish Projects Publications.

Siracusano, G. in press. Preliminary results of the faunal analysis from the site of Zeytinli Bahçe, in Tuna, N., Özturk, J. and Velibeyoglu, J. (eds), *Salvage Project of the Archaeological Heritage of the Ilisu and Carchemish Dam Reservoirs. Activities in 2001*. Ankara: Middle East Technical University (METU) Keban Project and Carchemish Projects Publications.

Tuna, N., Özturk, J. and Velibeyoglu, J. 2001. *Salvage Project of the Archaeological Heritage of the Ilisu and Carchemish Dam Reservoirs. Activities in 1999*. Ankara: Middle East Technical University (METU) Keban Project and Carchemish Projects Publications.

Tuna, N., Özturk, J. and Velibeyoglu, J. 2002. *Salvage Project of the Archaeological Heritage of the Ilisu and Carchemish Dam Reservoirs. Activities in 2000*. Ankara: Middle East Technical University (METU) Keban Project and Carchemish Projects Publications.

van Wijngaarden-Bakker, L. H. 1989. The animal remains from Tell Sabi Abyad – Square P14, pp. 301–323 in Akkermans, P. M. M. G. (ed.), *Excavations at Tell Sabi Abyad* (BAR International Series 468). Oxford: British Archaeological Reports.

Wattenmaker, P. and Stein, G. 1986. Early pastoral production in Southeast Anatolia: faunal remains from Kurban Höyük and Gritille Höyük, pp. 90–6 in Algaze, G., Ataman, K., Ingraham, M., Marfoe, L., McDonald, M., Miller, M., Snow, C., Stein, G., Verhaaren, B., Wattenmaker, P., Wilkinson, T. and Tener, A. (eds) The Chicago Euphrates Archaeological Project 1980-1984: an interim report. *Anatolica* **13**, 37–148.

Weiss, H., Courty, M. A., Wellerstrom, W., Guichard, F., Senior, L., Meadow, R. and Currow, A. 1993. The genesis and collapse of third millennium north Mesopotamian civilization. *Science* **291**, 995–1088.

Zeder, M. A. 1991. *Feeding Cities: Specialized Animal Economies in the Ancient Near East*. Washington: Smithsonian Institution Press.

Zeder, M. A. 1998a. Pigs and emergent complexity in the Ancient Near East, pp. 109–22 in Nelson, S. M. (ed.), *Ancestors for the Pigs* (MASCA Research Papers in Science and Archaeology 15). Philadelphia: University of Pennsylvania Museum of Archaeology and Anthropology.

Zeder, M. A. 1998b. Environment, economy, and subsistence on the treshold of urban emergence in northern Mesopotamia, pp. 55–67 in Fortin, M. and Aurenche, O. (eds), *Espace Naturel, Espace Habité en Syrie du Nord (10e-2e Millénaire av. J.-C.)* (Bulletin of the Canadian Society for Mesopotamian Studies 33 and Travaux de la Maison de l'Orient 28). Québec: The Canadian Society for Mesopotamian Studies ; Lyon: La Maison de l'Orient Méditerranéen.

Giovanni Siracusano
Missione Archeologica Italiana nell'Anatolia Orientale,
Università di Roma La Sapienza, Via Palestro 63,
00185 Roma, Italy
E-mail: sirgianni@tin.it

9th ICAZ Conference, Durham 2002
Behaviour Behind Bones, (ed. Sharyn Jones O'Day et al.) pp. 198–202

22. Food for the dead, the priest, and the mayor: looking for status and identity in the Middle Kingdom settlement at South Abydos, Egypt

Stine Rossel

The Middle Kingdom site at South Abydos, Egypt, features the mortuary temple of the pharao Senwosret III and an associated settlement in which several elite houses have been excavated, including the local officeholder's household. Iconographic and textual records of the ancient periods in Egypt suggest that animal products were ranked and that food choices were to some extent related to status. This proposition is tested archaeologically by studying the relationship between the composition of the faunal assemblage from South Abydos and its archaeological context. This paper reports how preliminary results of the faunal analysis show overall correspondence with the impressions derived from texts and depictions. It is suggested that the study of ancient Egypt fauna is a potentially useful tool for the identification of social status.

Introduction

It has been proposed that food choices can help define and cement the identity of social groups (Samuel 1999; Miracle and Milner 2002). Differences in status, affluence, and power can be reflected in food procurement strategies, including those invested in animal exploitation, and accordingly they may translate into the diet of past communities. Many archaeological studies of status have focused on the identification of variation between socio-economic strata, *e.g.* high versus low status. One method relies on the quantification of valuable items from settlements and burials, with value defined by variables relating to extraction, processing, and transportation costs. High-status objects are employed as emblems of economic wealth and social status of the owners. This method was applied by Richards (1997) to argue that the social fabric of Egyptian society comprised more variation than previously believed and included a so-called middle class. The textual record suggests that Egyptian society during the dynastic periods comprised at least two tiers, namely an elite of government officials, and people that were not directly linked to the state administration. Barry Kemp (1989) proposes that to understand the socio-economic structure of ancient Egyptian society it is essential to explore the most affluent segment of the population for which sources are most abundant.

The Middle Kingdom settlement at South Abydos provides an opportunity to examine how faunal remains can be used to identify and characterize variation in status within a privileged, single socio-economic stratum of dynastic Egyptian society. The primary question that is addressed below is whether it is possible to detect a correlation between food choices and social rank within elite households during the Middle Kingdom. The method employed focuses on daily consumables because the waste of day-to-day activities accumulated over time by a household can provide a measure for the wealth of that household and possibly identify differences in affluence among households. The analytical unit is consequently the household with its members, including associated attendants. Waste deposits, comprising habitually used materials, such as faunal remains and pottery, are available for study regardless of the status of a particular household. Consequently, exploration of any socio-economic group within a community is made possible by comparing and contrasting living spaces and consumption habits as they are reflected in floor deposits and refuse areas.

Consumption habits reflected in four types of contexts at South Abydos were examined with taxonomic frequencies for each presented below. That the deposits were produced by people of the same socio-economic segment of society, is inferred from architecture, artifacts, and seal

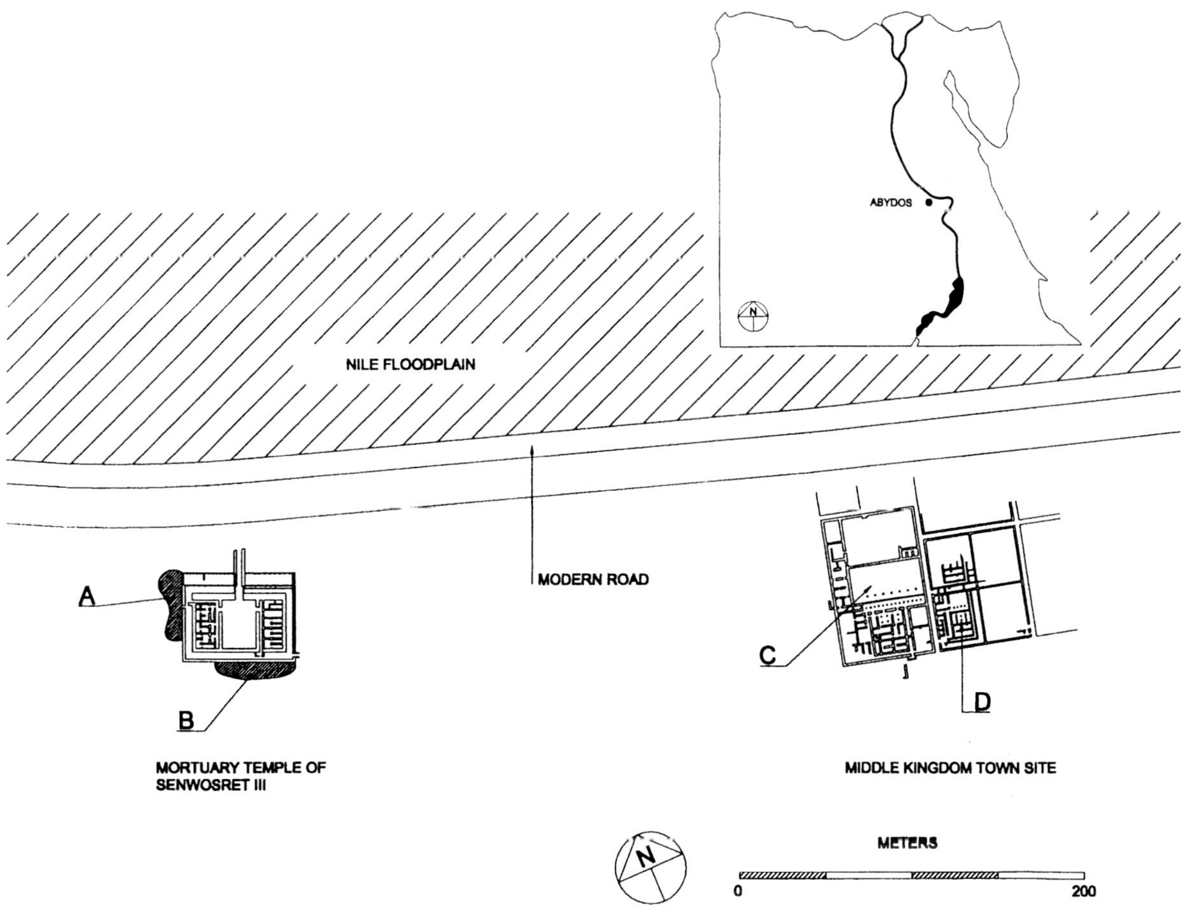

Fig. 1. Location of Abydos and map with the various studied contexts. A, West Block deposit; B, East Block deposit; C, Mayor's house; D, large residence (after Wegner 2001).

impressions (Wegner 2000; Wegner 2001). Variation in taxon frequencies between the different areas is discussed in the light of the contextual information.

South Abydos: four spatial contexts

The ancient structures at Abydos are located on the western edge of the Nile Valley, approximately 80 kilometers northwest of ancient Thebes (Fig.1). The Middle Kingdom settlement at South Abydos was inhabited from the late 12th dynasty into the late 13th dynasty, corresponding roughly to 1850–1700 BC (Wegner 2000). The site comprises the mortuary temple of Senwosret III and the planned settlement associated with the maintenance of the cult for the deceased king. Five seasons of excavations have yielded detailed information on the temple itself as well as exposed a series of large households at the nearby town site. The buildings uncovered to this point seemingly housed an elite segment of society judging from their sizes, titles on seal impressions, and artifactual inventory.

Faunal remains from four spatially distinct contexts were chosen for this study (Fig.1). These comprise the household of the local office-holder (the so-called Mayor), a large residence next to the Mayor's house, and two undisturbed refuse areas at the temple that we can associate with specific groups of temple staff. One refuse deposit (the West Block deposit) is located directly behind the priestly and administrative quarters at the temple whereas the other deposit (the East Block deposit) is located in the proximity of facilities for housing cultic objects and for final preparation of offerings – the area where the non-priestly temple staff performed their daily duties (Wegner 2001).

Data presentation, summary

This analysis is preliminary and includes only mammal and fish remains; bird and reptile bones are still in the process of being analyzed. The 8508 bones that have been surveyed at this point from four contexts have provided a number of identified specimens to genus (or more specific: NISP) of 2291, or 27 % of the total number of finds. Fig. 2 shows the overall distribution of the primary domesticates from all four analyzed contexts. The local animal

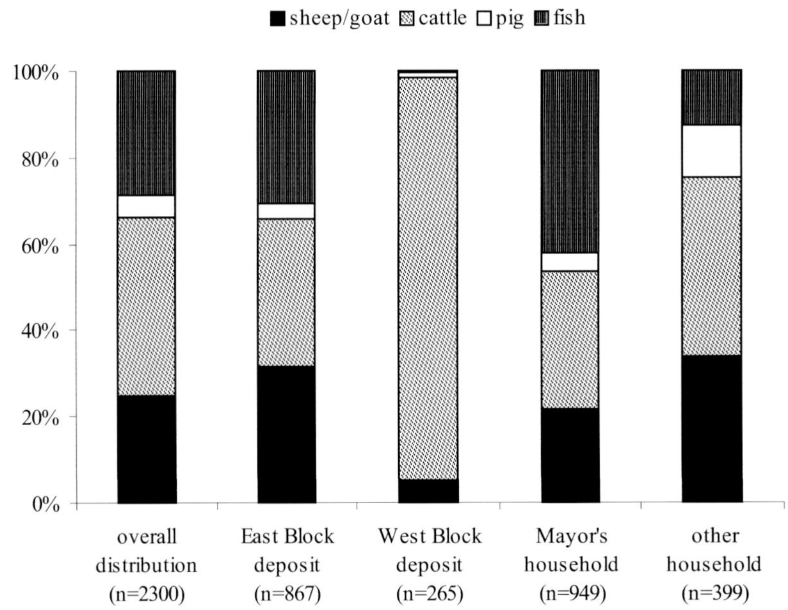

Fig. 2. Relative frequencies of major food animals at South Abydos.

Taxon	**NISP**
Hyperopisus bebe	1
Distichodus sp.	1
Citharinus sp.	4
Citharinus sp./*Distichodus* sp.	1
Auchenoglanis occidentalis	1
Auchenoglanis sp.	4
Bagrus bajad	1
Bagrus sp.	33
Schilbe mystus	1
Clarias gariepinus	7
Clariidae	357
Malapterurus electricus	1
Synodontis schall	6
Synodontis sp.	41
Synodontis sp./Clariidae	1
Lates niloticus	21
Tilapiini	18
Mugil cephalus	1

Fig. 3. Absolute frequencies of fish remains from South Abydos.

economy relied heavily on products from cattle, sheep, goats, pigs, and fish. Cattle dominate most contexts by counts, and certainly by weight, followed by ovicaprines in a ratio of two goats to one sheep. This pattern differs from most other dynastic settlements in Egypt where sheep are more common, but is similar to the earlier settlement at Abydos during the Old Kingdom through the First Intermediate Period (Redding, *pers. comm.*). In contrast, pig is much less commonly represented but was still an important source of meat. Fish remains are numerous in most of the analyzed areas of the site suggesting that fish was a staple food. The vast majority of the fish bones come from the family Clariidae (Fig. 3) perhaps because the cranial plates preserve very well.

The distribution of taxa by NISP can be used for comparison of areas that are characterized by the same geochemical environment and exposed to similar taphonomic processes – as seems to be the case at South Abydos. The West Block deposit located behind the priestly residence in the temple deviates significantly from the overall distribution. Here the faunal refuse is composed almost entirely of cattle remains with very few pig and ovicaprine bones, and only a single fish specimen. Bird and rodent bones, yet unidentified, preserve well in all four contexts and the physical appearance of bones from all contexts is similar suggesting that this is not a reflection of differential preservation. The East Block deposit contains almost equal amounts of cattle, ovicaprines, and fish, and a much smaller number of pig specimens, suggesting that pig was only occasionally consumed there. The faunal profile from the Mayor's house is fairly similar to that from the East Block deposit with the difference that there is a higher concentration of fish remains. The other household indicates less fish consumption but a heavier reliance on ovicaprines and pig.

Discussion

How can the relative distribution of the major food animals help us understand aspects of social status within this portion of dynastic Egyptian society? In order to use food as an indicator of the status level of a group it is important to establish that there is a relationship between food choice and access and social rank. The archaeological context at South Abydos in addition to textual and iconographic records, are informative on aspects of food choices. Contemporary accounts of offerings, as well as accounts from adjacent periods, unanimously report that the sacrificial animals consumed were cattle and fowl. Offering lists are known in two formats, namely idealizing stelae listing one thousand each of sacrificed cattle and fowls, symbolizing infinite amounts, and actual administrative accounts that list specific quantities of food.

The textual record also contains useful indicators of how different animals and their products were ranked. Janssen's seminal study (1975) of the ostraca from the New Kingdom community of workmen at Deir el-Medina, provides several records of animal transactions. Price indices show that cattle were the most expensive of the major domesticates with prices ranging between 20–50 or 100–120 *deben* of copper, with oxen being highest in cost. Sheep and goats were cheaper with prices ranging between 1–3 *deben*. Pigs were slightly more expensive than the ovicaprines with prices from 3–5 *deben*. Janssen notes that most fish at Deir el-Medina were supplied by the government, and only one example documents how 50 ungutted fish came to cost 2 *deben*. He concludes that fish were the cheapest food along with bread and beer (Janssen 1975, 353). An interpretation of these figures should take into consideration a taxon's utility for secondary products and the body size of the animal and its volume. Despite weight-differences between cattle and ovicaprines, the prices suggest that cattle would have been an expensive investment, especially draught-animals.

At South Abydos the temple deposits are more informative than household contents in two ways. Firstly, they consist of undisturbed refuse areas, and secondly seal impressions from the refuse areas as well as from adjacent room-structures provide detailed information on the group and the activities involved in the making of the dump areas. The West Block deposit is believed to have housed the top hierarchy of temple staff, i.e. the priests, based on the concentration of seal impressions found inside the rooms as well as in the refuse deposit itself (Wegner 2000). These include a series of titles and names of priests as well as of other staff that we can associate with the administrative tasks of daily maintenance of the temple. According to textual sources priests received the largest share of the temple offerings in the so-called "reversion of offerings" (Helck 1975; Pierce 1976), which would have included substantial amounts of cattle. The remaining staff members received little of the offerings and accordingly must have been provisioned from alternative sources acquired by the temple. There is thus good correspondence between the expectations that can be generated based on the archaeological context and textual sources, and the actual outcome of the faunal analysis at the West Block.

Textual evidence on animal keeping is important to keep in mind when discussing the contexts within the settlement. The East Block deposit consists of roughly equal counts of bones from cattle, ovicaprines, and fish with a small amount of pig. Although sheep, goats, and fish do not generally appear in the renditions of offering tables, it cannot be excluded that these taxa occasionally formed part of the offerings but were not worthy of mention in the idealizing representations. It is clear, however, that the regular temple staff discarded ovicaprine remains as well as fish leftovers, and – based on element distribution – it also appears that the room-structure connected with the refuse area was used for processes of preparation, storage and consumption.

The faunal remains from the West Block deposit suggest that a very specialized consumption pattern took place inside the West Block whereas the East Block deposit is composed in a way that is more similar to the distributional patterns from the settlement. The ceramics from the two temple-discard areas show the opposite pattern (Wegner 2000). The pottery of the West Block is domestic in nature and similar to the pottery from the settlement while the East Block deposit contains mostly ceramics associated with cultic activities and storage, namely bread molds, beer jars and serving plates. This may suggest, however, that the lower status temple staff received a large portion of plant-based offerings supplemented by temple provided fish. Priests lived in the temple on a rotational basis, the so-called *phyle*-system, which would have required a ceramic selection relatively similar to that of any household, even if meat was provided by the offerings. Looking at the information extracted from the ceramics accordingly can be combined with the faunal data to suggest how food resources circulated and were distributed inside the temple and document how textual ideas were actually played out inside the temple compound.

At the settlement, the faunal remains were recovered from floors, courtyards, and refuse areas as well as from room fill and can be expected to represent a very general picture of consumption inside this larger household. The Mayor's house apparently contained public administrative facilities, such as grain storage installations, as well as domestic residential areas. The remains reflect the food of employer as well as of the employee, high and lower status. The expected picture based on the previous discussion of the rank of animal products would be one characterized by the animals of most extreme high and low cost values, cattle for the Mayor and fish for his employees. This does in fact correspond well with the distribution of taxa at the Mayor's house, although we cannot say for sure that only the staff ate the fish and the ruling family consumed beef, sheep, goats and pig.

The dataset from the second elite household is fairly similar to that from the Mayor's residence. The most significant difference is the relatively small amount of fish bone. In contrast ovicaprines and pig seemingly played a slightly more important role here than in the Mayoral household. Since the other household would presumably not have employed as many people as the Mayor's house and also likely was economically less privileged, the pattern of more middle-ranked animals fits nicely into the proposition that there is a strong correlation in the settlement between food resources and social rank.

Conclusion

At the cenotaph-temple of Senwosret III at South Abydos, the distribution of faunal remains and the content of textual records and images of offering tables all suggest that there was a strong correlation between status and diet. The distribution of faunal remains from the settlement is far more ambiguous. This ambiguity may derive from the fact that the faunal remains here come from an amalgamation of contexts from inside the households as opposed to the situation at the temple where the bones were recovered from refuse areas only.

Based on the discussion above it is argued that the interpretation of faunal data is dependent on its archaeological context. The example from South Abydos shows how faunal data can support other types of data, *e.g.* pottery, as well as help address very basic economic and social questions on the distribution and circulation of resources within a seemingly homogeneous economic stratum. The same basic questions will be explored in more detail in future analyses, looking into variables of age, sex, and element distribution.

Acknowledgements
I would like to thank Wim Van Neer (Royal Museum for Central Africa, Belgium) for his help with the identification of the fish remains. I am particularly grateful to Richard Meadow for helpful comments.

References

Helck, W. 1975. *Wirtschaftsgeschichte des Alten Ägypten im 3. und 2. Jahrtausend vor Chr.* (Handbuch der Orientalistik 1.1.5). Leiden: Brill.

Janssen, J. 1975. *Commodity Prices from the Ramessid Period. An Economic Study of the Village of Necropolis Workmen at Thebes*. Leiden: Brill.

Kemp, B. 1989. *Ancient Egypt: the Anatomy of a Civilization*. London: Routledge.

Miracle, P. and Milner, N. 2002. *Consuming Passions and Patterns of Consumption*. McDonald Institute Monographs. Oxford: Oxbow books.

Pierce, R. 1976. Land use, social organization and temple economy. *RAIN* **0(15)**, 15–7.

Richards, J. 1997. Ancient Egyptian mortuary practice and the study of socioeconomic differentiation, pp. 33–42 in Lustig, J. (ed.), *Anthropology and Egyptology: a Developing Dialogue* (Monographs in Mediterranean Archaeology 8). Sheffield: Sheffield Academic Press.

Samuel, D. 1999. Bread making and social interactions at the Amarna workmen's village, Egypt. *World Archaeology* **31**, 121–44.

Wegner, J. 2000. The organization of the temple *NFR-K3* of Senwosret III at Abydos. *Ägypten und Levante* **10**, 83–125.

Wegner, J. 2001. The town of *Wah-sut* at South Abydos: 1999 excavations. *Mitteilungen des Deutschen Archäologischen Instituts Abteilung Kairo* **57**, 281–308.

Stine Rossel
Department of Anthropology,
Harvard University,
Peabody Museum,
11 Divinity Avenue,
Cambridge, MA 02138, USA
E-mail: Rossel@fas.harvard.edu

9th ICAZ Conference, Durham 2002
Behaviour Behind Bones, (ed. Sharyn Jones O'Day et al.) pp. 203–214

23. Remains of traded fish in archaeological sites: indicators of status, or bulk food?

Wim Van Neer and Anton Ervynck

*The presence on archaeological sites of fish species that have been imported from distant areas has often been regarded as an indicator of high status. Typical examples include the Spanish mackerel (*Scomber japonicus*) and the fish sauces found at Roman sites from Middle and Western Europe, or the Nilotic fish found at Chalcolithic to Crusader period sites in the Eastern Mediterranean area. The available evidence in the literature for trade in those species is reviewed and also the herring trade in medieval Europe is discussed. The factors that have an influence on the price of those traded food items include the quality of the fish products, production costs, and transport, but are difficult to evaluate. It appears, moreover, that the number of sites with exotic fish is growing rapidly now that better recovery techniques are used and the quality of the reference collections used during study is improving. Traded fish seem to turn up in numerous settlement types of varying status, thus showing that caution is needed when using these remains as indicators of high status or purchasing power.*

Introduction

Fish traded over long distances have often been considered, be it in an explicit way or not, to be luxury products. Consequently, they are used as indicators of high status or purchasing power. In an attempt to classify food consumption, Ervynck *et al.* (2003) have proposed a fourfold definition that is useful within the context of this theme. They defined as level 1 the food that is physiologically necessary to survive, be active and grow, and as level 2 the food that is considered to be a basic need, above the level of strictly physiological needs. Affluence is placed at the third level and includes the consumption of level 1 and 2 ingredients beyond basic and considered needs, whereas luxury (level 4) is defined as the consumption, beyond the level of affluence, of goods that are special, limited in supply, difficult to procure, or very expensive due to other reasons. According to this scheme, it is mainly the costs involved with the procurement of the product that define whether it belongs to category 1 and 2, 3 or 4. Traded fish is often regarded as a typical ingredient of a level 4 menu.

Within the whole range of zoological groups represented by their remains on archaeological sites, fish are often easily subdivided into imported and locally caught specimens. Fish can be considered as imported food-stuffs when they belong to species that do not occur in the region of a given site. Marine fish found at inland sites are the typical example, but trade in freshwater fish has also been documented in the past, using the geographical distribution of fish species (*e.g.*, Van Neer *et al.* 1997; Forstenpointner *et al.* 2002). Besides indicators of former commercial relationships, remains of traded fish have often been considered as possible indicators of high status or purchasing power especially when they occurred on sites that were located far from the area where the fish originated.

However, the situation is often more complex than that. The distances over which fish can be transported depend not only on the state in which they are traded, but also on the efficiency of transport. Fresh fish can only be traded over short distances and need to be transported swiftly to prevent them from spoiling before their arrival at the consumer site. In the case of fish that have been preserved by drying, smoking or salting, larger distances can be covered, although a decrease in quality can occur when transport times are long. Moreover, the taste and nutritional value of fresh and preserved fish differ markedly, and thus also their market price and their position on the scale from

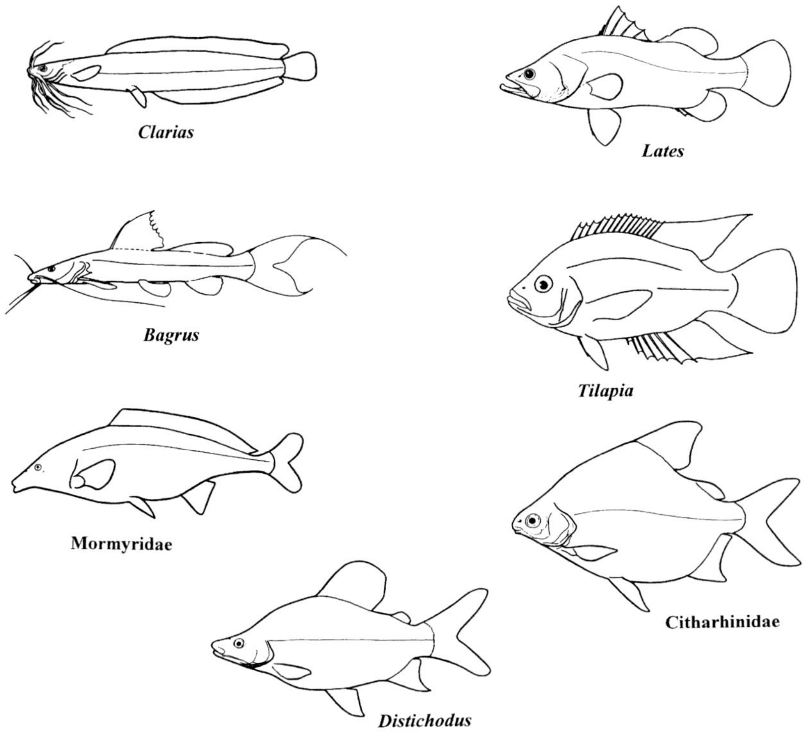

Fig. 1. Nilotic fish that have been found outside their natural range on archaeological sites in the Eastern Mediterranean.

bulk food to luxury product. Other factors that influence the status of the imported fish are, for example, the volume of import (making the difference between a common product or a rarity), or the familiarity of the public with the product (well known food-stuff versus novelty).

A completely different factor, possibly biasing any interpretations of the culinary status of the imported fish species, is the development of ichthyological research within archaeozoology, and the growing integration of data on fish remains within mainstream archaeological and historical interpretations. For example, it will be demonstrated below that the number of sites with fish traded over long distances is increasing rapidly as more and more archaeo-ichthyological studies are published. Additionally, archaeozoological data are increasingly interpreted within a wider anthropological framework, thus yielding better understanding of the reasons why certain species were present on, or absent from, the menu.

Generally, it must thus be questioned whether imported fish species, thus far considered as special food items, are still valid as indicators of a higher status for the households that consumed them. The examples that will be elaborated to address this question are chosen from different regions and represent different historical periods. The themes discussed include the trade in Nile fish in the Eastern Mediterranean area from the Chalcolithic period onwards, the commerce in salted fish (*salsamenta*) and in fish sauces within the Roman Empire, and the herring trade in medieval Europe.

Nilotic fish in the Eastern Mediterranean

The trade of fish from the Nile throughout the Eastern Mediterranean area has a long tradition. The major taxa that were exported from Egypt were a catfish (*Clarias* sp.) and Nile perch (*Lates niloticus*), although other fish such as *Synodontis* and *Bagrus* catfish, tilapia, Mormyridae, Citharhinidae and *Distichodus* sometimes also occur (Fig. 1). The earliest indication for such an import dates from a Late Chalcolithic site in Israel, but from the Middle Bronze Age onwards, the number of sites with Nile fish is considerably higher and this trade remains attested until Crusader times (Fig. 2). A total of 55 sites for the Eastern Mediterranean, and even two sites in Italy, exist with imported Nilotic fish, according to the literature and yet unpublished data (Van Neer *et al.*, *in prep.*) (Fig. 3). It is striking that the fish could be transported over such long distances, which indicates that they must have been sun dried or smoked. Already in the Egyptian Late Palaeolithic smoking was used to cure fish, as is indicated by the 12,000 year old finds from Makhadma (Van Neer *et al.* 2000). For the Pharaonic period iconographic evidence is available for drying and, to a lesser extent, smoking (*e.g.*, van Elsbergen 1997). Salting was apparently not carried out frequently.

A review of mainly Bronze Age sites in the eastern Mediterranean region was previously made by Rose (1994), who states that some of the traded fish were luxury or prestige foods. This supposition is based on the finds

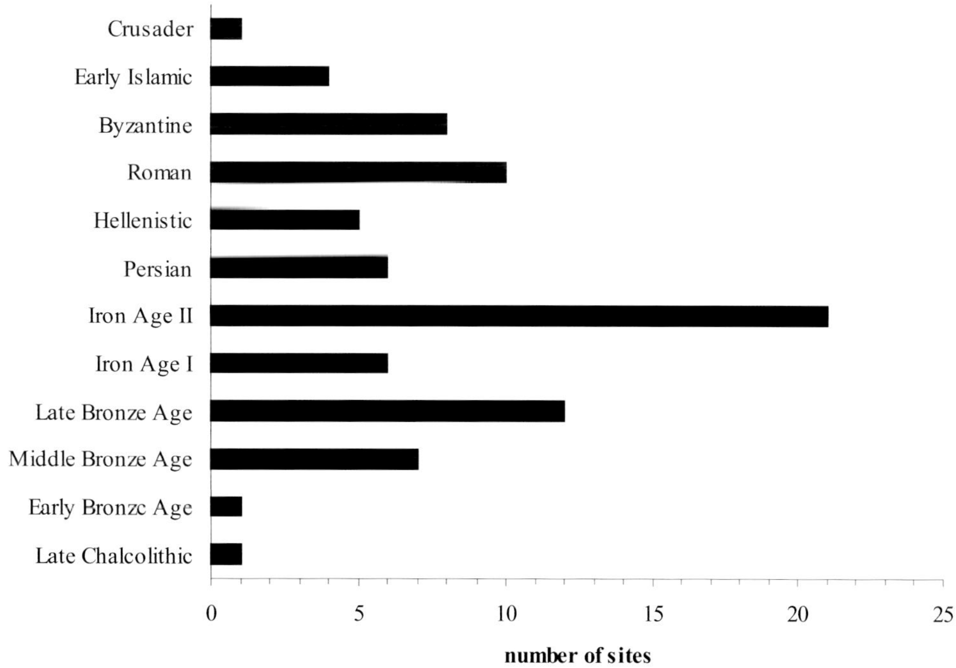

Fig. 2. Number of sites in the Eastern Mediterranean region with imported Nile fish, according to period.

of *Clarias* from an 8th c. BC and archaic cemetery at Salamis, Cyprus (Greenwood and Howes 1973), and from a Middle Bronze Age tomb at Sasa, Israel (Horwitz 1987). Since both finds are derived from rich burials it was concluded that this was prestige food. The find of *Clarias* in a temple context at Kommos, Crete (Rose 1994, 461), dated to ca. 375 BC–AD 200, was seen as an additional indication for luxury food, as was literary evidence dated to about 1075 BC. In the story of Wenamun (Erman 1966), the shipment of Nilotic fish to the Levant is documented and it is listed together with luxury goods, suggesting that it was also a prestigious item. It should be mentioned, however, that lentils were also listed among the shipped goods, possibly showing that the dried fish may have been a common food item taken along on such voyages.

Since the review of Rose (1994) the number of sites in the eastern Mediterranean area has increased considerably, albeit that many of them, mainly from Israel, remain unpublished thus far. The new picture that emerges from this enlarged database (Van Neer *et al.*, *in prep.*) (Fig. 3) is that Nilotic fish were quite common trade goods in the Levant from the Bronze Age onwards. However, the published data may be biased because sites with large architectural remains are often studied preferentially, resulting in the fact that only the upper level of society is documented. Additionally, the information on the exact find context of the archaeozoological material is usually too general to make inferences about the status of the people responsible for the disposal of the food remains. Accordingly it remains a major problem to establish the relationship between the presence of Nilotic fish and status. One of the few indications that Nile fish may not be as unusual as previously thought is the fact that such fish occur in Late Bronze Age Lachich in both simple dwellings and in palaces and temples (Lernau and Golani, *in press*). An additional argument is that Nilotic fish seem to appear on numerous Bronze Age and later sites in the Levant for which a sufficient fish bone sample is available, as long as those sites are situated near to the coast or along caravan routes.

It will perhaps become possible to better evaluate the meaning of the presence of Nilotic fish within the diet of a Mediterranean site when more sites have been studied, including those representing households with lower purchasing power. In doing this, however, it would be preferable to establish this purchasing power independently, on the basis of other finds categories. Of course, there is a danger in such an approach because, within the same household, different categories of consumption often suggest different levels of purchasing power, according to the efforts people wanted to put into the acquirement of certain goods. An example of this can be found in the family that obtained status by a luxurious decoration of the house, but that has no money left to buy expensive food products. Perhaps a solution to this problem can be found in an independent approach, by evaluating the costs involved in the transport between producers and consumers. Given the present state of the database, imported Nilotic fish could be regarded as belonging to an affluent (level 3) diet, being something extra but not really special. This interpretation is thus clearly based on the idea that the product was wanted, but also consumed by somewhat

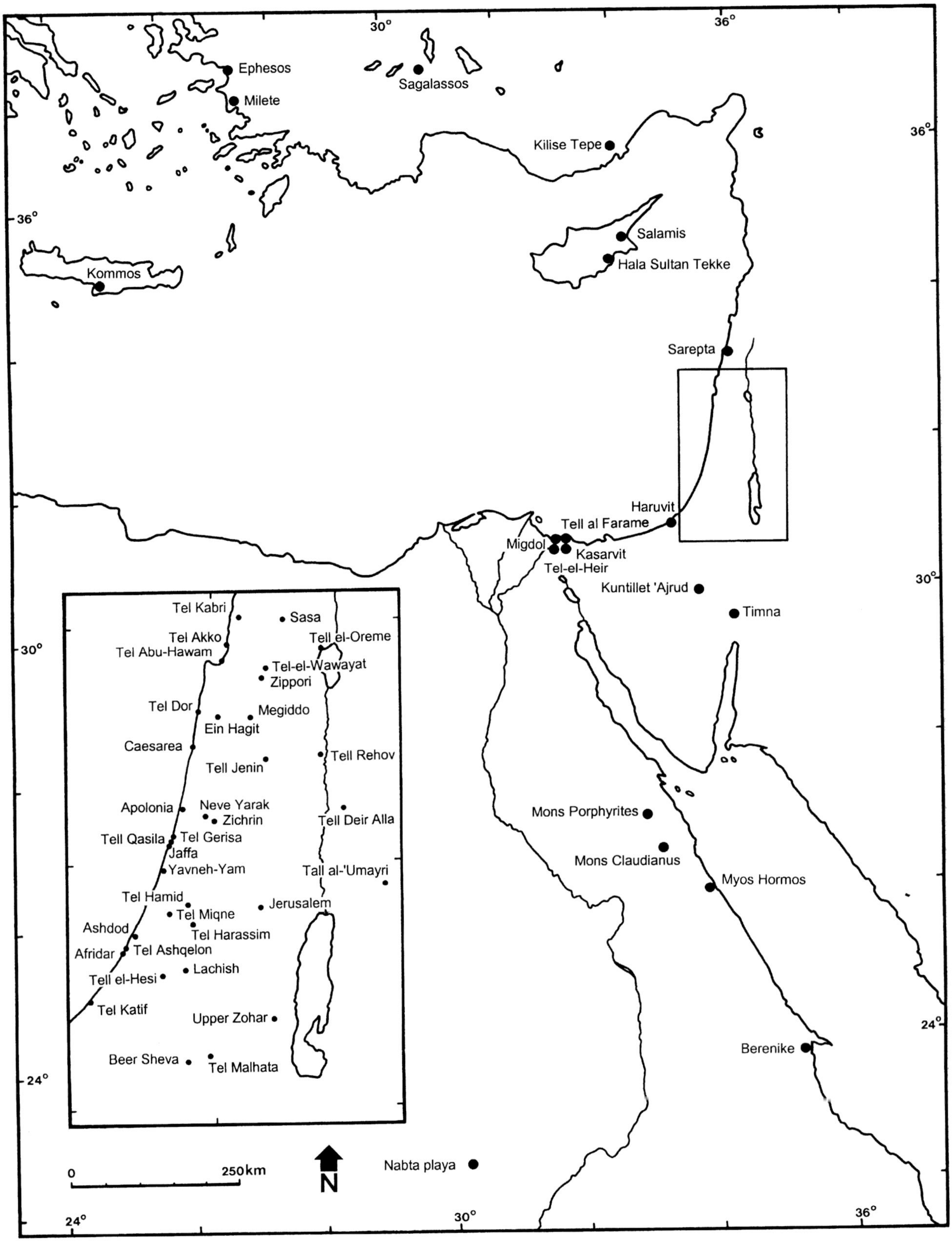

Fig. 3. Location of sites in the Eastern Mediterranean where imported Nile fish were found.

'poorer' households. However, it remains possible that the imported fish only represented a level 1 or 2 ingredient, if, for example, the costs involved with the transport were minimal. Perhaps, when ships or caravans were travelling from the Nile area to places where they wanted to acquire products, they took with them goods that were not the main aim of their trade but that made a little profit, in any case a better option than travelling without goods?

Salsamenta in the Roman world

The large-scale production and trade of salted fish (*salsamenta*) is well documented, certainly for parts of the Roman Empire. Salting installations are known, mainly from excavations in southern Spain and North Africa, and the trade in these products is documented by amphora studies (Curtis 1991). A methodological problem inherent to amphora studies is that these containers can be re-utilised, but in some instances there are still fish remains associated with the amphorae that can be studied by archaeozoologists. It appears that the major species encountered are sardines (*Sardina pilchardus*) and Spanish mackerel (*Scomber japonicus*), although other taxa are sometimes also identified, such as seabreams (Sparidae), horse mackerel (*Trachurus*), mullets (Mugilidae), wrasses (Labridae), picarel (*Spicara smaris*), tuna (*Thynnus thunnus*) and amberjack (*Seriola dumerilii*) (Delussu and Wilkens 2000). The type of fish, the reconstructed size, the skeletal elements present, and the presence or absence of anatomical articulation are criteria that can be used to determine whether the fish product represents salted fish or fish sauce (Desse-Berset and Desse 2000).

The Spanish mackerel is one of the best-represented salted fish found in Roman archaeological contexts. There are a few 1st and 2nd century AD finds from shipwrecks off the coast of Italy, Sardinia (Delussu and Wilkens 2000) and Corsica (Desse-Berset 1993), with amphorae containing mainly *Scomber japonicus*. Amphorae with this salted fish reached areas outside the Mediterranean, as shown by a 1st c. AD find from Nijmegen, The Netherlands (Lauwerier 1993), a contemporaneous example from Winchester Palace, Southwark, London (Yule 1989, Locker 1994) and a find from Mainz, Germany (Ehmig 2001) dated to about AD 100 (Ehmig, *pers. comm.*). Moreover, a number of bones of this species – not in connection with amphorae – are known from 2 sites each in Germany and Britain, 3 sites in the Netherlands, 4 sites in Belgium, and 12 sites in Switzerland (Van Neer *et al.*, *in prep.*). Most of the finds date from the period of the late 1st century BC to the 2nd century AD. The finds of this salted fish, so far from the source area, were usually considered as an indication of relatively high status (*e.g.*, Vanderhoeven *et al.* 1993; Hüster-Plogmann 1999a, 462; Murphy *et al.* 2000). In some cases the contextual evidence does indeed seem to indicate the special nature of this food item. The only Spanish mackerel bone found in an early Roman cemetery in Augusta Raurica, Switzerland, occurred in a grave with the highest diversity in species and the richest archaeological grave goods (Veszeli 1996). The species was also found amongst the leftovers of a ritual, sumptuous, meal held within a temple for Mithras at Tienen, Belgium (Vanderhoeven *et al.* 2001; Lentacker *et al.*, *this volume*) and the finds from Martigny in Switzerland also come from a *mithraeum*. However, given the recent increase in ichthyological studies, Spanish mackerel occurs in more and more Roman sites of different function: the species is now known from Roman towns, *vici*, *villae* and military settlements (Hüster-Plogmann 1999b). Although contextual evidence, that could possibly give more detailed information on the precise status from the find contexts, is usually lacking, the impression arises that this salted fish was not a very special commodity, and that it was probably quite easy to procure from the late 1st century BC to the 2nd century AD.

Were the *salsamenta* then merely a level 3 ingredient, or have they still to be regarded as a luxury? A major problem in the interpretation is that the functional description of a site does not say much about the consumers of the *salsamenta* found there. There have been large and small *villae*, important and minor military settlements. Moreover, within a site, there were people of higher and lower status, whose diet differed considerably. It will thus again be better to evaluate independently the costs involved in bringing the salted products from the Mediterranean to northern Gaul. Was there a specially organised transport of southern marine products to the northern provinces, or did the *salsamenta* arrive there with ships that came to collect the grain harvests (merely to prevent them from sailing with empty holds)? The consumption of salted fish was not a characteristic of the Iron Age diet in Gaul, suggesting that the import, under Roman occupation, was directed towards the upper part of society. However, how popular salted products became, and whether this was a slow or rapid process, remains unknown. Perhaps, in the earliest days of Roman occupation, the consumption of Mediterranean products was needed to demonstrate a Romanised way of living, while the *salsamenta* became an appreciated but not necessary product in later times? Perhaps they became a level 1 or 2 ingredient comparable to the Nilotic fish discussed earlier?

An interesting aspect of the mentioned interpretation problems is the observation that, in Roman times, as far as is known, the production of salted fish was never organised along the North Sea coast. It is clear that mackerel, and many other species, could be fished off the coasts of Northern Gaul, and salt was available. Nevertheless, the incentive was apparently not there to develop a local variant of the Mediterranean production. Does this mean that there was no major return to be expected from developing this production line? Perhaps the prices paid for *salsamenta* were not high enough to justify the establishment of a local production? Must it be concluded, therefore, that, certainly after the initial occupation period,

Roman *salsamenta* were not a luxury product in Gaul, and thus did not yield any status when served during a banquet?

Fish sauces in the Roman world

Fish sauces were produced in large quantities during the Roman period and arrived at consumer's places throughout the empire, thanks to a well-organised transport (Ponsich and Tarradell 1965; Curtis 1991). Fish sauces are a condiment rather than a food stuff, although they can contribute significantly to the diet, as is still the case today in poorer populations in the Far East where fish sauces form a major supplement of proteins and vitamins (Curtis 1991, 22). In the Roman world fish sauces were regarded an absolute necessity according to the 'original' Mediterranean Roman culinary taste. Therefore, the analysis of fish sauces can be of interest for the study of trade mechanisms, consumer behaviour, purchasing power and, ultimately, status. Traditionally, the limited number of Roman period sites, where the remains of fish sauce were found, was seen as an indicator for the luxury status of the product. However, as with the aforementioned Spanish mackerel, the number of archaeological finds of fish sauce has recently increased dramatically, as a result of better sampling techniques and a growing attention towards the potential of the analysis of animal remains. It remains to be verified whether fish sauces should continue to be considered as indicators of high status, or not.

Garum, *allec*, *muria* and *liquamen* were distinguished in antiquity and, depending on the additions to these basic products (spices, olive oil, wine, etc.), numerous additional names were applied (Grimal and Monod 1952; Corcoran 1963; Curtis 1991; Martínez Maganto 1992). In his discussion on the prices of salted fish products, Curtis (1991, 170–5) stresses that fish sauces were affordable to all classes, and that only particular types of sauce were expensive. Not only the ingredients and the general quality of the products determined the price, but also the transportation costs, including payment of tariffs. The impression that fish sauces were of great expense results from the interpretation, by certain modern scholars, of literary evidence from classical writers such as Martialis, Horacius, and Plinius who exaggerated the prices in Rome in order to criticise individuals or the society in general. The accessibility of fish sauce to all classes is demonstrated by the analysis of amphorae at Pompeii and Herculaneum, showing that this commodity was found throughout the cities, in wealthy and poor or average households, as well as in commercial establishments (Curtis 1991, 171–4). However, these observations were made in a Mediterranean city close to the production centres of Mediterranean fish sauce. Whether they can be extrapolated towards more northern parts of the Roman Empire, remains to be investigated.

Archaeozoological analysis can only document the occurrence of fish sauce when bones are preserved, which explains why thus far only *allec*, a less expensive variety of fish sauce with bones still present, has been reported. Bone finds recovered from amphora bottoms have been described from the Mediterranean area in Spain, the Italian mainland, Elba, Sicily, Sardinia, and Greece (Williams 1978; von den Driesch 1980; Wheeler and Locker 1985; Morales and Roselló 1989; Curtis 1991; Desse-Berset and Desse 2000; Delussu and Wilkens 2000). North of the Mediterranean area, faunal assemblages with the 'genuine' Mediterranean fish sauce are rare. There is one find from Austria (Lepiksaar 1986) and possibly two additional finds from Great Britain (Kenward *et al.* 1986; O'Connor 1988). Because of the distance from the place of production, and probably also because *allec* remains are relatively rare (or rarely recognised), these finds have received a lot of attention. However, does rarity equal luxury here? On the one hand, it is often overlooked that ceramic studies indicate that the trade in these products was not an uncommon practice in Roman times, certainly until at least the 2nd c. AD (Martin-Kilcher 1990). On the other hand, remains of fish sauce produced in Spain have been found as far as Masada, Israel (Cotton *et al.* 1996). The cost of transport towards that site must have been high and it appears that there are several indications that king Herodus, for whom this sauce is supposed to be meant, tried to emulate the tastes of the Roman aristocracy. This is shown not only by the fish sauce but also by wine imported from Italy, and by apples from Cumae (close to present-day Naples). Generally, it must be asked how these supposed high costs of transport can be married with the large scale of the trade mentioned earlier, an interpretation problem further complicated by the observation that, apparently, it was worthwhile to transport the second class version of the product (*allec* versus *garum*). The question thus remains: was Mediterranean fish sauce an expensive product, or a common product shipped throughout the empire?

Besides the typical Mediterranean fish sauces, there is now growing archaeozoological evidence for the production of fish sauces in other parts of the Roman empire, using local fish species. The fish that were used were mainly clupeiforms (herring, sprat and anchovy-like species), which are characterised by a high fat content. This was the case in Jordan in a late 4th – early 5th context at Petra (Studer 1994) and, in Egypt, in several late Roman contexts at Berenike (Van Neer and Ervynck 1998; Van Neer and Ervynck 1999) and in early Roman Quseir al-Qadim (Hamilton-Dyer and Van Neer, *unpublished*), where Red Sea clupeiforms were used. In a late Roman or early post-Roman context at York (Jones 1988), and in a mid-3rd c. AD context at London (Bateman and Locker 1982), local fish sauce with mainly herring and sprat was found. Herring and sprat also dominate four Belgian examples, although small flatfish were sometimes also frequent. These finds are from mid-2nd c. AD Tienen (Vanderhoeven *et al.* 2001), from 2nd c. AD Tongeren (Van Neer and Ervynck, *unpublished*), from 3rd c. AD

Braives (Van Neer and Lentacker 1994), and from 3rd c. AD Tienen (Lentacker *et al.*, *in press*). Sometimes even freshwater species are used, as is shown by two finds from Egypt and one from Israel. A 2nd c. AD fish sauce made from Nilotic species was found at Mons Claudianus, a quarry site in the Eastern Desert (Hamilton-Dyer 1994) and in the Nile valley itself such a fish sauce was discovered in a 5th c. AD context at Shenhur, north of Luxor (Van Neer, *unpublished*). In a 5th–7th c. AD context at Horvat Karkur, Israel, concentrations of small cichlids have also been considered as possible remains of fish sauce (Lernau, *pers. comm.*). Possibly assemblages of small freshwater fish found in Western Europe may also be considered as representing fish sauces. Concentrations mainly of small cyprinids were found in Belgium in a 2nd c. AD deposit at Tongeren, Veemarkt (Vanderhoeven *et al.* 1993), and also in Switzerland, Roman assemblages with small freshwater fish have been reported that are considered as possibly representing fish sauces (Hüster-Plogmann 1999b). Based on the distribution of these finds over various settlements, Hüster-Plogmann (1999b) postulates that these local fish sauces may not have been meant for the higher social classes, but she also does not exclude that the observed tendencies are related to chronology. Indeed, the picture that emerges for the moment is that in Switzerland these fish sauces appear only from the second half of the 2nd c. AD onwards, in Belgium and Great Britain the non-Mediterranean fish sauces are also not early Roman. The production of local fish sauces, both those made from marine and from freshwater species, may have started or increased from the late 2nd c. AD onwards to continue satisfying the Roman taste once the import of the 'genuine' Mediterranean variant slowed down, as suggested by the amphora studies (Martin-Kilcher 1990). Alternatively, it could be that, at some point, the local production started a fierce competition with the south-European products because of cheaper transport costs. In fact, the complete lack of economic context makes it difficult to evaluate the consumption of the Roman fish sauces. Was a freshwater fish sauce a cheap replacement for the 'genuine' thing, or a top class gastronomic achievement? Could people tell the difference between North Sea *garum* and the Mediterranean version? Are amphora studies a good method to evaluate the changes in the market, given the fact that in later periods more and more products were perhaps traded in barrels?

Without sufficient contextual information it will remain difficult to decide what the differences in quality and price may have been between the variants of Roman fish sauce. Possibly, the product was a common ingredient that could be either imported from the Mediterranean or from the less distant Atlantic coast, once it was firmly established within culinary practices. If local freshwater species were used, it would possibly have been no problem to obtain this ingredient at a reasonable price. Probably, particular variants may have been affordable only to the higher classes, because they were expensive due to the elevated transport costs, or because of the ingredients. In general, it will still need a lot of zooarchaeological and archaeological research before Roman fish sauce can safely be used as an indicator of status or not.

Herring trade in medieval Europe

The herring trade is well documented by historical sources, from late medieval times onwards (*e.g.*, Degryse 1944; Unger 1978; Unger 1980; Cushing 1988; Lampen 2000; Hoffmann 2001a; Hoffmann 2001b). Moreover, evidence from faunal analyses is also plentiful, certainly when compared to our knowledge about the trade in fish species in other parts of the world. The historical and archaeozoological information provided may, therefore, represent a good background against which the other fish species dealt with in this paper can be analysed, despite the fact that those examples deal with other periods and regions. Common knowledge would state that the herring was, and always has been, a cheap product, making it an interesting (conflicting) case study within the present context of searching for luxury items yielding status. However, a thorough analysis of the data illustrates that the picture is much more complex than that.

Both historical and archaeozoological data (for summaries see mainly Lampen 2000, and Hoffmann 2001a and 2001b) show that herring consumption in early medieval times was limited to the coastal inhabitants of Picardy (northern France) to the Baltic, and that those populations were starting to supply fish to nearby inland consumers only from around AD 1000. From the 12th century onwards, herring production increased significantly and became an important food supply, especially for the developing towns. The available historical data come from royal, noble and urban charters documenting commerce of herring, and from documents concerning taxes allowing statements about fish prices. Accounts of monastic and noble households also give an idea of the evolution of prices through time (*e.g.*, Lampen 2000). In late medieval times, barrelled herring was available in large quantities and its price was relatively low, although the interaction between supply and demand played an important role. The plague epidemics around AD 1350, for instance, resulted in a price decrease because of lower demand, but towards the end of the 14th century prices rose again as a result of political difficulties in Denmark, which had an effect on the export of herring from Schonen (Lampen 2000, 39; Hoffmann 2001a). In addition, the distance over which herring had to be transported also influenced its price. Herring in 15th century Flanders (northern Belgium) was a rather cheap food item, but in central Europe it was more expensive. Historical records on fish prices in 15th century Konstanz (southern Germany) show that the cheapest fish were the local whitefish (genus *Coregonus*), whereas all the other freshwater fish and imported herring and stockfish were more expensive. Herring was almost as pricey as the most expensive freshwater fish, which were pike and carp (Lampen 2000, 40).

Until the middle of the 15th century, Scandinavian herring dominated the markets of north-western Europe as a result of an efficient processing technique (*i.e.* gutting, see below), combined with protective trade regulations stipulating the exclusive right to commercialise herring in barrels. Later on, fishermen operating in the more southern and western parts of the North Sea also started gutting the herring they caught, making it possible to carry out longer journeys at sea and to start the preservation of herring while on board. Prior to this period, the herring needed to be brought ashore more rapidly and the preservation technique consisted of sprinkling salt on the complete or beheaded fish (Degryse 1966). The improvement of preservation techniques resulted in larger quantities of caught herring and also made the product more suitable for transport over long distances.

Since historical information does not inform us about the purchasing power of all social classes, and because the record is fragmentary both spatially and chronologically, it is vital to also include archaeozoological information in the discussion on herring trade and consumption. Herring remains are usually found in large numbers on late medieval sites of countries bordering the north-eastern part of the Atlantic, on the condition that sieving is practised. The species is found in monasteries, castles and urban sites and this in both rich and poor households (Van Neer and Ervynck 1994; Bødker Enghoff 1999; Bødker Enghoff 2000; Clavel 2001; Locker 2001). This indicates that in the region mentioned, herring can be considered as a staple food as suggested by the historical information. However, although it is clear from the foregoing that herring was generally a rather cheap commodity, it remains to be verified whether this was also the case before mass production occurred and well-organised transport existed. The onset of intensive herring fishing is impossible to reconstruct from the historical sources but can be documented by a diachronic analysis of ichthyofaunas from a given region. It appears from the archaeological record in Belgium that herring appears relatively late on inland sites compared to flatfish and gadids (Ervynck *et al.*, *in press*). A similar late appearance of herring in England (Jones 1981) has been linked to a technical innovation, *i.e.* the widespread introduction of floating nets since AD 1000. Herring finds from the 11th and 12th century are relatively rare in, for instance, Flanders, although this could be partly a result of the number of available find contexts, and the less detailed recovery techniques used in older excavations. The rare finds of early date from Flanders come from sites located along the river Scheldt, *i.e.* the AD 1000 *portus* site at Ename (Van Neer and Ervynck, *unpublished*), and 11th century early urban household refuse at Gent (Ervynck and Van Neer, *unpublished*). Farther upstream, the first occurrence of herring is dated around AD 1000 at Tournai (Pigière and Van Neer, *in prep.*). This town was a bishopric at that time, but the type of settlement to which the deposit belongs is unknown. From Luxembourg, early herring remains dated to the 10th–12th centuries AD are known from the Château du Bock (Ervynck *et al.* 1996). Herring consumption still farther away from the coast is documented by historical data, for instance, for the 11th century AD at the monastery of Sankt Gallen in Switzerland (Lampen 2000, 45). A few archaeozoological cases possibly illustrating early inland trade exist for Poland and Germany (Benecke 1982; Bødker Enghoff 1999; Hoffmann 2001b), but the dating of the contexts is not very precise and the status of the inhabitants that produced the refuse is unclear.

It could be tempting to consider the early inland finds of herring as indicators of higher status, because of their relatively early date, and because of their provenance from pre-urban settlements, a feudal castle, a rich monastery, or an episcopal centre. However, in the case of the finds from Ename and Gent, it is too obvious to assume a direct relationship between an early- or pre-urban settlement, dating from around 1000 AD, and a population of rich inhabitants enjoying high status. On the contrary, the overall archaeological context of both sites suggests that the refuse deposits from which these herring remains originated were generated by merely common people. At Ename, it is now hypothesised that the pre-urban portus site was a settlement of craftsmen, servants, or agricultural labourers servicing the castle (Ervynck, *unpublished*). The contexts excavated at Gent, although illustrating the oldest development phase of the town, could well represent a rural consumption pattern (Ervynck and Van Holmen, *unpublished*). For Tournai, contextual information is lacking. The herring from the castle site in Luxembourg and the monastery in Switzerland are generally speaking from a higher status settlement, but are located farther away from the coast. Additional data on sites predating the 13th century and with sufficient contextual information will be necessary to elucidate this matter. Could it be that the first herring arriving inland was a luxury item for sites away from the sea, but immediately a common commodity for people living close to the coast, as for example in the Scheldt basin? Alternatively, are we (archaeologically and historically) missing an early phase of herring import within the Scheldt basin, during which the fish was a luxury ingredient?

Before this last question can even be addressed, a further complicating factor must be taken into account, *i.e.* the state in which the herring was traded (fresh, dried, smoked, salted, etc.). Historical sources (from the late Middle Ages) (Lampen 2000) reveal that fresh herring was considerably more expensive than the salted product. This is logical as herring, once caught, goes off readily, which implies that a quick transport (involving extra costs) was needed to bring the product to the inland markets. Furthermore, due to the varying physical condition of the fish, the herring caught in different seasons within the year yielded different prices. Herring, of which the remains are found in rich households, can thus have been a completely different product (both physically and market-wise) compared to herring found amongst the consumption remains of common people. Archaeozoological data sometimes indicate the effects of market practices that are related to the large-scale commercialisation of herring. Both in Germany (Benecke 1987) and Belgium (Van Neer and Pieters 1997) a comparison of reconstructed fish sizes from herring at coastal and inland sites demonstrated that a size selection of fish was carried out in view of the export towards consumer sites inland. The traded herring are of a more or less standardised body length, whereas in coastal villages smaller specimens were also consumed. These fish sizes are not documented in the historical sources, and, from the archaeological data, a size difference cannot be observed from the herring remains consumed at different medieval sites, certainly not for the earliest period of herring import. More important, however, is the fact that, in some cases, archaeological evidence is found of the type of herring that was consumed at a site. The intra-skeletal distribution sometimes shows an absence, or severe under representation, of bones from the pectoral girdle and gill apparatus, which indicates the consumption of gutted herring (Seeman 1986). This processing technique involved the removal with a knife of the gills and of the anterior part of the digestive tract, after which the fish were packed in salt. In accordance with the available historical data, the earliest archaeozoological evidence for this practice is dated to around AD 1340 (Bødker Enghoff, *pers. comm.*) and comes from the site of Selsø-Vestby, Denmark, where an accumulation of gutting waste was found (Bødker Enghoff 1996; Bødker Enghoff 1999). The only example of consumed gutted herring known thus far in the region considered is from a 15th – 16th century context at the monastery of Ename, Belgium (Ervynck and Van Neer 1992). It should be stressed, however, that the number of sites allowing a statement on the type of consumed herring is limited and that, in the case of gutted herring, it has been impossible thus far to distinguish between fish imported from Schonen or from other areas. Since prices were also dependent on transport costs, the impossibility of inferring the precise origin when using archaeozoological data can sometimes be a drawback for the interpretation of the purchasing power. Moreover, it must always be taken into account that the herring remains from a certain site can represent a mixture of preserved or fresh fish, caught off the local coast, or farther away.

The interpretation problems considering the meaning of herring remains within the diet of medieval communities are thus manifold and complex. From the foregoing, it became clear that the state in which herring arrived must have, at least partially, defined its place on the gastronomic scale. The fact that the trade routes and the processing techniques, and certainly the freedom to use them, considerably evolved during the Middle Ages, poses further questions. Did the first herrings consumed at inland Flanders consist of preserved or fresh fish? Were the first herrings traded towards inland Flanders caught by Flemish fishermen or by Danish colleagues? Was it the 'invention' of gutting that made herring popular, or was it already very cheap before that time? Was fresh herring still a luxury product in late medieval times, when salted herring dominated the fish markets, and if so, to what extent?

The conclusion must be, therefore, that herring could have been a luxury food in certain periods and certain forms, but that this interpretation will always be difficult to make on the basis of the archaeological remains. Generally, in terms of luxury, this commodity must have varied between a level 1 ingredient (gutted herring as staple food) and a level 4 product (fresh herring).

Concluding remarks

From the examples discussed, it is clear that traded fish cannot be used as an indicator of purchasing power, and thus of status, without caution. Regardless of the historical and contextual information available, each case study has proven to evoke interpretation problems that cannot be solved easily. The fish sauces and salted fish from the Roman period may have been considered as essential cooking ingredients by the Romans and Romanised populations within the empire, but whether these products were expensive, in a certain period, remains to be established. Whether, in the case of fish sauce, the expensive nature of the product explains why local variants of the Mediterranean product were made, is also questionable. Perhaps we are dealing here with a competition between cheap products? Consumption of the genuine Mediterranean fish sauce may be considered as a sign of affluence when it needed to be transported over a longer distance, or even as a luxury as in the example from Masada (Cotton *et al.* 1996), but the costs involved in the production and transport are difficult to evaluate, and it was perhaps difficult even for the consumers to discriminate between local and imported versions of the product.

As far as the Nilotic fish is concerned, these may simply have been considered a basic need in some cases. An example is provided by the settlement of the Egyptian workmen at Timna, a Late Bronze Age to Early Iron Age gold mining site north of the Gulf of Aqaba (Lernau 1988). On this site no fish was consumed from the Red Sea, despite the proximity of those waters. Fish from the Nile and the Mediterranean Sea were eaten instead, both of which had to be transported over a longer distance than Red Sea fish, but to which the workmen were accustomed. It remains difficult, however, to classify all the other finds of Nilotic fish. Preserved Nile fish were not limited in supply; mass production of dried fish was already taking place in the Late Palaeolithic. From the Middle Bronze Age onwards, there was a well developed trade network between Egypt and the Syro-Palestinian area whereby goods went in both directions; the ships and caravans coming from Egypt seem to have often carried dried fish, perhaps at minimal costs. Possibly the earliest finds, such as the Late Chalcolithic Nile perch from Tel Katif (Lernau 1996), should be regarded as a luxury as the means of transport were still poorly developed. In the other cases, however, these dried fish may simply have represented bulk food that was a standard component of the traded or exchanged goods.

The case of the medieval herring consumption is perhaps even more complex. In certain sites, such as the late medieval monastery of Ename in Flanders, a luxurious fish consumption seems to have been maintained. A wide variety of locally caught freshwater fish and cultured carp were consumed, but on top of that there were 13 species of imported marine fish, most of which must have arrived in a fresh state (Van Neer and Ervynck 1996). Although the variety points to luxury, the problem remains what the place of the individual species was within this culinary spectrum. In other words, did the herring consumed at the site add to the luxurious nature of the menu, or not? Furthermore, what part was consumed fresh or preserved?

Finally, there are two additional factors that seriously affect all interpretations. Typical of all the fish taxa considered is that they occur seasonally in massive quantities and that the surplus catch can be kept and traded for future consumption once they are dried, smoked or salted. The costs of production can thus have been limited, but are generally difficult to evaluate, just as the costs of transport. Secondly, as has been mentioned before, the general trend appearing from the examples elaborated in this paper is that finds of exotic fish in distant sites increase as sampling methods improve, the number of ichthyofaunal studies increases, and collections of modern reference specimens are becoming more numerous and complete. In the case of the Spanish mackerel, for instance, almost all finds have been reported only since the 1990's. Similarly, the identification of Nilotic fish outside Egypt has been hampered in the past by inadequate reference collections, and in the case of fish sauces, it is obvious that it is mainly the more rigorous sampling techniques that are responsible for the increase in finds.

The identification of the status of the people responsible for the deposition of the refuse analysed by archaeozoologists thus suffers from various drawbacks. A rather crude method consists of considering the finds coming from various settlement types: the comparison of the late medieval animal remains from castles, monasteries, towns and rural settlements sometimes shows certain patterns within the food consumption (Ervynck, *this volume*), but it is obvious that people of different social status may have lived together within one settlement. The refuse deposited in and around a castle, for example, can consist not only of food remains of the noble, but also of their servants (Ervynck 1992). The hierarchy existing in monasteries is also reflected in the food consumption (Van Neer and Ervynck 1996), and within towns poor and rich households existed (IJzereef 1989). The analysed animal remains are usually confronted with the architectural remains to which they were associated, but it should be taken into account that the filling found in a prestigious building need not necessarily have been deposited by the well-to-do original inhabitants, but that it can represent refuse deposited once the structure was abandoned or taken over by other people. It is therefore vital that the type and quality of other archaeological objects such as ceramics, glass, and metal is also analysed as they can give an indication of the wealth of the inhabitants responsible for the refuse disposal. This kind of contextual analyses has, however, only rarely been practised thus far and it seems that the interdisciplinary approach risks ending in circular reasoning if the identification of status is not done in an independent way prior to confrontation of the data from various disciplines. The present contribution certainly suggests that, when fish are used within this context, caution must be taken.

Acknowledgements

The contribution of Wim Van Neer to this paper presents research results of the Interuniversity Poles of Attraction Programme – Belgian State, Prime Minister's Office – Federal Office for Scientific, Technical and Cultural Affairs. The following colleagues are acknowledged for sharing unpublished data with us: Inge Bødker Enghoff, Ulrike Ehmig, Sheila Hamilton-Dyer, Heide Hüster-Plogmann, Omri Lernau, Jacqueline Studer, Margarethe Uerpmann and Angela von den Driesch.

References

Bateman, N and Locker, A. 1982. The sauce of the Thames. *The London Archaeologist* **4**, 204–7.

Benecke, N. 1982. Zur frühmittelalterliche Heringsfischerei im südlichen Ostseeraum – ein archäozoologischer Beitrag. *Zeitschrift für Archäologie* **16**, 283–90.

Benecke, N. 1987. Die Fischreste aus einer frühmittelalterlichen Siedlung bei Menzlin, Kreis Anklam. *Bodendenkmalpflege in Mecklenburg Jahrbuch* **1986**, 225–39.

Bødker Enghoff, I. 1996. A medieval herring industry in Denmark and the importance of herring in eastern Denmark. *Archaeofauna* **5**, 43–7.

Bødker Enghoff, I. 1999. Fishing in the Baltic region from the 5th century BC to the 16th century AD: evidence from the fish bones. *Archaeofauna* **8**, 41–85.

Bødker Enghoff, I. 2000. Fishing in the southern North Sea region from the 1st to the 16th century AD: evidence from the fish bones. *Archaeofauna* **9**, 59–132.

Clavel, B. 2001. L'animal dans l'alimentation médievale et moderne en France du Nord (XIIe–XVIIe siècles). *Revue Archéologique de Picardie, Numéro Spécial* **19**, 1–204.

Corcoran, T. 1963. Roman fish sauces. *Classical Journal* **58**, 204–10.

Cotton, H., Lernau, H. and Goren, Y. 1996. Fish sauces from Herodian Masada. *Journal of Roman Archaeology* **9**, 223–38.

Curtis, R. I. 1991. *Garum and Salsamenta. Production and Commerce in Materia Medica.* Leiden: Brill.

Cushing, D. H. 1988. *The Provident Sea.* Cambridge: Cambridge University Press.

Degryse, R. 1944. *Vlaanderens Haringbedrijf in de Middeleeuwen*. Antwerpen: De Nederlandse Boekhandel.

Degryse, R. 1966. De laatmiddeleeuwse haringvisserij. *Bijdragen voor de Geschiedenis der Nederlanden* **21**, 82–121.

Delussu, F. and Wilkens, B. 2000. Le conserve di pesce. Alcuni dati da contesti italiani. *Mélanges de l'Ecole Française de Rome* **112**, 53–65.

Desse-Berset, N. 1993. Contenus d'amphores et surpêche: l'exemple de Sud-Perduto, pp. 341–6 in Desse, J. and Audoin-Rouzeau, F. (eds), *Exploitation des Animaux Sauvages à travers le Temps*. Juan-les-Pins: APDCA.

Desse-Berset, N. and Desse, J. 2000. Salsamenta, garum et autres préparations de poisson. Ce qu'en disent les os. *Mélanges de l'Ecole Française de Rome, Antiquité* **112**, 73–97.

Ehmig, U. 2001. Naturwissenschaftliche Untersuchungen an Tömischen Amphoren aus Mainz und ihre kulturhistorische Interpretation, pp. 85–100 in Frey, M. and Hanel, N. (eds) *Archäologie – Naturwissenschaften – Umwelt. Beiträge der Arbeitsgemeinschaft "Römische Archäologie" auf dem 3. Deutschen Archäologenkongress in Heidelberg 25.5 – 30.5. 1999*. (BAR International Series 929). Oxford: British Archaeological Reports.

Erman, A. 1966. *The Ancient Egyptians. A Source Book of their Sources*. New York: Harper & Row.

Ervynck, A. 1992. Medieval castles as top-predators of the feudal system: an archaeozoological approach. *Château Gaillard. Etudes de Castellologie médiévale* (Caen) **15**, 151–9.

Ervynck, A. and Van Neer, W. 1992. De voedselvoorziening in de Sint-Salvatorsabdij te Ename (stad Oudenaarde, prov. Oost-Vlaanderen). 1. Beenderen onder een keukenvloer (1450–1550 AD). *Archeologie in Vlaanderen* **2**, 390–403.

Ervynck, A., Lentacker, A. and Van Neer, W. 1996. Les restes fauniques du Château du Bock à Luxembourg-ville (G.-D. de Luxembourg), pp. 221–8 in Zimmer, J. (ed.), *Die Burgen des Luxemburger Landes*. Luxembourg: Editions Saint-Paul.

Ervynck, A., Van Neer, W., Hüster-Plogmann, H. and Schibler, J. 2003. Beyond affluence: the zooarchaeology of luxury, pp. 428–41 in van der Veen, M. (ed.), *Luxury Foods* (World Archaeology 34/3). London & New York: Routledge.

Ervynck, A., Van Neer, W. and Pieters, M. in press. How the North was won (and lost again). Historical and archaeological data on the exploitation of the North Atlantic by the Flemish fishery, in Housley, R. A. and Coles, G. M. (eds), *Atlantic Connections and Adaptations: Economies, Environments and Subsistence in Lands bordering the North Atlantic*. Oxford: Oxbow Books.

Ervynck, A. this volume. Orant, pugnant, laborant. The diet of the three orders in the feudal society of medieval north-western Europe, pp. 215–23 in O'Day, S. J., Van Neer, W. and Ervynck, A. (eds), *Behaviour behind Bones. The Zooarchaeology of Ritual, Religion, Status and Identity*. Oxford: Oxbow Books.

Forstenpointner, G., Weissengruber, G. and Galik, A. 2002. Banquets at Ephesos; archaeozoological evidence of well stratified Greek and Roman kitchen waste, pp. 282–304 in Buitenhuis H., Choyke A. M., Mashkour M. and Al-Shiyab A. H. (eds) *Archaeozoology of the Near East V. Proceedings of the Fifth International Symposium on the Archaeozoology of Southwestern Asia and Adjacent Areas*. (ARC-Publicaties 62). Groningen: ARC.

Greenwood, P. H. and Howes, G. 1973. Fish remains, pp. 259–68 in Karageorghis, V. (ed.) *Salamis V. Excavations in the Necropolis of Salamis III*. Nicosia: Zevallis Press.

Grimal, P. and Monod, Th. 1952. Sur la véritable nature du "garum". *Revue d'Etudes anciennes* **54**, 27–38.

Hamilton-Dyer, S. 1994. Preliminary report on the fish remains from Mons Claudianus, Egypt. *Offa* **51**, 275–8.

Hoffmann, R. C. 2001a. A longer view: is industrial metabolism really the problem? *Innovation* **14**, 143–55.

Hoffmann, R. C. 2001b. Frontier foods for late medieval consumers: culture, economy, ecology. *Environment and History* **7**, 131–67.

Horwitz, L. K. 1987. Animal offerings from two Middle Bronze Age tombs. *Israel Exploration Journal* **37**, 251–5.

Hüster-Plogmann, H. 1999a. Ergebnisse der Untersuchungen von Knochenresten aus den Schlämmproben, pp. 457–63 in Rychener, J. (ed.), *Der Römische Gutshof in Neftenbach* (Monographien der Kantonsarchäologie Zürich 31/1). Zürich und Egg: Kantonsarchäologie Zürich.

Hüster-Plogmann, H. 1999b. Fischreste aus Schlämmproben aus dem Bereich der Herdstelle in Raum B6, pp. 214–219 in Sütterlin, H. (ed.) *Kastelen 2. Die Älteren Steinbauten in den Insulae 1 und 2 von Augusta Raurica*. (Forschungen in Augst 22). Augst: Kantonsarchäologie Baselland.

IJzereef, F. G. 1989. Social differentiation from animal bone studies, pp. 41–53 in Serjeantson, D. and Waldron, T. (eds) *Diets and Crafts in Towns. The Evidence of Animal Remains from the Roman to the Post-medieval Periods* (BAR International Series 199). Oxford: British Archaeological Reports.

Jones, A. K. G. 1981. Reconstruction of fishing techniques from assemblages of fish bones, pp. 4–5 in Bødker Enghoff, I., Richter, J. and Rosenlund, K. (eds), *Fish Osteo-Archaeology Meeting. Copenhagen 28th-29th August 1981*. Copenhagen.

Jones, A. K. G. 1988. Fish bones from the excavations in the cemetery of St Mary Bishophill Junior. *The Archaeology of York* **15**, 126–30.

Kenward, H. K., Hall, A. R. and Jones, A. K. G. 1986. Environmental evidence from a Roman well and Anglian pits in the legionary fortress. *The Archaeology of York* **14**(5), 41–288.

Lampen, A. 2000. *Fischerei und Fischhandel im Mittelalter. Wirtschafts- und Sozialgeschichtliche Untersuchungen nach Urkundlichen und Archäologischen Quellen des 6. bis 14. Jahrhunderts im Gebiet des Deutschen Reiches* (Historische Studien 461). Husum: Matthiesen Verlag.

Lauwerier, R. C. G. M. 1993. Twenty-eight bird briskets in a pot: Roman preserved food from Nijmegen. *Archaeofauna* **2**, 15–9.

Lentacker, A., Ervynck, A. and Van Neer, W. in press. The symbolic meaning of the cock. The animal remains from the mithraeum at Tienen in Martens, M. and De Boe, G. (eds), *Roman Mithraism: the Evidence of the Small Finds* (Archeologie in Vlaanderen Monografie 5). Zellik and Tienen: Institute for the Archaeological Heritage of the Flemish Community.

Lentacker, A., Ervynck, A. and Van Neer, W. this volume. Gastronomy or religion? The animal remains from the Mithraeum at Tienen (Belgium), pp. 77–94 in O'Day, S. J., Van Neer, W. and Ervynck, A. (eds), *Behaviour behind Bones. The Zooarchaeology of Ritual, Religion, Status and Identity*. Oxford: Oxbow Books.

Lepiksaar, J. 1986. Tierreste in einer römischen Amphore aus Salzburg (Mozartplatz 4). *Bayerische Vorgeschichtsblätter* **51**, 163–85.

Lernau, H. 1988. Fish remains, pp. 241–6 in Rothenberg, B. (ed.), *The Egyptian Mining Temple at Timna*. London: Institute of Archaeology, University College London.

Lernau, O. 1996. Fish remains from Tel Harassim, pp. 14–23 in Givon, S. (ed.), *The Sixth Season of Excavation at Tel Harassim (Nahal Barkai) 1995*. Tel Aviv: Bar-Ilan University.

Lernau, O. and Golani, D. in press. The osteological remains (aquatic), in Ussishkin, D. (ed.), *The Renewed Archaeological Excavations at Lachish (1973–1994)*. (Tel Aviv University, Publications of the Institute of Archaeology). Tel Aviv.

Locker, A. 1994. The fish bones from excavation at Winchester Palace, Southwark, 1983. Unpublished report for Southwark and Lambeth Excavation Committee.

Locker, A. 2001. *The Role of Stored Fish in England 900–1750 AD; the Evidence from Historical and Archaeological Data*. Sofia : Publishing Group Limited.

Martin-Kilcher, S. 1990. Fischsauces und Fischkonserven aus dem römischen Gallien. *Archäologie der Schweiz* **13**, 37–44.

Martínez Maganto, J. 1992. Las salazones romanas: aportes historiográficos de interés en arqueozoología. *Archaeofauna* **1**, 73–81.

Morales Muñiz, A. and Roselló, E. 1989. Informe sobre la fauna ictiològica recuperada en una àmfora tardo-romana del Tipus Keay XXVI (Spatheion), pp. 324–8 in Ted, A. (ed.), *Un Abocador del Segle V D.C. en el Fòrum Provincial de Tàrraco* (Memòries d'Excavació 2). Tarragona: Ajuntament de Tarragona.

Murphy, P., Albarella, U., Germany, M. and Locker, A. 2000. Production, import and status: biological remains from a late Roan farm a Great Holts Farm, Boreham, Essex, UK. *Environmental Archaeology* **5**, 35–48.

O'Connor, T. P. 1988. Bones from the General Accident site, Tanner Row. *The Archaeology of York* **15**(2), 61–136.

Ponsich, M. and Tarradell, M. 1965. *Garum et Industries Antiques de Salaison dans la Méditerranée Occidentale* (Bibliothèque de l'Ecole des Hautes Etudes Hispaniques 36). Paris: Presses universitaires de France.

Rose, M. J. 1994. *With Line and Glittering Bronze Hook: Fishing in the Aegean Bronze Age*. Unpublished Ph.D. thesis, Indiana University.

Seeman, M. 1986. Fish remains from Smeerenburg, a 17th century Dutch whaling station on the westcoast of Spitsbergen, pp. 129–39 in Brinkhuizen D. C. and Clason A. T. (eds), *Fish and Archaeology* (BAR International Series 294). Oxford: British Archaeological Reports.

Studer, J. 1994. Roman fish sauce in Petra, Jordan, pp. 191–6 in Van Neer, W. (ed.) *Fish Exploitation in the Past. Proceedings of the 7th Meeting of the ICAZ Fish Remains Working Group* (Annales du Musée Royal de l'Afrique Centrale 274). Tervuren.

Unger, R.W. 1978. The Netherlands herring fishery in the late Middle Ages: the false legend of Willem Beukels of Biervliet. *Viator* **9**, 335–56.

Unger, R.W. 1980. Dutch herring, technology, and international trade in the seventeenth century. *Journal of Economic History* **40**, 253–80.

Vanderhoeven, A., Ervynck, A. and Van Neer, W. 1993. De dierlijke en menselijke resten, pp. 177–86 in Vanderhoeven, A., Vynckier, G. and Vynckier, P. (eds), Het oudheidkundig bodemonderzoek aan de Veemarkt te Tongeren (prov. Limburg). Eindverslag 1988. *Archeologie in Vlaanderen* **3**, 127–205.

Vanderhoeven, A., Martens, M., Ervynck, A., Cooremans, B. and Van Neer, W. 2001. Interdisziplinäre Untersuchungen im römischen Vicus von Tienen (Belgien). Die Integration von ökologischen und archäologischen Daten, pp. 13–31 in Frey, M. and Hanel, N. (eds) *Archäologie – Naturwissenschaften – Umwelt. Beiträge der Arbeitsgemeinschaft "Römische Archäologie" auf dem 3. Deutschen Archäologenkongress in Heidelberg 25.5 – 30.5.1999*. (BAR International Series 929). Oxford: British Archaeological Reports.

van Elsbergen, M. J. 1997. *Fischerei im Alten Ägypten. Untersuchungen zu den Fischfangdarstellungen in den Gräbern der 4. bis 6. Dynastie*. (Abhandlungen des Deutschen Archäologische Instituts Kairo. Ägyptologische Reihe, Band 14). Berlin: Achet Verlag.

Van Neer, W., De Cupere, B. and Waelkens, M. 1997. Remains of local and imported fish at the ancient site of Sagalassos (Burdur Prov., Turkey), pp. 571–86 in Waelkens, M. and Poblome, J. (eds), *Sagalassos IV. Report on the Survey and Excavation Campaigns of 1994 and 1995* (Acta Archaeologica Lovaniensia Monographiae 9). Leuven: Leuven University Press.

Van Neer, W. and Ervynck, A. 1994. New data on fish remains from Belgian archaeological sites, pp. 217–29 in Van Neer, W. (ed.), *Fish Exploitation in the Past. Proceedings of the 7th Meeting of the ICAZ Fish Remains Working Group* (Annales du Musée Royal de l'Afrique Centrale, Sciences Zoologiques 274). Tervuren.

Van Neer, W. and Ervynck, A. 1996. Food rules and status: patterns of fish consumption in a monastic community (Ename, Belgium). *Archaeofauna* **5**, 155–64.

Van Neer, W. and Ervynck, A. 1998. The faunal remains, pp. 349–88 in Sidebotham, S. and Wendrich, W. (eds), *Berenike '96. Report of the Excavations at Berenike (Egyptian Red Sea Coast) and the Survey of the Eastern Desert*. (CNWS Publications, Special Series 3). Leiden: CNWS.

Van Neer, W. and Ervynck, A. 1999. The faunal remains, pp. 325–48 in Sidebotham, S. and Wendrich, W. (eds), *Berenike '97. Report of the 1997 Excavations at Berenike and the Survey of the Egyptian Eastern Desert, including Excavations at Shenshef*. (CNWS Publications, Special Series 4). Leiden: CNWS.

Van Neer, W. and Lentacker, A. 1994. New archaeozoological evidence for the consumption of locally-produced fish sauce in the northern provinces of the Roman Empire. *Archaeofauna* **3**, 53–62.

Van Neer, W., Paulissen, E. and Vermeersch, P. M. 2000. Chronology, subsistence and environment of the Late Palaeolithic fishing sites of Makhadma-2 and 4, pp. 271–87 in Vermeersch, P. M. (ed.), *Palaeolithic Living Sites in Upper and Middle Egypt*. Leuven: Leuven University Press.

Van Neer, W. and Pieters, M. 1997. Evidence for processing of flatfish at Raversijde, a late medieval coastal site in Belgium. *Anthropozoologia* **25–26**, 571–78.

Veszeli, M. 1996. Tierische Speisebeigabe in den Brandgräbern, pp. 246–52 in Haeffelé, C. (ed.), *Die römischen Gräber an der Rheinstrasse 46 des Nordwestgräberfeldes von Augusta Raurica*. (Jahresberichte aus Augst und Kaiseraugst 17). Augst: Kantonsarchäologie Baselland.

von den Driesch, A. 1980. Osteoarchäologische Auswertung von Garum-Resten des Cerro del Mar. *Madrider Mitteilungen* **21**, 151–4.

Wheeler, A. and Locker, A. 1985. The estimation of size in sardines (*Sardina pilchardus*) from amphorae in a wreck at randello, Sicily. *Journal of Archaeological Science* **12**, 97–100.

Williams, Ch. K. 1978. Corinth, 1978: Forum Southwest. *Hesperia* **48**, 104–44.

Yule, B. 1989. Excavations at Winchester Palace, Southwark. *The London Archaeologist* **6**(2), 31–9.

Wim Van Neer
Royal Museum of Central Africa, IUAP 05/09,
Leuvensesteenweg 13, B-3080 Tervuren, Belgium, and
Katholieke Universiteit Leuven,
Laboratory of Comparative Anatomy and Biodiversity,
Ch. de Bériotstraat 32, B-3000 Leuven, Belgium
E-mail: vanneer@africamuseum.be

Anton Ervynck
Institute for the Archaeological Heritage of the Flemish
Community, Koning Albert II-laan 19, Box 5
B-1210 Brussel, Belgium
E-mail: lentacker.ervynck2@yucom.be

9th ICAZ Conference, Durham 2002
Behaviour Behind Bones, (ed. Sharyn Jones O'Day et al.) pp. 215–223

24. *Orant, pugnant, laborant*. The diet of the three orders in the feudal society of medieval north-western Europe

Anton Ervynck

'On earth, there are people who pray, people who fight and people who work. These three groups form a unity and do not tolerate separation'. *This statement, written down explicitly for the first time in the 11th century, describes the essence of the European feudal society. The self-image of a society consisting of a divinely created trinity of orders was only completely abandoned during the French revolution, and thus dominated European social history for almost a millennium. In a world of monks, knights and peasants, the three orders distinguished themselves by a different social behaviour, not only based on their specific economic position, but also upon ideology. As part of their characteristic behaviour, the three orders also ate differently. Indeed, the differences between the diet within a castle, an abbey or another type of household, are not the result of culinary preferences; they are a reflection of their position within society. This pattern can be illustrated, and interpreted, by a survey of zooarchaeological data from medieval and post-medieval sites from Belgium.*

Introduction: the three orders

Historical sources reveal that shortly after AD 1000, for the first time, the philosophical concept was written down that European society was subdivided in three orders, each with a different occupation and function. One of the texts described these functions using the terms *'orant'*, *'pugnant'* and *'laborant'*, which can be translated as 'those who pray', 'those who fight' and 'those who work'. The first order thus was the clergy; its members did not work, were not occupied with earthly matters, but by praying they acted as mediators between God and the other two orders. The second order consisted of the feudal nobility, which protected the other two groups by fighting, thus consolidating the structure of society. The third order, the people, laboured and produced food, and thus provided the basis for all activities of the two other groups (Duby 1978). It should be noted that the equality between 'fighting' and 'nobility' is perhaps an over-simplification; noble men may have preferred to see themselves as those who controlled fighting (or power) instead of those who fought themselves. Moreover, mere fighting did not *per se* result in a noble status (R. Hoffmann, *pers. comm.*) (see also the mention about mercenaries at the end).

The philosophical theory of the three orders was presented as a divine concept (Fig. 1), in which each order benefited from the activities of the others, and it was continuously reinforced in everybody's mind through the Catholic Church. The whole system was, in its original form, also based upon inheritance. At least for the people and the nobility, one stayed within the order in which one was born, while the members of the clergy (certainly the upper hierarchical levels) came out of the nobility. This 11th century static image of a tripartite society has been very important within European history; and it has had a strong impact until the late 18th century, when the *ancient regime* was finally abolished by the French Revolution. Only then did the last remnants of a society based upon inherent status and prestige slowly start to disappear. Meanwhile, through colonialism and cultural influence, this concept of three orders had spread in many parts of the world outside Europe, giving it a global relevance.

The key observation within this contribution is that the zooarchaeological analysis of medieval sites in Belgium (and in other countries) reveals that food was different for the three orders. The consumption refuse excavated in different categories of medieval dwelling places clearly shows specific characteristics, which will first be described. Within the context of this paper, it will not be

Fig. 1. The three orders represented as a divine concept: clergy (left), nobility (right) and people (bottom) (Lichtenberger Prognosticatio, Mainz 1492, Kupferstichkabinett Staatliche Museen Preussischer Kulturbesitz, Berlin) (after von Bieberstein 1989, 13).

possible to give an elaborate description of the sites, their stratigraphy and excavation history. Nevertheless, it has been taken into account that the taphonomy of certain contexts, the sometimes small sample sizes, or the problems of sampling and recovery methodology, may have biased the datasets. However, this is not considered to be a major problem, since it is assumed that stochastic variation, and even occasional aberrant results, will compensate for each other as long as the dataset is large enough. Moreover, what follows is a synthesis, and thus necessarily also a broad generalisation. The approach taken here is more that of an essay, than of a 'normal' zooarchaeological article.

The study area considered is present-day Belgium, roughly coinciding with parts of the medieval feudal entities of Flanders, Brabant, Hainaut, Namur, Liège, Limburg and Luxemburg (Fig. 2). Although strong parallels could be drawn between the present analysis and that in neighbouring countries (*e.g.*, Clavel 2001 or Lauwerier 2002 and the references therein), such an approach would be beyond the scope of this short paper. Only when the archaeological observations had to be put into their historical context, have broader interpretative works (of value for the whole of north-western Europe) been used.

Consumption patterns

The noble diet

When starting with the second order, the nobility, it is evident that their consumption behaviour can best be investigated on the basis of consumption refuse excavated from their typical residences: the castles. These sites initially present themselves as wooden fortifications on artificial mounds (mottes) or natural elevations, but evolve into stone castles during the late medieval period. A list of 19 sites investigated zooarchaeologically is given by Ervynck (1992b), but must now be updated with new studies by Ervynck (1991), Ervynck *et al.* (1994), Fiers (2000) and Pigière *et al.* (this volume).

Typical for the Belgian castles is the presence of the remains of large game such as red deer (*Cervus elaphus*), wild boar (*Sus scrofa*) and brown bear (*Ursus arctos*). These species are not found in other types of contemporaneous sites. The same exclusive pattern is true for large wild birds, such as heron (*Ardea cinerea*), swan (*Cygnus* species) or crane (*Grus grus*), which were apparently also only hunted and eaten by castle inhabitants. Another exclusive characteristic for the castle sites is the presence of extravagant food products, such as sperm whale (*Physeter catodon*).

Turning towards the common domestic meat providers (the trio of pig, sheep and cattle), castles identify themselves by a dominant frequency of pig remains. Only in the coastal area, with its salty meadows and salt-marshes, is this not the case, but even there significantly more pigs are consumed within the castle's walls compared to rural sites (Ervynck and De Meulemeester 1996, Fig. 2). In general, the remains of the domestic meat providers outnumber those of wild animals at all castle sites. Hunting was important for many reasons (see further) but did not yield a quantitatively significant addition to the daily food supply.

The ecclesiastic diet

Considering the clergy, the situation is rather complex because a distinction must be made between bishops, priests, monks and nuns. Moreover, within a monastic community there were clear hierarchical differences, and there were differences between monastic orders. Nevertheless, during the High Middle Ages, monastic life was the most important part of the religious society, both in spiritual and economic terms. Therefore, the analysis will concentrate upon this group.

The first observation that can be made from a survey of zooarchaeological material from Belgian monastic sites (seven male communities and two nunneries; see Ervynck

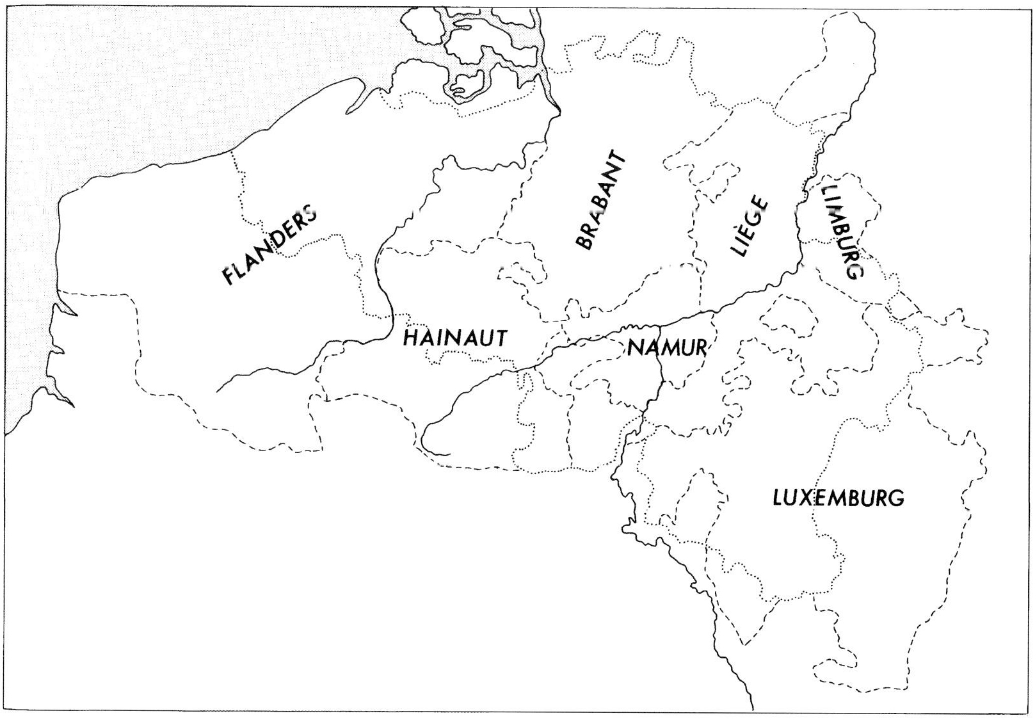

Fig. 2. Map of the study area (dotted line: present-day Belgium) with the location of the main feudal territories (broken lines).

1997) is that the remains of large game are always absent. When archaeological contexts are sieved using fine mesh sizes, one can clearly see another difference between monastic and secular contexts: in the former group, the frequency of fish remains is more elevated (Ervynck 1997, Fig. 3). At the same time, one can see that, apparently, within the abbeys, the meat of mammals was only occasionally consumed. When meat was eaten within the abbey walls, this was almost exclusively beef or mutton. Pig remains are always extremely rare, with the only exception being the late post-medieval material from the infirmary of the Sint-Pieter Abbey at Gent (Ervynck 1997, Fig. 4). These consumption patterns, however, did not imply that the diet at an abbey was necessarily austere. Even when the meat of mammals was not eaten, very rich banquets could be served, consisting, for example, of a wide variety of fish species (Van Neer and Ervynck 1996).

The people

The third order, that of the peasants, is not yet well documented within the zooarchaeology of Belgium. However, the scarce information available (Ervynck and De Meulemeester 1996; Ervynck, *unpublished*) indicates a meat consumption mainly based upon cattle or sheep (depending upon the region), and, again, an absence of game.

Explanations

Having summarised these data, the search for explanations can now begin. The main question is, of course, why and how the differences in food pattern originated? Culinary preferences can certainly not be the sole explanation, and purchasing power is also not a valid reason. The system of the three orders is not based upon money, and, indeed, large and small castles, important and less important abbeys all show the same basic patterns described.

Thinking in terms of ecology and economy, the theory can be put forward that the nobility's diet focused upon the meat of pig because these animals were herded in the forest, which was the exclusive territory of the lord of the castle. Indeed, the privileged use of the forest and other uncultivated lands by the nobility is at the very heart of the origin of the medieval feudal society. After the collapse of the Roman Empire and the invasion of people from Germania, the cultivated land came under the control of independent large landowners. They had to pay taxes to the functionaries of the central authority (king or emperor) but these functionaries did not own the territories they taxed. The functionaries also did not own, but only supervised, the wastelands and forests, which were still royal or imperial property. However, when central power weakened at the end of the Carolingian era, the functionaries took possession of the uncultivated land themselves, began the usurpation of the cultivated areas, pushing the independent large landowners aside, and became the feudal

Fig. 3. The Money Devil, in an armour of silver coins, dancing upon a chest filled with gold coins, is attacked by the clergy (left) and the nobility (right) (anonymous late 16th century drawing, Prentenkabinet Rijksmuseum, Amsterdam) (after van Beek et al. *1984).*

lords of the High Middle Ages (for the situation in Flanders, see Berings 1985). The rural people had only limited access to the forests; they had to pay taxes for herding pigs and for collecting wood. The abbeys were given land by the feudal lords, or, earlier, by the early medieval kings or emperors, and they mainly used it for agricultural purposes. All this could explain the dominant frequencies of pig remains in castles, as opposed to their limited presence in ecclesiastic sites and in the dwelling places of the common people.

The privileged use of the forests could also explain why hunted species such as red deer, wild boar or brown bear (which lived in woodlands) only ended up at the castle's table (Chastel 1990). Similarly, the exclusive consumption by the nobility of crane, heron and swan could be explained by their rights over uncultivated lands, in this case wetlands. However, land use is not the only factor of importance. During the early medieval period, the nobility also acquired hunting rights (Jarnut 1985). Within the Celto-Germanic tradition, hunting was free for everybody, but during Roman times it became possible to forbid hunting activities on the land one owned. Moreover, after an initial return to the pre-Roman situation at the beginning of the Middle Ages, this restriction on the basis of land ownership was again established and, most importantly, expanded to rights upon the animals themselves. This means that species such as red deer, wild boar, brown bear, and large birds, became the exclusive hunting prey for the nobility, regardless of where they were found. The castle's lords could thus hunt on everybody's fields and meadows for the animals that fell under their privileges, while the peasants were not allowed to kill these species even when they were seen on their fields (Smit 1911). A comparable situation was valid for marine animals stranded on the North Sea coast. Already in early medieval times, the waterways became considered as 'uncultivated lands', of which the use was exclusive to the feudal elite. Consequently, the animals that stranded upon the coast, or the monetary value of their sale, had to go (at least partly) to the count or duke (De Groote 1999). This could explain the exclusive presence of sperm whale remains in the castles within our study area.

Must the conclusion thus be that different ownership of land and rights explains why, at rural sites mainly sheep and cattle were slaughtered, why at castles one finds a lot of pigs, game and unusual animals, and why within the monasteries pork was such a rare commodity? The fact alone that land ownership cannot explain why monks rarely ate meat but did consume a lot of fish, suggests there is more at hand. Indeed, the food pattern of a monastic community cannot be studied without referring to food rules. During the initial, early medieval, development of monasticism many systems of behavioural rules were followed (Miccoli 1991). However, during the Carolingian period, the rule of Saint Benedict became adopted by all monasteries in Western Europe and after that period it stayed the most influential one within medieval Europe (Milis 1992). With it, came the prohibition against the consumption of the meat of four-legged terrestrial animals, for monks that were not weak or ill (Schmitz 1945). This explains the near absence of mammal bones at the monastic sites and implies why, with regard to animal proteins, fish became an especially important commodity.

The monastic diet thus seems to be largely based upon ideological grounds. Must this factor then also not be taken into account concerning the diet of the nobility? Indeed, those who fought were constantly concerned about their status. They had to reaffirm it regularly and could do this by showing the benefits of their privileges, such as the use of the forest and other wild lands, and, on top of all this, the exclusive hunting rights to certain species. Banquets were ideal events to do this, by serving animals from the forest (game and pigs), large birds from the swamps or stranded sea mammals. Moreover, serving and eating large quantities of meat fitted within the original Germanic tradition and brought with it a connotation of power, well-being, virility and martiality (Montanari 1994). In addition, because pig is the only large domestic mammal that was solely kept as a meat producer, it was the most 'meaty' food product one could serve and eat. The pig, although a domestic animal, also retained a connotation with the forest (because it was herded there) and with animals living in a wild condition (because the pig then still resembled the wild boar, its wild conspecific). Therefore, the symbolic value of the forest, as part of the Germanic cultural tradition (Smit 1911), and the high esteem attributed to the wild boar as the most dangerous hunting prey (Hainard 1948), were transferred to the noble host by serving pork at the castle's table.

Concerning status, it is also known that the Merovingian, Carolingian and high medieval nobility regarded the *Imperium Romanum* as the ideal, heavenly, empire they wanted to re-establish (Vones 1990). Their dress code, burial habits, and culinary culture imitated the Roman way of life (Plouvier 1990). From the zooarchaeological record of Roman sites (King 1984) and the texts on cooking (André 1981), we know that the Romans preferred pork above all other meat. Moreover, during the occupation of northern Gaul, a Roman diet had always been most prominent in military camps (Davies 1971; King 1984), again associating pork with a martial aspect of culture. Could it be possible that this association between pig, Roman culture and army also played a role in the composition of the medieval noble diet?

Returning to the monastic diet, it can now also be interpreted as a reaction against the behaviour of the nobility. The Mediterranean monastic model, which eventually became adopted throughout Europe, originated within the classical Roman culture, in which, regarding food consumption, modesty and refinement was highly esteemed (indeed, the infamous orgies of the Roman upper class must be seen as a violation against this cultural pattern). The monastic food rules were not based upon any biblical concept but did refer to the diet of the common people in the Mediterranean, mainly consisting of vegetable products and the occasional fish, and only very rarely of meat (Jungclaussen 1980). However, the north-western European nobility came out of the Germanic culture, in which, as has been stated, dining copiously, and even gluttony, was highly appreciated (Montanari 1994). The differences between the monastic and the noble diet can thus be seen as a clash of cultures, through which each order tried to identify themselves. Generally, the clergy considered itself not only to be different but also to be a superior order within the structure of society, focusing on heavenly matters as an antidote to the worldly attitude of those who fought. By introducing the abstinence from meat, they showed they stood above all temptation towards carnal pleasures, be it in the form of sexual acts or of eating meat. One could even state that they promoted the concept of Lent, or of fasting in general, as a way of imposing their diet upon the other orders. In that way, it makes sense that, when monks ate meat, as a disobedience to their food rules, they almost never choose pork. The consumption of pigs, being the most 'meaty' domestic animals and being only bred for meat production, could have been regarded as a greater sin than eating mutton or beef (except for sick monks that did not have to follow the food rules, hence the consumption of pork at the post-medieval infirmary of the Sint-Pieter Abbey at Gent). Within the Catholic conceptual framework, the pig even became the symbol of all carnal pleasures, thus also of sexuality. Sex was desired but wicked; the pig was an attractive but dirty animal. According to the psychology of Lacan, this symbolism is still very much alive today (Declercq 1999).

Complicating the model

The concept of the three orders has thus far only been described in its most simplified form. However, our image of the medieval world, and its consumption patterns, becomes more complicated when sites or groups of people are taken into account that do not fit easily within the categories of castle, Benedictine abbey, or rural household. Firstly, during the late medieval period, some agricultural sites became so wealthy that they tried to acquire the status of the nobility and to detach themselves from the feudal framework. In the western part of (medieval) Flanders, such sites can easily be recognised by their large, single or multiple moats, and the raised central 'living' island, imitating the defence works of a noble site (Verhaeghe 1981). Reliable studies of zooarchaeological material are lacking from these Flemish moated sites but it would be interesting to see whether the general tendency to imitate the nobility was also reflected within the diet of the inhabitants.

Another group within society for which the situation was not that clear-cut, is formed by the members and functionaries of the noble courts. Again zooarchaeological information is virtually absent, although a recent study from Namur indicates that court members (who lived in the town at the foot of the castle) could profit from the privileges owned by their lord, making 'noble' food ingredients appear in civilian households (Pigière *et al.*, *this volume*). Indeed, a lot of food officially meant for the highest elite ended up being consumed well down the social scale in the same households or even local communities (R. Hoffmann, *pers. comm.*). On the other hand, within a castle, of course not everybody enjoyed the pleasures of the noble banquet. When analysing the zooarchaeological finds from a noble site, a careful intra-site analysis must thus be made before conclusions can be drawn. It has, for example, been shown that meat consumption in the servants' quarter of the castle at Sugny included significantly less pork than was the case in the central donjon (Ervynck 1992b).

Considering the monasteries, only a very general view on food consumption has been obtained through archaeological analysis (Ervynck 1997). At present, although it is known that other rules than that of Saint Benedict existed, the differences between monastic orders have not yet been evaluated. Interesting cases will be those of the military orders, such as the Knights Templar or the Knights Hospitallers, the members of which were both soldiers and ecclesiastics. Another phenomenon that needs more attention is the hierarchy within an abbey, for example because the abbot was often exempted from the food rule when he was receiving guests in his private quarters. It must never be forgotten that in many abbeys those high within the hierarchy came out of noble families, which in terms of diet composition could have caused an internal conflict between noble background and monastic life. Another aspect of the obedience to the food rules is the wave-like trend of observance versus neglect that is apparent throughout the medieval period, and the final gradual loosening of the rules that certainly manifested itself towards the end of the Middle Ages and during the post-medieval period (Milis 1982). Again, differences between monastic groups must be taken into account here. Cistercians, for example, during their early 12th-century development, claimed that they were restoring the proper observance of Saint Benedict's Rule and followed a more austere rule compared to the contemporaneous Benedictines (R. Hoffmann, *pers. comm.*). Finally, outside of the monasteries, the position of the secular priests or of the bishops (and their court) is also not yet clarified in terms of dietary composition. Bishops were clerical leaders but often lived as noble lords. This was especially true in the eastern parts of present-day Belgium, for example in the medieval entities of Brabant and Liège, where some bishops were really feudal lords, thus possessing hunting rights or privileges upon land use.

The fourth order

Another complicating factor is that the model of three orders is essentially a rural one (Duby 1985), developed within a society in which towns were not yet important. Thus, when towns started to develop, the concept needed to be enlarged by the inclusion of a fourth order: the urban population. This process took considerable time, illustrated by the fact that, although towns such as Gent were already beginning to develop during the early 10th century, the first representation of the fourth order is only to be found in a codex dating from around AD 1420 (the *Livre des bonnes moeurs* by Jacques Legrand, see Smeyers 1998). This slow acceptance of the urban population within the medieval world view is probably related to the limited political power of the town populations during their first phase of growth. Only when the late medieval merchants and craftsmen acquired considerable wealth with their activities in towns, and when the latter groups organised themselves politically into guilds, did towns begin to challenge the nobility in a struggle for power. From then on, their presence could no longer be overlooked. However, only the richer part of the urban society benefited from this trend. The major part of the urban population consisted of labourers and other groups of less wealthy people, and they never gained recognition within the concept of orders.

The zooarchaeological record for Belgian medieval towns is still rather incoherent (Ervynck 1992a and the references there), although new data are slowly appearing for high and late medieval Ieper (Ervynck 1998), Oudenaarde (Ervynck and Van Neer, *unpublished*), Aalst (Pieters *et al.* 1993), Antwerpen (Ervynck 2002 and the references there), Mechelen (Ervynck and Van Neer, *unpublished*), Brussel (Van Neer 2001, and the references there), Leuven (Ervynck and Van Neer 1998) and Namur (De Cupere and Van Neer 1993; Lentacker *et al.* 1997; Pigière *et al.*, *this volume*). Most informative, however, is a gradual accumulation of consumption refuse excavated at Gent, dating from the early 10th to the late 12th century, and thus covering the initial development phase of the town. The recent analysis of this material has revealed that the labourers who came to live in town retained the rural diet of the surrounding countryside. Their diet for example does not include game and shows low frequencies of pig (Ervynck and Vanholme, *unpublished*). Only when a rich class eventually established itself within the town's walls, was a departure made from this rural consumption pattern. The zooarchaeological remains from many late medieval cesspits belonging to the houses of the richer tradesmen and craftsmen illustrate that the *nouveaux riches* tried to imitate the nobility. They did this, not through the acquisition of rights or privileges but, through the power of their money. The occasional game eaten came not from their feudal lands but must have been bought or acquired through contacts with the nobility. The frequency of pig remains can also be rather high within these cesspits, again not because these townsmen had the privileged use of forests but mostly because they bought the animals. However, it must also be taken into account that people started to breed pigs in towns (in the process turning the pig from a forest animal into a garbage eater), which could account for at least part of the pork consumption within the late medieval towns.

Interestingly, most towns represent a micro-cosmos illustrative for the whole of the medieval society. Within the urban centres, abbeys founded *refugia* while some feudal castles were located within the town's walls. Alternatively, feudal lords living in the countryside started to establish urban residences. For the urban monastic sites no meaningful data are available yet but considering the castles the case study of the Castle of the Counts at Gent has shown that this site, located within the town, displayed a completely different dietary pattern compared to the urban households. For the most part, this difference was not due to higher purchasing power but to the fact that the castle relied completely upon its properties in the countryside in terms of meat provisioning (Ervynck 1992b).

Clearly, the rise of the fourth category marked the beginning of the end for the whole static, divine, concept of a society divided into orders. The upper part of the urban society had established itself not through inheritance, status, or any appeal to a divine or religious concept. Instead, they had made their way through their economic importance, thus, in a way, through money. The rise of these urban 'capitalistic entrepreneurs' resulted in a loss of importance of both nobility and clergy (Fig. 3). At the same time, urbanisation started the continuous shrinking of the rural part of society. During the late Middle Ages, regarding meat supply, the situation also changed for the nobility in the countryside, for other reasons. Firstly, there was an almost complete depletion of game, a process visible in the zooarchaeological record (Ervynck *et al.* 1999). Furthermore, deforestation and conflicting options regarding the use of the remaining forest forced pigs to become farm animals, which must have changed their economic and symbolic meaning (Ervynck *et al., in prep.*). Finally, at the same time, the whole ideology of the nobility changed. The ideas about knighthood and chivalry faded away and gradually privileges were abolished. Meanwhile, in the abbeys, the food rules became more and more relaxed and in some cases, the economic function of the site gained more importance than the spiritual one. In general, purchasing power thus became the main factor for the composition of the European diet, instead of religion, privileges or inherent status. This implies that post-medieval zooarchaeological collections must be interpreted within a completely different conceptual frame work than medieval ones. In the former case, the differentiation between those with money and those without must be kept in mind; in the latter case, the system of three orders described here.

Evaluation and future prospects

The model of the three orders is certainly a useful approach in order to structuralise the interpretations of food patterns in high and late medieval sites. It pushes the analysis beyond the level of descriptive statistics, and reveals how differences within the medieval societies were reflected by differences in diet. In doing this, however, we do not yet reach the interpretative level at which we can answer the question why medieval society became subdivided in static orders, and whether food provisioning was a causal factor or only a symptom of the discriminating processes. Such an investigation must of course be carried out by historians; zooarchaeologists can be of help but cannot take the lead. On the other hand, zooarchaeologists should realise that, when they are studying contexts from historical periods, they are partly doing historical research. The texts available simply cannot be ignored. In that way, in the case of medieval and other historical sites, zooarchaeology has more links with history than with its most cited counterpart, cultural anthropology. Consequently, it must be regretted that in many zooarchaeological reports for historical periods references to texts are only anecdotal, or only factual instead of interpretative.

In the future, historical archaeology, in general, must benefit from the methodological developments within the field of history (Bintliff 1991), and this is of course equally true for historical zooarchaeology. Our search for the explanations 'why' could, for example, be stimulated by emphasising the *longue durée* approach of the historical school of the *Annales* (Bintliff 1991). Indeed, the food economy of the system of the three orders will never be fully understood without a profound analysis of how and why it developed from a Roman economy (with a society divided in *liberi* and *servi*, and with, for example, free hunting rights), via the Germanic, Merovingian and Carolingian economic paradigms, into the model of feudal Europe. Unfortunately, the early medieval zooarchaeological record for Belgium is still insufficient, but it would be interesting to see what the dietary pattern was in abbeys in which the abbots were noble lords that stayed laymen, or in the households of the functionaries of the central authority (Merovingian or Carolingian king or emperor) that had not yet become feudal lords, or in the large farmhouses of the early medieval independent landowners before their properties became usurped by the rising nobility. One of the few examples of sites that have already shed some light upon the transition from the Carolingian to the feudal period is provided by the 'Hoge Andjoen' motte at Werken (western Flanders) which evolved from a 9th century rural site to a high medieval feudal stronghold. Within the first phases of the occupation sheep and cattle dominate the domestic livestock while pigs take over during the phase of the motte castle (Demandt 1997)!

Another benefit of adopting a historical approach within medieval zooarchaeology, and more specifically that of the structural history of the *Annales* school, is that it provides a way to overcome the processual versus post-processual schism within theoretical archaeology (was human behaviour ruled by the environment or by cognitive aspects?). Structural history highlights the complementary nature of both approaches (Bintliff 1991, note 1) and the analysis presented here for medieval Belgium has also, be it implicitly, argued that both environment and ideology were important controlling factors for consumption behaviour.

Future zooarchaeological research of medieval sites must of course no longer try to prove, over and over again, that a tripartite subdivision occurred within the high to late medieval society. The archaeological information must challenge the historical data in order to refine the general model and to reveal patterns that are not documented within history. The potentially special position of moated sites, bishops' palaces, the households of secular priests, monastic military orders or abbeys not following the Benedictine food rules, has already been mentioned, just as the danger of neglecting hierarchical differentiation within sites. Within this context, it must also not be forgotten that the model of three orders is an ideological one (Duby 1985) while archaeology deals with reality. The animal remains can thus reveal behavioural patterns that have not been documented sufficiently within the written sources. It must further be realised that the model may never become restrictive for the interpretation of data. Indeed, even before the rise of the urban populations, many people within the north-west European medieval society were neglected within the concept of the three orders. Examples are Jews, gypsies, heretics, mercenaries, marginal figures (prostitutes, lepers, beggars, thieves, performing artists), pilgrims, crusaders, physicians and even seamen (Duby 1985; Le Goff 1991). Moreover, the model of a tripartite structure of society must not prevent the analysis of data along other dividing lines, such as 'woman versus man' or 'child versus adult'. After all, the concept of the three orders has a clear (adult) male, and thus sexist aspect to it.

Finally, it will always be useful to analyse the system of the three orders for comparison with societies that are organised in other ways, especially to avoid the pitfall of ethnocentrism, i.e. looking towards former (pre-contact) non-European human societies through the framework of medieval and post-medieval Europe.

Acknowledgements

Many thanks go to Prof. Dr. Richard Hoffmann (York University, Toronto, Canada) for very valuable historical comments and remarks.

References

André, J. 1981. *L'Alimentation et la Cuisine à Rome*. Paris: Les Belles Lettres.

Berings, G. 1985. Het oude land aan de rand van het vroegmiddeleeuws overstromingsgebied van de Noordzee. Landname en grondbezit tijdens de middeleeuwen. *Handelingen van de Maatschappij voor Geschiedenis en Oudheidkunde te Gent s.n.* **39**, 37–84.

Bintliff, J. 1991. The contribution of an *Annaliste* / structural history approach to archaeology, pp. 1–33 in Bintliff, J. (ed.), *The Annales School and Archaeology*. London, New York: Leicester University Press.

Chastel, A. 1990. *Le Château, la Chasse et la Forêt*. Luçon: Sud-Ouest.

Clavel, B. 2001. L'animal dans l'alimentation médievale et moderne en France du Nord (XIIe-XVIIe siècles). *Revue Archéologique de Picardie, Numéro Spécial* **19**, 1–204.

Davies, R. W. 1971. The Roman military diet. *Britannia* **2**, 122–42.

Declercq, F. 1999. De eeuwenoude ambivalentie tegenover het varken: Freudiaans-Lacaniaanse kanttekeningen. Het varken in de psychologie, pp. 69–83 in Dhaene, S. (ed.), *Hoge Hakken, Roze Billen*. Gent: Museum voor Volkskunde.

De Cupere, B. and Van Neer, W. 1993. La faune du site de l'Hospice Saint-Gilles à Namur: résultats préliminaires, pp. 87–92 in Corbiau, M. H. and Plumier, J. (eds), *Actes de la Première Journée d'Archéologie Namuroise*. Namur: Service des Fouilles de la Région Wallonne.

De Groote, M. 1999. De baljuwsrekening: venster op de praktijk van het middeleeuws strandrecht in Vlaanderen. *Madoc. Tijdschrift over de Middeleeuwen* **13**, 258–67.

Demandt, I. 1997. *Dierenresten uit de Vroegmiddeleeuwse Motte 'Hoge Andjoen': Tafonomie, Paleo-ecologie en Paleo-economie*. Unpublished Master's thesis, University of Leuven.

Duby, G. 1985. *De Drie Orden. Het Zelfbeeld van de Feodale Maatschappij. 1025–1225* (translation of *Les Trois Ordres ou l'Imaginaire du Féodalisme*. Paris: Gallimard. 1978). Amsterdam: Elsevier.

Ervynck, A. 1991. 'De beer die woonde op de burg...': mens en dier in een vroegmiddeleeuwse versterking, pp. 170–80 in De Witte, H. (ed.), *De Brugse Burg. Van Grafelijke Versterking tot Moderne Stadskern*. Brugge: VZW Archeo-Brugge.

Ervynck, A. 1992a. The economy of food in medieval towns: getting rid of the simple explanations, pp. 133–8 in *Medieval Europe 1992. Pre-printed Papers of the Conference on Medieval Archaeology in Europe, York, 1992. Volume I*. York.

Ervynck, A. 1992b. Medieval castles as top-predators of the feudal system: an archaeozoological approach. *Château Gaillard. Etudes de Castellologie médiévale* (Caen) **15**, 151–9.

Ervynck, A. 1997. Following the rule? Fish and meat consumption in monastic communities in Flanders (Belgium), pp. 67–81 in De Boe, G. and Verhaeghe, F. (eds), *Environment and Subsistence in Medieval Europe. Papers of the 'Medieval Europe Brugge 1997' Conference Volume 9* (I.A.P. Rapporten 9). Zellik: Instituut voor het Archeologisch Patrimonium.

Ervynck, A. 1998. Wool or mutton? An archaeozoological investigation of sheep husbandry around late medieval Ypres, pp. 77–88 in Dewilde, M., Ervynck, A. and Wielemans, A. (eds), *Ypres and the Medieval Cloth Industry in Flanders. Archaeological and Historical Contributions* (Archeologie in Vlaanderen Monografie 2). Zellik: Instituut voor het Archeologisch Patrimonium.

Ervynck, A. 2002. Natuurwetenschappelijk onderzoek en de Antwerpse stadsarcheologie: een historische schets, pp. 25–31 in Veeckman, J. and Bellens, T. (eds), *Vijftig Jaar Stadsarcheologie in Antwerpen. Wat nu?*. Antwerpen: Dienst Stadsarcheologie.

Ervynck, A. and De Meulemeester, J. 1996. La viande dans l'alimentation seigneuriale et la variété des terroirs: l'exemple des Pays-Bas méridionaux, pp. 36–41 in Colardelle, M. (ed.), *L'Homme et la Nature au Moyen Age. Paléoenvironnement et Sociétés Occidentales*. Paris: Editions Errance.

Ervynck, A., Lentacker, A., Müldner, G., Richards, M. and Dobney, K. (in prep.). Paradise Lost. The late medieval evolution from forest dwelling pigs to farm animals. A case study from Flanders, Belgium. In: *Proceedings of the Pigs and Humans' Workshop at Walworth Castle, 26th–28th September 2003*.

Ervynck, A. and Van Neer, W. 1998. Het archeologisch onderzoek van de voedseleconomie van laatmiddeleeuwse steden. Mogelijkheden en eerste resultaten voor Leuven, pp. 79–94 in Bessemans, L., Honoré, I., Smeyers, M., Vandekerchove, V and Van Uytven, R. (eds), *Leven te Leuven in de Late Middeleeuwen*. Leuven: Peeters.

Ervynck, A., Van Neer, W. and Lentacker, A. 1999. Introduction and extinction of wild animal species in historical times: the evidence from Belgium, pp. 399–407 in Benecke, N. (ed.), *The Holocene History of the European Vertebrate Fauna. Modern Aspects of Research* (Archäologie in Eurasien 6). Berlin: Deutsches Archäologisches Institut, Eurasien-Abteilung.

Ervynck, A., Van Neer, W. and Van der Plaetsen, P. 1994. Dierlijke resten, pp. 99–170 in Ervynck, A. (ed.), *'De Burcht' te Londerzeel. Bewoningsgeschiedenis van een Motte en een Bakstenen Kasteel* (Archeologie in Vlaanderen Monografie 1). Zellik: Instituut voor het Archeologisch Patrimonium.

Fiers, E. 2000. *Archeozoölogisch Onderzoek van de Kasteelsite Montaigle (Provincie Namen)*. Unpublished Master's thesis, University of Gent.

Hainard, R. 1948. *Les Mammifères Sauvages d'Europe. I: Insectivores, Chiroptères, Carnivores*. Neuchâtel: Delachaux et Niestlé.

Jarnut, J. 1985. Die frühmittelalterliche Jagd unter rechts- und sozialgeschichtlichen Aspekten. *Settimane di studio del Centro Italiano di studi sull'Alto Medioevo* **31(1)**, 765–98.

Jungclaussen, E. 1980. *Worte der Weisung*. Freiburg im Breisgau: Herder.

King, A. C. 1984. Animal bones and the dietary identity of military and civilian groups in Roman Britain, Germany and Gaul, pp. 187–217 in Blagg, T. F. C. and King, A. C. (eds), *Military and Civilian in Roman Britain. Cultural Relationships in a Frontier Province* (BAR British Series 136). Oxford: British Archaeological Reports.

Lauwerier, R. C. G. M. 2002. De archeozoölogie van de Middeleeuwen, pp. 225–33 in Woltering, P. J., Verwers, W. J. H. and Scheepstra, G. H. (eds), *Middeleeuwse Toestanden Archeologie, Geschiedenis en Monumentenzorg*. Amersfoort: Rijksdienst voor het Oudheidkundig Bodemonderzoek.

Le Goff, J. 1991. Inleiding, pp. 7–42 in Le Goff, J. (ed.), *De Wereld van de Middeleeuwen*. (translation of *L'Uomo Medievale*. Bari: Laterza. 1987).

Lentacker, A. Van Neer, W. and Plumier, J. 1997. Historical and archaeozoological data on water management and fishing during medieval and post-medieval times at Namur (Belgium), pp. 83–94 in De Boe, G. and Verhaeghe, F. (eds), *Environment and Subsistence in Medieval Europe. Papers of the 'Medieval Europe Brugge 1997' Conference Volume 9* (I.A.P. Rapporten 9). Zellik: Instituut voor het Archeologisch Patrimonium.

Miccoli, G. 1991. De monniken, pp. 43–79 in Le Goff, J. (ed.), *De Wereld van de Middeleeuwen*. (translation of *L'Uomo Medievale*. Bari: Laterza. 1987).

Milis, L. 1992. *Angelic Monks and Earthly Men*. Woodbridge. Rochester NY.

Montanari, M. 1994. *Honger en Overvloed*. (translation of *La Fame et l'Abbondanza: Storia dell' Alimentazione in Europa*. Roma: Laterza. 1993).

Pieters, M., Cooremans, B., Ervynck, A. and Van Neer, W. 1993. Van akkerland tot Heilige Geestkapel. Een kijk op de evolutie van de bewoningsgeschiedenis in de Kattestraat te Aalst (prov. Oost-Vlaanderen). *Archeologie in Vlaanderen* **3**, 299–329.

Pigière, F., Boone, I., Udrescu, M., Van Neer, W. and Vanpoucke, S. this volume. Status as reflected in food refuse of late medieval noble and urban households at Namur (Belgium), pp. 233–43 in O'Day, J. S., Van Neer, W. and Ervynck, A. (eds), *Behaviour behind Bones. The Zooarchaeology of Ritual, Religion, Status and Identity*. Oxford: Oxbow Books.

Plouvier, L. 1990. De "Belgische" keuken in de vroege middeleeuwen. *De Brabantse Folkore en Geschiedenis* **267–8**, 205–35.

Schmitz, P. 1945. *Sancti Benedicti Regula Monachorum*. Maredsous: Editions de Maredsous.

Smeyers, M. 1998. *Vlaamse Miniaturen van de 8ste tot het Midden van de 16de Eeuw*. Leuven: Davidsfonds.

Smit, J. P. W. A. 1911. *Het Brabantse Jachtrecht voor de Regering van Karel den Stouten*. Amsterdam: van Langenhuysen.

van Beek, B., Jacobi, H. and Scharloo, M. 1984. *Geld door de Eeuwen heen. Geschiedenis van het Geld in de Lage Landen*. Amsterdam: Pampus Associates.

Van Neer, W. 2001. Restes de poissons provenant de contextes médiévaux et post-médiévaux dans l'îlot Villers-Dinant. *Archéologie à Bruxelles/Archeologie in Brussel* **4**, 316–23

Van Neer, W. and Ervynck, A. 1996. Food rules and status: patterns of fish consumption in a monastic community. *Archaeofauna* **5**, 155–64.

Verhaeghe, F. 1981. Moated sites in Flanders. Features and significance, pp. 98–121 in Hoekstra, T. J., Janssen, H. L. and Moerman, I. W. L. (eds), *Liber Castellorum. 40 Variaties op het Thema Kasteel*. Zutphen: Walburg Pers.

von Bieberstein, J. 1989. *Adelsherrschaft und Adelskultur in Deutschland*. Frankfurt am Main: Peter Lang.

Vones, L. 1990. Monarchy and nobility: on the delicate balance of power in the Middle Ages, pp. 40–65 in Toman, R. (ed.), *The High Middle Ages in Germany*. Köln: Benedikt Taschen.

Anton Ervynck
Institute for the Archaeological Heritage of the Flemish Community, Koning Albert II-laan 19, Box 5,
B-1210 Brussel, Belgium
E-mail: lentacker.ervynck2@yucom.be

9th ICAZ Conference, Durham 2002
Behaviour Behind Bones, (ed. Sharyn Jones O'Day et al.) pp. 224–232

25. Dietary habits of a monastic community as indicated by animal bone remains from Early Modern Age in Austria

Alfred Galik and Günther Karl Kunst

Rescue excavations in the former Carthusian monastery Mauerbach (Lower Austria) yielded a series of animal bone samples that had accumulated during a rebuilding phase of the first half of the 17th century. The five assemblages studied were retrieved from different sections of the monastery, including levelling layers inside rooms and yards, and backfilled cesspits. The faunal content of all samples is dominated by the remains of fish and other aquatic animals such as terrapins, beavers, waterfowl and molluscs, while bones of the major domesticates are almost absent. This pattern is obviously related to the food rules of the monastic community: the monks had to abstain from the meat of warm-blooded animals. As a consequence, they adopted special dietary habits relying on fish and other cold-blooded or aquatic species regarded as a substitute for fish. Some of the assemblages studied clearly correspond to single dumping episodes, and variations in the frequency of the animal groups between the samples may be due to seasonality, type of context and different taphonomic pathways within the monastery. The increasing consumption of fish-substitute species probably developed in the wake of the Counter-Reformation.

Historical context and archaeological setting

The monastery of Mauerbach (Lower Austria) was founded by King Frederick the Fair as the first Carthusian convent in Austria in 1316, an era corresponding to an early expansion phase of this order (Jaritz 1999). Although situated just outside the western border of today's district of Vienna, the site, lying within a small valley of the Vienna woods, still retains a sense of remoteness and solitude, which was certainly once decisive for its location. Mauerbach, however, also became the first monastery dissolved in the wake of the ecclesiastical reforms under Emperor Joseph II in 1782. Serving as a charitable hospital and asylum during the 19th and the first half of the 20th century, the remaining parts of it are now administrated by the National Monument Office (BDA), and substantial renovation works are undertaken.

Since 1996, the rebuilding is supervised by a team of archaeologists of a social project (ASINOE), carrying out salvage excavations in the interior and at the precincts of the former monastic buildings (Kreitner 1999; Kreitner 2001). A major task is the investigation of the architectural development of the medieval monastery. It must be kept in mind that the Charterhouse, as it remains today, is mainly a product of the so-called baroque building mania (*Bauwut*) of the 17th century, clearly related to the Counter-Reformation of the Hapsburg Empire. During the era of the Counter-Reformation, governmental and ecclesiastical authorities succeeded in wiping out Protestantism, which had been adopted by large parts of the population by the middle of the 16th century, from most parts of the German-speaking regions of the Austrian territory, leading to the eventual ousting of corporative power and to the introduction of absolutism. This process was assisted or carried out by religious orders (Jesuits, Franciscans, Capuchins), partly originating from South European countries unaffected by Protestantism, and found its architectural expression by the adoption of Italian baroque style. In the decades after 1616, mainly in the years 1620–1640, the greater part of the medieval fabric was demolished and the whole monastery was almost totally erected anew. Parts of the medieval church, of some adjacent buildings, and of the southern cloister, were used as foundations for the equivalent baroque structures, whereas the cloister itself was significantly enlarged. The buried remnants of the north-eastern corner of the medieval cloister have only recently been discovered in the middle

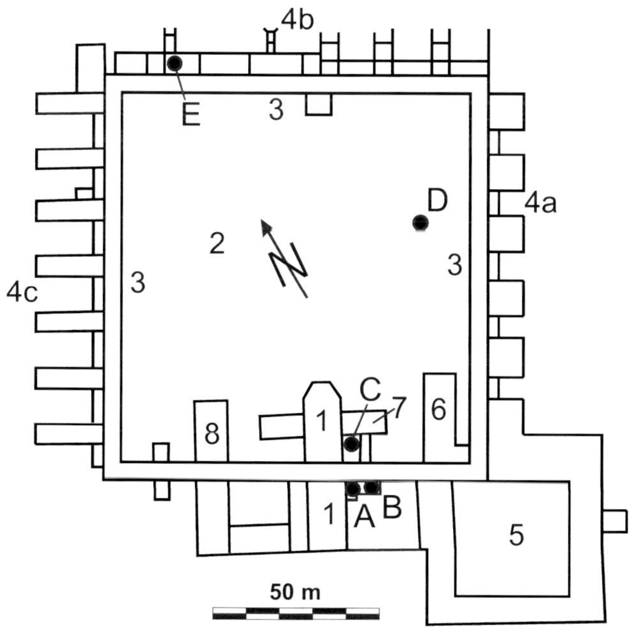

Fig. 1. Simplified ground plan of the baroque Carthusian monastery at Mauerbach and positions of samples A-E; numbers indicate main conventual buildings: 1 – church; 2 – central garden (cloister garth); 3 – great cloister; 4 a – east, 4 b – north, 4 c – west row of cells; 5 – prelate's (imperial) section; 6 – refectory; 7 – chapter house; 8 – sacrist's cell.

of the extant cloister garden (near point D in Fig. 1). An important point is that the 17th century reconstruction stage entailed a lot of soil movement and that a considerable amount of rubbish was either deliberately dumped, or was allowed to accumulate, when room floors were levelled or structures like latrines were abandoned. In fact, most of the animal bone assemblages of the monastic period are in some way related to the rebuilding, if not directly caused by it. Similar processes occurred during the post-dissolution period, when the building had to be adapted to its new functions. Thus, the history of the monastery is mirrored, if rather irregularly, in the archaeological record of the site.

The Carthusian way of life

The Carthusians sought to reconcile the two traditions of Christian monasticism: the eremitic (hermits) and the coenobitic (communities), resulting in a distinct and successful form of monastery (Barber and Thomas 2002, 1).

In fact, the way of life and the monastic rules of the Carthusians are already transparent from the baroque ground plan of Mauerbach (Fig. 1), or any other charterhouse: a row of detached cells for the individual monks are more or less regularly arranged around a central garden, and connected by a cloister. On the periphery, a secluded garden is attached to each cell. In the case of Mauerbach, the main conventual buildings, including the church and the prelate's section, are centred around the southern cloister and the south-eastern corner of the monastery.

Some time after the foundation of the Carthusian Order by S. Bruno in 1084, certain rules regulating a monk's daily routine became established, amongst which the *consuetudines* from the first half of the 12th century form the Order's first written rule (Parisse 1994, 416; Barber and Thomas 2002, 61). During the week, personal contacts are kept to a minimum and the monks meet each other only for prayer and during mass. Meals, handed through a serving hatch arranged aside each cell door, are taken in solitude except on Sundays and feast days, when the whole community assembles for dinner and supper in the refectory. On these days, there is also a common recreation with talking during a common walk.

Judging from the number of cells, the baroque monastery of Mauerbach was originally intended for about twenty monks. Historical sources indicate numbers between 12 and 21 monks (including the prior) for the 17th century. In 1782, the year of dissolution, the prelate, 18 clericals and 6 lay brothers lived in the buildings.

The food rules comprised a strict abstinence from meat of warm-blooded vertebrates. They forbade the consumption of all major domesticates, leaving only fish and fish substitutes, and were applied more rigorously during periods such as Lent or Advent. The day-to-day dietary pattern of a Carthusian community should therefore broadly correspond to the one obligatory for the catholic laity when "fasting", i.e. abstaining from meat of mammals and birds. Apart from fish, it was allowed to eat all animals that either resemble fish in anatomical (*e.g.*, being covered with scales), physiological (having "cold blood"; Mayerhofer and Pirquet 1923, 867–72) or ecological (aquatic habits) features. It is frequently noted and well known from anecdotal evidence, that certain mammal and bird species became to be regarded as "fish" as well. For certain religious communities water-living mammals like beavers, otters, seals and dolphins, and water birds putatively feeding on fish, like coots and diving ducks, were considered as part of the class of *Aquatilia* (Mayerhofer and Pirquet 1923; De Grossi Mazzorin and Minniti 1999) whose consumption was allowed on days of abstinence from meat. The adoption of these species as fish-substitutes during periods of fasting developed out of customary law during the Middle Ages concerning disputes about the nature and the systematic affiliation of the species mentioned. As regards the acceptance of certain species and species groups, a considerable variation can be observed according to religious order and geographical region. Chelonia, which apparently had a special significance for the Mauerbach monks, were allowed as food without breaking the rule of Lenten abstinence by Pope Hadrian VI (1522–1523), by *vivae vocis oraculo* ("word

	sample A		sample B		sample D		sample E	
	NISP	W	NISP	W	NISP	W	NISP	W
Mammals indet.	3	13.6	4	4.7	–	–	–	–
Domestic cat	–	–	–	–	9	3	17	3.2
Micromammals	–	–	–	–	7	0.2	28	0.52
Beaver	1	3.4	–	–	13	37.8	2	8.7
Birds indet.	–	–	–	–	1	0.7	–	–
Coot	–	–	32	12.7	–	–	–	–
Chelonians	535	1077.1	71	43.1	92	59.3	5	0.6
Fish	11	31.2	359	332.3	124	8.6	800	59.25
Total	550	1125.3	466	392.8	246	109.6	852	72.3

Fig. 2. Mauerbach: species composition in samples A, B, D and E; NISP = Number of identified specimens; W = Bone weight (g); sample C contained several thousand remains of Emys orbicularis.

of the mouth", oral decree; De Grossi Mazzorin and Minniti 1999, 48). In Catholic regions of Southern Germany and Austria, fasting practices involving the consumption of some of these animal groups continued until the 19th century, well documented by contemporaneous market reports and by cooking literature (Kunst and Gemel 2000).

Animal remains from the monastery

Considering the archaeozoological analysis, a serious methodological problem is caused by the fact that the vast majority of the animal bones recovered so far, mostly butchering waste of cattle, belongs to the post-dissolution period of the 19th and early 20th century. Certain context types, such as levelled room floors, are not always amenable to a refined phasing because of condensation (palimpsest type of deposits), or where a detailed chronology has yet to be worked out. When looking through the complete animal bone material of the excavation seasons 1996–99, it transpired that those samples containing both common livestock species and the elements considered to be typical of the monastic period were mostly from reworked horizons, or were from contexts disturbed by later activities. On the other hand, there are a few samples, from selected features securely dated within the monastic period of the site through their artefact content or stratigraphic position, composed almost exclusively of the remains of fish and other aquatic species. This study will therefore focus on this latter group. Also included are the faunal contents of two "closed" pre-baroque structures backfilled in the 17th century, which were completely wet-sieved during archaeobotanical investigations (Fig. 2).

Sample A – number 801, rubble layer

The first sample comes from a rubble layer within a staircase close to the eastern wall of the church, overlying a medieval horizon. According to pottery sherds and glass fragments, the rubble layer was deposited some time after

	A	B	D	E
Acipenserinae (sturgeon)	7	44	–	–
Cyprinidae	–	–	3	16
Barbus barbus (barbel)	–	–	2	–
Cyprinus carpio (carp)	–	40	5	29
Tinca tinca (tench)	–	–	1	–
Rutilus rutilus (roach)	–	–	–	1
Leuciscus idus (orf)	–	5	1	–
Leuciscus leuciscus (dace)	–	–	1	–
Gobio gobio (gudgeon)	–	–	–	1
Phoxinus phoxinus (minnow)	–	–	1	–
Salmo trutta f. fario (brook trout)	–	–	–	1
Esox lucius (pike)	1	44	1	46
Silurus glanis (catfish)	–	76	1	–
Clupeidae (herring?)	–	–	–	1
Gadus morhua (cod)	–	70	–	–
Pleuronectidae (flatfish)	–	–	3	3
Pleuronectes platessa (plaice)	–	–	–	1
Soleidae (sole)	–	–	–	2
Psetta maxima (turbot)	–	–	–	5

Fig. 3. Mauerbach: Number of identified fish remains from samples A, B, D and E; sample C yielded no fish remains.

AD1600. This sample, containing about 550 identifiable specimens, was rather carefully collected by hand, but parts of the small fraction may be lost (Figs. 2 and 3). Apart from a fragmented scapula of beaver (*Castor fiber*), three unidentified splinters of mammal long bones and eleven remains of fish, the assemblage is largely dominated by bones of Chelonia, the majority of which belong to the European pond terrapin (*Emys orbicularis*). About a hundred remains are attributable to a tortoise of the genus *Testudo*, most likely *T. hermanni*, Hermann's tortoise. Although the bony shields prevail, there are elements from other parts of the skeleton as well. Among the fish bones, skull parts of large-sized specimens of pike (*Esox lucius*) and beluga sturgeon (*Huso huso*) could be observed.

Sample B – number 848, sewage run

Derived from the same trench as sample A, this hand-

Fig. 4. Aspect of the bone assemblage from a filled up sewer (sample B) with the remains of fish, turtles, coots and marine bivalves.

Fig. 5. Densely packed layer of remains of the European pond terrapin, Emys orbicularis, *from sample C. Some of the bone plates of the plastron and the carapace are still connected.*

collected sample was taken from a sewer filled up some time after AD 1600. As for composition, fish remains are the prevailing group, both by number (over 360 out of 466 identified specimens) and by weight (Fig. 4). They contain a high proportion of larger, readily recognizable elements and fragments like bone scutes of acipenserids, probably all from beluga sturgeon. There are several dozens of vertebrae of catfish (*Silurus glanis*), parts of which belong to the same individuals. A series of rod-shaped fragments, obviously cut to similar length, could be identified as cleithra and other elements of cod (*Gadus morhua*). Consistency regarding skeletal part representation is also indicated by 24 bird tarsometatarsals and some tibiae, all from coot (*Fulica atra*). About 70 chelonian remains come from different parts of the skeleton. Only four unidentifiable fragments of mammal bones are present. 13 valves, some of them matching pairs, belong to the marine bivalve *Tapes decussatus*, the chequered carpet shell.

Sample C – number 540, levelling layer with Emys orbicularis

A levelling layer inside a small yard, bordered by a tower containing a spiral staircase, was deposited during, or shortly before, the rebuilding phase of the 17th century, when this area was still open ground. It was overlain by pieces of copper sheets and 19th century rubbish layers, all sealed by a stone-slabbed floor, also from the 19th century. A densely packed bone horizon, consisting exclusively of several thousand skeletal parts of *Emys orbicularis*, was discovered inside the 17th century levelling layer and was taken out as a bulk sample (Fig. 5). A subsample of this assemblage was analysed according to intra-skeletal distribution, corroborating the visual impression that anatomical elements are represented

Fig. 6. Water-sieved residue of sample C, containing phalanges and vertebrae of Emys orbicularis.

according to their natural frequency. Even tiny bones like caudal vertebrae and phalanges are present (Fig. 6).

Sample D – number 2161, cesspit in cloister garth

An excavation trench in the north-eastern part of the central garden revealed important subsurface features: the remains of medieval or early post-medieval fabric, notably the north-eastern corner of the pre-baroque cloister and parts of a monk's cell. The backfill of a sunken, square brick-walled 15th century structure (measuring about 1,5 by 1 m), readily identifiable as the cesspit of a latrine belonging to the excavated cell, was completely wet-sieved. As the pre-baroque cloister, the exact configuration of which is still unknown, was considerably smaller than its successor from the 17th century, point D (Fig. 1) is in no direct relationship to any of the structures shown in the ground plan. A sample of animal remains was collected from the residue of flotation samples according to standard archaeobotanical procedure, using a mesh size of 1 mm. The cesspit yielded, besides potsherds, glass fragments and some charred grains and wood particles, an assemblage of 14 almost complete ceramic vessels (jugs, pans and plates) from the beginning of the 17th century. This pottery assemblage indicates household activity and could well correspond to the personal equipment of a single monk, dumped into the latrine during the rebuilding period between 1620 and 1620/40, when this part of the pre-baroque monastery was broken down. Similar use and abandonment cycles for monastic cesspits are discussed by Van Neer and Ervynck (1996, 157) and Roselló Izqierdo *et al.* (1994, 325). The animal bone sample retrieved from the latrine (246 specimens) only weighs just over 100 g, owing to the predominance of small fragments. Mammals are represented by a few heavily butchered bones of beaver, whereas the skull fragments of a very young cat and the bones of small rodents are probably not to be regarded as consumption refuse. Both small-sized remains of fish (124 specimens) and turtle (92) prevail in this assemblage. As the fish bone assemblage consists mainly of fragmented ribs and fin rays, only a minor portion of it could be identified, including one marine group (flatfish).

Sample E – numbers 2413 to 2415, basement of cell 2, northern section

Within the basement of one of the baroque cells aligned alongside the northern cloister, a pit-like structure of unknown function was fully excavated and the sediment water-sieved. Just as for assemblage D, a sample of animal remains was collected from the residue of flotation samples according to common archaeobotanical procedure, using a mesh size of 1 mm. This recently discovered structure belongs to a yet largely unknown part of the pre-baroque building and may have been a cesspit. The backfill from the pit can be roughly dated into the first half of the 17th century by fragments of Venetian glassware. The sample produced some mineralised weed seeds and a high number of tiny fragments of skeletal elements of fish, including otoliths. The species list (Fig. 3), containing both freshwater and marine taxa, must be regarded as rather preliminary, leaving a great part of potentially identifiable small fragments for future study. There are some larger remains too, of the freshwater species carp and pike. As regards terrapins, there are only a few vertebrae and small-sized fragments of long bones and from the pectoral and pelvic girdles. The mammal bones provide a similar picture as in sample D, with micro-mammals, butchered beaver bones, and several elements of a very young cat, probably from a single skeleton. Eggshells are an important component of the residue.

Taphonomic analysis

The samples investigated show a high degree of variability, which could have been controlled by the very type of context, the location within the site and various taphonomic factors. The spatial organisation of waste disposal within the monastery certainly played a crucial role. In all instances, structures of the pre-baroque building were opportunistically used as refuse dumps. It seems problematic to assign the assemblages, which were all generated during the transitory era of the rebuilding phase, to functional features of the standing baroque monastery. Nevertheless, the samples A-C were collected near the main conventual buildings and the church, an area probably already serving common functions in the pre-baroque era, containing both the refectory and the kitchen after the rebuilding. Contexts D and E, located towards the periphery of the monastery, are more likely to belong to the sphere of individual monks.

The formation of assemblage B could correspond to a single depositional event. This is indicated by the simultaneous presence of fragments belonging to the same individual animals (catfish, carpet shells) or modified in exactly the same way (cod cleithra), and by the accumulation of identical waste parts (tarsometatarsals of coot). These factors, including the presence of fish individuals of particularly large dimensions, point to the organized provisioning of many people, in the course of which highly structured kitchen refuse, or a combination of kitchen and table refuse (consumption refuse s.s. *sensu* Gautier 1987), was produced. It is tempting to relate the making of sample B to a special event, conceivably a feast of high-status members of the monastic community.

The chelonian-dominated samples A and C need some special comments. Compared with mammals and fish, it is more difficult to demonstrate that chelonian bones represent remains of animals actually eaten, because little butchering is needed in the preparation of the smaller species. Dorsal and ventral shields will easily fall apart during cooking. As a consequence, modifications are rare, but skeletal part representation provides a clue. In sample A, the bony plates of carapace and plastron (430) largely exceed the elements of the locomotory apparatus (97) and even more so the ones of the skull and the axial skeleton (8). As it seems justified to regard the bony plates as waste parts, accumulated during processing, this assemblage could also be classified as kitchen refuse, again indicating a larger group of people being involved in consumption. The well-balanced element frequency of sample C and its monospecific composition make it likely that these pond terrapins were not eaten at all. Conceivably, a large delivery of animals perished during transport, or storage, or remained unconsumed for whatever reason, leading to the synchronous deposition of a large amount of carcasses.

The faunal assemblages of the water sieved samples D and E are heavily biased towards small fragments, which might not be a consequence of recovery techniques alone. The absence of large-sized bone debris is especially surprising in the case of cesspit D, a voluminous structure well suited for the disposal of bulky domestic rubbish. It appears that the biological remains may correspond to a period when this structure still served its original function and that the influx of bulky material was avoided. In this case, the pottery assemblage recovered would only provide a latest possible date for the backfill. The largest bones are meat-bearing elements (vertebrae, ribs, scapula, innominate, femur) of beaver, butchered by a chopping tool in a way one would expect in other mammals of comparable size. Likewise, the collected fish remains contain mainly fragment types (ribs, vertebrae) likely to stay in prepared food, and this holds true, to some extent, for the chelonian remains as well (shoulder and pelvic girdle). It seems reasonable, therefore, to interpret these two assemblages taphonomically as consumption refuse s.s. *sensu* Gautier (1987), probably as slowly accumulating day-to-day debris. At least they appear closer to consumption processes than A and B. There are no clear signs that any of the bones recovered were digested.

Among archaeologists working at the site, it is an open question if, during the week, food preparation and provisioning was organized centrally, or if cooking was carried out by each monk individually in its own cell. The pottery found in assemblage D, probably belonging to a single monk, contained cooking devices, and the pre-baroque cell fabric excavated nearby yielded clear signs of a hearth. No hearth structures are present in the standing 17th century cells.

Food economy and rules

Obviously, the monastic rules are bound to have influenced composition, accumulation and spatial distribution of food refuse in a charterhouse. Although in principle no bones of ordinary livestock animals are to be expected among the food refuse of the monks, a small number of them have been found within the contexts studied. However, for interpretation, several elements of uncertainty remain. Amongst them figure both the divergence from prescribed alimentary practices and the presence of people possibly not tied to the strict rules. According to a visitation report from 1656, forty paid servants lived inside the convent (Jaritz 1999). The diet of lay brothers probably did not entirely exclude occasional meat, and lay servants and guests can be assumed to have been completely free of dietary restrictions (Barber and Thomas 2002, 61).

Thus, the possible role of livestock animals like cattle, ovicaprines and pig during the monastic period in general remains unclear, awaiting a more detailed analysis of the associated artefacts. Moreover, the limited number of strictly medieval contexts investigated so far produced only a few animal bones, making a diachronic comparison with the baroque contexts impossible. Most of the written documents from the monastery were either lost in the course of the dissolution or were distributed among different archives and have not yet been investigated, leaving the biological remains and their respective contexts as the most reliable source of information on the former Carthusian diet.

The five faunal samples share the prevalence of fish or "fish-substitute animals" as a common feature, but the differences regarding relative frequency of species are great (Fig. 2). Although not contemporaneous in the strict sense, these assemblages were created during roughly the same period, the first half of the 17th century. Aside from taphonomic factors, the intra-site variability may be due to seasonal or other temporal variations, social differentiation within the monastery, or the availability of certain species. Marked seasonal differences in the relative importance of species groups became clear from an analysis of the account books of the Carthusian monastery of Santa Maria degli Angeli (Rome) for the years 1801–1809. There were reciprocal trends in the amount of animals purchased, *e.g.*, coots, otters, tortoises and frogs (De Grossi Mazzorin and Minniti 1999).

Some of the taxa identified from samples A-D are uncommon in Central European post-medieval settings and justify a comment on their zoological and possible cultural background. The remains of European pond terrapins are known from many prehistoric sites of low-lying areas of eastern Austria, whereas the autochthonous status of recently found animals is questionable. In theory, the terrapins eaten in Mauerbach could have been taken from local populations. If these still existed in the 17th century, they were probably not capable of sustaining the intensive exploitation, indicated by both archaeological evidence and written sources of this period, which might have led to a regional extinction of this species in parts of Central Europe. Documents report a large-scale import of this species from Eastern and Southern Europe into Catholic areas (Kunst and Gemel 2000). Terrestrial tortoises were certainly imported from the Mediterranean area, and their use as food in Central Europe has been apparently ignored because written documents do not discriminate between the two species. However, in Mauerbach, man-chelonian relationships went beyond the mere consumption of purchased animals, as a whole section of the outer garden was reserved for the keeping of these animals. This chelonian garden, which included several ponds, is clearly visible in contemporaneous engravings. According to the 17th century Austrian agronomist Hohberg (1701), terrapin ponds, serving both culinary and decorative purposes, were a widespread feature on rural estates of the local gentry and some efforts were undertaken in the breeding of these animals. A preliminary biometric analysis of the terrapin remains from the monastery points at a rather high size variability of the population consumed, but very young specimens are rare. It is doubtful if the monks succeeded in effectively breeding terrapins. For the lack of comparative data, this question cannot be solved at present. The chelonian garden of Mauerbach probably served for temporal storage or hatching. Its walls and outline could still be identified on aerial photographs from the early 1960's, but it is now mostly destroyed and not amenable to archaeological investigations. A document from the may of 1686 reports the delivery of 472 Chelonia, and 60 animals were still present in 1782, the year of dissolution.

The presence of marine and wild-caught (*e.g.*, sturgeon) freshwater species among the identified fish remains demonstrates that the Mauerbach monks partially depended on a market economy for their fish supply (Fig. 3). However, at least in the Middle Ages, the monastery possessed its own fishing ground on the river Danube, lying just 10 km north of the monastery. Carp and some other freshwater species were probably reared in nearby ponds, one does still exist today. The relative frequency of species from different origins varies among the samples and may be, just like overall species composition, mainly related to taphonomic and contextual causes and sampling techniques. Typically, the single clupeid vertebra, apparently the first one found from an Austrian archaeological site, was collected from the flotation sample E. However, a bias of the size spectra towards large specimens, as observed for several species in samples A and B, has also been reported from other Carthusian sites and may be related to both raised status and logistics in provisioning a large community (Roselló Izquierdo *et al.* 1994; De Grossi Mazzorin and Minniti 1999).

Inter-site comparison

Animal bone assemblages from medieval and post-medieval monastic sites have been enjoying a raised interest by archaeozoologists during the last decade. They provide the rare opportunity to study the culinary habits of a defined group of people whose social and mental background may be, at least to some degree, known. In many cases, an interdisciplinary approach, making use of both written accounts and archaeological evidence, is taken.

A central issue is to consider the ways in which food rules might potentially influence the composition of refuse contexts, and, as a consequence, if any differences from contemporaneous secular sites can be detected through animal bone analyses. As already mentioned, most of the monastic food rules entail an either permanent or temporary abstinence from the meat of warm-blooded terrestrial vertebrates, if not from meat at all. In their most pronounced version, dietary prescriptions might lead to the total absence of animal bones from a context. On the other hand, the remains of animal groups like fish, which were allowed to be eaten during periods such as Lent, may be hopelessly diluted among ordinary food refuse accumulated during the rest of the year. This situation is to be expected if we are dealing with temporarily fasting communities, a situation mandatory for most orders, but also for the Catholic laity until fairly recent times. This was indeed the picture met in the basement of the "Alte Aula", a section of the old university building of Vienna, backfilled at almost exactly the same time period and under similar circumstances as the samples from Mauerbach (Adam and Kunst 1999). Here, isolated finds of fish and terrapin were found scattered amongst huge masses of bones of the major domesticates (Kunst and Gemel 2000, 50). Thus, the mere presence of bones of "Lenten diet animals", or even of raised percentages of fish, may not be specific for a monastic situation at all. In the end, everything depends on the degree of spatial and temporal resolution offered by the sample studied. Ideally, the number of potential food consumers should be limited, and the time period of bone accumulation short. In a regional overview, Ervynck (1997) extensively discusses methodological problems in discriminating the food remains from monastic sites from those of other parts of the society.

A well-defined contextual framework and a reliable dating of the bone assemblages analysed seems therefore indispensable, even in apparently clear-cut settings like Carthusian communities. Medieval and post-medieval animal remains from monasteries of the Carthusian order have been described from La Cartuja, Sevilla (Roselló Izquierdo *et al.* 1994), Santa Maria degli Angeli, Rome (De Grossi Mazzorin and Minniti 1999) and Charterhouse, London (Barber and Thomas 2002). All mentioned studies take an interdisciplinary approach insofar as written documents, relating to the building history of the sites or to the development of culinary habits, are integrated. One further paper by Bielza (1996) on the role of fish within the Carthusian community of El Paular (Spain) is based entirely on archival material. Some influence of the food rules upon the formation of bone assemblages could be observed for all sites mentioned.

As regards the animal species represented, the bone midden from the monastery of Trinità dei Monti (Minims of Saint Francis of Paola), another site in Rome (De Grossi Mazzorin and Minniti 1999), seems of special relevance here. Apparently, apart from Mauerbach, the two samples from Rome are the only other known examples of bone collections from monastic sites containing a high percentage of vertebrate fish-substitute species. The assemblage from Santa Maria degli Angeli, dating from the 17th to the 18th century, yielded 216 identified specimens of vertebrates and shellfish. The remains of otter and fish (53% and 22%, respectively) are the main components, followed by the bones of common domesticated animals (15%). Chelonia and coot are present in low numbers. The preservation pattern of more than 100 otter remains, mainly limb elements with butchering marks, corresponds to the picture observed in the beaver bones of samples D and E from Mauerbach. The sample from Trinità dei Monti, dated into the second half of the 16th century, totalling 1245 bones, is clearly dominated by the remains of Chelonia (mostly *Testudo* and *Emys*), fish and molluscs. The respective frequencies are 38,4%, 17,2% and 42.3%. Tortoises were kept, and probably also bred naturally, inside the garden.

In La Cartuja, an assemblage from a 16th-century cesspool contained 426 fish remains, mostly marine species, associated with an unspecified number of bird and mammal bones (Roselló Izquierdo *et al.* 1994). The special needs of high-status residents or guests are held responsible for the high diversity of this ichthyocenosis. Consumption of Chelonia is known from historical sources but not evidenced archaeologically. In 18th century El Paular, terrapins were kept in a special pond and served to guests of the monastery (Bielza 1996).

In the London Charterhouse, as far as remains from the monastic period (about 1370–1538) are concerned, the relative percentages of fish and of the major domesticates vary according to the areas where the samples were taken. There is evidence both for primary refuse associated with the diet of the monks, where fish is more common, and for the presence of other elements of the Charterhouse community who were not under the same strict dietary regime (Barber and Thomas 2002, 65).

As for the Austrian situation, bone reports from other Carthusian sites are lacking. Short surveys have been carried out on assemblages of monasteries of different orders, as in the case of the Cistercian abbey of Neuberg (Styria), where a few terrapin and fish remains occurred within a 17/18th century rubble layer inside the former kitchen. The material is, however, dominated by domestic mammals. The account books of this monastery have been studied recently, indicating a large-scale supply of terrapins, marine and terrestrial molluscs, and marine and freshwater crayfish throughout the 17th and 18th century (Zeiringer 1991; Kunst and Gemel 2000).

Conclusion

It must be stressed that even five years of continuous excavation, however restricted by the needs of the ongoing renovation works, produced only a limited amount of suitable bone samples and context types from the monastic period, most of which cover only a comparatively short time span. For example, it has not yet been possible to localize and excavate the latrines of the baroque cells. Moreover, the foregoing showed that the process of site formation within, or in close vicinity of, the buildings occurred in a rather irregular, discontinuous, fashion. It is not yet possible to address diachronic, long-term, developments within the dietary regime of the monastery or, for example, to investigate biometrical changes of reared fish like carp, as carried out by Van Neer and Ervynck (1996) for the Flemish Benedictine site of Ename. The selection of the samples A-E, covering a short but important period of the history of Mauerbach, should be understood as a demonstration of the possible ways an animal bone assemblage might look like under the given circumstances.

Fig. 7. Putto *with turtle and crayfish from the main entrance of the Carthusian monastery of Mauerbach; 18th century.*

Generally, the consistent presence of animal resources taken as fish substitutes may be regarded as the most conspicuous feature of the food refuse from Mauerbach. Apparently, this dietary habit was far more widespread than can be evidenced archaeologically, leading to questions about its origin and evolution. Interestingly, De Grossi Mazzorin and Minniti (1999) argue that particular conditions of poverty drove the monks to adopt substitutes for fish in the form of molluscs and Chelonia. Indeed, today, freshwater terrapins and tortoises are valued as a rather low-prestige food throughout large parts of the Mediterranean. However, this attitude is probably not comparable to the one prevailing in 17th century Mauerbach and other Central European settings, where marine molluscs and fish and tortoises, but probably terrapins as well, had to be imported and can be regarded as luxury food. The time period covered by the samples A-E corresponds to an era of unprecedented wealth and prosperity for the monastery, which is indicated by the baroque rebuilding itself. Independent testimony for the sentiment of the monks towards their culinary practices is also provided by stone figures above the main entrance (Fig. 7). It suggests that the adoption of certain animal groups as a food resource was nothing to be ashamed of but, rather, contributed to the social identity of the monastic community. As long as there is no reliable archaeological and documentary evidence for the excessive consumption of Chelonia and *Aquatilia* other than fish significantly predating the time around 1600 AD, it seems reasonable to relate the spreading of these (Mediterranean?) dietary habits among both the clergy and the laity to the process of Counter-Reformation, the decisive religious and political movement in Austria and in other Catholic areas of the Holy Roman Empire throughout the 17th century. It is to be hoped that ongoing excavations will provide additional information regarding the regional diachronic development of dietary practices, both in monastic contexts and beyond.

Acknowledgements
Grateful thanks go to Thomas Kreitner (Vienna) from the excavation team at Mauerbach for providing essential information on the archaeological context of the bone assemblages studied, and to Wim Van Neer (Tervuren) who identified the cod remains from sample B as to their true identity. Both Wim van Neer and Anton Ervynck (Zellik) provided valuable refereeing comments. Figs 4 and 6 were photographed by Rudolf Gold (Vienna), Fig. 5 by ASINOE.

References

Adam, A. and Kunst, G. K. 1999. Aspekte der Tierknochenauswertung in einem urbanen Milieu am Beispiel der Grabung Wien/Alte Aula. *Beiträge zur Mittelalterarchäologie in Österreich* **15**, 157–75.

Barber, B. and Thomas, C. 2002. *The London Charterhouse* (MoLAS Monograph 10). London: Museum of London Archaeology Service.

Bielza, Mª. J. 1996. El pescado en la dieta de los Cartujos de El Paular en el siglo XVI. *Archaeofauna* **5**, 65–77.

De Grossi Mazzorin, J. and Minniti, C. 1999. Diet and religious practices: the example of two monastic orders in Rome between the XVIth and XVIIIth centuries. *Anthropozoologica* **30**, 33–50.

Ervynck, A. 1997. Following the rule? Fish and meat consumption in monastic communities in Flanders (Belgium), pp. 67–81 in De Boe, G. and Verhaeghe, F. (eds), *Environment and Subsistence in Medieval Europe, Papers of the 'Medieval Europe Brugge 1997' Conference Volume 9*. (I.A.P. Rapporten 9). Zellik.

Gautier, A. 1987. Taphonomic groups: how and why ? *ArchaeoZoologia* **1(2)**, 47–52.

Hohberg, W. H. 1701. *Georgica curiosa. Adeliges Land- und Feldleben* (reprint of 1682 edition). Nürnberg: Martin Endtner.

Jaritz, G. 1999. Die Kartäuser von Mauerbach und ihre Geschichte: Spirituelles Leben auf materieller Basis. *Österreichische Zeitschrift für Kunst und Denkmalpflege* **53**, 375–84.

Kreitner, T. 1999. Ein Bericht zum gegenwärtigen Stand der bauarchäologischen Untersuchungen 1996 bis 1999 in der Kartause Mauerbach. *Österreichische Zeitschrift für Kunst und Denkmalpflege* **53**, 411–30.

Kreitner, T. 2001. Ein Vorbericht über die archäologischen Ausgrabungen im Jahr 2000 in der Kartause Mauerbach, NÖ, pp. 207–13 in Krenn, M. (ed.), Bericht zu den Ausgrabungen des Vereins ASINOE im Projektjahr 2000. *Fundberichte aus Österreich* **39**, 181–337.

Kunst, G. K. and Gemel, R. 2000. Zur Kulturgeschichte der Schildkröten unter besonderer Berücksichtigung der Bedeutung der Europäischen Sumpfschildkröte, *Emys orbicularis* (L.) in Österreich. *Stapfia* **69**, 21–62.

Mayerhofer, E. and Pirquet, C. (eds) 1923. *Lexikon der Ernährungskunde*. Wien: Rikola.

Parisse, M. 1994. Zwischen actio and contemplatio – die Orden im 12. Jahrhundert, pp. 391–433 in Vauchez, A. (ed.), *Machtfülle des Papsttums, 1054–1274* (German edition edited by O. Engels). *Die Geschichte des Christentums 5*. Freiburg: Herder.

Roselló Izquierdo, E., Morales Muñiz, A. and Morales, D.C. 1994. La Cartuja/Spain: anthropogenic ichthyocenosis of culinary nature in a paleocultural context. *Offa* **51**, 323–31.

Van Neer, W. and Ervynck, A. 1996. Food rules and status: Patterns of fish consumption in a monastic community (Ename, Belgium). *Archaeofauna* **5**, 155–64.

Zeiringer, J. 1991. *Nahrung und Speisen im Cistercienserstift Neuberg/Mürz. Eine volkskundliche Untersuchung anhand der Küchenrechnungen des 17. und 18. Jahrhunderts*. Unpublished Ph.D thesis, Karl-Franzens University of Graz.

Alfred Galik
Institute of Anatomy,
University of Veterinary Medicine Vienna,
Veterinärplatz 1, A-1210 Vienna, Austria
E-mail: alfred.galik@vu-wien.ac.at

Günther Karl Kunst
VIAS – Vienna Institute for Archaeological Science,
c/o Institute for Palaeontology, University of Vienna,
Althanstr. 14, A-1090 Vienna, Austria
E-mail: guenther.karl.kunst@univie.ac.at

9th ICAZ Conference, Durham 2002
Behaviour Behind Bones, (ed. Sharyn Jones O'Day et al.) pp. 233–243

26. Status as reflected in food refuse of late medieval noble and urban households at Namur (Belgium)

Fabienne Pigière, Ides Boone, Mircea Udrescu, Wim Van Neer and Sofie Vanpoucke

The archaeozoological database available for the late medieval city of Namur (Belgium) allows the analysis of consumption patterns of different social groups and of the expression of their status through diet. Two sites, situated at the confluence of the rivers Sambre and Meuse, are considered for this analysis, i.e., the feudal settlement Château des Comtes *and the middle-class houses at the* Grognon. *Additionally, historical data giving information about some inhabitants of these sites are confronted with the archaeozoological results. The study of the dietary patterns, through the 11th – 15th centuries, shows the potential of documenting status using the proportions of the domestic mammals and their age distribution, as well as the proportion of game and the type of wild species represented. The type and abundance of domestic and wild birds also appear to be linked to status. Amongst the fish, the variety and the proportion of imported marine species seems significant as well as the sizes of the local freshwater species consumed.*

Introduction

The excavations conducted since 1990 within the city of Namur, Belgium (Fig. 1), by the Service de l'Archéologie du Ministère de la Région wallonne en Province de Namur, and the Société Archéologique de Namur, documented several occupation phases in the medieval centre of Namur. These excavations yielded faunal assemblages associated with various social classes in town, covering the time period from the 11th to the 15th century AD. Thanks to a historical study carried out in parallel with the archaeological study, the status of some inhabitants could be documented precisely. In this paper, the data are summarized of several faunal studies that will be published in detail in monographs dedicated to the individual archaeological sites. The fauna from various assemblages is described and used to characterize the food habits of different social groups within the medieval town of Namur.

The archaeological sites and the faunal material

The sites under review are situated at the confluence of two important rivers, the Sambre and the Meuse (Fig. 1). The first site, the *Château des Comtes* is the feudal settlement of the counts of Namur, the other is located in a residential urban area called the *Grognon*.

The Château des Comtes

The castle *Château des Comtes* (castle of the counts) is located on a rocky outcrop and overlooks the *Grognon*. The fortress was the political, military, administrative and juridical centre of the county for four and a half centuries (from the 10th century AD till 1429). It consisted of a complex of private, public and religious buildings, of which little remains today.

A written source of 1198 mentions a chapter of thirteen canons connected to the *Saint-Pierre-au-Château* church, built on the castle hill. They were in charge of the Lord's Prayer services in the castle. In the beginning these canons shared a dormitory. From 1330 onward, however, they lived in separate houses near the church. Eventually, in the second half of the 16th century, the canons were included in the chapter of the *Saint-Aubain* Cathedral in the city, after which their houses on the castle hill were demolished (Antoine 1997; Antoine 1998).

The *Château des Comtes* excavation took place between 1996 and 2000 (Antoine 1997; Antoine 1998;

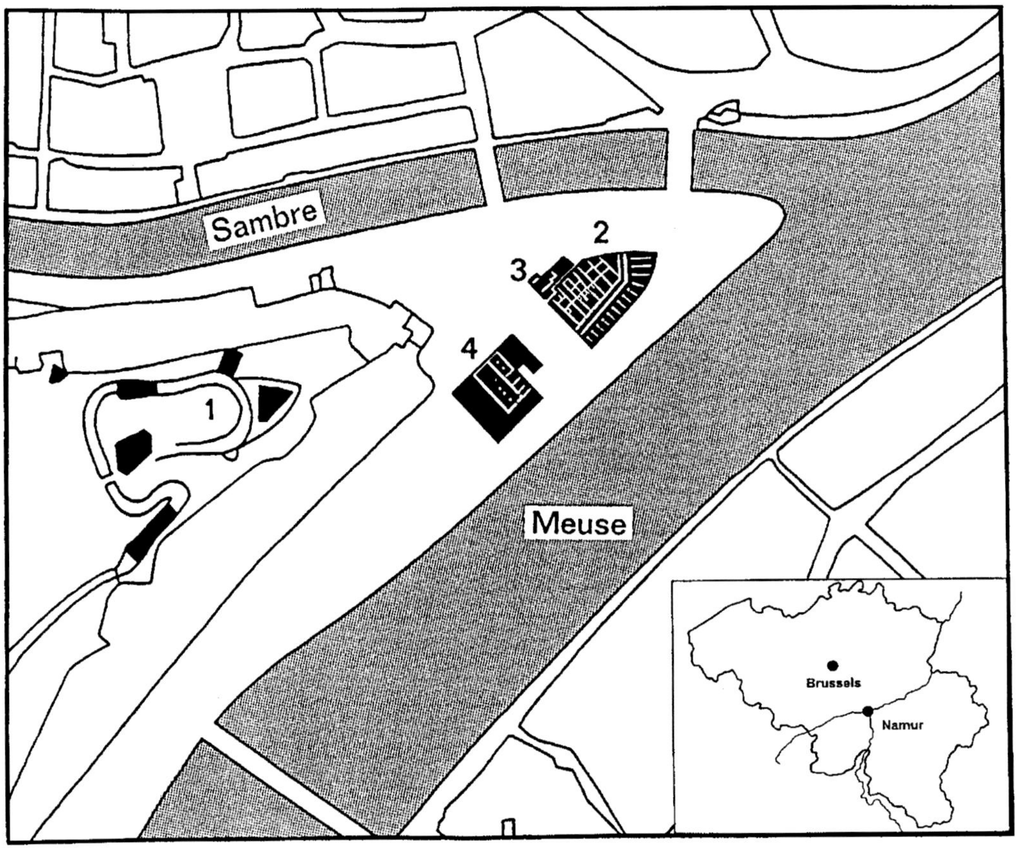

Fig. 1. Namur: the medieval urban area at the confluence of the Sambre and Meuse Rivers. Indicated are the analysed sites Château des Comtes *(1) and the* Grognon *(2), as well as the location of the chapel* Saint-Hilaire *(3) and the* Grand Hôpital *(4).*

Antoine 2000). The area yielded three faunal assemblages that can be associated with the feudal occupation (Fig. 2). The first collection consists of an 11th century garbage dump situated outside the defensive walls (PDC). The other two collections are from filling layers, inside the enclosure of the castle. One of these is dated to the 11th–12th century (CVC), the second originates from the foundation trench of the northern tower of the castle and is dated to the end of the 13th – beginning of the 14th century (TAF). A further assemblage (TDG), dating from between the 14th and the 15th century, has been recovered from a latrine that was probably filled up by the canons living in individual houses near the castle.

The Grognon

Due to its strategic location at the confluence of the rivers Sambre and Meuse, the site of the *Grognon* has been occupied since the middle Mesolithic period and was continuously inhabited from the first century AD onwards. The *Grognon* was the harbour district of the medieval city of Namur, at that time the modest capital of the county with the same name. The harbour was located at the head of the confluence (Fig. 1). The local fish market, called *l'Applé,* was located along the right bank of the river Sambre. Other characteristic topographical elements of the *Grognon* were the chapel *Saint-Hilaire*, built in the centre of the district, and the *Grand Hôpital*, later called the *Hospice Saint-Gilles*, established along the river Meuse and first mentioned in 1270. Finally, some houses were built on both sides of the road *Saint-Hilaire*, which crossed the confluence from west to east (Plumier *et al.* 1997; Bodart 2002). This district has been protected by a wall since the end of the 10th century (Vanmechelen *et al.* 2001).

Rescue excavations, carried out between 1994 and 2000, documented the residential zone located behind the wall alongside the river Meuse (Vanmechelen 1996; Plumier *et al.* 1997; Mees *et al.* 2001–2002). A first occupation phase, dating from the 11th to the beginning of the 12th century, is represented by a stone house (S 25) situated on a large plot. Three faunal collections belong to this phase (Fig. 2). The first one is likely to be associated with the inhabitants of house 25 and is from rubbish layers in the open area of parcel 25. The other two bone collections have been collected from filling layers in house 25 and from rubbish layers close to the urban wall. Both assemblages are to be considered as general urban

Sites	Date (c. AD)	Locus	Type of deposit	Sampling	Number of remains
Château des Comtes					
feudal settlement	11th	PDC	dump	hand	6790
				4, 2, 1 mm (P)	8280
	11th–12th	CVC	filling layers	Hand	7493
	end 13th–early 14th	TAF	filling layers	5 mm	2303
canons	14th–15th	TDG	cesspit	5 mm	2417
Grognon	11th–early 12th	Parcel 25	refuse layers	hand	3735
	2nd half 11th–early 12th	House 25	filling layers	hand	1322
	11th–12th	Urban wall	refuse layers	hand	255
				4, 2, 1 mm (P)	171
	2nd half 12th–13th	C 62	cesspit	5 mm	2812
				4, 2 mm (P)	554
	end 13th–mid 14th	Parcel 5–6	filling layers	hand	290
	end 13th–mid 14th	Cellar 6	filling layers	hand	1936
	13th–14th	C 57	cesspit	hand	1028
				4, 2, 1 mm (P)	1735
	14th–15th	C 40	cesspit	5 mm	9582
				4, 2, 1 mm (P)	18896
	15th	C 40	filling layers	hand	494
	15th	C 31	cesspit	5 mm	613
				4, 2 mm (P)	736

Fig. 2. Overview of the contexts studied from the Château des Comtes *and from the* Grognon. *P, partial.*

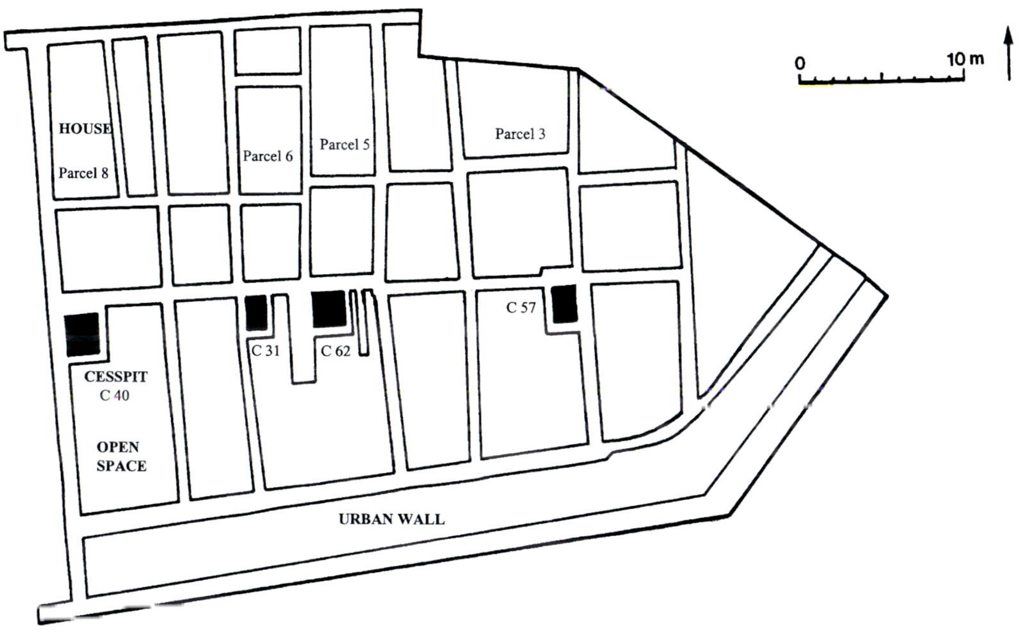

Fig. 3. Plan of the residential area of the Grognon. *Indication of the parcels with a stone house at the front and an open space with a cesspit behind.*

consumption refuse, because they probably come from a wider area of the *Grognon*. The second occupation phase, started in the first half of the 12th century, is represented by nine plots that were discovered during the excavation campaign (Fig. 3). All these parcels present a similar plan: at the front stood a house made of stone, behind which there was an open space, which could have been a garden or a yard. Most of these houses had their private cesspit built against the rear wall. The archaeozoological material from four cesspits (C 62, C 57, C 40 and C 31), belonging to four different houses, has been analysed. The utilisation layers of these latrines cover the period from the second half of the 12th century till the 15th century (Fig. 2). Again, the other assemblages are to be considered as

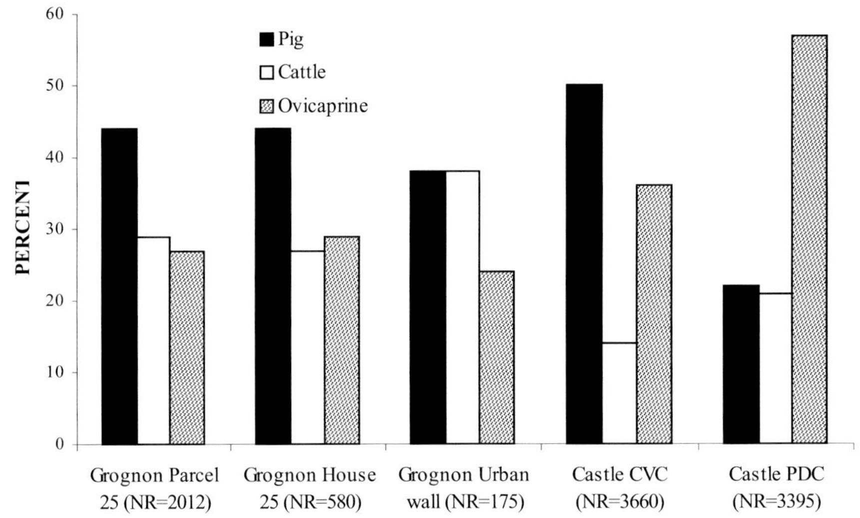

Fig. 4. Relative frequency of the remains of the three main meat suppliers in the 11th – 12th c. AD at the castle and the Grognon.

consumption refuse from the *Grognon*, since the faunal material cannot be associated with a specified house or plot. Two bone collections have been excavated from filling layers (Cellar 6 and Parcel 5-6), both dated to the end of the 13th - first half of the 14th century. Finally, one collection consists of 15th century filling layers from a cesspit (Filling layers C 40). The filling layers of plot 5–6 could also contain some refuse linked with the end of the occupation of houses 5–6.

According to archaeological data and written sources (Mees *et al.* 2001–2002; Bodart 2002), the inhabitants of these houses were middle-class people that practised independant professions. Several of them were merchants, bargemen or master fishers. Moreover, written sources from the 14th century mention that some of the owners rose to the level of the lower nobility as a result of social promotion. One of the inhabitants of the *Grognon*, probably a financier, is known as belonging to the count's entourage. Canons connected to the *Saint-Pierre-au-Château* have been mentioned as owners of houses of this sector during the 15th century (Bodart 2002). The faunal material from cesspits C 62, C 40 and C 31 as well as contexts TAF and TDG from the *Château des Comtes* were collected by sieving the sediment through a large, 5 mm, mesh (Fig. 2). The material from the other contexts has been collected by hand. Samples taken from the garbage dump of the castle (PDC), the rubbish layers close to the urban wall, and from all of the cesspits from the *Grognon* were sieved accurately using a 4 mm, 2 mm and 1 mm mesh.

Consumption at the *Château des Comtes* and at the *Grognon*

The livestock

During the 11th – 12th century, the remains of pigs were predominant amongst the three main meat suppliers at both the *Grognon* and the castle (Fig. 4). Only the garbage dump from the castle (PDC) is characterized by a preponderance of remains of sheep/goat. This can be explained by the specific nature of this context, as it consists mainly of sheep/goat skulls and mandibles, perhaps the result of a primary butchering process. Within the remains of pigs, we also found a preponderance of cranial fragments, while scapulae are over-represented amongst the cattle remains. In general, the differences between the castle and the *Grognon* are to be found in the two other meat suppliers, namely the cattle and the sheep/goat-group. The proportion of sheep/goat exceeds that of cattle in the filling layers of the castle (CVC), whereas at the Grognon the percentages of the two species are more or less equivalent.

At the end of the 13th – beginning of the 14th century, pigs are still predominant at the castle, while the proportions of cattle and sheep/goat are almost identical (Fig. 5). There is also an increase of cattle remains in the material from the filling layers of the *Grognon*. It reaches a proportion identical to that of pigs (approximately 40%), whereas the remains of sheep/goat represent only 15%. As for the cesspits at the Grognon, it appears that, between the second half of the 12th century and the 15th century, the remains of pigs were predominant among the three main meat suppliers (Fig. 6). Compared to the contemporary material from the filling layers of the *Grognon* (Fig. 5), the proportion of cattle bones is less important (roughly 20%), favouring that of pig remains. The under-

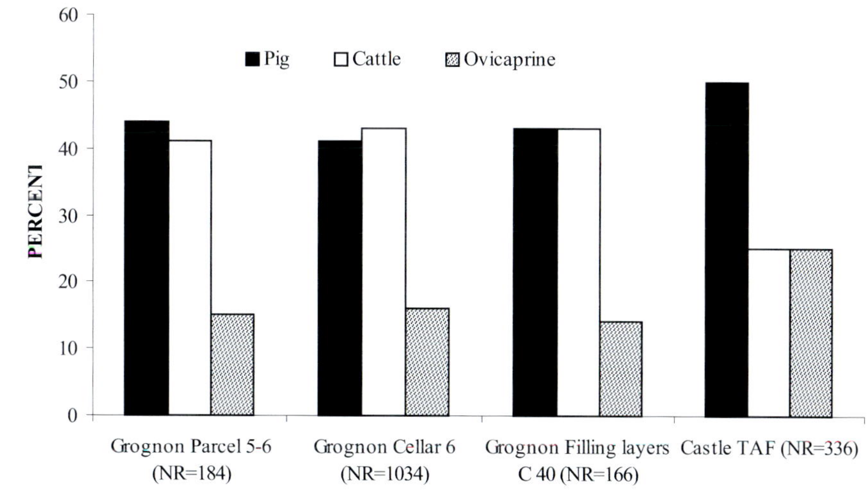

Fig. 5. Relative frequency of the remains of the three main meat suppliers in the 13th – 15th c. AD at the castle and the Grognon.

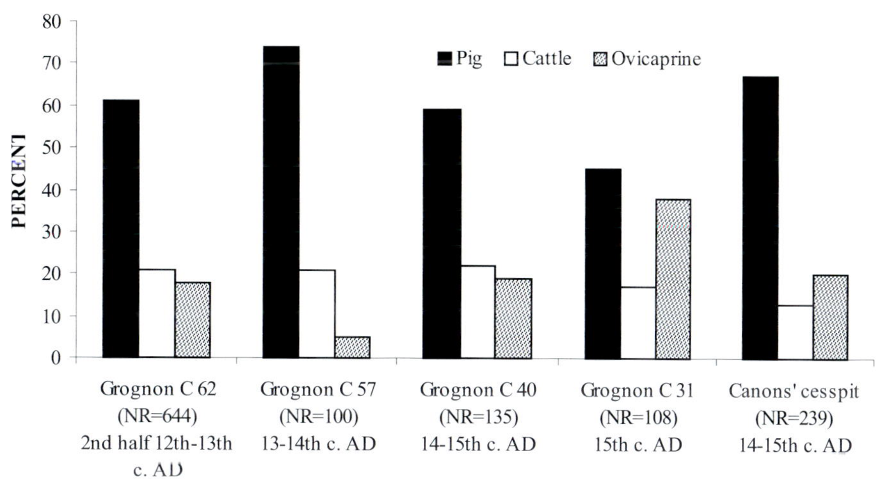

Fig. 6. Relative frequency of the remains of the three main meat suppliers from the Grognon*'s latrines and from the canons' cesspit in the 2nd half of the 12th – 15th c. AD.*

representation of cattle bones could be related to pre-depositional selection, connected with the context's nature, aiming to reduce the rapid infilling of the cesspits with large bones. All skeletal elements of the main meat suppliers are represented. The contents of the cesspits seemed not, therefore, to be exclusively composed of one particular type of refuse (*e.g.*, only table refuse). In order to evaluate the selection process of the cesspits' refuse, contexts of different nature are needed, for example a garbage dump *versus* cesspits, which could be associated with the same occupants.

The diet of the canons could only be evaluated with material from a 14th – 15th century cesspit. In this context pig remains are also largely predominant (Fig. 6).

During the whole period considered for this paper, the small proportions of skull fragments, metapodia and phalanges of cattle (less than 10 % of cattle remains) and of sheep/goat (less than 5 %) in the faunal material from the filling layers of the castle seem to indicate selective consumption towards meat-bearing elements. A comparison of the age distribution by means of epiphysial fusion of the long bones, between the filling layers of the castle and those of the *Grognon*, indicates a higher proportion of juveniles at the castle. This is particularly clear for the remains of pig, which is the predominant species (Fig. 7). During the whole period, the percentage of bones non-fused at one year is much higher for material from the castle.

	Grognon												Castle							
	Parcel 25 (NR=158)				House 25 (NR=46)				Cellar 6 (NR=81)				CVC (NR=371)				TAF (NR=41)			
	F	%	NF	%	F	%	NF	%	F	%	NF	%	F	%	NF	%	F	%	NF	%
1 year	48	**92**	4	**8**	4	**57**	3	**43**	18	**82**	4	**18**	48	**41**	68	**59**	4	**57**	3	**43**
2–2 1/4 years	32	**39**	51	**61**	2	**7**	28	**93**	9	**18**	41	**82**	19	**10**	165	**90**	2	**7**	28	**93**
3 1/2 years	–	–	23	**100**	–	–	9	**100**	3	**21**	11	**79**	6	**8**	65	**92**	1	**25**	3	**75**

Fig. 7. Epiphyseal fusion age data for pig. Number of remains and proportion of fused bones (F) and non-fused bones (NF) at the Grognon versus *the* Château des Comtes. *The table shows that the percentage of bones not fused at one year is much greater at the castle during the whole period.*

Poultry

The presence of peacock bones during the 11th – 12th centuries is remarkable. We found some bones at the castle, but also at the *Grognon* where a single bone has been found in the layers of rubbish of parcel 25. During the Middle Ages, peacocks were definitely considered as prestigious. These birds were kept and reared for the banqueting tables of the upper-class. They were not always eaten but often served decorated with their own plumage, to impress guests (Ervynck 1992, 153; Lauwerier and Laarman 1996, 9–10; Clavel 2001, 110–1). Chicken was the dominant species at the castle and at the *Grognon*, both in the filling layers and in the cesspits. In contrast with the *Grognon* and the castle, the canons' cesspit shows a high proportion of ducks within the poultry. The same phenomenon has been observed in religious contexts in northern France (Clavel 2001).

Game

In the 11th – 12th centuries, remains of game are found in very small quantities at both the castle and the *Grognon*. At the *Grognon*, in the rubbish layers of parcel 25, bones of red deer, roe deer and hare have been identified. Although the ratio of game reached only 1% in the castle, its species richness is obvious compared to the *Grognon*. In addition, many species were considered as noble game, in particular grey heron, capercaillie, red deer, roe deer and – among the most prestigious animals – wild boar and brown bear (Dam 1953; Ervynck 1991).

Two prestigious birds, namely swan and heron, have been discovered in the castle's faunal assemblage dated to the end of the 13th – beginning of the 14th century. Moreover, hare and rabbit were also consumed at the castle. Remarkably, in the contemporary contexts of the *Grognon*, namely the filling layers of cellar 6 and of parcel 5–6, remains of game are responsible for as much as 6% and 3% of all mammals identified. These contexts contained bones of hare but also noble game: red deer, roe deer and wild boar. Among the cesspits of the *Grognon*, only the 14th – 15th century cesspit C 40 yielded a moderate percentage of game within its utilisation and filling layers (3% and 4%). The species present comprise hare, rabbit and red deer.

Fish

Concerning the consumption of fish, it appears that freshwater species were predominant at the castle as well as at the *Grognon*. These were mainly composed of cyprinids, typical of the Meuse basin. However, the average size of the fish consumed by the nobility and the canons was larger than at the *Grognon*. Fig. 8 illustrates the differences in the reconstructed sizes of salmonids, pike and perch. These three species reached larger sizes at the castle than at the *Grognon*. Once more, it appears that the average size of the freshwater fish found in the canons' cesspit was larger than those from the latrines of the *Grognon* (Fig. 9). Furthermore, a higher proportion of perch has been found at the castle. This tasty fish was appreciated, as mentioned in a 15th century written source of Namur (Thomas 1994). Sturgeon remains have been discovered only in the 11th – 12th century context of the *Château des Comtes*. This rare and precious fish was highly esteemed by the elite and sometimes offered to impress high-ranking guests (Lampen 2000).

Marine fish already occur in the 11th century levels, but consisted only of herring at both sites. However, the relative frequencies differ between both sites (Fig. 10). At the castle, 24% of all the fish remains were identified as herring. One context of the *Grognon* (Urban Wall) contained 31% herring bones, but the sample is very small and the ratio may therefore not be significant. In the two later dated contexts of *Grognon*, the marine species, represented by herring, played virtually no role.

Since the faunal assemblage of the 13th – 14th century context from the castle (TAF) has only been sieved through the large 5 mm mesh, small-sized fish are underrepresented. On the other hand, the samples from the cesspits of the *Grognon* were accurately sieved through fine meshes. These contexts yielded mainly remains of fish with an average size smaller than 10 cm standard length (SL), mostly from the 2 mm sieved samples. The almost exclusive presence of small-sized fish probably cannot be explained in relation to the excrements from the cesspits. On the contrary, the excellent state of preservation of the fish bones suggests that few of them were digested. In addition, the other consumption remains confirm that the cesspits were used as garbage pits.

Consumption of domestic carp has been attested for

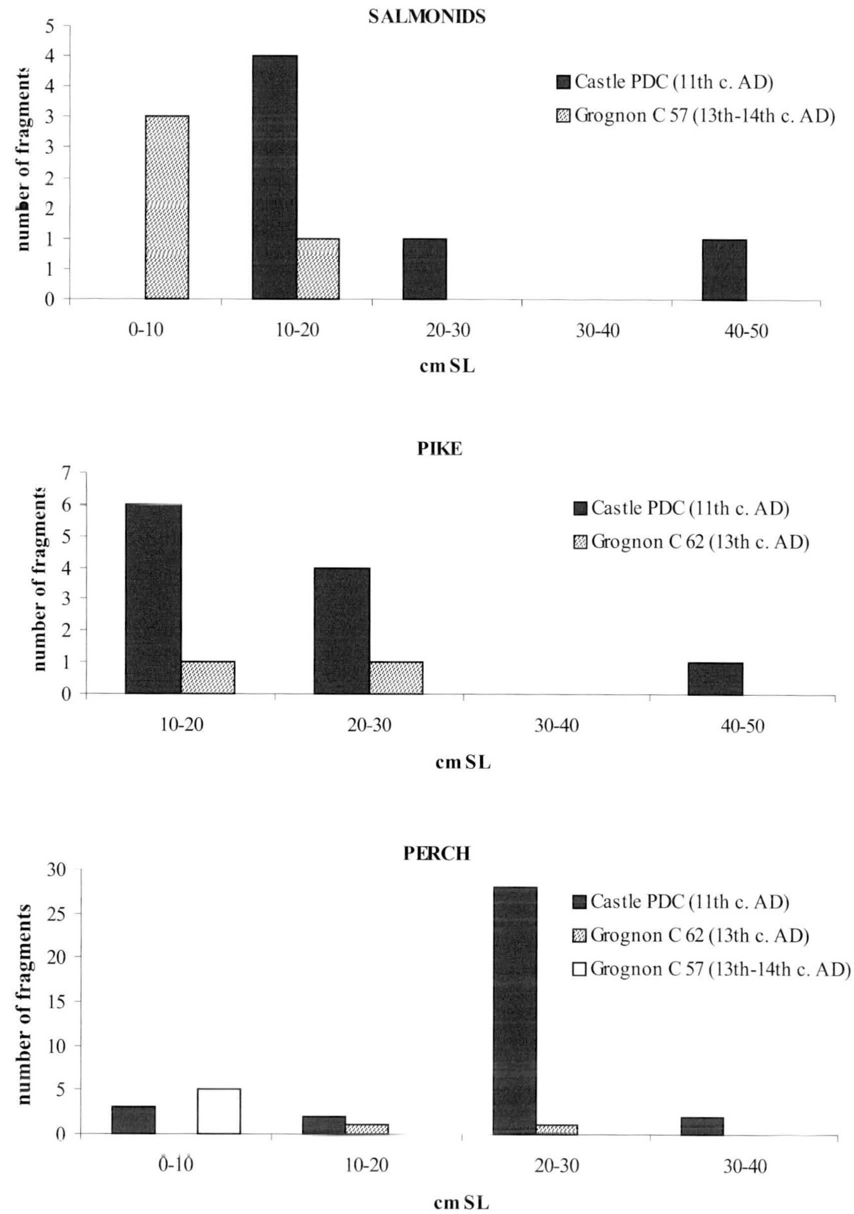

Fig. 8. Comparison of the reconstructed size of salmonids, pike and perch at the castle and at the Grognon *in the 11th – 14th c. AD (SL: Standard length in cm).*

the first time in filling layers of the castle dated to the end of the 13th – beginning of the 14th century (Fig. 11). The oldest remains of carp in Belgium have been discovered at the castles of Laarne and Londerzeel in Flanders (Van Neer and Ervynck 1994), both contemporary with the castle context of Namur. No remains of this fish were found at the *Grognon* in this early period. Even for the later periods carp remains are scarce in the assemblages of the *Grognon*. A 15th century historic source of Namur mentions it as quite expensive, worth half the daily salary of a watchman (Thomas 1994).

At the end of the 13th – beginning of the 14th century, at least five different marine species were consumed at the castle: herring, three gadids (cod, haddock and whiting) and flatfish (plaice/flounder/dab) (Fig. 11). The canons' cesspit contained a similar diversity of marine species. At the *Grognon*, however, herring was the main marine fish until the 16th century, and gadids and flatfish played virtually no role in the overall marine fish consumption. Moreover, the amount of marine fish does not exceed 5%, except for cesspit C 31, where a proportion of 34% of marine fish was observed (Fig. 12). The latter cesspit indicates the rise of marine fish consumption at the *Grognon* from the 15th century onwards. The castle and

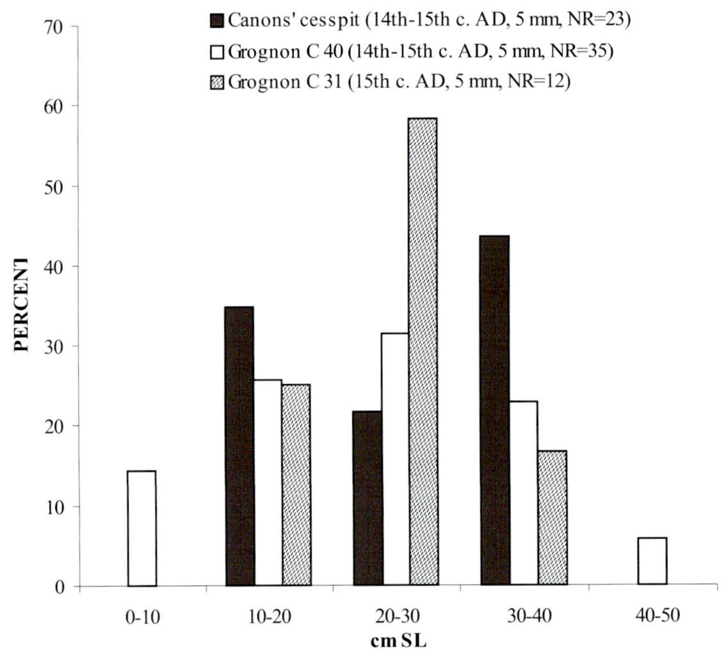

Fig. 9. Reconstructed size of the freshwater fish from the canons' cesspit and from the cesspits of the Grognon in the 14th – 15th c. AD (SL: Standard length in cm).

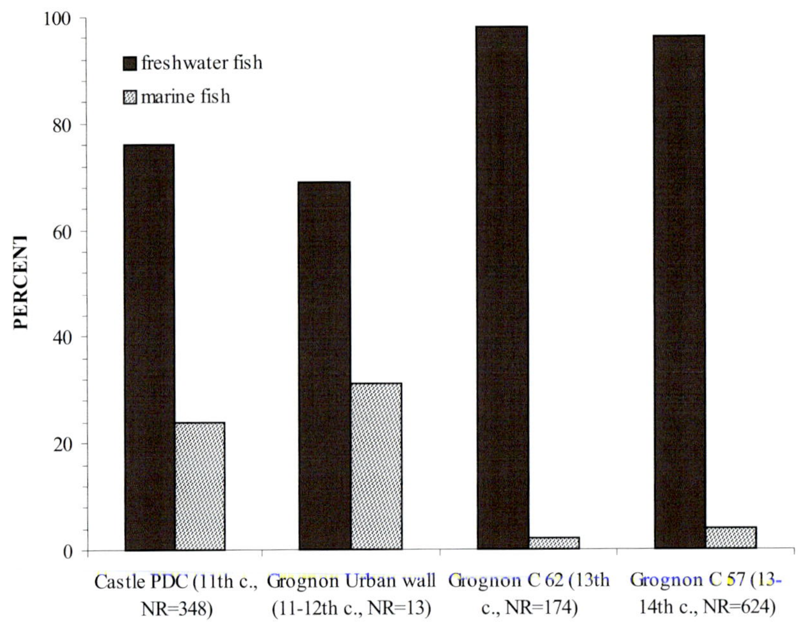

Fig. 10. Proportions of freshwater fish and marine fish at the castle versus *the* Grognon *in the 11th – 14th c. AD (remains from the 2 mm collections).*

the canons' cesspit contained proportions of approximately 40% and 50% marine fish. This fish bone material was retrieved by sieving the sediment through the large 5 mm mesh. In Fig. 11 and Fig. 12, it was not possible to confront the proportions of marine and freshwater fish from contexts of the castle *versus* those from the *Grognon*, based only on the collections from the sediment sieved at 5 mm, since the *Grognon* yielded few fish bones in the 5 mm samples.

	Grognon				Castle	
	C 62 (NR=200) 5+4+2 mm 13th c. AD	C 57 (NR=626) 5+4+2 mm 13th–14th c. AD	C 40 (NR=4672) 5+4+2 mm 14th–15th c. AD	C 31 (NR=497) 5+4+2 mm 15th c. AD	TAF (NR=315) 5 mm 13th–14th c. AD	Canons (NR= 92) 5 mm 14th–15th c. AD
freshwater fish	98	95	98	66	58	49
carp	–	–	P	P	3	–
herring	2	5	2	34	5	1
cod	–	P	–	P	4	7
haddock	–	–	–	–	13	1
whiting	–	–	–	–	1	–
gadids	–	–	P	–	3	3
flatfish	–	P	P	–	13	39

Fig. 11. Proportions of freshwater fish, carp and different species of marine fish at the castle and the Grognon *in the 13th – 15th c. AD. The differences in the volume of the sediment sieved for the same context were taken into consideration. P indicates species that are represented by only few bones.*

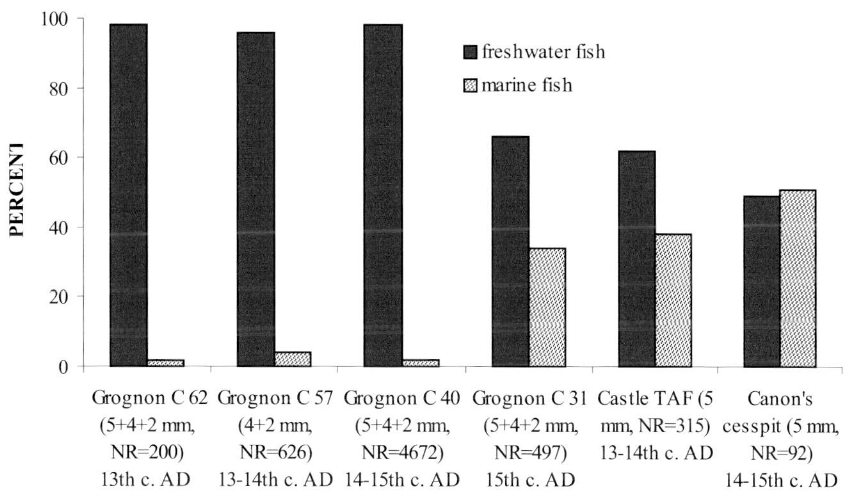

Fig. 12. Proportions of freshwater fish and marine fish at the castle and the Grognon *in the 13th – 15th c. AD. The differences in the volume of the sediment sieved for the same context were taken into consideration.*

The relative proportions from the cesspits of the *Grognon* have been calculated, therefore, by adding the 5, 4 and 2 mm sieving samples together, after being previously brought back to a comparable volume.

In addition, it is worth mentioning that many mussels were identified from the canons' cesspit of the 14th – 15th century. At the *Grognon*, finds of these marine molluscs are scarce until the 15th century.

Consumption refuse and social status

The dietary patterns of the nobility reflect their privileged social status. This pattern is particularly illustrated by those goods obtained by long-distance trade: the nobility consumed more marine fish, composed of varied species. At the *Grognon*, the households consumed only small quantities of herring, a cheap fish that middle-class people could easily afford (Lampen 2000). In general, the quality of the freshwater fish was also higher at the castle; the species spectrum is richer and the reconstructed size of the freshwater fish is larger. The presence of prestigious birds such as peacock again indicates the high standard of living of the nobility.

The variety of wild birds and mammals attested in the faunal assemblages from the castle reflects the exclusive hunting rights of the feudal lords. This was especially the case for large game. Written sources, in relation to hunting rights in postmedieval times in the county of Namur, mention this noble privilege and the gravity of offences against the game-act. These offences were assimilated with the worst crimes because they were considered as offences against the public authority represented by the feudal lords (Lelièvre 1870). Moreover, the upper-class status of the nobility from the *Château des Comtes* can be demonstrated thanks to the presence of wild boar and brown bear. Hunting these dangerous animals required a great number of men and trained hounds. The lower nobility probably could not afford this kind of hunting (Dam 1953; Ervynck 1991). It is noteworthy, however, that the inhabitants of the *Grognon* also had access to game, and even to noble game, during the entire period considered in this paper. Maybe those people obtained this game as a gift? Another possibility is that these wild animals were poached. Besides, the ennobled bourgeois, attested at the *Grognon* since the 14th century, could have obtained game by hunting them on their estates, at least if they were land-owners. Furthermore, the settling of the nobility in the town, related to the presence of the count, is not completely ruled out (Bodart, *pers. comm*).

Finally, concerning the consumption of the three main meat suppliers, it appears that the refuse associated with the nobility corresponds mainly to the best meat-bearing parts of the animals and that they had the opportunity to eat juvenile animals more frequently than at the *Grognon*. Nevertheless, since the 11th century, pig is predominant in the food remains of both the nobility and the inhabitants of the *Grognon*. Yet, it appears that in several regions a high level of pork consumption is characteristic of the feudal elite (Ervynck 1992; Audoin-Rouzeau 1995; Clavel 2001). This could be explained as a result of the privileges of the feudal lords who had a monopoly on the exploitation of the forests or also by the preference of the nobility for the prestigious pork meat (Ervynck and De Meulemeester 1993). At Namur, the reason for the high level of pork consumption among the middle-class people of the *Grognon* is maybe an ecological one. Indeed, the wooded environment around Namur is very favourable for herding pigs. Pig keeping within the city of Namur should also be taken into account. Written sources of the 14th and 15th centuries mention the presence of pigsties at Namur (Bodart, *pers. comm.*). However, the archaeozoological data of the neighbouring site, the *Hospice Saint-Gilles* (De Cupere and Van Neer 1993), suggest that the consumption of pigs in such high proportions at the *Grognon* could be related to the social status of its inhabitants. The consumption refuse from the *Hospice Saint-Gilles* could not be associated with one particular social class. The faunal material consists of numerous small assemblages, which had to be regrouped by period. In spite of that, this material gives a global indication of the consumption at Namur. It appears that between the 11th and the 15th centuries, the remains of cattle are predominant (approximately 40%), while pig bones represent about 30%.

Concerning the canons' diet, as it appears through the analysis of the contents of one cesspit, we also observed a high level of pork consumption among the three main meat suppliers. The food supply of the canons differs chiefly from that of the *Grognon* by its superior quality of marine and freshwater fish, comparable to the diet of the nobility.

Conclusion

The archaeozoological analysis of the food remains provides interesting information concerning the reflection of status through the diet of various social groups within the medieval city of Namur. The consumption patterns of the nobility clearly reflect their privileges. The high level of pig consumption, thoughout the entire period considered, is not however a characteristic of the nobility alone, unlike the situation observed in many other regions. The data available seem to indicate that the high level of pig consumption at the *Grognon* could be related to a certain wealth of the inhabitants, or to environmental conditions. In addition, consumption of game is probably indicative of inhabitants with a high social status at the *Grognon*, since the 11th – early 12th century. Differences of status between the middle-class people of the *Grognon* and the nobility from the *Château des Comtes* have been clearly pointed out through differences in fish consumption.

Namur is the first city in southern Belgium where a comparative study of the dietary patterns of several social classes has been carried out. However, more adequately sampled material will be necessary to confirm and refine the trends described here. This will also reduce the possible weakness of this study due to the variable taphonomic characteristics of the faunal assemblages. Future analyses of faunal collections from other social classes of the city should allow a more precise evaluation of the different factors that have an impact on the food supply of the inhabitants of Namur, and may help to document additional indicators of social status.

Acknowledgements

The site of the Grognon has been excavated under the supervision of J. Plumier, by N. Mees, C. Robinet and R. Vanmechelen (*Service de l'Archéologie du Ministère de la Région wallonne*). The excavation of the *Château des Comtes* has been carried out by J.-L. Antoine (*Musée archéologique de Namur*). We would like to thank Wim Wouters of the Royal Museum of Central Africa for sieving and sorting out enormous quantities of sediment. This contribution presents the research results of a Belgian Programme on Interuniversity Poles of Attraction initiated by the Belgian Government, Prime Minister's Office, Science Policy Programming.

References

Antoine, J.-L. 1997. Le château des comtes à Namur. Recherches de 1996, pp. 109–13 in Plumier, J., Plumier-Torfs, S. and Duhaut, C. (eds), *Actes de la Cinquième Journée d'Archéologie Namuroise*. Namur.

Antoine, J.-L. 1998. Le château des comtes à Namur. Recherches de 1997, pp. 111–6 in Plumier, J., and Duhaut, C. (eds), *Actes de la Sixième Journée d'Archéologie Namuroise*. Namur.

Antoine, J.-L. 2000. Le château des comtes à Namur. Recherches de 1999, pp. 7–9 in Plumier, J., Plumier-Torfs, S. and Duhaut, C. (eds), *Actes de la Huitième Journée d'Archéologie Namuroise*. Namur.

Audoin-Rouzeau, F. 1995. Compter et mesurer les os animaux. Pour une histoire de l'élevage et de l'alimentation en Europe de l'Antiquité aux Temps Modernes. *Histoire et Mesure* **X- 3/4**, 277–312.

Bodart, E. 2002. Namur. Grognon : Rapport d'enquête historique. Unpublished Région Wallonne Report, Archéologie namuroise.

Clavel, B. 2001. L'animal dans l'alimentation médievale et moderne en France du Nord (XIIe-XVIIe siècles). *Revue Archéologique de Picardie, Numéro Spécial* **19**, 1–204.

De Cupere, B. and Van Neer, W. 1993. La faune du site de l'Hospice Saint-Gilles à Namur : résultats préliminaires, pp. 87–92 in Corbiau, M. H. and Plumier, J. (eds), *Actes de la Première Journée d'Archéologie Namuroise*. Namur.

Dam, J. H. 1953. *Het Jachtbedrijf in Nederland en West-Europa*. Zutphen: Thieme & Cie.

Ervynck, A. 1991. 'De beer die woonde op de burg ...': mens en dier in een vroegmiddeleeuwse versterking, pp. 170–80 in De Witte, H. (ed.), *De Brugse Burg. Van Grafelijke Versterking tot Moderne Stadskern* (Archaeo-Brugge 2). Brugge: Die Keure NV.

Ervynck, A. 1992. Medieval castles as top-predators of the feudal system: an archaeozoological approach. *Château Gaillard* **15**, 151–9.

Ervynck, A. and De Meulemeester, J. 1993. La viande dans l'alimentation seigneuriale et la variété des terroirs : l'exemple des pays-Bas méridionaux, pp. 36–41 in Colardelle, M (ed.), *L'Homme et la Nature au Moyen Age. Paléoenvironnement des Sociétés Occidentales*. Paris: Errance.

Lampen, A. 2000. *Fischerei und Fischhandel im Mittelalter. Wirtschafts- und Sozialgeschichtliche Untersuchungen nach Urkundlichen und Archäologischen Quellen des 6. bis 14. Jahrhunderts im Gebiet des Deutschen Reiches* (Historische Studien 461). Husum: Matthiesen Verlag.

Lauwerier, R. C. G. M. and Laarman, F. J. 1996. Slachtafval en haute cuisine. Bot uit een beerput van het kasteel van Valkenisse (begin 15de eeuw). *Nehalennia* **108**, 8–11.

Lelièvre, X. 1870. Droit de chasse au comté de Namur. *Annales de la Société archéologique de Namur* **11**, 359–76.

Mees, N., Robinet, C. and Vanmechelen, R. 2001–2002. Namur. Grognon: Dossier d'étude archéozoologique. Unpublished Région Wallonne Report.

Plumier, J., Mees, N. and Vanmechelen, R. 1997. Elements de topographie urbaine au Bas Moyen Age, pp. 487–91 in Corbiau M. H. (ed.), *Le Patrimoine Archéologique de Wallonie*. Namur: Ministère de la Région Wallonne D.G.A.T.L.P.

Thomas, F. 1994. Hygiène, approvisionnement en eau et gestion hydrographique à Namur au XVème siècle. *Annales de la Société Archéologique de Namur* **68**, 235–305.

Vanmechelen, R. 1996. Aisances et dépotoirs: structures domestiques d'un quartier namurois (Le Grognon, 12e–18e siècle), pp. 89–104 in Plumier, J. and Corbiau, M. H. (eds), *Actes de la Quatrième Journée d'Archéologie Namuroise*. Namur.

Vanmechelen, R., Mees, N., Robinet, C. and Plumier, J. 2001. De la berge à l'enceinte: évolution du bord de Meuse au Grognon à Namur (5e–11e siècle) (Nr). *Archaeologia Medievalis* **24**, 125–8.

Van Neer, W. and Ervynck, A. 1994. New data on fish remains from Belgian archaeological sites, pp. 217–29 in Van Neer, W. (ed.), *Fish Exploitation in the Past. Proceedings of the 7th Meeting of the ICAZ Fish Remains Working Group* (Annales du Musée Royal de l'Afrique Centrale, Sciences Zoologiques 274). Tervuren.

Fabienne Pigière, Ides Boone, Mircea Udrescu, Wim Van Neer
Royal Museum of Central Africa, IUAP 05/09
Leuvensesteenweg 13, B-3080 Tervuren, Belgium
E-mail: fpigiere@africamuseum.be
mudrescu@africamuseum.be

Wim Van Neer
Katholieke Universiteit Leuven,
Laboratory of Comparative Anatomy and Biodiversity,
Ch. de Bériotstraat 32, B-3000 Leuven, Belgium
E-mail: vanneer@africamuseum.be

Sofie Vanpoucke
Katholieke Universiteit Leuven,
Department of Archaeology, Erasmushuis,
Blijde-Inkomststraat 21, B-3000 Leuven, Belgium
E-mail: sofie.vanpoucke@africamuseum.be

9th ICAZ Conference, Durham 2002
Behaviour Behind Bones, (ed. Sharyn Jones O'Day et al.) pp. 244–251

27. Food, status and formation processes: a case study from medieval England

Jonathan C. Driver

Zooarchaeological studies of social differentiation typically examine the composition of assemblages, and link variation in assemblage composition to social status or ethnicity. Faunal assemblages excavated from a single medieval property in Southampton (U.K.) are quite variable in species composition. The context of the assemblages suggests that such variability was not caused by differences in the social status of the people who deposited the assemblages. Instead, it appears that there were changes in the preparation, consumption and disposal of food from earlier to later medieval periods. As a result, earlier medieval assemblages from different contexts tend to be quite homogenous, whereas later medieval assemblages from different contexts on the same property are heterogenous. More extensive comparative studies would be required to establish whether these changes were widespread, or whether they reflect the behaviour of a wealthy middle class.

Introduction

Zooarchaeological studies of social organization generally rely on a form of spatial analysis known as compositional patterning (Ferring 1984). In this approach, the analyst examines assemblages from spatially discrete locations or contexts that are thought to have been used by identifiable groups or individuals of a particular social status. The assemblages are described in terms of their composition. Differences in composition may be interpreted as a reflection of different diets or differential access to resources, and are therefore often regarded as evidence for different social status or ethnic affiliation. When considering social status it is usually assumed that social rules govern the access of individuals to certain resources. When considering ethnicity it is assumed that individuals conform to certain customs or prohibitions concerning the preparation and consumption of food.

The description of assemblage composition may be based on relative frequency of species, but could also include discussion of relative frequency of skeletal elements, frequency of butchery techniques, or age at death profiles of particular species. For example, it is widely recognized that in market economies differential prices restrict consumption of certain species or carcass portions to people with adequate incomes. In some hunter-gatherer societies there are rules about who receives certain parts of an animal when it is divided and shared. Animals of different ages may be accorded different value – for example, young animals are often regarded as a delicacy, and may therefore be restricted to certain people.

Recently published textbooks of zooarchaeology make minimal reference to the use of animal remains as markers of social processes (*e.g.*, Reitz and Wing 1999, 332; O'Connor 2000, 167–69). This is understandable, because zooarchaeological studies are still dominated by considerations of diet and economy. However, there is a growing body of literature that not only demonstrates the potential of this "sociological" approach, but contributes significantly to understanding the nature of past social organization.Crabtree (1990) reviewed much of the pioneering literature concerned with deducing social organization from zooarchaeological data in complex societies (mainly chiefdoms and states). Many of the examples she cited utilized compositional patterning. Since then, zooarchaeologists have tried to incorporate social and ideological dimensions in their analyses of faunal remains from societies with a wider range of sociopolitical circumstances. Examples in which the analysts used compositional patterning include Marshall (1993), Jackson and Scott (1995), Potter (1997 and 2000), and Muir and Driver (2002).

Obviously, a potential problem with compositional

analysis is that many factors contribute to the composition of animal bone assemblages (Lyman 1994). Therefore, differences between assemblages of the same time period are not necessarily due to social or ethnic differences between the people who deposited the assemblages. One obvious confounding variable is the post-depositional history of assemblages – well-preserved assemblages may have a different composition from those that have been subject to numerous destructive processes, simply because fragile bones survive better within the former group. However, even if assemblages have similar taphonomic histories, one cannot assume that significant differences in the composition of contemporaneous assemblages necessarily relate to social or ethnic factors.

This paper considers a case study from medieval England that provides an example of how different waste disposal practices create strikingly different assemblages of animal bones, even though different assemblages were apparently deposited by the same people. In this example, differences between assemblages from different contexts probably do not reflect social differences between people depositing the assemblages. However, the very fact that assemblages from the same time period are differentiated may indicate a change in economic and social organization from the preceding period when homogeneity of assemblage composition was normal (see further).

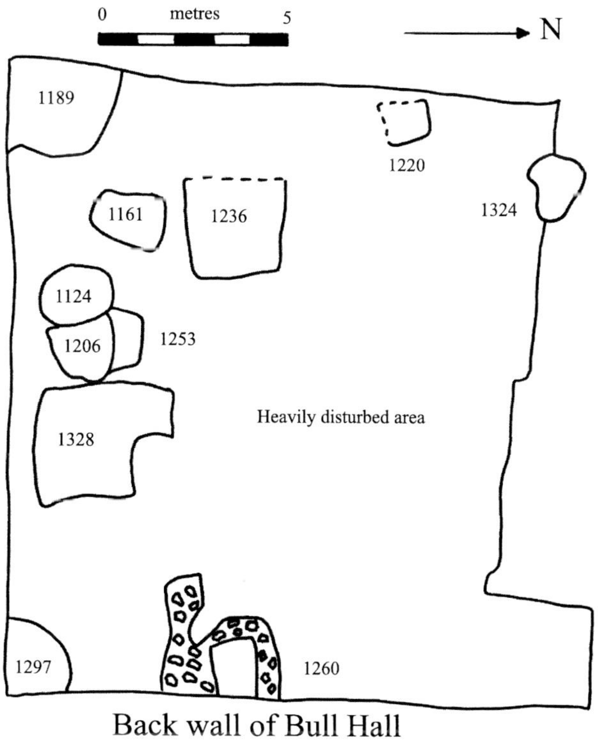

Fig. 1. Schematic plan of property and features associated with Bull Hall

Bull Hall and the Westgate Site

The example discussed here is from the medieval period in Southampton, a port city on the southern coast of England. During later medieval times, in the 13th and 14th centuries, wealthy merchants built substantial stone houses in this city (Platt and Coleman-Smith 1975). In 1979 archaeologists from Southampton Archaeological Research Committee (SARC) had the opportunity to excavate much of the property of one of these houses, known as Bull Hall. The house itself had been excavated by a different archaeological project some years before. The SARC excavations on SOU 25 (also known as SARC 25) are still unpublished, but a simplified diagram of the excavated area is provided in Fig. 1. The excavations behind Bull Hall were part of a larger series of excavations in the area of Southampton known as Westgate. Other data from the Westgate excavations will be discussed later.

Although some later disturbance to the site probably destroyed some contexts, excavations revealed a number of features associated with Bull Hall, and almost certainly all of the excavated features were contained within the property boundary of the land associated with the house. One feature (1260) was a stone-lined latrine or cess pit attached to the back wall of the house. Another feature (1328) was relatively shallow, and exhibited traces of burning. This feature has been interpreted as the hearth of a kitchen located behind the main house. Other features seem to have been refuse pits, and the excavators suggested that 1124, 1206 and 1253 were used for disposal of refuse from the kitchen. Contemporaneity of the features was demonstrated by the dates of pottery and by cross-fitting sherds of the same vessels between features. Unpublished analyses of pottery suggest that these features were utilized in the early to mid-14th century.

The composition of the faunal assemblages from all features except the latrine is shown in Fig. 2. Faunal remains were divided into three categories: (1) cattle, sheep and pig; (2) fowl and goose; (3) domestic cat. Other species were recovered, but were relatively rare and are not considered in detail here. As can be seen, there is considerable variation in the contents of different features. Some are dominated by the major domestic mammals, others by domestic birds, and two features contain a high percentage of cat bones. These features were divided into three groups. The first two groups have highly distinctive assemblages, and the third group is less homogenous. Group A (Fig. 3) is clearly dominated by the major domestic mammals. Group C (Fig. 4) is dominated by cat bones. Group B (Fig. 5) is less easy to characterize and the assemblages are more variable. Its most distinctive characteristic is that its features contain more domestic bird than Group A. Even though all the features are of similar construction, even though they date to the same time period and even though they are on the same property, the assemblage composition is quite varied.

Although we cannot explain all of the variability, it

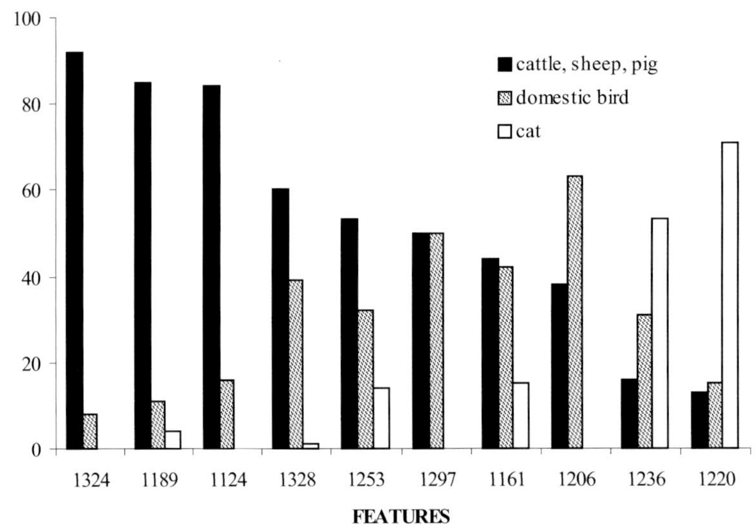

Fig. 2. Relative frequencies (%NISP) of common taxa for all later medieval features at Bull Hall, except feature 1260.

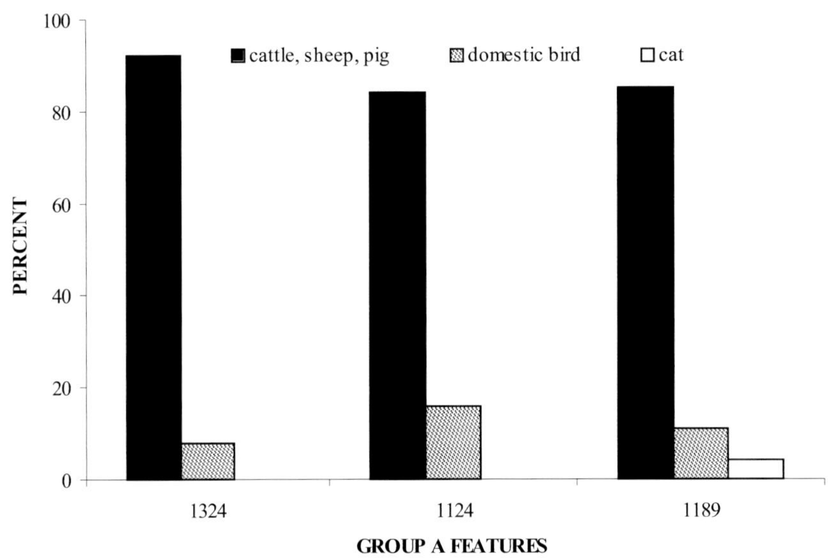

Fig. 3. Relative frequencies (%NISP) of common taxa in Group A features.

seems to be correlated in some way with the distance of the feature from the main residence. The Group A and Group C features are usually located further from Bull Hall than the Group B pits (see Fig. 1). Bearing in mind that Group B features tend to have more domestic birds and are usually located closer to the house, one can plot the overall percentage of domestic birds against the distance from the house, and show a fairly clear negative correlation between percentage of birds and distance of the feature from the back of the house (Fig. 6).

Here we are probably dealing with different methods of waste disposal, rather than different preservational characteristics of the various features. All features were cut into similar natural gravel deposits, and all had similar fills. It seems unlikely that post-depositional taphonomic processes can account for the differences between assemblages, and therefore the original mix of specimens probably varied from one feature to another. Features closer to the house tend to receive more refuse from animals that are cooked with meat on the bone, such as domestic chickens. In contrast, features further away from the house were used to dispose of uncooked bones and the carcasses of uneaten animals such as cats. Zooarchaeologists have recognized for some time that refuse disposal practices within households may create assemblages of variable composition. For example, Halstead *et al.* (1978) studied

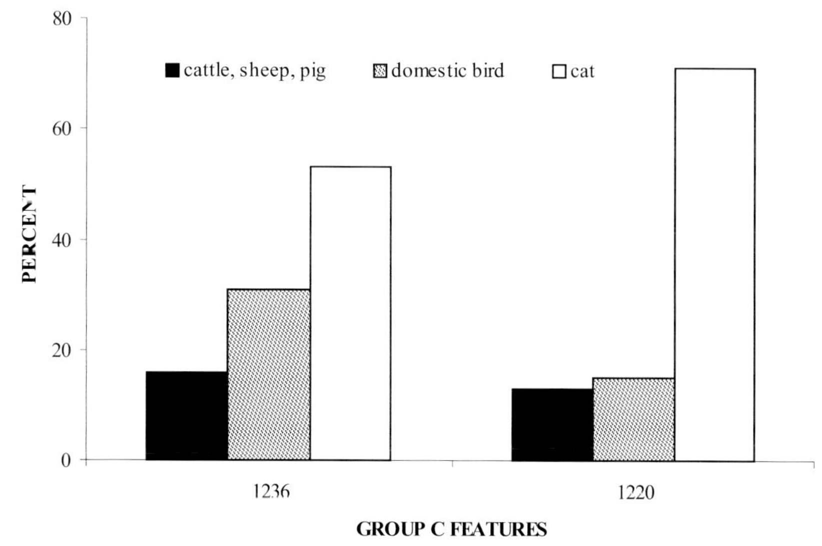

Fig. 4. Relative frequencies (%NISP) of common taxa in Group C features.

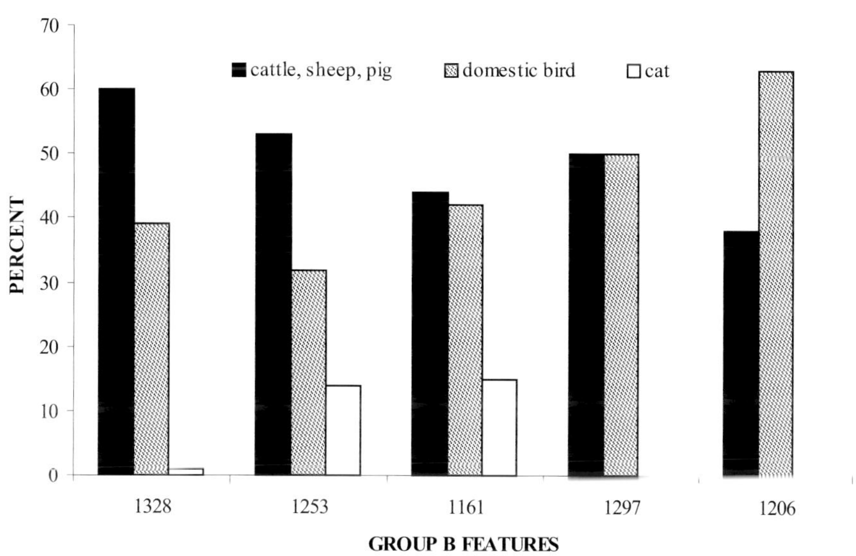

Fig. 5. Relative frequencies (%NISP) of common taxa in Group B features.

distribution of fauna at an Iron Age site. They noted that different sized animals may have been butchered and disposed of in different ways, and suggested that difference between "kitchen" and "table" waste deposits were spatially separated. Gautier (1987) used similar terminology (kitchen refuse and consumption refuse) in a discussion of taphonomy. Serjeantson (2000) has drawn attention to the need to consider the classification systems that may have been employed by people who actually created refuse deposits, and her examples demonstrate how this could affect the composition of animal bone assemblages.

At this point it is instructive to consider the stone-built latrine (Feature 1260) and its contents. As this feature is attached to the rear wall of the house one would expect it to have the characteristic assemblage of the Group B features – it should be dominated by domestic birds. However, this is not the case. About 70% of the fauna consists of the large domestic animals, making the assemblage quite similar to the Group A features. However, a closer look at the assemblage reveals why this is the case. Unlike the Group A features, many of the pig bones from the latrine are from very young individuals – suckling pigs. Suckling pigs are typically treated in the kitchen like goose or chicken – they are cooked on the bone. Thus, while suckling pig may be classified as pork, in culinary

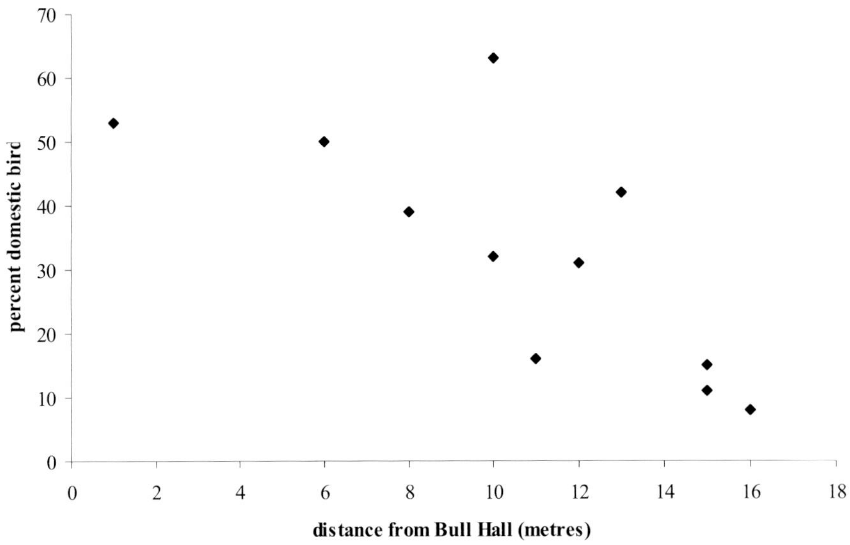

Fig. 6. Relationship between percentage of domestic birds in features and distance of features from Bull Hall.

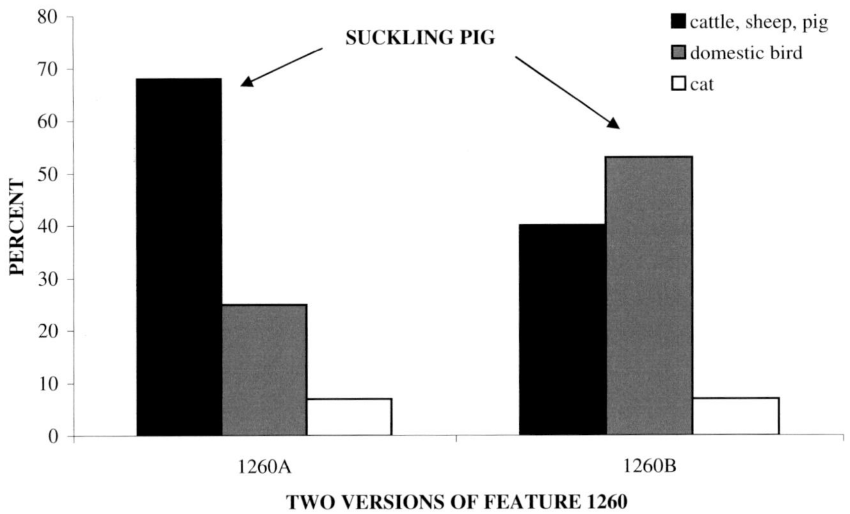

Fig. 7. Relative frequencies (%NISP) of common taxa in feature 1260. In 1260A suckling pig is counted in the cattle/sheep/pig assemblage. In 1260B suckling pig is counted with domestic birds.

terms it resembles fowl or goose. We can therefore present the assemblage composition data from feature 1260 in two ways (Fig. 7). First, we can classify the pig bones zoologically, and the assemblage resembles those from Group A features. Second, we can consider suckling pig to be the culinary equivalent of birds, in which case the assemblage now resembles the other Group B assemblages found near the house, as can be seen in Fig. 8.

Discussion

This example suggests that one must be cautious in using assemblage composition as a simple guide to social status. Assemblages produced in a single household can be quite variable, and this variability can be preserved if people dispose of different types of refuse in spatially discrete locations. It can be seen that one household created assemblages with quite different compositions, depending on the activities being undertaken. At least three different kinds of assemblages were represented on a single property. It is interesting to note that there was some spatial

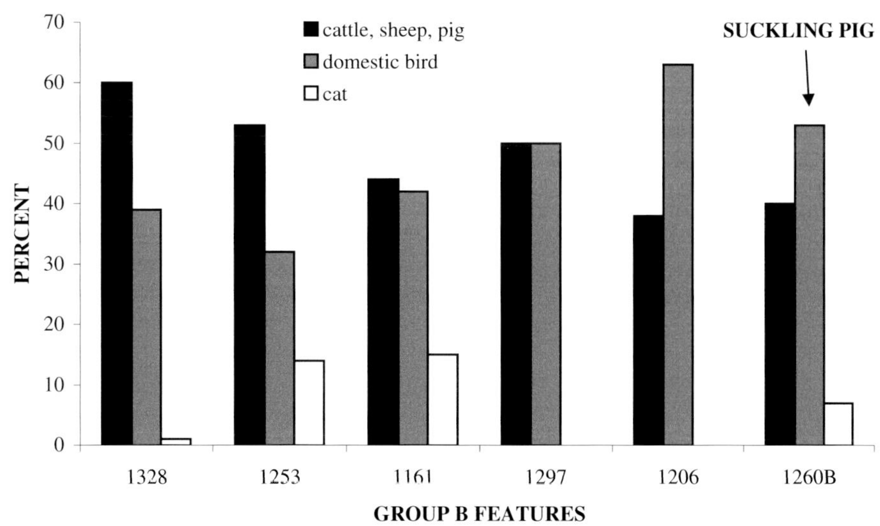

Fig. 8. Relative frequencies (%NISP) of common taxa in all Group B features and feature 1260.

	cattle		sheep		pig		cat		fowl		goose	
	NISP	%	NISP	%	NISP	%	NISP	%	NISP	%	NISP	%
11–12th centuries	867	44	446	22	435	22	23	1	194	10	18	1
13–14th centuries	252	17 (22)	204	14 (18)	208	14 (19)	338	23	409	28 (36)	48	3 (4)
Group A features	82	37	69	32	40	18	2	1	20	9	6	3

Fig. 9. Comparison of earlier and later assemblages from pit features at the Westgate site (SOU 25), Southampton. For the 13–14th centuries percentages in brackets indicate proportions excluding cat.

patterning to these assemblages, because spatial variation in assemblage composition has been used in other studies as possible evidence of social differentiation.

Faunal assemblages from features of the same age as those associated with Bull Hall were analyzed from other areas of Southampton by Noddle (1975) and Bramwell (1975). Samples from Cuckoo Lane and High Street were selected by the excavator (Platt 1975). At these sites, excavations revealed substantial stone buildings and associated features similar to Bull Hall. However, the samples submitted for faunal analysis were selected from a few of the many features encountered. At High Street most features were unlined refuse pits, as at Bull Hall. But of the four pit features selected from fifty for analysis, three were stone-lined cess pits and two were attached directly to the house, similar to Feature 1260 at Bull Hall. At Cuckoo Lane three features were selected: a well, a stone-lined cess pit and a regular refuse pit located very close to a house. The results of the analysis are difficult to compare to Bull Hall, because birds were identified separately and bone counts were not provided (Bramwell 1975). Nevertheless it is obvious from Platt's comments that substantial quantities of bird bone were recovered. He suggested that there was a significant change in the economy of medieval Southampton in the 13th and 14th centuries. Some of the changes he noted include: a reduction in "waste" bone; a reduction in the frequency of cattle, sheep and pig; an increase in smaller mammals (such as suckling pig); an increase in birds. These are the same trends that we see at Bull Hall when we compare features located close to the house with features located further away. As Platt selected unusual features from High Street and Cuckoo Lane, and as the Bull Hall assemblages demonstrate considerable variation across a single property, one cannot be sure that the economy changed as significantly as earlier reports suggested. Nevertheless, I suspect that Platt and his colleagues were correct in identifying important trends in the diet of later medieval merchants' households, and other data from the Westgate excavations support this view.

Prior to the construction of the merchant's house (Bull Hall) the Westgate area was the site of less substantial buildings inhabited mainly during the 11th and 12th centuries. Like the later period occupation, most fauna was recovered from refuse pits. However, there was no evidence for structured variation in faunal assemblages between the pits. Most features were dominated by cattle, sheep and pig, accounting for almost 90% of the assemblage, in contrast to about 50% in the later period. So some of the pits associated with Bull Hall (the Group A features) have similar assemblages to refuse pits in the earlier medieval period (Fig. 9). But in later medieval times, as Platt (1975) and Noddle (1975) suggested, there were changes to the diet and cuisine, at least in wealthier households. These can be summarized as follows:

1. the proportion of cattle, sheep and pig declines
2. cattle declines more than sheep or pig
3. domestic bird increases
4. suckling pig increases
5. assemblages are no longer homogenous, but are spatially structured in relation to the house

There may be a number of changes taking place over this period, and they are not simply linked to social status, but they may all involve social organization. The following hypotheses would have to be explored further in the context of larger samples from a greater variety of medieval properties in a single urban area:

1. The nature of the urban provisioning economy changed, so that smaller animals that could be raised in or near the town were favoured. This would account for increases in suckling pig, fowl and goose.
2. Wealthier households were more deeply committed to purchasing food through a market system, rather than producing food themselves. This would account for the decrease in cattle (assuming that more beef was purchased after de-boning) and possibly for the increased reliance on smaller animals in sites such as Bull Hall.
3. The nature of food preparation and cooking changed, at least in wealthier households.
4. Methods of refuse disposal changed as people changed their views about the nature of waste.

It may well be that wealthy urban households in the 13th and 14th centuries were distinguished by their ability to obtain, prepare and serve certain types of food that were not available to everyone and had not been common in earlier times. The contrast between the earlier and later assemblages at the Westgate site certainly suggests this, as did earlier studies of faunal assemblages in medieval Southampton. However, the detailed comparison of faunal assemblages from the Bull Hall property suggests that the socioeconomic transition was more complex. Some of the features at Bull Hall contained animal bone assemblages that are virtually indistinguishable from those of an earlier time period. In order to demonstrate that the shift in animal bone assemblages was related to the emergence of a wealthy merchant class, one would need samples of fauna taken from comparable features associated with a variety of households. For example, we would need to know whether only the merchant class participated in these changes, or if similar dietary changes occurred in households of lower or higher socioeconomic status than the merchants.

This study also suggests that one must be cautious when interpreting the composition of faunal assemblages from spatially discrete contexts. The differences between the Bull Hall assemblages are pronounced. It is likely that the differences between the assemblages from different features at Bull Hall are due to changing methods of food preparation, consumption and disposal. However, without understanding that the assemblages had been deposited by one household, it would be easy to ascribe the differences between the assemblages to differences in status and ethnicity of those who deposited them. This example demonstrates the importance of understanding the behavioural context of assemblage formation, and the need to consider alternative hypotheses when attempting to account for spatial variation in assemblage composition.

References

Bramwell, D. 1975. The bird bones, pp. 340–1 in Platt, C. and Coleman-Smith, R. (eds), *Excavations in Medieval Southampton 1953–1969*. Leicester: Leicester University Press.

Crabtree, P. J. 1990. Zooarchaeology and complex societies: some uses of faunal analysis for the study of trade, social status, and ethnicity, pp. 155–204 in Schiffer, M. B. (ed.), *Archaeological Method and Theory. Volume 2*. Tucson: University of Arizona Press.

Ferring, C. R. 1984. Intrasite spatial patterning: its role in settlement-subsistence systems analysis, pp. 116–26 in Hietala, H.J. (ed.), *Intrasite Spatial Analysis in Archaeology*. Cambridge: Cambridge University Press.

Gautier, A. 1987. Taphonomic groups: how and why? *Archaeo-Zoologia* **I(2)**, 47–52.

Halstead, P., Hodder, I. and Jones, G. 1978. Behavioural archaeology and refuse patterns: a case study. *Norwegian Archaeological Review* **11**, 118–31.

Jackson, H. E. and Scott, S. L. 1995. The faunal record of the southeastern elite: the implications of economy, social relations, and ideology. *Southeastern Archaeology* **14**, 103–19.

Lyman, R. L. 1994. *Vertebrate Taphonomy*. Cambridge: Cambridge University Press.

Marshall, F. 1993. Food sharing and the faunal record, pp. 228–46 in Hudson, J. (ed.), *From Bones to Behavior* (Southern Illinois University at Carbondale, Occasional Paper 21). Carbondale: Center for Archaeological Investigations.

Muir, R. J. and Driver, J. C. 2002. Scale of analysis and zooarchaeological interpretation: Pueblo III faunal variation in the northern San Juan region. *Journal of Anthropological Archaeology* **21**, 165–99.

Noddle, B.A. 1975. The animal bones, pp. 332–40 in Platt, C. and Coleman-Smith, R. (eds), *Excavations in Medieval Southampton 1953–1969*. Leicester: Leicester University Press.

O'Connor, T. 2000. *The Archaeology of Animal Bones*. Gloucestershire: Sutton Publishing

Platt, C. 1975. The excavations, 1966–1969, pp. 232–330 in Platt, C. and Coleman-Smith, R. (eds), *Excavations in Medieval Southampton 1953–1969*. Leicester: Leicester University Press.

Platt, C. and Coleman-Smith, R. (eds) 1975. *Excavations in Medieval Southampton 1953–1969*. Leicester: Leicester University Press.

Potter, J. 1997. Communal ritual and faunal remains: an example from the Dolores Anasazi. *Journal of Field Archaeology* **24**, 353–64.

Potter, J. 2000. Pots, parties, and politics: communal feasting in the American Southwest. *American Antiquity* **65**, 471–92.

Reitz, E. J. and Wing, E. S. 1999. *Zooarchaeology*. Cambridge: Cambridge University Press.

Serjeantson, D. 2000. Good to eat *and* good to think with: classifying animals from complex sites, pp. 178–89 in Rowley-Conwy, P. (ed.), *Animal Bones, Human Societies*. Oxford: Oxbow Books.

Jonathan C. Driver
Department of Archaeology,
Simon Fraser University,
Burnaby BC, Canada V5A 1S6
E-mail: driver@sfu.ca

9th ICAZ Conference, Durham 2002
Behaviour Behind Bones, (ed. Sharyn Jones O'Day et al.) pp. 252–261

28. Animal bones as indicators of *kosher* food refuse from 14th century AD Buda, Hungary

László Daróczi-Szabó

*Documentary and architectural evidence of medieval Jewish settlers in Buda has been available since the 13th c. AD establishment of the royal capital. This paper, however, is the first ever study of animal remains linked to the earliest settlement of this community. The fragment of a wooden plate decorated with a star of David and a glass fragment with Hebrew inscription were found in the fill from "Well 8" within the former Teleki Palace, near the so-called Jewish Gate. The deposit under discussion here dates to the late 14th century when these old quarters were abandoned. Deposits in this well were subdivided into an upper and lower section in relation to the 148 m level asl. While 100 (*c. *4%) pig bones were found in the upper section of the well, only three (*c. *0.2%) came to light from the lower levels. While sieved samples from the lower section associated with Jewish occupation contain numerous remains from scaled,* kosher *fish (esp. cyprinids and pike), the single bone from (non-scaled) catfish occurred at a level of 153.3 m, above the 148 m line of separation. In contrast to this taxonomic indicator of Christian meat consumption in the upper section, a typical trait in* kosher *diet, avoidance of meat from the hindquarters of domestic ruminants could not be observed in the lower, "Jewish" section of the deposit where several bones of the hind leg were recovered.*

Introduction

Excavations of the remains of the Teleki Palace in the Buda Castle district, the medieval royal capital of Hungary, were directed in 1999–2000 by Dorottya B. Nyékhelyi. Well 8 at this site was unusually rich in finds. The altitude of 148 m asl (above Adriatic Sea level) was chosen as an arbitrary cut point to distinguish between two major periods of the well. Layers above this level contained an admixture of post-Anjou Period medieval finds.

Deposits below the altitude of 148 m asl, were more clearly stratified and revealed, among other things, a fragment of a wooden plate decorated with a star of David as well as a glass sherd with traces of Hebrew inscription at 147 m. These objects are of paramount importance, since the location of the first medieval Jewish quarters were known only from written sources with no archaeological artefacts to support this information. In addition to the two finds, additional evidence was sought to reconfirm the ethnic/religious character of the assemblage. Analysing the relatively rich archaeozoological assemblage found in the well seemed a promising way of testing the hypothetical presence of a Jewish community, given the marked differences between Jewish dietary regulations and meat consumption by the Christian majority that inhabited medieval Buda. Methodologically, this meant a comparison between the archaeozoological finds from above and below the archaeologically estimated 148 m level.

A historical review

The chief aim of this paper, establishing patterns in the animal bone material which are characteristic of ethnic/religious identity, cannot be pursued without briefly reviewing the history of the early Jewish settlement at Buda. The Buda Castle was established by King Béla IV in 1247 on what is known today as Castle Hill, on the right bank of the Danube (Fig. 1). The foundation of the castle followed the devastating 1241–1242 Mongol Tartar invasion in Hungary. The new royal seat was fortified and its "burghers" recruited, sometimes by force, from as far away as Székesfehérvár (the former royal capital),

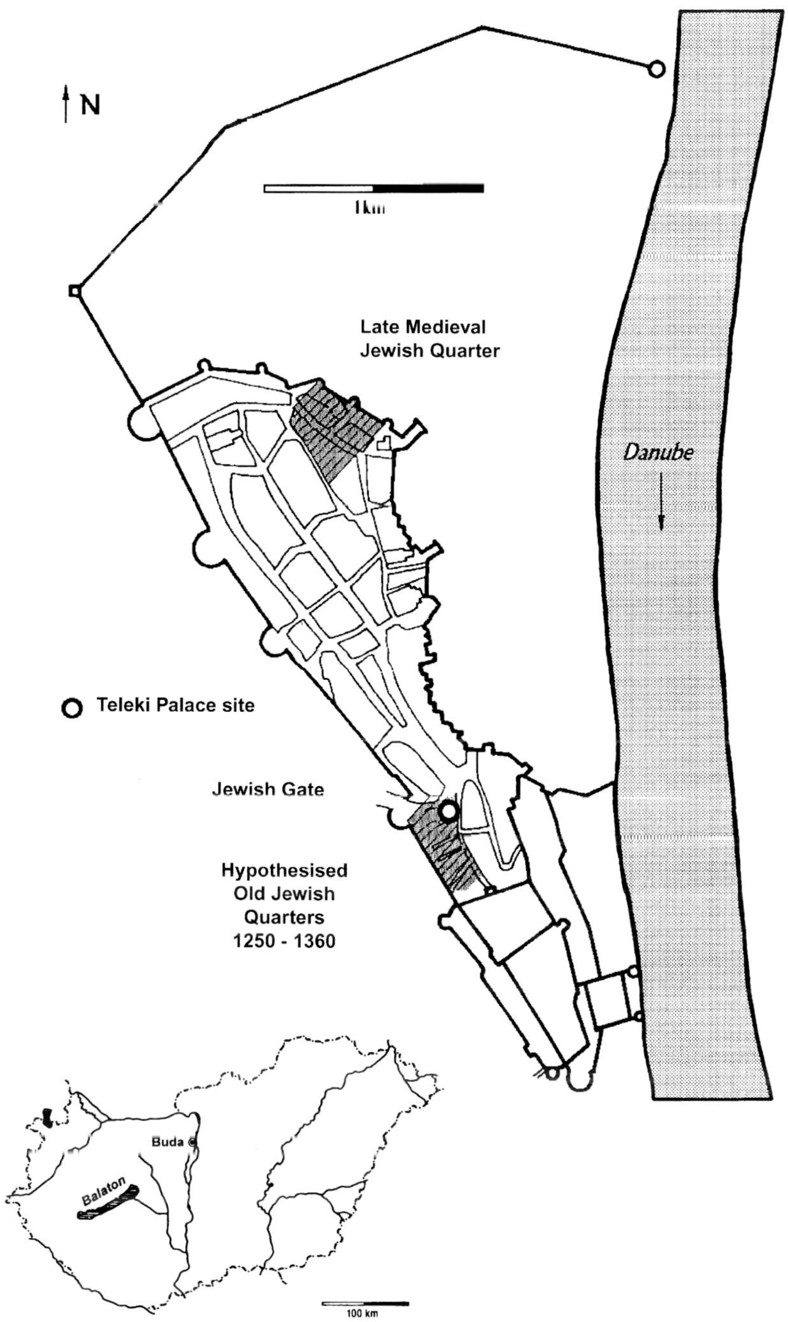

Fig. 1. The location of medieval Buda along the Danube within the borders of modern-day Hungary and the location of the site within the Buda Castle District.

Esztergom and even Zagreb (Bartosiewicz 2001, 36). At that time, also the first Jewish settlers arrived, between 1245 and 1250 (Zolnay 1987, 25). Both parties benefited from this move: the royal treasury, bankrupt after the Tartar invasion, was in need of both liquid capital and financial expertise. In turn, Jewry, persecuted throughout Europe, found temporary refuge in the emerging new capital. In 1251, King Béla IV granted basic rights to the community including religious freedom as well as the right to elect judges and clergy. In major legal cases, the Jewish community was directly responsible to the king. These rules and regulations remained in force practically until the end of the Middle Ages. Jewish leaders had their bill of rights recognised and signed by subsequent kings.

The Jewish community of Buda, however, came under attack in 1360, and were even expulsed *en bloc* by King Louis the Great. The exact reasons for this move remain unknown. According to coeval sources, Jews had refused to convert to Catholicism, thereby infuriating the king. The expulsion, however, lasted only four years. When the

taxon	upper layers	lower layers		total
	hand-collected	hand-collected	water-sieved	
mammal	2205	768	13	2986
bird	861	371	134	1366
fish	4	2	175	181
non-identifiable	636	95	1357	2088
total	3706	1179	1736	6621

Fig. 2. The composition of the assemblage by gross stratigraphic units and methods of recovery.

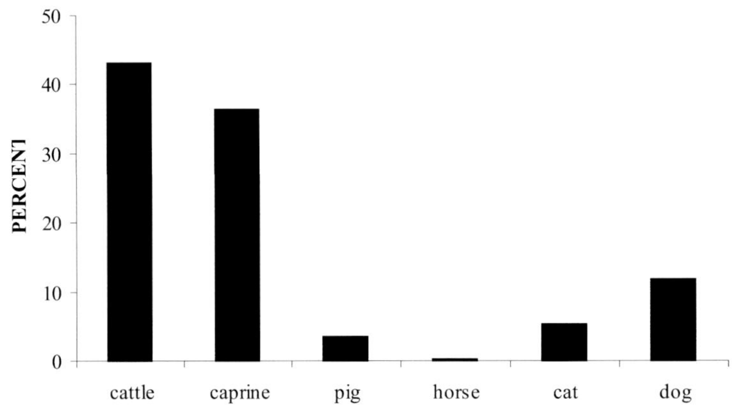

Fig. 3. Mammalian remains in the pooled assemblage (NISP=2973).

Jews of Buda returned, however, their former houses had all been occupied by Christians. Therefore a new area was assigned to the community in the northern section of the Castle district (Fig. 1).

Until now, the location of the early settlement (dating to before 1360) was unknown, as only a few written sources exist referring to it. A recurrent item in these sources was the medieval name of the so-called Székesfehérvár Gate, opening on the western side of Castle Hill. This western exit was called Jewish Gate in early sources. In addition, the sources indicate that there was also a synagogue operating near this gate.

The faunal assemblage

Material from the over 12 m deep "Well 8" was collected between the altitudes of 156.2 m asl down to 144.9 m asl. On the basis of the archaeological observations, the *c.* 11 m deep stratigraphy was sub-divided into an upper (*c.* 8 m deep) and a lower (*c.* 3 m deep) section at the level of 148 m asl. Hypothetically, these two sections corresponded to the earlier Jewish settlement (lower section) and subsequent Christian occupations (upper section). The volume of animal bones from the upper section was *c.* 740 l (58% of the total collection from the well, including over 3000 identifiable bones), while only *c.* 540 l (42% of the collection, including over 1000 identifiable bones) of material was available from the lower portion of the well. A total of 6621 bones were studied, 4533 of which could be identified to species. The remaining 2088 fragments could be assigned only to size categories, corresponding to small or large ungulates in terms of size (Fig. 2).

Owing to the sudden occurrence of special finds and more clearly stratified deposits, layers below the 148 m level were also water-sieved. In the case of identifiable mammalian remains, finer recovery resulted largely in the recovery of small bones (vertebrae, phalanges) from dogs and cats, as well as 12 bones from rodents and the mandible of a shrew (*Sorex* sp.). Among the bird remains, it was also the vertebrae and phalanges whose representation was improved by sieving. In these two groups of animals, however, the hand-collected and sieved samples could be merged with no risk of major distortion. The situation was different with fish, a group of animals that could not have been evaluated on the basis of the few hand-collected remains alone. A great number of small splinters could not be identified owing to the high degree of fragmentation. In quantitative terms, however, the great number of such small or large ungulate remains would have represented only a small amount of meat anyway. They would not have contributed to the estimation of the proportions between species either, since most of them were evidently fragments of larger, identifiable bones. The percentual distribution of mammalian remains in the pooled assemblage (NISP=2973, excl. rodents and shrew) is shown in Fig. 3.

The numerical dominance of cattle (*Bos taurus* L. 1758) bones (almost half of the mammalian NISP) may be considered a common phenomenon, since the importance

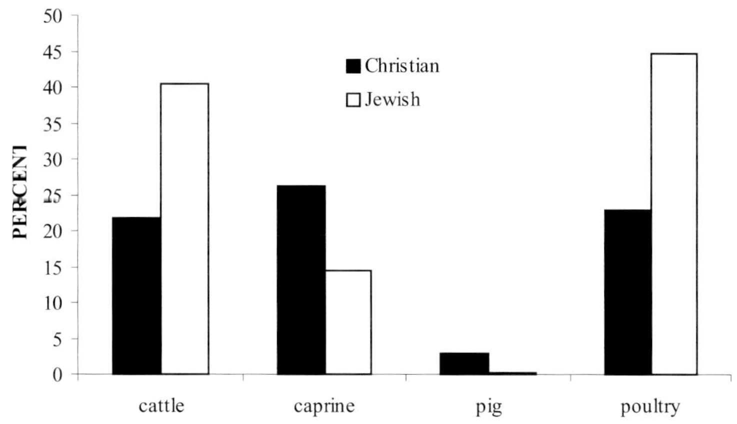

Fig. 4. Summary of differences in the contribution by major domesticates to the two sub-assemblages from Well 8.

of this multi-purpose animal was outstanding in many periods in Hungary (Bökönyi 1974). Moreover, the high degree of natural and artificial fragmentation tends to increase the number of bone fragments from large animals. In comparison with other medieval urban sites from Hungary, however, the percentual contribution of cattle remains to this assemblage is relatively low (Bartosiewicz 1999). Sheep and goat (Caprinae subfamily) were also evidently important, owing to their exploitation for wool (sheep) and possibly milk. Their remains comprised approximately a third of the mammalian NISP. No goat bones could be unambiguously identified. Pig (*Sus domesticus* Erxl. 1777) bones make up only 3.4% of the pooled mammalian assemblage, they are however, of outstanding interest from the viewpoint of this paper. Their absence may be diagnostic in possibly identifying Jewish layers on a zoological basis. The five horse (*Equus caballus* L. 1758) bones recovered make up only 0.2% of the identifiable material. Both Christians and Jews avoided horse flesh. Thus, similarly to the bones of dogs and cats, whose meat was similarly avoided during the Middle Ages, horse bones were not likely to have been thrown into the well as food refuse. Evidently, the often articulated remains of dog and cat (12% and 5.3% of the mammalian NISP) did not belong to kitchen refuse. These animals must have played traditional roles in guarding and pest control. Their bones also show that the well at one point was used as a dump for dead animals, beyond the deposition of food refuse. The significance of domestic hen (*Gallus domesticus* L. 1758) and goose (*Anser domesticus* L. 1758) is shown by the great number of bones from poultry, although the meat output of these animals is difficult to compare to those of domestic mammals. Considering the proximity of the Danube river, fish seems relatively underrepresented in this assemblage. The almost 200 bones comprise only a fraction of the total NISP, evidently related to the lack of sieving in the upper layers of the deposit.

It is difficult to tell, however, whether the well was first polluted with the bodies of dead animals and then abandoned, or was in general used as a garbage dump in which dead dogs and cats were thrown as well. The latter alternative looks to be the more likely, given the broad distribution of dog and cat bones across the stratigraphy.

Fig. 4 shows the contribution of the major domesticates to the two sub-assemblages, whereas the detailed anatomical distribution of mammalian and bird bones, subdivided by the upper and lower sections is shown in Figs. 5 and 6. The vertical distribution of hand-collected remains of economically important meat-purpose mammals within the stratigraphy is shown in Fig. 7. Cutmarks could be identified on 6.8% of all animal bones from the well. The most commonly occurring marks include diagonal cuts on ribs and the splitting of vertebrae. The first may be associated with secondary butchering in the kitchen, while the second form is more typical of primary carcass processing following slaughter. The majority of cuts were inflicted on the bones of cattle and caprines as is shown in Fig. 8. Marks of defleshing and possibly skinning occur on all cattle bones of major size. This is consonant with the more intensive butchering of bones from large animals. Marks of apparently non-professional primary butchery seem to occur on a cattle metacarpus (Fig. 9). In addition to randomly placed, deep hacking/chopping marks on the side of the diaphysis, a few precise cuts were inflicted on the palmar surface of this bone. In some cases, marks of defleshing and skinning cannot be distinguished. Additional cutmarks could be identified on the bones of smaller animals, but their distribution seems to be less patterned. Further culinary processing is difficult to identify, for example, owing to the lack of burnt bones. Only eight such bones were found.

Typically for urban food refuse, the distribution of bones by Uerpmann's (1973) meat value categories shows the dominance of high quality cuts in both sub-assemblages. This phenomenon, however, must be interpreted

	upper section							lower section						
	cattle	caprine	pig	horse	cat	dog	total	cattle	caprine	pig	cat	dog	total
atlas	13	10	1	–	–	2	26	9	1	–	–	1	11
axis	7	10	–	–	–	1	18	2	2	–	–	–	4
vert. cervicalis	72	67	8	–	–	4	151	40	16	–	7	–	63
vert. thoracalis	62	122	5	–	–	20	209	49	14	–	16	3	82
vert. lumbaris	39	67	1	–	–	3	110	9	1	–	6	7	23
vert. indet.	–	–	–	–	–	–	–	–	–	–	2	–	2
sacrum	4	8	–	–	–	–	12	–	–	–	1	–	1
scapula	76	64	5	–	4	8	157	27	11	–	8	–	46
humerus	50	52	5	–	8	11	126	24	8	–	7	1	40
pelvis	**28**	**66**	**5**	**–**	**–**	**10**	**109**	**4**	**2**	**–**	**6**	**3**	**15**
femur	**50**	**40**	**5**	**–**	**–**	**19**	**114**	**9**	**3**	**–**	**5**	**4**	**21**
patella	**9**	**–**	**1**	**–**	**–**	**–**	**10**	**–**	**–**	**–**	**–**	**2**	**2**
Category A	***410***	***506***	***36***	–	***12***	***78***	***1042***	***173***	***58***	–	***58***	***21***	***310***
radius	35	35	2	–	1	9	82	30	8	–	23	9	70
ulna	26	17	1	–	1	15	60	9	7	1	5	5	27
radius+ulna	3	–	–	–	–	–	3	1	–	–	–	–	1
tibia	**31**	**48**	**6**	**–**	**5**	**12**	**102**	**8**	**4**	**–**	**3**	**6**	**21**
fibula	**–**	**–**	**–**	**–**	**–**	**12**	**12**	**–**	**–**	**–**	**2**	**12**	**14**
tibia + fibula	**–**	**–**	**–**	**–**	**–**	**1**	**1**	**–**	**–**	**–**	**1**	**–**	**1**
cranium	25	–	4	–	4	6	39	–	–	–	4	1	5
mandibula	31	24	16	–	1	8	80	11	11	1	5	3	31
costa	139	199	15	–	–	96	449	72	41	–	–	18	131
sternum	2	7	–	–	–	–	9	–	1	–	–	–	1
Category B	*292*	*330*	*44*	–	*12*	*159*	*837*	*131*	*72*	*2*	*43*	*54*	*302*
proc. cornualis	2	–	–	–	–	–	2	–	1	–	–	–	1
maxilla	2	2	7	–	–	3	14	–	–	–	–	–	–
dentes	17	8	–	1	–	1	27	3	1	–	1	–	5
vert. caudalis	3	4	–	–	–	–	7	1	1	–	2	1	5
carpus	16	4	–	–	–	–	20	19	–	–	–	4	23
metacarpus	27	8	1	1	–	7	44	13	5	1	4	3	26
phalanges	50	39	2	2	–	–	93	34	2	–	18	12	66
os naviculare	–	–	–	1	–	–	1	–	–	–	–	–	–
astragalus	**9**	**9**	**3**	**–**	**–**	**–**	**21**	**2**	**–**	**–**	**–**	**–**	**2**
calcaneus	**12**	**6**	**–**	**–**	**–**	**–**	**18**	**–**	**1**	**–**	**2**	**–**	**3**
tarsus	**5**	**3**	**–**	**–**	**–**	**–**	**8**	**1**	**–**	**–**	**–**	**–**	**1**
metatarsus	**30**	**12**	**4**	**–**	**–**	**2**	**48**	**9**	**2**	**–**	**6**	**2**	**19**
baculum	–	–	–	–	–	1	1	–	–	–	–	–	–
Category C	*173*	*95*	*17*	*5*	–	*14*	*304*	*82*	*13*	*1*	*33*	*22*	*151*
long bone	11	7	–	–	–	–	18	1	2	–	–	–	3
flat bone	3	1	–	–	–	–	4	2	–	–	–	–	2
Total	*889*	*939*	*97*	*5*	*24*	*251*	*2205*	*389*	*145*	*3*	*134*	*97*	*768*

Fig. 5. The anatomical distribution of mammalian remains (hind leg elements are marked in bold).

carefully: NISP values may be biased by the fact that valuable meat bearing bones ("A" category, *cf.* Fig. 5) were exposed to more intensive butchering. Small, meatless bones in the "C" category are evidently less frequently cut into pieces.

Gnawing marks, presumably by dogs, were found only on three bones. In light of the nature of this deposit, however, this is not surprising. Much of the kitchen refuse may have ended up in the abandoned well before dogs could get to it.

Aspects of ethnic-religious identity

In order to interpret the faunal material as relevant to the hypothesised Jewish settlement, dietary regulations first have to be reviewed. Jewish settlers in Hungary tended to be Ashkenazim of German-East European descent. This distinction is made in contrast to Sephardic Jews from the Mediterranean region (Roden 1996, 14). Ashkenazi Jews are strictly observant even today. This somewhat facilitates the degree to which religious laws could be reflected in the animal remains from medieval Buda.

According to Jewish tradition, the fundamentals of dietary restrictions were laid down by Moses, as God's

	upper section			**lower section**		
	domestic hen	goose	**total**	domestic hen	goose	**total**
mandibula	3	11	14	4	1	5
atlas	1	1	2	2	–	2
axis	–	3	3	1	1	2
vert. cervicalis	17	47	64	8	7	15
vert. thoracalis	7	14	21	4	5	9
vert. lumbaris	6	–	6		2	2
synsacrum	8	11	19	4	5	9
clavicula	14	6	20	3	5	8
coracoideum	17	11	28	5	4	9
scapula	29	16	45	8	9	17
humerus	36	40	76	24	45	69
radius	–	2	2	10	–	10
ulna	30	28	58	22	18	40
carpometacarpus	21	12	33	4	10	14
phalanx 1 ant.	–	1	1	2	3	5
phalanx 2 ant.	–	9	9	1	1	2
phalanx 3 ant.	1	16	17	1	2	3
pelvis	16	4	20	6	8	14
femur	28	32	60	22	17	39
tibiotarsus	5	82	87	4	68	72
fibula	10	–	10	7	1	8
tarsometatarsus	47	44	91	44	15	59
phalanx 1 post.	–	14	14	8	8	16
phalanx 2 post.	–	9	9	9	5	14
phalanx 3 post.	–	15	15	3	9	12
costa	–	44	44	11	3	14
sternum	29	18	47	7	18	25
long bone	–	–	–	4	5	9
non–identifiable	22	24	46	2	–	2
total	**347**	**514**	**861**	**230**	**275**	**505**

Fig. 6. The anatomical distribution of bird remains in Well 8.

commandments (Roden 1996, 17). Although a few minor points have been re-interpreted over time, basic prohibitions have remained unchanged until today. Since this paper is concerned with animal remains, regulations concerning plant foods will not be discussed. Also, animal meat irrelevant to the Hungarian faunal record (*e.g.*, ostrich) will not be mentioned.

The most important dietary laws observed by Buda Jewry during the Middle Ages were summarised in the *Kashruth*, the system of Jewish dietary laws. The meat may be considered *kosher* if it meets three levels of requirements:

1. Taxonomic criterion: a *kosher* mammal must have cloven hooves and chew its cud. In zoological terms, this means ruminants. Other species, including pig, horse, rabbit and all carnivores are considered ritually unfit, i. e. *tref*. Raptor birds and scavengers also have to be avoided. Only fish with scales and fins are *kosher*, and no other creatures (frogs, snails, turtles, insects etc.) may be eaten.
2. Mode of slaughter: in order to yield *kosher* meat, animals must be in good health, killed in the least painful manner and the carcass must be washed down. Since animals that died a natural death, that were hunted, or that died of disease, were not killed and bled by the rules of *sheitah*, the ritual slaughter, their meat is not *kosher* (this excludes wild ruminants from the menu).
3. Anatomical criterion: certain body parts are prohibited, such as the fat of and by the tail, the fat on the kidneys and the blood that was in the veins or arteries at the time of death. Kidneys may also contain such residual blood, while bloodless brain and sweetbread (the pancreas or the thymus glands) are considered *kosher*. Of special interest is the sinew that surrounds the sciatic nerve of the hind leg. While most body parts in this list are made up by soft tissue and thus cannot be traced in the archaeological record, the hindquarters of domestic ruminants below the "sinew of Jacob" cannot be used. This would exclude bones of the hind leg from *kosher* food refuse.

These regulations always had to be observed, while on special occasions additional restrictions were imposed on the consumption of the *kosher* food. In addition to dietary laws, customs also played an important role. The biblical passage banning the consumption of the "sinew" is interpreted differently by the Sephardi and Ashkenazi branches of Orthodox Judaism, with Sephardim permitting consumption if the hind quarters are porged, that is "deveined properly", but Ashkenazim prohibiting consumption of hind quarters altogether (Encyclopaedia Judaica 1971, 27–45). Although this was not prescribed by law, fish was

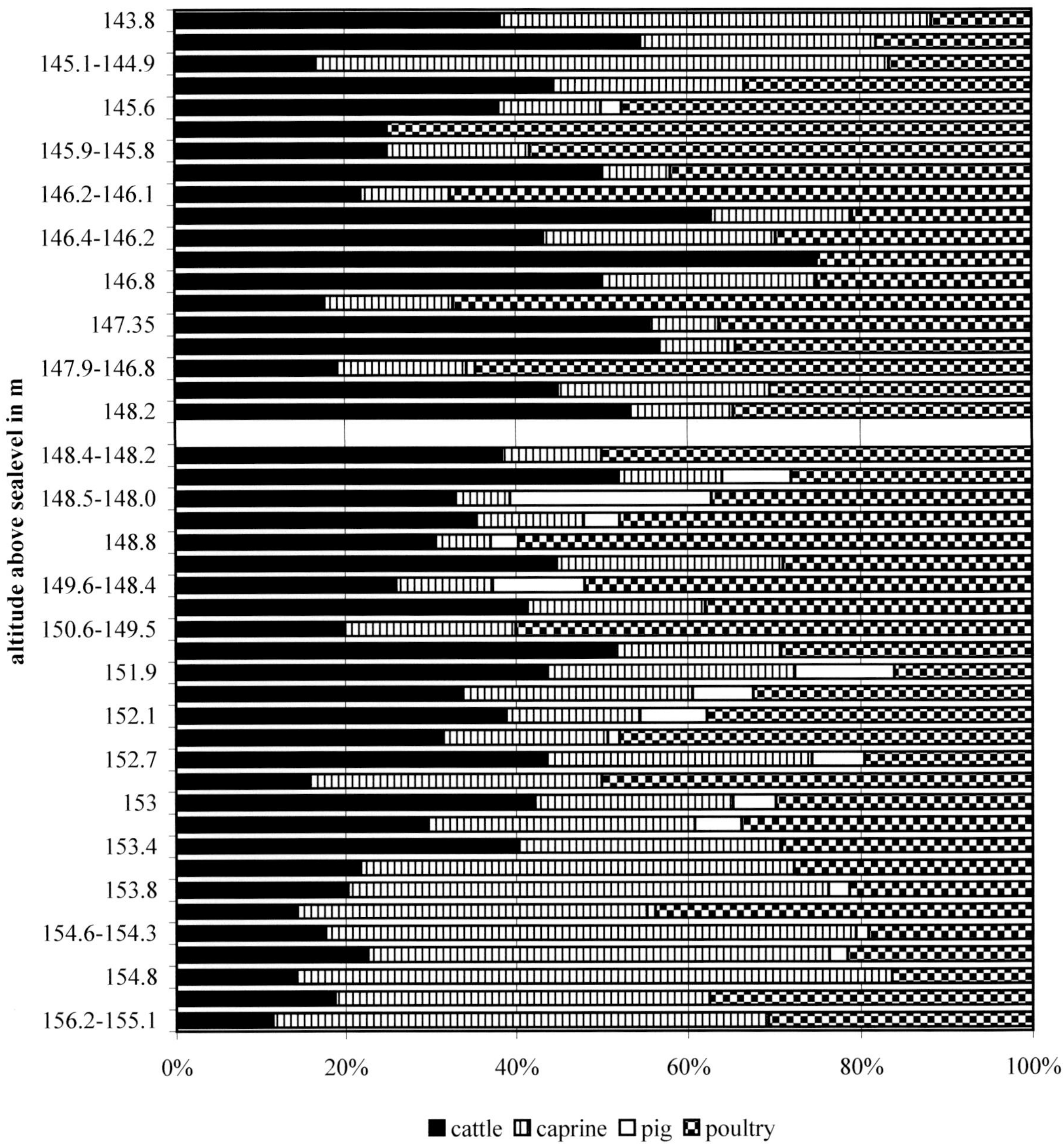

Fig. 7. The distribution of hand-collected animal remains within the stratigraphy of Well 8.

often consumed on Fridays before the Sabbath. The food rules neither were responsible for the popularity of domestic fowl, but chickens were cheaply available and easily slaughtered by a trained ritual butcher.

It is worth mentioning, that some of these regulations were also followed spontaneously by the early Christians. The only exception is hippophagy, that was only explicitly banned by Pope Gregory III in the 8th century (Bartosiewicz 1998, 162). Locally, this taboo was probably reinforced during the 11th century conversion of the Hungarians. Otherwise, the meat of dogs, cats, raptors, scavengers of any sort, reptiles and other "disgusting" animals was avoided with occasional exceptions such as famine.

Owing to overlaps between the diets of the two cultures, differentiation between their respective archaeological deposits can best be studied on the basis of pig bones, the avoidance of hind legs and remains of fish that had no scales or fins. In the case of the assemblage under discussion, it must be pointed out, that most pig and most ruminant extremity bones are large enough to be discovered without sieving, thus, their cultural historical interpretation is less biased by the recovery methods used than in the case of fish.

	upper section							lower section				
	cattle	caprine	pig	goose	large ungulate	small ungulate	**total upper**	cattle	caprine	goose	large ungulate	**total lower**
mandibula	2	2	–	–	–	–	4	7	–	–	–	7
atlas	–	–	–	–	–	–	–	1	–	–	–	1
axis	–	7	–	–	–	–	7	–	1	–	–	1
vert. cervicalis	15	30	1	–	–	–	46	5	2	–	–	7
vert. thoracalis	7	46	1	–	–	–	54	7	5	–	–	12
vert. lumbaris	9	36	–	–	–	–	45	2	–	–	–	2
sacrum	2	1	–	–	–	–	3	–	–	–	–	–
vertebra	1	–	–	–	–	–	1	–	–	–	–	–
costa	37	2	2	–	38	4	83	35	–	–	42	77
scapula	11	2	–	–	–	–	13	4	1	–	–	5
humerus	8	2	1	–	–	–	11	5	2	–	–	7
radius	8	5	–	–	–	–	13	2	3	–	–	5
ulna	2	–	–	–	–	–	2	1	–	–	–	1
radius + ulna	1	–	–	–	–	–	1	–	–	–	–	–
metacarpus	8	–	–	–	–	–	8	4	–	–	–	4
phalanx 1 ant.	1	–	–	–	–	–	1	–	–	–	–	–
pelvis	**4**	**7**	**1**	–	–	–	**12**	**1**	–	–	–	**1**
femur	**10**	**1**	–	–	–	–	**11**	**4**	–	**1**	–	**5**
tibia/tibiotarsus	**7**	–	–	**1**	–	–	**8**	**1**	–	–	–	**1**
calcaneus	**3**	–	–	–	–	–	**3**	–	–	–	–	–
astragalus	**2**	–	**2**	–	–	–	**4**	–	–	–	–	–
metatarsus	**3**	**1**	–	–	–	–	**4**	**4**	–	–	–	**4**
phalanx 1 post.	**1**	–	–	–	–	–	**1**	–	–	–	–	–
phalanx 3 post.	–	–	–	–	–	–	–	**1**	–	–	–	**1**
long bone	–	–	–	–	3	1	4	–	–	–	1	1
flat bone	42	9	3	–	5	4	63	–	–	–	–	–
total	184	151	11	1	46	9	402	84	14	1	43	142

Fig. 8. The anatomical distribution of cutmarks (hind leg elements are marked in bold).

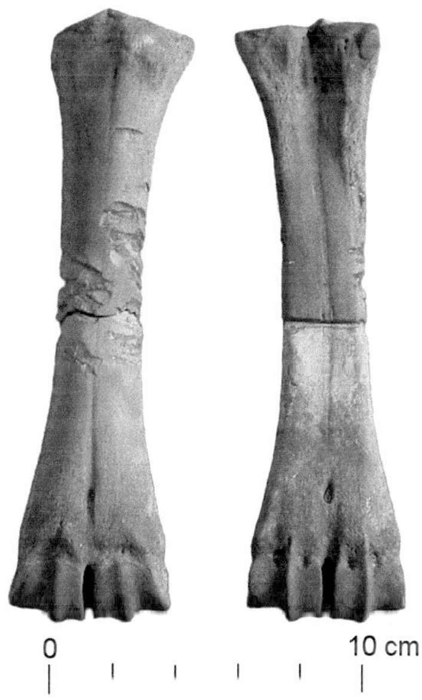

Fig. 9. Cattle metacarpus with irregular and regular butchering marks from the upper section.

Diagnostic differences between mammalian remains

Pig is virtually missing from the lower layers (3 pieces, as opposed to 97 in the upper section) and this seems to reflect a real difference in consumption traditions. Considering that the avoidance of pork is one of the strictest and most characteristic laws of *kosher* diets, this drastic decline in the number of pig bones in the well's stratigraphy is of indisputable diagnostic value. In addition, the three pig bones found in this lower section are of small size, i.e. prone to movement within the deposit. The ulna fragment of a subadult and the proximal phalanx of an adult pig were found right below the 148 m level, chosen *a priori* as a hypothetical cut-off point between the two sections of the well. Their presence may be attributed to the lack of precision in exactly defining the layer of transition between the two major deposits. It is only the single mandible fragment of an adult pig, however, that was found deep within the "Jewish" deposit (145.6 m asl).

Another potential difference, the avoidance of meat from the hindquarter of domestic ruminants is less clearly manifested in the material under discussion here. According to the Book of Genesis (32: 25–26, 32–33), during the nightlong fight between the "stranger" and Yaakov "the ball of Yaakov's thighbone became dislocated as he wrestled .... Therefore, the children of Israel are not to eat the displaced sinew on the ball of the thighbone ...". Following this event, all Jews are forbidden from eating the *gid hanasheh*, meaning "sinew of the sciatic". In veterinary terms, this is the sciatic nerve (*n. ischiadicus*; Bartosiewicz 1995, 38), reserved for the Temple sacrifice and thus forbidden as food. In principle, the hindquarters below this nerve known as the "sinew of Yaakov" cannot be used. During several millennia, however, this custom developed in a variety of ways in different Jewish communities, hence it is only of vague relevance to the little known 14th c. Jewry in Buda. Today in Israel and other

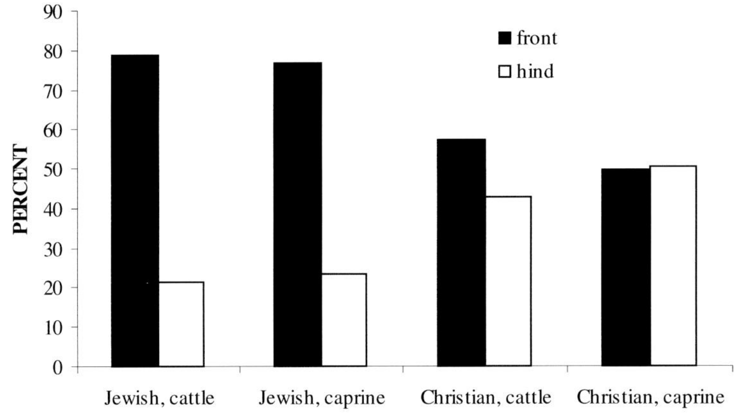

Fig. 10. The distribution of front and hind leg elements in domestic ruminants.

countries in the Middle East porging is practised as it was common in Victorian England as well (Roden 1996, 19). Today, the thigh is simply avoided by religious Jews in the US, as it is not cost effective to employ experts who can remove all the branches of the prohibited "tendon", as the nerve is called. IJzereef (1989, 47) points to the absence of hind limb bones in food refuse from Sephardic households in 17th–18th c. Amsterdam.

The difference is very visible in the material from Well 8 at Teleki Palace. The study of 407 cattle and 364 caprine limb bones from the Christian deposit and the comparison to the same 156 cattle and 51 caprine bones from the Jewish section show that the distribution of these elements is far more homogeneous in what looks like Christian food refuse (Fig. 10). In addition to long bones, elements of the shoulder and pelvic girdles (scapula and pelvis) were also included in these calculations, partly owing to the relevance of the latter to porging. The presence of a relatively small portion of hind leg elements (45 pieces) in the Jewish deposit may, in part, be attributed to contamination from the upper section (as with pig bones). Alternatively, the presence of hind leg bones of cattle and caprine among the *kosher* food refuse may indicate porging. The number of cutmarks present in the sample is too small to test whether porging is reflected in the Jewish layers by a higher ratio of traces in the hind limb compared to those from the fore limb.

Fish remains

The quantitative analysis of fish remains from this site were strongly biased by the fact that excavators felt encouraged to actually sieve the material, once a very special piece of Anjou Period draping was found near the bottom of the upper, "Christian" section. Consequently only 3 identifiable fish bones came to light from the layers above 148 m. The great number of bones from chicken and goose also indicate how important fine recovery is in appraising the economic significance of smaller animals (Fig. 8).

The 175 fish bones recovered from the sieved samples of the lower layers can be seen as characterising Jewish diets. Aside from commonly occurring carp (*Cyprinus carpio* L. 1758) and small cyprinids such as tench (*Tinca tinca* L. 1758) and bream (*Abramis brama* L. 1758), more luxurious food is represented by the remains of pike (*Esox lucius* L. 1758) and pikeperch (*Stizostedion lucioperca* L. 1758).

Fish played a prominent role in feasting, especially on Friday nights (according to a Jewish proverb, "there is no Sabbath without fish"), at New Year and during Shavuot, the celebration of the Torah. Fish also has symbolic meanings in Judaism as a sign of fertility (Yaakov blessed his sons to be as prolific as fish in the sea) as well as an impersonation of the Messiah as a huge fish.

Kosher fish are defined in the Torah as "all that have fins and scales..." (Leviticus XI, 9 12). This law has become a subject of vivid debate with the invention of the microscope that made all sorts of small scales visible on fish (Roden 1996, 18). The scales, however, must be true scales that can be removed without damaging the skin.

The number of hand-collected fish bones (6) from the upper and the sieved material from the lower sections (175) of the well cannot be compared in numerical terms (Fig. 11). In spite of the meagre representation of fish in the hand-collected assemblage from the upper section, one of the three bones recovered was a cleithrum fragment of a relatively small catfish (sheathfish or wels, *Silurus glanis* L. 1758). This delicious species falls in the category of fish which only have fins but no scales, and are therefore not *kosher*. Hence it confirms the hypothesis concerning the "Christian" origins of the upper section (153.7 m level) in the well. Fish remains in the lower section are mostly those of carp and pike, the latter species being represented by a number of rather large individuals.

	pike	pikeperch	catfish	carp	cyprinid	fish indet.	fish total
hand–collected							
upper section	–	–	1	1	–	2	4
lower section	1	–	–	–	–	1	2
total hand	1	–	1	1	–	3	6
water–sieved							
lower section	37	2	–	39	41	56	175

Fig. 11. The distribution of fish remains in hand-collected and sieved samples.

Conclusions

The archaeological sub-division between Christian and Jewish layers was largely supported by the vertical distribution of animal remains. The presence/absence of pig remains and the differential representation of cattle and caprine leg bones are consonant with known Jewish dietary laws. Meanwhile, some noise is evident in the overall picture that may result from the arbitrary definition of the 148 m cut point in the stratigraphy and the inevitable vertical movement of finds owing to a variety of taphonomic factors.

In summary, the absence of pig bones in the lower "Jewish" layers of Well 8 at Teleki Palace, as well as the presence of scaleless catfish in the upper, "Christian" section both confirm the hypothesis that these two sub-assemblages belonged to the respective religious communities. This means that the earliest Jewish community in Buda must have inhabited the area of or near Teleki Palace. The presence of some hind leg bones from cattle and caprines in the apparently *kosher* food refuse raises the question of whether one is dealing with coeval stratigraphic contamination in the deposit or a historical difference in rendering the hind leg edible by removing only the "sinew" from it.

As far as poultry is concerned, hens and geese were not subject to particular dietary laws, but they had a special significance for Jews. In fact, while the percentual contribution of beef was comparable between the two communities, in relative terms the consumption of both mutton and (naturally) pork was more important in the Christians of Buda. Although in addition to ethnic affiliation, social status may have influenced the composition of these assemblages, this paper does not discuss the relative wealth of the consumers, represented by the upper and lower deposit. At this point, there is no comparative material from verified Jewish sites in Budapest. Future research should shed more light on the presumed richness of the Jewish community and on the economic status of the Christians who took over their houses. It may be hypothesized that both types of households were of comparable purchasing power in urbanized medieval Buda. Testing this research hypothesis, however, will take additional historical, archaeological as well as zoological data.

Acknowledgements

Grateful thanks are due to Dorottya B. Nyékhelyi, the excavator of the Teleki Palace site, who provided me with the osteological material and supported its archaeozoological analysis. Help offered by György Terei should also be acknowledged. Ágoston Schmelowszky kindly supplied information on medieval Jewish dietary laws. Further help with the identification of fish bones, archaeozoological evaluation and English translation was offered by László Bartosiewicz. Special thanks are due to Omri Lernau, Wim Van Neer and Anton Ervynck for their important remarks and corrections that have significantly improved the original manuscript. The English text was revised by Alice M. Choyke.

References

Bartosiewicz, L. 1995. *Animals in the Urban Landscape in the Wake of the Middle Ages* (BAR International Series 609). Oxford: British Archaeological Reports.

Bartosiewicz, L. 1998. Mobile pastoralism and meat consumption: an archaeozoological perspective, pp. 157–78 in Laszlovszky, J. (ed.), *Tender Meat under the Saddle*. Krems: Medium Aevum Quotidianum.

Bartosiewicz, L. 1999. Animal husbandry and medieval settlement in Hungary: a review. *Beiträge zur Mittelalterarchäologie in Österreich* **15**, 139–55.

Bartosiewicz, L. 2001. Buda, pp. 36–8 in Crabtree, P. J. (ed.), *Medieval Archaeology: An Encyclopedia*. New York and London: Garland Publishing Inc.

Bökönyi, S. 1974. *History of domestic mammals in Central and Eastern Europe*. Budapest: Akadémiai Kiadó.

Encyclopaedia Judaica 6. 1971. New York: MacMillan Publishing Co.

IJzereef, G. F. 1989. Social differentiation from animal bone studies, pp. 41–53 in Serjeantson, D. and Waldron, T. (eds), *Diet and Crafts in Towns. The Evidence of Animal Remains from the Roman to the Post-Medieval Periods* (BAR British Series 199). Oxford: British Archaeological Reports.

Roden, C. 1996. *The Book of Jewish Food*. London: Penguin Books.

Uerpmann, H. P. 1973. Animal bone finds and economic archaeology. *World Archaeology* **4/3**, 308–9.

Zolnay, L. 1987. *Buda középkori zsidósága és zsinagógáik* (The medieval Jews of Buda and their synagogues). Budapest: Statisztikai Kiadó Vállalat.

László Daróczi-Szabó
Budapest Historical Museum,
Szent György Square 2,
H-1014 Budapest, Hungary.

9th ICAZ Conference, Durham 2002
Behaviour Behind Bones, (ed. Sharyn Jones O'Day et al.) pp. 262–270

29. Ethnic traditions in meat consumption and herding at a 16th century Cumanian settlement in the Great Hungarian Plain

Éva Ágnes Nyerges

During the mid-14th century the Cumanians, an ethnic group of Central Asian origin, settled in Szentkirály, central Hungary. These "nomads" traditionally kept sheep, cattle and horse. Animal remains available from the 16th century phase of this settlement show, however, that pork already played a major part in the diet. This change took place against the backdrop of 16th–17th century Ottoman Turkish occupation in the area. In addition to the numerous pig bones from Szentkirály, written documents and Turkish tax rolls also attest the relative importance of this species as opposed to the "nomadic" stereotypes. Comparison between archaeozoological evidence and written sources reveals an interesting interplay between economy and cultural change in the daily life of the settlement. In the absence of an explicit taboo, Cumanians pragmatically took up and adhered to pork consumption, in spite of Islamic rule that eventually tolerated and taxed pig keeping. In addition to the possible effects of ethnicity and religion, social and economic circumstances are also considered by this study. In spite of the established economic role of pig keeping, sheep retained its role in the economic life of this 16th century community.

Introduction

Archaeozoological studies are often carried out with ethnicity and "culture" in mind. In the best case, *ethnicity*, as a feature of a human group sharing racial, religious, linguistic, and certain other characteristics, however, can best be traced at archaeological sites where a coeval written record exists. Manifestations of *culture*, the total of the inherited ideas, beliefs, values, and knowledge providing a shared base for social action are usually limited to material finds in archaeology. *Tradition* adds the dimension of time to these two related concepts, as ethnicity and culture change diachronically at varying rates in the process of being handed down from generation to generation. A fundamental assumption in archaeozoology is that animal remains are potential indicators of these complex changes. In this study, a 16th century example from Hungary is presented to show the relationship between traditional elements and contemporary effects in animal exploitation by the Cumanians.

From the AD 1060s, the Cumanian tribal alliance thrived in the Pontic steppe south of the Russian Principalities. Their way of life is stereotypically described as nomadic (Matolcsi 1983), combined with military raids that supported complementary branches of economy such as slave trade. The importance of mobile pastoralism (a term preferred to "pure" nomadism in this paper owing to the complexity of Cumanian society) is shown in coeval Russian annals as well as in the *Codex cumanicus* compiled for Italian settlers in the Crimea at the turn of the 13th–14th centuries. Equestrian skills, indispensable in herding, were also fundamental in warfare for peoples of the steppe (Fig. 1).

Historical background

The Cumanian tribal alliance was defeated during the 1222–1223 European campaign by Genghis Khan. Some Cumanian groups fled west and in 1239 a major contingent reached the Carpathian Basin. They were granted freedom, but they had to submit themselves to the Hungarian King and convert to Christianity. According to Rogerius, "because of their great multitude, and because their people were hard and crude and did not know subordination ...[King Béla IV] nominated one of his own leaders to guide them into the centre of his country" (Pálóczi Horváth

Fig. 1. A reconstruction drawing of a mid-13th century Cumanian military leader from Csengele, SE Hungary (Horváth 2001; artwork by Márta Lacza and Ágoston Dékány).

1989, 47). Peaceful integration, however, was disrupted by the Mongol Tartar invasion in 1241. Even in peacetime, however, traditional Cumanian pastoralism would have been doomed in Hungary owing to the lack of space. Sedentism would have reduced the productivity of mobile pastoralism and agricultural tradition was lacking. Moreover, Cumanians were settled in the largely sandy Danube-Tisza interfluve, ill-suited to cultivation. Finally, a 1279 royal decree defined a contiguous Cumanian homeland in the desolated center of the Great Hungarian Plain. It prescribed that Cumanians leave their felt houses (*yurts*) and take up a Christian, sedentary way of life. From a legal document as late as 1347, however, we learn that a dozen Cumanians living in felt houses (*filtreas domus habentes*) moved onto a neighbouring Hungarian estate (Pálóczi Horváth 1989, 110).

Meanwhile, by 1280 all documented Cumanian trade had been cash based (Györffy 1990a, 286; although less formal bartering may not have been recorded) and Cumanian auxiliary forces remained instrumental in the royal army until the turn of the 14th–15th centuries. Leading strata of Cumanian society emulated Hungarian nobility and tried to privatise tribal lands. The 15th–16th centuries witnessed a process of spontaneous assimilation: permanent settlements with solid structures became increasingly common and Christian Cumanians even erected churches.

After 1541, the Cumanian homeland fell under Ottoman Turkish occupation. The settlement network was destroyed during the 16th–17th centuries, forcing large-scale internal migrations that blended Cumanians within the Hungarian population. Szentkirály village belonged to the Kecskemét District of the Buda *sanjak*, a subdivision of the Buda province (Pálóczi Horváth 1976, 278), a hierarchy that mirrored Turkish military occupation (Bertényi and Gyapay 1999, 216).

Problem statement

Cumanians arrived to Hungary as a "barbarian" steppe people of war-like persuasion. They were different from the local population in a racial, religious and, linguistic sense as well as in their material culture. This sharp contrast disappeared within 40–50 years, i.e. within two generations. In 200–250 years Cumanians caught up with the sedentary Hungarian population although they retained certain aspects of their traditional way of life (Pálóczi Horváth 1985, 71).

This hypothesis has been tested in a major Ottoman Turkish Period archaeozoological assemblage from the Cumanian settlement of Szentkirály. Special emphasis was laid on the role played by pigs, compared against the presence of bones of domesticates more typical of mobile pastoralism, such as horse and caprines. Archaeozoological evidence was used as the centrepiece of a three-stage diachronic evaluation:

1) Early stage: stereotypical descriptions of "nomadic" Cumanians in medieval Hungary.
2) Middle stage: osteological evidence of early 16th century animal exploitation in a well established Cumanian village.
3) Late stage: Turkish Period tax rolls from the second half of the 16th century reflecting "sedentary" animal keeping.

This three-tier reasoning was aimed at outlining ethnic identity as reflected in meat consumption and animal keeping.

The history of Szentkirály

Szentkirály lays in the proximity of the right bank of the Tisza river on the Great Hungarian Plain, near the intersect of 20°E and 47°N (Fig. 2). The core area of the settlement stretched some 900 m along a northwest-southeast elevation and covered 20–22 hectares. The village was flanked by the marshy floodplain so that during floods the Tisza river sometimes reached its periphery. Excavations at this site were conducted by András Pálóczi Horváth between 1969 and 1990 (Pálóczi Horváth 1996, 12).

This archaeological site was especially suitable for study since its history has been unusually well documented.

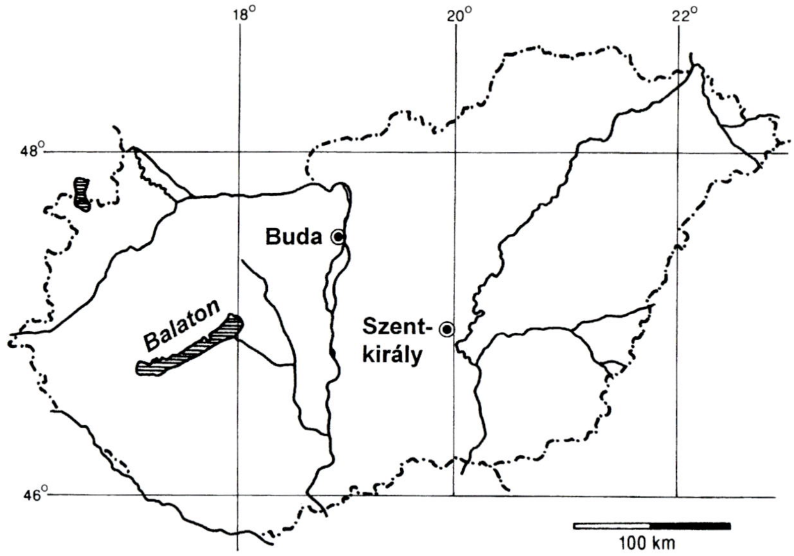

Fig. 2. The location of Szentkirály relative to Buda in present day Hungary.

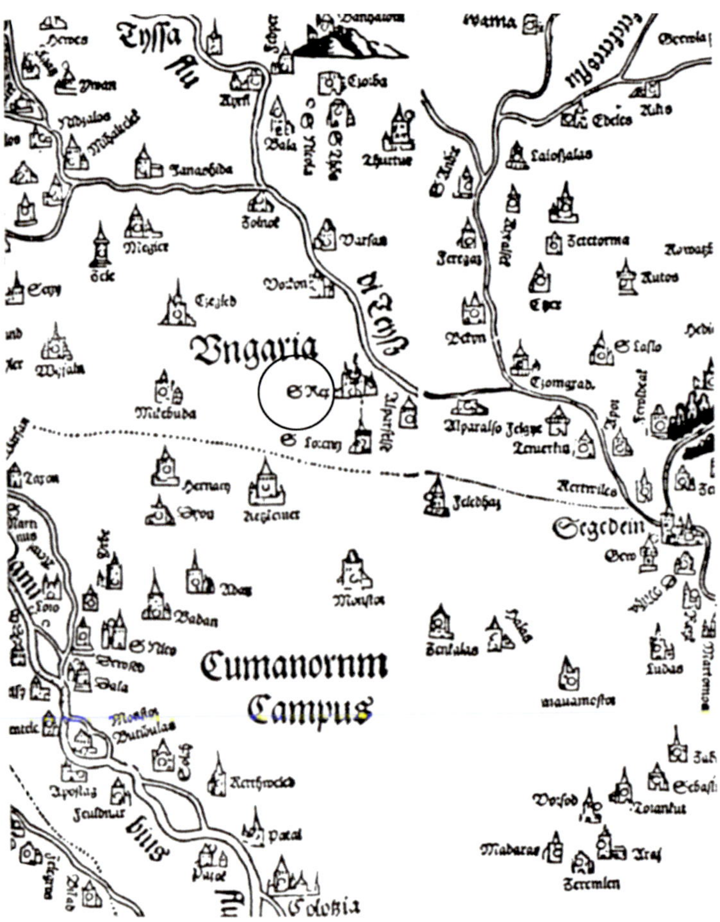

Fig. 3. The first appearance of Szentkirály ("S. Rex"; encircled) in Central Hungary on the map by Dean Lázár from 1528.

Vernacular name	Latin name	Note	NISP	%
Cattle	*Bos taurus* L. 1758	631 + 2 skulls + 1 bone belt buckle	634	52.1
Sheep	*Ovis aries* L. 1758	45 + 1 skull	46	3.8
Goat	*Capra hircus* L. 1758		10	0.8
Sheep/goat	Caprinae		196	16.1
Pig	*Sus domesticus* Erxl. 1777	109 + 1skull	110	9.0
Horse	*Equus caballus* L. 1758	77 + 2 skulls + 2bone skates	81	6.7
Dog	*Canis familiaris* L. 1758		25	2.2
Cat	*Felis catus* L. 1758		13	1.1
Domestic hen	*Gallus domesticus* L. 1758	63 + 2 portions of eggshell fragments	65	5.3
Goose	*Anser* sp.	6 + 1 portion eggshell fragments	7	0.6
Red deer	*Cervus elaphus* L. 1758		3	0.2
Brown hare	*Lepus europaeus* Pall. 1778		4	0.3
Mole	*Erinaceus europaeus* L. 1758		1	0.1
Pike	*Esox lucius* L. 1758		1	0.1
Carp	*Cyprinus carpio* L. 1758		20	1.6
Identifiable			**1216**	**100**
Large ungulate	Mammalia indet.		68	
Small ungulate	Mammalia indet.		32	
Bird	Aves		21	
Egg shell		2 portions of fragments	2	
Fish	Pisces indet.	15 + 2 portions of scales	17	
Non–identifiable			140	
Total			**1496**	

Fig. 4. Taxonomic composition of the material from Szentkirály.

The village was first destroyed in the 1241–1242 Tartar invasion. Later, Szentkirály (*Zenthkiral*) was donated to Cumanian elders by King Lajos the Great on March 30 in 1354, along with another desolated royal estate in Szolnok county, under the condition that they would observe Christian customs and rites. By the turn of the 15th–16th centuries, Szentkirály occupied a prominent position in the countrywide road network (Pálóczi Horváth 2000, 123). The settlement is marked as *S. Rex* on the first detailed map of Hungary published by Dean Lazar in 1528 (Fig. 3).

With its 50–60 habitation units, Szentkirály was one of the largest villages around the market town of Kecskemét prior to Ottoman Turkish occupation. The ownership of this area changed several times in the mid-16th century as Turkish forces advanced into the area. At this time, the Hungarian land owners still collected taxes but Turkish authorities also had tax revenues from the village. As may be seen in Turkish tax rolls (Aszt 2000), people in the village were evidently pauperised by Hungarian/Turkish dual taxing.

In the 1579 map by János Zsámboki the area of Szentkirály is covered only by an idyllic drawing of a herd of cattle, a reminder that the village lay well within the catchment area of large scale livestock exports that peaked during that time (Bartosiewicz 1995, 83, Fig. 53). The village reappears as *S. Kyral* in a 1585 map (by Mercator) as well as others drawn after the devastating Fifteen Years War (1593–1606) in 1626 (by John Speede) and 1635/1642 (by Guiljelmus and Johannes Blaeu). Owing to dwindling security, however, inhabitants moved into nearby Kecskemét during the Turkish Period. Szentkirály gradually became a deserted *puszta* of this latter market town and was abandoned for good in 1692. It was re-established only ten generations later, in 1952, under the name Lászlófalva.

Some thirty of the documented 50 to 60 houses could be identified at the site, twenty of which were excavated. House remains were discovered at distances of 50–70 m from each other, with structures indicative of animal keeping in between. Almost 300 settlement features and 400 graves also came to light (Aszt 2000,12–3).

The animal bone assemblage

Small sub-assemblages of animal bones from Szentkirály were studied by several experts (Matolcsi, 1983; Takács 1988; Takács 1989; Takács 1994; Somhegyi 1998). Excavations at this site were carried out in five major areas. Animal bones from Area III (House 4–4a and its lot, excavated in 1978–1980) were analysed by the author of this study. Coin finds date this material to the first half of the 16th century, possibly directly pre dating Turkish occupation (Pálóczi Horváth 2002, 191).

Sample characteristics

The taxonomic composition of this assemblage is summarised in Fig. 4. While during the late 1970s (with the exception of a special feature, a well; Takács 1988) finds at this site were hand-collected, the occurrence of small bones is indicative of relatively precise recovery. Since preservation is good, the majority of the animal remains could be identified to species. Aside from the remarks listed in Fig. 4, animals were presented in terms of the

	cattle	sheep	goat	caprine	pig	horse	dog	chicken	goose	red deer	brown hare	bird	large ungulate	small ungulate
atlas	4	–	–	–	2	–	–	–	–	–	–	–	–	–
axis	5	–	–	–	–	–	1	–	–	1	–	–	–	–
vert. cervicalis	17	–	–	5	4	2	–	–	–	–	–	–	–	–
vert. thoracalis	18	–	–	6	1	–	–	–	–	–	–	–	1	1
vert. lumbalis	10	–	–	9	–	–	1	–	–	–	–	–	–	–
os sacrum	–	–	–	1	–	–	–	2	1	–	–	1	–	–
sternum	–	–	–	–	1	–	–	5	–	–	–	1	–	–
scapula	41	4	–	9	3	3	–	2	–	–	–	–	5	2
os coracoideum	–	–	–		–	–	–	2	–	–	–	–	–	–
humerus	33	3	–	20	10	–	6	5	1	–	1	1	4	1
pelvis	14	3	–	9	2	1	1	–	–	–	1	1	2	–
femur	34	–	–	8	–	2	–	5	–	–	–	1	3	–
Category A	**176**	**10**	**–**	**67**	**23**	**8**	**9**	**21**	**2**	**1**	**2**	**5**	**15**	**4**
neurocranium	20	5	–	3	5	3	1	1	–	–	–	–	–	2
mandibula	57	2	–	30	25	–	6	–	–	–	–	–	2	1
costa	36	–	–	24	1	–	–	12	–	–	–	2	7	4
radius/ulna	55	3	1	18	7	5	4	9	1	–	–	2	1	1
tibia/fibula	51	9	3	19	3	5	–	10	1	–	–	3	8	–
Category B	**219**	**19**	**4**	**94**	**41**	**13**	**11**	**32**	**2**	**–**	**–**	**7**	**18**	**8**
viscerocranium	15	1	–	5	13	1	–	1	–	–	–	–	2	–
maxilla	5	–	–	–	6	–	–	–	–	–	–	–	–	–
dentes	27	1	–	16	12	12	1	–	–	–	–	–	1	1
carpalia	11	–	–	2	3	–	–	–	–	–	–	–	–	–
metacarpalia	45	7	6	6	1	3	2	3	–	–	–	1	–	–
ph. proximalis	35	1	–	–	4	33	–	–	1	–	–	1	–	–
ph. media	19	–	–	–	3	3	–	–	–	1	–	–	–	–
ph. distalis	8	–	–	1	1	3	–	–	–	–	–	–	–	–
calcaneus	7	1	–	3	–	–	–	–	–	–	–	–	–	–
astragalus	15	2	–	1	1	1	–	–	–	–	–	–	–	–
centrotarsale	4	–	–	–	–	–	–	–	–	–	–	–	–	–
metatarsalia	47	4	–	1	2	2	2	6	1	1	2	–	–	–
Category C	**238**	**17**	**6**	**35**	**46**	**58**	**5**	**10**	**2**	**2**	**2**	**2**	**3**	**1**
A+B+C total	**633**	**46**	**10**	**196**	**110**	**79**	**25**	**63**	**6**	**3**	**4**	**14**	**36**	**13**
long bone fragment	–	–	–	–	–	–	–	–	–	–	–	7	22	11
flat bone fragment	–	–	–	–	–	–	–	–	–	–	–	–	10	7
identifiable by size only	–	–	–	–	–	–	–	–	–	–	–	**21**	**68**	**32**

Fig. 5. The anatomical distribution of the material by Uerpmann's (1973) categories, excluding manufactured bones (buckle and skates)

number of identifiable specimens (NISP): estimating the numbers of individuals would have been distorted by the fact that only a few bones were found in an articulated (or even primary) position. The great number of datable but heavily disturbed secondary deposits would have made estimating the minimum number of individuals a completely theoretical exercise.

Cattle provided half of the material (NISP), the remains of sheep and goat approached 20%, while pig bones almost attained 10%. Horse, an animal associated with "nomadic" Cumanians was only fifth in this list (7%). This is all the more interesting, since both pig and horse were represented by high percentages in the small sample of 140 (!) bones studied by Takács (1989) from the same area, creating a peculiar picture of an "equestrian" culture with "extensive" pig keeping. Evidently, this phenomenon was random: such small samples are difficult to relate to ethnicity on a quantitative basis.

The anatomical distribution of bones by animal species (Fig. 5) shows that while meat bearing bones (Uerpmann 1973, Category A) form one third of all remains in ruminants, they are less common in pig and horse. Dry limb bones (Uerpmann 1973, Category C in Fig. 5) of negligible meat value are most common in the case of horses (Fig. 6). The bones of pig and even the non-meat purpose dog display a more balanced anatomical distribution. While dogs were not eaten, the "dry limbs" of pig, i.e. pig knuckles, are highly appreciated as food (Bartosiewicz 1997, 176), which may account for their presence in food refuse.

Pork consumption as a possible indicator of ethnic change

Food customs are often diagnostic for ethnic groups. Within the present case study, of the domesticates of economic significance, pigs may be considered the best indicators of cultural attitudes. These prolific animals kept exclusively for meat, are subject to religious taboos (Judaism, Islam) and are widely associated with a sedentary way of life. Traditionally, the complementary roles of pigs and caprines may be observed in many cultures (Bartosiewicz 1984). Although, in addition to meat, sheep also provided wool and milk, pig yielded, on average, twice as much meat and fat per animal.

The limited role of pigs in Cumanian tradition is

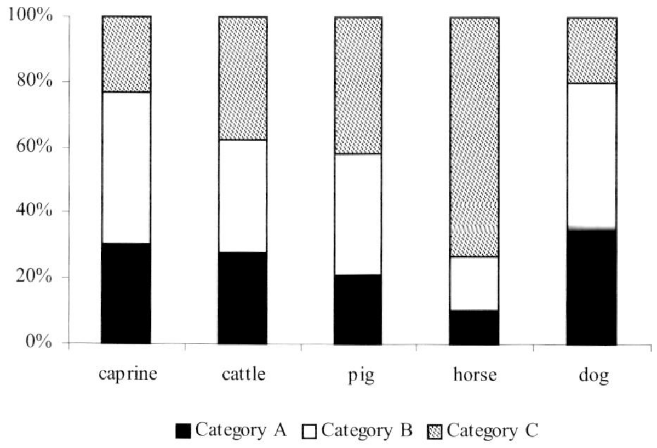

Fig. 6. The percentual distribution of skeletal elements by meat value categories (Uerpmann 1973).

Settlement type	Number of sites	Intersect	Exponent	Coefficient of determination
Urban	16	6.258	0.497	0.353**
High status	22	13.297	0.511	0.340**
Rural	23	14.640	0.457	0.324**

Fig. 7. Statistical parameters of the relationship between NISP values of caprines (x) and pig (y).

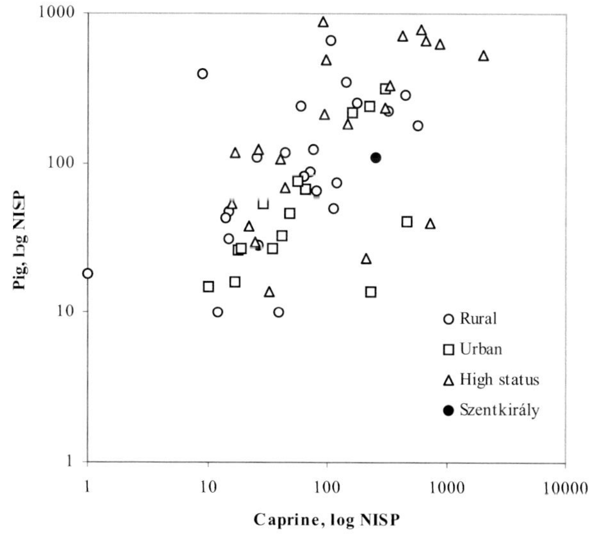

Fig. 8. The configuration of medieval sites in the plane defined by the number of caprine (x) and pig (y) remains.

reflected in the 1294–1295 Latin-Persian-Cumanian dictionary of the *Codex cumanicus* (written in the Crimea). It is especially rich in terms related to horse-, sheep- and cattle herding, showing the importance of these stocks. Although pigs are also sporadically mentioned in this text, no linguistic distinction is made between boars and sows, suggesting that pig breeding must have played at most a minor role in the economy of Cumanian pastoralists in the Pontic Region (Györffy 1990b, 244).

The less than 10% NISP calculated for pig remains at Szentkirály equals only half the proportion of caprine bones at this site (Fig. 4). NISP values of pig from 61 medieval and Turkish Period sites were plotted against those of caprine per settlement type, that is rural, urban and high status sites. Differences of several orders of magnitude between assemblage sizes were balanced by plotting the decimal logarithms of NISP values (cf. Bartosiewicz 1999, Fig. 2). Only assemblages in which at least 10 bones of both pigs and caprines could be identified were included in this analysis. Fig. 8 shows that medieval urban sites contained relatively few pig bones in contrast to rural and high status settlements. This pattern is in accordance with the hypothesis that pigs were the animal of choice in self-reliant (if not actually subsistence) economy in villages and high status centres such as forts, castles and monasteries. This tendency is clearly expressed when the statistical parameters of trend lines fitted to data points are compared (Fig. 7). In Fig. 8, the full circle standing for *rural* Szentkirály falls exactly on this lower, "urban" trendline; mutton consumption was important in many medieval cities in Hungary (Bartosiewicz 1996). At Szentkirály sheep remained fundamental to the meat diet, as is shown by the lamb *decima* (1/10 of all animals) tax data, which is comparable to the contribution of pigs.

Other rural sites in this position in Fig. 8 may be identified as villages from the 11th–13th centuries of the period of the Árpád Dynasty, inhabited by Hungarians shortly after their 10th century conquest of the Carpathian Basin. It is important to point out, that these latter villages show no regional trend (cf. Vörös 2000, 80, Tab. 2, Fig. 2), but are located across the Carpathian Basin. Early Hungarians were also pastoralists from the steppe region, adopting their animal keeping to local circumstances! Among the urban and high status assemblages in this figure, only a few Turkish Period settlements are characterised by even lower contributions of pig remains.

In contrast to the modest evidence of early 16th century pig bones from Szentkirály, Turkish tax rolls attest to the steadily increasing importance of these animals. Factors behind this change include the marshy, floodplain habitat near the settlement that could be best exploited by pigs by a "steppe" people who had no religious restrictions on pork consumption. By the 20th century Tálasi (1977, 171 cited by Aszt 2000, 43) even mentioned an extinct "Cumanian breed of swine", although no further details on this breed are provided.

"The circumstances of compiling [Turkish tax] registers, naturally, determine the reliability of these sources. Almost in any form of taxing, there was a conflict of interest when the data were recorded." (Káldy-Nagy 1985, 11). In the case of Szentkirály it is fortunate that tax collectors favoured the population, sometimes understating

	Unit	1546	1559	1562	1580	1590
Tax items						
Decima in barley	*Kile*	16	–	–	–	–
Decima in hay	*Akche*	200	–	–	–	–
Price of straw	*Akche*	–	–	–	150	150
Price of stray beasts	*Akche*	–	–	250	300	100
Slaughter dues	*Akche*	–	–	–	100	100
Decima in lamb	Animal	50	–	198	100	85
Number of pigs	**Animal**	**25**	**–**	**255**	**150**	**>150**
Sheep herding						
Number of sheep	Animal	765	1300	1928	1260	–
Number of sheep owners	Head	6	8	11	5	–
Mean flock size	**Animal**	**153**	**163**	**175**	**252**	–
Typical flock size	Animal	150–200	–	250–300	300–400	–
Census data						
Number of family heads		41	–	51	58	66
Number of people registered		53	–	84	99	119
Mean family size, head		1.3	–	1.6	1.7	1.8

Fig. 9. Some relevant economic data based on Turkish tax rolls (Káldy-Nagy 1985). 1 Kile *is* c. *35 l, depending on local measurement standards within the Empire,* Akche *are Turkish monetary units.*

the wealth there. Therefore, one should not be concerned about too high values in these records. In spite of evident bias, some animal keeping data contribute valuable information to the analysis of animal bones from this site (Fig. 9). It is of special interest that as sheep keeping starts prospering during the Ottoman Turkish Period, pig keeping also increases. It is possible that in this situation, owning pigs was another way to further increasing one's wealth. The number of pigs in tax rolls also suggests that during the first year available for study (1546) only a few (25) pigs may have been kept around the house. Fattening pigs on the household level seems to have become popular during the 16th century in Hungary. In addition to barley, Siberian buckwheat (*Fagopyrum tataricum* L. 1758) was often used as fodder (Dám 1995, 190–1). The idea of pig fattening is consonant with *decima* dues paid in barley in Szentkirály at a time when pigs begin to appear in small numbers. It seems that during the years to follow, the number of pigs kept was far greater (255 in 1562) than would have been typical for small scale, household rearing.

There is another interface between the osteological material and written sources. Taxes were levied only on pigs older than 1 year. In the studied bone assemblage, half of the pig remains originated from juvenile/subadult individuals (age <1 year), while the other came from adult/mature pigs (age >1 year). As for the anatomical distribution of skeletal parts, no significant difference was found between these two gross age categories (Fig. 10).

Non meat-purpose animals

While emphasis in this paper has been laid on the increased importance of pork in the diet of Cumanians, the remains of at least two other animals may reflect ancient pastoral traditions. The aforementioned overrepresentation of meta- and autopodium bones from horse is commonly interpreted as a sign of skinning, since foot bones tend to be left attached to raw hides. This hypothesis is supported by skinning marks, only one of which was found in a young animal whose deformed proximal phalanx was cut. This may have been a lame animal culled at an unusually young age for horses. Six of the 19 cut marks identified on the bones of mature horses also unambiguously originated from skinning while the rest are difficult to categorise. Neither anatomical patterns nor cutmarks, however, would indicate hippophagy, as would be expected in the case of many pastoral communities from the Eurasian steppe region (*e g*., Sarmatians, Bartosiewicz 2003). On the other hand, Takács (1994) suggested that the two large skull remains originating from young horses had an apotropaic function as was observed with several pastoral peoples in Eurasia as well as in the Hungarian ethnographic record (Bökönyi 1978).

Another animal of interest is dog. Dog bones were scattered across the research area, many of them possibly belonging to the same individual. A cutmark occurs on the pelvis (*corpus ossis ilii*) from a large dog, in the size range of modern kuvasz-góral-Pyrenese group (such dogs were used in guarding, rather than herding). Another cut was found on the coronoid process of a similarly large mandible, and may be the result of decapitation. These disarticulated bones came to light in two adjacent squares. Although they were not found in a primary position their mere presence is a reminder that dog sacrifices were part of the Cumanian tradition. Especially the ilium (covered by the mass of rump musculature) could not have been cut if the animal was simply skinned. The mutilation of dog carcasses is known from written sources as well. In 1247, the first son of King Béla IV, king of Hungary, married the daughter of the new Cumanian sovereign. According

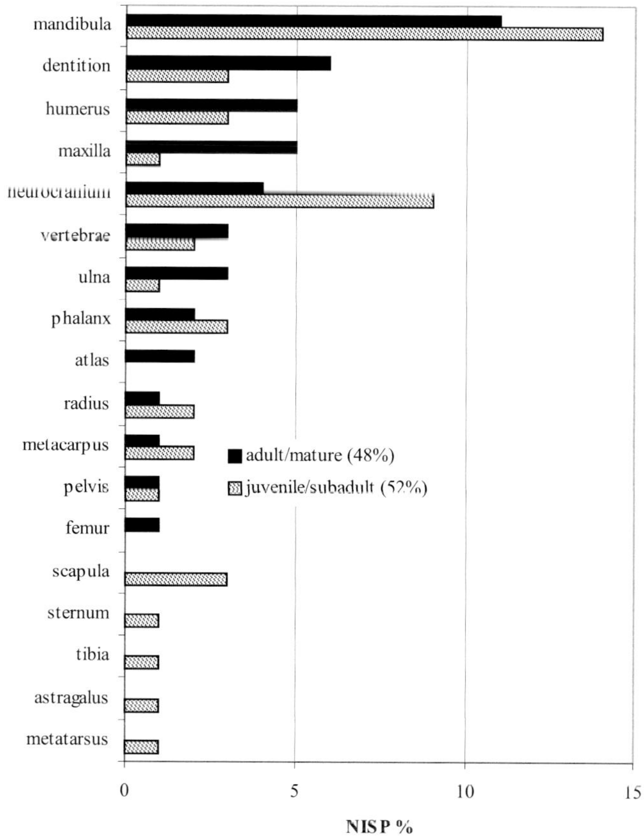

Fig. 10. Similarities between the anatomical distribution of 110 pig bones from younger and old individuals.

to an account by John of Plano Carpini, during the wedding celebration "ten of the Cumans came together and made an oath according to their custom, with their swords on a dog that had been sundered in two, that they would defend the lands of the Magyars as would the king's own supporters against the Tartars and barbarian peoples" (Pálóczi Horváth 1989, 47). Possible archaeological evidence of dog sacrifice may be found in the form of an articulated dog skeleton found within a fireplace at the settlement of Csengele (Horváth 2001, 122).

Conclusions

This paper was built on three sources of evidence, *i.e.* the stereotypical medieval view of "nomadic" Cumanians in Hungary, osteological evidence of early 16th century meat consumption in a well established Cumanian village at the onset of the Ottoman Turkish occupation, and Turkish Period tax rolls from the second half of the 16th century.

In the faunal assemblage, pig is the third most frequently occurring species, possibly indicating that Cumanians in Szentkirály had just began keeping pig relatively recently. The comparatively higher values of caprine remains are similar to those observed at villages of newly settled 11th–13th century Hungarians, mobile pastoralists with a comparable steppe tradition.

The Turkish Period tax rolls suggest that a form of "sedentary" animal keeping, characterised by a major contribution of pig (in addition to the traditional sheep and cattle), started prospering in Szentkirály during the Turkish Period. A comparison between osteological and taxing data from Szentkirály seem to confirm that whether pigs had been (a marginal) part of the Asiatic Cumanian heritage or their keeping was adopted only after the Cumanians came to the Carpathian Basin, they continued to be raised by individual households (Andrásfalvy 1997, 87) at this settlement.

Aside from this diachronic review of pig keeping by the Cumanians, animal remains from Szentkirály revealed dog and horse finds of possibly ritual significance, although these remains originate from secondary deposits and therefore need to be interpreted prudently.

Acknowledgements

Grateful thanks are due to Dr. András Pálóczi Horváth, the excavator of Szentkirály, who entrusted me with the analysis of the material and offered guidance throughout its archaeological interpretation. Special thanks go to Márta Lacza and Ágoston Dékány the artists who released copyright for Fig. 1, prepared on the basis of scholarly reconstructions by Dr. Ferenc Horváth. Dr. Gyöngyi Kovács offered information on the immense complexity of interpreting *Kile* units in Fig. 9. Further help with the archaeozoological evaluation and English translation was provided by Drs. László Bartosiewicz and Alice M. Choyke.

References

Andrásfalvy, B. 1997. Nyájszervezet – csordatartás. (Herd organization and flock keeping), pp. 85–94 in Kovács, L. and Paládi-Kovács, A. (eds), *Honfoglalás és néprajz. A honfoglalásról sokszemmel IV*. Budapest: Balassi Kiadó.

Aszt, Á. 2000. *Állattartó objektumok a középkori magyarországi falvakban*. Unpublished MA Thesis, Loránd Eötvös University, Budapest.

Bartosiewicz, L. 1984. Az állatállomány faji összetételének összefüggése a lakosság étrendjének energiatartalmával néhány fejlődő országban (Correlations between herd composition and the per capita energy consumption of the human population in developing countries). *Állattenyésztés és Takarmányozás* **33/3**, 193–203.

Bartosiewicz, L. 1995. *Animals in the Urban Landscape in the Wake of the Middle Ages* (BAR International Series 609). Oxford: British Archaeological Reports.

Bartosiewicz, L. 1996. Közép- és török kori állatmaradványok Segesdről (Medieval and Turkish Period animal remains from Segesd). *Somogyi Múzeumok Közleményei* **12**, 183–222.

Bartosiewicz, L. 1997. This little piggy went to market... An archaeozoological study of modern meat values. *Journal of European Archaeology* **5**, 170–182.

Bartosiewicz, L. 1999. Animal husbandry and medieval settlement in Hungary: a review. *Beiträge zur Mittelalterarchäologie in Österreich* **15**, 139–55.

Bartosiewicz, L. 2003. A millennium of migrations: Protohistoric mobile pastoralism in Hungary, pp. 101–30 in King, F. W. and Porter, C. M. (eds), *Zooarchaeology: Papers to Honor Elizabeth S. Wing*. Bulletin of the Florida Museum of Natural History vol. **44**.

Bertényi, I. and Gyapay, G. 1999. *Magyarország rövid története* (A short history of Hungary). Budapest: Maecenas Könyvek.

Bökönyi, S. 1978. Árpád-kori szokás analógiája: kiaggatott lókoponyák a Közel-Keleten (The analogy of a Hungarian custom from the Árpád Period: horse skulls on display in villages of the Middle East). *Archeológiai Értesítõ* **104**, 91–4.

Dám, L. 1995. *Magyar népi állattartás és pásztorodás* (Hungarian traditional animal keeping and pastoralism) (Néprajz egyetemi hallgatóknak 19). Debrecen: Kossuth Lajos Tudományegyetem.

Györffy, Gy. 1990a. A magyarországi kun társadalom a XIII-XIV. században (Cumanian society in Hungary during the 13–14th c.), pp. 274–304 in Györffy, Gy. (ed.), *A magyarság keleti elemei*. Budapest: Akadémiai Kiadó.

Györffy, Gy. 1990b. A kipcsak kun társadalom a Codex Cumanicus alapján (Kipchak and Cumanian society on the basis of the Codex Cumanicus), pp. 242–73 in Györffy, Gy. (ed.), *A magyarság keleti elemei*. Budapest: Akadémiai Kiadó.

Horváth, F. 2001. *A csengelei kunok ura és népe* (The lord and people of the Cumanians in Csengele). Budapest: Archaeolingua.

Káldy-Nagy, Gy. 1985. *A budai szandzsák 1546–1590 évi összírásai. Demográfiai és gazdaságtörténeti adatok* (The 1546–1590 census data of the Buda *sanjak*. Data on demography and economic history). Budapest: Pest Megye Múltjából.

Matolcsi, J. 1983. A középkori nomád állattenyésztés kelet-európai jellegzetességei (East-European characteristics of medieval nomadic animal keeping), pp. 281–306 in Tőkei, F. (eds), *Nomád társadalmak és államalakulatok* (Körösi Coma Kiskönyvtár 18). Budapest: Gondolat.

Pálóczi Horváth, A. 1976. A Lászlófalván 1969–1974-ben végzett régészeti ásatások eredményei (Results of the 1969–1974 excavations at Lászlófalva). *Cumania Archeologia* **4/1976**, 275–309.

Pálóczi Horváth, A. 1985. A kunok bejövetele és letelepedése. *Forrás* **17**, 65–71.

Pálóczi Horváth, A. 1989. *Pechenegs Cumans, Iasians. Steppe Peoples in Medieval Hungary*. Budapest: Corvina.

Pálóczi Horváth, A. 1996. *Élet egy középkori faluban. 25 év régészeti kutatása a 900 éves Szentkirályon* (Life in a medieval village. 25 years of archaeological research at the 900 years old village of Szentkirály). Budapest: Magyar Mezőgazdasági Múzeum.

Pálóczi Horváth, A. 2000. Lakóház és telek rekonstrukciója Szentkirályon, egy alföldi késő középkori faluban I. (The reconstruction of a house and a lot at Szentkirály, a medieval village in the Great Hungarian Plain I), pp. 121–49 in Bende, L. and Lőrinczy, G. (eds), *A középkori magyar agrárium*. Ópusztaszer: Tudományos ülésszak Ópusztaszeren.

Pálóczi Horváth, A. 2002. Szentkirály továbbélése a török korban (The continuity of Szentkirály in the Turkish Period), pp. 189–94 in Gerelyes, I. and Kovács, Gy. (eds), *A hódoltság régészeti kutatása* (Opuscula Hungarica 3) Budapest: Magyar Nemzeti Múzeum.

Somhegyi, T. 1998 A húsfeldolgozás és –fogyasztás jelei a középkori Szentkirályon (Evidence of meat processing and -consumption at medieval Szentkirály). *Magyar Mezőgazdasági Múzeum Közleményei* **1995–1997**, 9–20.

Takács, I. 1988. Collecting biological finds from the well of a medieval village, pp. 275–82 in Járó, M. and Költõ, L. (eds), *Archaeometrical Research in Hungary*. Budapest: National Centre of Museums.

Takács, I. 1989. Szentkirály középkori falu zoológiai leletei (Zoological finds from the Medieval village of Szentkirály). *Magyar Mezőgazdasági Múzeum Közleményei* **1988–1989**, 95–110.

Takács, I. 1994. Cumanian animal keeping at Szentkirály, a 15th–16th century site in central Hungary. *Magyar Mezőgazdasági Múzeum Közleményei* **1992–1994**, 15–23.

Tálasi, I. 1977. *Kiskunság*. Budapest: Gondolat.

Uerpmann, H.-P. 1973. Animal bone finds and economic archaeology. *World Archaeology* **4**, 308–309.

Vörös, I. 2000. Adatok az Árpád-kori állattartás történetéhez, pp. 71–119 in Bende, L. and Lőrinczy, G. (eds), *A középkori magyar agrárium*. Ópusztaszer: Tudományos ülésszak Ópusztaszeren.

Vörös, I. 2002. Török kori állatcsontleletek Magyarországon, pp. 340–52 in Gerelyes, I. and Kovács, Gy. (eds), *A hódoltság régészeti kutatása* (Opuscula Hungarica 3). Budapest: Magyar Nemzeti Múzeum.

Éva Ágnes Nyerges
Institute of Archaeological Sciences,
Loránd Eötvös University,
Múzeum körút 4/B,
H-1088 Budapest, Hungary
E-mail: nyergeseva@hotmail.com

9th ICAZ Conference, Durham 2002
Behaviour Behind Bones, (ed. Sharyn Jones O'Day et al.) pp. 271–284

30. Rich, poor, shaman, child: animals, rank, and status in the 'Gran Coclé' culture area of pre-Columbian Panama

Richard Cooke

In pre-Columbian burials at eight sites distributed across the Pacific littoral of the 'Gran Coclé' culture area of Panama, artifacts made out of animal remains and biomorphic icons depicted on mortuary arts provide information about the social rank, status and occupation of the deceased. Some adult males, who were presumably of high rank, owned very large numbers of animal products, such as shark, domestic dog and peccary teeth, and smaller numbers of exquisite crafts made of bone and whale ivory, these often overlain with gold-copper foil. People of lower rank or status, including women and children, were buried with fewer and simpler objects, e.g., *necklaces of teeth, shell and bone, and household tools. In some cemeteries, associations of animal products with other artifacts (for example, aprons sewn with felid teeth,* Spondylus *shell and gold) and clusters of sting-ray spines and stone and bone tools, allude to specific occupations, such as that of shaman, curer, chanter, warrior and hunter. The differential distribution of well-made* Spondylus *shell pendants shaped like animals may point to some kind of ranking among young people although evidence for unusually rich children, adolescents and adult women is absent. Some animal species commonly used for food, like the white-tailed deer, were depicted on craft objects. Other species apparently tabooed as food, eg., sea turtles and crocodiles, were painted and modelled frequently. The teeth of domestic dogs were used to make necklaces, but their flesh does not seem to have been consumed. It is likely, though impossible to prove with current data, that some animal icons represented specific sectors of the regional population, such as clans and descent-groups. One of these – a crocodilian – appears in high rank or high status graves dressed like an important human whereas in poorer peoples' burials it retains its unadorned animal guise.*

Geography, time and the social environment

The socio-cultural focus of this paper is the 'Gran Coclé' culture area of central Panama (Fig. 1; Sánchez 2000), which is well known for its 'semiotic tradition' of geometric and biomorphic icons inter-linked by repetitive artistic conventions (i.e., standardized claws, tongues, frets, and scrolls) (Lothrop 1937; Lothrop 1942; Linares 1977; Helms 1995; Labbé 1995; Helms 2000). These icons are wrought in several media: hammered and cast gold-copper alloys, clay, stone, marine shell, bone, and wood. Since depictions of animals are quite realistic, but not necessarily biologically rational, they can be evaluated with observational taxonomic criteria (Helms 1977; Linares 1977; Cooke 1984; Helms 1977; Benson 1992; Helms 1995; Cooke 1998; Helms 2000). This paper focuses on the relationship between social inequality and mortuary arts made of animal products (*e.g.*, bone, teeth and shell), and comments briefly on some iconographic details relevant to this theme. The religious significance of 'Gran Coclé' imagery is dealt with in another paper (Cooke, *this volume*).

The 'Gran Coclé' semiotic tradition is manifest around 200 cal BC (Cooke *et al.*, 2003) and subsequently underwent continual stylistic and technical transformations until Spanish conquest and settlement between AD 1502 and 1550 (Ladd 1964; Ichon 1980; Cooke 1985; Labbé 1995; Sánchez 1995; Sánchez 2000; Cooke and Sánchez 2003, Fig. 2). During this 1800-year period, regional native peoples lived in hamlets and villages of palm-and-thatch houses whose maximum population likely did not exceed 2000 people *per* community. They buried their dead in earthen graves and low earth mounds, the structural simplicity of which belies, not only the impressive material wealth accumulated by a few people, but also the variety

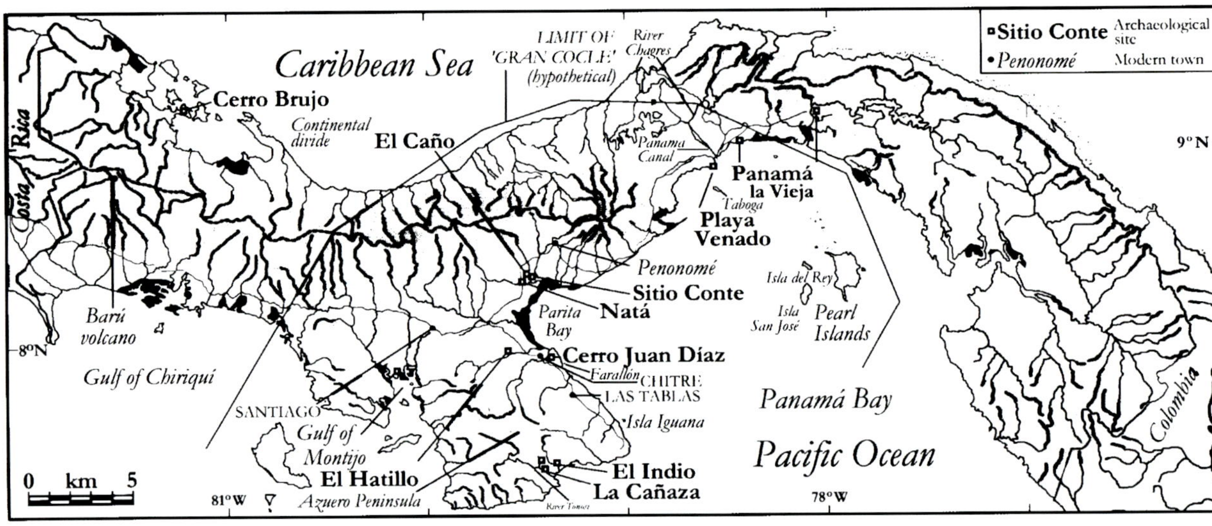

Fig. 1. Map of Panama, which shows the location of archaeological sites and geographical features mentioned in the text.

and complexity of mortuary treatments and placements of the dead, even those of low rank or status (Lothrop 1937; Ladd 1964; Briggs 1989; Hearne and Sharer 1992; Cooke and Sánchez 1998; Díaz 1999; Cooke *et al.* 2000). Undoubtedly, 'Gran Coclé' society was very concerned with the well being of ancestors in the after-life and with their continuing contact with the living (Díaz 1999; Cooke 2001).

At the time of Spanish contact 'Gran Coclé' was divided into many political units, which generally comprised one or more river valleys (or sections thereof) situated between less fertile land or natural barriers. Settlement nucleation is less pronounced on the Caribbean slopes where it appears to be correlated as much with the proximity of valuable 'point resources,' such as gold and copper ores and volcanic rocks for axes and grinding stones, as with ecological carrying capacity (Griggs 1998; Griggs *et al.* 2002; Cooke *et al.*, 2003). It is customary for anthropologists to uncritically accept assumptions made by Spanish soldiers and *literati* that each of these political units constituted a territorially stable 'chiefdom' ruled by a 'chief' who was a member of a hereditary elite (Helms 1979; Drennan 1991; Drennan 1996). There are good reasons for inferring that social groups on the isthmus were ranked in some way and that political leaders were drawn from top-ranked polities. Even so, although it is clear from the following discussions that a few individuals were considerably richer, in a material sense, than the majority of the population, it is difficult to determine with current archaeological data whether their wealth was due to inherited rank, acquired status or position, or occupation (Linares 1977; Linares 1979). As in several chiefdoms elsewhere in the world, heredity and ascription probably went hand-in-hand. An additional problem for interpretations of the data summarized herein is the likelihood that, in pre-Columbian times, social relations were governed by the same kinds of complex behaviours that are described by, or can be inferred from contact-period documents and the ethnography of lowland Neotropical societies, such as multilingualism, constant group fissioning, complex marriage and residence rules and fickle descent group politics. My use of the term 'high rank' and 'high status', then, should not be construed as affirming that the power or wealth of the individual(s) concerned was necessarily inherited. The absence of unusually rich women and children in 'Gran Coclé' is an apposite *caveat* against this assumption (Cooke *et al.*, *in press*).

Environment

In spite of its small area, 'Gran Coclé' comprises a wide range of coastal and terrestrial habitats. The Atlantic and Pacific coasts, less than 100 km apart, are ecologically and physically very different (Jackson and D'Croz 1997). Some animal resources that are unique to the Caribbean were used by communities on the opposite coast (*e.g.*, manatee bone and some marine shells). The salient feature of the Pacific coastline are two estuarine systems, Parita Bay and the Gulf of Montijo, and several coral-fringed islands, some small and inshore (*e.g.*, Isla Iguana, Farallón de Chitré and Taboga), others large and a long canoe ride offshore (*e.g.*, Rey and San José islands in the Pearl island archipelago). Hard substrates fringing these islands probably provided most of the clear-water marine shells, which were used for making personal ornaments (i.e., *Spondylus*, *Oliva* and *Pinctada*).

The east-west central *cordillera* is narrow and steep, and rises to over 2000 m. Another north-south mountain chain dissects the Azuero Peninsula rising to 1600 m at its southern extremity. This T-shaped orography, in tandem with the annual migration of the Inter-Tropical Convergence Zone across the isthmus, leads to contrasting climatic regimes on the Caribbean and Pacific slopes and on the eastern and western side of the Azuero Peninsula. The steepest parts of the Caribbean are extremely wet (3000–5000 mm rainfall annually) and lack prolonged dry periods that allow fallen vegetation to be burnt effectively. Over most of the Pacific side, dry seasons last for 3–5 months and bring strong adiabatic winds, which desiccate the landscape quickly, especially where lowered elevations in the central *cordillera* accelerate wind velocity, as in the area bordering Parita Bay.

After several millennia of pre-Columbian agricultural activities (Cooke *et al.* 1996; Piperno and Pearsall 1998) by 200 cal BC much of the Pacific side would have been under grassy and wooded savannas interspersed with gallery forests along watercourses and wooded tracts on hills and in the mountains (Cooke and Ranere 1992). Extensive and minimally impacted forests would have remained only in the steepest and rainiest areas. The taxonomic composition and abundance of animal taxa recorded in archaeological sites throughout 'Gran Coclé' are logically conditioned by this longevally anthropogenic vegetation history.

One might expect external contacts and long-distance trade to have figured prominently in pre-Columbian isthmian life. Empirical evidence for it is, in fact, rather thin.

The archaeological sites to which I refer are all located within 30 km of Parita Bay. Archaeozoological research demonstrates that these sites' inhabitants could have obtained all or most of the animal taxa they used for food or for making utilitarian artifacts within a day's walking or canoing distance. In samples of animal bones from kitchen middens, the white-tailed deer (*Odocoileus virginianus*) is by far the most abundant species, so much so, in fact, that its ease of capture probably influenced the low representation or absence of other widespread Neotropical mammals. Iguanas (Iguanidae) were also regularly consumed at all times and sites. Bird and turtle species reflect opportunistic hunting in the nearest available habitats, i.e., mud-flats, mangroves, swamps, and savannas. Fishing focused on abundant in-shore marine species, especially marine catfish (Ariidae), small, shoaling taxa (*e.g.*, thread herrings [*Opisthonema*], lookdowns [*Selene*] and grunts [*Orthopristis*]), and large predators on the nektonic fauna [*e.g.*, corvinas [*Cynoscion*], snook [*Centropomus*] and jacks [*Caranx caninus*]). Littoral food resources, such as dried or salt fish, were transported to sites located further inland. At inland sites, a few freshwater fish taxa, such as catfish (*Parauchenipterus* and *Rhamdia*), gymnotid eels (*Sternopygus*) and characids (*Hoplias*), were important food sources. Spanish captains describe markets at large villages such as Natá. Although some of the non-food animal products which I discuss probably originated from beyond the normal subsistence catchment of the territories where they were found, all could have been obtained within the 'Gran Coclé' culture area through regular, short-distance exchange channels (Ranere and Hansell 1978; Cooke 1979; Cooke 1992a; Cooke and Ranere 1992; Cooke 1993b; Zohar and Cooke 1997; Cooke and Ranere 1999; Jiménez 1999; Jiménez and Cooke 2001).

Mortuary samples and social hierarchies

The number of mortuary sites in 'Gran Coclé', which have provided samples of animal materials relevant to the research question, is very small (8) – so small, in fact, that it complicates interpretations of social inequality. Six sites are located around Parita Bay, in the driest area of the Pacific watershed, and two in the central part of Panama Bay, near modern Panama City, an area which represents the waxing and waning eastern boundary of the 'Gran Coclé' tradition (Fig. 1; Sánchez and Cooke 2000). The archaeologists who excavated these sites had different attitudes towards the value of faunal materials, which is why taxonomic precision, quantification and contextual associations are uneven.

One site – Sitio Conte (cal AD 750–950) – stands out from the others because it is only here that categorical wealth differences among graves, burials and individuals can be demonstrated. The wealthiest people – always adult males – were patently of high rank or status. Ethnohistoric evidence suggests that this was a society in which prowess in war was an important determinant of social worth and that the acquisition and display of booty were an integral part of this bellicose behavior. The placement of artifacts at Sitio Conte supports this relationship (Lothrop 1937; Helms 1977; Linares 1977; Helms 1979; Briggs 1989; Cooke *et al.* 2000). It is also likely that social rank was related to affiliation to hierarchized descent groups, which, to judge from recent ethnographic evidence (*e.g.*, Stone 1961), may have had eponymous animal 'ancestors'.

Sitio Conte is a unique locality in the 'Gran Coclé' culture area. If we assume that it constitutes a single, very large site – in conjunction with contiguous El Caño and Cerro Zuela – it is the only one, which contains monumental features suggestive of paramountcy, i.e., rows of un-carved and carved basalt columns, cobble stone pavements and stone-faced terraces (Fig. 2a; Lothrop 1937, 39–43, Figs. 16 and 17; Cooke *et al.* 2000). This suggests that Sitio Conte's importance transcended the limits of the 'chiefdom' in whose territory it was located. I propose that it functioned as a pan-regional meeting-place used for affirming common ancestry and for celebrating ritual games and the funerary rites of esteemed people, such as the members of high-ranked descent groups. In recent times, the Ngöbére-speaking Guaymí of central and western Panamá used a ritual game, the *balsería,* for similar ends (Young 1976). *Balserías* were held at maize and peach palm (*Bactris gasipaes*) harvest peaks and were

Fig. 2. El Caño, Coclé, Panamá. a: rows of columnar basalt that delineate a ceremonial space (photo: Carlos Fitzgerald B.), b: stone column of a human with a monkey on his back, perhaps a pet or a stuffed animal used in a ritual game like the modern balsería *(from Museum Rieteberg 1984), c: human male torso carved in stone, which wears a frog figurine presumably of gold-copper alloy (drawing: Arcadio Rodaniche).*

sponsored by individuals who had both good political connections and the wherewithal to amass sufficient food and fermented drinks to satisfy the invitees. Successful players were well respected: usually they wore stuffed animals on their backs, which they considered to be symbolic of their prowess (Johnson 1948, plate 45). At a huge *balsería* celebrated at Canquintú, in the lower Cricamola valley (Caribbean, Bocas del Toro province), in November 1974, I observed *balseros* wearing the following animals: cayman (*Caiman fuscus*), giant anteater (*Myrmecophaga tridactyla*), tayra (*Eira barbara*), ocelot (*Leopardus pardalis*) and jaguar (*Panthera onca*). A stone statue found at the El Caño section of the Sitio Conte complex depicts a person who wears an animal that appears to be a monkey on his back (Fig. 2b). On another statue (Fig. 2c), a thick-set male displays a frog figurine dangling from a cord on his chest, which surely represents a golden image as a status symbol (albeit of unreal size). The monkey could, of course, be a pet! But, in view of the regional uniqueness of the stone-lined ceremonial space at El Caño, a possible parallel with the modern Ngöbé *balseros'* use of animals should be noted.

Another site at which lavish burials have been discovered is El Hatillo or Finca Calderón (He-4) (cal AD 750–1520). It was excavated by a Smithsonian Institution/National Geographic team in the 1940s and '50s (Ladd 1964) and subsequently ransacked by 'amateurs,' who discovered the richest burials (Biese 1967). El Hatillo is a very large site with evidence for dwelling zones as well as tumuli with burials (Ladd 1964; Haller, *in preparation*). It was probably the chief village of a confederation centered on the lower valleys of the La Villa and Parita rivers, which, at Spanish contact (AD 1516–17), was under the aegis of a regionally influential chieftain, 'Parita' ('París' or 'Antatará') (Lothrop 1937, 46; Cooke and Bray 1985; Cooke *et al.*, 2003). It lacks, however, the monumental features of the Sitio Conte centre (Cooke 1993a).

Playa Venado is another site whose mortuary arts suggest that high rank or high status people were buried there. Some of the 360 burials were excavated by a professional archaeologist (S. K. Lothrop) in the 1950s, but most were "studied" by amateurs in search of gold ornaments to sell. No monograph exists although some grave descriptions have been published (Lothrop 1954; Feriz 1956; Lothrop 1956; Bull 1958; Sander *et al.* 1958; Sander *et al.* 1959; Bull 1961). Most of the painted mortuary pottery found is comparable with the 'Cubitá' and 'Early Conte' styles of the 'Gran Coclé' polychrome pottery sequence, which were in vogue between cal AD 550 and cal AD 850 (Sánchez and Cooke 2000). Some fine shell ornaments and necklaces of animal teeth (Fig. 5c; Lothrop *et al.* 1957) cannot be related to particular individuals, but their quality alludes to people of high rank or status. Several finely worked cast gold-copper pendants and embossed hammered plaques were found here (Lothrop 1956; Lothrop *et al.* 1957; Cooke and Bray 1985, Figs. 10 and 15; Bray 1992, Fig. 3.7).

The five other sites that I will consider – Sitio Sierra, Cerro Juan Díaz, El Indio, La Cañaza and Panamá la Vieja – have cemeteries in which distinctions among grave goods appear to be symptomatic of age, sex and occupation, rather than of social rank or status. These cemeteries received children, adolescents and adults, men and women. Some people were buried with unusual artifacts, however, which probably relate to a special position or occupation (*e.g.*, that of 'shaman', 'curer' or 'chanter'). Archaeological survey shows that Sitio Sierra and Cerro Juan Díaz were quite large settlements, with a maximum extent of about 45 and 200 hectares, respectively. Particular areas within all these sites were set aside for burials. At Cerro Juan Díaz, it is likely that families, clans or similar social groups maintained particular sections of the intentionally levelled graveyard for several generations because it is evident that tombs were kept open and burials were continually being added to them (Ichon 1975; Cooke 1979; Ichon 1980; Briggs 1989; Isaza 1993; Cooke and Sánchez 1998; Cooke *et al.* 2000).

Animals and rich people at Sitio Conte

The universities of Harvard and Pennsylvania excavated about 100 graves at Sitio Conte in the 1930s and '40s. Briggs (1989, 73) estimated that 94% of the 93 skeletons that could be aged were adult; 77.4% were males and 22.6% females. Only two children's skeletons were found. According to Briggs (1989, 138–39), the placement of mortuary arts at Sitio Conte is characteristic of an "additive" pattern of status recognition: the more important the individual, the more objects and the more categories of artifacts he or she possessed. The richest people are adult males. Not only can they be identified on the basis of the quality and quantity of their mortuary goods, but also their graves are larger and have more people buried in them, often in layers with the primary occupant in the middle, *buried in a seated position* – as in two of the most splendid graves, 26 and 74 (Lothrop 1942, Fig. 31; Briggs 1989; Hearne 1992, Figs. 1.9 and 1.10). Some skeletons in graves, which have several people buried in them (particularly those placed face-down in an extended position) may be prisoners executed for inhumation with dead chiefs – a custom described by Spanish soldiers in the early 16th century AD. Alternatively, some may be the embalmed corpses of ancestors preserved in special mortuary houses – another historically recorded funerary custom in 'Gran Coclé', which has recently been identified archaeologically at Cerro Juan Díaz (Cooke 2001).

Certain artifacts accompany only the richest individuals (quantitatively defined [Briggs 1989]), *e.g.*, gold-copper disks, plaques, pendants, greaves, cuffs and helmets, which usually bear embossed or modelled animal designs (Fig. 3 c-f), and carved whale teeth and manatee bone, sometimes decorated with gold leaf (Fig. 3a-b; Briggs 1989; Hearne and Sharer 1992, Plates 22–28). Pre-1550 Spanish chronicles describe prominent warriors wearing such apparel to war (Cooke *et al.*, 2003). A Spanish captain happened upon the dead chieftain 'Parita' wrapped in fine mantles bedecked with similar golden apparel near the site of El Hatillo in AD 1517 (Lothrop 1937, 43; Cooke and Bray 1985). Lothrop identified the whale teeth as coming "from the lower jaw of a female sperm whale" (*Physeter macrocephalus*) (Lothrop 1937, 170). Sperm whales are frequently seen in Panama Bay, especially during the dry season, and around the Pearl Islands (Fig. 1). There is no evidence for whaling, and pre-Columbian teeth were surely collected from stranded animals. Lothrop (*op. cit.*) proposed that artisans chose the proximal ends of manatee ribs because of their similarity to the shape of whale teeth. Manatee bone must have been traded in from the Caribbean coast since there are no Quaternary records of this taxon from the eastern tropical Pacific. Manatees were hunted in pre-Columbian times on the Caribbean (Wing 1980) and are recorded as common in early historical documents as for example, in the Chagres river (Oviedo 1853, 148).

One of the three burials in a large, very rich grave (number 32), had more carved bone and ivory objects than any other at Sitio Conte, including five spear-throwers, a comb, a set of six bone crocodile pendants with gold-copper overlay, four carved sperm whale teeth (one with gold-copper overlay) and a set of carved deer vertebrae (Fig. 4; Lothrop 1937, 283). Perhaps the deceased was a bone-carver. Or perhaps the cluster of spear-thrower parts alludes to his having been a hunter or warrior. One item depicts a realistic spotted cat (Fig. 4j). Felid images are notably few in number at this site (Cooke 1998).

Some top-ranked graves possess very large numbers of the products of ubiquitous animal taxa, which, as we shall see, also occur in much lesser quantities in the graves of lower ranked or poorer people (Fig. 5; Cooke 1998, Table 4.3). For example, one individual at Sitio Conte was buried with 200 peccary (*Tayassu*) tusks arranged as eight necklaces (Lothrop 1937, Fig. 130); another with a necklace or apron made of more than 300 domestic dog teeth (*Canis familiaris*) (Lothrop 1937, Fig. 33). The peccary tusks have not been identified to species. Peccaries (*Tayassu tajacu* and *T. pecari*) were apparently rare around Parita Bay settlements during the time period under consideration: surprisingly few peccary bones are present in dietary archaeofaunas that date after 200 cal BC (Cooke and Ranere 1989; Cooke and Ranere 1992; Jiménez 1999; Jiménez and Cooke 2001; Cooke and Ranere). A necklace with so many tusks, then, points to acquisition from afar or very good trade contacts – and an object of considerable value and prestige.

Worthy of note is the clustering in some Sitio Conte graves of items of marine origin: in the upper level of grave 74, for example, one individual (number 5) was associated with two carved whale teeth, about 200 sting ray spines and 100 shark teeth (Briggs 1989, 112 and 202–3). One individual in grave 32 owned 248 perforated shark teeth (Lothrop 1937, Figs 32 and 34 top, Fig. 132a). The ray-shark association may have a military connotation. Spanish documents record wooden sword-clubs, called *macanas*, which had shark teeth fixed to their edges (Jopling 1994, 57). Lothrop's (1937, 98–9) proposal that sting-ray spines were used as projectile points is in line with Spanish chronicler Fernández de Oviedo's (1853, 129) reference to "fish bones" serving this purpose. A grave at Panama la Vieja, which contained a single adolescent aged between 13 and 15, presents an interesting association of 82 *Spondylus* shell ornaments, at least 46 sting-ray spines and 30 well-made stemmed unifacial chalcedony blades (Fig. 6 b-d; Proyecto Arqueológico Panamá la Vieja 1998). The position of the fish spines and stone blades at the side and feet of the deceased suggests that each group of objects was buried in individual bags. (At nearby Playa Venado, Bull [1958] reported finding a robust adult male buried with a pouch decorated with shell and gold beads). As far as I can tell, all the spines are from rays of the genus *Dasyatis*. The largest are comparable in size and robustness to spines from two *Dasyatis longus*, housed in the Smithsonian Tropical

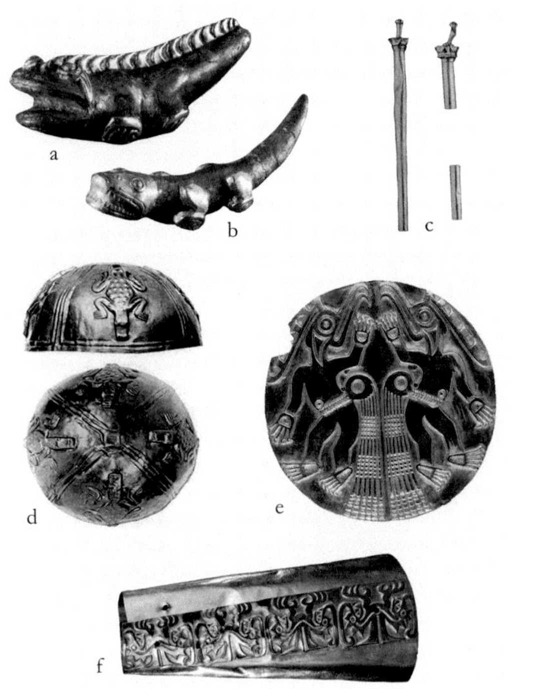

Fig. 3. Primary symbols of rank and power in 'Gran Coclé.' a-b: sperm whale ivory with gold-copper leaf, c: ear-rods, d: helmet, e: plaque embossed with humanized bird figures, f: cuff embossed with crocodile figures. a-c, e-f, from Hearne and Sharer 1992, reproduced with permission. d, from Lothrop 1937, reproduced with permission of the Trustees of Harvard University.

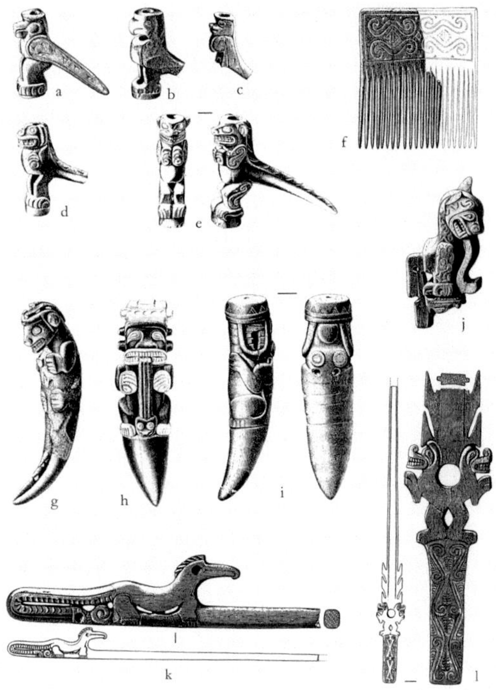

Fig. 4. Bone artifacts found in the middle layer of burials in grave 32 at Sitio Conte. a-e: carved deer vertebrae, f: comb, g-i: sperm whale teeth, j-l: parts of spear throwers. From: Lothrop 1937, reproduced with permission of the Trustees of Harvard University.

Taxon	Article	Maximum number in a single grave	Grave number
Emydid turtle	Carapace (whole)	9	1
Whale	Carved teeth	7	32
Stingray	Caudal spines	>205	74
Rabbit	Incisors	176	74
Shark	Perforated teeth	248	32
Peccary	Tusks	186 = 4 necklaces	1
Deer	Carved vertebrae	16	32
Manatee	Carved ribs	8	1
Dog (domestic)	Perforated canine teeth	321	74

Fig. 5. Maximum numbers of animal remains, which were found only in the richest graves at Sitio Conte or were present in them in unusually large numbers. "Richest" means graves in clusters 1–4 of Briggs' (1989) analysis; see Cooke (1998) for a list of all animal products found at Sitio Conte.

Research comparative skeleton collection, which weighed 28.9 and 50.9 kg, respectively (Fig. 7e-f). Some of the archaeological examples have been thinned and tapered at their proximal ends and scored with a sharp instrument, presumably for hafting (Fig. 7a-d). This modification would, of course, be appropriate for using them as spear points. It is also possible that the spines and stone blades represent another kind of activity, such as blood-letting: Oviedo (1853, 138) mentions that stone flakes were used on the isthmus for bleeding legs when people "felt tired". Sting-ray spines were used for elite blood-letting rituals among the Maya and figure prominently on Maya art, as they do on 'Gran Coclé' polychrome plates (Benson 1988). A bizarre ritual using a "particular fish spine" was recorded by Fray Adrián de Ufeldre, a Flemish priest who worked among the Ngöbé in the seventeenth century: piercing the foreskin with the spine, running a cotton cord through the hole and then having two people pull at either end in order

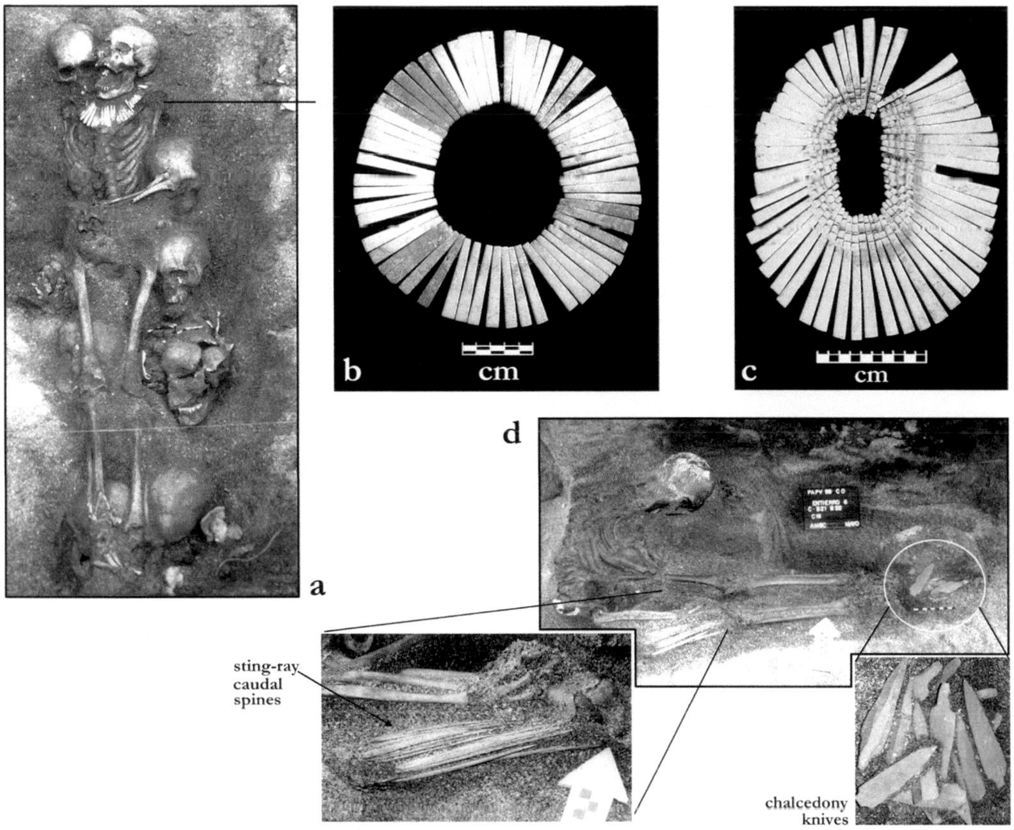

Fig. 6. Two burials at Panamá la Vieja. a: a young adult woman who wears a Spondylus *pendant necklace and was buried with seven human skulls, b: close-up of* Spondylus *necklace, c:* Spondylus *necklace found at Playa Venado, d: an adolescent whose mortuary arts consisted of* Spondylus *beads and pendants, 30 stone blades (circled) and about 46 caudal spines of sting-rays (*Dasyatis*). Photos: a: Richard Cooke, b: Tomás Mendizábal, c: Luís A. Sánchez, reproduced courtesy of the Dumbarton Oaks Museum, Washington, D.C., d: Tomás Mendizábal, published with the approval of the Fundación Panamá Viejo, Panamá.*

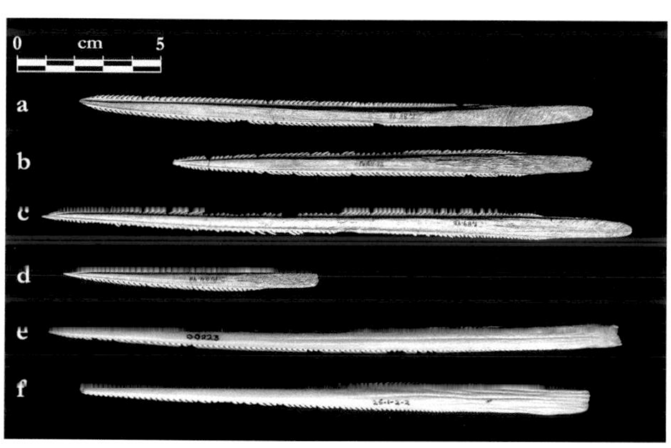

*Fig. 7. a-d: Sting-ray spines (*Dasyatis*) found in a grave occupied by a single adolescent at Panamá la Vieja (see Fig. 5d). They have been tapered and thinned distally. e: caudal spine from* Dasyatis longus, *total length: 2.78 m, weight: 58.86 kg, f: caudal spine from* Dasyatis longus, *total length: 1.6 m, weight: 20.88 kg.*

to bleed the penis. This ceremony was apparently practiced by groups of males in order to placate a "lightning deity" when a bolt fell to earth (Torres de Araúz 1965).

Manatee bone at El Hatillo

In the 1960s a looter excavated a grave at El Hatillo in which he found about 30 cast gold effigies of humanized animal warriors who wear the headdresses and belts of high rank, and brandish spear-throwers and sword-clubs of palm wood, some of whose blades are made of manatee bone (Biese 1967; Bray 1992, Fig. 3.11; Cooke *et al.*, 2003 Fig. 8). Professional archaeologists found beautifully carved manatee bones in a grave whose pottery suggests a contact-period date (around AD 1500). It contained two urns with human bone and a necklace of 737 perforated human teeth (Ladd 1964, 245 and Plate 1a-c) – raw numbers that allude that the person who wore them had special status. Another set of carved bones (not identified taxonomically) found by amateurs at El Hatillo (apparently in a single grave) is of exquisite workmanship. It includes a spear-thrower hook (Torres de Araúz 1972, 73). These finds reaffirm the importance of quality crafting to the 'Gran Coclé' value system (Helms 1993).

Cemeteries with mostly more modest mortuary goods

In contrast to Sitio Conte and El Hatillo, no one stands out as being unusually wealthy in two cemeteries at Cerro Juan Díaz, two at Sitio Sierra and three located at the southern tip of the Azuero Peninsula (two at El Indio and one at La Cañaza). Some people, however, were buried with special artifacts or unusual combinations of objects, which allude to particular occupations. At some of these sites, small ornaments of animal teeth, bone and shell appear to be proportionally more frequent with children and adolescents, than with adults.

At Cerro Juan Díaz, the oldest burial horizon was located underneath a circular arrangement of stone-lined ovens (Cooke and Sánchez 1998, Fig. 3; Cooke *et al.* 2000, Fig. 8.6). They represent the period cal AD 150–550. One grave contained the primary extended skeleton of an adult, one a flexed adult woman laid on top of a disturbed earlier burial, and two graves, several packets of secondary burials wrapped in bark cloth. The extended adult male (Operation 3, Feature 1) was buried with a special group of goods: two pottery incense burners, two gold plaques with raised spirals, 400 *Spondylus* beads, and 25 perforated canines of puma (*Puma concolor*) and jaguar (*Panthera onca*). The non-ceramic objects were probably part of a composite apron or shirt (Cooke and Sánchez 1998, Fig. 4; Cooke *et al.* 2000, Figs. 8.1, l-m and 8.6; Cooke 2001, Fig. on p. 58).

An earlier grave (Feature 16) received at least 18 individuals buried in secondary bundles. One of these contained an adult and an adolescent, whose funerary goods are unique in this feature: two polished stone bar-pendants, a copper ring, and two necklaces, one of puma and the other of ocelot (*Leopardus pardalis*) and raccoon (*Procyon lotor*) canines (Cooke 1998; Cooke and Sánchez 1998, Figs. 4, 5 and 8; Cooke *et al.* 2000, Fig. 8.6 and 8.7). These individuals did not receive special burial treatment, unlike that afforded the central occupants of the richest Sitio Conte graves, but were rather wrapped in a bark cloth bundle like everyone else in the grave.

The mortuary associations of Features 1 and 16 at Cerro Juan Díaz allude to some ritualistic occupation, such as that of shaman or curer – a ubiquitous office in Neotropical societies, and one that is not necessarily commensurate with political power. The adult and adolescent buried together are perhaps evidence for a shaman (= "puma") and his younger apprentice (= "ocelot"). Given the fact that the gold-work found in Feature 16 is the oldest yet radiocarbon-dated in Central America (cal AD 130 [250] 370) (Cooke *et al.*, 2003), it lends support to the idea that the initial significance of gold-copper artifacts in isthmian society was ceremonial or arcane, and that only later did it acquire a manifestly political association with power, influence and showing-off wealth that is so evident at Sitio Conte between cal AD 750 and 950. Confirmation of a metal – marine shell – animal tooth necklace association is provided by one of the four individuals buried in Grave 1 at Sitio Conte (one of the oldest at the site) who owned several exquisite gold pieces and hundreds of perforated dog teeth interspersed with hundreds of elongate beads, which are described as bone by Lothrop (1937, Fig. 33), but are in fact the same kinds of *Spondylus* beads that were found in Feature 1 at Cerro Juan Diaz.

In a grave in the second cemetery at neighbouring Sitio Sierra (dated to around cal AD 980-1270), an unusually small male aged between 20 and 35 had few burial goods, but among them was a greenstone necklace, a pelican bone flute *(Pelecanus occidentalis)* and the skeleton of a macaw – probably the scarlet macaw *(Ara* cf. *macao)*, once common in Panamanian dry forests (Cooke 1984). A materially poor person, then, who, in spite of being buried face-down, seemingly had a special job, i.e., that of curer or chanter. At Sitio Conte, extended skeletons that are arranged around principal occupants, and may have been executed captives, are generally placed face-down. Therefore being buried in this position may have been a dishonor. The only other animal remains in this sample of 13 adults were two perforated teeth of two species of inshore shark (*Carcharhinus* cf. *limbatus* and *C. leucas*), each one representing a single amulet worn around the neck.

At Cerro Juan Díaz, hundreds of graves were placed over the stone-lined ovens between cal AD 650 and 1350. These were mostly shallow tombs, less than 1.5 m deep, which received bodies prepared in several ways (primary, secondary, cremated, and in urns) – often in the same grave. Some burial units are exact contemporaries of the later, richest graves at Sitio Conte (numbers 5, 24 and 26) and have the same style of polychrome pottery albeit in much smaller numbers (Cooke *et al.* 2000, Fig. 89 a-e). Claudia Díaz (1999) has aged and sexed a sample of 115 skeletons (see also Cooke 2001, Fig. on p. 58; Cooke *et al.* 2000, Fig. 8.5). Forty-nine (43%) belonged to children of which 10 were neonates and 11 infants under 5 years. Twenty-nine (74%) of the 35 adult skeletons, which could be aged and sexed, were women. Animal-derived paraphernalia represent many of the same taxa that adorned the rich dead at Sitio Conte – *but in much more modest numbers*. One adult owned a necklace of sixteen counterfeit big cat claws made of a very ordinary mud-flat bivalve (*Anadara grandis*), twelve mother-of-pearl beads, three *Spondylus* beads and a necklace of teeth: two bull shark (*Carcharhinus leucas*), nineteen domestic dog, one white-tailed deer, three raccoon (*Procyon lotor*), one tayra (*Eira barbara*), one jaguar and one howler monkey (*Alouatta*) (Fig. 7l). Very similar counterfeit claws were reported by Ichon (1980, Plate LV) from La Cañaza (see below). The monkey tooth is one of only three occurrences of primate materials in all 'Gran Coclé' sites. Another is that of a

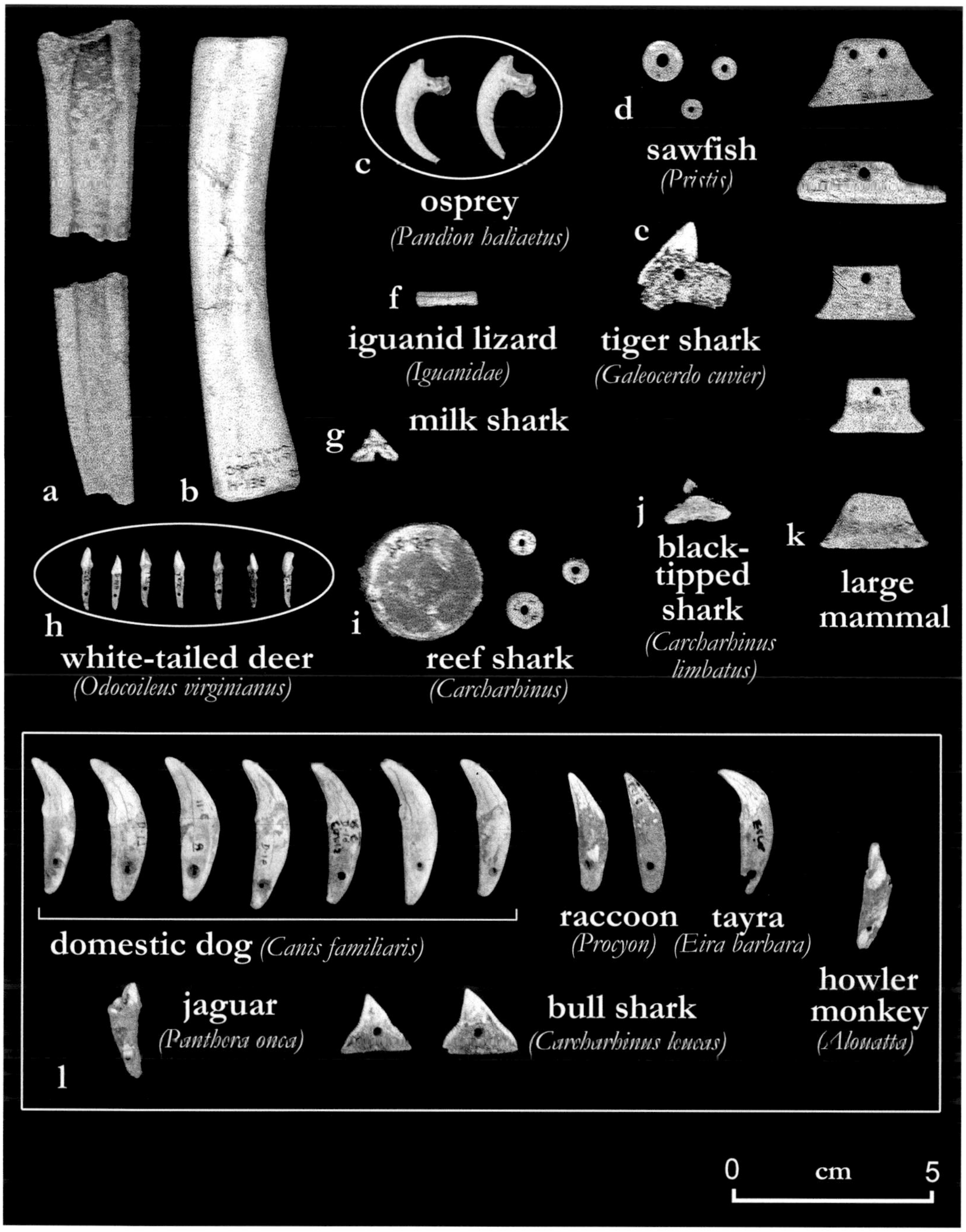

Fig. 8. Modest mortuary artifacts from the second cemetery at Cerro Juan Díaz, Panamá. a: split and polished metatarsal, b: split and polished humerus, c: distal phalanges, d: vertebrae, e: tooth, f: long bone, g: tooth, h: perforated incisors, i: vertebrae, j: burnt and polished tooth, k: long bone fragments, l: teeth belonging to a single necklace.

skeleton of a small monkey found complete alongside a burial at the El Caño section of Sitio Conte by Institute of Culture personnel. It was subsequently lost. I identified it without reference to a comparative skeleton as a white-faced capuchin (*Cebus capucinus*). Was it a pet or a *balsero's* embalmed animal avatar (cf. Fig. 2b)? The third report refers to a six year old child buried in a lidded urn at Playa Venado, whose funerary offerings were limited to *Spondylus* pendants and a necklace of "monkey" and "dog" teeth (archaeozoologically unconfirmed) with a solitary gold spacer bead.

Other animal products buried with the few people who possessed them in the second Cerro Juan Díaz cemetery were functional items, such as a deer metapodial tool and extremely simple personal adornments, *e.g.*, bone tubes, between 1 and 14 perforated teeth used as amulets or necklaces, two osprey (*Pandion haliaetus*) claws, perforated dog teeth, a few perforated and unperforated

*Fig. 9. Representations of sawfish (*Pristis*) on polychrome pottery from 'Gran Coclé'. From Labbé 1995, Figs. 88 and 89, reproduced with permission of Bowers Museum of Cultural Art, Los Angeles.*

elasmobranch centra and teeth, and a single polished and burned tubular bead made out of an iguana (Iguanidae) femur! (Fig. 8a-k).

Sawfish *(Pristis)* were once common in inshore tropical waters (Fig. 9); they have a bizarre appearance, bear live young, live in fresh and salt water, and grow to a very large size. Their rostral spikes were often hafted, perhaps as projectile points and awls (Lothrop 1937, Fig. 65). But the abundant use of their vertebrae for beads suggests they possessed a significance that transcended their practical usage. That this may have had a cognitive basis is suggested by tiny sawfish spikes from a single rostrum small enough to have been an embryo, which were found alongside an 8-year old child at Cerro Juan Díaz (Carvajal, *personal information*).

According to Briggs (1989, 62–3), three cemeteries excavated by Alain Ichon in the Tonosí valley at the southern tip of the Azuero Peninsula – El Indio I (cal AD 250–550), El Indio II (cal AD 750–950) and La Cañaza (cal AD 550–950) – point to an 'egalitarian' society in which the acquisition of sumptuary goods nevertheless increases through time. None of the individuals buried in the two cemeteries contemporary with Sitio Conte (El Indio II and La Cañaza) approach the richness of this site's alpha males. The scatter of modest animal bone amulets, necklaces and functional items, *e.g.*, shell and bone beads and pendants and a few perforated dog and shark teeth, is similar to that of the second cemetery at Cerro Juan Díaz. Twenty-five graves out of a total of 29 that had artifacts made of animal products contained the remains of children and adolescents. The largest numbers of teeth in a necklace were 32 perforated dog's teeth buried with an adult at El Indio (II) – modest by Sitio Conte's standards. In the earlier cemetery at Sitio Sierra, with 25 graves (300 cal BC – cal AD 500), only two had animal products – a group of seven sting-ray caudal spines from the spotted eagle ray (*Aeteobatus narinari*) and thornback ray (*Dasyatis*), buried in a group alongside an adult, and a few tear-shaped shell beads with a child (Isaza 1993).

Spondylus shell pendants: special artifacts?

Beautiful *Spondylus* shell pendants carved like animals (anurans, sea turtles, crocodiles, 'monkeys', 'dogs', 'felines' etc.) were recovered in several graves at Cerro Juan Díaz, El Indio (II), La Cañaza, Playa Venado and Panamá la Vieja (Lothrop *et al.* 1957; Ichon 1980, Figs. 86–89; Briggs 1989, Figs. 4 and 5; Cooke and Sánchez 1998; Cooke *et al.* 2000, Fig. 8.7 l-q). At Cerro Juan Díaz, many of these beads were found in Feature 16 (Cooke *et al.* 2000, Fig. 8.6) although the jumbled nature of the secondary packages jammed into this grave make it difficult to associate specific groups of artifacts with individual skeletons. Some shell beads were found inside the package, which held the remains of the purported 'shaman' and his 'apprentice,' and could have been the property of the latter. In the second cemetery at El Indio, Briggs (1989, 34–54) noted that shell pendants and beads were most prominent in children's graves, of which one (grave 42) had an unusual number of very large and beautiful *Spondylus* ornaments. Consequently, he inferred that the differential distribution of such artefacts pointed towards some kind of rank or status *among young people*. It is relevant that the beautiful necklace of club- or claw-shaped pendant from Panamá la Vieja adorned the neck of a 13–15 year old adolescent (Fig. 7a-b). This person was buried with seven human skulls placed carefully around the body – a custom that is also frequent at Cerro Juan Díaz, and, in my opinion, is more likely to represent the intentional hoarding of ancestors' remains, than the taking of trophy heads in battle. That *Spondylus* artefacts, however, were not the *exclusive* possessions of sub-adult people is indicated by finds of hundreds of long beads with the purported 'shamans' at Cerro Juan Díaz (Operation 3, Feature 1) and Sitio Conte (Grave 1).

Animal icons, food, and human society

The very small and geographically unrepresentative sample of professionally excavated mortuary sites in pre-Columbian 'Gran Coclé' makes it frustratingly difficult to identify and interpret the nature of wealth and social hierarchy and its correlation with the placement of animal products and icons in individual graves. Current knowledge suggests that high rank or high status people – always adult males, it seems – were not necessarily identified by a particular animal or group of animals, but by the biggest

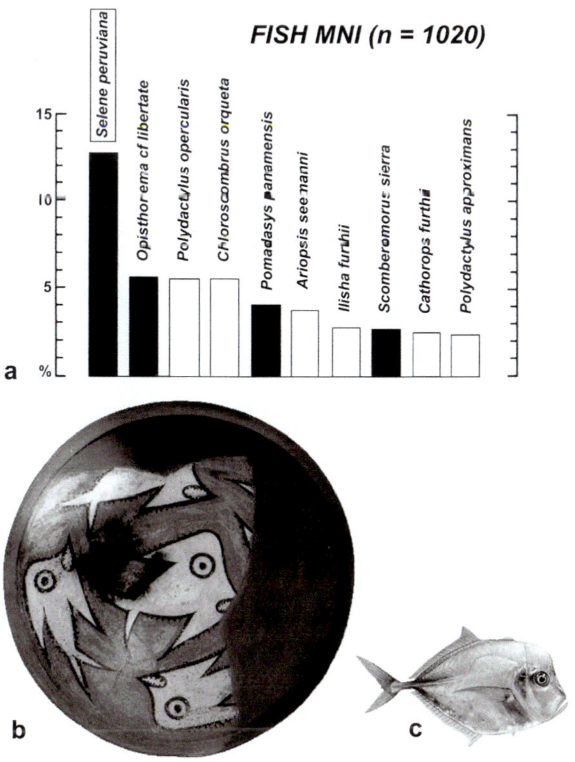

Fig. 10. a: Abundance (% minimum number of individuals [MNI]) of marine fish species in five middens at Cerro Juan Díaz, Panamá (after Jiménez and Cooke 2001). b: Red painted plate of the Early Conte style of the 'Gran Coclé' polychrome pottery sequence, which depicts a shoal of fish of the genus Selene *(of which there are three species in the eastern Tropical Pacific). c:* Selene peruviana, *from Allen and Robertson 1986, reproduced with permission of Crawford House Press, Bathurst.*

and best necklaces, belts and aprons, by the most exotic or scarcest materials (such as sperm whale teeth), and by the most elegantly crafted objects. It is also possible that necklaces made out of large numbers of felid, dog and peccary teeth; sting-ray caudal spines; macaw remains; *Spondylus* beads sewn to garments or used as pectorals with gold and felid teeth; and carved bone spear-thrower parts may identify people with a special *occupation*, such as that of shaman, curer or chanter, warrior and hunter, and their apprentices or offspring (Cooke and Ranere 1989; Cooke 1992b; Cooke and Ranere 1992; Cooke and Ranere, *in press*).

In 'Gran Coclé,' even the most mundane food species were painted and modelled, such as the Pacific lookdown or moonfish *(Selene peruviana)* depicted on the trichrome plate in Fig. 10b. This is the commonest fish taxon in Cerro Juan Díaz kitchen middens and the second or third most abundant at other sites around Parita Bay (Fig. 10a; Cooke 1992a; Cooke and Ranere 1999). The white-tailed deer, by far the most frequent mammalian species in all sampled 'Gran Coclé' middens, is often depicted on polychrome pottery, recognizable by its branched antlers (Cooke 1992b; Labbé 1995, Figs. 63 and 114; Helms 2000). We have seen that its vertebrae were occasionally exquisitely carved (Fig. 4 a-e). Conversely, some animal taxa, whose bones are absent or extremely rare in Pacific-side middens, are iconographically prominent, especially crocodilians and sea turtles, and, to a lesser degree, monkeys and domestic dogs. Presumably these animals were tabooed as food because they possessed a special cognitive significance. Yet other animal species, whose flesh was regularly consumed, such as the abundant Neotropical rodent, the paca (*Agouti paca*), never appear on 'Gran Coclé' art in recognizable forms (Linares 1976). Dog teeth were frequently displayed as necklaces and amulets by rich and poor. In an area well endowed with animal protein, their value as guardians and hunting aids presumably outweighed their dietary potential.

Many authors (*e.g.*, Helms 1977; Linares 1977; Helms 1979; Briggs 1989; Cooke 1998) have noted that some animal icons are more frequent in some graves, sites and regions in 'Gran Coclé' than in others. One trend in iconography that appears to be diachronic is the reduction in diversity of animal icons depicted on 'Gran Coclé' art and, concomitantly, the increasing iconographic prominence of a few taxa. Some of these, such as the king vulture (*Sarcorhamphus papa*), probably had a religious significance (Cooke, *this volume)*. Others may have been mythical figures well-known to the local audience *(e.g.*, Helms 1977; Helms 2000). Another possibility is that they were the animal counterparts of specific human groups. It is known that historic tribes in the isthmian region belonged to ranked clans with animal identifiers (Stone 1961). Many of the beautifully crafted artifacts in the top-ranked graves at Sitio Conte depict a humanized crocodile, which is dressed like *important* humans: with a headdress, baton or sword-club, ear rods and belt (Hearne and Sharer 1992, Plate 1). It is interesting that the crocodile is one of only two animal icons found in the second cemetery at Cerro Juan Díaz, where so many people of modest means were buried at the same time as Sitio Conte's rich and influential males. But none of these icons sports human attire. The hypothesis that, as time went on, the social group led in the supernatural world by the crocodile became progressively more dominant in the region bordering Parita Bay, accords with the increasing concentration of wealth, which is apparent from the archaeological record after around cal AD 750 at sites like Sitio Conte and El Hatillo. Sadly, although it is theoretically testable with mortuary remains, I doubt whether any high-rank cemeteries remain sufficiently intact in 'Gran Coclé' to provide the necessary data to evaluate this hypothesis.

Acknowledgements

The author expresses his gratitude for support granted to the Cerro Juan Díaz project (1991-2001) from the following institutions: Instituto Nacional de Cultura (Panama), National Geographic Society, Smithsonian Institution ('Scholarly Studies' and 'Latino Initiatives' programmes), Smithsonian Tropical Research Institute, Universidad de Costa Rica, Universidad de Panamá, Centro Regional Universitario de Azuero (Panama), Instituto Enrico Fermi (Panama), London University Institute of Archaeology, Université de Montréal, Arcillas de Chitré. He also acknowledges gratefully the dedication and skills in field and laboratory of the following persons who were involved in the recovery, conservation and analysis of the materials presented in these two papers: Adrián Badilla, Diana Carvajal, Benoit Desjardins, Claudia Díaz, Claudia Espejel, John Griggs, Ilean Isaza, Máximo Jiménez, Luís Alberto Sánchez, Olman Solís, Conrado Tapia, and Aureliano Valencia.

References

Allen, G. R. and Robertson, D. R. 1986. *The Complete Divers' and Fishermens' Guide to Fishes of the Tropical Eastern Pacific*. Bathurst: Crawford House Press.

Benson, E. P. 1988. A knife in the water: the sting-ray in Mesoamerican ritual life. *Journal of Latin American Lore* **14**, 12–24.

Benson, E. P. 1992. Motifs, meanings and myths: the iconography of Sitio Conte gold plaques, pp. 22–31 in Hearne, P. and Sharer, R. J. (eds), *River of Gold: Precolumbian Treasures from the Sitio Conte*. Philadelphia: University of Pennsylvania Museum of Archaeology and Anthropology.

Biese, L. 1967. The gold of Parita. *Archaeology* **20(5)**, 202–8.

Bray, W. M. 1992. Sitio Conte metalwork in its pan-American context, pp. 33–6 in Hearne, P. and Sharer, R. J. (eds), *River of Gold: Precolumbian Treasures from the Sitio Conte*. Philadelphia: University of Pennsylvania Museum of Archaeology and Anthropology.

Briggs, P. S. 1989. *Art, Death and Social Order: the Mortuary Arts of Pre-Conquest Central Panama*. (BAR International Series 550). Oxford: British Archaeological Reports.

Bull, T. 1958. Excavations at Venado Beach, Canal Zone, Panama. *Panama Archaeologist* **1**, 6–14.

Bull, T. 1961. An urn burial at Venado Beach, Canal Zone. *Panama Archaeologist* **4:** 42–47.

Cooke, R. G. 1979. Los impactos de las comunidades agricolas precolombinas sobre los ambientes del Trópico estacional: datos del Panamá prehistórico, pp. 919–73 in *Actas del IV Simposio de Ecología Tropical,* Vol. 3. Panama City: Instituto Nacional de Cultura and Smithsonian Tropical Research Institute.

Cooke, R. G. 1984. Birds and men in prehistoric central Panama, pp. 243–81 in Lange, F. W. (ed.), *Recent Developments in Isthmian Archaeology*. (BAR International Series 212). Oxford: British Archaeological Reports.

Cooke, R. G. 1985. Ancient painted pottery from central Panama. *Archeology* **July/August**, 33–9.

Cooke, R. G. 1992a. Prehistoric nearshore and littoral fishing in the eastern tropical Pacific: an ichthyological evaluation. *World Archaeology* **6**, 1–49.

Cooke, R. G. 1992b. Preliminary observations on vertebrate food avoidance by the Precolombian Amerinds of Panama, with comments on the relevance of this behaviour to archaeozoology and palaeoenvironmental reconstruction, pp. 59–107 in Ortiz-Troncoso, O. and van der Hammen, T. (eds), *Archaeology and Environment in Latin America*. Amsterdam: Instituut voor Pre-en Protohistorische Archeologie Albert Egges van Giffen, Universiteit van Amsterdam.

Cooke, R. G. 1993a. Alianzas y relaciones comerciales entre indígenas y españoles durante el período de contacto: el caso de Urracá, Esquegua y los vecinos de Natá. *Revista Nacional de Cultura (Panamá)* **25**, 111–22.

Cooke, R. G. 1993b. The past and present distribution of sea catfishes (Ariidae) in a small estuarine embayment in Panama: relevance to Precolumbian fishing practices, pp. 57–74 in Lanata, J. L. (ed.), *Explotación de Recursos Faunísticos en Sistemas Adaptativos Americanos* (Arqueología Contemporánea 4, Edición Especial). Buenos Aires: Universidad de la Plata.

Cooke, R. G. 1998. The Felidae in Pre-Columbian Panama: a thematic approach to their imagery and symbolism, pp. 77–121 in Saunders, N. J. (ed.), *Icons of Power: Felid Symbolism in the Americas*. London: Routledge.

Cooke, R. G. 2001. Cuidando a los ancestros: rasgos mortuorios precolombinos en cerro Juan Díaz, Los Santos, pp. 54–62 in Heckadon-Moreno, S. (ed.), *Panamá: Puente Biológico*. Panamá: Smithsonian Tropical Research Institute.

Cooke, R. G. this volume. Observations on the religious content of the animal imagery of the 'Gran Coclé' semiotic tradition of pre-Columbian Panama, pp. xx-xx in O'Day, S. J., Van Neer, W. and Ervynck, A. (eds), *Behaviour behind Bones. The Zooarchaeology of Ritual, Religion, Status and Identity*. Oxford: Oxbow Books.

Cooke, R. G. and Sánchez, L. A. 2003. Alain Ichon in Panama (1967–1970): A reappraisal of the Tonosí Research Project in the light of new research, pp. 13–27 in Arnauld, C., Breton, A., Fauvet-Bertholet, M. F. and Valdés, J. A. (eds), *Misceláneas en Honor a Alain Ichon*. Guatemala: Impresora Caudal.

Cooke, R. G. and Bray, W. M. 1985. The goldwork of Panama: an iconographic and chronological perspective, pp. 35–49 in Jones, J. (ed.), *The Art of Precolumbian Gold: the Jan Mitchell Collection*. London: Weidenfield and Nicholson.

Cooke, R. G., Isaza, I., Griggs, J., Desjardins, B. and Sánchez, L. A. 2003. Who crafted, exchanged and displayed gold in pre-Columbian Panama? pp. 91–158 in Quilter, J. and Hoopes, J. W. (eds), *Gold and Power in the Intermediate Area*. Washington D.C.: Dumbarton Oaks.

Cooke, R. G., Norr, L. and Piperno, D. R. 1996. Native Americans and the Panamanian landscape: harmony and discord between data sets appropriate for environmental history, pp. 103–26 in Reitz, E. A., Newsom, L. A. and Scudder, S. J. (eds), *Case Studies in Environmental Archaeology*. New York: Plenum Press.

Cooke, R. G. and Ranere, A. J. 1989. Hunting in prehistoric Panama: a diachronic perspective, pp. 295–315 in Clutton-Brock, J. (ed.), *The Walking Larder: Patterns of Domestication, Pastoralism and Predation*. London: Unwin Hyman.

Cooke, R. G. and Ranere, A. J. 1992. Human influences on the zoogeography of Panama: an update based on archaeological and ethnohistorical evidence, pp. 21–58 in Darwin, S. P. and Welden, A. P. (eds), *Biogeography of Mesoamerica. Proceedings of a Symposium (Mérida, Yucatán, México, October 26-30, 1984)*. Tulane: Special Publication of the Mesoamerican Ecology Institute, Tulane University.

Cooke, R. G. and Ranere, A. J. 1999. Precolumbian fishing on the Pacific coast of Panama, pp. 103–22 in Blake, M. (ed.), *Pacific Latin America in Prehistory: the Evolution of Archaic and Formative Cultures*. Pullman: Washington State University Press.

Cooke, R. G. and Ranere, A. J. *in press*. Influencias humanas sobre la vegetación y fauna de vertebrados de Panamá: actualización de datos arqueozoológicos y su relación con el paisaje antrópico, in Leigh, E. G. (ed.), *Ecología y Conservación en Panamá*. Panama: Smithsonian Tropical Research Institute.

Cooke, R. G. and Sánchez, L. A. 1998. Coetaneidad de metalurgia,

artesanías de concha y cerámica pintada en cerro Juan Díaz, Panamá. *Boletín Museo del Oro* **42**, 57–85.
Cooke, R. G., Sánchez, L. A. and Udagawa, K. 2000. Contextualized goldwork from 'Gran Cocle', Panama: an update based on recent excavations and new radiocarbon dates for associated pottery styles, pp. 154–76 in McEwan, C. (ed.), *Precolumbian Gold: Technology, Style and Iconography*. London: British Museum Press.
Díaz, C. P. 1999. *Estudio Bio-Antropológico de Rasgos Mortuorios de la Operación 4 del Sitio Arqueológico Cerro Juan Díaz, Panamá Central*. Unpublished graduation thesis, Universidad de los Andes, Santa Fé de Bogotá, Colombia.
Drennan, R. D. 1991. Pre-Hispanic chiefdom trajectories in Mesoamerica, Central America, and northern South America, pp. 263–87 in Earle, T. K. (ed.), *Chiefdoms, Power, Economy and Ideology*. Cambridge: Cambridge University Press.
Drennan, R. D. 1996. Betwixt and between in the Intermediate Area. *Journal of Archaeological Research* **4**, 95–131.
Feriz, H. 1956. Bericht über eine Ausgrabung an der "Venado beach", Panama Canal Zone. *Wiener Völkerkündliche Mitteilungen Jahr* **4–5**, 191–7.
Griggs, J. C. 1998. Un estudio preliminar arqueológico de la Concesión Minera de Petaquilla, Provincia de Colón, República de Panamá. Reporte del Proyecto para Investigaciones Realizados en la Concesión Minera de Petaquilla. Vancouver: Teck Corporation.
Griggs, J. C., Sánchez, L. A., Cooke, R. G., Díaz, C. P. and Carvajal, D. R. 2002. Recopilación y Presentación de Datos Ambientales y Culturales en la Región Occidental de la Cuenca del Canal de Panamá. Tarea 6: Inventario de Sitios de Recursos Culturales y Evaluacón del Potencial de Sitios Adicionales. Volumen 2: Informe de los Sitios de Recursos Culturales Fuera de las Areas de Impacto Directo y Sitios de Recursos Culturales dentro de las Areas de Impacto Directo en las Cuencas de los Ríos Caño Sucio e Indio. Panama: Autoridad del Canal, Panamá.
Hearne, P. 1992. The story of the river of gold, pp. 1–21 in Hearne, P. and Sharer, R. J. (eds), *River of Gold: Precolumbian Treasures from the Sitio Conte*. Philadelphia: University of Pennsylvania Museum of Archaeology and Anthropology.
Hearne, P. and Sharer, R. J. (eds) 1992. *River of Gold: Precolumbian Treasures from the Sitio Conte.1:* Philadelphia: University of Pennsylvania Museum of Archaeology and Anthropology.
Helms, M. W. 1977. Iguanas and crocodilians in tropical American mythology and iconography with special reference to Panama. *Journal of Latin American Lore* **3**, 51–132.
Helms, M. W. 1979. *Ancient Panama: Chiefs in Search of Power*. Austin: University of Texas Press.
Helms, M. W. 1993. *Craft and the Kingly Ideal: Art, Trade, and Power*. Austin: University of Texas Press.
Helms, M. W. 1995. *Creations of the Rainbow Serpent: Polychrome Ceramic Designs from Ancient Panama*. Albuquerque: University of New Mexico Press.
Helms, M. W. 2000. *The Curassow's Crest. Myths and Symbols in the Ceramics of Ancient Panama*. Gainesville: University of Florida Press.
Ichon, A. 1975. *Tipos de Sepultura Precolombina en el Sur de la Península de Azuero*. Panama City: Instituto de Cultura.
Ichon, A. 1980. *L' Archéologie du Sud de la Péninsule d' Azuero, Panama*. Etudes Mésoamericaines – Serie II. México D. F.: Mission Archéologique et Ethnologique Française au Méxique.
Isaza, I. I. 1993. *Desarrollo Estilístico de la Cerámica Pintada del Panamá Central con Énfasis en el Período 500 a.C.–500 d.C.* Unpublished graduation thesis, Universidad Autónoma de Guadalajara, México.
Jackson, J. B. C. and D'Croz, L. 1997. The ocean divided, pp. 38–71 in Coates, A. G. (ed.), *Central America: a Natural and Cultural History*. New Haven: Yale University Press.
Jiménez, M. 1999. *Explotación de Vertebrados Acuáticos y Terrestres por los Indígenas Precolombinos en Cerro Juan Díaz, Los Santos, durante el Periodo 300–700 d.C.* Unpublished graduation thesis, Universidad de Panamá.
Jiménez, M. and Cooke, R. G. 2001. *La pesca en el borde de un estuario neotropical: el caso de Cerro Juan Díaz (Bahía de Parita, costa del Pacífico de Panamá)* (Noticias de Arqueología y Antropología). Buenos Aires: Grupo NaYa, CD-ROM.
Jonson, F. W. 1948. The Caribbean lowland tribes: the Talamanca division, pp. 231–51 in Steward, J. H. (ed.), *Handbook of South American Indians, Vol. 4, The Circum-Caribbean Tribes*. Washington D.C.: United States Government Printing Office.
Jopling, C. (ed.) 1994. *Indios y Negros en Panamá en los Siglos XVI y XVII: Selecciones de los Documentos del Archivo General de Indias*. Woodstock V.T.: C. Lutz.
Labbé, A. J. 1995. *Guardians of the Life Stream: Shamans, Art and Power in Prehispanic Central Panamá*. Los Angeles: Bowers Museum of Cultural Art.
Ladd, J. 1964. *Archaeological investigations in the Parita and Santa María zones of Panama* (Smithsonian Institution Bureau of the American Ethnology, Bulletin 193). Washington D.C.: Smithsonian Institution.
Linares, O. F. 1976. Animals that were bad to eat were good to compete with: an analysis of the Conte style from ancient Panama, pp. 1–20 in Young, P. D. and Howe, J. D. (eds), *Ritual and Symbol in Native South America* (University of Oregon Anthropological Papers 9). Eugene: University of Oregon Press.
Linares, O. F. 1977. *Ecology and the Arts in Ancient Panama: on the Development of Rank and Symbolism in the Central Provinces* (Studies in Precolumbian Art and Archaeology 17). Washington D.C.: Dumbarton Oaks.
Linares, O. F. 1979. What is lower central American archaeology? *Annual Review of Anthropology* **8**, 21–43.
Lothrop, S. K. 1937. *Coclé: an Archaeological study of Central Panama, Part 1* (Memoirs of the Peabody Museum of Archaeology and Ethnology 7). Cambridge: Harvard University Press.
Lothrop, S. K. 1942. *Coclé: an Archaeological Study of Central Panama, Part 2* (Memoirs of the Peabody Museum of Archaeology and Ethnology 8). Cambridge: Harvard University Press.
Lothrop, S. K. 1954. Suicide, sacrifice and mutilations in burials at Venado Beach, Panama. *American Antiquity* **19**, 226–34.
Lothrop, S. K . 1956. Jewelry from the Panama Canal zone. *Archaeology* **9**, 34–40.
Lothrop, S. K., Foster, W. F. and Mahler, J. (eds) 1957. *The Robert Woods Bliss Collection of Precolumbian Art*. New York: Phaidon.
Museum Rieteberg 1984. *Die Kunst des Indianischen Amerika*. Zurich: Atlantis Verlag
Oviedo y Valdés, G. Fernández de 1853. *Historia General y Natural de las Indias, Islas y Tierra Firme del Mar Océano* (edited by J. Amador de los Ríos). Madrid: Real Academia de Historia.
Piperno, D. R. and Pearsall, D. M. 1998. *The Origins of Agriculture in the Lowland Tropics*. San Diego: Academic Press.
Proyecto Arqueológico Panamá la Vieja 1998. Informe de la temporada de excavación mayo-junio, presentado por Alvaro Brizuela.
Ranere, A. J. and Hansell, P. 1978. Early subsistence patterns along the Pacific coast of Panama, pp. 43–59 in Stark, B. L. and Voorhies, B. (eds), *Prehistoric Coastal Adaptation*. New York: Academic Press.
Sánchez., L. A. 1995. *Análisis Estilístico de Dos Componentes Cerámicos de Cerro Juan Díaz: su Relación con el Surgimiento de las Sociedades Cacicales en Panamá*. Unpublished graduation thesis, Universidad de Costa Rica.
Sánchez, L. A. 2000. Panamá: arqueología y evolución cultural,

pp. 115–45 in *Artes de los Pueblos Precolombinos de América Central*. Barcelona: Institut de Cultura and Museu Barbier-Mueller.

Sánchez, L. A. and Cooke, R. G. 2000. Cubitá: un nuevo eslabón estilístico en la tradición cerámica del "Gran Coclé, Panamá. *Precolombart* **3**, 5–20.

Sander, D., Mitchell, R. H. and Turner, R. G. 1958. Report on Venado Beach excavations, Canal Zone. *Panama Archaeologist* **1**, 26–31.

Sander, D., Mitchell, R. H. and Turner, R. G. 1959. Report on fabric and figurine, Venado Beach, Canal Zone. *Panama Archaeologist* **2**, 52–3.

Stone, D. Z. 1961. *The Talamanca Tribes of Costa Rica* (Peabody Museum Papers 43). Cambridge: Harvard University Press.

Torres de Araúz, R. 1965. Dimensión etnohistórica del misionero Fray Adrián de Santo Tomás. *Hombre y Cultura (Panama)* **1**, 66–122.

Torres de Araúz, R. 1972. *Arte Precolombino de Panamá*. Panama City: Instituto Nacional de Cultura y Deportes.

Wing, E. S. 1980. Aquatic fauna and reptiles from the Atlantic and Pacific sites, pp. 194–215 in Linares, O. F. and Ranere, A. J. (eds), *Adaptive Radiations in Prehistoric Panama* (Peabody Museum Monographs 5). Cambridge: Harvard University Press.

Young, P. D. 1976. The expression of harmony and discord in Guaymí ritual, pp. 77–90 in Young, P. D. and Howe, J. (eds), *Ritual and Symbol in Native North America* (University of Oregon Anthropological Papers 9). Eugene: University of Oregon Press.

Zohar, I. and Cooke, R. G. 1997. The impact of salting and drying on fish skeletons: preliminary observations from Parita Bay, Panama. *Archaeofauna* **6**, 59–66.

Richard Cooke
Smithsonian Tropical Research Institute,
Apdo. 2072, Balboa,
Ancón, Panama,
República de Panamá
E-mail: cooker@naos.si.edu

9th ICAZ Conference, Durham 2002
Behaviour Behind Bones, (ed. Sharyn Jones O'Day et al.) pp. 285–292

31. Hunting and social differentiation in the late prehispanic American Southwest

James M. Potter

Social differentiation is examined within two sequentially occupied late prehispanic villages in the Zuni region of the American Southwest. Faunal data suggest that, after about 1275, communal hunting practices increase in importance in the Zuni region. The effect was two-fold: more large game was brought into the village, and the distribution of large game within the village became more uneven. It is suggested that as communal hunting increased in importance, those in control of the ritual related to this practice were allowed primary access to the resources procured in such hunts prior to the distribution of these resources within the community. It is further suggested that these individuals were legitimized and empowered in their position within the community through organizing successful communal hunts.

Introduction

Recent zooarchaeological analyses in the Americas have identified a clear pattern of faunal distributional differences within and among prehispanic communities in the Southeastern United States. Zooarchaeological data from Cahokia for instance indicates a classic chiefdom pattern, with elites being provisioned with highly valued hunted resources. Social ranking is evidenced in stark differences in access to large game (Kelly 2000) and ritually important species (Kelly 1997) and these differences are mirrored in unequal access to other resources (Pauketat 1994).

If we define social differentiation as the differential ability of people to produce intended social effects and to secure access to material, social, and ideological resources, then there are very strong indications of social differentiation in the Southeast that can be explained with a fairly straightforward chiefdom model.

In the American Southwest, it is a different story. Currently it appears that there were no provisioned chiefs and even the most "complex" systems in the Southwest, such as the so-called Chaco Phenomenon, were not ranked in the way that a classic chiefdom was. This, however, is not to say that there was no social differentiation, that communities were simply "egalitarian", but rather that it was subtler in its material and social manifestation than in the Southeast.

This subtlety of expression makes social differentiation not only difficult to detect archaeologically, but also difficult to model, particularly because of the myriad sources of social differentiation, including clan affiliation, membership in societies, exchange relationships, household size and status, gender, and access to ritual knowledge. In this paper I suggest that, at certain times, the organization of hunting was a particularly potent source of social differentiation within and among Southwestern communities and that this makes faunal analysis an important component of any investigation of social differentiation in the Southwest.

The American Southwest in the late prehispanic period

The particular time of focus is the Pueblo III to Pueblo IV period transition from about AD 1250–1375. In the Zuni region, this transition was marked by extraordinary changes in community structure, social organization, and ritual practices. Communities reorganized themselves from loosely planned aggregates of room blocks to very large, often highly planned, inward focused, plaza-oriented communities.

There was also a major shift in ritual practice at this time. Prior to 1275 unroofed and roofed great kivas were the predominant locus of communal ritual activity. After 1275, large, enclosed plazas replaced great kivas as communal ritual space, which allowed entire villages to witness and participate in communal rituals (Adams 1989). At the same time, kivas, or ceremonial rooms, became more specialized and proportionately less numerous. Rather than each individual household having access to a separate kiva, Pueblo IV period villages might contain fewer than five ceremonial rooms. Thus, while some aspects of ritual became more inclusive and visible, others became more restricted, exclusionary, and secretive. This dualistic aspect of ritual is the foundation of social differentiation in Pueblo IV villages. The co-occurrence of public, plaza-oriented ritual and restricted-access kiva-based rituals established a system that was at once integrative and hierarchical. And this hierarchy was based on differential access to rituals and the knowledge necessary to properly conduct these rituals (Potter 1997).

All of these changes are a reflection of increased *communalism* at this time and an organizational shift from individual, fairly autonomous, competing households to a system of villages composed of highly integrated households. This communalism manifested itself in a number of ways. In addition to the plaza becoming the focus of communal rituals, and as a consequence allowing larger more inclusive participation in these rituals, the actual planning and construction of houses and villages became much more of a communal task than in the Pueblo III period. In the Zuni region, prior to about 1275, houses were planned and constructed separately, and over time villages would grow and contract according to the domestic cycles of individual households and through migration. Communities often comprised loosely aggregated room blocks arranged according to little perceptible plan. Thus, there appear to have been few constraints on the size and configuration of the residential unit, and in consequence residential units were quite variable.

After 1275, communities were highly planned in the Zuni region. Bonding and abutting data from Pueblo de los Muertos, for example, indicate that the outer wall of this Pueblo IV community was continuous except for one break on the west side. Once the shape of the settlement was defined by this outer wall, the community developed toward the interior. Entire rows of rooms were built at once, eventually defining the plaza. The end result was several hundred uniform cells, each measuring 3 × 2.5 m. Residential units were highly uniform, consisting of one or two rows of rooms that extended from the outer wall to the plaza (Watson *et al.* 1980).

I suggest that a third aspect of puebloan life that became more communal after 1275, in addition to public ritual and the planning and construction of pueblos, was hunting. While it is well known that hunting in modern and historic pueblos exhibit a strong pattern of communal cooperation, it has not yet been demonstrated when this pattern came about, or even if the organization of hunting was substantially different prior to Spanish contact. However, there are different methods used to hunt various large game species in the Southwest, specifically mule deer (*Odocoileus hemionus*) and pronghorn antelope (*Antilocapra americana*), that require varying levels of communal participation, cooperation, and organization, and thus a shift from a focus on one species to the other should reflect specific organizational changes.

Hunting large game in the Southwest

Deer are browsers that range to high elevations but are also found in lower, pinyon-juniper and desert habitats. They were primarily hunted in the fall and early winter, when they were the fattest. Deer tend to herd in small family groups, and are thus effectively hunted by one or a few hunters using hunting blinds.

Pronghorn antelope are a gregarious species and tend to herd in large groups. Pronghorn have a cruising speed of about 50 km an hour and can achieve short bursts that reach speeds of 100 km an hour. Due to their gregariousness and speed, cooperation among several or many hunters was the preferred and most effective method for hunting pronghorn by Pueblo Indians. Leslie White (1932) recounts an antelope hunt that involved as many as 74 Puebloan hunters and reputedly killed 744 antelope in one day.

The most effective method for hunting pronghorn was the use of the corral drive. Stephen (1936) presents a detailed description of a Hopi antelope-corral drive. The corral, called an antelope house, was an elliptical stockade of tightly interwoven tree trunks and branches built on a large mesa. At one end of the stockade was a 4 to 6 m opening, and for a distance of 1000 m two wing fences of brush and branches extended from this opening. Piles of brush and tree limbs stuck in the ground were placed in the ground every so often for a distance of 6 to 20 km, still gradually diverging. When antelope were spotted grazing in the vicinity of such a structure, eight young men would surround the animals on three sides. At the same time other Hopi men would space themselves along both sides of the corridor. When the eight men were in place, the two behind started fires of cedar bark and yelled like wolves. The other six then, with their own smoking cedar bark and wolf yells, tried to scare the antelope into the corridor between the wings. The men outside the wing fences kept the antelope from escaping. Once the antelope were inside the corral, two men closed the entrance with piles of brush and branches and they and several more hunters proceeded to shoot the antelope with arrows.

Zuni tradition refers to capturing antelope in winged corrals of tens of square kilometres in size. The corral, hidden from the animal's view around a bend in a curving valley, were made of logs and brush. Once the animals were inside, the corral entrance was closed off with hide ropes and the animals dispatched with arrows (Anell 1969, 60–1). A shift then from a focus on deer to a focus on pronghorn that cannot be accounted for by environmental factors should represent an increase in the importance of communal hunting over small-group hunting as the dominant technique.

The El Morro Valley: a case study

The following presents a short case study that suggests that such a shift in focus on antelope occurred at about 1275 in the Zuni region. The El Morro Valley in west-central New Mexico is in the heart of the American Southwest, at the base of the Zuni Mountains just east of the Arizona border (Fig. 1). It is a high, broad valley, that lies at about 2130 m above sea level. Water is available in the form of springs and natural catchments, and the broad, flat plains are well suited to runoff agriculture. Game, especially deer, is plentiful in and around the Zuni Mountains. The primary limiting subsistence variable is the short growing season, which averages about 113 days. Nevertheless, several very large villages were occupied in the valley from about AD 1250 to 1375 and one of the most striking archaeological patterns in the El Morro Valley is the sudden appearance of nucleated, inward-focused, plaza-oriented villages at about AD 1275. At this time aggregated settlements comprising unplanned, dispersed room blocks were abandoned, and large, highly planned, plaza-oriented communities were established. The Scribe S site is an example of a late Pueblo III village in the El Morro Valley composed of 15 contemporaneously occupied room blocks of varying sizes and configurations. At 1275 this site was abandoned and a large, plaza-oriented nucleated pueblo, Pueblo de Los Muertos, was constructed 100 m to the west (Fig. 2). These sites were intensively excavated in the late 1970s and 1980s and over 12,000 faunal specimens were recovered from a variety of randomly sampled contexts within each village.

Chronology

Another notable aspect of the archaeological record of this area is the systematic and rapid stylistic change of decorated pottery types. Decorated red ware and white ware styles that are common in 1250, such as Tularosa Black-on-white and St. Johns Polychrome, are, by 1300, largely replaced by Heshotauthla, Pinedale, and Kwakina Polychrome styles. Ceramic chronologies coupled with dendrochronological dates thus allow for a very fine grained seriation and dating of deposits within and among villages in this area (see Duff 1996; Potter 1997, 126, 128).

The decorated ceramic data from Scribe S and Pueblo de Los Muertos allows for the clustering of deposits into an early, middle, and late phase. The early phase corresponds to the Scribe S phase, the middle phase is the early phase of occupation at Pueblo de Los Muertos, and the late phase represents the final occupation of Pueblo de Los Muertos, immediately after which the valley was depopulated. This fine-grained temporal sequence then allows for a well controlled study of change over time.

Taphonomy, depositional variation, and sampling

The excavation of the Scribe S site and Pueblo de los Muertos represents one of the original attempts at stratified random sampling of large village sites in the Southwest (Watson *et al.* 1980). A sampling strategy that ensured representative samples from all depositional contexts within a site, including rooms, plazas, and extramural areas, makes these assemblages some of the most appropriate to intrasite statistical analyses. For this particular study, comparisons were either at the site level (or temporal component level) or only room assemblages were compared. Thus, differential deposition and taphonomic processes among various contexts (*e.g.*, rooms versus plaza areas) was avoided.

Gross comparisons among sites are relatively unproblematic as well. The sites are in close proximity to each other so that access to hunting grounds was the same. Additionally, sampling and recovery techniques were comparable at each site; trowels and shovels were used and the majority of excavated sediment was passed through 6 mm mesh screen. However, comparisons between room assemblages from Scribe S and Pueblo de los Muertos pose a particular challenge. Scribe S room assemblages represent primary and de facto refuse associated with structure floors, while those from Pueblo de los Muertos consist mostly of secondary refuse in the form of trash. Nonetheless, I argue that the room assemblages from each site are, in their own way, representative of the behaviors associated with domestic architecture of these particular sites, and are thus comparable in a general sense. Room assemblages from Scribe S, for example, appear to be, in some sense, "living assemblages". Floor assemblages from this site included many "in use" items, such as whole pots, some of which were still filled with corn. The faunal assemblages also indicate some degree of "de-facto-ness". Several rooms at CS 12 (a large room block at Scribe S), for example, exhibit a disproportionate number of unfragmented deer ulnae and radii associated with their floors, as if these bones were being stored for tool manufacture or for some ritual use. In addition, the presence of "ritual" fauna (wild bird and carnivore bones) at each room block indicates that items of potentially high value were left behind at this site. And even if some cleaning behavior occurred at the site prior to abandonment, the room assemblages from Scribe S represent, to a certain extent, "in use" assemblages.

The room deposits at Pueblo de los Muertos, in contrast, are secondary deposits, and were accumulated as trash. Thus, although the artifacts in these contexts were not generated by behaviors conducted in the structures them-

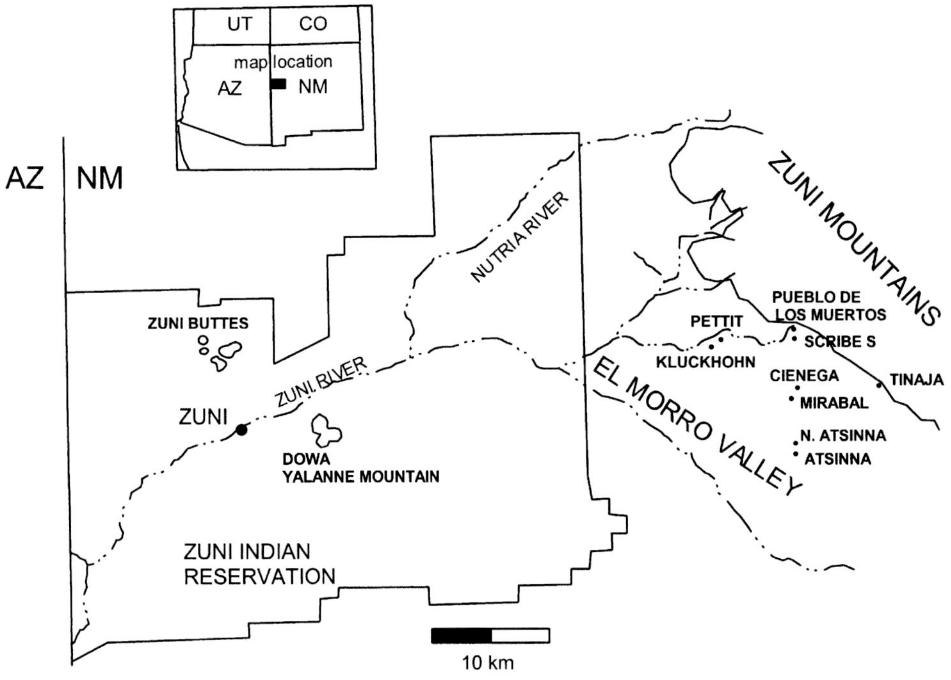

Fig. 1. Map of El Morro Valley with major sites plotted.

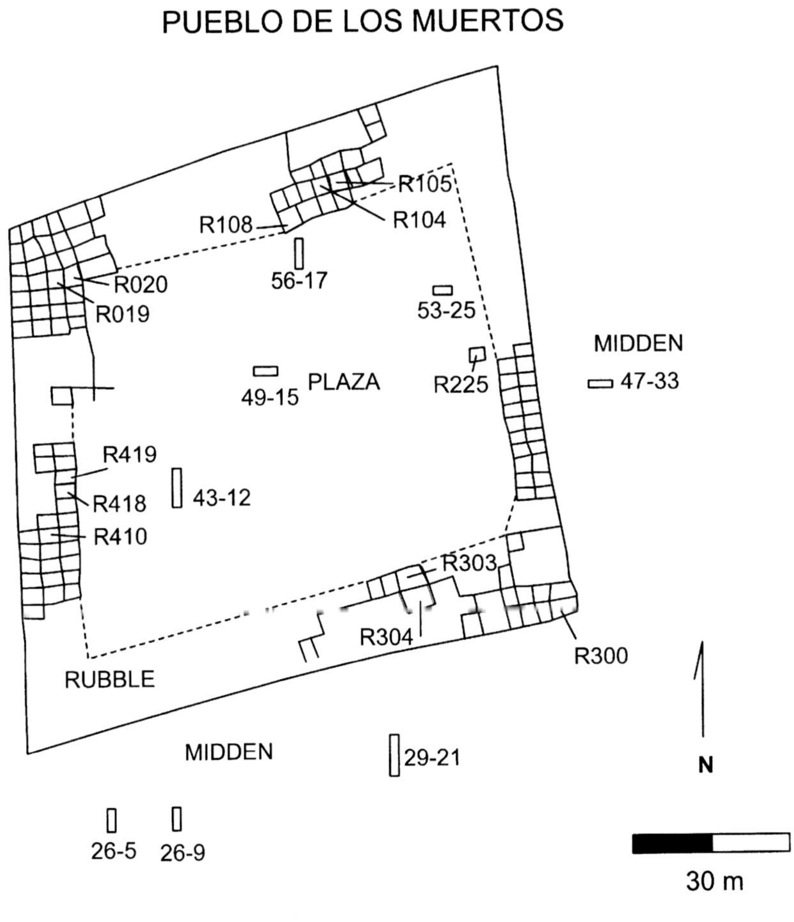

Fig. 2. Map of Pueblo de los Muertos showing excavated areas.

selves, it is highly probable that the room assemblages represent behaviors conducted in close proximity to the structures. Consequently, I argue that general comparisons between *room block* (a series of isolated multiple conjoined rooms) assemblages at Scribes S and *room* assemblages (each of which contained the refuse from activities conducted in adjacent rooms) from Pueblo de los Muertos can be made.

The frequency data

NISP data from Scribe S and Pueblo de Los Muertos indicate that ratios of pronghorn antelope to mule deer increased slightly at 1275 (from 0.10 to 0.12 [pronghorn divided by deer]), and rise dramatically during the late Pueblo IV period (0.20). This pattern suggests that communal hunting of large game increased in importance through time. This trend had several important results. The first was an increase in the amount of large game being brought into the site during the late period. Ungulate proportions (deer + antelope divided by assemblage total) were 0.11 in the early period, 0.04 in the middle, and finally 0.13 in the late period.

The second potential consequence of increased communal hunting was the disproportionate distribution of these resources within the community. Fig. 3 presents the range estimates for ungulate proportions from Scribe S and early and late period room deposits at Pueblo de los Muertos. These estimates were based on total assemblages that included only those taxa that were identifiable to the taxonomic level of species and were most likely not post-depositional in origin. These were ungulates, lagomorphs, birds (including turkey), and carnivores (including dog). This was done to minimize the effects of unidentified bone (*e.g.*, "small mammal" bones) and potential post-occupational bones (*e.g.*, rodent bones) on assemblage proportions. 95% confidence intervals within Scribe S and early period assemblages from Pueblo de los Muertos overlap considerably and no single assemblage stands out as being outside the range of other assemblages. Late period assemblages at Pueblo de los Muertos, on the other hand, include two (Rooms 108 and 418) that were well outside the range of most of the other assemblages, and one assemblage, Room 108, that was outside the range of all but one late period assemblage, which has a wide range due to its small sample size. In addition, the variance of the ungulate proportions (ungulates divided by total assemblage) in the late period assemblages at Pueblo de los Muertos was more than twice that of the early period assemblages (F-ratio=2.47, p-value=0.25), indicating the greater variability in access among residence units in the later period. It is important to point out that even if Room 108 were excluded from the analysis, the ungulate proportions in late period room assemblages would be more variable than early period Pueblo de los Muertos assemblages (F-ratio=1.43). The Scribe S assemblages, while fewer in number, overlap considerably with each other. This may have to do with the relatively small size of some of the samples (*e.g.*, Roomblock 9), which broaden the range estimates. However, one would expect greater variation from smaller samples, and thus the greater variation exhibited by the late Pueblo de los Muertos assemblages, which are more numerous and generally larger, is all that much more significant.

The spatial distribution of various anatomical portions of ungulate within Pueblo de los Muertos elements was also informative in terms of understanding behavioral/access differences within the site (Fig. 4). Ungulate elements were divided here into two categories: "high-utility" and "low-utility". High-utility elements were those that are associated with a high amount of meat and marrow (Binford 1978; Metcalf and Jones 1988; Potter 1995). These include axial elements, such as ribs, vertebrae, and innominates, and upper limb bones, including femora and humeri. Low-utility elements include those that have relatively little usable meat and marrow associated with them, such as lower limb and foot bones, and crania. As Fig. 4 shows, late period assemblages with relatively high ungulate frequencies (*e.g.*, Rooms 108, 418, and 419) exhibited a preponderance of high-utility elements, indicating that those individuals or groups who had better than average access to large game, also had fairly consistent access to the better portions of those animals.

Social differentiation in the historic record

Historically, communal large-game hunts in Puebloan society were organized and enacted only by the leading men or specific societies within a Puebloan community and required specific ritual knowledge that only certain individuals possessed (Anell 1969, 61; White 1974, 302; Gnabasik 1981, 45, 48, 103). The spoils of such hunts were often differentially apportioned according to the level of an individual's involvement in the hunt. In addition, the primary ritual specialist in the community, or the head of the society organizing the hunt, or both, also received a sizable portion of the game killed in the hunt. For example, at Zia,

> a deer is divided into two parts, one for the *cacique* (ritual specialist), the other for the hunter who killed it. The hunter gets the head, the skin and part of the backbone and the chest from the neck down to and including the fourth rib, and a part of the belly. The rest goes to the *cacique* (White 1974, 303).

Similarly, at Taos, after a deer hunt,

> The first two deer to be killed went to the Hunt chief who had 'made the talk,' i.e. prayed before the hunt began, asking the deer not to be afraid to give themselves to the hunters (Gnabasik 1981, 48, quoting Parsons 1920, footnote 37 on page 19).

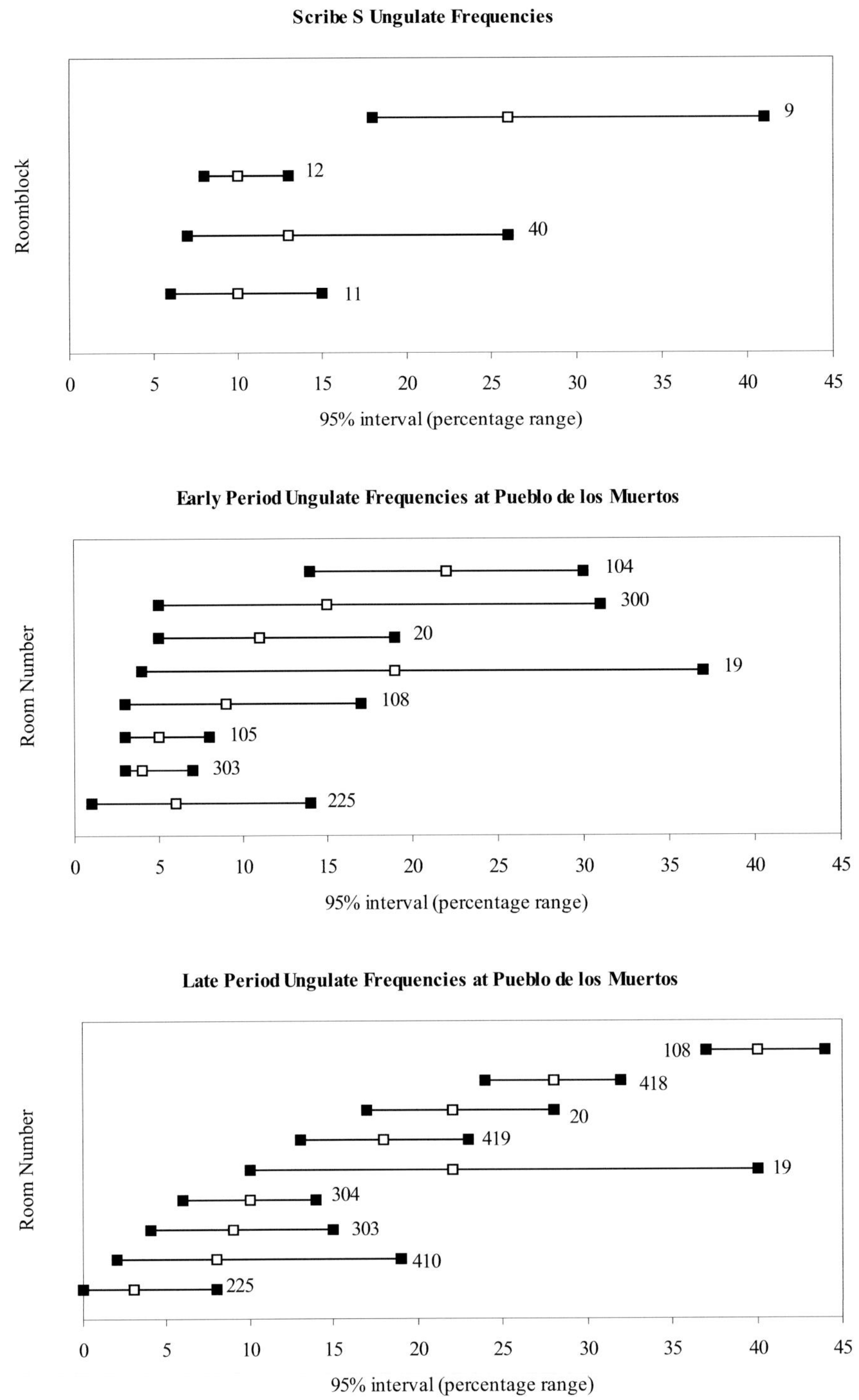

Fig. 3. Binomially derived range estimates for ungulate frequencies in Scribe S room block assemblages and early and late room assemblages from Pueblo de los Muertos. The black squares indicate the 95% probability distribution and the blank square the actual assemblage percent.

Here it is evident that by virtue of the ritual knowledge that the Hunt chief possesses he is entitled to a sizable portion of the hunt spoils. Moreover, in the example above from Zia, the hunter ends up with some fairly low-utility elements (head and skin) while the ritual specialist receives more of the high utility portions (*e.g.*, the haunch).

Communally hunted antelope are also distributed by and among the most powerful ritual specialists within many

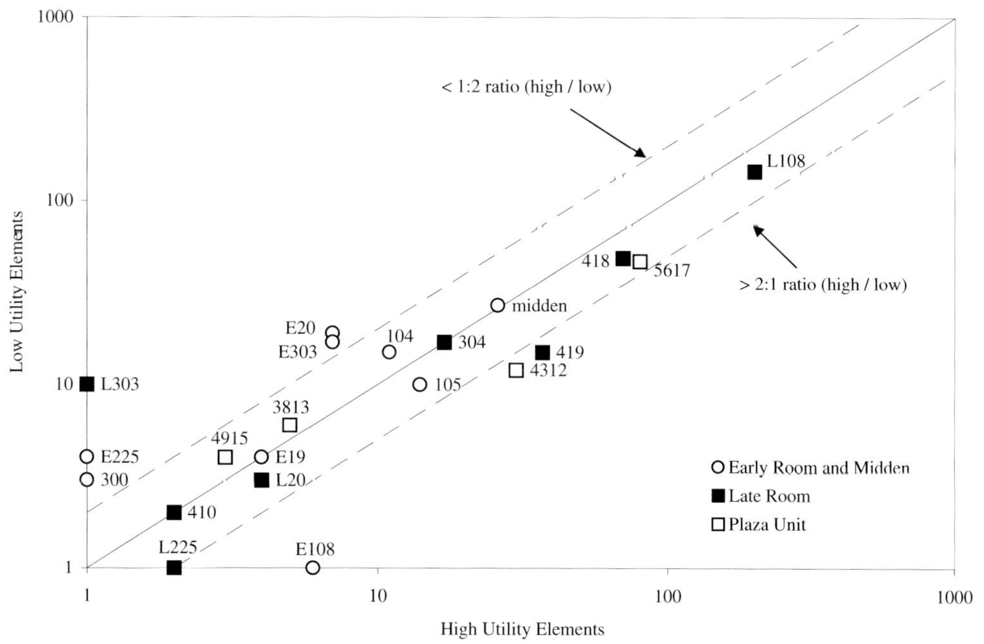

Fig. 4. Plot of high versus low utility ungulate elements within Pueblo de los Muertos. Note that axes are logged.

Puebloan groups. At Zia, for example, after a successful hunt, the antelope meat is brought to the *cacique's* office, where the war chief decides when it is to be divided among the heads of the curing and weather control societies, be it one, two, or even three days after the hunter's return (Stevenson 1894, 119–21).

The organization of successful communal large-game hunts, while bringing more large-bodied game into the community, would have simultaneously legitimized and empowered those in control of the ritual related to hunting and allowed them primary access to the resources procured in such hunts prior to the distribution of these resources within the community. The reinforcing relationship between ritual-based status, the organization of communal activities, and access to resources may ultimately have translated into more enduring expressions of social differentiation within the community. This would have been the case, for instance, in situations in which certain individuals controlled aspects of ritual that were related to effectively dealing with the impacts of large sedentary populations on the local environment, such as the depletion of large game over time due to over hunting.

Conclusions

To conclude, although communal hunting certainly occurred in earlier time periods, it appears, at least in the Zuni area, to have become more institutionalized and formalized as a practice in the early Pueblo IV period, allowing leaders to control not only the organization of hunts, for instance when they occurred and who could participate, but also the distribution of the game once it was brought back to the village. The American Southwest offers an interesting distinction from classic anthropological models of social organization. Social differentiation is not the product of competitive feasting or other status enhancing behaviors as in big man systems, nor is it inherited directly through lineage as in a chiefdom. It derives instead from a variety of sources, especially through initiation into societies and the control of ritual knowledge through membership in societies, which can be both an achieved and ascribed status. For instance, membership in some societies is related to clan or moiety affiliation, while others require specific accomplishments on the part of initiates.

This multidimensional aspect of social differentiation makes it difficult to detect and model archaeologically. Indeed, in early Pueblo IV villages the only material realm that even suggests economic distinctions among households is faunal distributions. I suggest that these inequities were the result of the increased importance of communal hunting and the organizational needs that arose from increased communalism in general, and that faunal analysis may be our best tool for not only detecting differences in resource access among households but also in modeling why those differences occurred in the first place.

References

Adams, E. C. 1989. Changing form and function in Western Pueblo ceremonial architecture from AD 1000 to AD1500, pp. 155–60 in Lipe, W. and Hegmon M. (eds), *The Architecture of Social Integration in Prehistoric Pueblos* (Occasional Papers No.1). Cortez: Crow Canyon Archaeological Center.

Anell, B. 1969. *Running Down and Driving of Game in North America* (Studia Ethnographica Upsaliencia 30). Lund: Berlingska Boktryckeri.

Binford, L. R. 1978. *Nunamiut Ethnoarchaeology*. New York: Academic Press.

Duff, A. I., 1996. Ceramic micro-seriation: types or attributes. *American Antiquity* **61**, 89–101.

Gnabasik, V. 1981. *Faunal Utilization by the Pueblo Indians*. Unpublished masters thesis, Eastern New Mexico University.

Kelly, L. S. 1997. Patterns of faunal exploitation at Cahokia, pp. 69–88 in Pauketat, T. and Erickson, T. (eds), *Cahokia: Domination and Ideology in the Mississippian World*. Lincoln: University of Nebraska Press.

Kelly, L. S. 2000. *Social Implications of Faunal Provisioning for the Cahokia Site: Initial Mississippian, Lohman Phase*. Unpublished PhD dissertation, Washington University.

Metcalf, D. and Jones , K. 1988. A reconsideration of animal body part indices. *American Antiquity* **53**, 486–504.

Parsons, E. C. 1920. Notes on Isleta, Santa Ana, and Acoma. *American Anthropologist* **23**, 149–69.

Potter, J. M. 1995. The effects of sedentism on the processing of hunted carcasses in the American Southwest: a comparison of two Pueblo IV sites in central New Mexico. *The Kiva* **60**, 411–28.

Potter, J. M. 1997. *Communal Ritual, Feasting, and Social Differentiation in Late Prehistoric Zuni Communities*. Unpublished PhD dissertation, Arizona State University.

Pauketat, T. R. 1994. *The Ascent of Chiefs: Cahokia and Mississippian Politics in Native North America*. Tuscaloosa: University of Alabama Press.

Stephen, A. 1936. *Hopi Journal* (edited in 1994 by Parsons, E.C.) (Columbia University Contributions to Anthropology 23). New York: Columbia University Press.

Stevenson, M. C. 1894. The Sia, pp. 3–157 in Powell, J. W. (ed.), *Eleventh Annual Report of the Bureau of American Ethnology, 1899–90*. Washington D.C.: Smithsonian Institution.

Watson, P. J., LeBlanc, S. and Redman, C. 1980. Aspects of Zuni prehistory: preliminary report on excavations and survey in the El Morro Valley of New Mexico. *Journal of Field Archaeology* **7**, 201–18.

White, L. 1932. *The Pueblo of San Felipe* (Memoirs of the American Anthropological Association 38). Millwood N.Y.: Kraus.

White, L. 1974 *Zia – the Sun Symbol Pueblo*. (originally published 1962 as "The Pueblo of Sia, New Mexico", *Bureau of American Ethnology, Bulletin* 184). Albuquerque: University of Albuquerque Press.

James M. Potter
SWCA, 208 Parker Drive, Suite C,
Durango, Colorado 81301, USA
E-mail: jpotter@swca.com

9th ICAZ Conference, Durham 2002
Behaviour Behind Bones, (ed. Sharyn Jones O'Day et al.) pp. 293–303

32. Zooarchaeological evidence for changing socioeconomic status within early historic Native American communities in Mid-Atlantic North America

Heather A. Lapham

Within decades of European settlement in eastern North America, an overseas demand for leather imported from the Americas led processed deerskins to become one of the most important commodities produced for trade by Native Americans in the greater Mid-Atlantic region. The deerskin trade, which flourished in the c. *AD 1600s–1700s, offered Native Americans new opportunities to acquire socially-valued goods which, in turn, provided the means for certain groups to enhance their socioeconomic status by trading processed hides for nonlocal goods that would have been symbols of wealth and prestige within the native community. Comparative analyses of faunal remains from southwest Virginia indicate that several changes occurred within some Native American villages in the early 1600s: deer use and venison consumption increased, deer harvests became more selective, and deer skinning techniques occasionally reflected an attention to detail geared toward preserving maximum hide size. These changes suggest that certain Native American settlements chose to alter deer harvest and processing activities in order to participate in a growing interregional trade in hides, furs, and nonlocal status goods.*

Introduction

Native American economies throughout the Mid-Atlantic region relied heavily on white tailed deer in the Late Woodland period (*c.* AD 900–1600). Deer meat contributed an important source of protein and fat to Native American diet and skin, bone, and tendon provided the raw materials needed to manufacture tools, clothing, and other life-essential items (Swanton 1946; Hudson 1976). Following settlement by the English in coastal Virginia in the early AD 1600s, Native Americans' use of deer in the Mid-Atlantic began to change. Colony merchants and traders sought processed deer hides to export overseas where extensive deforestation in Great Britain and other areas of western Europe had devastated indigenous deer populations (Thompson 1975). The highest quality deerskins were shipped to England where they were used to manufacture book covers, fine leather gloves, and other apparel and accessories (Clarkson 1960; Parrish 1972). Lower quality hides, those that did not meet foreign industry standards, were sold to local leather manufacturers in Virginia or routed to the northern American colonies (Crane 1928, 111). Government inventories that recorded the annual number of deerskins shipped to Great Britain from the Virginia and Carolina colonies provide compelling documentation of this trans-Atlantic trade. Between 1698 and 1710, the earliest years for which such figures exist, the two colonies combined exported an average of 72,000 deerskins per year, a total of more than 800,000 hides in little more than a decade (Crane 1928, 328).

By the mid-to-late 1600s, many Native American groups in the Mid-Atlantic were involved in a lucrative trade in processed deer hides. In exchange for these deerskins, colonial traders supplied Native Americans with valued European-manufactured and native-made goods. Traders brought goods to the region in varieties and quantities tailored to meet the demands of their Native American trading partners (Tingling 1977, 30, 31, 41, 57, 63, 66). Appropriate trade stock grew in diversity to include everything from glass beads to iron knives, English guns, cotton cloth, and eventually items such as coffee and sugar (Ewan and Ewan 1970, 385; Waselkov and Braund 1995, 157). Over time this new Native American economy, one that emphasized hunting for commercial hide production, along with increased competition for European commodities and the broader impacts of European colonization redirected exchange networks (Usner 1992), altered political alliances and gave rise to the development of political factions and middlemen (Ward and Davis 1993; Waselkov 1993; Martin 1994), reshaped gender relations (Perdue 1998), and transformed cultural belief systems (Martin 1978). Yet many questions remain regarding the initial trajectories of these changes. What archaeological evidence exists, for example, for deerskin production for the purpose of commercial trade in the early Historic period? And, how did socioeconomic relationships change within Native American communities participating in this trade?

In this paper I use faunal remains to track participation in the deerskin trade. I am then able to monitor interregional exchange in deerskins and access to nonlocal status goods which, in turn, allows me to evaluate continuity and change in socioeconomic status in historic Native American communities. My research provides several lines of evidence for intensified interregional trade and shifting social relations within seventeenth century Native American settlements in the southern Mid-Atlantic region. The zooarchaeological data indicate that, in the 1600s, venison gained dietary importance as deer were being increasingly hunted for their hides and the processing of these hides intensified. The distribution of the nonlocal status goods received in return for processed deerskins suggests that certain individuals within Native American villages were more active participants in the trade and that they did indeed benefit from this new commerce. In the following paper I first consider the cultural significance and archaeological distribution of three main nonperishable prestige materials (copper, shell, and glass) that moved through Mid-Atlantic exchange spheres. I then turn to the zooarchaeological data to present evidence for the commercial production and trade of deerskins. I discuss the faunal remains from three perspectives: general subsistence practices, deer hunting strategies, and deer butchering techniques. Assemblages from three Native American settlements (Crab Orchard, Hoge, and Trigg) located in southwest Virginia provide the basis for this study. Comparative data from other southern Virginia sites are also considered, following which I discuss the broader implications of participation in an interregional trade in hides, furs, and status goods.

Prestige materials in Mid-Atlantic trade

In the greater Mid-Atlantic region and elsewhere in eastern North America, interregional exchange in nonperishable prestige goods prior to European colonization consisted largely of marine shell and copper. Status-marking goods circulated within trade networks at varying temporal and spatial levels throughout the Woodland period (*c.* 1000 BC–AD 1600) (Stewart 1994, 88; Hantman and Gold 2002). The Late Woodland period (*c.* AD 900–1600) in the southern Mid-Atlantic witnessed a significant increase in the distribution of marine shell artifacts (Stewart 1989, 63). Inland Native American groups obtained marine shell in both unmodified and finished product form from coastal peoples. Shell beads regularly appear in mortuary contexts. Shell pendants and gorgets are also found in burials, but less frequently than beads. Marine shell is considered by many archaeologists to have been a "standard of wealth" in past Native American societies (Knight and Steponaitis 1998, 11), although research has also shown that access to such items was not limited solely to central leaders (Muller 1997, 18). Marine shell, as well as copper, continued to be valued by many Native Americans following the introduction of European commodities in the AD 1600s. The early historic grave of a high-status individual, possibly a petty chief, located along the Potomac River in northeastern Virginia, contained an inventory of local and nonlocal goods including five shell mask gorgets engraved with stylized human faces, plain shell gorgets, tubular shell beads, and several complete whelk shells along with numerous copper items, glass beads, and a metal cross (Potter 1993, 213–8).

Copper, in particular, gained importance in prestige trade relations late in the Late Woodland period, beginning around AD 1400 (Hantman 2001). Copper was used in mortuary ceremonialism during this time in the form of gorgets, geometric-shaped pendents, and rolled beads interred with high-status individuals (Potter 1993, 213–20; Hantman and Gold 2002). Chiefly leaders among the Algonquian-speaking Powhatan considered copper to be a key symbol of political power and religious authority (Potter 1989, 153; Hantman 1990, 685). The importance placed upon copper in the Mid-Atlantic led English colonists to comment in the early 1600s on the great value bestowed upon this metal by coastal Algonquian societies (Rountree 1989, 86; Mallios 1998). Although copper obtained considerable symbolic significance in most Eastern Woodland societies (Hantman 1993, 109), some argue that it never reached the social importance in the Late Woodland period in southwest Virginia as it did elsewhere to the north and east (Barber *et al.* 1996, 17).

The social significance and meaning conferred by upon glass beads, which first arrived in Virginia along with European colonists in the early AD 1600s, is slightly more ambiguous than that of copper and marine shell. Substantial archaeological and ethnohistoric evidence suggests, however, that many Native American groups in the greater Mid-Atlantic region identified the earliest glass beads as prestige items (Hamell 1983, 25; Rountree 1989, 73; Potter 1993, 218–9; Gleach 1997, 57). Glass beads have been found in association with high-status burials (Potter 1993, 218–9) and ritual paraphernalia (Hamell 1983, 24, 27). Virginia trader William Byrd I complained, in the late seventeenth century, that a "want of beads" often resulted in unsuccessful exchanges if his trade good stock did not contain the specific types and colors of glass beads that his Native American trading partners had requested (Tinling 1977, 14, 64).

Archaeological distributions in southwest Virginia

In southwest Virginia, marine shells were distributed through well-defined, broad-based exchange networks. The abundance of buscycon, marginella, and olivella shell ornaments further suggests that some individuals within certain communities also hoarded these valued objects (Stewart 1989, 64). Village sites often contain significant amounts of marine shell (Egloff 1992), however the intrasite distribution of these artifacts within burial contexts varies (Boyd and Boyd 1992). Detailed mortuary studies that consider the type, frequency, and context of marine shell artifacts specifically, and grave goods generally, are lacking for the study region. One recent study of mortuary practices in southwest Virginia indicates that the distribution of marine shell as well as other types of grave goods varied by age class, sex, and grave location (Lapham 2002). Further research into the distribution of grave goods within burial populations is greatly needed in order to elucidate local and regional patterns of prestige goods use.

The distribution patterns for copper or copper alloys, which consistently occur in much lower quantities than shell, suggest either focused exchange or hoarding (Stewart 1989, 64). Copper artifacts are rare finds in the Late Woodland period in southwest Virginia, and few sites in general have yielded copper items. At the Shannon site (44My8) in northcentral Montgomery County, the burial of an older adult male contained a copper object of undetermined form (Benthall 1969, 93) and a burial at the Richlands Hospital site (44Tz51) in southern Tazewell County produced a copper effigy-like pendent and a rolled cone "tinkler" (Egloff and Turner 1988, 19). Spectrographic analysis identified the copper from both sites as North American rather than European in origin (*ibid.*), however elsewhere in the eastern North America the tinkler ornament form has been associated almost exclusively with historic contexts. The Crab Orchard site (44Tz1) contained more than twenty-five copper/brass ornaments, mostly rolled beads found in burial contexts (Egloff and Reed 1980; MacCord and Buchanan 1980). Spectographic tests conducted on one of the twenty-some metal artifacts from the Crab Orchard site revealed it to be of a North American source (Egloff 1992, 203).

Native American trade networks throughout eastern North America witnessed profound changes in the early Historic period. Exchange spheres began to circulate larger quantities and a greater variety of goods, due largely to an increasing supply of marketable European commodities in the seventeenth century and an overseas demand for hides and furs imported from the Americas. By the early-to-mid 1600s, European trade brought glass beads, non-native copper, and the occasional iron artifact to southwest Virginia inhabitants (MacCord 1977; Geier and McFee 1981; Geier and Warren 1982; Whyte and Geier 1982; Barber 1989; Barber *et al.* 1996). Glass beads and ornaments made from European copper alloy have been found at only a few known early-to-mid seventeenth century Native American sites in the region, including Hurt Powers Plant (44Py144), Thomas Sawyer (44Rn39), and Trigg (44My3). Edged iron tools and guns were introduced to the region by the mid-to-late 1600s, as evidenced by the find of a trigger from an English-manufactured snaphaunce firearm at the *c.* AD 1660–1680 Graham-White site (44Rn21) (Klatka and Klein 1993).

The international market for North American deerskins offered Native Americans new opportunities to acquire socially-valued goods. As a result, some Native Americans may have been able to enhance their socioeconomic status by trading processed hides for nonlocal goods that would have been symbols of wealth and prestige within the native community. In the following section, I consider the evidence for continuity and change in subsistence practices, deer harvests, and deer butchery between Late Woodland and Early Historic Native American settlements in southwest Virginia which provides insights into changing socioeconomic status in the seventeenth century.

Tracking the skin trade

Faunal material from three Native American villages provide the basis for this study: Crab Orchard (44Tz1), Hoge (44Tz1), and Trigg (44My3). The sites are located in southwest Virginia within the physiographic province known as the Ridge and Valley. Characterized by longitudinal ridges separated by valleys that vary in width, the Ridge and Valley extends for more than fifteen hundred kilometers from the northern Hudson Valley to central Alabama in eastern North America. In Virginia, the Ridge and Valley comprises the portion of the state that lies west of the Blue Ridge Mountains with the exception of a small section of Appalachian Plateau in the southwestern-most corner of the state. Ridge and Valley topography represents some of the most varied in Virginia. Elevations range from 300 to 1400 meters above sea level. Oak and chestnut tree forests dominate the region, although mixed mesophytic communities (basswood, maple, tuliptree, hemlock, beech, etc.) commonly grow along ravine slopes (Braun 1964, 231–3). Five major watersheds drain the Ridge and Valley in Virginia. These include the Shenandoah, James, Roanoke, New, and Tennessee rivers.

The study sites are located within an eighty kilometer radius of one another in the southern Ridge and Valley region in Virginia (Fig. 1). All three villages are characterized by settlement patterns of circular-shaped domestic structures encircling a central plaza area and surrounded by a palisade, subsistence strategies of maize horticulture supplemented by hunting and gathering, the presence of limestone-tempered pottery, a preference for flexed interments, and marine shell acquired in significant quantities through interregional exchanges (MacCord 1989). The architectural and artifactual diversity seen throughout

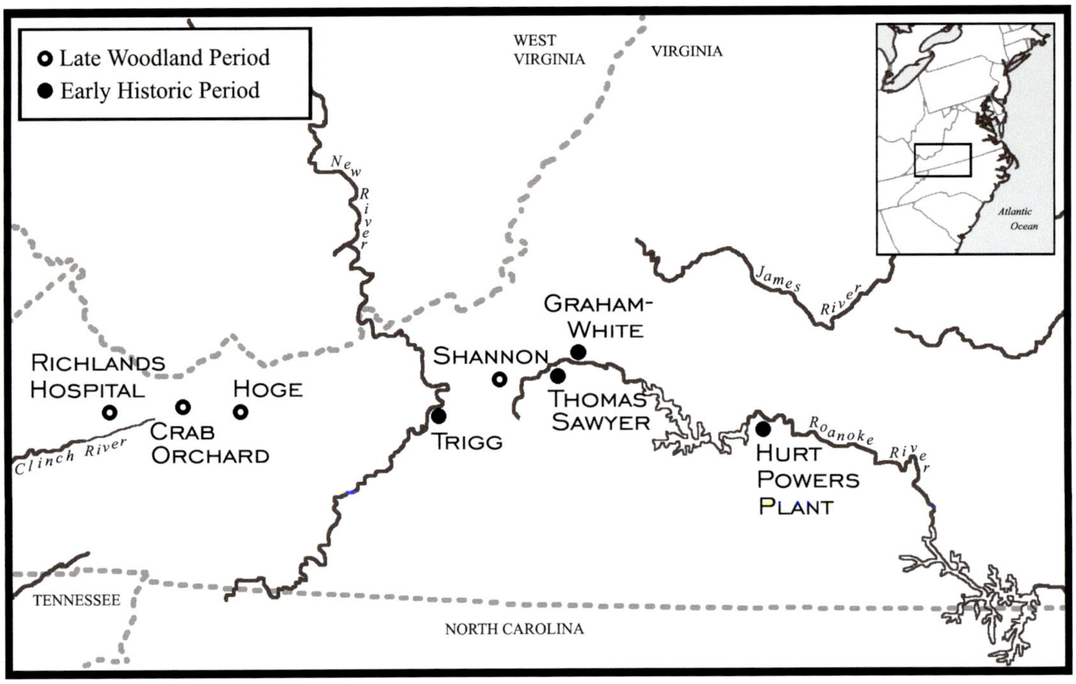

Fig. 1. Study area with archaeological sites mentioned in the text identified.

southwest Virginia suggests the regionalization of peoples who shared common cultural and linguistic bonds (Egloff 1992, 204).

Two of the villages considered in this study, the Crab Orchard and Hoge sites, were occupied during the Late Woodland period in the general range of *c.* AD 1450–1600 based on radiocarbon dates and regional ceramic sequences (Egloff and Reed 1980; MacCord and Buchanan 1980), directly prior to permanent European settlement in Virginia. The third village, the Trigg site, was occupied in the Early Historic period *c.* AD 1600–1650 based on radiocarbon dates, regional ceramics sequences, and regional glass bead chronologies (MacCord 1977; Buchanan 1984; Egloff 1992; Boyd 1993). Data from these three sites therefore provide a baseline for assessing continuity and change in socioeconomic status in historic Native American communities in the region.

The Late Woodland Crab Orchard site is located along the Clinch River, a tributary of the Tennessee River, in the Pisgah Valley about six kilometers west of the town of Tazewell in Tazewell County. The site is the largest of the three settlements discussed here, totaling slightly more than one hectare in area with the palisade measuring about 125 meters in diameter (MacCord and Buchanan 1980). My research focuses on materials recovered from features in the northwest section of the village, an area excavated by archaeologists from the Virginia Research Center for Archaeology in 1978 (Egloff and Reed 1980). The Crab Orchard faunal assemblage contains more than 15,500 animal bones from twenty-four features and a midden area. The second site, Hoge, is located about sixteen kilometers east of the Crab Orchard settlement near Burkes Garden Creek in the town of Burkes Garden in Tazewell County. This Late Woodland village is the smallest of the three study sites, totaling less than one-half hectare in area. Emory Eugene Jones excavated the Hoge site on an intermittent basis over a period of twenty years, beginning in 1976 (Jones and MacCord 2001). The Hoge site assemblage consists of more than 5,500 animal bones from fifty-one features. Faunal remains from all excavated storage/refuse pits at both the Crab Orchard and Hoge sites were studied.

The third study site, Trigg, is situated approximately seventy-two kilometers east of Hoge and eighty-eight kilometers east of Crab Orchard. This Early Historic village lies on the southeast bank of the New River within the present day city limits of Radford. The palisade is about 100 meters in diameter, which makes it slightly smaller than the Crab Orchard village. The Archaeological Society of Virginia excavated the Trigg site in a 1974 and a 1975 field season (Buchanan 1984). Excavations at Trigg unearthed European-manufactured glass beads and copper artifacts which indicates that the Trigg site occupants engaged in interregional exchanges with European traders or, more likely, with their Native American neighbors who had access to European goods. More than 11,500 animal bones from a sample of one hundred and eleven storage/refuse pit features were analyzed for this study. The one hundred and eleven feature sample included all storage and refuse pits where the fauna had retained its provenience information following a post-excavation fire at the storage facility that housed the artifacts. This

	Crab Orchard	Hoge	Trigg
Number of features analysed	24	51	111
Total NISP	15,729	5,564	11,525
Total MNI	81	79	264
Total biomass in kg	212	333	437
Number of species identified	26	24	40

Fig. 2. Basic composition of the faunal assemblages.

	Crab Orchard	Hoge	Trigg
Butchery marks	0.1	1	1.3
Carnivore gnawing	0.6	3.7	3.1
Rodent gnawing	<0.1	0.1	0.1
Digested bone	0.1	0.2	0.3
Burning			
burnt	0.4	0.3	0.1
carbonized	22.6	2.3	0.5
calcined	22.7	2.4	1

Fig. 3. Bone modifications (percent of total NISP).

collection constitutes 26% of the total number of storage/refuse pit features, and comprises approximately 22% of the total fauna recovered from the site. The features studied are well-distributed throughout all areas of the village.

General subsistence

General subsistence practices in southwest Virginia were evaluated from the analysis of more than 32,000 animal bones collected from the three study sites. A total of forty-six different species were identified in the faunal assemblages, including twenty-one mammals, fifteen birds, two fishes, six reptiles, and two amphibians. Fig. 2 provides summary statistics for each assemblage, including the number of features analyzed, number of identified specimens (NISP), minimum number of individuals (MNI), total biomass, and the number of species identified. MNI was calculated on a site-wide basis from paired elements with age of identified taxon taken into account. Biomass, computed for this study using bone weight, follows Reitz and Cordier (1983), Reitz *et al.* (1987), and Reitz and Wing (1999).

The field recovery methods implemented during excavations were similar at each of the three sites. Excavators dry-sifted feature fill through standard-size, one-quarter inch mesh hardware. At the Crab Orchard site, excavators also water screened select samples of feature fill through fine mesh hardware, a technique that generally increases the recovery of small bones and bone fragments. Only one third of the unidentified bone in the Crab Orchard assemblage came from such contexts, however. Two thirds of the bone not specified to class or a lower taxonomic level were recovered from dry-screened proveniences. I suspect that the high recovery rate of unidentified specimens resulted largely from an effort on the part of the excavators to collect even the smallest artifacts, the effects of which can be seen in the high NISP relative to the low biomass estimate when compared to the same calculations for the Hoge and Trigg sites (Fig. 2).

Modifications to bone represent additional factors that can potentially bias the composition of archaeofaunas. Animal bones from all three sites exhibit excellent preservation and little to no weathering or root damage was noted on the bone from the three assemblages. Butchered bone is represented in similar proportions at the Hoge (1.0%) and Trigg (1.3%) sites (Fig. 3). The Crab Orchard assemblage contains a much lower frequency of butchered bones (0.1%). This pattern holds even when all unidentified fragments are excluded from the equation. Deer bones comprise 91% of all butchered bones at Crab Orchard, 83% at Hoge, and 93% at Trigg. Butchery of deer will be discussed in greater detail in a following section. Carnivore gnawed bone was observed in all three assemblages. Carnivore gnawing, which most often signifies the presence of domestic dogs in a village, accounts for about one half of one percent of all bone at the Crab Orchard site. The Hoge and Trigg settlements exhibit slightly higher frequencies of chewed bone (3.7% and 3.1%, respectively). Domestic canid is also represented at Crab Orchard in the form of two dog burials (MacCord and Buchanan 1980, 68-70) and isolated dog bones were found in five features at Trigg. Digested bone and rodent gnawing is present in low frequencies at each site.

The proportion of burnt bone in each assemblage varies significantly. Burning occurs most frequently in the Crab Orchard assemblage where almost half (45%) of the fauna recovered exhibits some degree of burning (Fig. 3). Burning occurs on fauna recovered from both dry- and wet-screened contexts. Burnt bones comprise 5% of the Hoge collection and less than 2% of the Trigg site assemblage. The high number of burnt bones at Crab Orchard may account, in part, for the low frequency of gnawed and butchered bones. Variation in burning and preservation has been shown to affect the frequency or visibility of chew marks. Gnawing tends to be less visible on burnt or poorly preserved bone (Klein and Cruz-Uribe 1984, 42). This is probably true for butchery marks as well. Most of the burnt bone at Crab Orchard is either carbonized (burnt black) or calcined (burnt gray/white). Both types of burning are caused by prolonged exposure to intense heat and would not have been a result of cooking activities (David 1990). The high number of carbonized and calcined specimens suggests that the Crab Orchard occupants may have incinerated bones as a form of refuse disposal, at least more so than their neighbors.

Species representation indicates that white-tailed deer (*Odocoileus virginianus*) comprised an important part of Native American diet at each settlement. Fig. 4 shows the proportions of the total biomass of animal bones identified to the taxonomic level of family or lower for each of the species indicated. Biomass estimates the total weight of an animal that the archaeological specimen, in this case

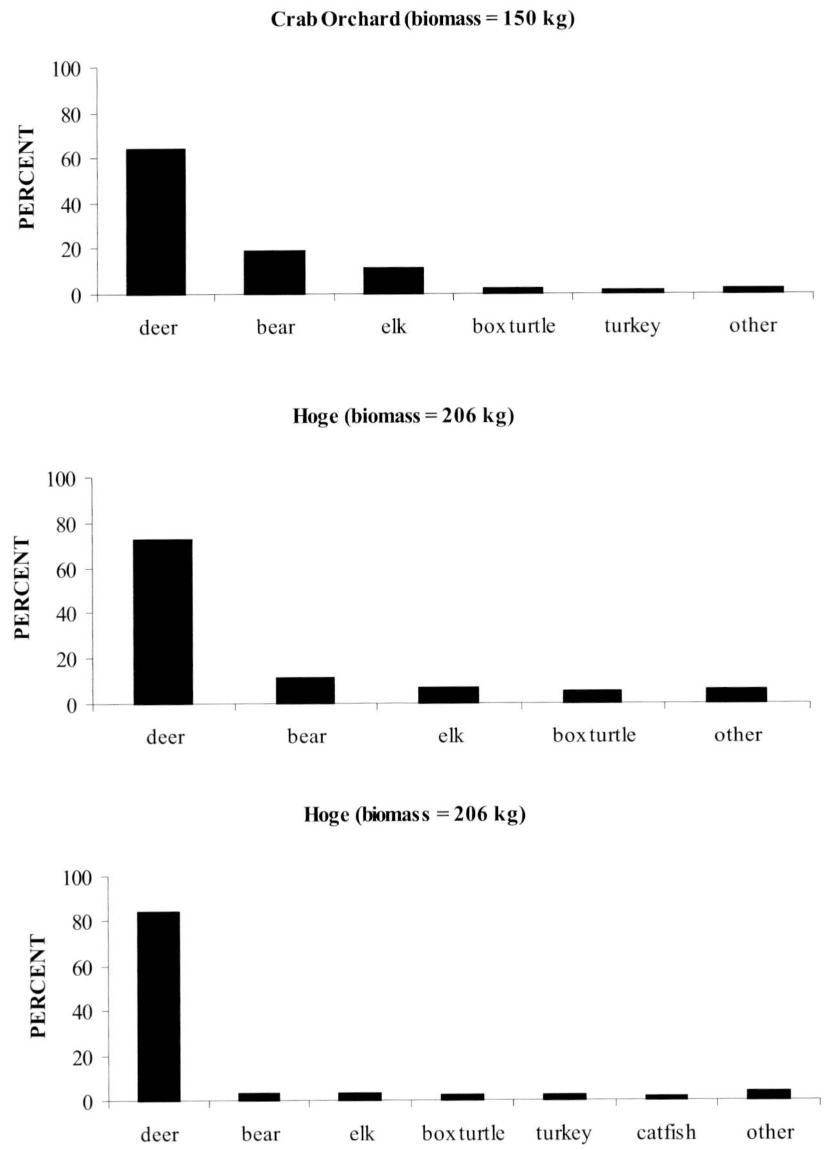

Fig. 4. Species representation in archaeological assemblages (percent of total biomass of identified fauna).

bone weight, may represent (Reitz and Wing 1999, 227). Four to five top biomass contributors were identified in each assemblage. The summed biomass represented by the indicated species varied, depending on the site, from 96–98% of the total biomass of animal bones identified to the taxonomic level of family or lower. In addition to deer, at the two Late Woodland settlements black bear (*Ursus americanus*) and wapiti (*Cervus elephus*) also contributed a significant amount of mass to vertebrate fauna subsistence. Wild turkey (*Meleagris gallapavo*) and eastern box turtle (*Terrapene carolina*) are present in lesser amounts at all three villages. Other taxa such as squirrel (*Sciurus carolinensis* and *Sciurus niger*), raccoon (*Procyon lotor*), Canada goose (*Branta canadensis*), ruffed grouse (*Bonasa umbellus*), and various fish provided diet diversity. All mammals and fish represented in the faunal assemblages occupied habitats adjacent to or nearby the three study sites. Birds identified range from permanent residents of the study region (i.e., turkey and grouse) to transient and winter visitors, such as the Canada goose.

Two important patterns emerge from these data. First, in each assemblage a limited number of taxa contribute the majority (between 96% and 98%) of the total biomass of identified specimens. This pattern is consistent with the subsistence model defined by Michael B. Barber (n.d.) and Eugene Barfield and Barber (1992) for the Mountain region of Virginia. They predict an animal resource base focused largely on deer with a substantial amount of meat supplied by bear and wapiti. Turkey and box turtles were also harvested in significant numbers. Differences among the three assemblages are evident in the species distribution of the top biomass contributors. Both Late Woodland sites show a more even representation of animal species compared to the Early Historic Trigg site. At the Trigg site, deer are by far the single most predominant biomass contributor. Compared to deer, other animal species appear in consistently low proportions (less than 3%) in the Trigg assemblage.

The second noteworthy pattern seen in these data is an increase in the utilization of white-tailed deer between the Late Woodland and Early Historic periods. Biomass for deer is greatest at the Early Historic Trigg site at 84% of identified taxa. Both Late Woodland sites contain deer in lesser amounts, 72% at Hoge and 64% at Crab Orchard. Regardless of the zooarchaeological measure used, deer are consistently represented in higher proportions in the Early Historic assemblage. This patterns holds for biomass, NISP, bone weight, and MNI. The increase in venison consumption seen at the Early Historic Trigg site suggests a shift to a greater utilization of white-tailed deer in the seventeenth century.

Deer hunting

Another line of evidence for shifting white-tailed deer utilization which provides further insights into changing socioeconomic status is apparent in changes in deer hunting strategies. Deer harvests were evaluated based on three different variables: age of deer at death (established from tooth eruption and occlusal wear patterns following Severinghaus 1949), sex of deer hunted (determined from an examination of the pelvic bone following Edwards *et al.* 1982), and season of deer kill (estimated from aged mandibles and antler development using regionally specific data on mean fawn-drop date and antler discard dates found in Linzey 1998; McGinnes and Downing 1973; Severinghaus 1949). The resulting harvest patterns have been described in detail elsewhere (Lapham, *in press*), therefore only a summary of the findings will be presented here.

Differences in Late Woodland and Early Historic period deer harvests are evident in all three categories (age, sex, and seasonality). At the two Late Woodland sites, kill off patterns closely resemble the natural structure of a white-tailed deer population in terms of the age and sex of the animals killed. Late Woodland hunters killed a significant number of fawns, some prime-age animals, and a few older adult deer, with male and female deer almost equally represented. Deer harvests occurred primarily from late summer to early winter. The Early Historic period Trigg site, in contrast, exhibits a kill-off pattern dominated by prime-age animals, more male than female deer killed, and deer hunted fairly regularly throughout the year.

These data indicate that a shift occurred in deer harvests from a more opportunistic deer procurement practice in the Late Woodland period to a hunting strategy in the Early Historic period that at times selected for prime-age deer, especially male deer. Deer exploitation may have also broadened from a seasonal activity to year-round harvesting. Such changes provide strong evidence that seventeenth century Native American hunters developed a selective predation strategy intended to procure deerskins specifically for commercial trade, and that some native hunters selected for large animals, or male deer, whose hides would bring the most competitive exchange rates on the commercial market.

Deer butchering

In addition to greater venison use and selective deer harvest strategies, butchery marks on deer bones also show that changes occurred in the seventeenth century in deer butchering techniques. A study of butchering type, length, frequency, orientation, and location revealed a general pattern of deer butchery consistent with those observed elsewhere in the eastern North America (*e.g.*, see Guilday *et al.* 1962; Guilday 1971; Waselkov 1977, 87–9; Binford 1981), with marks intended to sever tendons and ligaments at limb joints prominent at all three study sites. Two main types of butchering marks were identified in the three assemblages: cuts (defined as narrow incised lines) and hacks (which appeared as slightly deeper, wedge-shaped scars). Less than five percent of all deer bones bore butchering scars (Fig. 5). The Early Historic Trigg site contained double the proportion of butchered deer bones than the Late Woodland Crab Orchard or Hoge sites. Cut marks made by metal tools, as differentiated by their hairline size and bone "shelf" (see Binford 1981, 105), were not noted in the Trigg site assemblage even though the possibly exists that metal knives could have been obtained in small numbers through interregional trade. All deer skeletal elements exhibited butchering marks, but the single most common scar in all three assemblages occurred in the form of cut marks on tarsal bones. Butchery was observed primarily on astragali, but several calcanei, central and fourth tarsals, and a second and third tarsal were also scarred. Cut marks of this sort have been associated with the dismemberment of the hindfoot from the lower hindlimb (Binford 1981, 199–20). Cuts on the distal humerus, characteristic of dismemberment and filleting, were also fairly common.

Butchering marks associated with skinning (i.e., removing the hide from the carcass) tend to be localized scars found on the lower jaw, on the skull encircling the antlers, and on the metapodials and phalanges (Guilday *et al.* 1962, 72–3; Binford 1981, 106–7; Lev-Tov 2000, 143). In his landmark study of animal bone modifications, Lewis Binford (1981, 103, 107) found that among the Alaskan Nunamiut skinning-for-skins differed from skinning as a stage of butchery. At multiple-kill sites where meat was the primary objective, skinning often began with a cut encircling the proximal metapodial or sometimes the distal lower limb bone (i.e., radius/ulna and tibia). During late summer and early fall hunts, when the skins of caribou yearlings and calves were most desired, Nunamiut butchers

	Crab Orchard	Hoge	Trigg
Deer NISP	2,063	2,112	3,898
Number of butchered deer bones	20	46	139
Butchery type			
Cut	15	41	119
Hack	5	5	19
Combination	–	–	1
% butchered bones of total deer NISP	1%	2%	4%
% skinning-for-skins cuts of total butchered deer bones	0%	2%	7%

Fig. 5: Butchered deer bones

often took greater care to begin skinning at the hoof of the animal in order to maximize the size of the hide removed intact from the animal. John Guilday *et al.* (1962, 72–3) identified a similar pattern to that noted by Binford at an Early Historic period Susquehannock site in south-central Pennsylvania where large quantities of European goods along with historical references signify an active participation in the deerskin trade by the village residents. Here, skinning a deer most often began on the distal shaft of the metapodial just above the dew-claws.

Cut marks consistent with a skinning-for-skins butchering technique are absent in the Late Woodland Crab Orchard assemblage and only two examples of scarred distal metapodials are present at the Late Woodland Hoge site (Fig. 5). Mandibles, distal metapodials, and phalanges exhibit scars at a significantly greater frequency at the Trigg site–more than three times the frequency–than at the Late Woodland sites combined. Although skinning-for-skins cuts are more prevalent in the Early Historic Trigg assemblage, the overall frequency of these marks in the butchered deer bone assemblage is quite low (7%). These data suggest that even though occasionally greater care was taken to ensure maximum hide removal in the Early Historic period, the main concern of the butcher continued to be to obtain meat and other usable parts from the carcass.

Changing socioeconomic status in the Early Historic period

A comparison of faunal remains from Late Woodland and Early Historic period contexts in southwest Virginia indicate that several changes occurred within some Native American communities in the early-to-mid 1600s: deer use and venison consumption increased, deer harvests became more selective, and deer skinning techniques occasionally reflected an attention to detail geared toward preserving maximum hide size. When these changes are considered within a regional context it becomes apparent that villages (and individuals and families within those villages) made different choices as to whether or not to participate in the deerskin trade and how much to alter their existing socioeconomic practices in order to do so. At the Hurt Powers Plant site (44PY144), for example, Native American lifeways changed little from Late Woodland times and no evidence exists to suggest that the deerskin trade influenced deer hunting (Barber *et al.* 1996, 369–70). Located on the Staunton River in the western Virginia piedmont about seventy miles east of the Trigg site, this village was occupied in the early-to-mid seventeenth century. Glass beads, copper alloy ornaments, and an iron knife fragment indicate contacts with European traders or with Native American groups who had access to European goods. At the mid-to-late seventeenth century Graham-White site (44RN21), a settlement located on the Roanoke River less than forty miles northeast of the Trigg village, deer exploitation increased significantly between the Late Woodland and Historic components (Moore 1994; Moore and Lapham 1997). Native American hunters also displayed a preference for harvesting prime-age deer. European artifacts recovered at Graham-White include a gun trigger, glass beads, copper alloy ornaments, and an iron object (Klatka and Klein 1993). By the time Native American peoples settled at Graham-White, the trade and commercial export of deerskins had become one the fastest growing industries in the southern colonies (Martin 1994, 310).

Changes evident in deer exploitation practices in the Early Historic period suggest that certain Native American settlements in southwest Virginia chose to alter deer harvest and processing activities in order to participate in a growing interregional trade in hides, furs, and nonlocal goods. How these changes affected socioeconomic status requires the data to be examined from a perspective that considers the broader implications of economic intensification. The intensified production of goods can provide an individual or social group with the material means to enhance social prestige and political authority through the public display, reciprocal gift-giving, and exchange of socially important objects. Among some emergent chiefdom polities in southeastern North America, for example, the local manufacture of valued craft items, particularly shell beads and beaded garments, intensified as individuals and kin-groups vied for power and prestige (Steponaitis 1986; Steponaitis 1991). These objects served to heighten social status when displayed as possessions and cultivate alliances or accrue the social obligations of others when given as gifts (*ibid.*). In the 1600s in southwest Virginia, deerskins may have served a somewhat similar purpose, or rather they may have provided the material resources to acquire the nonlocal status goods that served these purposes. I argue that certain individuals or families choose to alter deer harvests and carcass processing in order to produce surplus hides that could be exchanged for valued goods whose acquisition would have ultimately increased personal or family wealth, prestige, or power.

As discussed at the beginning of this paper, in the Late Woodland period two nonlocal materials, marine shell and copper, were the primary prestige goods obtained by southwest Virginia inhabitants through interregional exchanges. This list expands in the Early Historic period to include European-manufactured items which, for the study area, are generally limited to copper alloy ornaments and glass beads. When these materials are compared between the Late Woodland and Early Historic study sites it becomes apparent that nonlocal prestige goods were acquired in greater quantities in the seventeenth century than they had been in prior times. Marine shell artifacts triple in number between the Late Woodland Crab Orchard and Hoge sites and the Early Historic Trigg site (Lapham 2002). Calculations based on the mean number of marine shells items per burial containing marine shell also indicate that these artifacts increase in quantity, but in this case Early Historic contexts contain only 25% more marine shell than do the Late Woodland period burials. Copper artifacts, which are rare finds in the study region prior to AD 1600, also show a dramatic increase in the Early Historic period. The raw count of copper artifacts at the Early Historic period Trigg site is more than twenty times greater than that seen at Late Woodland Crab Orchard and Hoge sites combined (*ibid.*). The substantial increase in copper at the Early Historic Trigg village seems to be a clear indication that these artifacts are the product of a new, European, source of copper and not derived from a North American source. In addition, more than 350 European-manufactured glass beads were recovered from the Early Historic Trigg site. A preliminary comparison of the Trigg site beads with glass bead assemblages from contemporaneous sites throughout eastern North America tentatively suggest that the Trigg site occupants may have been drawing off several interregional exchange networks over an extended period of time (Lapham 1998; Lapham 2002).

This marked increased in prestige goods and the potential variety in the sources of these items suggests that competition growing among certain social groups at the Early Historic Trigg village led to intensified deerskin production as a way for individuals to secure the resources needed to obtain socially-valued goods, and in doing so heighten their own socioeconomic status within Native American society. The zooarchaeological data presented in this paper identify several signatures of commercial hide production and trade in the seventeenth century southern Mid-Atlantic. At some, but not all, Native American settlements venison gained dietary importance as deer were being increasingly hunted for their hides and the processing of these hides intensified. In trade for processed deerskins, prestige goods were acquired. New sources of prestige goods offered new opportunities to gain and define socioeconomic status. Further research on the intra-site level is needed in order to better understand precisely who benefited from these new status items and how access to prestige goods varied within a single village.

References

Barber, M. B. n.d. Late Woodland vertebrate faunal utilization patterns in the South Ridge and Valley of Virginia: changes through time. Unpublished report, Preservation Technologies, Inc., Salem, Virginia.

Barber, M. B. 1989. Preliminary vertebrate faunal analysis: the Hall site, Montgomery County, Virginia. Unpublished report, Department of Historic Resources, Richmond, Virginia.

Barber, M. B., Solberg, T. N., and Barfield, E. B. 1996. The composition of copper recovered from contact Native American sites in the Lower Piedmont and Southern Ridge and Valley, Virginia. *Virginia Journal of Science* **47**, 9–18.

Barfield, E. B. and Barber, M. B. 1992. Archaeological and ethnographic evidence of subsistence in Virginia during the Late Woodland period, pp. 225–48 in Reinhart, T.R. and Hodges, M. E. N. (eds), *Middle and Late Woodland Research in Virginia: A Synthesis* (Special Publication 29). Richmond: Archaeological Society of Virginia.

Benthall, J. L. 1969. *Archeological Investigation of the Shannon Site, Montgomery County, Virginia.* Richmond: The Virginia State Library.

Binford, L. R. 1981. *Bones: Ancient Men and Modern Myths.* New York: Academic Press.

Boyd, C. C. 1993. The Trigg site (44My3) and other Late Woodland/Contact period sites along the New River Valley in Southwest Virginia. Unpublished paper presented at the Southeastern Archaeological Conference, Raleigh, North Carolina.

Boyd, D. C. and Boyd, C. C. 1992. Late Woodland mortuary variability in Virginia, pp. 249–75 in Reinhart, T. R. and Hodges, M. E. N. (eds), *Middle and Late Woodland Research in Virginia: A Synthesis* (Special Publication 29). Richmond: Archaeological Society of Virginia.

Braun, E. L. 1964. *Deciduous Forests of Eastern North America.* New York: Hafner Publishing Company.

Buchanan, W. T. 1984. *The Trigg Site, City of Radford, Virginia.* Richmond: The Archeological Society of Virginia.

Clarkson, L. A. 1960. The organization of the English leather industry in the late sixteenth and seventeenth centuries. *The Economic History Review* **13**, 245–56.

Crane, V. W. 1928. *The Southern Frontier, 1670–1732.* Durham: Duke University Press.

David, B. 1990. How was this bone burnt?, pp. 65–79 in Davidson, I. and Solomon, D. W. S. (eds), *Problem Solving in Taphonomy: Archaeological and Palaeontological Studies from Europe, Africa, and Oceania* (Anthropology Museum Tempus Vol. 2). St. Lucia, Brisbane: University of Queensland.

Edwards, J. K., Marchinton, L. R. and Smith, G. F. 1982. Pelvic girdle criteria for sex determination of white-tailed deer. *Journal of Wildlife Management* **46**, 544–7.

Egloff, K. T. 1992. The Late Woodland period in southwestern Virginia, pp. 187–223 in Reinhart, T. R. and Hodges, M. E. N. (eds), *Middle and Late Woodland Research in Virginia: A Synthesis* (Special Publication 29). Richmond: Archaeological Society of Virginia.

Egloff, K. T. and Reed, C. 1980. Crab Orchard site: a Late Woodland palisaded village. *Archeological Society of Virginia Quarterly Bulletin* **34**, 130–48.

Egloff, K. T. and Turner, E. R. 1988. Archaeological investigations at the Richlands Hospital site (44Tz51), *Archeological Society of Virginia Quarterly Bulletin* **43**, 14–23.

Ewan, J. and Ewan, N. 1970. *John Banister and his Natural History of Virginia, 1678-1692*. Urbana: University of Illinois Press.

Geier, C. R. and McFee, M. 1981. *The Harry Jaeger Site (44Ba489): A Multicomponent Protohistoric Indian/Eighteenth Century Historic Site, Bath County, Virginia* (Archaeological Research Center Occasional Papers in Anthropology No. 7). Harrisonburg, Virginia: James Madison University.

Geier, C. and Warren, J. C. 1982. *The Noah's Ark Site (44Ba15): A Late Woodland/Protohistoric Site on the Jackson River, Bath County, Virginia* (Archaeological Research Center Occasional Papers in Anthropology No. 14). Harrisonburg, Virginia: James Madison University.

Gleach, F. W. 1997. *Powhatan's World and Colonial Virginia: A Conflict of Cultures*. Lincoln: University of Nebraska Press.

Guilday, J. E. 1971. *Biological and Archeological Analysis of Bones from a 17th Century Indian Village (46PU31), Putnam County, West Virginia* (Report of Archaeological Investigations No. 4). Morgantown: West Virginia Geological and Economic Survey.

Guilday, J. E., Parmalee, P. W. and Tanner, D. P. 1962. Aboriginal butchering techniques at the Eschelman site (36La12), Lancaster County, Pennsylvania. *Pennsylvania Archaeologist* **32**, 59–83.

Hamell, G. R. 1983. Trading in metaphors: the magic of beads, pp. 5–28 in Hayes, C. F. (ed.), *Proceedings of the 1982 Glass Trade Bead Conference* (Research Records No. 16). Rochester: Rochester Museum and Science Center.

Hantman, J. L. 1990. Between Powhatan and Quirank: reconstructing Monacan culture and history in the context of Jamestown. *American Anthropologist* **92**, 676–90.

Hantman, J. L. 1993. Powhatan's relations with the Piedmont Monacans, pp. 94–111 in Rountree, H. (ed.), *Powhatan Foreign Relations*. Charlottesville: The University of Virginia Press.

Hantman, J. L. 2001. Monacan archaeology of the Virginia interior, AD 1400–1700, pp. 107–23 in Cowan, C. W., Mainfort, R. C. and Brose, D. S. (eds), *Societies in Eclipse: Archaeology of the Eastern Woodlands Indians, AD 1400–1700*. Washington, D.C.: Smithsonian Institution Press.

Hantman, J. L. and Gold, D. L. 2002. The Woodland in the Middle Atlantic: ranking and dynamic political stability, pp. 270–91 in Anderson, D. G. and Mainfort, R. C. (eds), *The Woodland Southeast*. Tuscaloosa: University of Alabama Press.

Hudson, Ch. M. 1976. *The Southeastern Indians*. Knoxville: The University of Tennessee Press.

Jones, E. E. and MacCord, H. A. 2001. The Hoge site, Tazewell County, Virginia, pp. 109–82 in *Contributions to the Archaeology of Southwestern Virginia: A Volume Honoring the Author Emory Eugene Jones, Jr., of Bluefield, West Virginia* (Special Publication No. 40). Richmond: The Archeological Society of Virginia.

Klatka, T. S. and Klein, M. J. 1993. Preliminary report on the Graham-White site: a late prehistoric and Contact period site in Western Virginia. Paper presented at the Southeastern Archaeological Conference, Raleigh, North Carolina.

Klein, R. and Cruz-Uribe, K. 1984. *The Analysis of Animal Bones from Archaeological Sites*. Chicago: University of Chicago Press.

Knight, V. J. and Steponaitis, V. P. 1998. *Archaeology of the Moundville Chiefdom*. Washington, D.C.: Smithsonian Institution Press.

Lapham, H. A. 1998. *Some thoughts on glass beads and exchange patterns in the interior Chesapeake Bay region*. Paper presented at the 63rd Annual Meeting of the Society for American Archaeology, Seattle, Washington.

Lapham, H. A. 2002. *Deerskin Production and Prestige Goods Acquisition in Late Woodland and Early Historic Southwest Virginia*. Unpublished Ph.D. dissertation, University of Virginia, Charlottesville.

Lapham, H. A. in press. "Their complement of deer-skins and furs": changing white-tailed deer exploitation patterns in the seventeenth century southern Chesapeake and Virginia hinterlands, in Blanton, D. B. and King, J. (eds), *Contact in Context: New Archaeological, Anthropological, and Historical Perspectives on Natives and Europeans in the Mid-Atlantic*. Gainesville: University Presses of Florida.

Lev-Tov, J. S. E. 2000. *Pigs, Philistines, and the Ancient Animal Economy of Ekron from the Late Bronze Age to the Iron Age II*. Ph.D. Dissertation, Department of Anthropology, University of Tennessee, Knoxville.

Linzey, D. W. 1998. *The Mammals of Virginia*. Blacksburg, Virginia: The McDonald and Woodward Publishing Company.

MacCord, H. A. 1977. Trade goods from the Trigg site, Radford, Virginia, pp. 60–8 in South, S A. (ed.), *The Conference on Historic Site Archaeology Papers, Vol. 10*. Columbia: The University of South Carolina Press.

MacCord, H. A. 1989. The intermontane culture: a Middle Appalachian Late Woodland manifestation. *Archaeology of Eastern North America* **17**, 89–108.

MacCord, H. A. and Buchanan, W. T. 1980. *The Crab Orchard Site, Tazewell County, Virginia*. Richmond: The Archeological Society of Virginia.

Mallios, S. W. 1998. *In the Hands of "Indian givers": Exchange and Violence at Ajacan, Roanoke, and Jamestown*. Unpublished Ph.D. Dissertation, Department of Anthropology, University of Virginia, Charlottesville.

Martin, C. 1978. *Keepers of the Game: Indian-Animal Relationships and the Fur Trade*. Berkeley: University of California Press.

Martin, J. W. 1994. Southeastern Indians and the English trade in skins and slaves, pp. 304–24 in Hudson, C. and Tesser, C. C. (eds), *The Forgotten Centuries: Indians and Europeans in the American South, 1521–1704*. Athens: The University of Georgia Press.

McGinnes, B. S. and Downing, R. L. 1973. Variation in peaks of fawning in Virginia, pp. 22–7 in Mitchell, A. L. (ed.), *Proceedings of the Twenty-Sixth Annual Conference, Southeastern Association of Game and Fish Commissioners*. Columbia, South Carolina: Southeastern Association of Game and Fish Commissioners.

Moore, E. A. 1994. *Native American hunting strategies in 17th-century Virginia*. Paper presented at the annual meeting of the Southeastern Archaeological Conference, Knoxville, Tennessee.

Moore, E. A. and Lapham, H. A. 1997. Vertebrate fauna from the Graham-White site, 44Rn21. Unpublished report, Virginia Department of Historic Resources, Salem Office, Salem, Virginia.

Muller, J. 1997. *Mississippian Political Economy*. New York: Plenum Press.

Parrish, S. V. 1972. *The Fur and Skin Trade of Colonial Virginia*. Unpublished M.A. thesis. Old Dominion University, Norfolk, Virginia.

Perdue, T. 1998. *Cherokee Women: Gender and Culture Change, 1700-1835*. Lincoln: University of Nebraska Press.

Potter, S. R. 1989. Early English effects on Virginia Algonquian Exchange and tribute in the Tidewater Potomac, pp. 151–72 in Wood, P. H., Waselkov, G. A. and Hatley, M. T. (eds), *Powhatan's Mantle*. Lincoln: University of Nebraska.

Potter, S. R. 1993. *Commoners, Tribute, and Chiefs: The Development of Algonquian Culture in the Potomac Valley*. Charlottesville: The University of Virginia Press.

Reitz, E. J. and Cordier, D. 1983. Use of allometry in zooarchaeological analysis, pp. 237–52 in Clutton-Brock, J. and Grigson, C. (eds), *Animals and Archaeology 2: Shell Middens, Fishes, and Birds* (BAR International Series 183). Oxford: British Archaeological Reports.

Reitz, E. J. and Wing, E. S. 1999. *Zooarchaeology*. Cambridge: Cambridge University Press.

Reitz, E. J., Quitmyer, I. R., Hale, H. S., Schudder, S. J. and Wing, E. S. 1987. Application of allometry to zooarchaeology. *American Antiquity* **52**, 304–17.

Rountree, H. 1989. *The Powhatan Indians of Virginia*. Norman: The University of Oklahoma Press.

Severinghaus, C. W. 1949. Tooth development and wear as criteria of age in white-tailed deer. *Journal of Wildlife Management* **13**, 195–216.

Steponaitis, V. P. 1986. Prehistoric archaeology in the southeastern United States, 1970–1985. *Annual Review of Anthropology* **15**, 363–404.

Steponaitis, V. P. 1991. Contrasting patterns of Mississippian development, pp. 193–228 in Earle, T. (ed.), *Chiefdoms: Power, Economy, and Ideology*. Cambridge University Press, Cambridge.

Stewart, R. M. 1989. Trade and exchange in Middle Atlantic Region prehistory. *Archaeology of Eastern North America* **17**, 47–88.

Stewart, R. M. 1994. Late Archaic through Late Woodland exchange in the Middle Atlantic Region, pp. 73–98 in Baugh, T. G. and Ericson, J. E. (eds), *Prehistoric Exchange Systems in North America*. New York: Plenum Press.

Swanton, J. 1946. *The Indians of the Southeastern United States* (Bureau of American Ethnology Bulletin 137). Washington, DC: Government Printing Office.

Thompson, E. P. 1975. *Whigs and Hunters: The Origin of the Black Act*. London: Allen Lane.

Tinling, M. 1977. *The Correspondence of the Three William Byrds of Westover, Virginia, 1684–1776*. Charlottesville: The University of Virginia.

Usner, D. H. 1992. *Indians, Settlers, and Slaves in a Frontier Exchange Economy*, Chapel Hill: University of North Carolina.

Ward, H. T. and Davis, R. P. S. 1993. *Indian Communities on the North Carolina Piedmont, A.D. 1000–1700* (Research Laboratories of Anthropology Monograph 2). Chapel Hill: University of North Carolina.

Waselkov, G. A. 1977. *Prehistoric Dan River Hunting Strategies*. M.A. Thesis, Department of Anthropology, University of North Carolina, Chapel Hill.

Waselkov, G. A. 1993. Historic Creek Indian responses to European trade and the rise of political factions, pp. 123–31 in Rogers, D. and Wilson, S. (eds), *Ethnohistory and Archaeology: Approaches to Postcontact Change in the Americas*. New York: Plenum Press.

Waselkov, G. A. and Braund, K. E. H. 1995. *William Bartram on the Southeastern Indians*. Lincoln: University of Nebraska Press.

Whyte, T. R. and Geier, C. 1982. *The Perkins Point site (44Ba3): A Protohistoric Stockaded Village on the Jackson River, Bath County, Virginia* (Archaeological Research Center Occasional Papers in Anthropology No. 11). Harrisonburg, Virginia: James Madison University.

Heather A. Lapham
Center for Archaeological Investigations,
Southern Illinois University Carbondale,
Carbondale, Illinois, 62901–4527, USA
E-mail: hlapham@siu.edu

9th ICAZ Conference, Durham 2002
Behaviour Behind Bones, (ed. Sharyn Jones O'Day et al.) pp. 304–317

33. Implications of risk theory for understanding nineteenth century slave diets in the southern United States

Justin S. E. Lev-Tov

Archaeologists working on 19th century southeastern U.S. plantations have focused their research primarily along the coastal areas and for the most part have confined their questions to a search for African-American ethnicity, and distinguishing the status differences between master and slave, or domestic and field slaves. Zooarchaeological studies have been a key within this research. Yet that plantation research paradigm is narrow, and is in need of expansion, both in geography and in the approach to slave diets. This paper will compare and contrast zooarchaeological data gathered from plantation sites not only in the Atlantic coastal region, but also in the economically distinct upland interior of the southeast. The overriding theme guiding this comparison is to explain the differences visible in slave diets using a theory of risk, i.e., the idea that slaves faced risks in not having enough or adequate food, and had to devise ways of minimizing it. The degree of risk for slaves varied according to the labor system used on the plantations; labor systems depended on the types of crops produced, which in turn was dependent upon the regional location of the plantation. One of the most important categories of dietary evidence employed here is the role of hunting and fishing played in slave communities – possibly one way to measure the need slaves had to supplement their supplied rations.

Introduction

The plantation economic system, which in North America developed in the southern United States and the Caribbean, could sustain itself economically only by having huge numbers of slaves laboring hard in agricultural fields. To this end, by the first half of the 19th century, millions of Africans had been stolen from their home areas to be slaves in the Americas. Plantations were organized as independent and privately-held estates, albeit very wealthy planters might own several such sprawling farms, and were never organized into local or regional units administered by a governmental body. The study of American slavery and the New World plantation system has for much of the 20th century held a prominent place within the field of history, but it is only within the last thirty years that the archaeological study of plantations gained legitimacy in the prehistoric-dominated Americanist archaeology paradigm.

Archaeological studies of slavery in the United States have often focused on the idea of isolating African or African-American ethnicity based on material culture correlates, as with trying to find African antecedents for slave house architecture, religion, ceramics, *etc.* (McKee 1987, 31–2). Animal bone assemblages from slave quarters at former plantations have also been used to trace African ethnicity in the New World. As with the archaeological pursuit of ethnic groups in general (cf. Shennan 1989), that research paradigm has not entirely borne the fruit which seemed so promising in the beginning.

Zooarchaeologists have attempted comparisons between slave and non-slave faunal assemblages, and in so doing distilled various characteristics which at first glance seemed to separate dietary remains of the two groups. The result of ethnicity paradigm research has been the generation of what might be called ethnic faunal indices. Two different proposals for such ethnic markers have both proven inconsistent markers of identity both within and between plantations. These proposals consist of Otto's (1984, 172–3) idea that a higher ratio of opossums to raccoons at slave cabins, and the opposite pattern at the homes of white plantation residents would serve as an ethnic marker for an African-American slave diet. As well, McKee (1987) proposed that the relative abundance of cleaver-made butchering marks as opposed to scars left by other tools might prove distinctive for the same population's dietary habits. Those characteristics instead appear correlated with poor and rural population elements of both European and African extraction (Otto 1984, 158 and 164–5). While African identity may have remained with slaves and can perhaps be traced through aspects of pottery, religion, or slave quarter architecture (Yentsch 1994), the diets of slaves varied due to factors other than what can safely be attributed to ethnic preference (cf. Scott 2001, 677–8).

Because attempts to isolate an ethnic African-American slave diet have proven unsuccessful, it is necessary to approach the dietary aspect of slave life in a different manner. From the historical record of planters' discussions of ways to manage their slaves, as well as archaeological explorations of 19th century and earlier slave cabins, we know that slaves were sometimes allowed to manage their own subsistence economies. That is, rather than – or in addition to – being supplied with rations supplied by the plantation owner, slaves on some plantations were allowed to raise their own livestock as well as crops (Breeden 1980, 84; Berlin and Morgan 1993, 30–1).

While the historic record supplies evidence of slaves having possessed their own livestock, especially fowl (Berlin and Morgan 1993, 30), it is less clear whether hunting, trapping and fishing were allowed, and to what extent slaves engaged in those activities. We know that some African-Americans hunted, fished, and trapped (cf. Berlin *et al.* 1998, 13 and 35 and 39). Slave owners were generally afraid of slave rebellions and retributional violence, and therefore would not have looked favorably on slaves' hunting with guns. And, while fishing and raising fowl were allowed activities by many slaveholders (Berlin *et al.* 1998, 39), other planters objected to the fact that this permitted slaves to generate an independent and rival economy where they bartered or sold their surplus (Breeder 1980, 270–5). In addition to documentary evidence, we have material culture remains in the form of both zooarchaeological evidence and recovered gun parts and fishhooks in slave quarters. These independent sources confirm that slaves did indeed hunt, trap, and fish in the South during the decades prior to the Civil War of 1860-1865, the era known as the antebellum period. We don't, however, know how important the additional food sources were in the diet of the enslaved population.

A geographical and labor-based approach to the problem of slave diet

The plantation systems of North America may be roughly divided into two categories, according to the types of crops grown. On the one hand there were the largely mono-cropping plantations which specialized in growing either inedible cash crops like tobacco, indigo, and cotton, or a solitary edible one like rice or sugar cane. Such plantations were located in the lowland areas of the southern coastal plain and in the Caribbean. On the other hand, there was another type of plantation, where diversified agriculture was practiced. These institutions, essentially very large farms, raised pigs as well as grew wheat and maize. Such plantations were located in the hilly piedmont and interior valley regions of the United States, where soils, climate, and topography prevented planters from having large slave populations producing a single cash crop. This area has been referred to as the 'interior' or 'upland' South as opposed to the 'coastal' or 'lowland' South (Fig. 1) (Singleton 1990; Sanford 1994). North American historical archaeology has maintained a focus on excavations of coastal plantations, such that some have called for a research program specifically aimed at exploring upland ones (Andrews and Young 1992).

The zooarchaeological record demonstrates that, from plantation to plantation, there was significant variation in the array and relative importance of animals consumed by slaves (cf. Reitz *et al.* 1985). The diversity of slave diets as revealed by an increasing number of zooarchaeological studies is in itself a piece of new information. We conceive of slave diets as having been monotonous, consisting primarily of pork and maize plus other planter-provided staple rations, and perhaps lacking either quantity or nutritional quality – or both. That idea is supported by some planters' discussions of appropriate food rations (Breeder 1980, 89–90) as well as by former slaves' own recollections (*e.g.*, Breeder 1980, 186).

At the same time, both documentary sources – planter and slave – make it clear that, although slave diets centered on the staples of maize meal and pork, many other items supplanted these. Sometimes planters saw the need for slaves to have variety or at least (19th century concepts of) nutritional balance, and so provided vegetables and fruits. Slaves were often able to plant gardens or diversify their diets through hunting and fishing (Morgan 1982; Yentsch 1994). Matching the documentary record, archaeological excavations of slave cabins have produced faunal assemblages in which domestic animals dominate (Reitz *et al.* 1985, 169 and 185). However, the relative contributions of wild species – from fish to birds and small mammals – varied greatly (cf. Reitz *et al.* 1985, 183–6). The question is, for what reason(s) did the amount of wild game and fish in slave diets vary so much?

This paper is divided into two sections, a review of zooarchaeological evidence from the upland South plantation of Locust Grove, and a broad-based comparison of slave diets from several upland and lowland plantations. Using this comparative approach, I wish to explore two hypotheses that in combination may explain variation in slave diets. The first hypothesis is one common in plantation archaeology, and adapted by Reitz *et al.* (1985), among others, to elucidate the major, interrelated, factors influencing slave foodways. The argument holds that it was the location of plantations in certain environmental

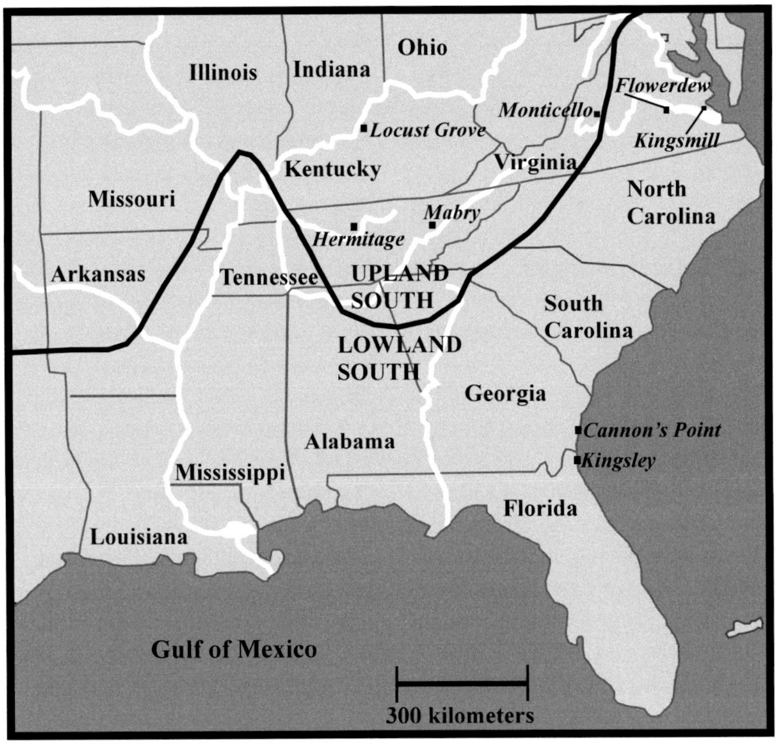

Fig. 1. Map of the Southeastern United States, showing the boundary between the coastal and upland South, and locations of plantations used in this comparative study.

zones, coastal or inland, that directly determined the crops grown and animals raised, and therefore to a certain extent dictated what foods slaves subsisted on. Slaves' access to wild game and fish may have differed according to the location of plantations, with slaves having much better access to wild game and especially fish in the flat coastal plain region as opposed to the hilly interior locales (*e.g*, Reitz *et al.* 1985, 184; McKee 1999, 231). The proof for this idea is still lacking however, given that the concentration of archaeological, and therefore zooarchaeological, research has been on plantations located in the coastal areas of the southeastern United States (Reitz *et al.* 1985, 184; Andrews and Young 1992). The uneven distribution of plantation excavations has resulted in a situation wherein archaeological dietary data from inland plantations are sorely lacking, and as a result slave diet has been written about mainly from a coastal perspective (cf. Reitz *et al.* 1985; McKee 1999). Therefore this explanation for variation in slave diets has to date not been fully explored, something which this essay will attempt to rectify with zooarchaeological information from the upland South. This and the following hypothesis will be explored through both analysis of a faunal assemblage from such a plantation as a case study, as well as a comparison of faunal information from several upland and coastal plantations' slave quarters.

The second hypothesis tested here is the power of risk theory (Wiessner 1982) to account for regional patterning in slave diets. Slaves naturally sought out the best possible diet in terms of variety and amount of food. That is, slaves – as in any human society – consciously or unconsciously made the effort to minimize the risk of starvation from an inadequate diet or simply improve a diet thought unsatisfactory in some other way (cf. Wiessner 1982, 172; McKee 1999, 233). A theory of risk reduction may lend a framework for interpreting African-American slave diets. Wiessner (1982, 172–3) has divided risk management into four categories:

- prevention of loss
- transfer of risk/loss to another party
- storage, and
- sharing of risk among a group.

Arguably, as Young (1997a) has demonstrated, all of these risk categories are applicable to the study of plantation slavery. The first risk category, prevention of loss, in this context may be understood as efforts by slaves to build a subsistence economy independent of the rations supplied by the master. Food was used as an instrument of power by slaveowners, and as punishments slaves sometimes were forced to make do with less or no rations, while at other occasions bonuses were given them (McKee 1999, 225–7). Establishment of an independent subsistence economy also provided the opportunity for slaves to transfer risk to the planter, as with the often successful efforts of African-Americans to market their poultry, hogs, game, thus at times competing with either their own master or other European-Americans (Campbell 1991; Yentsch 1994). Surreptitious storage of self-produced food and food-sharing among other slaves may have been means of lessening the real risks of food shortage and other harsh realities of slavery (cf. Young 1997a, 14).

Species	NISP	Percent	MNI	Percent
Scalopus aquaticus	6	<1	2	3
Rattus norvegicus	107	10	13	17
Rattus sp.	10	<1	3	4
Sciurus carolinensis	5	<1	2	3
Sciurus sp.	2	<1	1	1
Marmota monax	6	<1	1	1
Didelphis virginianus	19	2	1	1
Sylvilagus floridanus	73	7	4	5
Procyon lotor	18	2	2	3
Felis catus	2	<1	1	1
Canis familiaris	2	<1	1	1
Ovis aries	6	<1	1	1
Ovis/Capra	39	3	2	3
Bos taurus	39	3	2	3
Sus scrofa	526	47	7	9
unidentifiable mammal	1196			
Anas platyrhynchos	1	<1	1	1
Anas sp.	2	<1	1	1
Branta canadensis	2	<1	1	1
Gallus gallus	184	16	13	17
Meleagris gallopavo	9	<1	1	1
Columbia livia	2	<1	1	1
Zenaida macroura	1	<1	1	1
Falco sparverius	1	<1	1	1
Melanerpes sp.	1	<1	1	1
Colaptes auratus	1	<1	1	1
Corvus brachyrynchos	1	<1	1	1
unidentifiable bird	206			
Chelydra serpentina	6	<1	1	1
unidentifiable reptile	14			
Acipenser fulvescens	15	1	1	1
Aplodinotus grunniens	29	3	3	4
Micropterus salmoides	1	<1	1	1
Ictalurus punctatus	1	<1	1	1
unidentifiable fish	47			
unidentifiable molluscs	9			
unidentifiable bones	76			
Total, identifiable bones	1117		73	
Total, all bones	2651			

Fig. 2. Species list for all Locust Grove slave cabins combined.

Faunal assemblage recovery and quantification methods

At Locust Grove Plantation, a substantial faunal assemblage was recovered from the three slave cabin cellars and areas immediately surrounding the cabin foundations (Fig. 2) (Young 1997b). The second half of this paper is a cross regional comparison of slaves' diets based on ten faunal assemblages excavated from eight different plantations, four Upland and four Coastal ones. The plantation assemblages used here include faunal collections from the Upland sites of the Hermitage (Breitburg 1976; Thomas 1998), Mabry (Young 1993), and Monticello (Crader 1990) as well as Locust Grove. The coastal assemblages were excavated from Cannon's Point (Otto 1984), Flowerdew (McKee 1999), Kingsley (Walker 1988), and Kingsmill Quarter (McKee 1987). Sites used for this study were chosen for several reasons. Archaeological, and especially subsistence data, are still lacking from upland South plantations. Therefore this article presents a forum to make available as yet incompletely published faunal data from one site (Locust Grove) as well as that from another site to date not published at all (Mabry), and forms a broader comparison by utilizing the well-published data from the Hermitage and Monticello plantations. Although many more coastal plantations have been excavated, publications incorporating faunal data from slave quarters are frequently contained within the vast, unpublished 'gray literature' produced by archaeological firms in the course of salvage work, and these reports are difficult to find. The four coastal plantation sites used here were chosen mainly by completeness of data presentation and faunal assemblage size, as well as their proximity to one another (see Fig. 1 for site locations).

The comparative nature of this study naturally must incorporate potential sources of error arising from different types of excavation contexts and recovery procedures, among other things. Four of the samples, including those from Locust Grove (Young 1997a), Mabry (Young 1993), the Hermitage (Thomas 1998), and Kingsley (Walker 1988) were recovered using 6.4 mm mesh to sieve excavated sediment, while at Cannon's Point 3 mm screens were employed (Otto 1984). Kelso (1984) does not discuss recovery methods used at Kingsmill, but Reitz *et al.* (1985, 176) conclude that he did not sieve at all, a practice unfortunately repeated during his later excavations at Monticello (Scholnick *et al.* 2001, 5). Although excavators do not specify sieving procedures used at Flowerdew, it is clear from excavation photographs that screens were used (cf. Deetz 1993, 121) but flotation may not have been, given that no shad or herring bones were recovered (Deetz 1993, 141–3). Differences in sieving methods have undoubtedly introduced at least a degree of bias to the comparison since it has been demonstrated that 3 mm sieves collect much more small mammal and fish bones than do 6.4 mm ones (cf. Shaffer and Sanchez 1994). The amount of bias should, however, be comparable between many of the sites, since most excavations used the same size sieves. More problematically, the lack of sieving at Monticello and Kingsmill no doubt resulted in truncated species lists, a bias taken into account as much as possible in the cross-regional comparison.

Faunal assemblages used for the comparison were excavated not only using different recovery methods, but also were collected from somewhat different depositional environments. Four of the faunal assemblages were recovered from various types of pit features under or close to the slave cabins, while three collections came mainly from sheet refuse middens in the areas surrounding the houses. Two additional collections originate from both cellars and midden refuse scattered elsewhere beneath and around the houses. Naturally, those assemblages scattered on the ground were subject to trampling and other taphonomic processes, while the bones deposited into pit features would have been more protected from such activities. Taphonomic processes like trampling should have affected assemblages from sheet midden contexts more than those from pit features, such that bones of small animals might have been disproportionately destroyed beyond identifiability. Nevertheless, the assemblage from Flowerdew, which McKee (1999, 228–9) describes as being severely fragmented, returned the second-highest diversity score, suggesting that trampling may not have impacted the collection as much as feared. Similarly, the bones from Monticello also came from an exposed deposit and, while that site produced the lowest diversity score, it seems more likely this is a result of a lack of sieving.

The Number of Identified Specimens (NISP) formed the basis for all comparisons, including the diversity analysis (see below), since it was the one quantification method published for all sites incorporated into the comparison. NISP was also employed because several of the samples were small enough such that using the Minimum Number of Individuals (MNI) statistic might have introduced errors such as exaggerating the importance of rare species (cf. Grayson 1978). MNI calculations for Locust Grove are, however, presented in Fig. 2. One of the tools used here for comparing the slave cabin faunal assemblages is diversity, a statistic used to measure distribution of nominal scale data among several categories, for example species inhabiting an ecosystem (Zar 1999, 40). In zooarchaeology diversity has been used to measure the number of taxa contributing to a faunal assemblage (Cruz-Uribe 1988, 130). Diversity actually has two components, richness (variety) and evenness (relative abundance) (Meltzer *et al.* 1992, 377); only richness scores were utilized for this research. Following Reitz and Scarry (1985, 20) and Reitz and Wing (1999, 234–5) the Shannon-Weaver index was used to obtain richness scores, registered on a scale of 0 to 4.99, with higher scores representing assemblages with a greater variety of fauna. Diversity calculations were derived from NISP values for the reasons stated above, but it is probable that MNI-based enumerations would have produced similar results. All identified species or higher taxonomic categories were used in these calculations when these did not have the potential to overlap with lower categories. Commensal fauna like rats and moles, potential pets like cats and dogs, in addition to horses were excluded from diversity calculations.

The faunal assemblages from each of the Locust Grove cabins returned similar richness values, between 1.4 and 1.5. These scores indicate, first of all, that this group of slaves had similar diets and essentially the same access to both domestic and wild animals; status differences between these slaves – as has sometimes been postulated from slave cabin faunal remains (Crader 1990) – seem not to have existed. The relatively low richness scores result from the fact that the faunal collections did not contain many different species of wild animals, consisting mostly of bones from just the same few species out of the twenty or so taxa identified in each of the cabin assemblages. Because the Locust Grove cabin collections were so similar in diversity and species make-up, the assemblages were combined in order to calculate a single diversity statistic for the site. Diversity scores calculated for Locust Grove and the other sites are used in this study as a tool to summarize any dietary divergences between slave diets in the upland and coastal regions. As recommended by Zar (1999, 156) the richness scores calculated were then tested for differences using the Student's *t*-test of significance.

A case study in slave diets: Locust Grove plantation, Kentucky

The concept of risk and 19th century slave management practices is applied here to a slave faunal assemblage from the upland South area of the United States. These data derive from the bone collection produced by excavation of three slave cabins at the Locust Grove plantation. The University of Louisville between 1987 and 1989 conducted excavations at Locust Grove. The plantation existed for somewhat more than half a century, between the late 18th and mid-19th centuries. Compromise measures between representatives from northern and southern U.S. states had worked out geographic limitations to slavery by the early 19th century, establishing among other boundaries the Ohio River as slavery's northern limit. Its location eight kilometers to the east of the city of Louisville, Kentucky on the south side of the Ohio River placed the plantation at the northernmost extent of slavery and plantation agriculture. The Locust grove estate originally covered an area of 282 hectares, with a slave population of 41 in the early 19th century. Major William Croghan and his family, the plantation's owners, used the slaves' labor to produce maize, wheat, hogs and sheep and as such could supply food for the resident population (Young 1997a, 5). By contrast many plantations on the coastal plain covered some 500 hectares, owned 100 or more slaves, and did not raise food crops or livestock (Young 1997a, 8).

Excavations at Locust Grove during the late 1980's revealed the foundations of three slave cabins (Fig. 3), aligned north to south and situated close together approximately 200 meters from the main house (Young 1997a, 9). For convenience when discussed individually, they are hereafter referred to as the north cabin, the central cabin, and the south cabin. The slave cabins were originally built of logs or wood frames, as suggested by nail analysis, and rested on limestone foundations (Young 1994; Young 1997a, 10–3). Especially intriguing about the architecture of the slave houses was that each of them possessed a small pit cellar excavated in front of the chimney pad

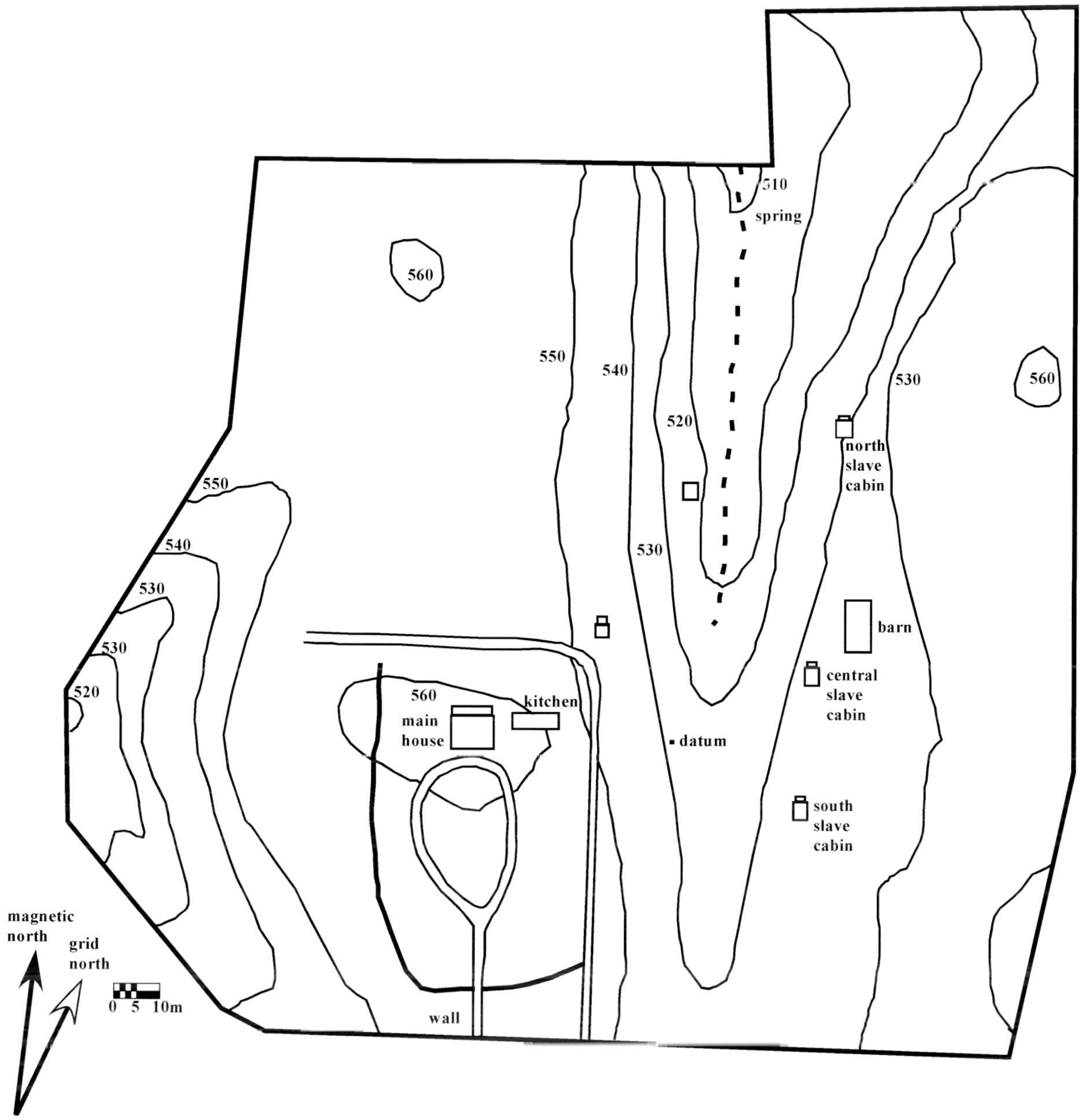

Fig. 3. Layout of Locust Grove plantation. Note locations of the three slave houses.

(Fig. 4) and originally concealed beneath the interior floor (Young 1997c, 96–8).

The cellars return us to the topic of risk management, specifically the principle of food storage. There remains a pervasive myth in Kentucky that slavery there had an unusually mild character to it, an argument based primarily on their material conditions compared to those elsewhere (Coleman 1940, 53 and 66–7). As Young (1997a, 8) put it, if "African-American slaves at Locust Grove . . . [were] capable of sustaining life, this was likely the result of . . .[their own] efforts rather than the . . . 'mildness' of the institution in Kentucky". Indeed, slaves there faced similar perils to those elsewhere, including physical abuse, being sold away from family, and work-related death or injury (Young 1997a, 14). Archaeological excavations of slave cabins have uncovered storage pits dug beneath the floors of slave cabins, a feature found mainly in the upland South region (Singleton 1995, 124). The existence of these pits is interesting since food storage is one way of coping with or minimizing dietary risks. Plantation owners during the 19th century frequently discussed what to do about these slave cabin cellars, and advocated that they be filled. The pits were viewed as challenges to their power, and therefore slave owners forbade their construction and use (McKee 1992, 197–8 and 203–4). DeBoer (1988, 3 and 14) has argued that storage pits, because they conceal

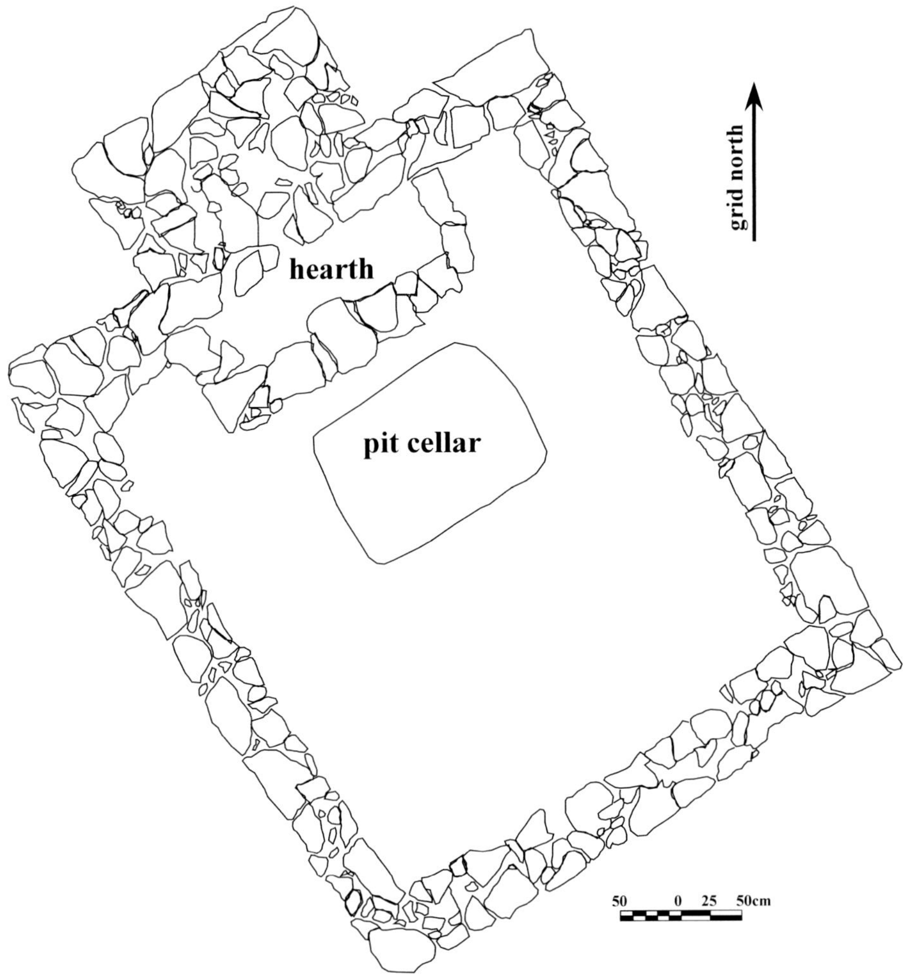

Fig. 4. Plan view showing the foundations and pit cellar of the south slave house at Locust Grove (drawing courtesy Amy Lambeck Young).

whatever is stashed inside them, are a form of resistance to social inequalities in which the distribution of goods is controlled by an elite class. This class conflict certainly existed between plantation owners and slaves. The two groups negotiated with each other to try to produce agreements on amount of permissable work, rest days, rations, and other issues. Slaves, despite internal class hierarchies (Coleman 1940, 50–1), often negotiated in unison and held common views on such issues as reasonable workloads and the fairness of taking the master's possessions for themselves (Berlin and Morgan 1993, 4–7; Berlin *et al.* 1998, 192–3 and 221–2).

As mentioned earlier, the practice of hunting by slaves is historically documented (*e.g.*, Schlotterbeck 1991), but what animals they hunted is not well known from such sources. At Locust Grove, the majority of these species were small mammals likely to have been trapped, songbirds, waterfowl, and fish (see Fig. 2). Though some plantation owners forbade slaves to hunt (Berlin and Morgan 1993, 24–25, 41) since it gave them a degree of economic independence, it is clear that the Locust Grove slaves did – though perhaps without consent of their master, given that hunting and fishing implements were all found concealed within the cellars. Corroborating discoveries of the gun parts, bullets, and fishhooks in the cabin cellars (Young 1997a, 17), excavations also produced a variety of wild animals' bones.

Hunting, trapping and fishing probably was important to slaves in the sense that these activities to some extent freed them from the control of their owner. As well, wild game and fish added diversity to the monotonous salted pork and cornmeal rations, the standard foods given to slaves as provisions. One can also frame the contest over hunting and fishing rights between slaves and masters in terms of Wiessner's (1982) concepts of risk management. On the one hand, planters attempted to prevent losses of authority and money by curtailing slaves' independent food provisioning and subsequent market of surpluses. At the same time, slaves, in pursuing a supplemented diet and perhaps an economy of their own, transferred economic risks to the plantation owner (Campbell 1991, 131).

Regardless of the social implications for hunting upon

either master or slave, wild animals did not form a major portion of the assemblages. In each cabin, no more than 20 percent of the bones were from wild fauna. Slaves in general may have secretly sought fish and wild game or demanded the right to hunt and fish. Those at Locust Grove either were not very interested in pursuing those activities, did not have the time to do so, or were effectively prevented from doing so by their owner. It seems more likely that the slaves at Locust Grove plantation simply did not possess the spare time to often pursue such activities, and as such concentrated their surplus food procurement activities on raising livestock or growing garden crops. After all, given the gun parts and fish hooks discovered, it seems that interest in hunting and fishing, along with the ability to do so did exist within the slave population. However, it is also possible they traded or sold much of their catches to plantation owners or at city markets, a common practice despite its having been frowned upon or even banned by many planters (cf. Berlin and Morgan 1993, 191; Yentsch 1994).

The data presented in Fig. 2 demonstrate that domestic animals formed one-third (according to MNI-based estimates) or even the major portion (according to NISP-based estimates) of the Locust Grove faunal assemblage. With regard to livestock, the question remains as to whether African-Americans were provided meat rations, or whether they raised livestock and poultry themselves. Chicken bones were very common in the assemblage, presumably because slaves raised the birds; this was a privilege commonly allowed by plantation owners (Campbell 1991). Indeed the assemblage contained bones of both immature and mature individuals, as well as tarsometatarsi with cockspurs, indicating slaves' access to a flock rather than selected birds. Neither sheep/goat nor cattle bones were frequent in the assemblage despite plantation records noting that these animals were raised there. Sheep were undoubtedly raised not for their meat but rather for their wool, while cattle probably were too valuable as dairy and draft animals to be sacrificed for regular slaughter. Interestingly, however, two of the cattle bones were professionally butchered 'slices' of upper limb bones (Fig. 5). These types of steak slices are common on mid-19th century U.S. sites, no doubt the result of the meat-packing industry's rise to prominence by that time (Clemen 1923). In fact, one of the industry's centers was in Cincinnati, Ohio, a city located not far up the Ohio River from Locust Grove. Packed meat was cheap, and plantations elsewhere in the South bought it in order to inexpensively feed slaves. Normally, barreled beef consisted of all saleable parts and excluded heads, legs and feet (Skaggs 1986). Given the presence of these professionally sawn bone slices, slaves likely received rations of beef purchased as barreled meat by the master. The leg bone slices are perhaps to be explained as contradictions to the recorded practice of meat packers, or else as a cheaper selection of beef cuts purchased for slaves' consumption. Identically butchered bones have been recognized in a slave context on a British West Indies plantation (Klippel 2001).

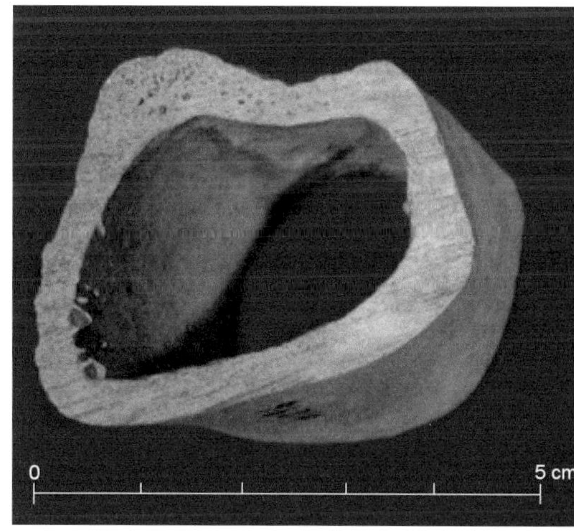

Fig. 5. Slice of a cattle bone, found in pit cellar of south cabin at Locust Grove.

A limited amount of beef may have found its way into slaves' diets via purchases of barreled beef, while excess male sheep may have been slaughtered on occasion without harming wool production (cf. Payne 1973). But what can we say about the role of pigs, given that this species' bones dominated the Locust Grove faunal assemblage? Element distributions provide a means of assessing the question of whether pork rations or pig herds were provided to the slaves. Element distributions for pigs (Fig. 6) demonstrate a general pattern where portions of the carcass are present in the same relative frequency as in an entire pig skeleton. If slaves had received master-supplied pork rations consisting of heads, feet, and other low-valued carcass portions (Crader 1990, 699), we would expect a pattern significantly different from the distribution of bones in an unmodified pig skeleton. A ´*t*-test comparing the percentages of body parts in the Locust Grove assemblage with those of a complete pig skeleton, demonstrates that the two suites of numbers are statistically indistinguishable ($p = .05$). At this particular plantation, slaves either received entire hogs raised on the plantation, or had their own herds. Masters sometimes banned slaves from having their own livestock herds (Schlotterbeck 1991, 174) as it granted them a degree of independence and a potentially rival economy, but nevertheless other slave communities were allowed such rights (Berlin and Morgan 1993, 31).

Mortality profiles provide another avenue to explore the question of whether slaves here received or raised their own livestock or had only master-controlled access to swine. Age estimation, based on the relatively small sample of ageable long bones in the collection, demonstrates that variously aged hogs – from neonatal to fully adult – were slaughtered (Fig. 7). The emphasis, however, was clearly on rather young pigs a year or less in

Body portion	Number of bones in assemblage	Percentage of bones in assemblage	Percentage of bones in a typical pig skeleton	Number of bones expected in a typical pig skeleton
Head	46	28	24	62
Axial	78	47	31	79
Upper Limbs	20	12	6	16
Lower Limbs/Feet	23	14	39	100

Fig. 6. Pig skeletal part distributions at Locust Grove. Following Klippel (2001, 1193), skeletal elements were divided into body portions as follows: Head: skulls, mandibles, hyoids, teeth; Axial: vertebrae, ribs, scapulae, pelves; Upper Limbs: humeri, radii, ulnae, femora, tibiae, lateral malleoli, patellae; Lower Limbs/Feet: carpals, tarsals, metacarpals, metatarsals, phalanges.

Plantation	≤ 1 year	1½ – 3 years	> 3 years
Locust Grove	57% (16)	25% (7)	18% (5)
Monticello	17% (3)	67% (12)	17% (3)

Fig. 7. Pig mortality patterns from Locust Grove and Monticello, based on unfused skeletal elements. The number of specimens in each age category is listed in parentheses next to the percentage.

age, a pattern different from that at Thomas Jefferson's Monticello Plantation in Virginia (Crader 1990, 696) as well as others (cf. Miller 1978; Tuma 1998). The usual practice was to slaughter pigs in the autumn, after the animals reached their full size, at around one and a half to two years of age (Crader 1990, 694). One possible explanation for the mortality profile is that slaves sometimes stole young pigs from plantation herds. Slave owners frequently complained about thefts of young animals (Genovese 1976, 599–601; McKee 1999, 227–8). Slaves viewed the practice as legitimate resistance, a way to slightly readjust the unfair power and wealth distribution (Genovese 1976, 603–8). The practice of stealing young animals transferred the risk to the master of either herd failure when too many young were stolen to guarantee herd security (Genovese 1976, 600), or of economic losses from stock lost before it could be fattened and sold. The pattern of pig mortality provides merely some intriguing data concerning the possibility of animal theft, but cannot prove that it occurred since "[a] rib from a stolen pig looks no different from . . . [one] distributed as rations" (McKee 1999, 232).

Faunal remains from Locust Grove present an intriguing set of data concerning the adaptation of slaves to a geographic area and economic system deviant from the more commonly known one which existed in the lowlands. As Young (1997a) has previously argued, slavery in Kentucky was not mild, only different from our 'Gone with the Wind' image of the antebellum South. The faunal remains suggest that slaves devised dietary, as well as other (cf. Young 1997a, 32–3) means of coping with and surmounting risks; stealing hogs may have been one such mechanism. The central thesis here is that slave diets were affected by the risks brought on by the types of crops produced and the labor system employed by plantations. A broader view, looking beyond Locust Grove plantation, is necessary in order to determine its applicability to an expanded area.

Variation in slave diets examined on a regional scale

The analysis of the faunal assemblage from Locust Grove begs the question of slave diet in general, and especially whether we can extend this analysis to encompass regional patterning. Regional patterns may have existed due both to the response of the enslaved and the economic systems of the enslavers. To examine these questions, I compared faunal reports from a number of excavated slave quarters from across the slaveholding areas.

Previous zooarchaeological research has already remarked upon the presence of a diverse array of wild fauna in slave assemblages from coastal, cash-cropping, plantations (Gibbs *et al.* 1980, 209–17; Otto 1984, 45–59). Archaeological research on slave diets from the upland South has been acquired more recently and therefore it remains to be seen how and why slave diets varied between the regions and their entrenched economic systems. I relied principally on diversity analyses to assess variation in slave diets. Statistical comparison of diversity test results calculated on a total of ten assemblages from eight different plantations (Fig. 8) demonstrate a significant (p = .01), difference between the two data sets in terms of richness. In essence, t-test results show that slaves living on coastal plantations generally had much more varied diets than their upland counterparts. This general difference seems largely unrelated to the aforementioned variations in recovery methods, since the faunal assemblages from most of the sites, both coastal and upland, were collected using 6.4 mm mesh. Further, differences in diversity are slight even between the assemblage collected using 3 mm mesh (Cannon's Point) and the close by site of Kingsley, where 6.4 mm sieves were employed. A greater problem than mesh size was the lack of sieving at least at Monticello and perhaps at Kingsmill as well. These faunal assemblages returned the lowest diversity values in the two sample sets, upland and coastal, but the t-test results are

Lowland plantations	Diversity scores	Upland plantations	Diversity scores
Cannon's Point	2.48	Hermitage (Breitburg)	1.24
Flowerdew	2.44	Hermitage (Thomas)	1.66
Kingsley, Cabin 3	2.15	Locust Grove	1.76
Kingsley, Cabin 6	2.21	Mabry	1.82
Kingsmill Quarter	1.79	Monticello	1.03

Fig. 8. Diversity scores from upland versus lowland plantations. Data for upland plantations derives from Locust Grove, Monticello (Crader 1990), Mabry (Young 1993) and the Hermitage (Breitburg 1976; Thomas 1998). Data for lowland plantations derives from Cannon's Point (Otto 1984), Flowerdew (McKee 1999), Kingsley (Walker 1988) and Kingsmill Quarter (McKee 1987). Student's t*-test (two sample equal variance): $p < .01$*

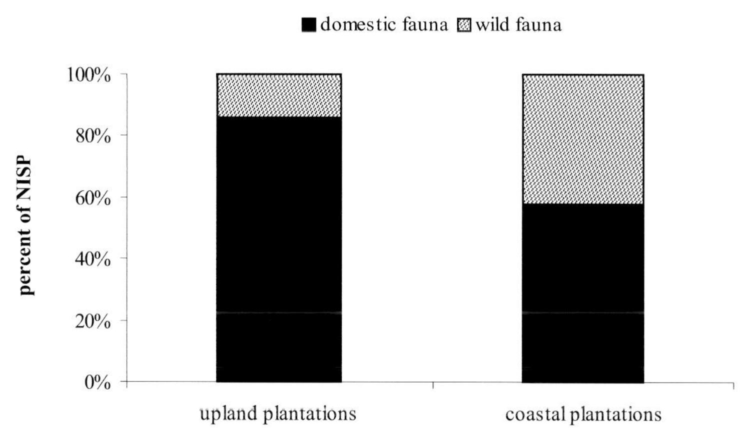

Fig. 9. Relative abundance of wild versus domestic fauna at upland and coastal plantations. Data from Brimstone Hill, Cannon's Point, Hermitage, Kingsmill, Locust Grove, Mabry, Monticello, and Saragossa plantations.

still significant ($p < .01$) even when the latter collections are not included in the diversity comparison.

The difference in dietary diversity is principally due to the prevalence of wild fauna, both in number and variety, which reflects the greater role that hunting, trapping and fishing played in the diets of coastal slaves (Fig. 9). Although there are certainly differences in environments between the upland and coastal South, environmental explanations alone cannot provide complete answers to the discrepancy. Wild fauna were as abundant and species diversity as rich inland as on the coast. In fact one of the published faunal assemblages from the Hermitage, a very large cotton plantation in an upland section of Tennessee, contained a great variety of wild game, fowl, and fresh-water fish (cf. Thomas 1998, 544). This finding leads to the hypothesis that the number and types of crops grown on a plantation, in addition to the related factors of geographic location and labor organization, may be another important component that affected slave dietary choices. With the introduction of a Mexican-derived short-staple variety of cotton in the early 19th century, upland plantation owners were able to at once grow a single cash crop while allowing their slaves free time in the evenings and on Sundays (Campbell 1991, 132–6). Slaves on many coastal plantations generally had considerable amounts of free time already from the start of plantation slavery in the 17th century. They, like upland slaves later, had negotiated with their masters to have Sundays and evenings off from work, an allowance made possible by the schedule of farming work and the early adoption of a flexible labor system (Campbell 1991, 132; McDonald 1991, 191–2; Berlin *et al.* 1998, 39). Free time may have presented the possibility for slaves to manage dietary risks by either supplementing rations with significant amounts of wild game and fish, and/or by raising domesticated animals for personal consumption as well as sale in the surrounding market economy.

Although there is a general divergence between upland and lowland slaves' diets with respect to the relative abundance and variety of wild fauna incorporated, a closer look at these dietary differences (Fig. 10) reveals a more specific divergence point. The divergence hinges upon the fact that fish and other aquatic animals constitute 50 percent or more of lowland assemblages, but a much smaller part of upland ones. Fishing is an activity which, depending on how it is practiced, can be both very profitable in terms of harvest, but also very time-consuming. So, time to pursue fishing more than the possibilities for fishing allowed by inland rivers as opposed to coastal bays and estuaries, may have been a key factor

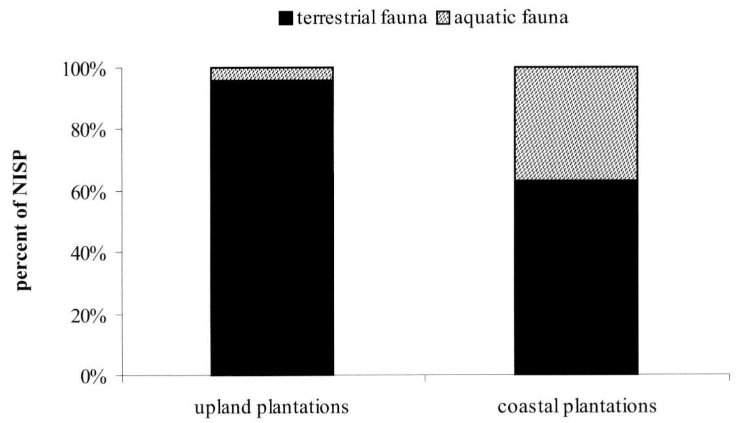

Fig. 10. Relative abundance of aquatic versus terrestrial fauna at upland and coastal plantations. Data from Brimstone Hill, Cannon's Point, Hermitage, Kingsmill, Locust Grove, Mabry, Monticello, and Saragossa plantations.

in dietary choices made by slaves. Zooarchaeological interpretations of slave diets at inland plantations (*e.g.*, Reitz *et al.* 1985, 184; McKee 1999, 225–6 and 228) have sometimes commented that the relatively small numbers of fish bones in such assemblages reflects the constricted range of fish species available in freshwater locales. McKee (1999, 226–228) for instance emphasizes the commercial, plantation-based fishing industry which formerly existed on estuarine plantations, and concludes that the virtual absence of fish bones at Monticello was due to the site's location "on a mountaintop in the foothills of Virginia's western mountain region, isolated from . . . riparian resources". In fact, Monticello sits on a sizeable tributary of the James River, whose tidal lower reaches were lined by wealthy plantations and early colonial settlements. Although it is generally true that upland plantation faunal collections typically contain few fish bones, it need not follow that this is because interior rivers were fish resource poor. In fact aquatic biologists (*e.g.*, Walsh *et al.* 1995) have characterized the southeastern United States' interior waterways as particularly diverse in fish species. These same rivers supplied southeastern Native Americans with a considerable array and quantity of freshwater fish (Limp and Reidhead 1979), as is apparent from the fish bones found on prehistoric sites in the region (*e.g.*, Bogan and Polhemus 1987).

Cobb (1976) as well as Mahaffey (1978) identified historic period fish traps in eastern Tennessee and Alabama, which suggests that freshwater fish were once both sought after and abundant, a conclusion further supported by Lund's (1995) documentation of the formerly thriving Ohio River fishing industry. The riverine fish weirs may have been exclusively rebuilt and re-used by slaves after their original construction by Native Americans, since the traps were evidently unknown in both Europe and West Africa (Mahaffey 1978; Yentsch 1992, 286–92). Even African-Americans on estuarine plantations caught and ate freshwater fish. One former slave recalled catching "mos'ly cats, an' perch, an' heaps an' heaps of suckers", as well as anadromous shad with either hook and line or basket traps (James Bolton, quoted in Berlin *et al.* 1998, 114). The possibilities for slaves to fish in the freshwater streams and rivers of the upland South were clearly great, but it is apparent from most faunal assemblages that they did not regularly or intensively pursue this activity.

Another point of hunting divergence between the upcountry and low country seems to have been slaves' ability to hunt deer, perhaps also a factor of time. While slave faunal assemblages from lowland plantations generally contain at least some deer bones, only one from an interior site (Monticello) has so far reported such remains (see Breitberg 1976; Wheaton *et al.* 1983; McKee 1987; Young 1993; Thomas 1998; Scott 2001; Young *et al.* 2001). Reitz *et al.* (1985, 185) observed that coastal faunal assemblages contain much smaller numbers of wild birds and small mammals as compared to upland sites. This observation caused them to speculate that upland plantations may have employed certain slaves as designated hunters, while coastal slaves used their free time to independently hunt and fish. The pattern observed by Reitz *et al.* (1985) is also visible in the current comparison (Fig. 10), as approximately 85 percent of faunal remains from wild species in upland assemblages came from small mammals and birds, as compared to only nine percent in coastal assemblages.

Although slaves in the upland South did make greater use of terrestrial and avian fauna, these animals still formed a comparatively small portion of their diet as compared to domestic animals. Even taking into account possible biasing factors such as reliance by coastal slaves on rations made up mainly of preserved pork containing no bones (cf. Reitz *et al.* 1985, 169), there nevertheless remains a large regional difference with respect to the importance of barnyard fauna. Historians (*e.g.*, Schlotterbeck 1991, 174) have discussed the fact that slaves in many places, but especially the interior regions, became the principal raisers and sellers of chickens and eggs. Zooarchaeological evidence demonstrates that the bird was of much greater relative importance to slaves in interior regions of the South. Chickens contributed an average of 14 percent to upland slave cabin faunal assemblages, but made up only one percent of coastal ones.

Labor systems and the dietary responses slaves invented to survive them may be part of the unifying explanation needed to understand the dietary variations in the world of slave-holding southern plantations. Because of the types of crops grown, plantations in the two subregions of the southern U.S. employed two different slave labor systems. Coastal plantations, due to their solitary focus on a single cash crop with limited growing seasons, mostly used the task system. This system made slaves work at a particular assignment until that job was finished, and from that point on what time left in the day was their own. The theory among masters employing that system was that slaves would work harder if more free time were available, while white owners would profit from the slaves' extra efforts in terms of harvest yields or other fruits of labor (Berlin and Morgan 1993, 9–10 and 14–5).

Due to the diverse array of crops grown in the uplands, slave owners there preferred gang labor, since that type of agriculture demanded constant day-long and year-round labor scheduling. There were always a myriad of tasks to perform on these diversified plantations; when one crop was harvested, the other could be planted. Even when there were no crops to manage slaves could be 'rented out' to neighboring plantations or, in evenings or bad weather, put to work indoors at cotton gins or spinning wool (Schlotterbeck 1991, 171–2). In essence, gang labor on diversified plantations forced slaves to work at broadly defined and closely supervised chores from dawn to dusk (Berlin and Morgan 1993, 14–5).

Because labor systems had such large impacts on slaves' free time, they undoubtedly in turn affected the slaves' abilities to hunt, fish, and trap. The previously observed divergence in diets between upland and lowland is, I believe, a reflection of these labor systems. Yet dietary diversity should not be read as a measure of the quality of slave life in task versus gang labor systems. If task labor gave slaves more latitude for producing their own food, the labors necessary to support those agricultural systems themselves took a great toll on slaves' bodies, especially the brutal sugar plantations of the southernmost United States (McDonald 1991, 185). In fact the successful efforts of coastal slaves to diversify and fortify their diets through their efforts off the plantation can be read as their means of minimizing the risk of food shortages or otherwise inadequate food supplies. The task labor system left slaves to be mostly responsible for acquiring their own food supplies beyond the provided rations of pork and cornmeal (Reidy 1993, 140). Indeed, Africans and African-Americans became quite proficient at hunting and fishing, so much so that Yentsch (1994, 249–53) has speculated that they may have been the primary ones to supply the diverse array of game sold in urban markets and served at slave owners' meals.

I assert that the reason for the difference in fishing intensity between the coast and the uplands was not environmental in nature, but rather a reflection of slaves' responses to the risks brought on by different plantation agricultural orientations and labor management systems. The gang labor system left upland slaves less free time for hunting and fishing. To some extent, this disparity was made up for by the general willingness among upland plantation owners to let slaves own livestock like hogs and poultry – animals which need minimal attention during the day – as well as tend gardens. That measure alone did not eliminate dietary risks for slaves, since masters frequently sited slaves' gardens and pastures in woods or distant areas of the plantation, where they were difficult to tend and away from optimal growing and foddering areas (Reidy 1993, 143–4). It is clear then that Africans and African-Americans enslaved into either labor system faced a variety of risks, especially those associated with poor or inadequate diets.

While slaves tasking on coastal cash-cropping plantations were allotted considerable free time to supplement their diets by either poultry-raising or pursuing wild game and fish, those toiling under gang labor systems had to resort to night hunting and trapping in order to supplement their diets. At first glance, the efforts made by upland slaves to catch wild animals is surprising given that they seemed to have had regular and ready access to either planter-provided meat rations or flesh from animals they themselves raised. Yet, as McKee (1999, 232) has argued, for slaves, food had not only nutritional but also social connotations. Whether or not slaves received nutritionally adequate rations misses the point. Both being able to raise animals as well as hunt or trap game gave slaves in the uplands a measure of control over their diets, just as extensive fishing did for slaves along the coast. Arguably, therefore, slaves attempted to ameliorate diet-related risks through a series of actions, including hunting, trapping, fishing, raising barnyard animals and selling or trading the fruits of these efforts for other goods. This behavior was in itself a risk if they hunted and fished when it was forbidden, as it sometimes was (cf. James Bolton, quoted in Berlin *et al.* 1998, 186) by their owners. The mere existence of slave cabin cellars, not to mention their contents of gun parts, fishing equipment, wild game, and even religious paraphernalia (cf. Young 1997c) – perhaps connected phenomena – demonstrates how slaves needed to find ways to cope with food-related uncertainties, and indeed balance the risks of illegal actions like stealing pigs or hunting against the need for a nutritionally adequate or sufficiently diverse and appealing diet (cf. McKee 1999).

Conclusions

What I have tried to do here is explore, first of all, the ways in which the diet of the bottom-most class varied in content and quality. Secondly, I have tried to find an explanation which better accounts for slave diet and dietary variation than do socioeconomic status and ethnicity paradigms. The comparisons across the two geographic/economic regions of the antebellum southern United States above all demonstrate that slave life, as viewed from diet, was certainly not uniform. Although the latter is a conclusion already reached by historians (*e.g.*, Berlin and Morgan 1993), the extent of dietary supplementation through slaves' own efforts, as well as the types of animals netted in the two areas' different subsistence strategies, have certainly not been as well documented. Beyond whether or not the regional comparison in itself reveals new aspects of slave life, the couching of the study in a risk theory framework arguably does contribute a fresh perspective.

The zooarchaeological record of slavery framed in terms of risk theory demonstrates that African-Americans could take advantage of the peculiarities of labor systems in order to minimize their risk of starvation or inadequate food supplies. Arguably, the risks involved with, and slaves' responses to, the owner-provided diet is understandable in terms of this theory. The two labor systems that prevailed in the South forced risks upon the slaves. Dietary analysis demonstrates how slaves met (1) the risk of loss by diversifying their diets, (2) that, at least at Locust Grove, they may have transferred risks to the plantation owners through stealing pigs, an action well-attested to in the documentary record (cf. Berlin *et al.* 1998, 192–4), and (3) stored food along with illegal hunting technology (guns, ammunition) in hidden cellars to ensure against shortfalls created when masters withheld rations as a form of punishment.

Possibly, slaves used kin-based networks on and off specific plantations to exchange or hide food, thus pooling the risks they may have faced by stealing livestock and owning firearms, possession of which could get slaves beaten, sold, or even killed (cf. Yentsch 1994, 250; Young *et al.* 2001). Although not attempted with the Locust Grove collection, food-sharing might be testable with faunal remains through refitting and element matching studies between assemblages from different cabins on the same plantation. Finally, risk theory applied to zooarchaeological evidence from the Locust Grove plantation slave cabin cellars, in addition to the regional comparison, helps to free the archaeological study of American slavery from a passive, victim-based understanding of this history. It instead allows for the possibility that even least powerful members of that society, the enslaved, played an active role in negotiating with their masters for or taking it upon themselves to create better living conditions. As well, the risk framework expands the study of slave diets from the limited viewpoint of whether or not food intake fulfilled their nutritional requirements (cf. Reitz *et al.* 1985); it also considers that African-Americans may have faced food-related risks from narrow, monotonous diets, such that they undertook other risks to remedy that situation.

References

Andrews, S. and Young, Amy Lambeck 1992. Plantations on the periphery: modeling a new approach for the Upper South. *Tennessee Anthropologist* **17**, 1–12.

Berlin, I. and Morgan, P. D. 1993. Introduction: labor and the shaping of slave life in the Americas, pp. 1–48 in Berlin, I. and Morgan, P. D. (eds), *Cultivation and Culture: Labor and the Shaping of Slave Life in the Americas.* Charlottesville: University Press of Virginia.

Berlin, I., Favreau, M. and Miller, S. F. 1998. *Remembering Slavery: African Americans talk about their Personal Experiences of Slavery and Freedom.* New York: The New Press.

Bogan, A. E. and Polhemus, R. R. 1987. Faunal analysis, pp. 971–1112 in Polhemus, R. R. (ed.), *The Toqua Site – 40MR6: A Late Mississippian, Dallas Phase Town* (Report of Investigations 41). Knoxville: Department of Anthropology, University of Tennessee.

Breeden, J. O. 1980. *Advice among Masters: the Ideal in Slave Management in the Old South.* Westport: Greenwood Press.

Breitburg, E. 1976. Faunal remains from the First Hermitage, pp. 249–69 in Smith, S. D. (ed.), *An Archaeological and Historical Assessment of the First Hermitage* (Research Series 2). Nashville: Tennessee Department of Conservation.

Campbell, J. 1991. As 'a kind of freeman'?: slaves' market-related activities in the South Carolina upcountry, 1800–1860, pp. 131–69 in Berlin, I. and Morgan, P. D. (eds), *The Slaves' Economy: Independent Production by Slaves in the Americas.* London: Frank Cass.

Clemen, R. A. 1923. *American Meat and Livestock Industry.* New York: The Ronald Press Company.

Cobb, J. E. 1976. Historic fish traps on the lower Holston River. *Tennessee Anthropologist* **3**, 31–58.

Coleman, J. W. 1940. *Slavery Times in Kentucky.* Chapel Hill: The University of North Carolina Press.

Crader, D. C. 1990. Slave diet at Monticello. *American Antiquity* **55**, 690–717.

Cruz-Uribe, K. 1988. The use and meaning of species diversity and richness in archaeological faunas. *Journal of Archaeological Science* **15**, 179–96.

DeBoer, W. R. 1988. Subterranean storage and the organization of surplus: the view from eastern North America. *Southeastern Archaeology* **7**, 1–20.

Deetz, J. 1993. *Flowerdew Hundred: the Archaeology of a Virginia Plantation, 1619-1864.* Charlottesville: University of Virginia Press.

Genovese, E. D. 1976. *Roll Jordan roll: the World the Slaves made.* New York: Random House.

Gibbs, T., Cargill, K., Lieberman, L. S. and Reitz, E. 1980. Nutrition in a slave population: an anthropological examination. *Medical Anthropology* **4**, 175–262.

Grayson, D. K. 1978. Minimum numbers and sample size in vertebrate faunal analysis. *American Antiquity* **43**, 53–65.

Kelso, W. M. 1984. *Kingsmill Plantations 1619-1800: Archaeology of Country Life in Colonial Virginia.* Orlando: Academic Press.

Klippel, W. E. 2001. Sugar monoculture, bovid skeletal part frequencies, and stable carbon isotopes: interpreting enslaved African diet at Brimstone Hill, St. Kitts, West Indies. *Journal of Archaeological Science* **28**, 1191–8.

Limp, W. F. and Reidhead, V. A. 1979. An economic evaluation of the potential of fish utilization in riverine environments. *American Antiquity* **44**, 70–8.

Lund, J. 1995. *Flatheads and Spooneys: Fishing for a Living in the Ohio River Valley.* Lexington: University Press of Kentucky.

Mahaffey, J. J. 1978. *A Typology and Functional Analysis of Fish Weir Configuration, R.L. Harris Reservoir.* Unpublished M. A. thesis, University of Alabama.

McDonald, R. A. 1991, Independent economic production by slaves on Antebellum Louisiana sugar plantations, pp. 182–208 in Berlin, I. and Morgan, P. D. (eds), *The Slaves' Economy: Independent Production by Slaves in the Americas*. London: Frank Cass.

McKee, L. W. 1987. Delineating ethnicity from the garbage of early Virginians: faunal remains from Kingsmill Plantation slave quarter. *American Archaeology* **6**, 31–9

McKee, L. W. 1992. The ideals and realities behind the design and use of 19th century Virginia slave cabins, pp. 195–214 in Yentsch, A. E. and Beaudry, M. C. (eds), *The Art and Mystery of Historical Archaeology: Essays in Honor of James Deetz*. Boca Raton: CRC Press.

McKee, L. 1999. Food supply and plantation social order: An archaeological perspective, pp. 218–39 in Singleton, T. A. (ed.), *"I, too, am America" Archaeological Studies of African-American Life*. Charlottesville: The University of Virginia Press.

Meltzer, D. J., Leonard, R. D. and Stratton, S. K. 1992. The relationship between sample size and diversity in archaeological assemblages. *Journal of Archaeological Science* **19**, 375–87.

Miller, H. M. 1978. Pettus and Utopia: a comparison of the faunal remains from two late seventeeth century Virginia households. *Conference on Historic Site Archaeology Papers* **13**, 158–79.

Morgan, P. D. 1982. Work and culture: the task system and the world of the lowcountry Blacks 1700 to 1880. *William and Mary Quarterly* **39**, 563–99.

Otto, J. S. 1984. *Cannon's Point Plantation 1794–1860: Living Conditions and Status Patterns in the Old South*. Orlando: Academic Press.

Payne, S. 1973. Kill-off patterns in sheep and goats: the mandibles from Asvan Kale. *Anatolian Studies* **23**, 283–303.

Reidy, J. P. 1993. Obligation and right: patterns of labor, subsistence, and exchange in the cotton belt of Georgia, 1790–1860, pp. 138–54 in Berlin, I. and Morgan, P. D. (eds), *Cultivation and Culture: Labor and the Shaping of Slave Life in the Americas*. Charlottesville: University Press of Virginia.

Reitz, E. J., Gibbs, T. and Rathbun, T. A. 1985. Archaeological evidence for slave subsistence on coastal plantations, pp. 163–87 in Singleton, T. A. (ed.), *The Archaeology of Slavery and Plantation Life*. San Diego: Academic Press.

Reitz, E. J. and Scarry, C. M. 1985. *Reconstructing Historic Subsistence with an Example from Sixteenth Century Spanish Florid*. (Special Publications Series 3). California, Pennsylvania: The Society for Historical Archaeology.

Reitz, E. J. and Wing, E. S. 1999. *Zooarchaeology*. Cambridge: Cambridge University Press.

Sanford, D. W. 1994. The archaeology of slavery and plantation life in piedmont Virginia: context and process, pp. 115–30 in Shackel, P. A. and Little, B. J. (eds), *Historical Archaeology of the Chesapeake*. Washington, D. C.: Smithsonian Institution Press.

Schlotterbeck, J. T. 1991. The internal economy of slavery in rural piedmont Virginia, pp. 170–81 in Berlin, I. and Morgan, P. D. (eds), *The Slaves' Economy: Independent Production by Slaves in the Americas*. London: Frank Cass.

Scholnick, J., Wheeler, D. and Neiman, F. 2001. *Mulberry Row Reassessment: the Building l Site* (Monticello Department of Archaeology Technical Report Series Number 3). Monticello, Virginia.

Scott, E. M. 2001. Food and social relations at Nina Plantation. *American Anthropologist* **10**, 671–91.

Shaffer, B. S. and Sanchez, J. L. J. 1994. Comparison of ⅛"-and ¼"-mesh recovery of controlled samples of small-to-medium-sized mammals. *American Antiquity* **59**, 525–30.

Shennan, S. 1989. Introduction: archaeological approaches to cultural identity, pp. 1–32 in Shennan, S. (ed.), *Archaeological Approaches to Cultural Identity*. London: Routledge.

Singleton, T. A. 1990. The archaeology of the plantation South: a review of approaches and goals. *Historical Archaeology* **24**, 70–7.

Singleton, T. A. 1995. The archaeology of slavery in North America. *Annual Review of Anthropology* **24**, 119–40.

Skaggs, J. M. 1986. *Prime Cut*. College Station: Texas A&M University Press.

Thomas, B. W. 1998. Power and community: the archaeology of slavery at the Hermitage Plantation. *American Antiquity* **63**, 531–52.

Tuma, M. 1998. Slave subsistence at Saragossa: preliminary report on faunal data. *Mississippi Archaeology* **33**, 125–38.

Walker, K. J. 1988. *Kingsley and his Slaves: Anthropological Interpretation and Evaluation*. Unpublished M. A. thesis, University of Florida.

Walsh, S. J., Burkhead, N. M. and Williams, J. D. 1995. Southeastern freshwater fishes, pp. 144–7 in LaRoe, E. T., Farris, G. S., Puckett, C. E., Doran, P. D., and Mac, M. J. (eds), *Our Living Resources*. Washington, D. C.: U.S. Department of the Interior, National Biological Service.

Wiessner, P. 1982. Beyond willow smoke and dogs' tails: a comment on Binford's analysis of hunter-gatherer settlement systems. *American Antiquity* **47**, 171–7.

Wheaton, T. R., Friedlander, A. and Garrow, P.H. 1983. *Yaughan and Curriboo Plantations: Studies in Afro-American Archaeology*. Marietta, Georgia: Soil Systems, Inc.

Yentsch, A. E. 1992. Gudgeons, mullet, and proud pigs: historicity, black fishing, and southern myth, pp. 283–314 in Yentsch, A. E. and Beaudry, M. C. (eds), *The Art and Mystery of Historical Archaeology: Essays in Honor of James Deetz*. Boca Raton: CRC Press.

Yentsch, A. E. 1994. *A Chesapeake Family and Their Slaves. a Study in Historical Archaeology*. Cambridge: Cambridge University Press.

Young, Amy Lambeck 1994. Nailing down the pattern. *Tennessee Anthropologist* **19**, 1–21.

Young, Amy Lambeck 1997a. Risk management strategies among African American slaves at Locust Grove. *International Journal of Historical Archaeology* **1**, 5–37.

Young, Amy Lambeck 1997b. Task and gang labor: work patterns at a Kentucky plantation. *North American Archaeologist* **18**, 41–66.

Young, Amy Lambeck 1997c. Cellars and African-American slave sites: new data from an upland south plantation. *Midcontinental Journal of Archaeology* **22**, 95–116.

Young, Amy Lambeck, Tuma, M. and Jenkins, C. 2001. The role of hunting to cope with risk at Saragossa Plantation, Natchez, Mississippi. *American Anthropologist* **103**, 692–704.

Young, Amy Lynne 1993. *Slave Subsistence at the Upper South Mabry Site, East Tennessee: Regional Variability in Plantation Diet of the Southeastern United States*. Unpublished M. A. thesis, University of Tennessee.

Zar, J. H. 1999. *Biostatistical Analysis* (fourth edition). Upper Saddle River: Prentice Hall.

Justin S. E. Lev-Tov
Department of Anthropology and Social Work,
1212 Ullman Building,
University of Alabama at Birmingham,
Birmingham, AL 35294–3350, USA.
E-mail: jlevtov@bama.ua.edu

9th ICAZ Conference, Durham 2002
Behaviour Behind Bones, (ed. Sharyn Jones O'Day et al.) pp. 318–326

34. Cultural identity and the consumption of dogs in western Africa

Veerle Linseele

The dog inhumations at Chin Tafidet (Niger), dating to 2600–1300 BC, represent the oldest evidence of the domestic dog in West Africa. There is a lack of archaeological dog remains between 1300 and 200 BC, but afterwards most sites contain dog bones among their faunal remains. These are often interpreted as consumption refuse. New, unequivocal, evidence for cynophagy is now available from Late Iron Age (AD 1000–1400) Northeast Burkina Faso, which at that time began to be influenced by North African people, Berbers and Islamic Arabs, through trans-Saharan trade. Berbers are notorious for their habit of eating dogs, while throughout Islam the dog, together with the pig, is considered as one of the most impure animals. Starting from archaeozoological data, confronted with historical and ethnographical information, this paper discusses the interaction in West Africa of several cultural groups with different attitudes towards dog eating.

Introduction

The ancestor of the domestic dog, the wolf (*Canis lupus*), does not occur in Africa, except in Libya and Egypt (Ferguson 1981), and domestic canids are therefore not native to Africa. A survey of the earliest dog finds given by Gautier (2001) indicates that they were probably introduced from the Middle East together with small livestock during the 5th millennium BC, most likely as shepherd dogs (Gallant 2002; Gautier 2001; Gautier 2002). Today, however, African dogs are rarely used for herding (Frank 1965), and in Saharan rock art the occurrence of dogs in pastoral scenes is rare, they are mostly depicted as hunting dogs (Cesarino 1997). Three major dog types occur in Africa today. The most common one is the pariah, a dog of a primitive generalised racial type and the descendant of domestic dogs that turned semi-feral (Epstein 1971). The term pariah, however, is also used for canids that are in an intermediate stage of domestication (Epstein 1971). Gallant (2002) only accepts this second definition and considers the first one as inaccurate. He argues that 'feral dog' is the only correct term in that sense. The second important type of African dog is the greyhound, which is a hunting dog (Epstein 1971). The third one, the africanis, has recently been described as a primitive hound endemic to southern Africa (Gallant 2002). In West Africa only the pariah occurs.

The first domestic animals in West Africa south of the Sahara for which there is osteological evidence are cattle, sheep and goats (MacDonald and MacDonald 2000). They appear in the late 3rd – early 2nd millennium BC. Three burials at Chin Tafidet (Paris 1984; Paris 2000) in Niger (Fig. 1) yielded the oldest West African remains of domestic dog in an area with inhumations of cattle and goat, dated to 2600–1300 BC. There are no indications that the dogs were buried ritually. Cranio-metric data and the morphological characters of the limbs and the vertebral column of these dogs indicate that they belong to a greyhound type (Paris 1984; Paris 2000). Remarkably, no complete dog inhumations or articulated finds are known for West Africa, apart from those at Chin Tafidet, and there are no other West African sites with secure dog identifications before 200 BC (MacDonald and MacDonald 2000). Their absence could just be a consequence of sample sizes, as dogs are usually only represented by a few bones. Their remains are, moreover, not easy to separate from jackals (*Canis aureus/Canis adustus*).

Most of the West African sites younger than 200 BC contain dog bones among their faunal remains. Generally, dogs represent between 1 and 20 % of the total amount of domesticated mammals. This proportion does not seem to show any trend in space or time. The dog bones are usually

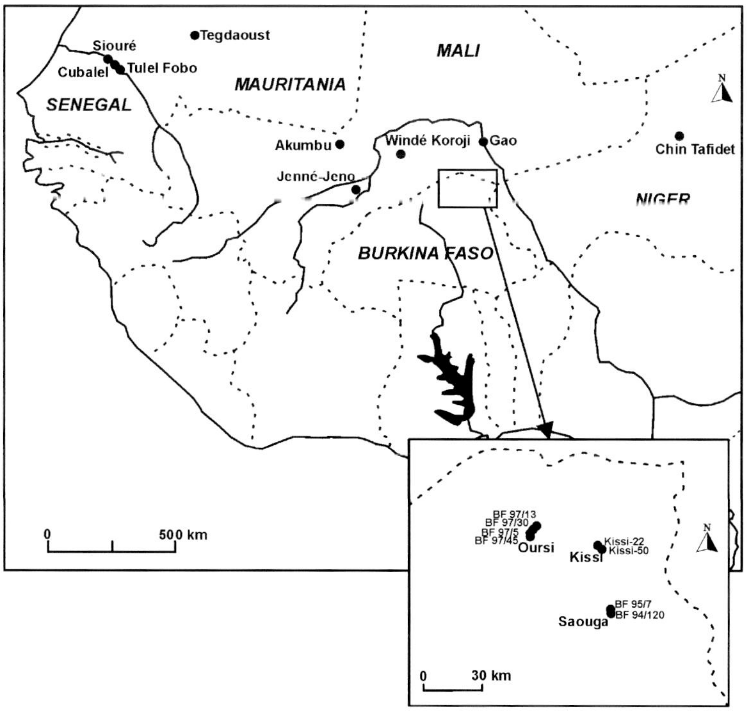

Fig. 1. Map of localities mentioned in the text. Cubalel and Siouré (McIntosh et al. *1992), Tulel Fobo (Bocoum 1986), Tegdaoust (Devisse and Robert 1993), Akumbu (Togola 1996), Jenné-Jeno (McIntosh 1995), Windé Koroji (MacDonald 1996), Gao (Insoll 1996), Chin Tafidet (Paris 1984), BF 97/13 and BF 94/45 (Hallier 1998), BF 97/30 (Hallier and Petit 2001), BF 97/5 (Breunig and Neumann 2002), BF 94/45 (Hallier 1998), Kissi-22 and Kissi-50 (Magnavita* et al. *2002), Saouga 94/120 and Saouga 95/7 (Vogelsang* et al. *1999).*

interpreted as consumption refuse, as at Akumbu (AD 400–1400) (MacDonald and Van Neer 1993), Cubalel and Siouré (AD 0–1400) (MacDonald and MacDonald 2000), Tulel Fobo (AD 300–900) (Van Neer and Bocoum 1991) and Tegdaoust (AD 600–1200) (Bouchud 1993). The conclusion that the dogs were eaten is based on one or more of the following arguments: presence of remains near living areas, body part distribution, pre-depositional fragmentation and charring. However, none of the sites mentioned yielded dog remains with traces of butchery, although bovid bones did bear such traces. New, unequivocal, evidence for cynophagy is now available from Iron Age sites in Northeast Burkina Faso. Starting from this example, different attitudes of human groups in West Africa towards dog eating will be discussed. A few reflections on cynophagy outside Africa will be added as well.

Evidence for cynophagy in Iron Age Northeast Burkina Faso

The Northeast of Burkina Faso is today part of the Sahel, which is the climatological-ecological transition zone between the arid Sahara and the semi-humid Sudan Zone. Geographically it is characterised by stabilised sand dunes, east-west trending, 1–10 km wide, alternated by plains. Because of the presence of water, in the form of springs and/or lakes, these dunes were, and still are, favourable for human habitation (Breunig and Neumann 2002).

Despite extensive archaeological survey, no sites older than 2200 BC have been discovered thus far. Apparently, the North of Burkina Faso, like many other regions in West Africa, was not or only very sparsely populated during the early Holocene. The oldest sites found date between 2200 and 1000 BC (Final Stone Age). These are small dune sites with few cultural remains and no settlement structures. Their inhabitants were probably immigrants coming from the Sahara (Breunig and Neumann 2002). The bone preservation on the Final Stone Age sites is generally very poor and no domestic animals have been attested thus far. Cocorba (BF 97/5), the only site with good preservation, contains mostly fish and some wild mammals (Van Neer and Lambrecht *unpublished data*). In contemporary sites in the nearby Windé Koroji region of Mali, domestic cattle and ovicaprines are present, although wild fauna is still dominant (MacDonald 1996). In the Northeast of Burkina Faso, no archaeological traces were found for the entire first millennium BC, but around 2000 years ago large Iron Age sites appear. These consist of groups of settlement mounds on top of the dunes, separated by flat areas, and are interpreted as villages or, at least as large hamlets. Most sites are large accumulations (up to 8 metres thick) of cultural remains, without much stratification and with only indirect evidence for settlement structures (Breunig and Neumann 2002). Excavations at Oursi hu-beero (BF 97/30), however, have revealed the burnt remains of a large two-storied house, on top of a settlement mound, consisting of at least 14 rooms (Hallier and Petit 2001; Petit and Hallier 2002). Mud brick debris sealed off this building and preserved it in the exact state it was left 1000 years ago. In the 14th century AD the settlement mounds were abandoned, probably as a consequence of political instability in the region, and an ethnical change took place whereby nomadic groups replaced the sedentary ones (Albert *et al.* 2000).

The Iron Age in North Eastern Burkina Faso has been divided into Early (AD 0–500), Middle (AD 500–1000) and Late Iron Age (AD 1000–1400) (Albert *et al.* 2000) based on changes in ceramic decoration. Faunal remains are abundant on the Iron Age sites in North Eastern Burkina Faso but they are usually not well preserved, as shown by the low identification rate (5–20 %). With the beginning of the Iron Age, domestic livestock (cattle, sheep, and goat) appeared in Northeast Burkina Faso. This was the most important source of animal proteins from that moment onwards, but hunting, fishing and fowling persisted throughout the entire Iron Age. Fig. 2 indicates the composition of the domesticated mammalian fauna of the studied Iron Age sites by phase. Dog remains become remarkably frequent from the Middle Iron Age onwards and especially during the subsequent Late Iron Age. Because skulls and complete long bones are not preserved, a statement on the type of dogs is difficult, although the measurements and the presence at Saouga 95/7 of two mandibles, significantly differing in size, suggest a large variation in the dog population. Judging from tooth eruption and fusion data, all remains are from adult animals, with only a few exceptions. Due to the poor preservation of the material, the skeletal distribution is distorted towards compact elements (phalanges, metapodials) and elements which are easily recognisable from small fragments (teeth, skull parts). A similar distortion exists for the other mammalian species present. The dog remains are scattered and found among consumption refuse, indicating that the animals were probably eaten. For the settlement mounds a possible objection is that the archaeological deposits could have been disturbed after deposition, but this objection does not hold for the site of Oursi hu-beero (BF 97/30) (cf. *supra*), where the finds are *in situ*. Several dog bones are burnt or charred. This is sometimes used as an argument for consumption, although Olsen (2000) does not accept burning or charring as evidence. According to him it is the result of grilling to a point where edible flesh has been completely destroyed. A distinction should be made, however, between bones that are burnt entirely and bones that are just charred at the ends. Only the latter can be used as possible evidence for food processing, but has not been attested for any of the species identified at Iron Age Northeast Burkina Faso. The best evidence for cynophagy in Burkina Faso is the presence of cut marks on dog bones from two different contexts, both dating to the Late Iron Age. An axis from Saouga 95/7 (Fig. 3) bears cut marks on all sides. These marks were most likely formed when the dog's head was removed from its body. A burnt talus (Fig. 4) from Oursi village (BF 97/13) has cut marks on the dorsal side, probably the consequence of the removal of the foot from the rest of the body. A second metacarpus (Fig. 5) from Oursi village (BF 97/13) shows cut marks on its dorsal side near the proximal end, resulting from skinning. No traces of butchery were found on long bones, but these are underrepresented due to the poor preservation at the sites. Apparently dog bones were sometimes worked as well, as suggested by a third metatarsal from Saouga 95/7 where the proximal end has been removed from the shaft by cutting or sawing it on all sides.

Cynophagy introduced in West Africa by North African Berbers?

Based on Arabic written sources, Lewicki (1974) has postulated the existence of cynophagy in medieval West Africa. He suggests that Berber influences were responsible for the introduction of this custom among certain groups. Berbers are the indigenous habitants of Northwest Africa and they are notorious for their habit of eating dogs (Simoons 1981). The persistence of cynophagy in Northwest Africa until recent times is illustrated by the lively debate on this practice in the Sahara during the 1950's in the *Bulletin de Liaison Saharienne* (Canard 1953; Bureau 1954; Chalumeau 1954; Thiriet 1954). In reaction to this debate a few short articles on instances of dog eating in western Africa were published in *Notes Africaines* (Mauny 1954; Holas 1955; You 1955).

From the 6th–7th century AD onwards there is good archaeological evidence for long distance trade from West to North Africa, as for example at the cemetery of Kissi in Burkina Faso (Thom 1998; Pelzer and Magnavita 2000). This trans-Saharan trade was controlled by Arabs and (Arabised) Berbers and intensified in the 10th century AD, with a significant additional rise in the subsequent two centuries (Insoll 1996). The presence of horse remains in the Late Iron Age sites in Northeast Burkina Faso and the find of what is interpreted as a slave-chain at Oursi hu-beero (BF 97/30) (Petit and Hallier, *pers. comm.*), point to the possibility that the inhabitants of these sites took part in the slave-trade economy across the Sahara, which is well attested by Arab authors by that time (Levtzion and Hopkins 1981). It is thus plausible that the observed (Late) Iron Age rise in cynophagy in Northeast Burkina Faso is a consequence of intensified contacts with North African Berbers. Excavations at a few trade centres on the routes to North Africa did not, however, point to a similar increase in dog eating. At Gao (AD 900–1250)

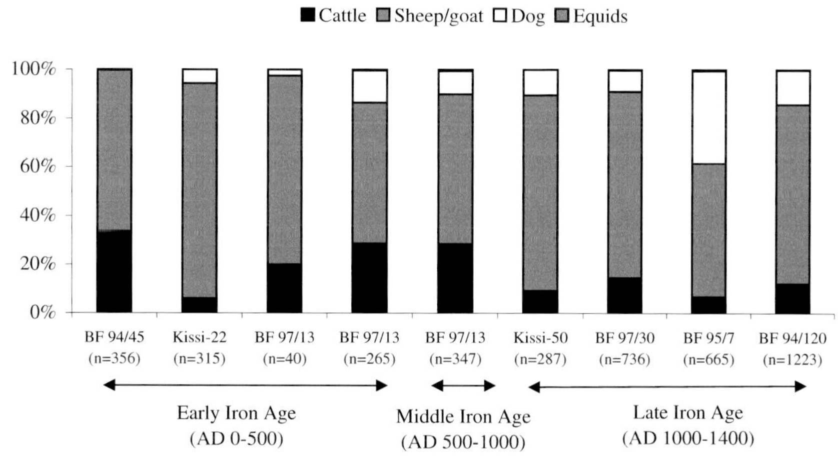

Fig. 2. Proportions of the domestic mammals by site and phase.

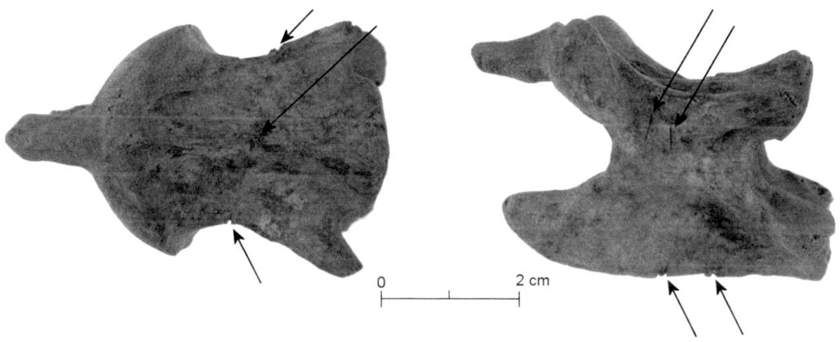

Fig. 3. Dog axis with cut marks from Saouga 95/7.

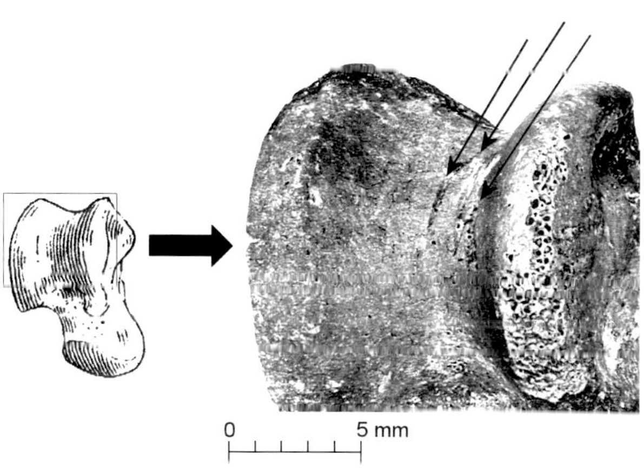

Fig. 4. Burnt dog talus with cut marks from Oursi village (BF 97/13).

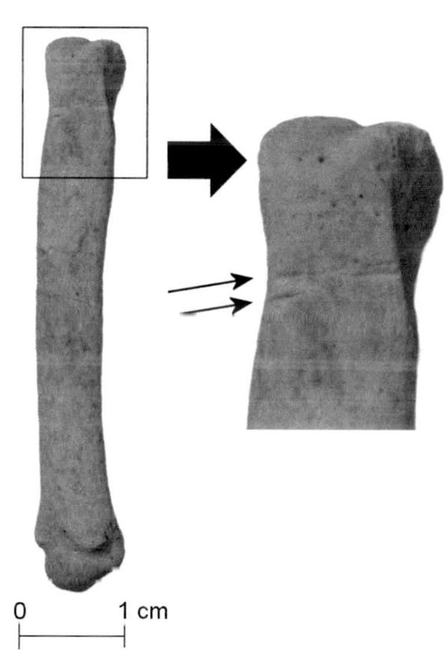

Fig. 5. Second metacarpus of dog with cut marks from Oursi village (BF 97/13).

(MacDonald and MacDonald 1996) and Jenné-Jeno (200BC – AD 1400) (MacDonald 1995), for example, dog bones were not very numerous and no conclusion was drawn as to the use of the animals, although the finding of disarticulated bones among consumption refuse does suggest that they were eaten. At 9th – 11th century Tegdaoust (Bouchud 1993) the dog remains were interpreted as food waste, but unfortunately no quantitative data are available.

Cynophagy may have developed in West Africa independently from North Africa, as there are indications that it already existed – albeit on a smaller scale – in the region before regular trans-Saharan trade. Contacts with North Africa date from before the 6th century AD, however, as suggested for example by the find of a camel bone in Siouré, dated to AD 250–400 (MacDonald and MacDonald 2000). There would already have been sporadic encounters between North and West Africans more than a thousand years before the advent of the Arabs in North Africa, although no important cultural influences spread through these early contacts (Masonen 1997). A possible explanation for cultural parallels between North and West Africa, older than trans-Saharan trade, is offered by Lam (1982). According to him ancient Egypt was a central diffusion centre from where some cultural traits were dispersed to both regions.

Islam and dogs

Islam is the oldest and most important anti-dog eating force in West Africa. The Koran divides food into two basic categories: halal, or permitted, and haram, or forbidden. Six foods or food categories are specifically forbidden: blood, carrion, pork, intoxicating beverages prepared from grapes, intoxicating drugs and foods previously dedicated or offered to idols. There is thus no Koranic prohibition against dog eating. However, in the Islamic law (shariah) dog meat is forbidden (The Encyclopaedia of Islam 1999). The Hadiths, or collected traditions and sayings attributed to the Prophet Mohammed, delineate quadrupeds that seize their prey with their teeth as animals to be avoided as food. However, only hyenas, foxes and elephants are specifically mentioned (Grivetti 2000).

Considered one of the most impure animals, together with the pig, the dog is generally if not despised, then at least avoided, throughout Islam. These negative feelings are probably a consequence of the Prophet's dislike of the dog, although they may have intensified when Islam was established in Iran, in a negative reaction to the unusual high regard in which the animal was held by the Zoroastrians. The reservation towards the dog is also linked with a superstitious caution arising from the belief that the animal is a demonic emanation belonging to the category of evil spirits (The Encyclopaedia of Islam 1999) and, according to the Hadiths *angels do not enter a house where there is a dog* (Dramani-Issifou 1988). Only useful dogs that obey a master, like trained hunting dogs and watchdogs, are freed from the Prophet's condemnation. Dogs are therefore generally better regarded among nomads and rural folk than in cities, but all dogs are unclean with respect to religious practices. Everything a dog touches or licks is rendered impure and the place where it has lain must be purified with water, following the practice of the Prophet on one occasion (The Encyclopaedia of Islam 1999). Islam also resolutely opposes the excessive manifestations of the keeping of dogs and in particular cynophagy (Dramani-Issifou 1988).

North Africa was invaded by Arabo-Islamic troops in the 7th century AD and through trans-Saharan trade Islamic people subsequently penetrated into western Africa (El Fasi and Hrbek 1988a). The North African traders brought the West African animistic societies into a first contact with Islam, but for the actual rooting of the religion lettered Muslims or 'men of religion' were required (Diallo 1990). Direct contacts with North Africans were mostly limited to the desert-edge cities of West Africa, where the North African traders lived in their own quarters (Perinbam 1989 *fide* Masonen 1997). Outside their trading colonies the Arab merchants had to follow local laws and customs (Masonen 1997).

The first wave of Islamisation in West Africa, from the 8th century AD onwards, happened through Ibadite traders (Dramani-Issifou 1988; El Fasi and Hrbek 1988b). The Ibadites are a branch of the Kharijites, the first schism from Islam (7th century AD) with its own government. Because Ibadite doctrines are tolerant and egalitarian they rapidly gained adherence among the Berber population of North Africa (Monès 1988). A second wave of Islamisation of the Sudanic belt came about in the 11th century AD with the rule of the Berber dynasty of the Almoravids (El Fasi and Hrbek 1988b; Hrbek 1988). The Almoravids preached orthodox Sunni Islam and they concerted the first major, rational, effort of conversion in West Africa (Hrbek and Devisse 1988). During their rule trans-Saharan trade rose to its peak (Devisse 1988). By the 16th century AD, Islam was well established all along the Sudanese belt from the Atlantic to Lake Chad, although it remained a cult of the courts and the trade centres for a long time (El Fasi and Hrbek 1988b; Masonen 1997). Furthermore, for most converts the acceptance of Islam did not mean a total abandonment of non-Islamic practices associated with the African traditional religion (El Fasi and Hrbek 1988b). The rural hinterland remained only slightly touched by Islam until the 19th and 20th century when large-scale conversions took place (de Benoist 1983).

The medieval Arabic accounts are silent on the question of whether Muslim actions against dogs in Northwest Africa ever led to friction with early non-Muslim Berber groups, as it did with the Zoroastrians of Iran (Simoons 1981). It is striking that a practice like cynophagy, considered as repugnant, persisted for so long in that area, especially since Islam generally has an impressive record at gaining adherence to its strictures. This is perhaps a too generalised view as the Ibadites (cf. *supra*), for example, are very tolerant and usually do not interfere with the persistence of non-Islamic practices like the (ritual) consumption of dogs (Bonte 2002). Cynophagy is presently spread in those zones of Northwest Africa that are under Ibadite influence. Initially the Islamic conquests did not aim at gaining new adherents and no forced conversions were made (El Fasi and Hrbek 1988b; Hrbek 1988). North Africa was possibly also protected from Islamisation by schisms in the east and, in the west, by the slowness of the process (Simoons 1981). In addition the Berbers of the Northwest African hinterland did stage a few famous revolts against the Arabic conquerors because, although the general Berber population had no objection to adopting Islam, their chiefs were unwilling to let themselves be absorbed into the empire of a foreign power (El Fasi and Hrbek 1988b; Monès 1988).

Despite Islamic presence in both Northwest and West Africa for more than a millennium, it is apparently only from the 19th century onwards that the opposition against cynophagy in the two regions has been strong enough to lead to a significant decline in the practice. Besides Islam, European colonists have been an important factor in this process, although specific actions against dog eating never seem to have been undertaken.

Dog eating in present-day sub-Saharan Africa

Frank (1965) discerns two core areas of dog eating in present sub-Saharan Africa: the Congo Basin and West Africa. In these two areas dog meat is usually not a regular source of protein, but it is associated with religion (Holas 1955; Frank 1965). Outside the core areas dogs are only eaten under extenuating circumstances (Simoons 1994). It is hard to make generalisations on cynophagy in sub-Saharan Africa as the customs surrounding it differ from one ethnic group to the other, suggesting that the practice has a long history. Recurrent is the fact that dog meat is prohibited to women, and sometimes also to other groups within a society (Frank 1965). Generally, no special types of dogs are kept for consumption (Frank 1965). To obtain tender meat animals are castrated (Frank 1965; Simoons 1994) and some groups believe that this can also be achieved by ill treating the dogs, as in the Congo Basin, where one breaks an animals legs and then leaves it for several hours before slaughtering (Frank 1965). The dogs are usually killed by beating them on the head with a stick (Frank 1965), so that their blood is not lost (Staffe 1938). In some cases dog meat is commercialised, as for example in the Mandara Mountains in North Cameroon (Thys and Nyssens 1983). Besides consumption, which is its most common use within Sub-Saharan Africa according to Boettger (1958), the dog has a wide variety of other functions. Frank (1965) states that hunting dogs generally receive special treatment. They are well fed and highly esteemed, whereas the other dogs are left to find their own food.

Cynophagy outside Africa

Other regions in the world where dogs are eaten nowadays are parts of Southeast Asia, North and South-America and the Pacific. As in Africa, their consumption is associated with more complex reflections than eating of other domestics (Serpell 1995). In Europe, dog consumption has been attested archaeozoologically in all prehistoric periods from the Mesolithic to the Iron Age. The brains seem to have been a favourite delicacy judging from the finds of crushed skulls (Bökönyi 1988). There is also historical evidence for dog eating from both ancient Rome and Greece (Simoons 1994) where it had mainly medicinal reasons. Geppert (1990) made an extensive study on the consumption of dogs in 19th and 20th century Munich (Germany) and concluded it was related to industrialisation and population growth. Dog meat was mainly eaten by the poor and working class people for whom regular meat was too expensive. In the centuries before, dog meat and fat was regularly used as a medicine. That dog eating persisted in Europe until very recently is also proven by the fact that several countries have a legislation that regulates dog slaughter and marketing (Geppert 1990).

While genetic research is now being carried out to help unravel the place and date of domestication of the dog (Savolainen *et al.* 2002), opinions on why the animal was domesticated still differ greatly. Nutritional value and food preferences are sometimes mentioned as possible reasons. Indeed, dog meat contains as much protein and less fat than, for example, pork (Simoons 1994), and it is furthermore often described as being very tasty. No Palaeolithic sites seem to exist, though, with dog remains among the consumption refuse. Gautier (1990) considers it very unlikely that prehistoric societies regularly used dogs as an important source of protein because of their small body size. Benecke (1994) argues that dogs and men are food competitors and that dogs will only be kept for food where they can live off abundant human refuse. He mentions for example Mesolithic Maglemose settlements where large numbers of dogs could survive on leftovers of fish

Different cultural attitudes towards dog eating

Within most human societies the dog occupies a special place among the domestic animals. This is expressed in several ways, for example in the burial of an animal after its death, or as contradictory as it may seem, in its consumption. The worship of an animal (or of a spirit through the animal that it is supposed to embody) may lead to its (cultic) consumption or to the opposite, a taboo on its meat. As already pointed out, dog meat is very rarely an ordinary source of protein, but is associated with religion, magic, or is used as a medicine. People also eat dogs in times of need or famine. Strikingly consistent among many human groups is that hunting dogs enjoy a higher esteem and are not eaten. It is not possible to assess which motivations were involved in the eating of dogs in Iron Age North Eastern Burkina Faso, although nothing points in the direction of food shortage.

A strong opposition against dog eating exists today among Hindus, Buddhists, westerners and, as already mentioned, Muslims. The most important reasons for anti-dog flesh feelings are the familiarity between humans and their pets and the opinion that a dog, as a scatophagous animal, carrion eater and scavenger, is unclean. For Hindus and Buddhists, it fits within a framework of general non-violence and a belief in the sanctity of all life (Simoons 1994).

The cultural background of a scholar might influence his results while doing research into cynophagy. Simoons (1981) has noted for example that ethnographic research in Northwest Africa is mostly carried out by people who do not have a tradition of cynophagy. Interviewed cynophagists would therefore not say that they eat dog flesh for its taste but rather suggest other reasons (religion, medicine, etc.), more acceptable to the interviewer. The conclusion that dogs are not a usual source of food may therefore be questionable in certain cases. Worth mentioning in that respect is that people in Burkina Faso today consume dog meat simply because they fancy it (Pelzer, *pers. comm.*).

Conclusions

The oldest dog remains in West Africa are the inhumated animals from Chin Tafidet. These animals were considered as greyhounds, although this dog breed does not occur in present-day West Africa. Faye (1984) proposed that they were either hunting- or shepherd dogs. However, greyhounds are nowadays mostly used for hunting and ethnographical data stress that hunting dogs are usually not eaten. The finds at Chin Tafidet suggest that this could also have been the case in West Africa around 2600–1300 BC. All younger dog remains, found at sites postdating 200 BC, are probably from consumed animals, which implies that cynophagy in West Africa is at least two millennia old. A great antiquity of cynophagy in this part of the African continent is also suggested by the diversity of customs associated with the practice today. This diversity makes it hard to put an interpretation forward for the dog eating recorded at the Iron Age sites in Burkina Faso, but there are no indications to assume that the animals were eaten because of food shortage. In Northeast Burkina Faso, cynophagy gained importance at a time when contacts with North Africa increased. It is therefore proposed that influences from North African Berbers were responsible for the observed rise in dog consumption. No other West African finds could corroborate this hypothesis, however. Considering the strong anti-dog eating feelings of Islam, it is remarkable that in Burkina Faso dog consumption rose in a period when Islamic people started to penetrate into western Africa. This can probably be explained by the fact that these people were mostly traders with no interest in imposing their religion on the local population. During a long period of West African history, Islam also remained restricted to royal courts and trade centres. The abandonment of dog eating in West Africa under the influence of foreign groups, Muslim and Western, is a phenomenon from the last two centuries.

Acknowledgements
This research was carried out within the framework of the project SFB 268 "Kulturentwicklung und Sprachgeschichte der Westafrikanische Savanne" at the Johann-Wolfgang Goethe University, Frankfurt-am-Main, Germany and sponsored by the Deutsche Forschungs Gemeinschaft (DFG). It will be incorporated in a doctoral thesis prepared by the author at the Royal Museum for Central Africa, Tervuren, Belgium with a research grant from The Belgian Federal Office for Scientific, Technical and Cultural Affairs (OSTC).

References

Albert, K-D., Hallier, M., Kahlheber, S. and Pelzer, C. 2000. Montée et abandon des collines d'occupation de l'Age de Fer au Nord du Burkina Faso. *Berichte des Sonderforschungsbereichs 268* **14**, 335–51.

Benecke, N. 1994. *Der Mensch und seine Haustiere. Die Geschichte einer Jahrtausendealten Beziehung*. Stuttgart: Theiss.

Bocoum, H. 1986. *La Métallurgie du Fer: Approche Archéologique, Technologique et Historique*. Unpublished Ph.D. thesis, University of Paris 1.

Boettger, C. R. 1958. *Die Haustiere Afrikas*. Jena: G. Fischer Verlag.

Bökönyi, S. 1988. *History of Domestic Mammals in Central and Eastern Europe*. Budapest: Akadémiai Kiadó.

Bonte, P. 2002. *Entre mythes et sacrifices. Le dossier inachevé de la cynophagie dans le monde berbère*. Paper presented at the "VIIe colloque international de l'association HASRI", 21–23 November 2002, Villeurbanne (France).

Bouchud, J. 1993. Paléofaune de Tegdaoust, pp. 355–63 in Devisse, J. and Robert, S. (eds), *Tegdaoust III. Recherches sur Aoudaghost: Campagnes 1960–1965*. Paris: Editions Recherche sur les Civilisations.

Breunig, P. and Neumann, K. 2002. From hunters and gatherers to food producers: new archaeological and archaeobotanical evidence from the West African Sahel, pp. 123–55 in Hassan, F. A. (ed.), *Droughts, Food and Culture. Ecological Change and Food Security in Africa's later Prehistory*. New York/Boston/Dordrecht/London/Moscow: Kluwer Academic/Plenum Publishers.

Bureau, R. 1954. Manger du chien! et du chat? *Bulletin de Liaison Saharienne* **5** (16), 15–6.

Canard, M. 1953. La cynophagie au Sahara. *Bulletin de Liaison Saharienne* **4** (15), 2–8.

Cesarino, F. 1997. I cani del Sahara. *Sahara* **9**, 93–113.

Chalumeau, P. 1954. Cynophagie. *Bulletin de Liaison Saharienne* **5** (17), 77–84.

de Benoist, J. R. 1983. Tendances actuelles de l'Islam en Afrique de l'Ouest. *Etudes* **359** (6), 667–82.

Devisse J. 1988. Trade and trade routes in West Africa, pp. 367–435 in El Fasi, M. and Hrbek, I. (eds), *General History of Africa III. Africa from the Seventh to the Eleventh Century*. Berkeley: University of California Press.

Devisse, J. and Robert, S. 1993. *Tegdaoust III. Recherches sur Aoudaghost: Campagnes 1960-1965*. Paris: Editions Recherche sur les Civilisations.

Diallo, H. 1990. Introduction à l'étude de l'histoire de l'Islam dans l'Ouest du Burkina Faso: des débuts à la fin du XIXe siècle. *Islam et Sociétés au Sud du Sahara* **4**, 33–45.

Dramani-Issifou, Z. 1988. Islam as a social system in Africa since the seventh century, pp. 92–118 in El Fasi, M. and Hrbek, I. (eds), *General History of Africa III. Africa from the Seventh to the Eleventh Century*. Berkeley: University of California Press.

El Fasi, M. and Hrbek, I. 1988a. *General History of Africa III. Africa from the Seventh to the Eleventh Century*. Berkeley: University of California Press.

El Fasi, M. and Hrbek, I. 1988b. Stages in the development of Islam and its dissemination in Africa, pp. 56–91 in El Fasi, M. and Hrbek, I. (eds), *General History of Africa III. Africa from the Seventh to the Eleventh Century*. Berkeley: University of California Press.

Epstein, H. 1971. *The Origin of the Domestic Animals of Africa* (Volume I). New York: Africana Publishing.

Faye, B. 1984. Etude ethnozoologique, pp. 72–4 in Paris, F. (ed.), *La Région d' In Gall – Tegidda N Tesemt (Niger): Programme Archéologique d'Urgence 1977–1981. III. Les Sépultures du Néolithique Final à l'Islam* (Etudes Nigériennes 50). Niamey: Institut de Recherches en Sciences Humaines.

Ferguson, W. W. 1981. The systematic position of *Canis aureus lupaster* (Carnivora: Canidae) and the occurrence of *Canis lupus* in North Africa, Egypt and Sinai. *Mammalia* **45**, 459–65.

Frank, B. 1965. *Die Rolle des Hundes in afrikanischen Kulturen*. Wiesbaden: Franz Steiner Verlag.

Gallant, J. 2002. *The Story of the African Dog*. Pietermaritzburg: University of Natal Press.

Gautier, A. 1990. *La Domestication. Et l'Homme créa ses Animaux*. Paris: Editions Errance.

Gautier, A. 2001. The Early to Late Neolithic archaeofaunas from Nabta and Bir Kiseiba, pp. 609–35 in Wendorf, F., Schild, R. (eds) *Holocene Settlement of the Egyptian Sahara. Volume 1. The Archaeology of Nabta Playa*. New York/Boston/Dordrecht/London/Moscow: Kluwer Academic/Plenum Publishers.

Gautier, A. 2002. The evidence for the earliest livestock in North Africa: or adventures with large bovids, ovicaprids, dogs and pigs, pp. 195–207 in Hassan, F. A. (ed.), *Droughts, Food and Culture. Ecological Change and Food Security in Africa's Later Prehistory*. New York/Boston/Dordrecht/London/Moscow: Kluwer Academic/Plenum Publishers.

Geppert, P. 1990. *Hundeschlachtungen in Deutschland im 19. und 20. Jahrhundert unter besonderer Berücksichtigung der Verhältnisse in München*. Unpublished Ph.D. thesis, Ludwig-Maximilians-University of München.

Grivetti, L. E. 2000. Food prejudices and taboos, pp. 1495–513 in Kiple, K. F. and Ornelas, K. C. (eds), *The Cambridge world history of food*. Cambridge: Cambridge University Press.

Hallier, M. 1998. Recherches archéologiques en hiver 1997/1998 au nord du Burkina Faso: les collines d'occupation de l'Age du Fer. *Nyame Akuma* **49,** 2–6.

Hallier, M. and Petit, L. 2001. Fouille d'une maison de l'Age du Fer dans le nord du Burkina Faso. *Nyame Akuma* **56,** 2–3.

Holas, B. 1955. De nouveaux cas de cynophagie en Afrique Occidental. *Notes Africaines* **65**, 22–3.

Hrbek, I. 1988. Africa in the context of world history, pp. 1–30 in El Fasi, M. and Hrbek, I. (eds), *General History of Africa III. Africa from the Seventh to the Eleventh Century*. Berkeley: University of California Press.

Hrbek, I. and Devisse, J. 1988. The Almoravids, pp. 336–66 in El Fasi, M. and Hrbek, I. (eds), *General History of Africa III. Africa from the Seventh to the Eleventh Century*. Berkeley: University of California Press.

Insoll, T. 1996. *Islam, archaeology and history. Gao region (Mali) ca. 900–1250* (Cambridge Monographs in African Archaeology 39. BAR International Series 647). Oxford: British Archaeological Reports.

Lam, A. M. 1982. Quelques similitudes culturelles entre l'Afrique du Nord et l'Afrique de l'Ouest. *Notes Africaines* **176**, 90–5.

Levtzion, N. and Hopkins, J. 1981. *Corpus of Early Arabic Sources for West African History*. Cambridge: Cambridge University Press.

Lewicki, T. 1974. *West African food in the Middle Ages*. London: Cambridge University Press.

MacDonald, K. C. 1995. Analysis of the mammalian, avian and reptilian Remains, pp. 291–318 in McIntosh, S. K. (ed.), *Excavations at Jenné-Jeno, Hambarketolo, and Kaniana (Inland Niger Delta), the 1981 Season* (Anthropology 20). Berkeley and Los Angeles: University of California Press.

MacDonald, K. C. 1996. The Windé Koroji complex: Evidence for the peopling of the Eastern Inland Niger Delta (2100–500 BC). *Préhistoire Anthropologie Méditerranéennes* **5**, 147–65.

MacDonald, K. C. and MacDonald, R. H. 1996. A preliminary report on the faunal remains recovered from Gao Ancien and Gao Saney (1993 season), pp. 124–6 in Insoll, T. (ed.), *Islam, Archaeology and History. Gao Region (Mali) ca. 900–1250* (Cambridge Monographs in African Archaeology 39. BAR International Series 647). Oxford: British Archaeological Reports.

MacDonald, K. C. and MacDonald, R. H. 2000. The origins and development of domesticated animals in arid West-Africa, pp. 127–62 in Blench, R. M. and MacDonald, K. C. (eds), *The Origins and Development of African Livestock. Archaeology, Genetics, Linguistics and Ethnography*. London: UCL Press.

MacDonald, K. C. and Van Neer, W. 1993. An initial report on the fauna of Akumbu (Mali). Appendix to Togola, T., *Archaeological Investigations of Iron Age Sites in the Méma Region, Mali (West Africa)*. Unpublished Ph.D. Thesis, Rice University of Houston.

Magnavita, S., Hallier, M., Pelzer, C., Kahlheber, S. and Linseele, V. 2002. Nobles, guerriers, paysans. Une nécropole de l'Age de Fer et son emplacement dans l'Oudalan pré- et protohistorique. *Beiträge zur allgemeinen und vergleichenden Archäologie* **22**, 21–64.

Masonen, P. 1997. Trans-Saharan trade and the West African discovery of the Mediterranean, pp. 116–42 in Sabour, M. and Vikør, K. S. (eds), *Ethnic Encounter and Cultural Change. Papers from the Third Nordic Conference on Middle Eastern Studies* (Nordic Research on the Middle East 3). London: Hurst and Co Publishers.

Mauny, R. 1954. La cynophagie en Afrique occidentale. *Notes Africaines* **64**, 114.

McIntosh, S. K. 1995 (ed.), *Excavations at Jenné-Jeno, Hambarketolo, and Kaniana (Inland Niger Delta), the 1981 Season* (Anthropology 20). Berkeley and Los Angeles: University of California Press.

McIntosh, S. K., McIntosh, R. J. and Bocoum, H. 1992. The Middle Senegal Valley project: preliminary results from the 1990–91 field season. *Nyame Akuma* **38**, 47–61.

Monès, H. 1988. The conquest of North Africa and Berber resistance, pp. 224–45 in El Fasi, M. and Hrbek, I. (eds), *General History of Africa III. Africa from the Seventh to the Eleventh Century*. Berkeley: University of California Press.

Olsen, S. J. 2000. Dogs, pp. 508–17 in Kiple, K. F. and Ornelas, K.

C. (eds), *The Cambridge World History of Food*. Cambridge: Cambridge University Press.
Paris, F. 1984 (ed.). *La région d' In Gall – Tegidda N Tesemt (Niger): Programme Archéologique d'Urgence 1977–1981. III. Les Sépultures du Néolithique Final à l'Islam* (Etudes Nigériennes 50). Niamey: Institut de Recherches en Sciences Humaines.
Paris, F. 2000. African livestock remains from Sahara mortuary contexts, pp. 111–26 in Blench, R. M. and MacDonald, K. C. (eds), *The Origins and Development of African Livestock. Archaeology, Genetics, Linguistics and Ethnography*. London: UCL Press.
Pelzer, C. and Magnavita, S. 2000. La nécropole de Kissi et ses implications historiques. *Berichte des Sonderforschungsbereichs 268* **14**, 367–73.
Perinbam, M. B. 1989. Soninke-Ibadiyya interactions in the Western Sudan c. ninth to c. eleventh century. *The Maghreb Review* **14**, 73–4.
Petit, L. and Hallier, M. 2002. *Tragedy in the dunes: the end of an Iron Age residence in northern Burkina Faso*. Paper presented at the 16th Biennial Conference of the Society of Africanist Archaeologists, University of Arizona, Tucson, Arizona (U.S.).
Savolainen, P., Zhang, Y., Luo, J., Lundeberg, J. and Leintner, T. 2002. Genetic evidence for an East Asian origin of domestic dogs. *Science* **298**, 1610–3.
Serpell, J. 1995. From paragon to pariah: some reflections on human attitudes to dogs, pp. 245–56 in Serpell, J. (ed.), *The Domestic Dog. Its Evolution, Behaviour and Interactions with People*. Cambridge: Cambridge University Press.
Simoons, F. J. 1981. Dogs as human food in northwest Africa. *Appetite: Journal for Intake Research* **2**, 253–66.
Simoons, F. J. 1994. *Eat not this Flesh. Food Avoidances from Prehistory to the Present* (2nd edition). Wisconsin: University of Wisconsin Press.
Staffe, A. 1938. Die Haustiere der Kosi (Rasse, Herkunft, Haltung). *Zeitschrift für Züchtung und Züchtungsbiologie* (Reihe B) **40** (2), 252–85 and **40** (3), 301–42.
The Encyclopaedia of Islam 1999. CD-ROM edition v. 1.0. Leiden: Koninklijk Brill NV.
Thiriet, C. 1954. La cynophagie à Ouargla. *Bulletin de Liaison Saharienne* **5** (16), 11–4.
Thom, S. 1998. *Die eisenzeitliche Nekropole von Kissi, Prov. Oudalan, Burkina Faso*. Unpublished MA thesis, Johann Wolfgang Goethe-University in Frankfurt am Main.
Thys, E. and Nyssens, O. 1983. Préparation et commercialisation de la viande canine chez les Vame-Mbreme, population animiste des monts Mandara (Nord-Cameroun) pp. 511–7 in *Proceedings of the International Colloquium on Tropical Animal Production for the Benefit of Man*. Antwerp: Prince Leopold Institute of Tropical Medicine.
Togola, T. 1996. Iron Age occupation in the Méma region, Mali. *The African Archaeological Review* **13**, 91–110.
Van Neer, W. and Bocoum, H. 1991. Etude archéozoologique de Tulel-Fobo, site protohistorique (IVe–Xe siècle) de la moyenne vallée du Fleuve Sénégal (République du Sénégal). *Archaeo-Zoologia* **4** (1), 93–114.
Vogelsang, R., Albert, K.-D. and Kahleber S. 1999. Le sable savant: les cordons dunaires sahéliens au Burkina Faso comme archive archéologique et paléoécologique du Holocène. *Sahara* **11**, 51–68.
You, R. 1955. Cynophagie. *Notes Africaines* **66**, 40.

Veerle Linseele
Royal Museum of Central Africa, IUAP 05/09,
Leuvensesteenweg 13,
B-3080 Tervuren, Belgium
E-mail: vlinseele@africamuseum.be

9th ICAZ Conference, Durham 2002
Behaviour Behind Bones, (ed. Sharyn Jones O'Day et al.) pp. 327–334

35. Hunting practices and consumption patterns in rural communities in the Rif mountains (Morocco) – some ethno-zoological notes

Marta Moreno-García

Ethnographic observations made on the role played by hunting among rural pastoral and farming communities in the western Rif mountains (Morocco) are presented. Until recently, before this activity was made illegal, wild animals were thoroughly exploited and provided an important supplement to the local economy. We discuss the traditional trapping techniques used and the way species such as wild boar, jackal and hedgehog are exploited. Results show that these self-sufficient communities do not waste any of the meat resources available, including those forbidden by their religious beliefs. Further, magico-religious and nutritional or medicinal significance is attributed to a diversity of products derived from these species.

Introduction

Most of the ethno-zoological studies carried out among traditional Mediterranean rural communities focus exclusively on domestic species (Lewthwaite 1983; Barker 1990; Halstead 1990; Barker and Grant 1991; Chang and Tourtellotte 1993; Moreno García 1999). These studies' main aim is to assess the role of domestic animal husbandry in the general economy of these often self-sufficient communities in order to create ethno-archaeological models that can be used when studying past societies. In particular, the study of production strategies related to the exploitation, on a small or large scale, of animal products, and its archaeological evidence, has received most of the attention. However, there has been little interest in studying the role of wild animals and especially, in assessing the significance of hunting among these agro-pastoral communities (Zvelebil 1992).

Between 1999 and 2001 I participated in an inter-disciplinary ethno-archaeological project set up in the Jebala Mountains, in the western part of the Rif mountain range, in northern Morocco (Fig. 1) (González Urquijo *et al.* 2001; Ibáñez *et al.* 2002; Ibáñez *et al., in press*). Our aim was to record ethnographic data from the traditional rural populations that inhabit this area in order to construct models for use in the study of archaeological self-producing societies (Peña-Chocarro *et al.* 2000). My area of research in this project was concerned with the exploitation of domestic and wild faunal resources.

In the attempt to fulfil part of the already mentioned gap that exists in the ethno-zoological literature, this paper presents some of the observations made on three wild species: the wild boar (*Sus scrofa algira*), the jackal (*Canis aureus*) and the hedgehog (*Erinaceus algirus*) that are or were, until the recent past, regularly hunted by the inhabitants of the Jebala Mountains. In order to understand the background in which hunting takes place a brief account of the social and economic situation of these rural communities as well as the role played by domestic animals is also included.

The Jebala Mountains rural communities: social and economic considerations

Our study area is located to the south of the northern Moroccan city of Chefchaouen, between the localities of Tanakoub, Mokrisset, Ksar-el Kébir and Ouzzane (Fig. 1). In most cases, villages are difficult to reach since they are situated on the slopes or at the valley bottoms. Very narrow and steep sand tracks, emerging from the main road, provide the only access. Some of these are impracticable for driving, so riding donkeys or walking are the usual means of moving around.

The number of families inhabiting each village is small and usually never exceeds thirty. However, family units are large, with grandparents, their children and

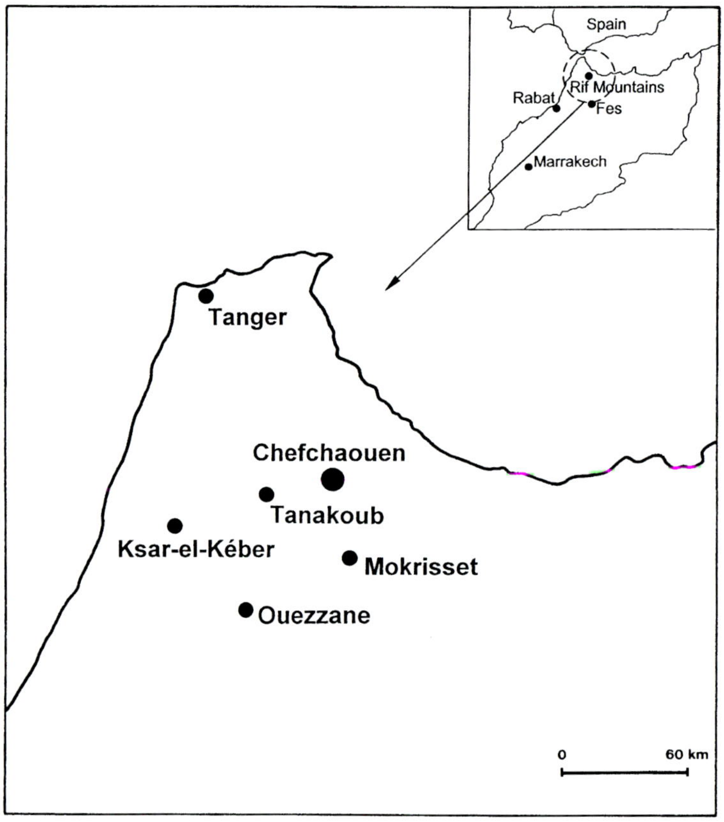

Fig. 1. Location of the area of research in the Rif Mountains.

grandchildren often sharing the same household. The economic status of many of these families is very poor. Interviewing the elderly members of the community was preferred (Fig. 2) because it was evident that the younger generations are rapidly abandoning traditional ways of life and show little interest in continuing to exploit local resources. This does not mean that agriculture or husbandry are being mechanised but that emigration and cultivation of hachis (*Cannabis sativa*) are seen as the only viable way of escaping poverty.

Cattle, donkeys, and occasionally horses are employed for traction in all agricultural tasks such as ploughing (Fig. 3) and threshing. These activities are performed exclusively by men, while women and especially young girls are in charge of the day to day husbandry of domestic animals.

Not all families can afford to have cattle. They are very useful animals that are highly valued for their so-called "secondary products". They remain stabled during the autumn and winter. Fresh fodder is brought to them daily by women who go into the woods to cut leaves and branches that are transported on their own backs (Fig. 4). Cattle are only released to pasture freely during the spring or the summer months. From their milk women make butter and yoghurt. Cow dung is equally precious. It is carefully collected and left to dry in the form of cakes to be used as fuel in domestic fires not only to cook food, but also for baking pottery (González Urquijo *et al.* 2001). Traditional local ceramists also use cow dung as material for making storage pots (Ibáñez *et al.*, *in press*).

Goats are more common livestock and their meat is frequently found in the butcher's shops that are located by the main roads or in the local urban markets. However, among these rural communities goat meat consumption is related to special occasions such as religious festivities or family celebrations. When animals are sold in the local markets, meat and skin can be sold separately. Skins can be kept by the owner of the animal to be cured at home or are handed in to the local tanner (Ibáñez *et al.* 2002). Like cattle, goat milk and dung are both highly valued. The latter is collected daily from the stables by women and stored in small enclosures near the houses until it is ready to be spread as manure in each family cultivation plot. Since there are no chemical fertilisers available, the dung of domestic animals is very important in maintaining soil fertility. Moreover, as recorded in classical Hispano-Arabic treatises of agriculture (Millás Vallicrosa 1943, 305–6; Millás Vallicrosa 1948, 363–7; Eguaras Ibáñez 1988, 208–9), farmers in the Jebala region are aware of the different qualities of manure depending on the animals from which it is derived.

Finally, poultry are the other "meat" animals that share

Fig. 2. Interviewing the Haddush family in the village of Briet.

Fig. 4. Woman transporting fodder on her back.

Fig. 3. Cattle ploughing a field.

the domestic space with people. They are more commonly eaten than livestock but still they are kept for their eggs, an important component of the daily diet.

Thus, one can conclude that for these traditional communities domestic animals are not primarily exploited for their meat. On the contrary, they are interested in maintaining them alive, since they provide them with products that are essential to their self-sufficient economy. Equally, they represent a banking resource that can be cashed outside the domestic environment whenever it is needed (Halstead 1993). It is against this background that one can attempt to assess the role of hunting.

Hunting practices and consumption patterns

Nowadays in Morocco, hunting is controlled by the Forestry Department, which means that certain regulations have to be observed. However, most of the men interviewed confessed that illegal hunting, using traditional techniques, is a relatively common activity in the Jebala region.

The interviews carried out focussed on five main subjects that are as follows:

1) identification of the species hunted;
2) reasons for hunting: are wild animals a threat to these agro-pastoral communities? Is hunting a leisure activity or a need?
3) hunting techniques: types of traps, where they are placed, use of baits, the role of dogs, etc. Is hunting an individual or a collective activity? Are there any species that can be hunted accidentally or is hunting always intentional?
4) processing of the carcass: what happens to the trapped animal? Is the carcass shared in any way by the hunters or the community? Where and how is the processing carried out?
5) and finally, are there any popular beliefs (i.e., superstitions, medicinal properties, rituals) associated with any of the locally hunted wild species?

The results obtained show that for each of the above topics not only was there a variety of answers but also that many of the circumstances surrounding hunting practices would leave no visible trace behind. Thus, assessing the role of this activity among past agro-pastoral communities appears to be more complex than could originally have been thought. To illustrate this point the observations made on the wild boar, the jackal and the hedgehog are presented below.

Wild boar (*Sus scrofa algira*)

Reasons for hunting

The wild boar is considered the most dangerous animal species in the region. It represents a great threat to the small crops of these communities since in a few hours it can eat or destroy what will support a whole family for months. Also when hurt it can attack man. At present, its hunting is forbidden by law, notwithstanding it is regularly practised in a clandestine way.

Hunting techniques

Wild boar always was and still is hunted in parties, in which men armed with sticks, ropes, knives and accompanied by several dogs work together. The sort of trap traditionally used is a guillotine iron toothed snare. Several of these traps are usually buried along the tracks of wild boar in the forest or near the water courses and swamps where they go to have their mud baths. Sometimes, to attract their attention to these points, men use baits, spread on top of the traps, such as dried grapes, figs, broad beans, etc.

The animal can get caught by its legs or by the snout, in such a way that often it is only injured but not killed. Until the recent past, it was common to take the trapped animal into the village to be exhibited as a trophy in front of all neighbours. Such practise has been abandoned since wild boar hunting is illegal and hunters can be fined. However, the feeling that men who participate in the hunting party are heroes still persists because they help the community by getting rid of a serious threat.

Traditionally, to diminish the fierceness and forces of the trapped wild boar, dogs were sent to attack it while all men threw stones at it or beat it with wooden sticks on its head. Then, to immobilise it the hunters used to put a stick across its mouth, from which ends two ropes came out that could be tied around the animal's neck to suffocate it. In order to reduce their danger, the tusks were cut with stones or an axe. If the animal was caught by a leg then they used to cut the tendons of the free legs with knives. Once paralysed, it could be tied with ropes to be killed on the spot or dragged down to the village.

Processing

Given that the Jebala rural communities are Moslem, to kill a wild boar they follow the usual procedure of cutting the animal's throat and leave it to bleed, although sometimes they can just abandon the trapped animal until it dies of starvation. Only the most religious and wealthy families interviewed referred to the processing of the wild boar carcass exclusively for feeding the dogs. The rest reported, as a general use, their own consumption of wild boar meat. They do not seem to extend the Islamic prohibition of pork consumption to wild boar flesh and therefore do not feel they are breaking any religious law by eating it. Indeed: "*The eating of wild and domestic pigs as food has survived here and there in the Islamic world*" (Simoons 1994, 35–36). Furthermore, De Planhol (1959), as quoted in Simoons (1994, 35), notes that "*formerly people raised swine in most of the mountain core of the Rif, a wooded environment eminently suited since antiquity to the practice*". Therefore, it appears that the eating of wild boar meat has a long tradition in the area.

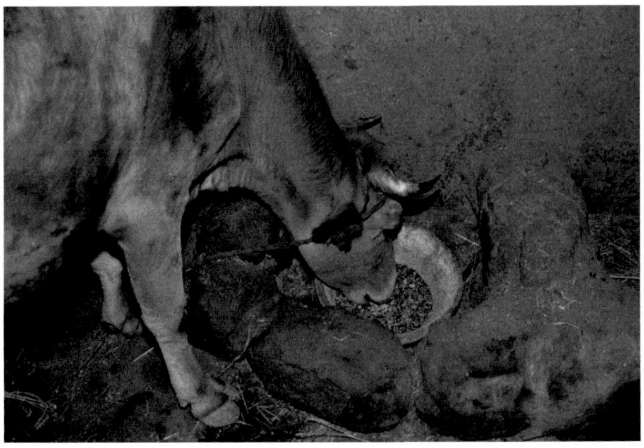

Fig. 5. Cow feeding on chaff and salted wild boar meat.

Primary processing can take place either in the forest or in the main square of the village, never in the households as is usually the case with domestic animals. They start by cutting the hind legs, from the internal side and from the lower part to the thigh. These are considered to be the portions with the best meat. Then they cut the loins and finally the front legs. Entrails, feet and head are discarded. If this primary processing occurs in the forest, the carcass is divided by all men who participated in the hunting party. Each of them has the right to take home a joint. However, if the animal is killed in the village, the meat is shared among all those neighbours who want to have some. In this case, secondary processing, i.e., filleting and cutting up of the meat into smaller pieces, takes place also in the public square and not in the houses. Then the bones are left for the dogs.

It seems evident that consumption of wild boar meat is done relatively quickly after the animal has been killed. However, an exception to this situation was reported in several villages. Some families admitted that occasionally wild boar meat is preserved by salting it. This is apparently done only with the hind legs and it is not consumed by people but by cattle and horses! The whole hind quarter is hung in the sun, covered in salt, for several weeks until its meat is dry. Once it is ready, it is cut in very small pieces and mixed with the cereals, chaff and fodder cattle and horses feed upon (Fig. 5). Since it is salted, they like it very much. After all meat has been consumed the bones are given to the dogs.

Popular beliefs

From the above observations it is evident that for these people the wild boar represents a very strong animal. Strength is a highly valued quality important for coping with the harsh living conditions in the Jebala Mountains. Thus, there is the local belief that the eating of wild boar meat helps gaining strength. It is highly recommended for children, weak people and during the winter cold. For the same reason, wild boar meat is fed to working animals – so that they grow strong and as powerful as the wild boar!

The nutritious significance of wild boar meat appears to be an old one, already noted in the historical Islamic literature. Avicenna, the tenth-century Persian doctor, "*in his classification of foodstuffs, included pork among foods that strengthen the blood*" (Simoons 1994, 35). An eleventh-century Arab manuscript found in the library of El Escorial (Madrid, Spain), translated by Ruiz Bravo-Villasante (1980, 23) points also to the rich quality of this kind of meat.

But it is not only the meat that in their opinion has some special properties. Thus, the de-fleshed bones of a wild boar hind leg are sometimes hung from the branches of a tree near the family cultivation plot in the belief that they frighten away other wild boars and prevent them from entering to eat the harvest. Also, the tusks are kept as a sort of amulet for protection against the evil eye. They can be perforated through the root and are worn on a string both by men and women. This latter custom has been noted among other Moslem populations such as the Arabs of Palestine (Macalister 1912 (2), 8, as quoted in Simoons 1994, note 165, 340)

Jackal (*Canis aureus*)

Reasons for hunting

This wild species represents a threat to livestock. It attacks goats and cattle and it can even come close to the villages and enter houses to kill poultry and rabbits. Almost all families have stories to tell about how their animals were killed by jackals. They believe that if a goat is injured by a jackal, it never recovers totally because of the "poison" it has received. As protection, most flocks and herds are accompanied by a shepherd dog. Therefore the hunting of this species is purely for protective reasons.

It seems there is not a particular season in the year when they are more abundant and particularly hunted. Females can sometimes be seen running down the mountain with their puppies while males are usually on their own.

Hunting techniques

Traditionally, as with the wild boar, jackal hunting is done in parties of ten-fifteen men, armed with rudimentary implements (i.e., sticks, stones and knives) and accompanied by dogs. The sort of traps used are similar to the iron snares described above, however they are not always employed. Hunting parties with several dogs can go into the woods making a noise to force the jackals out of their dens. Once they are surrounded, hunters send the dogs to attack them. It is not unusual to locate the den and kill the puppies while the mother is away.

When a jackal is caught in a trap and is still alive, it can be brought to the village to be exhibited in front of the community or left in the forest to die of starvation. When brought to the village it is used for training the hunting dogs. To facilitate its transport they put a muzzle, tied by ropes behind the neck around its mouth, and the legs are also tied up.

Processing

Skins of jackals have been traditionally used for decoration, as carpets or to make the hachis pipe bags. They are highly-priced in the markets and outsiders sometimes come to buy them. Skinning of the animal is usually done in the forest and the procedure they follow is as for a goat or a cow. They start from the inner part of the lower hind legs and proceed to the trunk area to reach the front legs and finally the head. Then the skin is washed, rolled up and taken to the village to be cured (Ibáñez *et al.* 2002). The skinned carcass of the animal is abandoned on the spot.

Although not a tradition shared by all interviewed families, it was recorded that jackal meat has occasionally been consumed, until the recent past. If the animal was caught in a trap, his throat could be cut hence there was no problem to eat it. The carcass was skinned, cleaned of its entrails and head and paws removed. In such a way, it was ready to be transported to the house where it was cut up and cooked like goat meat in a stew with vegetables. They reported that it smells strong but tastes good. Particularly, the meat of puppies, cooked in a soup, is considered to be very good and nutritious for children. Some of the interviewees admitted having eaten it when small.

Popular beliefs

In some instances, consuming jackal flesh is an act of revenge. The jackal becomes the victim in the hands of men after having been the predator of their livestock. In addition, as noted for the wild boar, some medicinal properties are recognised in this kind of meat. Firstly, it strengthens weak people such as children or the sick. Secondly, they believe that it helps to get rid the blood of sugar and so it is recommended for diabetics. According to the treatment, the patient should eat small pieces of meat, boiled with salt and pepper, for a week; rest for forty days and eat this meat again for another week.

Besides the meat, different parts and organs of the jackal have traditionally been used as healing remedies or as amulets by these communities. Some examples are:

- against the evil eye one should inhale the smoke produced after burning any **bone** of the jackal skeleton;
- wearing a jackal **skin** protects against sickness and sleeping on top of one prevents an attack by other animals, such as snakes;
- hanging the right **eye** of a jackal in the corner of a house prevents entry by other animals;

- its **testicles** and **gall bladder** wrapped in paper, on which a prayer has been written, must be burnt and its smoke inhaled by the sick patient;
- the liquid obtained from boiling the dried **intestine** of jackal in water cures heart maladies and it is good for stomach problems. They said that it is still possible to find this product in certain markets in Tanger;
- the cooked **liver** of jackal eliminates intestine worms.

These magical-medicinal uses of the jackal seem to be part of the traditional knowledge that has been transmitted from generation to generation. Some of these and others are described in detail in the above mentioned medieval Arab manuscript on the uses of animals (Ruiz Bravo-Villasante 1980). Simoons (1994, 228) also reports on the nutritional and medicinal significance attributed, in general, to dogflesh in Morocco.

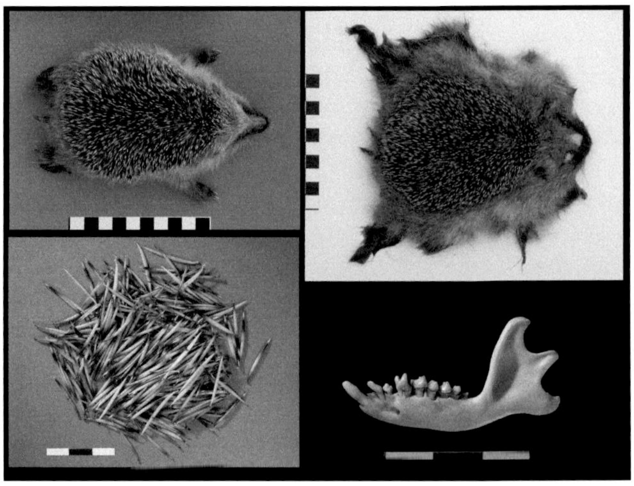

*Fig. 6. Hedgehog (*Erinaceus algirus*). Skin, spines and mandible.*

Hedgehog (*Erinaceus algirus*) (Fig. 6)

Reasons for hunting

This species is neither a predator nor a threat, but very abundant in the area. The men interviewed appear to be well acquainted with its behaviour and feeding habits. It is hunted for eating, although often it is hunted unintentionally. It falls accidentally into traps set for rabbits or small carnivores.

Hunting techniques

It is a very easy animal to catch since it moves around slowly. It can be hunted by hand, especially at night when they roll themselves up.

Processing

Hedgehog meat is widely consumed among the peoples of the Jebala. After the throat of the animal has been cut and it has bled three different methods are used to skin it:

1) grilling the animal on hot coals so that the burnt spines can be removed by scraping with a knife;
2) boiling in hot water for several minutes frees the spines as one plucks the feathers of a bird;
3) making a cut along or across the belly and pulling out the whole skin with the spines.

The next step, after skinning, is to eliminate the entrails, paws and snout. Head, heart and liver are kept and cooked. The carcass can be divided into smaller portions and stewed with vegetables, potatoes and couscous. Or it can be roasted whole.

Popular uses and beliefs

Despite its small size, its insignificant appearance and its harmless condition, the hedgehog is an animal of multiple uses in the popular and traditional culture of the Jebala peoples. The following examples illustrate this point:

- farmers make the calves wear the **skin with the spines** as a muzzle so that when they approach their mothers to suckle, they prick them and get rejected. This method for weaning livestock is apparently a traditional common practise used also in other areas beyond Morocco. The Portuguese ethnographer Leite de Vasconcellos (1982, 402) reports that in Nisa (Alto Alentejo, Portugal) "*para desmamar os vitelos atam ao focinho da vaca – enfiado pela cabeça e preso com um cordel- um pedaço de pele de ouriço-cacheiro com os picos, para que, quando as crias vão mamar e a mãe as lambe, se piquem e se afastem*" [to wean the calves, they tie a piece of hedgehog skin with its spines around the cow's mouth so that when the calf approaches to suckle and the mother licks it, it gets pricked and flees];
- isolated **spines** (Fig. 6) are used by women as implements to paint their eyes with kohl or their hands and feet with henna;
- **spines** are also sold in the market together with medicinal plants to make amulets;
- the **ulna** and the **spines** are used as toothpicks;
- the **mandible** (Fig. 6) is a well-known amulet against the evil-eye. Women wear it in the fold of their skirts, at the waist height, and men hang it on a string from their necks;
- the powder of a **burnt hedgehog** mixed with honey is used as a remedy against asthma;
- its **blood** is a good remedy for skin infections;
- the dried **gall bladder** is also good for pimples and helps stop bleeding after a cut.

Discussion

These observations made on wild boar, jackal and hedgehog hunting among the agro-pastoral rural communities of the Jebala Mountains (northern Morocco) reveal that this activity has traditionally had, until the recent past, an important and complementary role in their self-sufficient economy. Not only is it a means of protecting their basic resources (*i.e.*, crops and livestock) from attack by animal predators but also it is an activity that provides additional supplies for free (*i.e.*, meat, skins, etc.). In my opinion, this is a key issue that needs to be tackled before trying to evaluate the 'social and symbolic implications' of hunting in farming societies (Kent 1989).

All our interviewees agreed that hunting is the only way that they have to defend their means of life. For this reason, they do not seem to understand why nowadays the local authorities forbid it and instead protect dangerous species such as the wild boar and the jackal. These animals compete with men for the same resources. They cannot find a better argument to justify them keeping up hunting. Their subsistence farming and husbandry systems are very precarious. Therefore, they need to control as much as possible any external threats as those represented by these two predators. Hunting in these rural mountain communities has functioned primarily as a "defence mechanism".

It was evident that it is not an activity in which they invest a lot of time. Hunting parties are organised only when their scarce properties are at risk. Trapping small carnivores (*i.e*, genets, foxes) or other small mammals such as hedgehogs usually occurs accidentally when they are caught in snares set for bigger animals. Neither special skill is required. Men in each family can hunt or be part of a hunting party; no special abilities or strength are necessary.

It is important to understand that among these mountain rural peoples any social, cultural or symbolic aspects related to hunting (*i.e.*, the fact of considering heroes men who participate in a hunting party) derive from the solidarity spirit that exists among members of the community. From the evidence collected it cannot be ascertained that here hunting serves as "power mechanism", as could have been the case in other farming societies (Hamilakis, *in press*). It is a collective activity in which a substantial number of neighbours participate, independently of their social or economic status. When it comes to guard basic living resources to the community, solidarity prevails over personal interests. The fact that the wild boar carcass may be brought to the village and its meat can be shared by all those who wish to get some, even if they have not participated in its hunting, illustrates this point.

As shown in the first section of this paper, domestic animals are exploited mainly for their "secondary products". Consumption of their meat only occurs in special occasions. Furthermore, interviewees referred to women's distaste of consuming animals that have been bred by them such as goats and cows. Meat does not constitute an essential part of their diets although they recognise it provides strength and therefore its consumption is recommended for children, the sick and men.

Against this background, meat derived from hunted animals appears to be a resource not to be wasted whenever available. However, none of the interviewees suggested that wild animals, with the exception of rabbits, are primarily hunted for their meat. One crucial aspect that cannot be forgotten is the fact that these peoples are Muslim and do not eat meat from animals that have not been bled. Thus, it is quite usual that if they find a trapped animal already dead they just skin it, if this is of any worth, and leave the carcass behind for the dogs.

Finally, it was intriguing and surprising to find out that cultural traditions related to the magico-religious and nutritional or medicinal significance attributed to a diversity of products derived from hunted animals have managed to survive for centuries (Ruiz Bravo-Villasante 1980). They are secretively commercialised outside the agro-pastoral communities in the local shops of the nearest towns. Thus, they appear to be recognised by all social groups independently of their social status.

Conclusion

This ethno-zoological study has proved interesting in showing the difficulties there are in assessing the social and cognitive aspects of the interaction between man and animals. The results indicate that there is no uniformity in that relationship. It is much more complex than initially can be thought. Even when it is supposed to occur, as in the present case, due to the Moslem condition of the rural communities in the Jebala region, unexpected observations were made (*i.e.,* meat consumption of "impure" species).

The social and symbolic implications relating the role of hunting in other present and past farming societies (Kent 1989; Hamilakis, *in press*) do not seem to exactly hold among the mountain agro-pastoral communities of the Moroccan Rif. Here defence and solidarity appear to be essential issues, which in turn, guarantee the survival of their means of life.

To conclude, it is evident that to recognise some of the topics this work deals with in the archaeozoological record seem difficult since many of the variables involved leave no traces. Thus, ethnographic research on traditional communities is presented as a worthwhile working methodology that allows one to be aware of the occurrence of some of those invisible variables.

Acknowledgements

This work is part of the project "*Las primeras sociedades campesinas. El aporte de la etnoarqueología en Marruecos*", supported by the Spanish Fundación Marcelino Botín. I am very grateful to Carlos M. Pimenta who was part of the team for the last two years and provided immense help in conducting the interviews. Since hunting is practised exclusively by men, his presence was essential in encouraging our interviewees to talk. I am thankful to Simon Davis for correcting my English, to Yannis Hamilakis for useful comments and for making available *in press* work. I would like to thank also José Paulo Ruas (IPA) for taking the photographs shown in Figure 6 and Abdullah Maknin for his excellent work as our Arabic interpreter.

References

Barker, G. W. W. 1990. Archaeological survey and ethnoarchaeology in the Cicolano Mountains, Central Italy. Preliminary results, in Maggi, R., Nisbet, R. and Barker, G. (eds), Archeologia della Pastorizia nell' Europa Meridionale, 1989. *Rivista di Studi Liguri* **56**, 109–21.

Barker, G. and Grant, A. 1991. Ancient and modern pastoralism in Central Italy: an interdisciplinary study in the Cicolano Mountains. *Papers of the British School at Rome* **59**, 15–88.

Chang, C. and Tourtellotte, P. A. 1993. Ethnoarchaeological survey of pastoral transhumance sites in the Grevena region, Greece. *Journal of Field Archaeology* **20**, 249–64.

de Planhol, X. 1959. *The World of Islam*. Ithaca, N.Y.: Cornell University Press.

Eguaras Ibáñez, J. 1988. *'Ibn Luyun: Tratado de Agricultura*. Volumen 2. Granada: Patronato de la Alhambra y Generalife.

González Urquijo, J. E., Ibáñez, J. J., Zapata, L. and Peña-Chocarro, L. 2001. Estudio etnoarqueológico sobre la cerámica Gzaua (Marruecos). Técnica y contexto social de un artesanato arcaico. *Trabajos de Prehistoria* **58**, 5–27.

Halstead, P. 1990. Present to past in the Pindhos: diversification and specialization in mountain economies, in Maggi, R., Nisbet, R. and Barker, G. (eds), Archeologia della pastorizia nell' Europa Meridionale, 1989. *Rivista di Studi Liguri* **56**, 61–80.

Halstead, P. 1993. Banking on livestock: indirect storage in Greek agriculture. *Bulletin on Sumerian Agriculture* **7**, 63–75.

Hamilakis, Y. in press. The sacred geography of hunting: wild animal, social power and gender in early farming societies, in Kotjabopoulou, E., Hamilakis, Y., Halstead, P., Gamble, C. and Elefanti, V. (eds), *Zooarchaeology in Greece: Recent Advances*. London: BSA Studies.

Ibáñez, J. J., González Urquijo, J. E., and Moreno, M. 2002. Le travail de la peau en millieu rural: le cas de la Jebala marocaine, pp. 79–97 in Audoin-Rouzeau, F. and Beyries, S. (eds), *Le Travail du Cuir de la Préhistoire à nos Jours. XXII Rencontres Internationales d'Archéologie et d'Histoire d'Antibes*. Antibes: Editions APDCA.

Ibáñez, J. J., González Urquijo, J. E., Zapata, L., Peña-Chocarro, L. and Moreno, M. in press. Les récipients en bouse de vache et argile non cuite (*tabtoba* et *tonna*) dans la région Jebala (Maroc). Exemple d'un processus technique de type domestique. *Techniques et Culture* **34**.

Kent, S. 1989. Cross-cultural perceptions of farmers as hunters and the value of meat, pp. 1–17 in Kent, S. (ed.), *Farmers as Hunters*. Cambridge: Cambridge University Press.

Leite de Vasconcellos, J. 1982. *Etnografia Portuguesa. Vol. 5*. Lisboa: Imprensa Nacional-Casa da Moeda.

Lewthwaite, J. 1983. The art of coarse herding: archaeological insights from recent pastoral practices on west Mediterranean islands, pp. 25–37 in Clutton-Brock, J. and Grigson, C. (eds), *Animals and Archaeology 3. Early Herders and their Flocks* (BAR International Series 202). Oxford: British Archaeological Reports.

Macalister, R. A. S. 1912. *The Excavation of Gezer 1902-1905 and 1907-1909*. 3 vols. London: John Murray.

Millás Vallicrosa, J. M. 1943. El tratado de agricultura de 'Ibn Wafid. *Al-Andalus* **8**, 281–332.

Millás Vallicrosa, J. M. 1948. La traducción castellana de "El tratado de agricultura" de 'Ibn Bassal. *Al-Andalus* **13**, 347–430.

Moreno-García, M. 1999. Ethnographic observations of transhumant husbandry practices in Spain and their applicability to the archaeological sample, in Bartosiewicz, L. and Greenfield, H. (eds.), Transhumant Pastoralism in Southern Europe. Recent Perspectives from Archaeology, History and Ethnology. *Archaeolingua, Series Minor* **11**, 159–77.

Peña-Chocarro, L., Zapata, L., González Urquijo, J. E. and Ibáñez, J. J. 2000. Agricultura, alimentación y uso del combustible: aplicación de modelos etnográficos en arqueobotánica. *Saguntum* extra **3**, 403–20.

Ruiz Bravo-Villasante, C. 1980. *Libro de las Utilidades de los Animales*. Madrid: Fundación Universitaria Española.

Simoons, F. J. 1994. *Eat not this Flesh. Food Avoidances from Prehistory to the Present*. Wisconsin: The University of Wisconsin Press.

Zvelebil, M. 1992. Hunting in farming societies – the prehistoric perspective, in Grant, A. (ed.), Animals and their Products in Trade and Exchange. *Anthropozoologica* **16**, 7–18.

Marta Moreno-García
Instituto Português de Arqueologia,
Av. da Índia 136,
P-1300-300 Lisboa, Portugal
E-mail: marta@ipa.min-cultura.pt